# Multigrid

Dedicated to Linde, Lukas, Sophia, Kristian, Eva, Filipp, Katharina,
Anasja, Wim, Agnes,
Annette, Sonja and Timo

# MULTIGRID

## U. Trottenberg
## C.W. Oosterlee
## A. Schüller

*Institute of Algorithms and Scientific Computing (SCAI),*
*GMD – German National Research Center for Information Technology,*
*Schloss Birlinghoven, D-53734 St Augustin, Germany*

with guest contributions by

## A. Brandt

*Department of Computer Science and Applied Mathematics,*
*The Weizmann Institute of Science, Rehovot 761000, Israel*

## P. Oswald

*Bell Laboratories, Lucent Technologies,*
*Rm 2C-403, 600 Mountain Avenue, Murray Hill,*
*NJ 07974, USA*

and

## K. Stüben

*Institute of Algorithms and Scientific Computing (SCAI),*
*GMD – German National Research Center for Information Technology,*
*Schloss Birlinghoven, D-53734 St Augustin, Germany*

**ELSEVIER**
ACADEMIC
PRESS

AMSTERDAM  BOSTON  HEIDELBERG  LONDON  NEW YORK  OXFORD
PARIS  SAN DIEGO  SAN FRANCISCO  SINGAPORE  SYDNEY  TOKYO

Elsevier Academic Press
525 B Street, Suite1900, San Diego, California 92101-4495, USA
http://www.elsevier.com

Elsevier Academic Press
84 Theobald's Road, London WC1X 8RR, UK
http://www.elsevier.com

**British Library Cataloguing in Publication Data**
A catalogue record for this book is available from the British Library

Library of Congress Catalog Number: 00-103940

ISBN 0-12-701070-X

Typeset by Newgen Imaging Systems (P) Ltd., Chennai, India

03  04  05  06  07  08  9  8  7  6  5  4  3  2

Transferred to digital printing in 2007.

# CONTENTS

# PREFACE

This book is intended for multigrid practitioners and multigrid students: for engineers, mathematicians, physicists, chemists etc.

We will give a systematic introduction to basic and advanced multigrid. Our intention is to give sufficient detail on multigrid for model problems so that the reader can solve further applications and new problems with appropriate multigrid techniques. In addition, we will cover advanced techniques (adaptive and parallel multigrid) and we will present applications based on these new developments.

Clearly, we would not have written this book if we had thought that there was a better book that fulfilled the same purpose. No doubt, there are a number of excellent multigrid books available. However, in our opinion all the other books emphasize multigrid aspects different from those that we are interested in and which we want to emphasize.

Mathematical results that can be rigorously proved may be formulated in different ways. Practically oriented mathematicians often prefer a presentation in which the assumptions and results are motivated by some typical examples and applications. These assumptions are usually not the most general ones, and thus the results may not be the strongest that can be obtained. We prefer such a presentation of results, as we want to provide as much motivation and direct understanding as possible. However, in many cases, we add some remarks about generalizations and extensions, and about weaker assumptions.

With respect to multigrid theory, we give (elementary) proofs, wherever we regard them as helpful for multigrid practice. All the theoretical tools which we use should be understood by mathematicians, scientists and engineers who have a general background in analysis, linear algebra, numerical analysis and partial differential equations (PDEs). Wherever more sophisticated mathematical tools are needed to derive practically relevant theoretical results, we cite the appropriate literature. However, we are not interested in theory as an end in itself.

This book has three authors and, in addition, three guest authors. While the guest contributions are supposed to be fully consistent with the contents of the book and fit with its general philosophy and message, they are still independent and self-contained. The guest authors are experts in the fields they present here and they use their own style to express their views and approaches to multigrid.

The three main authors, on the other hand, have agreed on all the material that is presented in this book. They did not distribute the chapters among themselves and did not distribute the responsibility. They also agreed on the way the material is presented.

Multigrid methods are generally accepted as being *the fastest numerical methods* for the solution of elliptic partial differential equations. Furthermore, they are regarded as *among the fastest methods* for many other problems, like other types of partial differential equations, integral equations etc. If the multigrid idea is generalized to structures other than grids, one obtains *multilevel*, *multiscale* or *multiresolution methods*, which can also be used successfully for very different types of problems, e.g. problems which are characterized by matrix structures, particle structures, lattice structures etc. However, the literature does not have a uniform definition of the terms multigrid, multilevel etc.

This book is devoted to PDEs and to the *"algebraic multigrid approach"* for matrix problems.

We assume that the reader has some basic knowledge of numerical methods for PDEs. This includes fundamental discretization approaches and solution methods for linear and nonlinear systems of algebraic equations. Implicitly, this also means that the reader is familiar with basics of PDEs (type classification, characteristics, separation of variables etc. see, for example, [395]) and of direct and iterative solvers for linear systems.

We will not, however, assume detailed knowledge about existence theories for PDEs, Sobolev spaces and the like. In this respect, the book is addressed to students and practitioners from different disciplines. On the other hand, in some sections, advanced applications are treated, in particular from computational fluid dynamics. For a full understanding of these applications, a basic knowledge of general PDEs may not be sufficient. In this respect, we will assume additional knowledge in these sections and we will give references to background material in the corresponding fields.

We do not assume that the reader works "linearly" with this book from the beginning to the end though this is suitable to obtain a good insight into multigrid and its relation to similar approaches. The multigrid beginner may well skip certain sections. We will lead the reader in this respect through the book, pointing out what can be skipped and what is needed.

The overall structure of the book is determined by its chapters. The first half of the book (Chapters 1–6) discusses standard multigrid, the second half (Chapters 7–10) deals with advanced approaches up to real-life applications. Accordingly, the style and presentation in the first half is more detailed. In addition to the basic techniques introduced in the first six chapters, we add many more specific remarks and algorithmical details. These may not be so interesting for beginners but should be helpful for practitioners who want to write efficient multigrid programs. Mistakes that are easily made are mentioned in several places.

The second part of the book (Chapters 7–10) is presented in a more condensed form, i.e. in a more research oriented style.

This structure of the book is also reflected by the nature of the equations and applications we deal with. There is no doubt about the fact that multigrid methods work excellently for "nicely" elliptic PDEs. This is confirmed by rigorous mathematical theory.

For typical real-life applications (PDE systems with nonelliptic features and nonlinear terms), however, such a theory is generally not available. Nevertheless, as we will see in this book, multigrid can be applied to such problems although they may not be "nicely" elliptic or even not elliptic at all. In answering the question "when does multigrid work?", we will give insight, based on 20 years of multigrid practice and multigrid software development.

## ACKNOWLEDGMENTS

We would like to thank the many friends, colleagues, collaborators and students who have helped us to write this book. First, we thank the three guest authors, Achi Brandt, Peter Oswald and Klaus Stüben for their contributions. Klaus Stüben was also the chief reader of the manuscript. His criticism and comments were extremely helpful.

Rudolph Lorentz also commented extensively on the manuscript and checked our English.

Achi Brandt, who, to our regret, never wrote a multigrid book himself, closely followed our progress and made many helpful comments.

Others who commented on our manuscript included: Tanja Füllenbach, Jürgen Jakumeit, Rene Redler, Roman Wienands and Kristian Witsch. We would like to thank them for their efforts and suggestions.

All colleagues working in the GMD–Institute for Algorithms and Scientific Computing (SCAI) who supported us are gratefully acknowledged. In particular, we thank Ingrid Filter, Wolfgang Joppich, Axel Klawonn, Johannes Linden, Hans-Günther Reschke, Hubert Ritz-dorf, Horst Schwichtenberg, Frauke Sprengel, Barbara Steckel, Clemens-August Thole and Monika Wendel.

Some of the colleagues and friends whom we thank for clarifying discussions, interesting cooperations and for pointers to the literature include: Wolfgang Dahmen, Francisco Gaspar, Michael Griebel, Norbert Kroll, Oliver McBryan, Takumi Washio, Pieter Wesseling, Olof Widlund, Jinchao Xu, Irad Yavneh and Christoph Zenger.

We would like to thank Eric Peters for the cover design.

The initiative to write this book was strongly influenced by Olof Widlund, who invited Ulrich Trottenberg to the Courant Institute of Mathematical Sciences of the New York University, New York City as a visiting professor to teach a multigrid course for graduate students.

Of course, this book could not have been written without the support and patience of our families, to whom we owe most.

# I

# INTRODUCTION

We start this chapter with a short introduction of some of the equations that we will treat in this book in Section 1.1 and with some information on grids and discretization approaches in Section 1.2. In Section 1.3 we will introduce some of our terminology. The 2D Poisson equation with Dirichlet boundary conditions is the prototype of an elliptic boundary value problem. It is introduced and discussed in Section 1.4. In Section 1.5 we will take a first glance at multigrid and obtain an impression of the multigrid idea. Some facts and methods on basic numerics are listed in Section 1.6.

## 1.1 TYPES OF PDEs

As we will see in this book, elliptic boundary value problems are the type of problem to which multigrid methods can be applied very efficiently. However, multigrid or multigrid-like methods have also been developed for many PDEs with nonelliptic features.

We will start with the usual classification of second-order scalar 2D PDEs. Generalizations of this classification to 3D, higher order equations or systems of PDEs can be found [150]. We consider equations $Lu = f$ in some domain $\Omega \in \mathbb{R}^2$ where

$$Lu = a_{11}u_{xx} + a_{12}u_{xy} + a_{22}u_{yy} + a_1u_x + a_2u_y + a_0u \quad (\Omega), \tag{1.1.1}$$

with coefficients $a_{ij}, a_i, a_0$ and a right-hand side $f$ which, in general, may depend on $x, y, u, u_x, u_y$ (the quasilinear case). In most parts of this book, $Lu = f$ is assumed to be a linear differential equation, which means that the coefficients and the right-hand side $f$ only depend on $(x, y)$. $L$ is called

- elliptic if $4a_{11}a_{22} > a_{12}^2$,
- hyperbolic if $4a_{11}a_{22} < a_{12}^2$,
- parabolic if $4a_{11}a_{22} = a_{12}^2$.

In general, this classification depends on $(x, y)$ and, in the nonlinear case, also on the solution $u$. Prototypes of the above equation are

- the Poisson equation $-\Delta u = -u_{xx} - u_{yy} = f$,
- the wave equation $u_{xx} - u_{yy} = 0$,
- the heat equation $u_{xx} - u_y = 0$.

Since multigrid methods work excellently for nicely elliptic problems, most of our presentation in the first chapters is oriented to Poisson's and Poisson-like equations. Other important model equations that we will treat in this book include

- the anisotropic model equation $-\varepsilon u_{xx} - u_{yy} = f$,
- the convection–diffusion equation $-\varepsilon \Delta u + a_1 u_x + a_2 u_y = f$,
- the equation with mixed derivatives $-\Delta u + \tau u_{xy} = f$.

All these equations will serve as model equations for special features and complications and are thus representative of a larger class of problems with similar features. These model equations depend crucially on a parameter $\varepsilon$ or $\tau$. For certain parameter values we have a singular perturbation: the type of the equation changes and the solution behaves qualitatively different (if it exists at all). For instance, the anisotropic equation becomes parabolic for $\varepsilon \to 0$, the equation with mixed derivatives is elliptic for $|\tau| < 2$, parabolic for $|\tau| = 2$ and hyperbolic for $|\tau| > 2$. All the model equations represent classes of problems which are of practical relevance.

In this book, the applicability of multigrid is connected to a quantity, the "$h$-ellipticity measure $E_h$", that we will introduce in Section 4.7. This $h$-ellipticity measure is not applied to the differential operator itself, but to the corresponding discrete operator. It can be used to analyze whether or not the discretization is appropriate for a straightforward multigrid treatment. Nonelliptic problems can also have some $h$-ellipticity if discretized accordingly.

The above model equations, except the wave and the heat conduction equations, are typically connected with pure *boundary conditions*. The wave and the heat conduction equations are typically characterized by *initial conditions* with respect to one of the variables ($y$) and by boundary conditions with respect to the other ($x$).

We will call problems which are characterized by pure boundary conditions *space-type problems*. For problems with initial conditions, we interpret the variable for which the initial condition is stated as the *time variable $t$ ($= y$)*, and call these problems *time-type*. Usually these problems exhibit a marching or evolution behavior with respect to $t$. Space-type equations, on the other hand, usually describe stationary situations. Note that this distinction is different from the standard classification of elliptic, hyperbolic and parabolic. Elliptic problems are usually space-type while hyperbolic and parabolic problems are often time-type. However, the stationary supersonic full potential equation, the convection equation (see Chapter 7) and the stationary Euler equations (see Section 8.9) are examples of hyperbolic equations with space-type behavior. (In certain situations, the same equation can be interpreted as space- or as time-type, and each interpretation may have its specific meaning. An example is the convection equation.)

> In this book, we will present and discuss *multigrid methods for space-type problems.*

For time-type problems, there is usually the option to discretize the time variable explicitly, implicitly or in a hybrid manner (i.e. semi-implicitly). The implicit or semi-implicit discretization yields discrete *space-type* problems which have to be solved in each time step. We will briefly consider multigrid approaches for the solution of such space-type problems which arise in each time step in Section 2.8. Typically, these problems have similar but more convenient numerical properties than the corresponding stationary problems with respect to multigrid.

**Remark 1.1.1** Some authors have proposed multigrid approaches which include the time direction directly. These approaches consider time to be just another "space-type" direction. Such approaches are discussed in [78, 175, 198, 199, 396] for so-called parabolic multigrid and multigrid-like methods based on "waveform relaxation".                    ≫

## 1.2 GRIDS AND DISCRETIZATION APPROACHES

In this book, we assume that the differential equations to be solved are *discretized on a suitable grid* (or, synonymously, mesh). Here, we give a rough survey (with some examples) of those types of grids which are treated systematically in this book and those which are only touched on. We also make some remarks about discretization approaches.

### 1.2.1 Grids

The general remarks in this section may not be so interesting to multigrid beginners. They may start with Section 1.4 on Poisson's equation and return to this section later.

Most parts of this book refer to: *Cartesian grids, boundary-fitted logically rectangular grids* and *block-structured boundary-fitted grids.* Figure 1.1 is an example of a Cartesian grid. For simple domains with simple boundaries, Cartesian grids are numerically convenient. We will use them for Model Problem 1 (see Section 1.3.2) and several other cases.

Figure 1.1 also gives an example of a boundary-fitted grid. Boundary-fitted grids will be used in more advanced examples in this book. A systematic introduction into boundary-fitted grids is given in [391]. We will discuss the *generation* of boundary-fitted grids in Section 10.3.

In the context of boundary-fitted grids, there are two different approaches. In the first, coordinate transformations are used to obtain simple (for example rectangular) domains, and correspondingly simple (rectangular) grids. Here the differential (and/or the discrete) equations are transformed to the new curvilinear coordinates. In the second approach, the computations are performed in the physical domain with the original (nontransformed) equations. In this book we concentrate on the second approach.

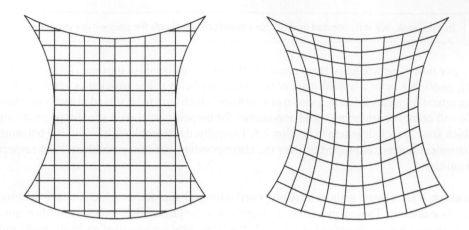

**Figure 1.1.** A square Cartesian grid (left) in a domain $\Omega$ and a boundary-fitted curvilinear (logically rectangular) grid (right).

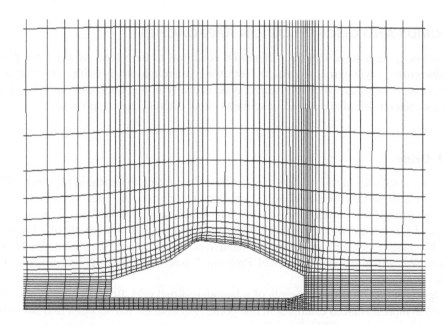

**Figure 1.2.** Boundary-fitted block-structured grid around a car.

Block-structured boundary-fitted grids are used if the given domain cannot (or cannot reasonably) be mapped to a rectangular domain, but can be decomposed into a finite number of subdomains each of which can be covered with a boundary-fitted grid. Quite complicated domains may be treated with this approach, as can be seen in Fig. 1.2. Block-structured boundary-fitted grids are discussed in Chapters 6 and 8–10.

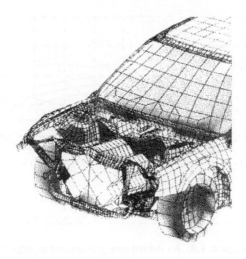

**Figure 1.3.** Unstructured grid around a car in a crash simulation application.

In many software packages, *unstructured, irregular grids* are used today. These grids have become quite popular because unstructured automatic mesh generation is much easier than the generation of block-structured grids for very complicated 2D and 3D domains.

An example of an unstructured grid is the grid around a car during a crash simulation. A part of this grid is presented in Fig. 1.3.

From the multigrid point of view, unstructured grids are a complication. For a given unstructured grid, it is usually not difficult to define a sequence of *finer* grids, but *it may be difficult to define a sequence of reasonable coarser grids.* (A hierarchy of coarse grids is needed for multigrid.)

The *algebraic multigrid* (AMG) method presented in Appendix A constructs a hierarchy of coarse grids automatically and is thus particularly well-suited for problems on unstructured grids.

Although unstructured grids are widely used, even complicated domains allow other than *purely* unstructured grid approaches. Often a *hybrid* approach, an unstructured grid close to the boundary and a structured grid in the interior part of the domain (or vice versa) is suitable for the treatment of such complicated domains.

More generally, all types of grids mentioned above can be used in the context of *overlapping grids*. A typical situation for the use of overlapping grids is when an overall Cartesian grid is combined with a local boundary-fitted grid (Chimera technique). An example of this approach is shown in Fig. 1.4. Such overlapping grids are also called composite grids.

Finally, we mention *self-adaptive grids*. They are constructed automatically during the solution process according to an error estimator that takes the behavior of the solution into account. Self-adaptive grids are very natural in the multigrid context. We will treat the self-adaptive multigrid approach systematically in Chapter 9. An example is given in Fig. 1.5.

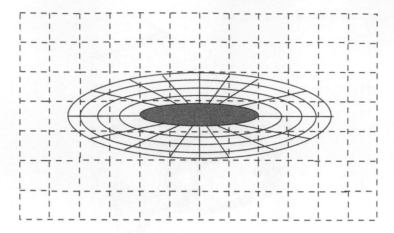

**Figure 1.4.** An overlapping grid around an object.

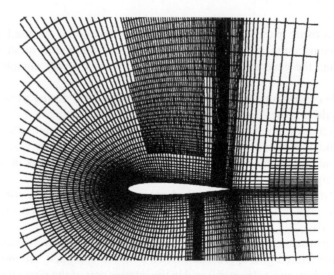

**Figure 1.5.** Adaptively refined grid around an airfoil.

## 1.2.2 Discretization Approaches

In principle, any type of grid can be used with any type of *discretization approach*. In practice, however, finite difference and finite volume methods are traditionally used in the context of Cartesian, logically rectangular and block-structured boundary-fitted grids, whereas finite elements and finite volumes are widely used in the context of unstructured grids.

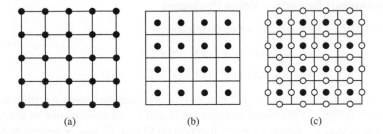

**Figure 1.6.** Three arrangements of unknowns in a Cartesian grid: (a) a vertex-centered grid; (b) a cell-centered grid; (c) a staggered grid.

In this book, we will focus on finite difference and finite volume discretizations on structured and block-structured grids.

Finally, another important choice to be made in discretization is the *arrangement of the unknowns within a grid*. The unknowns can be defined at the vertices of a grid (*vertex-centered location of unknowns*). Another option is the choice of the unknowns at cell centers (*cell-centered location of unknowns*).

For systems of PDEs it is possible to choose different locations for different types of unknowns. A well-known example is the *staggered grid* for the system of the incompressible Navier–Stokes equations discussed in Chapter 8, in which pressure unknowns are placed at the cell centers and velocity components at cell boundaries. Examples of a vertex-centered, a cell-centered and a staggered Cartesian grid are sketched in Fig. 1.6. Often, the discrete equations are defined at the same locations as the unknowns. It is hard to say which location of unknowns and which location of the discrete equations is best in general. Often these choices depend on the type of boundary conditions and on the application. In the following chapters we mainly present results for vertex-centered locations of unknowns. Multigrid components for cell-centered arrangements of unknowns are presented in Section 2.8.4.

## 1.3 SOME NOTATION

We start with some general notation. In this book, we use the classical formulation of differential equations with differential (and boundary) operators rather than a weak formulation. For discretization, we use the terminology of *discrete differential operators*. This also means that we use *grid functions* rather than vectors and *grid operators* rather than matrices. In our opinion, the grid-oriented notation emphasizes more clearly the correspondence between the discrete and continuous problems, and between the discrete formulations on different grids of the multigrid structure. In that respect, we regard it as a natural formulation for the multigrid approach. If, for example, the discrete differential operator can be described by a *difference stencil* (see Section 1.3.4) this is clearly a more condensed formulation than a matrix notation. On the other hand, there are situations in which the matrix notation has some advantages and is more general. Then we will not hesitate to use it.

### 1.3.1 Continuous Boundary Value Problems

We denote scalar linear boundary value problems by

$$
\begin{aligned}
L^\Omega u(x) &= f^\Omega(x) \quad (x \in \Omega) \\
L^\Gamma u(x) &= f^\Gamma(x) \quad (x \in \Gamma := \partial\Omega).
\end{aligned}
\tag{1.3.1}
$$

Here $x = (x_1, \ldots, x_d)^T$ and $\Omega \subset \mathbb{R}^d$ is a given open bounded domain with boundary $\Gamma$. $L^\Omega$ is a linear (elliptic) differential operator on $\Omega$ and $L^\Gamma$ represents one or several linear boundary operators. $f^\Omega$ denotes a given function on $\Omega$ and $f^\Gamma$ one or several functions on $\Gamma$. We always denote solutions of (1.3.1) by $u = u(x)$. For brevity, we also simply write $Lu = f$ instead of (1.3.1). Most concrete considerations refer to the cases $d = 2$ and $d = 3$. (Multigrid is usually not needed for $d = 1$, in which case it degenerates to direct or other simple well-known solvers, see Section 6.4.1.)

In the case $d = 2$ or $d = 3$, we will usually write $(x, y)$ instead of $(x_1, x_2)$ and $(x, y, z)$ instead of $(x_1, x_2, x_3)$.

This and the following chapters essentially refer to scalar equations. Chapter 8 and other parts of the book, however, will refer to systems of PDEs. The above notation is also used in that case. $\boldsymbol{L}^\Omega$ then stands for a vector of differential operators and $\boldsymbol{u}, \boldsymbol{f}$ etc. are vector-valued functions.

### 1.3.2 Discrete Boundary Value Problems

The following considerations are formulated for the 2D case. They can, of course, be directly generalized to higher dimensions. The discrete analog of (1.3.1) is denoted by

$$
\begin{aligned}
L_h^\Omega u_h(x, y) &= f_h^\Omega(x, y) \quad ((x, y) \in \Omega_h) \\
L_h^\Gamma u_h(x, y) &= f_h^\Gamma(x, y) \quad ((x, y) \in \Gamma_h).
\end{aligned}
\tag{1.3.2}
$$

Here $h$ is a (formal) discretization parameter. Using the infinite grid

$$
\boxed{\mathbf{G}_h := \{(x, y): x = x_i = i h_x, \ y = y_j = j h_y; \ i, j \in \mathbb{Z}\}}
\tag{1.3.3}
$$

where $\boldsymbol{h} = (h_x, h_y)$ is a vector of fixed mesh sizes, we define $\Omega_h = \Omega \cap \mathbf{G}_h$ and $\Gamma_h$ as the set of discrete intersection points of the "grid lines" with $\Gamma$. In the special case of square grids, we simply identify $h = h_x = h_y$.

The discrete solution $u_h$ is a function defined on the grid $\Omega_h \cup \Gamma_h$, i.e., a grid function, and $f_h^\Omega$ and $f_h^\Gamma$ are discrete analogs of $f^\Omega$ and $f^\Gamma$, respectively, where $f^\Omega, f^\Gamma$ are restricted to the grid. Instead of $u_h(x, y) = u_h(x_i, y_j) = u_h(i h_x, j h_y)$, we sometimes simply write $u_{i,j}$.

$L_h^\Omega$ and $L_h^\Gamma$ are grid operators, i.e., mappings between spaces of grid functions. ($L_h^\Omega$ is also called a *discrete* (differential) or *difference operator*, $L_h^\Gamma$ a *discrete boundary operator*.)

Clearly the concrete definitions of $\Omega_h$, $\Gamma_h$ etc. also depend on the given PDE, the domain $\Omega$, the boundary conditions, the grid approach and the discretization.

For ease of presentation, we will first assume that discrete boundary equations are eliminated from (1.3.2). (In general, a proper multigrid treatment of boundary conditions needs an explicit consideration of *noneliminated* boundary conditions. This is discussed in more detail in Section 5.6.)

In the case of second-order equations with Dirichlet boundary conditions, which means $u_h = f_h^\Gamma$, $(x, y) \in \Gamma_h$, eliminated boundary conditions can be considered as a natural approach. We then simply write

$$L_h u_h = f_h \quad (\Omega_h). \tag{1.3.4}$$

Here $u_h$ and $f_h$ are grid functions on $\Omega_h$ and $L_h$ is a linear operator

$$L_h : \mathcal{G}(\Omega_h) \to \mathcal{G}(\Omega_h), \tag{1.3.5}$$

where $\mathcal{G}(\Omega_h)$ denotes the *linear space of grid functions on $\Omega_h$*. Clearly, (1.3.4) can be represented as a system of linear algebraic equations. However, we usually consider it as one *grid equation*.

Even if the boundary conditions are not eliminated, we may use the notation (1.3.4) for the discrete problem. The notation then represents an abstract equation (in a finite dimensional space).

### 1.3.3 Inner Products and Norms

For convergence considerations and many other purposes, we need to use norms in the finite dimensional space $\mathcal{G}(\Omega_h)$. Most of our considerations (and many of our measurements) will be based on the *Euclidean inner product*

$$\langle u_h, v_h \rangle_2 := \frac{1}{\#\Omega_h} \sum_{x \in \Omega_h} u_h(x)\overline{v_h(x)}, \tag{1.3.6}$$

where $\#\Omega_h$ is the number of grid points of $\Omega_h$. The scaling factor $(\#\Omega_h)^{-1}$ allows us to compare grid functions on different grids and also the corresponding continuous functions on $\Omega$. The induced norm is $||u_h||_2 = \sqrt{\langle u_h, u_h \rangle_2}$. The corresponding operator norm for discrete operators $L_h$ on $\mathcal{G}(\Omega_h)$ is the *spectral* norm

$$||B_h||_S = \sqrt{\rho(B_h B_h^*)},$$

where $B_h$ denotes any linear operator $B_h : \mathcal{G}(\Omega_h) \to \mathcal{G}(\Omega_h)$, and where $\rho$ is the spectral radius.

For $L_h$, symmetric and positive definite, we also consider the *energy inner product*

$$\langle u_h, v_h \rangle_E := \langle L_h u_h, v_h \rangle_2 \tag{1.3.7}$$

and the corresponding operator norm $|| \cdot ||_E$, which is given by

$$||B_h||_E = ||L_h^{1/2} B_h L_h^{-1/2}||_S = \sqrt{\rho(L_h B_h L_h^{-1} B_h^*)}. \tag{1.3.8}$$

For practical purposes the *infinity norm*

$$||u_h||_\infty := \max\{|u_h(x)|: x \in \Omega_h\} \tag{1.3.9}$$

and the corresponding operator norm $|| \cdot ||_\infty$ are commonly used.

### 1.3.4 Stencil Notation

For the concrete definition of discrete operators $L_h^\Omega$ (on a Cartesian or a logically rectangular grid) the stencil terminology is convenient. We restrict ourselves to the case $d = 2$. The extension to $d = 3$ (and to general $d$) is straightforward (see Section 2.9). We first introduce the stencil notation for the infinite grid $\mathbf{G}_h$ (defined in Section 1.3.2). On $\mathbf{G}_h$, we consider grid functions

$$w_h : \mathbf{G}_h \longrightarrow \mathbb{R} \quad (\text{or } \mathbb{C})$$
$$(x, y) \longmapsto w_h(x, y).$$

A general stencil $[s_{\kappa_1 \kappa_2}]_h$

$$[s_{\kappa_1 \kappa_2}]_h = \begin{bmatrix} & \vdots & \vdots & \vdots & \\ \cdots & s_{-1,1} & s_{0,1} & s_{1,1} & \cdots \\ \cdots & s_{-1,0} & s_{0,0} & s_{1,0} & \cdots \\ \cdots & s_{-1,-1} & s_{0,-1} & s_{1,-1} & \cdots \\ & \vdots & \vdots & \vdots & \end{bmatrix}_h \quad (s_{\kappa_1 \kappa_2} \in \mathbb{R})$$

defines an operator on the set of grid functions by

$$\boxed{[s_{\kappa_1 \kappa_2}]_h w_h(x, y) = \sum_{(\kappa_1, \kappa_2)} s_{\kappa_1 \kappa_2} w_h(x + \kappa_1 h_x, y + \kappa_2 h_y)}. \qquad (1.3.10)$$

Here we assume that only a *finite number of coefficients* $s_{\kappa_1 \kappa_2}$ *are nonzero.*

Many of the stencils we will consider will be *five-point* or *compact nine-point* stencils

$$\begin{bmatrix} & s_{0,1} & \\ s_{-1,0} & s_{0,0} & s_{1,0} \\ & s_{0,-1} & \end{bmatrix}_h, \quad \begin{bmatrix} s_{-1,1} & s_{0,1} & s_{1,1} \\ s_{-1,0} & s_{0,0} & s_{1,0} \\ s_{-1,-1} & s_{0,-1} & s_{1,-1} \end{bmatrix}_h. \qquad (1.3.11)$$

The discrete operators $L_h^\Omega$ are usually given only on a finite grid $\Omega_h$. For the identification of a discrete operator $L_h^\Omega$ with "its" stencil $[s_{\kappa_1 \kappa_2}]_h$, we usually have to restrict the stencil to $\Omega_h$ (instead of $\mathbf{G}_h$). Near boundary points the stencils may have to be modified. In square or rectangular domains $\Omega$, which are the basis for the examples in Chapter 2, this modification of $[s_{\kappa_1 \kappa_2}]_h$ to $\Omega_h$ is straightforward (see, e.g., Example 1.4.1 below). Let us finally mention that the coefficients $s_{\kappa_1 \kappa_2}$ will depend on $(x, y)$:

$$s_{\kappa_1 \kappa_2} = s_{\kappa_1 \kappa_2}(x, y)$$

if the coefficients of $L^\Omega$ and/or $L_h^\Omega$ depend on $(x, y)$.

## 1.4 POISSON'S EQUATION AND MODEL PROBLEM 1

In this section, we introduce Model Problem 1, the classical model for a discrete elliptic boundary value problem. Every numerical solver has been applied to this problem for comparison.

**Model Problem 1** *At many places in this book, we will study in detail the discrete Poisson equation with Dirichlet boundary conditions*

$$-\Delta_h u_h(x, y) = f_h^\Omega(x, y) \quad ((x, y) \in \Omega_h)$$
$$u_h(x, y) = f_h^\Gamma(x, y) \quad ((x, y) \in \Gamma_h = \partial\Omega_h)$$

(1.4.1)

*in the unit square $\Omega = (0, 1)^2 \subset \mathbb{R}^2$ with $h = 1/n, n \in \mathbb{N}$. Here, $L_h = -\Delta_h$ is the standard five-point $O(h^2)$ approximation (explained below) of the partial differential operator L,*

$$Lu = -\Delta u = -u_{xx} - u_{yy}$$

(1.4.2)

*on the square grid $\mathbf{G}_h$.*

$O(h^2)$ here means that one can derive consistency relations of the form

$$Lu - L_h u = O(h^2) \quad \text{for } h \to 0$$

for sufficiently smooth functions $u (u \in C_4(\bar{\Omega})$, for example).

To illustrate exactly what the elimination of boundary conditions means, we consider the following example.

**Example 1.4.1** The discrete Poisson equation with eliminated Dirichlet boundary conditions can formally be written in the form (1.3.4). For $(x, y) \in \Omega_h$ not adjacent to a boundary this means:

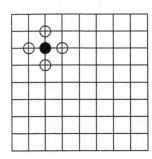

$$f_h(x, y) = f^\Omega(x, y)$$
$$L_h u_h(x, y) = -\Delta_h u_h(x, y)$$
$$= \frac{1}{h^2}[4u_h(x, y) - u_h(x - h, y) - u_h(x + h, y)$$
$$- u_h(x, y - h) - u_h(x, y + h)]$$
$$= \frac{1}{h^2}\begin{bmatrix} & -1 & \\ -1 & 4 & -1 \\ & -1 & \end{bmatrix}_h u_h(x, y)$$

The notation

$$-\Delta_h = \frac{1}{h^2}\begin{bmatrix} & -1 & \\ -1 & 4 & -1 \\ & -1 & \end{bmatrix}_h$$

is a first example of the stencil notation (1.3.11).

For $(x, y) \in \Omega_h$ adjacent to a (here: west) boundary, $L_h$ (1.3.4) reads

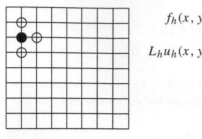

$$f_h(x, y) = f^\Omega(x, y) + \frac{1}{h^2} f^\Gamma(x - h, y)$$

$$L_h u_h(x, y) = \frac{1}{h^2}[4u_h(x, y) - u_h(x + h, y)$$

$$- u_h(x, y - h) - u_h(x, y + h)]$$

$$= \frac{1}{h^2} \begin{bmatrix} & -1 & \\ 0 & 4 & -1 \\ & -1 & \end{bmatrix}_h u_h(x, y).$$

For $(x, y) \in \Omega_h$ in a (here: the north-west) corner we have

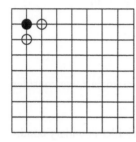

$$f_h(x, y) = f^\Omega(x, y) + \frac{1}{h^2}[f^\Gamma(x - h, y) + f^\Gamma(x, y + h)]$$

$$L_h u_h(x, y) = \frac{1}{h^2}[4u_h(x, y) - u_h(x + h, y) - u_h(x, y - h)]$$

$$= \frac{1}{h^2} \begin{bmatrix} & 0 & \\ 0 & 4 & -1 \\ & -1 & \end{bmatrix}_h u_h(x, y). \qquad \triangle$$

In this example, elimination of the boundary conditions is simple. Often, elimination of boundary conditions may be complicated and is not preferred. In such cases, a particular treatment of the boundary conditions in the multigrid process may be needed (see Section 5.6).

### 1.4.1 Matrix Terminology

Discrete operators $L_h$ are often represented by *matrices* $A_h$. Each matrix row then represents connections of one unknown in the discretization of a PDE to its neighbor unknowns. Which of the matrix entries is different from 0, depends on the ordering of grid points, i.e., the ordering of the components of the vector of unknowns.

As an example we consider Model Problem 1. For a column- or row-wise ordering of grid points (also called *lexicographical* ordering, see Fig. 1.7(a), starting with points at the left lower corner) and eliminated Dirichlet boundary conditions, the resulting matrix is a block tridiagonal matrix with a regular sparsity pattern:

$$A_h = \frac{1}{h^2} \begin{pmatrix} T & -I & & \\ -I & T & -I & \\ & -I & T & -I \\ & & -I & T \end{pmatrix}, \qquad (1.4.3)$$

| 13 | 14 | 15 | 16 |  | 15 | 7 | 16 | 8 |
|----|----|----|----|--|----|---|----|---|
| 9 | 10 | 11 | 12 |  | 5 | 13 | 6 | 14 |
| 5 | 6 | 7 | 8 |  | 11 | 3 | 12 | 4 |
| 1 | 2 | 3 | 4 |  | 1 | 9 | 2 | 10 |
| (a) | | | | | (b) | | | |

**Figure 1.7.** (a) Lexicographic; (b) red–black ordering of grid points.

where

$$
I = \begin{pmatrix} 1 & & & \\ & 1 & & \\ & & 1 & \\ & & & 1 \end{pmatrix} \quad \text{and} \quad T = \begin{pmatrix} 4 & -1 & & \\ -1 & 4 & -1 & \\ & -1 & 4 & -1 \\ & & -1 & 4 \end{pmatrix}.
$$

Due to the elimination of Dirichlet boundary points, every matrix row corresponding to a grid point near a boundary has only three or two entries connecting them to neighbor grid points. This can be seen, for example, in the first rows of the matrix, where only two or three $-1$ entries can be found instead of four for interior points.

The dependence of the matrix structure on the ordering of the grid points can be seen when we write the matrix down for a *red–black ordering* of the grid points (see Fig. 1.7(b)). First all unknowns at odd (red) grid points are considered, then the unknowns at the even (black) points. The corresponding matrix $A_h$ is now a block matrix with blocks $A_{rr}$, representing the connections of the red grid points to red grid points, $A_{rb}$ the connections of the red points to the black points, $A_{br}$ the connections of black points to red points, and $A_{bb}$ the connections of black points to black points. So:

$$
A_h = \begin{bmatrix} A_{rr} & A_{rb} \\ A_{br} & A_{bb} \end{bmatrix}. \tag{1.4.4}
$$

For Model Problem 1, the blocks $A_{rr}$ and $A_{bb}$ are diagonal matrices with $4/h^2$ as the diagonal elements. The resulting block matrix $A_{rb}$ ($= A_{br}^T$) of the above example in Fig. 1.7(b) is

$$
A_{rb} = \frac{1}{h^2}
\begin{pmatrix}
-1 & 0 & -1 & & & & & \\
-1 & -1 & 0 & -1 & & & & \\
-1 & 0 & -1 & -1 & -1 & & & \\
 & -1 & 0 & -1 & 0 & -1 & & \\
 & & -1 & 0 & -1 & 0 & -1 & \\
 & & & -1 & -1 & -1 & 0 & -1 \\
 & & & & -1 & 0 & -1 & -1 \\
 & & & & & -1 & 0 & -1
\end{pmatrix}.
\tag{1.4.5}
$$

**Remark 1.4.1** As we will see below (Section 1.5), in a multigrid algorithm it is usually not necessary to build up the matrix $A_h$ coming from the discretization. The multigrid components are based on "local" operations; multiplications and additions are carried out grid point by grid point. The storage that is needed in a multigrid code mainly consists of solution vectors, defects and right-hand sides on all grid levels.　　　　　　　　≫

### 1.4.2 Poisson Solvers

Table 1.1 gives an overview on the complexity of different solution methods (including fast Poisson solvers) applied to Model Problem 1. Here direct and iterative solvers are listed. For the iterative solvers, we assume an accuracy (stopping criterion) in the range of the discretization accuracy. This is reflected by the $\log \varepsilon$ term. The full multigrid (FMG) variant of multigrid which we will introduce in Section 2.6 is a solver up to discretization accuracy.

It is generally expected that the more general a solution method is, the less efficient it is and vice versa. Multigrid is, however, a counter example for this pattern—indeed multigrid

**Table 1.1.** Complexity of different solvers for the 2D Poisson problem ($N$ denotes the total number of unknowns).

| Method | # operations in 2D |
|---|---|
| Gaussian elimination (band version) | $O(N^2)$ |
| Jacobi iteration | $O(N^2 \log \varepsilon)$ |
| Gauss–Seidel iteration | $O(N^2 \log \varepsilon)$ |
| Successive overrelaxation (SOR) [431] | $O(N^{3/2} \log \varepsilon)$ |
| Conjugate gradient (CG) [194] | $O(N^{3/2} \log \varepsilon)$ |
| Nested dissection (see, for example, [9]) | $O(N^{3/2})$ |
| ICCG [264] | $O(N^{5/4} \log \varepsilon)$ |
| ADI (see, for example, [403]) | $O(N \log N \log \varepsilon)$ |
| Fast Fourier transform (FFT) [112] | $O(N \log N)$ |
| Buneman [93] | $O(N \log N)$ |
| Total reduction [342] | $O(N)$ |
| Multigrid (iterative) | $O(N \log \varepsilon)$ |
| Multigrid (FMG) | $O(N)$ |

will turn out to be a general principle. On the other hand, Table 1.1 shows that the most efficient version of multigrid yields an algorithm that is at least as efficient as the highly efficient and highly specialized fast Poisson solvers. Here, the total reduction method and the Buneman algorithm are typical fast Poisson solvers, very efficient but essentially designed and tailored exclusively for elliptic *model* problems.

## 1.5 A FIRST GLANCE AT MULTIGRID

### 1.5.1 The Two Ingredients of Multigrid

In this section, we introduce the multigrid idea heuristically for Model Problem 1. We use the grid function oriented notation introduced in Section 1.3.2.

The iteration formula of the classical *lexicographical* Gauss–Seidel method (GS-LEX) for Poisson's equation reads

$$u_h^{m+1}(x_i, y_j) = \tfrac{1}{4}[h^2 f_h(x_i, y_j) + u_h^{m+1}(x_i - h, y_j) + u_h^m(x_i + h, y_j)$$
$$+ u_h^{m+1}(x_i, y_j - h) + u_h^m(x_i, y_j + h)], \tag{1.5.1}$$

where $(x_i, y_j) \in \Omega_h$. Here $u_h^m$ and $u_h^{m+1}$ are the approximations of $u_h(x_i, y_j)$ before and after an iteration, respectively.

If we apply (1.5.1) to Poisson's equation, we recognize the following phenomenon. After a few iteration steps, the *error* of the approximation becomes *smooth*. It doesn't necessarily become small, but it does become smooth. See Fig. 1.8 for an illustration of this error smoothing effect. Looking at the error

$$v_h^m(x_i, y_j) = u_h(x_i, y_j) - u_h^m(x_i, y_j),$$

Formula (1.5.1) means

$$v_h^{m+1}(x_i, y_j) = \tfrac{1}{4}[v_h^{m+1}(x_i - h, y_j) + v_h^m(x_i + h, y_j)$$
$$+ v_h^{m+1}(x_i, y_j - h) + v_h^m(x_i, y_j + h)]. \tag{1.5.2}$$

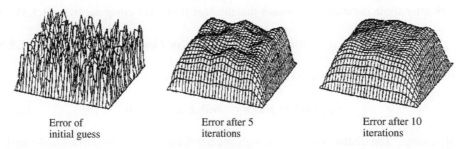

Error of
initial guess

Error after 5
iterations

Error after 10
iterations

**Figure 1.8.** Influence of lexicographic Gauss–Seidel iteration (1.5.1) on the error (Model Problem 1).

Obviously, the iteration formula can be interpreted as an error averaging process. Error smoothing is one of the two basic principles of the multigrid approach.

> **Smoothing principle**  Many classical iterative methods (Gauss–Seidel etc.) if appropriately applied to discrete elliptic problems have a strong smoothing effect on the error of any approximation.

The other basic principle is the following: a quantity that is smooth on a certain grid can, without any essential loss of information, also be approximated on a coarser grid, say a grid with double the mesh size. In other words: if we are sure that the error of our approximation has become smooth after some iteration steps, we may approximate this error by a suitable procedure on a (much) coarser grid.

Qualitatively, this is the coarse grid approximation principle.

> **Coarse grid principle**  A smooth error term is well approximated on a coarse grid. A coarse grid procedure is substantially less expensive (substantially fewer grid points) than a fine grid procedure.

As this principle holds for error or "correction" quantities, we also speak of the *coarse grid correction* (CGC) principle.

Let us illustrate these considerations and explain them heuristically by looking at the Fourier expansion of the error. In our model problem the error $v_h = v_h^m(x, y)$ considered as a function of the discrete variables $x$ and $y$ can be written as

$$v_h(x, y) = \sum_{k,\ell=1}^{n-1} \alpha_{k,\ell} \sin k\pi x \sin \ell\pi y. \tag{1.5.3}$$

For $(x, y) \in \Omega_h$, the functions

$$\varphi_h^{k,\ell}(x, y) = \sin k\pi x \sin \ell\pi y \quad (k, \ell = 1, \ldots, n-1) \tag{1.5.4}$$

are the discrete eigenfunctions of the discrete operator $\Delta_h$. The fact that this error becomes smooth after some iteration steps means that the *high frequency components* in (1.5.3), i.e.

$$\alpha_{k,\ell} \sin k\pi x \sin \ell\pi y \quad \text{with } k \text{ or } l \text{ large} \tag{1.5.5}$$

become small after a few iterations whereas the *low frequency components*, i.e.

$$\alpha_{k,\ell} \sin k\pi x \sin \ell\pi y \quad \text{with } k \text{ and } l \text{ small} \tag{1.5.6}$$

hardly change. The distinction between high and low frequencies is important in the multigrid context. In the following subsection, we will give a first definition of high and low frequencies and show how this concept is related to the *coarse* grid under consideration.

### 1.5.2 High and Low Frequencies, and Coarse Meshes

We again consider Model Problem 1 on a grid $\Omega_h$ with mesh size $h = 1/n$. Additionally, we consider Model Problem 1 on a coarser grid $\Omega_H$ with mesh size $H > h$. Assuming that $n$ is an even number, we may, for instance, choose $H = 2h$, which is a very natural choice in the multigrid context. This choice of the coarse grid is, therefore, called *standard coarsening*.

For the definition of the high and low frequencies, we return to the eigenfunctions $\varphi^{k,\ell} = \varphi_h^{k,\ell}$ in (1.5.4). For given $(k, \ell)$, we consider the (four) eigenfunctions

$$\varphi^{k,\ell}, \quad \varphi^{n-k,n-\ell}, \quad \varphi^{n-k,\ell}, \quad \varphi^{k,n-\ell}$$

and observe that they coincide on $\Omega_{2h}$ in the following sense:

$$\varphi^{k,\ell}(x, y) = -\varphi^{n-k,\ell}(x, y) = -\varphi^{k,n-\ell}(x, y) = \varphi^{n-k,n-\ell}(x, y) \quad \text{for } (x, y) \in \Omega_{2h}.$$

This means that these four eigenfunctions cannot be distinguished on $\Omega_{2h}$. (For $k$ or $l = n/2$, the $\varphi^{k,\ell}$ vanish on $\Omega_{2h}$.) This gives rise to the following definition of low and high frequencies:

---

**Definition** (in the context of Model Problem 1)  For $k, \ell \in \{1, \ldots, n - 1\}$, we denote $\varphi^{k,\ell}$ to be an eigenfunction (or a *component*) of

$$\text{low frequency} \quad \text{if } \max(k, \ell) < n/2,$$
$$\text{high frequency} \quad \text{if } n/2 \leq \max(k, \ell) < n.$$

---

Obviously, only the low frequencies *are visible* on $\Omega_{2h}$ since all high frequencies coincide with a low frequency on $\Omega_{2h}$ (or vanish on $\Omega_{2h}$). The fact that high frequencies coincide with low ones is also called *aliasing* of frequencies. For the 1D case with $n = 8$, we illustrate the above definition in Fig. 1.9. We summarize this consideration:

---

The low frequency components also represent meaningful grid functions on a grid $\Omega_{2h}$ with double the mesh size, whereas the high frequency components do not. Their high frequencies are not "visible" on the $\Omega_{2h}$ grid.

---

If we apply this distinction of low and high frequencies to the representation of $v_h(x, y)$ in (1.5.3), we can decompose the sum in (1.5.3) into corresponding partial sums:

$$\sum_{k,\ell=1}^{n-1} \alpha_{k,\ell}\varphi^{k,\ell} = \sum\nolimits^{\text{high}} \alpha_{k,\ell}\varphi^{k,\ell} + \sum\nolimits^{\text{low}} \alpha_{k,\ell}\varphi^{k,\ell}$$

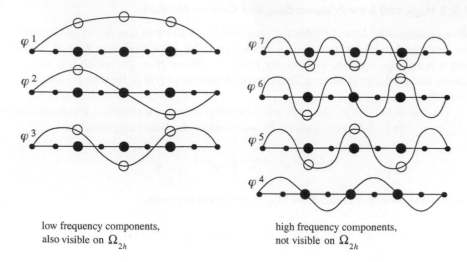

low frequency components,
also visible on $\Omega_{2h}$

high frequency components,
not visible on $\Omega_{2h}$

**Figure 1.9.** Low and high frequency components for a 1D example ($n = 8$).

where

$$\sum{}^{\text{low}} \alpha_{k,\ell}\varphi^{k,\ell} = \sum_{k,\ell=1}^{n/2-1} \alpha_{k,\ell}\varphi^{k,\ell}$$

and

$$\sum{}^{\text{high}} \alpha_{k,\ell}\varphi^{k,\ell} = \sum_{\substack{k,\ell \\ n/2 \leq \max(k,\ell)}}^{n-1} \alpha_{k,\ell}\varphi^{k,\ell}.$$

**Remark 1.5.1 ($H = 4h$ and other choices of $H$)** From our definition, it immediately becomes clear that the terms "high frequency" and "low frequency" are related to both the fine grid $\Omega_h$ *and* the coarse grid $\Omega_H$ that we consider. (If we want to emphasize this dependence on the grids $\Omega_h$ and $\Omega_H$, we will speak of $(h, H)$-low and $(h, H)$-high frequencies.) If, for example, we would choose

$$H = 4h$$

(assuming that $n$ is a multiple of 4), our definition of high and low frequencies would have to be modified in the following way:

$\varphi^{k,l}$ is a $(h, 4h)$-low frequency component    if $\max(k, l) < n/4$.

$\varphi^{k,l}$ is a $(h, 4h)$-high frequency component    otherwise.

The choice $H = 4h$ is not very practical in the multigrid context since it usually does not lead to the most efficient algorithms, but it is not out of the question, either.

Other, more practical choices of coarse grids, different from standard coarsening, are introduced in Section 2.3.1.                                              ≫

### 1.5.3 From Two Grids to Multigrid

In the following, we will continue our heuristic considerations a little further extending them from two-grid levels $\Omega_h$, $\Omega_{2h}$ to a sequence of levels. In this setting, we will also be able to explain the *multigrid* (not only the two-grid) idea. Fig. 1.10 shows such a sequence of grids.

We assume additionally that $n$ is a power of 2 meaning that $h = 2^{-p}$. Then we can form the grid sequence

$$\Omega_h, \Omega_{2h}, \Omega_{4h}, \ldots, \Omega_{h_0} \tag{1.5.7}$$

just by doubling the mesh size successively. We assume that this sequence ends with a coarsest grid $\Omega_{h_0}$ (which may well be the grid consisting of only one interior grid point $(x, y) = (1/2, 1/2)$, i.e. $h_0 = 1/2$, see Fig. 1.10).

We now consider a decomposition of the error representation (1.5.3) into partial sums which correspond to the grid sequence (1.5.7), having the following idea in mind. In the same way as we have distinguished low and high frequencies with respect to the pair of grids $\Omega_h$ and $\Omega_{2h}$ in the previous section, we now make an additional distinction between high and low frequencies with respect to the pair $\Omega_{2h}$ and $\Omega_{4h}$. We continue with the pair $\Omega_{4h}$ and $\Omega_{8h}$ and so on.

As discussed above, by using Gauss–Seidel type iterations on the original $\Omega_h$ grid, we cause the $(h, 2h)$-high frequency error components to become small rapidly. The remaining low frequency error components are visible and can thus be approximated on $\Omega_{2h}$. (Of course, an equation which determines the low frequency error has to be defined in a suitable way on $\Omega_{2h}$.) Performing Gauss–Seidel-type iteration not only on the original $\Omega_h$ grid, but also on $\Omega_{2h}$ and correspondingly on $\Omega_{4h}$ etc., causes the $(2h, 4h)$-high frequency, the $(4h, 8h)$-high frequency components etc. to decrease rapidly. Only on the coarsest grid might it be necessary to do something special, which is trivial if $\Omega_{h_0}$ consists of only one (or few) point(s). Altogether, this leads to a very fast overall reduction of the error.

**Summary** What we have described and explained so far is the basic idea of multigrid. Our description is not at all an algorithm. It is not even a clear verbal description of an algorithmic principle. For example, we have neither said precisely what a Gauss–Seidel iteration on $\Omega_{2h}$, $\Omega_{4h}$ etc. means algorithmically and to which grid functions it is to be applied, nor how to go from one level to the other etc. However, the flavor of multigrid is in it.

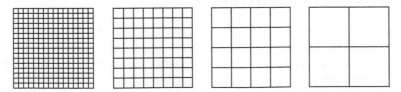

**Figure 1.10.** A sequence of coarse grids starting with $h = 1/16$.

> **Flavor of multigrid**　Gauss–Seidel iteration (or more generally, an appropriate iterative scheme) on different grid levels gives rapid reduction of the corresponding high frequency components and as this process covers all frequencies, a rapid reduction of the overall error can be achieved.

### 1.5.4 Multigrid Features

The multigrid idea is very fundamental and can also be applied in other contexts. Accordingly, there are also several different ways to view multigrid. In order to clarify our understanding of multigrid, we want to briefly discuss different multigrid aspects and to point out those multigrid features which we regard as the most significant ones.

**Multigrid as an iterative linear solver**　The most basic way to view multigrid is to consider it as an iterative linear solver for a discrete elliptic boundary value problem. Here we assume the problem, the grid and the discretization to be given and fixed. A characteristic feature of the iterative multigrid approach is that the multigrid convergence speed is independent of the discretization mesh size $h$ and that the number of arithmetic operations per iteration step is proportional to the number of grid points. The multigrid principle allows us to construct very efficient linear solvers, and, in that respect, the iterative approach is important and fundamental. This approach is described in detail in Chapter 2 of this book. On the other hand, this view is somewhat restricted and does not exploit the full potential of the multigrid idea. For example, even the direct application of multigrid to *nonlinear problems* (discussed in detail in Section 5.3) is not covered by this approach.

**Multigrid as a solver for the differential problem**　From a more sophisticated point of view, multigrid can be regarded as a solution method for the (continuous) differential problem. In this view it is not appropriate to separate the discretization and the solution of the discrete problem, but rather to regard both processes as interdependent: the solution process can, according to this view, be performed the more efficiently the more the continuous differential background is exploited.

One simple, but very natural way of looking at multigrid as a "differential solver" is represented by the *full multigrid method* (FMG, see Section 2.6). Here, the method is oriented to minimizing the differential error rather than to minimizing the algebraic error (corresponding to the linear system). Self-adaptive grid refinements and related approaches, which are very natural in the multigrid context, also belong to the "differential view" of multigrid (and will be presented in Chapter 9).

**Efficiency**　Multigrid methods are highly efficient. Efficiency here relates to both a theoretical feature and a practical one. The theoretical efficiency is more precisely expressed by the term "optimality" in a complexity theory sense. If interpreted appropriately, (full) multigrid methods are typically optimal in the sense that the number of arithmetic operations needed to solve a (discrete) problem is proportional to the number $N$ of unknowns in the problem considered.

Efficiency in the practical sense means that the proportionality constants in this $O(N)$ statement are small or moderate. This is indeed the case for multigrid: if designed well, the $h$-independent convergence factors can be made very small (in the range 0.1–0.2 or even less) and the operation count per unknown per iteration step is also small. If a concrete multigrid algorithm has this property, we speak of the *typical multigrid efficiency* (sometimes also called top multigrid efficiency).

**Generality**  The second striking property of multigrid is its generality. Multigrid methods are as efficient as the so-called fast elliptic solvers [95, 342] but are less limited in their range of application. Multigrid methods can be applied with full efficiency

- to general elliptic equations with variable coefficients,
- in general domains,
- for general boundary conditions,
- in 2D, 3D and higher dimensions (trivially also for 1D problems),
- to scalar equations and to systems of equations.

Very importantly, multigrid can also be applied *directly*, i.e. without global linearization, to *nonlinear elliptic problems*.

Furthermore, multigrid is not restricted to a certain discretization approach. In principle, multigrid can be used in connection with any type of grid-based discretization: finite differences, finite volumes and finite elements. (Collocation, variational, spectral or particle-based discretization methods can also be combined with the multigrid principle. In this book, however, we will concentrate on grid-type approaches.)

**Optimality versus robustness**  Multigrid methods are characterized by their so-called *components*. We will introduce them systematically in Sections 2.1–2.4. The components are the *smoothing procedure*, the *coarsening strategy*, the *coarse grid operators*, the *transfer operators from fine grids to coarse and from coarse to fine and the cycle type*. These components have to be specified for each concrete problem. Although it is well known how to choose suitable multigrid components for large classes of problems, it may be (very) difficult to define the right or reasonable ones in complicated new applications. This is still an "art" requiring theoretical insight, experience and numerical experiments. There are two trends with respect to the choice of multigrid components.

In *optimized multigrid* algorithms, one tries to tailor the components to the problem at hand in order to obtain the highest possible efficiency for the solution process. This optimized approach clearly makes sense if a very large scale problem has to be (repeatedly) solved or if a smaller core problem needs to be solved many times every day, like, for example, the 3D Helmholtz equation in daily weather prediction [251].

On the other hand, the idea of *robust multigrid* algorithms is to choose the components independently of the given problem, uniformly for as large a class of problems as possible. The robust approach is often used in software packages where the highest efficiency for a single problem is not so important. The AMG method in Appendix A is an example of a robust multigrid method.

Optimization of multigrid is in many cases limited by practical circumstances. For instance, important large software packages are based on certain grid structures, discretization techniques and solution methods, which are not oriented to multigrid requirements. In such situations, the users and developers of the software packages may be very interested in *accelerating* the program by introducing some multigrid features. It may, however, be impossible or too costly to change the data structures in such a way that optimal multigrid features can be integrated. From a puristic multigrid point of view, such an accelerating approach may be very unsatisfactory because much higher efficiency is achievable in principle. From a practical point of view, this approach can, however, be a good compromise. A simple modification of the program may give a significant reduction in computing time.

**Adaptivity**    Defining a global grid for the discretization of a given problem independently of the solution process is often insufficient. Only during the solution process may certain features of the solution, like shocks, singularities, oscillations, turbulent behavior and the like be recognized. In such cases, the local discretization error is of different size in different parts of the domain and therefore it would be natural to adapt the grids (and perhaps also the discretization) to the behavior of the solution. This is one of the main reasons for using adaptive grids that are dynamically constructed within the solution process. For 3D problems, this is particularly necessary since a grid that is globally as fine as is needed at some crucial parts of the domain is not affordable.

For these reasons, adaptivity of grids is one of the major trends in today's numerical simulation and scientific computing.

Adaptivity can be combined with the multigrid principle in a very natural way. In the adaptive multigrid process finer and finer grids are not constructed globally. They are only constructed in those parts of the domain where the current discretization error is significantly large. Essentially, all other multigrid components are maintained as usual. What is specifically needed for this approach, are criteria to measure or estimate the current local discretization error. Adaptive multigrid will be discussed in Chapter 9.

**Parallel features**    A promising and challenging trend in numerical simulation and scientific computing is the use of *parallelism* in numerical algorithms. The background to this trend is the fact that most high performance computers are now parallel systems.

In order to solve any problem on a parallel computer, an algorithm with an appropriate degree of parallelism has to be made available. In addition, the data being processed in the algorithm should be suitably organized. This is particularly important if the parallel computer has *distributed memory*. In this case, the "communication overhead" has to be limited, i.e. the internal transfer of data should cost only a (small) fraction of the overall effort for solving the given problem.

In Chapter 6, we deal mainly with the practical aspects of *parallel multigrid*, focusing on the *grid partitioning approach*, in which the given grid is partitioned into a number of subgrids. Each processor works on "its" subgrid but has to communicate with the other processors. In this approach all multigrid components that have to be specified should be parallel, or as parallel as possible, but at the same time as efficient as possible.

**1D problems**    The multigrid idea leads to optimal algorithms for many PDEs. Since ordinary differential equations (ODEs) are a special case of PDEs, the question arises whether multigrid methods are also useful for ODEs. Indeed, the multigrid principle can also be applied to ODEs and, of course, leads to optimal (i.e. $O(N)$) algorithms in 1D. However, in 1D other optimal methods are available and multigrid methods typically coincide with (and degenerate to) well-known optimal solvers. In that respect, the multigrid principle is also applicable to the 1D case, but not really needed there, at least not for standard problems.

For example, if linear ordinary boundary value problems are discretized with standard discretization methods (for example, finite differences) band matrices are obtained, where the bandwidth is independent of the mesh size $h$. In particular, if a three-point discretization is used for the differential operator, typically the discrete problem is characterized by a linear system *with a tridiagonal* $(N \times N)$ *matrix*. Such tridiagonal systems, and more general band matrices with a fixed small bandwidth, can be solved in $O(N)$ operations by Gaussian elimination-type methods.

### 1.5.5 Multigrid History

The forerunners of multigrid are the ideas of *nested iteration, error smoothing by relaxation* and *total reduction*. Nested iteration (see also Section 2.6) has been used for a long time to obtain first approximations (initial guesses) on fine grids from coarse grids, for instance in the context of Newton's method. Also, the fact that in many cases relaxation processes have an error smoothing property has been known for a long time [367, 368, 374]. The total reduction method by Schröder and Trottenberg [342–344] has the complete multigrid structure. The main difference from standard multigrid is, however, that on the coarse grids equations are constructed, which are equivalent to the fine grid equations for the respective unknowns; smoothing plays no role then.

The first studies investigating *multigrid methods in a strict sense* were those by Fedorenko [139, 140] (1962, 1964) and that of Bakhvalov [16] (1966). While Fedorenko [140] restricted the convergence investigation to a discrete boundary value problem of second order with variable coefficients (in the unit square), Bakhvalov also indicated the possibility of combining multigrid methods with nested iteration.

Though the studies published by Fedorenko and Bakhavalov had shown the asymptotic optimality of the multigrid approach (and to a certain extent its generality), their *actual efficiency* was first recognized by Brandt in the early 1970s. Studying adaptive grid refinements and their relation to fast solvers, Brandt had been led to the papers of Fedorenko and Bakhvalov by Widlund. In his first two papers [56, 57] (1973, 1976) and then summarized in the systematic 1977 work [58], Brandt showed the actual efficiency of multigrid methods. His essential contributions (in the early studies) include the introduction of a nonlinear multigrid method (FAS) and of adaptive techniques (MLAT), the discussion of general domains, the systematic application of the nested iteration idea (FMG) and, last but not least, the provision of the local Fourier analysis tool for theoretical investigation and method design.

The following papers, which we would like to mention as being "historically" relevant contributions, are representative of the early development of multigrid.

In 1971 Astrakhantsev [6] generalized Bakhvalov's convergence result to general boundary conditions. Like Bakhvalov he used a variational formulation in his theoretical approach.
After a first study of multigrid methods for Poisson's equation in a square, Nicolaides [278]
(1975) discussed multigrid ideas in connection with *finite element* discretizations systematically in [279] (1977). In 1975, Frederickson [142] introduced an approximate multigrid-like
solver which can be regarded as a forerunner of the MGR methods [141, 319].

In 1975–1976, Hackbusch developed the fundamental elements of multigrid methods
anew. It was again Widlund who informed Hackbusch about the studies which were already
available. Hackbusch's first systematic report [169] (1976) contained many theoretical and
practical investigations which have been taken up and developed further by several authors.
So one finds Fourier analysis considerations, the use of "red–black" and "four-colour"
relaxation methods for smoothing, the treatment of nonrectangular domains and of nonlinear problems etc. Hackbusch then presented a general convergence theory of multigrid
methods [170–172].

Since the early 1980s, the field has been exploding and many researchers have con-
tributed to the field. Two series of conferences dedicated to multigrid were set up. The
European Multigrid Conferences (EMG) have been held at Cologne (1981, 1985), Bonn
(1991), Amsterdam (1993), Stuttgart (1996) and Ghent (1999). And in the US the Copper
Mountain Conferences on multigrid have been held bi-annually since 1983. Proceedings
of the European meetings have appeared in [129, 174, 177, 178, 181, 189] and of the
Copper Mountain Conferences in special issues of journals: *Applied Numerical Mathematics* (Vol. 13, 1983; Vol. 19, 1986), *Communications in Applied Numerical Methods* (Vol. 8,
1992), *SIAM Journal of Numerical Analysis* (Vol. 30, 1993), *Electronic Transactions on
Numerical Analysis* (Vol. 6, 1996). Another rich source of information on multigrid is the
MGNet website maintained by C.C. Douglas: `http://www.mgnet.org`. This website includes a large multigrid bibliography with more than 3000 entries. Some multigrid
textbooks and monographies are [54, 66, 91, 176, 206, 262, 332, 351, 378, 415].

## 1.6 INTERMEZZO: SOME BASIC FACTS AND METHODS

This section contains some general numerical material which is needed at several places
of the book. One reason for presenting this material here is to clarify terminology. The
reader who has a general idea of the material presented here may have a quick look at it and
return to it for details later. Those readers for whom the material is new may read it now or
postpone its study. (We will refer these readers back to these sections later.)

### 1.6.1 Iterative Solvers, Splittings and Preconditioners

Since the facts listed in this section are valid for general matrices and are not restricted to
discrete differential operators, we use linear algebra notation in this section, i.e. matrices
$A_h$ or $A$ instead of discrete operators $L_h$.

Let

$$Au = f \qquad\qquad (1.6.1)$$

be a linear system with an invertible matrix $A$. The simplest iterative scheme for this equation is the *Richardson iteration*

$$u^{m+1} = u^m + \tau(f - Au^m) = (I - \tau A)u^m + \tau f \quad (m = 0, 1, \dots) \tag{1.6.2}$$

with some acceleration parameter $\tau \neq 0$.

A more *general iteration* is

$$u^{m+1} = Mu^m + s \quad (m = 0, 1, \dots). \tag{1.6.3}$$

Here, $M$ is the *iteration matrix*. We assume that the original equation $Au = f$ is equivalent to the fixed point equation $u = Mu + s$. For Richardson's iteration, we have $M = I - \tau A, s = \tau f$.

The convergence (and the asymptotic convergence speed) of Richardson's iteration and of the general iteration are characterized by the spectral radii $\rho(I - \tau A)$ and $\rho(M)$, respectively. The spectral radius of a matrix $M$ is defined as

$$\rho(M) = \max\{|\lambda|: \lambda \text{ eigenvalue of } M\}. \tag{1.6.4}$$

The spectral radius is the asymptotic convergence factor of the iteration. Asymptotically (i.e., for $m \to \infty$) we have $||u - u^{m+1}|| \leq \rho(M)||u - u^m||$.

There are many ways to specify $M$ leading to different iterative solvers. (In Chapter 2 we will specify $M$ in such a way that standard multigrid methods are obtained.) Here we present three different but equivalent ways to formulate or to introduce the iteration (1.6.3). These three approaches differ only in their motivation (their point of view), not mathematically. All three points of view will be used at different places in this book.

### Approximate solution of the defect equation

If $u^m$ is any approximation of $u$ and

$$d^m = f - Au^m \tag{1.6.5}$$

is its *defect*, then the defect equation $Av^m = d^m$ is equivalent to the original equation. By solving for the *correction* $v^m$, we obtain the solution $u = u^m + v^m$. However, if we use an *approximation* $\hat{A}$ *of* $A$, such that

$$\hat{A}\hat{v}^m = d^m \tag{1.6.6}$$

can be solved more easily, we obtain an iterative process of the form

$$d^m = f - Au^m, \quad \hat{A}\hat{v}^m = d^m, \quad u^{m+1} = u^m + \hat{v}^m \quad (m = 0, 1, 2 \dots). \tag{1.6.7}$$

This process is obviously equivalent to (1.6.3) where

$$M = I - (\hat{A})^{-1}A.$$

Vice versa, if $M$ is given, this yields an approximation $\hat{A}$ of $A$ according to

$$\hat{A} = A(I - M)^{-1}.$$

## Splitting

An equivalent way of constructing $M$ is to start with a *splitting*

$$A = \hat{A} - R$$

and to use the iteration

$$\hat{A}u^{m+1} = Ru^m + f. \tag{1.6.8}$$

Here

$$M = (\hat{A})^{-1}R = I - (\hat{A})^{-1}A.$$

## Preconditioning

A third, also equivalent, approach is based on the idea of *preconditioning*. Here the original equation $Au = f$ is replaced by an equivalent equation

$$CAu = Cf \tag{1.6.9}$$

where $C$ is an invertible matrix. $C$ is called a *(left) preconditioner* of $A$. The identification with the above terminology is by

$$(\hat{A})^{-1} = C.$$

In other words, the inverse $(\hat{A})^{-1}$ of any (invertible) approximation $\hat{A}$ is a left preconditioner and vice versa.

Furthermore, we see that Richardson's iteration for the preconditioned system (1.6.9) (with $\tau = 1$)

$$u^{m+1} = u^m + C(f - Au^m) = (I - CA)u^m + Cf \tag{1.6.10}$$

is equivalent to the general iteration (1.6.3) with $M = I - CA$. This also means that any iteration of the general form (1.6.3) is a Richardson iteration (with $\tau = 1$) for the system that is obtained by preconditioning the original system (1.6.1).

**Remark 1.6.1** One speaks of a *right preconditioner* $C$, if the original equation is replaced by

$$ACz = f, \qquad u = Cz. \tag{1.6.11}$$

Richardson's method based on right preconditioning (with $\tau = 1$) would result in

$$z^{m+1} = (I - AC)z^m + f. \tag{1.6.12}$$

Since

$$(I - AC) = A(I - CA)A^{-1},$$

we have

$$\rho(I - AC) = \rho(I - CA). \qquad\qquad \gg$$

So far, we have stated relations between the defect interpretation, the splitting and the preconditioning approaches. We would like to add some remarks, which further motivate the term *preconditioning*. The idea behind preconditioning is that the condition of the system (1.6.1), measured by the condition number

$$\kappa(A) = ||A|| \ ||A^{-1}|| \tag{1.6.13}$$

(in some appropriate norm), is to be improved by multiplying $A$ by $C$, in the form (1.6.9) or (1.6.11). The condition number, on the other hand, is relevant for the convergence speed of certain iterative approaches, for instance Richardson's iteration and conjugate gradient-type methods. We summarize some well-known facts here. For that purpose, we assume that the matrix $A$ is *symmetric and positive definite* (s.p.d.) with maximum and minimum eigenvalues $\lambda_{max}, \lambda_{min} > 0$, respectively. We consider the Euclidean norm in $\mathbb{R}^N$ and the corresponding spectral matrix norm $||A||_S = \lambda_{max}$.

**Remark 1.6.2**   If $A$ is s.p.d., the Richardson iteration converges for

$$0 < \tau < 2||A||_S^{-1}.$$

Then the optimal $\tau$ (for which the spectral radius $\rho(I - \tau A)$ becomes minimal) is $\tau_{opt} = 2/(\lambda_{max} + \lambda_{min})$, and one can prove

$$\rho(I - \tau_{opt}A) = ||I - \tau_{opt}A||_S = \frac{\lambda_{max} - \lambda_{min}}{\lambda_{max} + \lambda_{min}} = \frac{\kappa_S(A) - 1}{\kappa_S(A) + 1}, \tag{1.6.14}$$

where $\kappa_S(A)$ is the spectral condition number of $A$ (for the proof, see for example [180]).
If we use a left preconditioner $C$ which is also s.p.d., we obtain

$$\frac{\kappa_S(CA) - 1}{\kappa_S(CA) + 1} \tag{1.6.15}$$

instead of (1.6.14).                                                                                           ≫

**Remark 1.6.3**   The term preconditioning is commonly used in the context of conjugate gradient type or, more generally, Krylov subspace methods [159, 337]. Here the spectral condition number $\kappa_S = \kappa_S(A)$ of $A$ also plays a role in error estimates for conjugate gradient iterants. Actually, under the same assumptions as above, instead of (1.6.14) and (1.6.15) the improved convergence factors

$$\frac{\sqrt{\kappa_S(A)} - 1}{\sqrt{\kappa_S(A)} + 1} \quad \text{and} \quad \frac{\sqrt{\kappa_S(CA)} - 1}{\sqrt{\kappa_S(CA)} + 1}, \tag{1.6.16}$$

respectively, appear in the corresponding estimates (see [180] for a proof and details).   ≫

# 2

# BASIC MULTIGRID I

The multigrid idea is based on two principles: error smoothing and coarse grid correction. In this chapter, we will explain how these principles are combined to form a multigrid algorithm. Basic multigrid will be described systematically.

In Section 2.1, we discuss the smoothing properties of classical iterative solvers. Sections 2.2, 2.4 and 2.6 give a systematic introduction to two-grid iteration, multigrid iteration and the full multigrid method, respectively. Some standard multigrid components are described in Section 2.3.

We prefer a presentation of the two-grid cycle in Section 2.2, which starts with the idea of an approximate solution of the defect equation and then brings together smoothing and coarse grid correction. The methods described in Sections 2.2–2.4 and 2.6 are general, although all concrete examples refer to Poisson's equation. Concrete fast multigrid Poisson solvers for the 2D and 3D cases are presented in Sections 2.5 and 2.9, respectively.

Some straightforward generalizations of the 2D method are discussed in Section 2.8. In Section 2.7, we resume the discussion on transfer operators and focus on some practical aspects.

## 2.1 ERROR SMOOTHING PROCEDURES

We have observed in Section 1.5 for Model Problem 1 that the usual Gauss–Seidel iteration has a remarkable smoothing effect on the error $v_h^m$ of an approximation $u_h^m$. As this property is fundamental for the multigrid idea, we discuss smoothing procedures in more detail here.

We will, in particular, consider two classical iteration methods: Gauss–Seidel-type and Jacobi-type iterations. We will see that these methods are suitable for error smoothing. The smoothing properties will, however, turn out to be dependent on the right choice of *relaxation parameters* and, in the case of the Gauss–Seidel iteration, also on the *ordering of grid points*.

All iterative methods which we discuss in this chapter, are *pointwise* iterations, line- or block-type iterations are *not* yet considered here. We start our discussion with Jacobi-type iterations since the analysis is particularly easy and illustrative.

> In general, however, appropriate Gauss–Seidel-type iterations turn out to be better
> smoothers than appropriate Jacobi-type iterations.

In the following we will speak of Jacobi- and Gauss–Seidel-type *relaxation* methods
rather than *iteration* methods.

> **Relaxation methods**   Classical iteration methods such as Gauss–Seidel-type and
> Jacobi-type iterations are often called *relaxation* methods (or smoothing methods
> or smoothers) if they are used for the purpose of error smoothing.

### 2.1.1 Jacobi-type Iteration (Relaxation)

For Model Problem 1, the iteration formula of the Jacobi iteration reads

$$z_h^{m+1}(x_i, y_j) = \tfrac{1}{4}\big[h^2 f_h(x_i, y_j) + u_h^m(x_i - h, y_j) + u_h^m(x_i + h, y_j)$$
$$+ u_h^m(x_i, y_j - h) + u_h^m(x_i, y_j + h)\big] \tag{2.1.1}$$
$$u_h^{m+1} = z_h^{m+1},$$

(with $(x_i, y_j) \in \Omega_h$), where $u_h^m$ denotes the old approximation and $u_h^{m+1}$ the new approximation of the iteration. The Jacobi iteration can be written as

$$u_h^{m+1} = S_h u_h^m + \frac{h^2}{4} f_h$$

with the iteration operator

$$S_h = I_h - \frac{h^2}{4} L_h$$

(where $I_h$ is the identity operator). We can generalize this iteration by introducing a relaxation parameter $\omega$

$$u_h^{m+1} = u_h^m + \omega(z_h^{m+1} - u_h^m), \tag{2.1.2}$$

which is called the $\omega$-(damped) Jacobi relaxation ($\omega$-JAC). Obviously, $\omega$-JAC and Jacobi
iteration coincide for $\omega = 1$. The iteration operator for $\omega$-JAC reads

$$S_h(\omega) = I_h - \frac{\omega h^2}{4} L_h = \frac{\omega}{4} \begin{bmatrix} & 1 & \\ 1 & 4(1/\omega - 1) & 1 \\ & 1 & \end{bmatrix}_h. \tag{2.1.3}$$

The *convergence properties* of $\omega$-JAC can be easily analyzed by considering the eigenfunctions of $S_h$, which are the same as those of $L_h$, namely

$$\varphi_h^{k,\ell}(x) = \sin k\pi x \sin \ell\pi y \quad ((x, y) \in \Omega_h; (k, \ell = 1, \dots, n-1)). \tag{2.1.4}$$

The corresponding eigenvalues of $S_h$ are

$$\chi_h^{k,\ell} = \chi_h^{k,\ell}(\omega) = 1 - \frac{\omega}{2}(2 - \cos k\pi h - \cos \ell\pi h). \tag{2.1.5}$$

For the spectral radius $\rho(S_h) = \max\{|\chi_h^{k,\ell}|: (k, \ell = 1, \dots, n-1)\}$, we obtain

$$\begin{aligned}
&\text{for } 0 < \omega \le 1: \ \rho(S_h) = |\chi_h^{1,1}| = |1 - \omega(1 - \cos \pi h)| = 1 - O(\omega h^2) \\
&\text{else}: \ \rho(S_h) \ge 1 \quad \text{(for } h \text{ small enough)}.
\end{aligned} \tag{2.1.6}$$

In particular, when regarding the (unsatisfactory) asymptotic convergence, there is no use in introducing the relaxation parameter: $\omega = 1$ is the best choice.

### 2.1.2 Smoothing Properties of $\omega$-Jacobi Relaxation

The situation is different with respect to the *smoothing properties* of $\omega$-Jacobi relaxation. In order to achieve reasonable smoothing, we have to introduce a parameter $\omega \ne 1$.

For $0 < \omega \le 1$, we first observe from (2.1.6) that the smoothest eigenfunction $\varphi_h^{1,1}$ is responsible for the slow convergence of Jacobi's method. Highly oscillating eigenfunctions are reduced much faster *if $\omega$ is chosen properly*. To see this, we consider the approximations before ($w_h$) and after ($\bar{w}_h$) one relaxation step and expand the errors before ($v_h$) and after ($\bar{v}_h$) the relaxation step, namely

$$v_h := u_h - w_h \quad \text{and} \quad \bar{v}_h := u_h - \bar{w}_h,$$

into discrete eigenfunction series

$$v_h = \sum_{k,\ell=1}^{n-1} \alpha_{k,\ell}\varphi_h^{k,\ell}, \qquad \bar{v}_h = \sum_{k,\ell=1}^{n-1} \chi_h^{k,\ell}\alpha_{k,\ell}\varphi_h^{k,\ell}. \tag{2.1.7}$$

As discussed in Section 1.5, in order to analyze the smoothing properties of $S_h(\omega)$ we distinguish between *low* and *high* frequencies (with respect to the coarser grid $\Omega_{2h}$ used).

In order to measure the smoothing properties of $S_h(\omega)$ quantitatively, we introduce the *smoothing factor* of $S_h(\omega)$ as follows:

---

**Definition**     Smoothing factor (of $\omega$-JAC for Model Problem 1)

The *smoothing factor* $\mu(h; \omega)$ of $S_h(\omega)$, representing the worst factor by which *high frequency* error components are reduced per relaxation step, and its supremum $\mu^*$ over $h$, are defined as

$$\begin{aligned}
\mu(h; \omega) &:= \max\{|\chi_h^{k,\ell}(\omega)|: n/2 \le \max(k, \ell) \le n-1\}, \\
\mu^*(\omega) &:= \sup_{h \in \mathcal{H}} \mu(h; \omega).
\end{aligned} \tag{2.1.8}$$

---

From here on, $\mathcal{H}$ denotes the set of admissible (or reasonable) mesh sizes. Below we will see, for example, that for $\Omega = (0, 1)^2$ the coarsest grid on which smoothing is applied is characterized by $h = 1/4$. In this case we would then define $\mathcal{H} = \{h = 1/n : n \in \mathbb{N}, n \geq 4\}$.

Inserting (2.1.5), we obtain from (2.1.8)

$$\mu(h; \omega) = \max\left\{\left|1 - \frac{\omega}{2}(2 - \cos k\pi h - \cos \ell\pi h)\right| : n/2 \leq \max(k, \ell) \leq n - 1\right\}$$

$$\mu^*(\omega) = \max\{|1 - \omega/2|, |1 - 2\omega|\}.$$

$$(2.1.9)$$

This shows that Jacobi's relaxation has no smoothing properties for $\omega \leq 0$ or $\omega > 1$

$$\mu(h; \omega) \geq 1 \quad \text{if } \omega \leq 0 \quad \text{or } \omega > 1 \quad \text{(and } h \text{ sufficiently small).}$$

For $0 < \omega < 1$, however, the smoothing factor is smaller than 1 and bounded away from 1, independently of $h$. For $\omega = 1$, we have a smoothing factor of $1 - O(h^2)$ only. In particular, we find from (2.1.9) that

$$\mu(h; \omega) = \begin{cases} \cos \pi h & \text{if } \omega = 1 \\ (2 + \cos \pi h)/4 & \text{if } \omega = 1/2 \\ (1 + 2\cos \pi h)/5 & \text{if } \omega = 4/5 \end{cases} \qquad \mu^*(\omega) = \begin{cases} 1 & \text{if } \omega = 1 \\ 3/4 & \text{if } \omega = 1/2 \\ 3/5 & \text{if } \omega = 4/5. \end{cases}$$

$$(2.1.10)$$

The choice $\omega = 4/5$ is optimal in the following sense:

$$\inf \{\mu^*(\omega) : 0 \leq \omega \leq 1\} = \mu^*(4/5) = 3/5.$$

With respect to $\mu(h; \omega)$, one obtains

$$\inf \{\mu(h; \omega) : 0 \leq \omega \leq 1\} = \mu\left(h; \frac{4}{4 + \cos \pi h}\right) = \frac{3\cos \pi h}{4 + \cos \pi h} = \frac{3}{5} - |O(h^2)|.$$

This means that one step of $\omega$-JAC with $\omega = 4/5$ reduces all high frequency error components by at least a factor of $3/5$ (independent of the grid size $h$).

The above consideration is a first example of what we call *smoothing analysis*.

### 2.1.3 Gauss–Seidel-type Iteration (Relaxation)

In Section 1.5.1 we introduced Gauss–Seidel iteration with a lexicographic ordering of the grid points. A different ordering is the so-called red–black ordering (see Fig. 1.7). If we use this red–black ordering for Gauss–Seidel iteration, we obtain the Gauss–Seidel red–black (GS-RB) method.

**Remark 2.1.1** The red–black ordering of grid points is also called odd–even ordering. This notation has the advantage that the two types of grid points are more clearly addressed (a grid point $(x_i, y_j)$ is odd/even if $i + j$ is odd/even) than with the term red–black. Since red–black is more often used in the literature, we will stay with red–black.                    ≫

The significance of using a relaxation parameter $\omega$ in Gauss–Seidel iteration is well known in the classical Gauss–Seidel *convergence* theory. For Model Problem 1, lexicographic Gauss–Seidel with a relaxation parameter is described by

$$z_h^{m+1}(x_i, y_j) = \tfrac{1}{4}\big[h^2 f_h(x_i, y_j) + u_h^{m+1}(x_i - h, y_j) + u_h^m(x_i + h, y_j)$$
$$+ u_h^{m+1}(x_i, y_j - h) + u_h^m(x_i, y_j + h)\big] \tag{2.1.11}$$
$$u_h^{m+1} = u_h^m + \omega(z_h^{m+1} - u_h^m).$$

The parameter $\omega$ not only enters explicitly in (2.1.11), but also implicitly in the "new values" $u_h^{m+1}(x_i - h, y_j)$ and $u_h^{m+1}(x_i, y_j - h)$. We will call this algorithm $\omega$-GS-LEX in the following. The corresponding algorithm with red–black ordering of the grid points and relaxation parameter $\omega$ is called $\omega$-GS-RB in this book.

We recall that, for Model Problem 1, the *convergence* of Gauss–Seidel iteration can be substantially improved by an overrelaxation parameter $\omega^*$. With

$$\omega^* = \frac{2}{1 + \sqrt{1 - \rho(\mathrm{JAC})^2}} = \frac{2}{1 + \sin \pi h}$$

we obtain

$$\rho(\omega^*\text{-GS}) = \omega^* - 1 = \frac{1 - \sin \pi h}{1 + \sin \pi h} = 1 - O(h)$$

instead of

$$\rho(\mathrm{GS}) = 1 - O(h^2) \quad (\text{for } \omega = 1).$$

This is the classical result on *successive overrelaxation* (SOR) [431]. Note that this result for Gauss–Seidel iteration is independent of the ordering of grid points.

Gauss–Seidel-type methods represent a particularly important class of smoothers. In the multigrid context the *smoothing properties* of Gauss–Seidel are much more important than the convergence properties. We will, however, not analyze the smoothing properties of Gauss–Seidel-type relaxations, here. Since, different from the Jacobi situation, the eigenfunctions of $L_h$ (2.1.4) are *not* eigenfunctions of the Gauss–Seidel operator, we need different tools for this analysis, which will be discussed in detail in Chapter 4. Here, we summarize the results of the smoothing analysis for Gauss–Seidel-type relaxations. For Model Problem 1, we obtain the smoothing factors

$$\mu(\text{GS-LEX}) = 0.50 \quad (\text{for } \omega = 1),$$
$$\mu(\text{GS-RB}) = 0.25 \quad (\text{for } \omega = 1).$$

(The factor of 0.25 for GS-RB is valid if only one or two smoothing steps are performed.) This result shows that the ordering of the grid points has an essential influence on the smoothing properties in the case of Gauss–Seidel relaxations. On the other hand, for Model Problem 1, the introduction of a relaxation parameter does *not* improve the smoothing properties of GS-LEX relaxation essentially (see Section 4.3).

The situation is somewhat different for GS-RB, for which an overrelaxation parameter can improve the smoothing properties (see Section 2.9 and [427, 428]).

### 2.1.4 Parallel Properties of Smoothers

Here, we compare the smoothing properties and the parallel features of $\omega$-JAC, GS-LEX and GS-RB. First, $\omega$-JAC is *"fully $\Omega_h$ parallel"*. By this we mean that the $\omega$-Jacobi operator (2.1.3) can be applied to *all* grid points $\Omega_h$ simultaneously; the new values do not depend on each other. We also say that the degree of parallelism (par-deg) is

$$\text{par-deg}(\omega\text{-JAC}) = \#\Omega_h.$$

If we use GS-LEX instead, we have dependences since we want to use the most recent values of $u_h$ wherever possible. Grid points lying on a diagonal in $\Omega_h$ (see Fig. 2.1) are independent of each other for five-point discretizations and can be treated in parallel. The degree of parallelism here clearly varies from one diagonal to the next and is restricted by

$$\text{par-deg}(\text{GS-LEX}) \leq (\#\Omega_h)^{1/2}.$$

In case of GS-RB each step of the Gauss–Seidel relaxation consists of two half-steps. In the first half-step, all red grid points ($\circ$) are treated simultaneously and independently (see Fig. 2.2). In the second half-step, all black grid points ($\bullet$) are treated, using the updated values in the red points. The degree of parallelism is

$$\text{par-deg}(\text{GS-RB}) = \tfrac{1}{2}\#\Omega_h.$$

Table 2.1 summarizes the properties of these relaxation schemes for Model Problem 1: $\omega$-JAC is fully parallel, but unfortunately not a really good smoother (not even with an

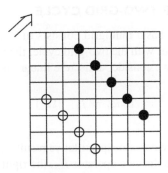

**Figure 2.1.** Diagonal grid points such as the $\bullet$ (or the $\circ$) can be treated in parallel in GS-LEX. Going from one diagonal to the next ($\nearrow$) is sequential.

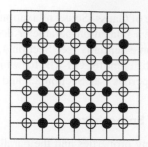

**Figure 2.2.** Red–black distribution of grid points in $\Omega_h$.

**Table 2.1.** Smoothing factors for various relaxation schemes. The smoothing factors marked * will be obtained from the analysis in Sections 4.3 and 4.5; the factor marked † is only valid if at most one or two smoothing steps are performed. (Here, $N$ denotes the number of grid points #$\Omega_h$ corresponding to the unknown grid values of $u_h$.)

| Relaxation | Smoothing factor | Smoothing | | Parallel degree |
|---|---|---|---|---|
| $\omega$-JAC, $\omega = 1$ | 1 | No | $N$ | Full |
| $\omega$-JAC, $\omega = 0.5$ | 0.75 | Unsatisfactory | $N$ | Full |
| $\omega$-JAC, $\omega = 0.8$ | 0.6 | Acceptable | $N$ | Full |
| GS-LEX, $\omega = 1$ | 0.5* | Good | $\leq \sqrt{N}$ | Square root |
| GS-RB, $\omega = 1$ | 0.25*† | Very good | $\frac{1}{2}N$ | Half |

optimized parameter $\omega$) whereas GS-LEX has reasonable smoothing properties but is not satisfactorily parallel. However, GS-RB is both a very good smoother (much better than $\omega$-JAC) and highly parallel.

## 2.2 INTRODUCING THE TWO-GRID CYCLE

As stated in Section 1.5, the basic multigrid consists of two ingredients: smoothing and coarse grid correction. We start with the two-grid cycle, the natural basis for any multigrid algorithm. For this purpose, we consider a discrete linear elliptic boundary value problem of the form

$$L_h u_h = f_h \quad (\Omega_h) \tag{2.2.1}$$

and assume that $L_h^{-1}$ exists.

As we are not going to give concrete quantitative results in this section, we do not make precise assumptions about the discrete operator $L_h$, the right-hand side $f_h$ or the grid $\Omega_h$. For a simple but characteristic example, one may always think of Model Problem 1 or some more general Poisson-like equation.

### 2.2.1 Iteration by Approximate Solution of the Defect Equation

One way of introducing the two-grid idea is to start from a general iteration based on an approximate solution of the defect equation.

For any *approximation* $u_h^m$ of the solution $u_h$ of (2.2.1), we denote the *error* by

$$v_h^m := u_h - u_h^m \tag{2.2.2}$$

and the *defect* (or *residual*) by

$$\boxed{d_h^m := f_h - L_h u_h^m} . \tag{2.2.3}$$

Trivially, the *defect equation*

$$L_h v_h^m = d_h^m \tag{2.2.4}$$

is equivalent to the original equation (2.2.1) since

$$u_h = u_h^m + v_h^m . \tag{2.2.5}$$

We describe these steps by the following procedural formulation:

$$u_h^m \longrightarrow d_h^m = f_h - L_h u_h^m \longrightarrow L_h v_h^m = d_h^m \longrightarrow u_h = u_h^m + v_h^m . \tag{2.2.6}$$

This procedure, however, is not a meaningful numerical process. However, if $L_h$ is approximated here by any "simpler" operator $\hat{L}_h$ such that $\hat{L}_h^{-1}$ exists, the solution $\hat{v}_h^m$ of

$$\hat{L}_h \hat{v}_h^m = d_h^m \tag{2.2.7}$$

gives a new approximation

$$u_h^{m+1} := u_h^m + \hat{v}_h^m . \tag{2.2.8}$$

The procedural formulation then looks like

$$\boxed{u_h^m \longrightarrow d_h^m = f_h - L_h u_h^m \longrightarrow \hat{L}_h \hat{v}_h^m = d_h^m \longrightarrow u_h^{m+1} = u_h^m + \hat{v}_h^m} . \tag{2.2.9}$$

Starting with some $u_h^0$, the successive application of this process defines an *iterative procedure*. The *iteration operator* of this method is given by

$$M_h = I_h - C_h L_h : \mathcal{G}(\Omega_h) \to \mathcal{G}(\Omega_h), \tag{2.2.10}$$

where $C_h := (\hat{L}_h)^{-1}$ and $I_h$ denotes the identity on $\mathcal{G}(\Omega_h)$. We have

$$u_h^{m+1} = M_h u_h^m + s_h \quad \text{with } s_h = (\hat{L}_h)^{-1} f_h \ (m = 0, 1, 2, \dots). \tag{2.2.11}$$

For the errors, it follows that

$$v_h^{m+1} = M_h v_h^m = (I_h - C_h L_h) v_h^m \quad (m = 0, 1, 2, \dots) \tag{2.2.12}$$

and for the defects that

$$d_h^{m+1} = L_h M_h L_h^{-1} d_h^m = (I_h - L_h C_h) d_h^m \quad (m = 0, 1, 2, \dots). \tag{2.2.13}$$

If we start the iteration with $u^0 = 0$, then we can represent $u_h^m$ $(m = 1, 2, \dots)$ as

$$\begin{aligned} u_h^m &= (I_h + M_h + M_h^2 + \cdots + M_h^{m-1})(\hat{L}_h)^{-1} f_h \\ &= (I_h - M_h^m)(I_h - M_h)^{-1}(\hat{L}_h)^{-1} f_h \\ &= (I_h - M_h^m) L_h^{-1} f_h. \end{aligned} \tag{2.2.14}$$

For general remarks on iterations such as (2.2.11), we refer to the discussion in Section 1.6.1. The asymptotic convergence properties of the above iterative process are characterized by the *spectral radius* (*asymptotic convergence factor*) of the iteration operator, i.e.

$$\rho(M_h) = \rho(I_h - C_h L_h) = \rho(I_h - L_h C_h). \tag{2.2.15}$$

If some norm $\| \cdot \|$ on $\mathcal{G}(\Omega_h)$ is introduced,

$$\| I_h - C_h L_h \|, \qquad \| I_h - L_h C_h \| \tag{2.2.16}$$

give upper bounds for the *error reduction factor* and the *defect reduction factor*, respectively, for one iteration.

**Remark 2.2.1**   Classical iterative linear solvers such as Jacobi or Gauss–Seidel iterations if applied to (2.2.1) can be interpreted as (iterated) approximate solvers for the defect equation. For $\omega$-JAC we have, for example,

$$\hat{L}_h = \frac{1}{\omega} D_h$$

where $D_h$ is the "diagonal" part of the matrix corresponding to $L_h$. Similarly, GS-LEX is obtained by setting $\hat{L}_h$ to be the "lower triangular" part of the matrix corresponding to $L_h$ including its diagonal part.                                                                 ≫

**Remark 2.2.2**   More generally, any of the classical iterative linear solvers of the form (2.2.11) can be interpreted as iterated approximate solvers for the defect equation if

$$C_h := (I_h - M_h) L_h^{-1}$$

is invertible. For then we can set $\hat{L}_h := C_h^{-1}$.                                                 ≫

### 2.2.2 Coarse Grid Correction

One idea to approximately solve the defect equation is to use an appropriate approximation $L_H$ of $L_h$ on a coarser grid $\Omega_H$, for instance the grid with mesh size $H = 2h$. This means that the defect equation (2.2.4) is replaced by

$$L_H \hat{v}_H^m = d_H^m. \tag{2.2.17}$$

Here we assume

$$L_H : \mathcal{G}(\Omega_H) \to \mathcal{G}(\Omega_H), \qquad \dim \mathcal{G}(\Omega_H) < \dim \mathcal{G}(\Omega_h) \tag{2.2.18}$$

and $L_H^{-1}$ to exist. As $d_H^m$ and $\hat{v}_H^m$ are grid functions on the coarser grid $\Omega_H$, we assume two (linear) transfer operators

$$I_h^H : \mathcal{G}(\Omega_h) \to \mathcal{G}(\Omega_H), \qquad I_H^h : \mathcal{G}(\Omega_H) \to \mathcal{G}(\Omega_h) \tag{2.2.19}$$

to be given. $I_h^H$ is used to *restrict* $d_h^m$ to $\Omega_H$:

$$\boxed{d_H^m := I_h^H d_h^m} \tag{2.2.20}$$

and $I_H^h$ is used to *interpolate* (or *prolongate*) the correction $\hat{v}_H^m$ to $\Omega_h$:

$$\boxed{\hat{v}_h^m := I_H^h \hat{v}_H^m}. \tag{2.2.21}$$

The simplest example for a restriction operator is the "injection" operator

$$d_H(P) = I_h^H d_h(P) := d_h(P) \quad \text{for } P \in \Omega_H \subset \Omega_h, \tag{2.2.22}$$

which identifies grid functions at coarse grid points with the corresponding grid functions at fine grid points. A fine and a coarse grid with the injection operator are presented in Fig. 2.3.

Altogether, one coarse grid correction step (calculating $u_h^{m+1}$ from $u_h^m$) proceeds as follows.

---

**Coarse grid correction** $\quad u_h^m \to u_h^{m+1}$

- –. Compute the defect $\qquad\qquad\qquad\qquad\qquad d_h^m = f_h - L_h u_h^m$
- – Restrict the defect (fine-to-coarse transfer) $\quad d_H^m = I_h^H d_h^m$
- – Solve on $\Omega_H$ $\qquad\qquad\qquad\qquad\qquad\quad L_H \hat{v}_H^m = d_H^m$
- – Interpolate the correction (coarse-to-fine transfer) $\quad \hat{v}_h^m = I_H^h \hat{v}_H^m$
- – Compute a new approximation $\qquad\qquad\quad u_h^{m+1} = u_h^m + \hat{v}_h^m$

---

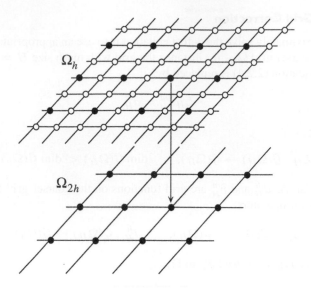

**Figure 2.3.** A fine and a coarse grid with the injection operator.

The associated iteration operator is given by

$$I_h - C_h L_h \quad \text{with } C_h = I_H^h L_H^{-1} I_h^H.$$  (2.2.23)

However:

---

**Remark 2.2.3** Taken on its own, the coarse grid correction process is of no use. It is not convergent! We have

$$\rho\left(I_h - I_H^h \, L_H^{-1} I_h^H L_h\right) \geq 1.$$  (2.2.24)

$$\gg$$

---

This remark follows directly from the fact that $I_h^H$ maps $\mathcal{G}(\Omega_h)$ into the lower dimensional space $\mathcal{G}(\Omega_H)$ and therefore $C_h = I_H^h L_H^{-1} I_h^H$ is not invertible. This implies that

$$C_h L_h v_h = 0 \quad \text{for certain } v_h \neq 0.$$

**Example 2.2.1** It may be illustrative to see what $C_h L_h v_h = 0$ means in practice. For the simple injection operator (2.2.22), for example, any error function $v_h \in \mathcal{G}(\Omega_h)$ with

$$L_h v_h(P) = \begin{cases} 0 & \text{for } P \in \Omega_H \\ \text{arbitrary} & \text{for } P \notin \Omega_H \end{cases}$$

is annihilated by $I_h^H$ and therefore by $C_h$. Such an error function $v_h$ will thus not be changed by a pure coarse grid correction.

As a more concrete example, consider Model Problem 1 ($L_h = -\Delta_h$), $h = 1/n$, and

$$v_h(x, y) = \sin\frac{n}{2}\pi x \, \sin\frac{n}{2}\pi y. \tag{2.2.25}$$

For standard coarsening, we find

$$v_h(P) = L_h v_h(P) = I_h^{2h} L_h v_h(P) = 0 \quad \text{for all } P \in \Omega_{2h}.$$

(This relation holds for *any* restriction operator $I_h^{2h}$ as long as the stencil of $I_h^{2h}$ is symmetric in $x$ and $y$.) Clearly, $v_h$ belongs to the high frequency part of the eigenfunctions of $L_h$.

$\triangle$

### 2.2.3 Structure of the Two-grid Operator

The above considerations imply that it is necessary to combine the two processes of smoothing and of coarse grid correction.

Each iteration step (cycle) of a two-grid method consists of a *presmoothing*, a *coarse grid correction* and a *postsmoothing* part. One step of such an iterative two-grid method (calculating $u_h^{m+1}$ from $u_h^m$) proceeds as follows:

---

**Two-grid cycle**   $u_h^{m+1} = \text{TGCYC}(u_h^m, L_h, f_h, \nu_1, \nu_2)$

(1) Presmoothing
- Compute $\bar{u}_h^m$ by applying $\nu_1 \, (\geq 0)$ steps of a given smoothing procedure to $u_h^m$:

$$\bar{u}_h^m = \text{SMOOTH}^{\nu_1}(u_h^m, L_h, f_h). \tag{2.2.26}$$

(2) Coarse grid correction (CGC)
- Compute the defect $\qquad \bar{d}_h^m = f_h - L_h \bar{u}_h^m.$
- Restrict the defect $\qquad \bar{d}_H^m = I_h^H \bar{d}_h^m.$
  (fine-to-coarse transfer)
- Solve on $\Omega_H \qquad\qquad L_H \hat{v}_H^m = \bar{d}_H^m. \tag{2.2.27}$

- Interpolate the correction $\qquad \hat{v}_h^m = I_H^h \hat{v}_H^m.$
  (coarse-to-fine transfer)
- Compute the corrected $\qquad u_h^{m,\text{after CGC}} = \bar{u}_h^m + \hat{v}_h^m.$
  approximation

(3) Postsmoothing
- Compute $u_h^{m+1}$ by applying $\nu_2 \, (\geq 0)$ steps of the given smoothing procedure to $u_h^{m,\text{after CGC}}$:

$$u_h^{m+1} = \text{SMOOTH}^{\nu_2}(u_h^{m,\text{after CGC}}, L_h, f_h). \tag{2.2.28}$$

---

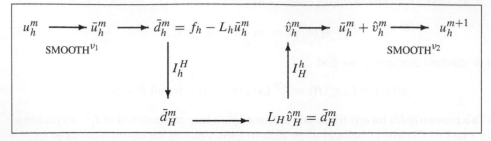

**Figure 2.4.** Structure of a two-grid cycle.

For the formal description of smoothing procedures we have used the notation

$$\bar{u}_h^m = \text{SMOOTH}^\nu\left(u_h^m, L_h, f_h\right).$$

With this notation we combine the advantages of a more mathematically oriented operator-like notation and a more computer science oriented formal procedural notation. In particular, the number $\nu$ of smoothing steps appears as an upper (power) index. Similar notation is also used for the two-grid and for the multigrid procedure.

The two-grid cycle is illustrated in Fig. 2.4.

From the above description, one immediately obtains the iteration operator $M_h^H$ of the $(h, H)$ two-grid cycle:

$$\boxed{M_h^H = S_h^{\nu_2} K_h^H S_h^{\nu_1} \quad \text{with } K_h^H := I_h - I_H^h L_H^{-1} I_h^H L_h}. \tag{2.2.29}$$

From Fig. 2.4, we see that the following individual *components of the two-grid cycle* have to be specified:

- the smoothing procedure SMOOTH $(u_h^m, L_h, f_h)$,
- the numbers $\nu_1$, $\nu_2$ of smoothing steps,
- the coarse grid $\Omega_H$,
- the fine-to-coarse restriction operator $I_h^H$,
- the coarse grid operator $L_H$,
- the coarse-to-fine interpolation operator $I_H^h$.

Experience with multigrid methods (and multigrid theory) shows that the choice of these components may have a strong influence on the efficiency of the resulting algorithm. On the other hand, there are no general simple rules on how to choose the individual components in order to construct optimal algorithms for complicated applications. One can, however, recommend certain choices for certain situations. The main objective of the *elementary multigrid theory* (see Chapter 3) and the *local Fourier analysis* (see Chapter 4) is to analyze the convergence properties of multigrid theoretically and to provide tools for the proper choice of multigrid components.

## 2.3 MULTIGRID COMPONENTS

In this section, we will introduce and list some important examples of how some of the multigrid components can be specified.

The idea of giving these specifications is to make our presentation more concrete and to introduce corresponding notations. The multigrid components specified here are needed in Section 2.5.1, where a specific multigrid Poisson solver is introduced. Therefore, readers who are more interested in the general structure of multigrid than in specific details may skip this section for the time being.

### 2.3.1 Choices of Coarse Grids

In this subsection, we will mention some possible and common choices for the grid $\Omega_H$. The simplest and most frequently used choice is *standard coarsening*, doubling the mesh size $h$ in every direction. Most of the results and considerations in this book refer to this choice. In $d$ dimensions, the relation between the number of grid points (neglecting boundary effects) is

$$\#\Omega_H \approx \frac{1}{2^d} \#\Omega_h.$$

We speak of *semicoarsening* in 2D if the mesh size $h$ is doubled in one direction only, i.e. $H = (2h_x, h_y)$ ($x$-semicoarsening, see Fig. 2.5) or $H = (h_x, 2h_y)$ ($y$-semicoarsening). This is especially of interest for anisotropic operators (see Section 5.1). Note that in this case

$$\#\Omega_H \approx \tfrac{1}{2} \#\Omega_h. \tag{2.3.1}$$

In 3D, we have additional types of semicoarsening: we can double the mesh size in one or in two directions (see Section 5.2).

We speak of *red–black coarsening* if the coarse grid points are distributed in the fine grid in a red–black (checkerboard) manner (see Fig. 2.5).

Also other coarsenings, like $4h$-coarsening, are sometimes of interest.

We mention that in the context of the AMG approach (see Appendix A), the coarse grid $\Omega_H$ is not formed according to such a fixed simple strategy. Using the algebraic relations in the corresponding matrix, $\Omega_H$ is determined by the AMG process itself in the course of calculation. We will see that the red–black coarsened grid is a standard choice for Model Problem 1 in AMG.

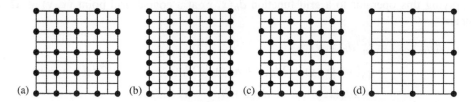

**Figure 2.5.** Examples of standard; $x$-semi-; red–black; $(h, 4h)$-coarsening in a square computational domain $\Omega$. The grid points of $\Omega_H$ are marked by dots.

### 2.3.2 Choice of the Coarse Grid Operator

So far, we have not described precisely how the coarse grid operator $L_H$ can be chosen. A natural choice is to use the direct analog of $L_h$ on the grid $\Omega_H$. For Model Problem 1, this means

$$L_H = \frac{1}{H^2} \begin{bmatrix} 0 & -1 & 0 \\ -1 & 4 & -1 \\ 0 & -1 & 0 \end{bmatrix}_H .$$

The direct coarse grid analog of the fine grid operator $L_h$ will be used in most parts of this book, in particular in all chapters on basic multigrid.

There are, however, applications and multigrid algorithms, which make use of a different approach. *The so-called Galerkin coarse grid operator is defined by*

$$L_H := I_h^H L_h I_H^h, \tag{2.3.2}$$

where $I_h^H$ and $I_H^h$ *are appropriate transfer operators*. We will return to the Galerkin approach in the context of problems with discontinuous coefficients in Chapter 7 and in Appendix A, where algebraic systems of equations without a grid-oriented background will be treated.

### 2.3.3 Transfer Operators: Restriction

The choice of restriction and interpolation operators $I_h^H$ and $I_H^h$, for the intergrid transfer of grid functions, is closely related to the choice of the coarse grid. Here, we introduce transfer operators for standard coarsening (see Fig. 2.5), i.e. the grid transfers between the grid $\Omega_h$ and the $2h$-grid $\Omega_{2h}$.

A *restriction operator* $I_h^{2h}$ maps $h$-grid functions to $2h$-grid functions. One restriction operator already discussed is the *injection* operator (2.2.22). Another frequently used restriction operator is the *full weighting* (FW) operator, which in stencil notation reads

$$\frac{1}{16} \begin{bmatrix} 1 & 2 & 1 \\ 2 & 4 & 2 \\ 1 & 2 & 1 \end{bmatrix}_h^{2h} . \tag{2.3.3}$$

Applying this operator to a grid function $d_h(x, y)$ at a coarse grid point $(x, y) \in \Omega_{2h}$ means

$$\begin{aligned} d_{2h}(x, y) &= I_h^{2h} d_h(x, y) \\ &= \tfrac{1}{16}[4d_h(x, y) + 2d_h(x + h, y) + 2d_h(x - h, y) + 2d_h(x, y + h) \\ &\quad + 2d_h(x, y - h) + d_h(x + h, y + h) + d_h(x + h, y - h) \\ &\quad + d_h(x - h, y + h) + d_h(x - h, y - h)]. \end{aligned} \tag{2.3.4}$$

Obviously, a nine-point weighted average of $d_h$ is obtained.

**Remark 2.3.1**   The FW operator can be derived from a discrete version of the condition

$$\int_{\Omega_{x,y}} w(\tilde{x}, \tilde{y}) \, d\Omega = \int_{\Omega_{x,y}} (I_h^{2h} w)(\tilde{x}, \tilde{y}) \, d\Omega \tag{2.3.5}$$

for $\Omega_{x,y} = [x-h, x+h] \times [y-h, y+h]$ where the midpoint rule is used to approximate the integral on the right-hand side of the equation and the trapezoidal rule is used to approximate the integral on the left-hand side.   $\gg$

**Remark 2.3.2**   The FW operator in $d$ dimensions can also be constructed as a tensor product of the one-dimensional FW operators

$$I_h^{2h} = \frac{1}{4}[1 \quad 2 \quad 1] \qquad I_h^{2h} = \frac{1}{4}\begin{bmatrix} 1 \\ 2 \\ 1 \end{bmatrix}.$$

These 1D restriction operators are of particular interest in combination with the semicoarsening approach discussed in the previous section. They are also commonly used in case of *noneliminated* boundary conditions (see Section 5.6.2).   $\gg$

A third restriction operator is *half weighting* (HW):

$$\frac{1}{8}\begin{bmatrix} 0 & 1 & 0 \\ 1 & 4 & 1 \\ 0 & 1 & 0 \end{bmatrix}_h^{2h}. \tag{2.3.6}$$

### 2.3.4 Transfer Operators: Interpolation

The *interpolation ( prolongation) operators* map $2h$-grid functions into $h$-grid functions. A very frequently used interpolation method is *bilinear interpolation* from $\mathbf{G}_{2h}$ to $\mathbf{G}_h$, which is given by:

$$I_{2h}^h \hat{v}_{2h}(x, y) = \begin{cases} \hat{v}_{2h}(x, y) & \text{for } \bullet \\ \frac{1}{2}[\hat{v}_{2h}(x, y+h) + \hat{v}_{2h}(x, y-h)] & \text{for } \square \\ \frac{1}{2}[\hat{v}_{2h}(x+h, y) + \hat{v}_{2h}(x-h, y)] & \text{for } \Diamond \\ \frac{1}{4}[\hat{v}_{2h}(x+h, y+h) + \hat{v}_{2h}(x+h, y-h) \\ \quad + \hat{v}_{2h}(x-h, y+h) + \hat{v}_{2h}(x-h, y-h)] & \text{for } \circ . \end{cases} \tag{2.3.7}$$

Figure 2.6 presents (part of) a fine grid with the symbols for the fine and coarse grid points referred to by (2.3.7).

**Remark 2.3.3**   Linear interpolation in $d$ dimensions can easily be constructed by a recursive procedure (over the $d$ dimensions) of 1D linear interpolations. Also $d$-dimensional higher order interpolations can be constructed and programmed very efficiently in this way.   $\gg$

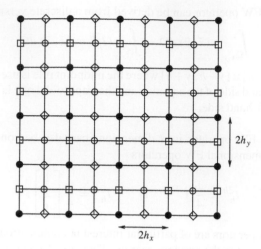

**Figure 2.6.** A fine grid with symbols indicating the bilinear interpolation (2.3.7) used for the transfer from the coarse grid (•).

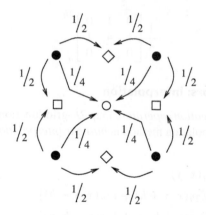

**Figure 2.7.** The distribution process for the bilinear interpolation operator. •, $G_{2h}$ grid; □, ◇ and ○ as in (2.3.7) and in Fig. 2.6.

In stencil notation we write the bilinear interpolation operator $I_{2h}^h$ (2.3.7) as

$$I_{2h}^h = \frac{1}{4} \left]\begin{array}{ccc} 1 & 2 & 1 \\ 2 & 4 & 2 \\ 1 & 2 & 1 \end{array}\right[_{2h}^h .$$  (2.3.8)

In this notation, the stencil entries correspond to weights in a *distribution process* as illustrated in Fig. 2.7. Therefore, the brackets are reversed.

For more general interpolation operators the stencils read

$$
I_{2h}^h \stackrel{\triangle}{=} ]t_{\kappa_1,\kappa_2}[_{2h}^h := 
\begin{bmatrix}
\cdot & \cdot & \cdot \\
\cdot & \cdot & \cdot \\
\cdot\cdot & t_{-1,1} & t_{0,1} & t_{1,1} & \cdot\cdot \\
\cdot\cdot & t_{-1,0} & t_{0,0} & t_{1,0} & \cdot\cdot \\
\cdot\cdot & t_{-1,-1} & t_{0,-1} & t_{1,-1} & \cdot\cdot \\
\cdot & \cdot & \cdot \\
\cdot & \cdot & \cdot
\end{bmatrix}_{2h}^h
\qquad (2.3.9)
$$

Again, only a finite number of coefficients $t_{\kappa_1,\kappa_2}$ is assumed to be nonzero. The meaning of this stencil terminology is that coarse grid values are distributed to the fine grid and the weights in this distribution process are the factors $t_{\kappa_1,\kappa_2}$.

**Remark 2.3.4** The formulas for linear interpolation of corrections can be applied near Dirichlet boundaries even if the boundary conditions have been eliminated. The corrections from a boundary point are assumed to be 0 in this case. $\gg$

## 2.4 THE MULTIGRID CYCLE

In Section 2.2 we have described the multigrid principle only in its two-grid version. In the multigrid context, two-grid methods are of little practical significance (due to the still large complexity of the coarse grid problem). However, they serve as the basis for the *multi*grid method.

**Remark 2.4.1** Methods involving only two grids (or a fixed, small, number of grids) are of some interest in other frameworks, e.g. in the context of certain domain decomposition and related methods [362]. $\gg$

---

**From two-grid to multigrid** The *multigrid idea* starts from the observation that in a well converged two-grid method it is neither useful nor necessary to solve the coarse grid defect equation (2.2.27) exactly. Instead, without essential loss of convergence speed, one may replace $\hat{v}_H^m$ by a suitable approximation. A natural way to obtain such an approximation is to apply the two-grid idea to (2.2.27) again, now employing an even coarser grid than $\Omega_H$.

---

This is possible, as obviously the coarse grid equation (2.2.27) is of the same form as the original equation (2.2.1). If the convergence factor of the two-grid method is small enough, it is sufficient to perform only a few, say $\gamma$, two-grid iteration steps (see Fig. 2.8) to obtain a good enough approximation to the solution of (2.2.27). This idea can, in a straight-forward manner, be applied recursively, using coarser and coarser grids, down to some coarsest grid.

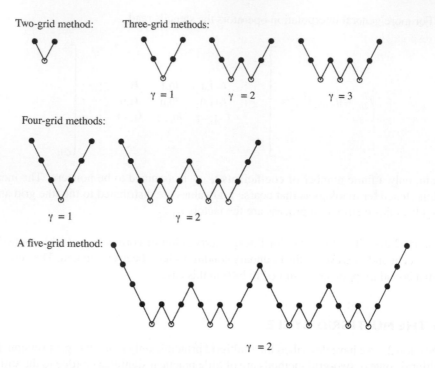

**Figure 2.8.** Structure of one multigrid cycle for different numbers of grids and different values of the cycle index $\gamma$ (●, smoothing; ○, exact solution; \, fine-to-coarse; /, coarse-to-fine transfer).

On this coarsest grid any solution method may be used (e.g. a direct method or some relaxation-type method if it has sufficiently good convergence properties on that coarsest grid). In ideal cases, the coarsest grid consists of just one grid point.

### 2.4.1 Sequences of Grids and Operators

Figure 2.8 illustrates the structure of *one iteration step* (*cycle*) of a multigrid method with a few pictures. Usually, the cases $\gamma = 1$ and $\gamma = 2$ are particularly interesting.

> For obvious reasons, we refer to the case $\gamma = 1$ as *V-cycles* and to $\gamma = 2$ as *W-cycles*. The number $\gamma$ is also called *cycle index*.

### 2.4.2 Recursive Definition

For a formal description of multigrid methods we now use a sequence of coarser and coarser grids $\Omega_{h_k}$, characterized by a sequence of mesh sizes $h_k$:

$$\Omega_{h_\ell}, \ \Omega_{h_{\ell-1}}, \ldots, \ \Omega_{h_0}.$$

The coarsest grid is characterized by the mesh size $h_0$ (index 0), whereas the index $\ell$ corresponds to the finest grid $\Omega_h$: $h = h_\ell$. For simplicity, we replace the index $h_k$ by $k$ (for grids, grid functions and grid operators) in the following. For each $\Omega_k$, we assume that linear operators

$$L_k : \mathcal{G}(\Omega_k) \to \mathcal{G}(\Omega_k), \qquad S_k : \mathcal{G}(\Omega_k) \to \mathcal{G}(\Omega_k),$$

$$I_k^{k-1} : \mathcal{G}(\Omega_k) \to \mathcal{G}(\Omega_{k-1}), \qquad I_{k-1}^k : \mathcal{G}(\Omega_{k-1}) \to \mathcal{G}(\Omega_k) \tag{2.4.1}$$

are given, where the $L_k$ are discretizations of $L$ on $\Omega_k$ for $k = \ell, \ldots, 0$, and where the original equation (2.2.1) reads

$$L_\ell u_\ell = f_\ell \quad (\Omega_\ell) \tag{2.4.2}$$

and is the discrete problem to solve. The operators $S_k$ denote the linear iteration operators corresponding to given smoothing methods on $\Omega_k$. As in the description of the two-grid cycle, performing $\nu$ smoothing steps (applied to any discrete problem of the form $L_k u_k = f_k$ with initial approximation $w_k$) resulting in the approximation $\overline{w}_k$ will also be denoted by

$$\overline{w}_k = \mathrm{SMOOTH}^\nu(w_k, L_k, f_k).$$

We now describe *a multigrid cycle* (multigrid iteration (MGI)), more precisely an $(\ell + 1)$-*grid cycle*, to solve (2.4.2) for a fixed $\ell \geq 1$. Using the operators $L_k$ $(k = \ell, \ell - 1, \ldots, 0)$ as well as $S_k, I_k^{k-1}, I_{k-1}^k$ $(k = \ell, \ell - 1, \ldots, 1)$, assuming the parameters $\nu_1, \nu_2$ (the number of pre- and postsmoothing iterations) and $\gamma$ to be fixed and starting on the finest grid $k = \ell$, the calculation of a new iterate $u_k^{m+1}$ from a given approximation $u_k^m$ to the solution $u_k$ proceeds as follows:

---

**Multigrid cycle**   $u_k^{m+1} = \mathrm{MGCYC}(k, \gamma, u_k^m, L_k, f_k, \nu_1, \nu_2)$

(1) Presmoothing
  – Compute $\bar{u}_k^m$ by applying $\nu_1$ $(\geq 0)$ smoothing steps to $u_k^m$

$$\bar{u}_k^m = \mathrm{SMOOTH}^{\nu_1}(u_k^m, L_k, f_k).$$

(2) Coarse grid correction
  – Compute the defect        $\bar{d}_k^m = f_k - L_k \bar{u}_k^m$ .
  – Restrict the defect       $\bar{d}_{k-1}^m = I_k^{k-1} \bar{d}_k^m$ .

- Compute an approximate solution $\hat{v}_{k-1}^m$ of the defect equation on $\Omega_{k-1}$

$$L_{k-1}\hat{v}_{k-1}^m = \bar{d}_{k-1}^m \qquad (2.4.3)$$

by

> If $k = 1$, use a direct or fast iterative solver for (2.4.3).
> If $k > 1$, solve (2.4.3) approximately by performing $\gamma (\geq 1)$ $k$-grid cycles using the zero grid function as a first approximation
>
> $$\hat{v}_{k-1}^m = \mathrm{MGCYC}^\gamma (k-1, \gamma, 0, L_{k-1}, \bar{d}_{k-1}^m, \nu_1, \nu_2). \qquad (2.4.4)$$

- Interpolate the correction        $\hat{v}_k^m = I_{k-1}^k \hat{v}_{k-1}^m$ .
- Compute the corrected
  approximation on $\Omega_k$        $u_k^{m,\,\text{after CGC}} = \bar{u}_k^m + \hat{v}_k^m$.

(3) Postsmoothing

- Compute $u_k^{m+1}$ by applying $\nu_2 (\geq 0)$ smoothing steps to $u_k^{m,\,\text{after CGC}}$:

$$u_k^{m+1} = \mathrm{SMOOTH}^{\nu_2}(u_k^{m,\,\text{after CGC}}, L_k, f_k).$$

In (2.4.4) the parameter $\gamma$ appears twice: once (as an argument of multigrid cycle (MGCYC)) for the indication of the cycle type to be employed on coarser levels and once (as a power) to specify the number of cycles to be carried out on the current coarse grid level.

**Remark 2.4.2** Since on coarse grids we deal with *corrections* to the fine grid approximation, this multigrid scheme is also called the *correction scheme* (CS).                    $\gg$

Let $M_k$ denote the iteration operator of the multigrid method described in the previous algorithm.

**Theorem 2.4.1**  *The multigrid iteration operator $M_\ell$ is given by the following recursion:*

$$
\begin{aligned}
&M_0 = 0\\
&M_k = S_k^{\nu_2}(I_k - I_{k-1}^k(I_{k-1} - (M_{k-1})^\gamma)(L_{k-1})^{-1}I_k^{k-1}L_k)S_k^{\nu_1} \quad (k = 1, \dots, \ell).
\end{aligned}
$$
$$(2.4.5)$$

An explicit *proof* of (2.4.5) can easily be given by means of induction on $k$. Implicitly, a proof is also contained in the following remark.

**Remark 2.4.3** The difference between the $(h_k, h_{k-1})$ two-grid iteration operator

$$M_k^{k-1} = S_k^{\nu_2}\big(I_k - I_{k-1}^k(L_{k-1})^{-1}I_k^{k-1}L_k\big)S_k^{\nu_1}, \tag{2.4.6}$$

and the above multigrid iteration operator $M_k$ is obviously that

$$(L_{k-1})^{-1} \text{ is replaced by } (I_{k-1} - (M_{k-1})^\gamma)(L_{k-1})^{-1}. \tag{2.4.7}$$

This reflects the fact that the coarse grid equation (2.4.3) is solved approximately by $\gamma$ multigrid steps on the grid $\Omega_{k-1}$ starting with zero initial approximation (compare with (2.2.14)). Thus, (2.4.5) follows immediately from (2.2.14). ≫

**Remark 2.4.4 (F-cycle)** It is convenient, but not necessary, that the parameters $\nu_1$, $\nu_2$ and the cycle index $\gamma$ are fixed numbers. In particular, $\gamma = \gamma_k$ may depend on $k$. Certain combinations of $\gamma = 1$ and $\gamma = 2$ are indeed used in practice. We mention here only the so-called *F-cycle* [66] which is illustrated in Fig. 2.9. The corresponding iteration operator $M_\ell^F$ is recursively defined by $M_1^F = M_1$ (as in (2.4.5)) and

$$M_k^F = S_k^{\nu_2}(I_k - I_{k-1}^k(I_{k-1} - M_{k-1}^V M_{k-1}^F)(L_{k-1})^{-1}I_k^{k-1}L_k)S_k^{\nu_1}$$
$$(k = 2, \ldots, \ell).$$

Here $M_{k-1}^V$ is the corresponding V-cycle iteration operator (i.e. (2.4.5) with $\gamma = 1$ and $k-1$ instead of $k$). ≫

**Remark 2.4.5** In *self-controlling* algorithms as proposed in [59, 64], variable cycles are used. Switching from one grid to another (to a finer or a coarser one) is controlled by suitable accommodative criteria. ≫

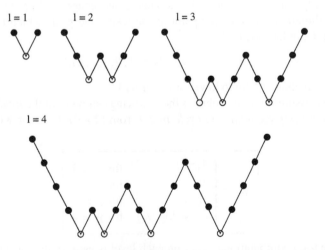

**Figure 2.9.** Structure of an F-cycle.

### 2.4.3 Computational Work

In Section 2.5.2 and Chapter 3, we will see that we can achieve a multigrid convergence factor that is independent of the mesh size: *The convergence speed does not depend on the size of the finest grid.* But the fact that a certain method has an *h*-independent convergence factor says *nothing* about its efficiency as long as the computational work is not taken into account. In the following, we will estimate the computational work of a multigrid method.

From the recursive definition of a multigrid cycle as given in Theorem 2.4.1 it immediately follows that the *computational work* $W_\ell$ per multigrid cycle on $\Omega_\ell$ is recursively given by

$$W_1 = W_1^0 + W_0, \qquad W_{k+1} = W_{k+1}^k + \gamma_k W_k \quad (k = 1, \ldots, \ell - 1). \tag{2.4.8}$$

Here $W_{k+1}^k$ denotes the computational work of one $(h_{k+1}, h_k)$ two-grid cycle *excluding* the work needed to solve the defect equations on $\Omega_k$, and $W_0$ denotes the work needed to compute the exact solution on the coarsest grid $\Omega_0$. By "computational work", we always denote some reasonable measure, typically the number of arithmetic operations needed. If $\gamma$ is *independent of k*, we obtain from (2.4.8)

$$W_\ell = \sum_{k=1}^{\ell} \gamma^{\ell-k} W_k^{k-1} + \gamma^{\ell-1} W_0 \quad (\ell \geq 1). \tag{2.4.9}$$

Let us first discuss the case of *standard coarsening* in 2D with $\gamma$ independent of $k$. Obviously,

$$N_k \doteq 4N_{k-1} \quad (k = 1, 2, \ldots, \ell) \tag{2.4.10}$$

where $N_k = \#\Omega_k$ (number of grid points on $\Omega_k$) and "$\doteq$" means equality up to lower order terms (boundary effects). Furthermore, we assume that the multigrid components (relaxation, computation of defects, fine-to-coarse and coarse-to-fine transfer) require a number of arithmetic operations per point of the respective grids which is bounded by a small constant $C$, independent of $k$:

$$W_k^{k-1} \dot{\leq} CN_k \quad (k = 1, 2, \ldots, \ell). \tag{2.4.11}$$

(As above, "$\dot{\leq}$" means "$\leq$" up to lower order terms.)

Under these assumptions, one obtains the following estimate for the *total computational work $W_\ell$ of one complete multigrid cycle in 2D* from (2.4.9), together with (2.4.10) and (2.4.11):

$$W_\ell \dot{\leq} \begin{cases} \frac{4}{3}CN_\ell & \text{for } \gamma = 1 \\ 2CN_\ell & \text{for } \gamma = 2 \\ 4CN_\ell & \text{for } \gamma = 3 \\ O(N_\ell \log N_\ell) & \text{for } \gamma = 4 \end{cases}. \tag{2.4.12}$$

For $\gamma = 4$ the total computational work on each level is essentially constant (up to lower order terms) and the number of grid levels is $O(\log N_\ell)$.

**Summary** This estimate of $W_\ell$ shows that the number of arithmetic operations needed for one 2D multigrid cycle is proportional to the number of grid points of the finest grid for $\gamma \leq 3$ and standard coarsening (under quite natural assumptions which are satisfied for reasonable multigrid methods). Together with the $h$-independent convergence, this means that multigrid methods achieve a fixed reduction (independent of $h$) of the error in $O(N)$ operations. The constant of proportionality depends on the type of the cycle, i.e. on $\gamma$, the type of coarsening and the other multigrid components. For reasonable choices of these components, the constants of proportionality are small.

**Remark 2.4.6** In practice, it is not necessary to choose a grid consisting of only one interior point as the coarsest grid. Instead, it is sufficient to choose the coarsest grid $W_0$ such that the amount of work (of the corresponding solution method) on $W_0$ is negligible.

$\gg$

**Remark 2.4.7** $W_k^{k-1}$ in (2.4.8) or, in other words, the constant $C$ in (2.4.11) is determined by the computational work needed for the individual multigrid components of the $(h_k, h_{k-1})$ two-grid method, namely

$$W_k^{k-1} \doteq (\nu w_0 + w_1 + w_2) N_k. \tag{2.4.13}$$

Here $\nu = \nu_1 + \nu_2$ is the number of smoothing steps used, $w_0$, $w_1$ and $w_2$ are measures for the computational work per grid point of $\Omega_k$ needed for the single components, namely

- $w_0$: one smoothing step on $\Omega_k$,
- $w_1$: computation of the defect and its transfer to $\Omega_{k-1}$,
- $w_2$: interpolation of the correction to $\Omega_k$ and its addition to the previous approximation.

$\gg$

More generally, in particular for other than standard coarsening, we can assume

$$N_k \doteq \tau N_{k-1} (k = 1, 2, \ldots, \ell) \quad \text{with } \tau > 1.$$

In that case we obtain for $\gamma$ independent of $k$

$$W_\ell \doteq \begin{cases} \tau/(\tau - \gamma) C N_\ell & \text{for } \gamma < \tau \\ O(N_\ell \log N_\ell) & \text{for } \gamma = \tau, \end{cases} \tag{2.4.14}$$

generalizing (2.4.12), where $\tau = 4$. If we consider, for example, *red–black coarsening* or *semicoarsening* (see Fig. 2.5), we have $\tau = 2$. In this case, we already see that W-cycles do not yield an asymptotically optimal multigrid method: for $\tau = 2$ and fixed $\gamma$, only $\gamma = 1$ yields a cycle for which $W_\ell$ is proportional to $N_\ell$.

**Remark 2.4.8** The computational work $W_F$ of an F-cycle as introduced in Remark 2.4.4 in combination with standard coarsening is also $O(N_\ell)$ since this cycle is more expensive than the V-cycle and less expensive than the W-cycle. In particular, we obtain

$$W_F \lesssim W_\ell^{\ell-1} \sum_{k=1}^{\infty} k \left(\frac{1}{4}\right)^{k-1} = W_\ell^{\ell-1} \sum_{k=1}^{\infty} \sum_{m=k}^{\infty} \left(\frac{1}{4}\right)^{m-1} = \frac{16}{9} W_\ell^{\ell-1} \qquad (2.4.15)$$

and thus

$$W_F \lesssim \frac{16}{9} C N_\ell. \qquad \gg$$

**Remark 2.4.9** For 2D semicoarsening ($\tau = 2$) we find that the computational work of an F-cycle still is of the form $O(N_\ell)$. This can be seen if we take into account that a grid $\Omega_k$ is processed once more often than grid $\Omega_{k+1}$ (see Fig. 2.9). Correspondingly, if $W_\ell^{\ell-1}$ is the amount of work spent on the finest grid, then the asymptotical amount of work (for $\ell \to \infty$) of the F-cycle in case of semicoarsening is

$$W_F \lesssim W_\ell^{\ell-1} \sum_{k=1}^{\infty} k \left(\frac{1}{2}\right)^{k-1} = 4 W_\ell^{\ell-1}. \qquad \gg$$

So far, we have estimated and discussed the computational work of multigrid cycles. In order to assess the numerical efficiency of such an iterative multigrid solver precisely, it is necessary to take into account both its convergence behavior and the computational effort per iteration step (cycle). For example, in order to decide how many relaxation steps $\nu = \nu_1 + \nu_2$ are appropriate per cycle and whether a V-, F- or W-cycle should be used, the effect of this decision on both the convergence speed *and* the computational work have to be analyzed. We will discuss these efficiency questions in the following section for a specific multigrid Poisson solver.

## 2.5 MULTIGRID CONVERGENCE AND EFFICIENCY

In this section we introduce the first specific multigrid algorithm. For that purpose we return to Model Problem 1, the discrete Poisson equation with Dirichlet boundary conditions in the 2D unit square. The algorithm presented here is a highly efficient Poisson solver. Its most characteristic multigrid component is the GS-RB relaxation for smoothing. We call it the red–black multigrid Poisson solver (RBMPS). In addition to its numerical efficiency, the algorithm is also highly parallel.

### 2.5.1 An Efficient 2D Multigrid Poisson Solver

The algorithm is characterized by the following multigrid components. In this characterization certain parameters and components still have to be specified:

$$- L_k = L_{h_k} = -\Delta_{h_k} = \frac{1}{h_k^2} \begin{bmatrix} & -1 & \\ -1 & 4 & -1 \\ & -1 & \end{bmatrix}_{h_k},$$

- smoother: GS-RB relaxation as presented in Section 2.1,
- restriction $I_k^{k-1}$: full weighting (2.3.4) (or half weighting (2.3.6), see Remark 2.5.1),
- prolongation $I_{k-1}^k$: bilinear interpolation (2.3.7),
- standard coarsening: $h_{k+1} = h_k/2$,
- size of coarsest grid: $h_0 = 1/2$.

**Remark 2.5.1**   This version of the red–black multigrid Poisson solver is a particularly simple one. There are further variants of this algorithm, some of which are even more efficient [319, 378], but less generally applicable. For example, the restriction operator FW can be replaced by HW (2.3.6) which is more efficient for certain choices of $\nu$ ($\nu \geq 3$, see Table 2.4 and Section 3.3.1).                                                          ≫

In the following we want to discuss the influence of the number of relaxations $\nu = \nu_1 + \nu_2$ and the cycle type (V, F, W) on the convergence speed of the algorithm and consequently the *numerical efficiency* of the algorithm.

We use the notation $V(\nu_1, \nu_2)$, $F(\nu_1, \nu_2)$ or $W(\nu_1, \nu_2)$ to indicate the cycle type and the number of pre- and postsmoothing steps employed. The finest grid is $h = 1/256$. Furthermore, we can compare theoretical convergence results (which we will obtain in Chapter 3) with the measured convergence results. In Fig. 2.10 the multigrid convergence of the V(0,1)-cycle (meaning 0 pre- and 1 postsmoothing), of the V(1,1)-, the W(0,1)- and

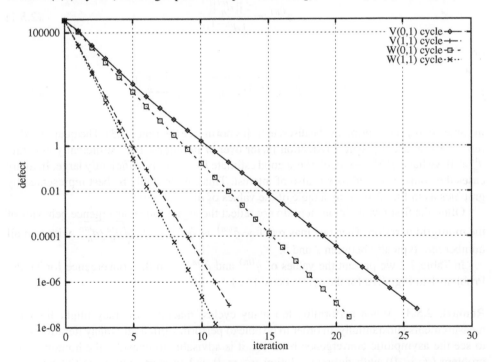

**Figure 2.10.** The convergence history of different RBMPS cycles for Model Problem 1.

the W(1,1)-cycle is presented, using FW as the restriction operator. The $l_2$ norm of the defect is plotted in a log scale along the $y$-axis. The $x$-axis shows the number of multigrid iterations. (The results obtained by the F-cycle and by the W-cycle are nearly identical.)

From Fig. 2.10, we observe rapid convergence of multigrid, especially for the V(1,1)- and W(1,1)-cycles: they reduce the defect by a factor of $10^{-12}$ within 12 multigrid iterations. Also the benefits of processing the coarse grid levels more frequently can be seen: the W-cycle (and the F-cycle) shows a better convergence than the V-cycle.

**Remark 2.5.2**  In practice, it is usually *not* necessary to reduce the defect by a factor of $10^{-12}$. Convergence to discretization accuracy $O(h^2)$ (see the discussion in Section 2.6) is sufficient in most cases, and is obtained much faster. Here, we reduce the defect further in order to illustrate the *asymptotic* convergence of the multigrid cycle.                    $\gg$

### 2.5.2 How to Measure the Multigrid Convergence Factor in Practice

In order to construct, evaluate and analyze a multigrid iteration one often wants to determine its convergence factor $\rho$ empirically (by measurement). In general, the only quantities that are available for the determination of $\rho$ are the defects $d_h^m$ ($m = 1, 2, \dots$). We can measure, for example,

$$q^{(m)} := \frac{\| d_h^m \|}{\| d_h^{m-1} \|} \tag{2.5.1}$$

or

$$\hat{q}^{(m)} := \sqrt[m]{q^{(m)} q^{(m-1)} \dots q^{(1)}} = \sqrt[m]{\frac{\| d_h^m \|}{\| d_h^0 \|}} \tag{2.5.2}$$

in some appropriate norm, say the discrete $\| \cdot \|_2$ norm (see Section 1.3.3). The quantity $\hat{q}^{(m)}$ represents an average defect reduction factor over $m$ iterations. For "sufficiently general" $d_h^0 \neq 0$ we have $\hat{q}^{(m)} \longrightarrow \rho$. $\hat{q}^{(m)}$ is a good estimate for $\rho$ if $m$ is sufficiently large. In many cases the *convergence history* is also of interest. This can probably be best represented by graphics as in Fig. 2.10 or by a table of the values of $q^{(m)}$.

Often the first few iteration steps do not reflect the asymptotic convergence behavior of the multigrid iteration. Then one may redefine $\hat{q}^{(m)}$ as $\hat{q}^{(m)} = \sqrt[m-m_0]{d_h^m / d_h^{m_0}}$ with a small number $m_0$, typically between 2 and 5.

In Table 2.2 we present the values of $q^{(m)}$ and $\hat{q}^{(m)}$ from the convergence for Model Problem 1 in Fig. 2.10 for the cycles considered.

**Remark 2.5.3**  When performing too many cycles, machine accuracy might limit the measurements substantially. In order to be able to perform sufficiently many iteration steps to see the asymptotic convergence behavior, it is advisable to consider the homogeneous problem ($f = 0$) with discrete solution $u_h \equiv 0$ and to start with an initial guess $u^0$ that is sufficiently large and general. In the case of our example, we then find asymptotic

**Table 2.2.** The quantities $q^{(m)}$ and $\hat{q}^{(m)}$ as a measure for the convergence of the RBMPS (with FW) for Model Problem I ($m_0 = 0$).

| | | |
|---|---|---|
| V(0,1): | $q^{(26)} = 0.343$ | $\hat{q}^{(26)} = 0.333$ |
| V(1,1): | $q^{(12)} = 0.101$ | $\hat{q}^{(12)} = 0.089$ |
| W(0,1): | $q^{(21)} = 0.243$ | $\hat{q}^{(21)} = 0.238$ |
| W(1,1): | $q^{(11)} = 0.063$ | $\hat{q}^{(11)} = 0.060$ |

**Table 2.3.** Measured convergence factors of the RBMPS (with FW) for Model Problem I on grids of different mesh size. In each case the coarsest grid has a mesh size of $h_0 = 1/2$.

| Cycle | $h = 1/512$ | $h = 1/256$ | $h = 1/128$ | $h = 1/64$ | $h = 1/32$ | $h = 1/16$ |
|---|---|---|---|---|---|---|
| V(1,1): | 0.10 | 0.10 | 0.10 | 0.10 | 0.11 | 0.12 |
| F(1,1): | 0.063 | 0.063 | 0.063 | 0.063 | 0.063 | 0.067 |
| W(1,1): | 0.063 | 0.063 | 0.063 | 0.063 | 0.063 | 0.067 |

convergence factors of 0.25 for the F(0,1)- or W(0,1)-cycle and of 0.074 for the F(1,1)- or W(1,1)-cycle after a large number of multigrid iterations.                                    ≫

---

**$h$-independent convergence of multigrid**   The numerically measured convergence of the RBMPS is essentially independent of the size of the finest grid in the multigrid cycle. This behavior is demonstrated by the results in Table 2.3. In Chapter 3, we will see that this behavior is also confirmed by multigrid theory.

---

### 2.5.3 Numerical Efficiency

In order to choose the most efficient multigrid solver, it is necessary to look at both, its convergence speed and its costs. In practice, the time needed to solve the problem is the most interesting quantity. Table 2.4 shows the wall clock times for different multigrid (cycle) iterations. The iterations stopped after the initial defects had been reduced by a factor of $10^{-12}$. The times were measured on a single workstation.

Table 2.4 throws a somewhat different light on the convergence results obtained before. The V(2,1)-cycle with HW is most efficient with respect to the wall clock time on the $256^2$ grid. Since W- and F-cycles have the same convergence speed for this model problem, the W-cycle clearly is less efficient than the F-cycle here.

---

**Remark 2.5.4**   Table 2.4 shows that it does not pay to use large values for $\nu_1$ and $\nu_2$. This is a general observation. Though the convergence factors become (somewhat) better if the number of smoothing steps is increased, it is more efficient not to smooth the error too much but rather carry out a few more multigrid cycles. In practice, common choices are $\nu = \nu_1 + \nu_2 \leq 3$. This observation can also be verified theoretically (see Section 3.3.1).                                    ≫

---

**Table 2.4.** Wall clock times and number of iterations for a defect reduction of a factor of $10^{-12}$ for different cycles and different restriction operators for Model Problem 1 on a $256^2$ grid.

| | FW | | HW | |
|---|---|---|---|---|
| Cycle | Iterations | Time (msec) | Iterations | Time (msec) |
| V(0,1): | 26 | 1290 | 167 | 7310 |
| V(1,1): | 12 | 759 | 13 | 740 |
| V(2,1): | 10 | 759 | 9 | 629 |
| V(2,2): | 9 | 799 | 8 | 669 |
| F(0,1): | 20 | 1270 | 34 | 1910 |
| F(1,1): | 10 | 819 | 10 | 740 |
| F(2,1): | 9 | 890 | 9 | 829 |
| F(2,2): | 8 | 930 | 8 | 880 |
| W(0,1): | 20 | 2269 | 34 | 3780 |
| W(1,1): | 10 | 1379 | 10 | 1379 |
| W(2,1): | 9 | 1450 | 9 | 1479 |
| W(2,2): | 8 | 1469 | 8 | 1460 |

**Remark 2.5.5** Although the V-cycle is obviously the most efficient one for Model Problem 1, we will see later that F- or W-cycles may be superior for more complicated applications. ≫

## 2.6 FULL MULTIGRID

An initial approximation for iterative solvers, like multigrid, can be obtained by *nested iteration*. The general idea of nested iteration is to provide an initial approximation on a grid $\Omega_\ell$ by the computation and interpolation of approximations on coarser grids. Within an arbitrary iterative process for the solution of a given discrete problem, this principle simply means that a lower (coarser) discretization level is used in order to provide a good initial approximation for the iteration on the next higher (finer) discretization level [226, 227].

The efficiency of iterative multigrid (MGI) can be improved if it is properly combined with the nested iteration idea. This combination is called the *full multigrid* (FMG) technique [58]. Typically, the FMG scheme is the most efficient multigrid version.

FMG has two fundamental properties:

(1) *An approximation $u_h^{\text{FMG}}$ of the discrete solution $u_h$ can be computed up to an error $\| u_h - u_h^{\text{FMG}} \|$ which is approximately equal to the discretization error $\| u - u_h \|$.*

(2) *FMG is an asymptotically optimal method, i.e the number of arithmetic operations required to compute $u_h^{\text{FMG}}$, is proportional to the number of grid points of $\Omega_h$ (with only a small constant of proportionality).*

The first property is explained in the following remark.

**Remark 2.6.1** Since the discrete solution $u_h$ approximates the continuous solution $u$ of the PDE only up to *discretization accuracy*, in many cases it does not make much sense to solve the discrete problem (1.3.2) exactly. We have to live with the discretization error $||u - u_h||$ anyway. That is why it is usually sufficient to compute an approximation $u_h^{FMG}$ only up to discretization accuracy. By this we mean that

$$||u_h - u_h^{FMG}|| \approx ||u - u_h||. \qquad (2.6.1)$$

More concretely we will derive estimates of the form

$$||u_h - u_h^{FMG}|| \leq \beta ||u - u_h|| \qquad (2.6.2)$$

in Section 3.2.2, which immediately implies

$$||u - u_h^{FMG}|| \leq (1 + \beta)||u - u_h||. \qquad (2.6.3)$$

We regard $\beta \approx 1$ as a sufficiently good value. In that case, "up to discretization accuracy" would mean that we allow a factor of 2 in comparing $||u - u_h^{FMG}||$ with the discretization error. If necessary, much smaller values of $\beta$ can also be achieved.

Generally, it is not worth investing more work in a more accurate approximation of $u_h^{FMG}$ than suggested here because it is more cost-effective to refine the grid once more than to solve the present problem more accurately. It should be kept in mind that the final goal is usually a good approximation to the differential problem, not an extremely good solution to the discrete problem. ≫

At this point, we add a remark, which refers to a question multigrid beginners sometimes have.

**Remark 2.6.2** In general, it is not sufficient to start the solution process on a very coarse grid, interpolate the approximation of the coarse grid solution to the next finer grid, smooth the visible error components and so on until the finest grid is reached. Actually, the interpolation of the approximation leads to nonnegligible high and low frequency error components on the fine grid (see Section 7.3 in [66] for a heuristic explanation) that can efficiently be reduced only by a subsequent smoothing of the error on all grid levels, i.e. by revisiting the coarse levels in multigrid cycles. ≫

### 2.6.1 Structure of Full Multigrid

As in Section 2.4, we consider (2.4.2), i.e. a sequence of discrete approximations to (2.2.1). As in Section 2.4.2, we use the notation

$$\text{MGCYC}^r(k + 1, \gamma, w_k, L_k, f_k, \nu_1, \nu_2) \ : \ \mathbf{G}(\Omega_k) \rightarrow \mathbf{G}(\Omega_k) \qquad (2.6.4)$$

for a procedure consisting of $r$ steps of a suitable iterative $(k + 1)$-grid cycle with cycle index $\gamma$ for solving (2.4.2) with initial approximation $w_k$ (using grids $\Omega_k, \ldots, \Omega_0$). The

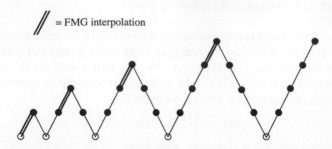

**Figure 2.11.** Structure of FMG with $r = 1$ and $\gamma = 1$ when using $l = 4$ (i.e. five grid levels).

right-hand sides $f_k$ on $\Omega_k$ can be defined recursively by $f_k = I_{k+1}^k f_{k+1}$ with $f_\ell = f|_{\Omega_\ell}$ or simply by $f_k = f|_{\Omega_k}$. The objective is to achieve discretization accuracy on each level within a few (typically $r = 1$ or $r = 2$) multigrid cycles. (In general $r$ may depend on $k$.) The structure of FMG (with $r = 1$ and $\gamma = 1$) is illustrated in Fig. 2.11.

*In contrast to the interpolation in the multigrid cycle $I_{k-1}^k$, which is applied to corrections, the FMG interpolation $\Pi_{k-1}^k$ from $\Omega_{k-1}$ to $\Omega_k$ transfers approximations of the solution to the fine grid.* Moreover, the operator $\Pi_{k-1}^k$ represents an interpolation procedure which usually is of higher accuracy than the interpolation used within the multigrid iteration.

The FMG algorithm proceeds as follows:

---

**Full multigrid**
For $k = 0$:
    Solve $L_0 u_0 = f_0$, providing $u_0^{\text{FMG}} = u_0$.

For $k = 1, 2, \ldots, \ell$:
$u_k^0 := \Pi_{k-1}^k u_{k-1}^{\text{FMG}}$
$u_k^{\text{FMG}} = \text{MGCYC}^r(k + 1, \gamma, u_k^0, L_k, f_k, \nu_1, \nu_2)$.

---

Here, $u_\ell^{\text{FMG}}$ denotes the resulting FMG approximation on grid $\Omega_\ell$.

In Section 3.2.2 we will see that FMG will deliver an approximation up to discretization accuracy if the multigrid cycle converges satisfactorily and if the order of the FMG interpolation is larger than the discretization order of $L_\ell$. A common FMG interpolation for second-order accurate discretizations is cubic (multipolynomial) interpolation. Cheaper interpolations especially suited for Model Problem 1 can also be constructed [378].

**Example 2.6.1** Often, bicubic interpolation is an appropriate FMG interpolation. If, for instance, $(x - 3h, y)$, $(x - h, y)$, $(x + h, y)$ and $(x + 3h, y)$ are points on the coarse grid, where an approximation $w_{2h}$ is known, cubic interpolation can be used to compute an approximation $w_h(x, y)$ at the fine grid point $(x, y)$:

$$w_h(x, y) = -\tfrac{1}{16}(w_{2h}(x - 3h, y) + w_{2h}(x + 3h, y))$$
$$+ \tfrac{9}{16}(w_{2h}(x - h, y) + w_{2h}(x + h, y))$$
$$= \tfrac{1}{16}[-1 \quad 9 \quad \cdot \quad 9 \quad -1]w_{2h}(x, y).$$

Cubic interpolation in the $y$-direction is analogous. Near the boundary, appropriate modifications can be used. $\triangle$

### 2.6.2 Computational Work

The computational work $W_\ell^{\mathrm{FMG}}$ needed for the FMG method can easily be estimated. By arguments similar to those in Section 2.4.3, one immediately obtains, for example:

$$W_\ell^{\mathrm{FMG}} \lesssim \begin{cases} \tfrac{4}{3}r W_\ell + \tfrac{4}{3} W_{\ell-1}^{\mathrm{INT}} & \text{for standard coarsening in 2D,} \\[2mm] 2r W_\ell + 2 W_{\ell-1}^{\mathrm{INT}} & \text{for semicoarsening in 2D,} \end{cases} \tag{2.6.5}$$

(again neglecting lower order terms). Here $W_{\ell-1}^{\mathrm{INT}}$ denotes the work needed for the FMG interpolation process from grid $\Omega_{\ell-1}$ to the finest grid $\Omega_\ell$ and $W_\ell$ is the work required for one multigrid cycle on the finest level $\Omega_\ell$.

---

**Remark 2.6.3** The number of operations required for FMG is governed by $W_\ell$ (and $W_{\ell-1}^{\mathrm{INT}}$). Under natural conditions, both of them are $O(N)$ (see (2.4.12) or (2.4.14)) with small constants. Thus, FMG only requires $O(N)$ operations. $\gg$

---

### 2.6.3 FMG for Poisson's Equation

We now give results for FMG with the RBMPS starting on the coarsest grid. As the FMG interpolation, we choose cubic interpolation. All other multigrid components are as introduced in Section 2.5.1. In particular we consider whether or not the discretization accuracy $O(h^2)$ is really achieved after FMG with $r = 1$.

In our model problem we choose the right-hand sides $f^\Omega(x, y)$ and $f^\Gamma(x, y)$ so that a known solution results from (1.3.1), for example,

$$u(x, y) = e^{xy}.$$

Thus, we can *observe* the second-order accuracy by comparing the discrete numerical solution with the analytical solution on finer and finer grids. Table 2.5 compares the accuracy obtained by FMG (for $r = 1$) with that of the exact discrete solution for a series of grids.

We see that the only cycle which does not achieve second-order accuracy for this example is the V(0,1)-cycle. For all the others, the discrete error is reduced by a factor of about four when the mesh size is reduced by a factor of two. Correspondingly, the errors are in the

**Table 2.5.** $||u - u_h^{\text{FMG}}||_\infty$ for different grid sizes, cycle types and number of relaxations ($r = 1$) (the exponent $E$ in $10^E$ is in brackets).

| Grid | $||u - u_h||_\infty$ | V(0,1) | V(1,1) | F(0,1) | F(1,1) |
|------|------|--------|--------|--------|--------|
| $32^2$ | 0.31 ($-5$) | 0.26 ($-4$) | 0.47 ($-5$) | 0.86 ($-5$) | 0.32 ($-5$) |
| $64^2$ | 0.77 ($-6$) | 0.83 ($-5$) | 0.12 ($-5$) | 0.13 ($-5$) | 0.77 ($-6$) |
| $128^2$ | 0.19 ($-6$) | 0.27 ($-5$) | 0.31 ($-6$) | 0.20 ($-6$) | 0.19 ($-6$) |
| $256^2$ | 0.48 ($-7$) | 0.87 ($-6$) | 0.78 ($-7$) | 0.48 ($-7$) | 0.48 ($-7$) |

**Table 2.6.** Computing times in milliseconds for FMG on a $256^2$ grid.

| V(0,1) | V(1,1) | F(0,1) | F(1,1) |
|--------|--------|--------|--------|
| 100 | 120 | 120 | 150 |

range of the discretization error. FMG thus proves to be a very efficient solution method, which costs less than two multigrid cycles. Table 2.6 shows the computing times obtained on a common workstation for the four FMG algorithms in Table 2.5 on the $256^2$ grid. The V(1,1)- and F(0,1)-cycles are thus the most efficient ones that achieve second-order accuracy for this test problem.

## 2.7 FURTHER REMARKS ON TRANSFER OPERATORS

In this section, we present some more background information on the choice of the transfer operators.

**Remark 2.7.1 (orders of restriction and interpolation of corrections)**   For Poisson's equation and for many other problems described by a second-order differential operator, the combination of full weighting and bilinear interpolation is a standard choice. For other problems, in particular those with higher order derivatives, different transfer operators may be required. The orders of these operators depend on the orders of the derivatives appearing in the PDE to be solved.

*The order of interpolation* is equal to $k + 1$ if an interpolation is exact for all polynomials of degree $k$. Bilinear interpolation, for example, has order 2. *The order of a restriction operator* is equal to the order of its transpose. Bilinear interpolation is the transpose of FW (see Remark 2.7.2), the order of FW is therefore also 2.

Let $m$ denote the order of the operator $L$ in the differential equation $Lu = f$. Let further $m_i$ denote the order of the restriction operator and $m^j$ denote the order of the interpolation operator. Then the orders of the transfer operators should fulfill

$$m_i + m^j > m \tag{2.7.1}$$

[64, 187]. This basic rule can be derived using the local Fourier analysis discussed in Chapter 4.

Note that the transpose of injection (2.2.22) does not even interpolate constant polynomials exactly since it does not give any information back to fine grid points which are not part of the coarse grid. Its order is therefore 0.

Thus, the combination of injection and bilinear interpolation does not satisfy (2.7.1) for Model Problem 1, whereas the combination of FW and bilinear interpolation does.    ≫

**Remark 2.7.2**    It can be shown that the bilinear interpolation operator corresponds to the FW restriction operator (2.3.4) in a natural way: these two operators are transpose to each other (see [173, 414]):

$$I_H^h = 2^d (I_h^H)^T.$$    (2.7.2)

≫

**Remark 2.7.3**    For five-point difference operators such as $\Delta_h$, HW coincides with half injection if it is preceded by one (or several) GS-RB relaxations. Half injection is injection with the weight 1/2

$$I_h^{2h} = \tfrac{1}{2}[1].$$

This can be seen very easily: the HW operator is a weighted average of defects at a coarse (red) grid point and its four direct neighbors. These four neighbors are all black points in the sense of GS-RB. After one GS-RB relaxation, however, the defects in all black points are zero.    ≫

**Remark 2.7.4 (parallelism of transfer operators)**    The multigrid components (calculation of defects, fine-to-coarse transfer and coarse-to-fine transfer) can typically be applied in parallel at all relevant grid points ($\Omega_h$ and $\Omega_{2h}$, respectively, see Section 6.1).    ≫

**Remark 2.7.5 (warning for beginners: correct scaling of the restriction of defects)**    We would like to point out that the discrete operators $L_h$ and $L_{2h}$ have the different factors $h^{-2}$ and $(2h)^{-2}$, respectively. In the practical implementation one often multiplies the original discrete equation by $h^2$. Multigrid beginners often make the mistake of also using this factor $h^2$ on the coarse grid instead of the correct factor $4h^2$. They forget the factor 4 in the transfer of defects from the fine to the coarse grid (by FW or other restriction operators).

If the fine grid equations are multiplied by $h^2$, the properly scaled injection and FW operators for the defects are

$$[4]_h^{2h} \quad \text{and} \quad \frac{4}{16} \begin{bmatrix} 1 & 2 & 1 \\ 2 & 4 & 2 \\ 1 & 2 & 1 \end{bmatrix}_h^{2h},$$

respectively.

The danger of making this mistake is even greater if the $h^2$ multiplied equation is the natural form of the discretization, as in the context of finite volumes and finite elements.

$\gg$

## 2.8 FIRST GENERALIZATIONS

Here, we consider first generalizations of Model Problem 1. We will discuss multigrid for Poisson-like equations in 2D, multigrid for time-dependent problems and give some results for multigrid on Cartesian grids in nonrectangular domains. Finally we give some details about multigrid methods for cell-centered discretizations.

### 2.8.1 2D Poisson-like Differential Equations

In this section, we will show that the multigrid method we have presented, the RBMPS, can be directly applied to more general elliptic boundary value problems of the form

$$
\begin{aligned}
-a(x, y)u_{xx} - b(x, y)u_{yy} &= f^\Omega(x, y) \quad (\Omega) \\
u(x, y) &= f^\Gamma(x, y) \quad (\Gamma = \partial\Omega)
\end{aligned}
\tag{2.8.1}
$$

for $\Omega = (0, 1)^2$ as long as the coefficients $a(x, y)$ and $b(x, y)$ are of about the same size and are smoothly varying. In stencil notation, the discrete equation then reads

$$
\frac{1}{h^2}
\begin{bmatrix}
 & -b(x_i, y_j) & \\
-a(x_i, y_j) & 2a(x_i, y_j) + 2b(x_i, y_j) & -a(x_i, y_j) \\
 & -b(x_i, y_j) &
\end{bmatrix}_h u_h(x_i, y_j)
$$
$$
= f_h(x_i, y_j).
\tag{2.8.2}
$$

We obtain the typical multigrid convergence in this case (presented in Example 2.8.1 below). Nonconstant (smooth) coefficients do not present problems to multigrid methods. This is different to other fast Poisson solvers such as FFTs for which *constant* coefficients in front of the derivatives are essential.

Another important problem to which the RBMPS can be applied directly is the *Helmholtz-like equation*

$$
\begin{aligned}
-\Delta u + c(x, y)u &= f^\Omega(x, y) \quad (\Omega = (0, 1)^2) \\
u(x, y) &= f^\Gamma(x, y) \quad (\Gamma = \partial\Omega).
\end{aligned}
\tag{2.8.3}
$$

Here, the function $c(x, y)$ is assumed to be *nonnegative* and smoothly varying. (This equation is a generalization of the Helmholtz equation where $c$ is constant.) Using the standard five-point discretization $\Delta_h$, we will obtain even better multigrid convergence than for Model Problem 1. The stencil of the corresponding discrete problem reads

$$
\frac{1}{h^2}
\begin{bmatrix}
 & -1 & \\
-1 & 4 + h^2 c(x_i, y_j) & -1 \\
 & -1 &
\end{bmatrix}_h,
$$

which means that we have additional nonnegative entries on the main diagonal of the corresponding matrix $A_h$ (as compared to the discrete Poisson equation). Therefore, diagonal dominance is amplified, which can be shown to result, here, in improved smoothing and convergence factors [378].

If we allow $c(x, y)$ to become negative, the multigrid convergence will essentially not deteriorate as long as $c(x, y)$ is smoothly varying and not too close to the first eigenvalue $\lambda_{1,1}$ of $-\Delta_h$,

$$c(x, y) \geq \text{ const } > \lambda_{1,1} = -\frac{1}{h^2}(4 - 4\cos(\pi h)).$$

If $c(x, y) \leq \lambda_{1,1}$, the problem may become indefinite and needs a more advanced multigrid treatment (see Section 10.2.4 and Appendix A).

**Example 2.8.1**  We will present the multigrid convergence obtained by the RBMPS applied to the following equation, with nonconstant coefficients:

$$\begin{aligned}
-a(x, y)u_{xx} - b(x, y)u_{yy} + c(x, y)u &= f^\Omega \quad (\Omega = (0, 1)^2) \\
u(x, y) &= f^\Gamma(x, y) \quad (\Gamma = \partial\Omega),
\end{aligned} \tag{2.8.4}$$

where $a(x, y) = 2+\sin(\pi x/2), b(x, y) = 2+\cos(\pi y/2)$. Three different choices of $c(x, y)$ in Table 2.7 show the (positive) effect of a large nonnegative $c$ on the multigrid convergence. Table 2.7 presents the measured multigrid convergence factors, $\hat{q}^{(m)}$ (see (2.5.2)), iteration index $m$ sufficiently large (see Remark 2.5.3), for V- and W-cycles with different numbers of pre- and postsmoothing steps on a $128^2$ grid. The restriction operators employed are FW, in the case of $v = 2$, and, for $v = 3$, HW. From Table 2.7 we observe that the typical multigrid convergence is obtained by the RBMPS. Furthermore, the convergence improves with increasing $c(x, y)$.                                                      △

### 2.8.2 Time-dependent Problems

One way to apply multigrid to time-dependent problems is to use an *implicit (or semiimplicit) time discretization* and to apply multigrid to each of the (discrete) problems that

**Table 2.7.** Measured convergence factors for a Poisson-like equation (see Example 2.8.1).

|            | FW      |         | HW      |         |
|------------|---------|---------|---------|---------|
| $c(x, y)$  | V(1,1)  | W(1,1)  | V(2,1)  | W(2,1)  |
| 0          | 0.15    | 0.12    | 0.081   | 0.067   |
| $x + y$    | 0.15    | 0.12    | 0.081   | 0.067   |
| $10^5(x + y)$ | 0.10 | 0.10    | 0.040   | 0.037   |

have to be solved in each time step. We explain this approach for two simple, but representative examples, a parabolic and a hyperbolic one. For both cases, we will discuss the discrete problems that arise and the consequences for their multigrid treatment.

**Example 2.8.2 (heat equation)** We consider the parabolic initial boundary value problem

$$
\begin{aligned}
u_t &= \Delta u & (\Omega, t > 0) \\
u(x, y, 0) &= u_0(x, y) & (\Omega, t = 0) \\
u(x, y, t) &= f^{\partial\Omega}(x, y, t) & (\partial\Omega, t > 0)
\end{aligned}
\tag{2.8.5}
$$

for the function $u = u(x, y, t)$, $(x, y) \in \Omega = (0, 1)^2$, $t \geq 0$. The simplest *explicit* discretization is the *forward Euler scheme*

$$
\frac{u_{h,\tau}(x, y, t + \tau) - u_{h,\tau}(x, y, t)}{\tau} = \Delta_h u_{h,\tau}(x, y, t),
$$

where $\tau$ is the step size in the $t$-direction and $h$ the grid size of the space discretization. $\Delta_h$ is again the usual five-point discretization of the Laplace operator.

Obviously, the values of $u_{h,\tau}$ at a new time step $t + \tau$ can be calculated immediately (explicitly) from the values of the previous time step. Such explicit time discretization schemes typically lead to a restrictive "*Courant–Friedrichs–Lewy (CFL) stability condition*" [113] of the form

$$
\tau \leq \text{const}\, h^2,
$$

where the constant is $1/2$ in our example. Implicit time discretization schemes, on the other hand, are *unconditionally stable* if arranged appropriately. In particular, there is no time step restriction.

We consider three implicit schemes, the *backward Euler scheme*

$$
\frac{u_{h,\tau}(x, y, t + \tau) - u_{h,\tau}(x, y, t)}{\tau} = \Delta_h u_{h,\tau}(x, y, t + \tau),
\tag{2.8.6}
$$

*the Crank–Nicolson scheme*

$$
\frac{u_{h,\tau}(x, y, t + \tau) - u_{h,\tau}(x, y, t)}{\tau} = \tfrac{1}{2}(\Delta_h u_{h,\tau}(x, y, t + \tau) + \Delta_h u_{h,\tau}(x, y, t))
\tag{2.8.7}
$$

and the so-called *backward difference formula BDF(2)* [154]

$$
\frac{3u_{h,\tau}(x, y, t + \tau) - 4u_{h,\tau}(x, y, t) + u_{h,\tau}(x, y, t - \tau)}{2\tau} = \Delta_h u_{h,\tau}(x, y, t + \tau).
\tag{2.8.8}
$$

The time discretization accuracy is $O(\tau)$ for the explicit and implicit Euler scheme, $O(\tau^2)$ for the Crank–Nicolson and the BDF(2) scheme. In order to calculate the grid function

$u_{h,\tau}(x, y, t + \tau)$ from $u_{h,\tau}(x, y, t)$ (and from $u_{h,\tau}(x, y, t - \tau)$ in the case of the BDF(2) scheme), one has to solve problems of the form

$$\frac{1}{h^2} \begin{bmatrix} & -1 & \\ -1 & 4 + \alpha(h^2/\tau) & -1 \\ & -1 & \end{bmatrix} u_{h,\tau}(x, y, t + \tau)$$

$$= F_{h,\tau}(u_{h,\tau}(x, y, t), u_{h,\tau}(x, y, t - \tau)) \tag{2.8.9}$$

in each time step, where $F_{h,\tau}\left(u_{h,\tau}(x, y, t), u_{h,\tau}(x, y, t - \tau)\right)$ contains only known values from previous time steps $t, t - \tau$. Here, we have $\alpha = 1$ for backward Euler, $\alpha = 2$ for Crank–Nicolson and $\alpha = 3/2$ for the BDF(2) scheme. In each case, this is obviously a discrete Helmholtz equation corresponding to a Helmholtz constant $c = \alpha/\tau > 0$. As was pointed out in Section 2.8.1, the positive contributions on the main diagonal of the corresponding matrix $A_h$ amplify the diagonal dominance so that the smoothing and convergence factors are improved (compared to the Poisson case). Obviously, the smaller the time steps, the stronger the diagonal dominance and the better are the smoothing and convergence factors.

$\triangle$

In many practical parabolic applications, however, one is interested in using time steps that are as large as possible. For $\tau \to \infty$, the discrete operator in (2.8.9) becomes $-\Delta_h$, so that all multigrid considerations for Poisson's equation apply.

Hyperbolic time-like problems have somewhat different features with respect to the multigrid treatment. Again, we consider a simple example.

**Example 2.8.3 (wave equation)**   We consider the hyperbolic initial boundary value problem

$$\begin{aligned} u_{tt} &= \Delta u & (\Omega, t \in \mathbb{R}) \\ u(x, y, 0) &= u_0(x, y) & (\Omega, t = 0) \\ u_t(x, y, 0) &= u_1(x, y) & (\Omega, t = 0) \\ u(x, y, t) &= f^{\partial\Omega}(x, y, t) & (\partial\Omega, t \in \mathbb{R}) \end{aligned} \tag{2.8.10}$$

for the function $u = u(x, y, t)$, $t \in \mathbb{R}$, $(x, y) \in \Omega = (0, 1)^2$. In such hyperbolic situations, the CFL stability condition for explicit schemes is much more moderate. It typically reads

$$\tau \leq \text{const}\, h.$$

Again, appropriate implicit time discretization schemes are unconditionally stable.

For the wave equation we discuss one common implicit discretization. Similar considerations apply to other implicit schemes. With respect to $t$, we use the approximation

$$u_{tt} = \frac{1}{\tau^2}[1 \quad -2 \quad 1]_\tau u(\cdot, t) + O(\tau^2),$$

whereas the following combination of three time levels is used for the space discretization

$$\Delta u = \Delta_h\left(\tfrac{1}{4}u(x, y, t + \tau) + \tfrac{1}{2}u(x, y, t) + \tfrac{1}{4}u(x, y, t - \tau)\right) + O(h^2) + O(\tau^2).$$

(2.8.11)

In this case, the resulting discrete problem that has to be solved in each time step is characterized by a discrete Helmholtz equation of the form

$$\frac{1}{h^2}\left[\begin{array}{ccc} & -1 & \\ -1 & 4 + h^2/\tau^2 & -1 \\ & -1 & \end{array}\right] u_{h,\tau}(x, y, t + \tau) = F_{h,\tau}(x, y, t),$$

(2.8.12)

which corresponds to the Helmholtz constant $c = h^2/\tau^2 > 0$. Note, that the diagonal dominance reflected by this Helmholtz constant is qualitatively stronger than in the parabolic case. If $\tau = O(h)$ (which is natural in the hyperbolic case) the constant is $c = \text{const}/h^2$ (const $> 0$) leading to a strong diagonal dominance constant, independent of $h$.    △

The considerations in the above examples are typical for more general time-dependent parabolic and hyperbolic problems. If a discretization $L_h$ of an operator $L$ can be treated efficiently by multigrid, this also holds for implicit discretizations of the problems $u_t = -Lu$ and $u_{tt} = -Lu$. The multigrid method for $L_h$ can also be used for the discrete problems that arise per time step. However, if the time steps become very small, even simple relaxation methods may have good convergence properties and may be competitive with (or even more efficient than) multigrid. The larger the time steps, the more will be gained by an appropriate multigrid treatment.

### 2.8.3 Cartesian Grids in Nonrectangular Domains

Here, we give some results for Poisson's equation on nonrectangular domains with Dirichlet boundary conditions. We assume that $\Omega$ is a bounded domain in $\mathbb{R}^2$. In particular, we consider the four examples of domains in Figure 2.12.

Although the domains are nonrectangular, we use again a square Cartesian grid $\Omega_h \cup \Gamma_h$ consisting of the set $\Omega_h$ of *interior grid points* and the set $\Gamma_h$ of *boundary grid points*. Here, $\Omega_h$ and $\mathbf{G}_h$ are defined as in Section 1.3.2. Again, the boundary grid points $\Gamma_h$ are just the discrete intersection points of $\partial\Omega$ with grid lines of $\mathbf{G}_h$.

We distinguish: *regular interior grid points* (all points of $\Omega_h$, whose neighbor grid points in northern, southern, western and eastern direction are also points of $\Omega_h$) and *irregular interior grid points* (all interior grid points which are not regular).

**Figure 2.12.** Four nonrectangular domains.

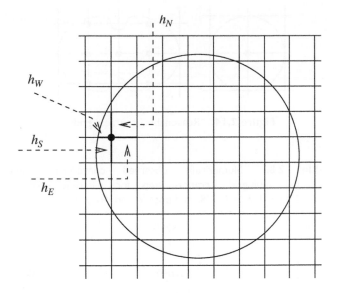

**Figure 2.13.** Irregular interior point with distances $h_i$ to neighbor grid points.

At regular interior grid points, the differential operator $L$ is approximated by the usual second-order five-point discretization (2.8.2). Near the boundary, i.e. at the irregular grid points, the five-point *Shortley–Weller approximation* [355] is used. This means, for example, that the differential operator in (2.8.1) is discretized by

$$
2 \left[
\begin{array}{cccc}
 & \dfrac{b(x_i, y_j)}{h_N(h_N + h_S)} & & \\[2ex]
\dfrac{a(x_i, y_j)}{h_W(h_W + h_E)} & -\dfrac{a(x_i, y_j)}{h_W h_E} - \dfrac{b(x_i, y_j)}{h_N h_S} & \dfrac{a(x_i, y_j)}{h_E(h_W + h_E)} \\[2ex]
 & \dfrac{b(x_i, y_j)}{h_S(h_N + h_S)} & &
\end{array}
\right]_h u_h(x_i, y_j).
$$

$$\tag{2.8.13}$$

Here, $h_N, h_S, h_W$ and $h_E$ denote the distances to the neighbor grid points in northern, southern, western and eastern directions, respectively (see Fig. 2.13).

We investigate the influence of the shape of the domains $\Omega$ shown in Fig. 2.12 on the multigrid convergence behavior. The multigrid algorithm consists of the following components:

- *Coarsening*: Coarse grids are obtained by doubling the (interior) mesh size in both directions, i.e. $\Omega_{2h} = \Omega \cap G_{2h}$. On all grids the discrete operators are constructed in the same way as on the finest grid. The coarsest grid may be "very coarse" (but should still be geometrically reasonable, see Fig. 2.14).
- *Smoothing*: GS-RB with $\nu_1 = 2, \nu_2 = 1$,

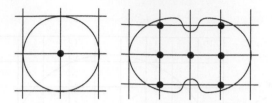

**Figure 2.14.** "Reasonable" coarsest grids.

**Table 2.8.** Numerically observed convergence factors per multigrid iteration step for different domains ($\nu_1 = 2, \nu_2 = 1, h = 1/128$) and both V- and W-cycles.

| Domain | V-cycle | W-cycle |
|---|---|---|
| ☐ | 0.059 | 0.033 |
| ◯ | 0.059 | 0.033 |
| ◇ | 0.063 | 0.032 |
| ∽∽ | 0.058 | 0.033 |
| ∽∽ | 0.088 | 0.033 |

- *Restriction $I_h^{2h}$*: HW,
- *Prolongation $I_{2h}^h$*: linear interpolation.

Here we employ HW instead of FW (according to Remark 2.5.1) which provides the better convergence for $\nu = \nu_1 + \nu_2 = 3$.

Table 2.8 shows numerically observed multigrid convergence factors $\hat{q}^{(m)}$ (as defined in Section 2.5.2, $m$ "large") for both V- and W-cycles for the above four domains. All these domains are comparable in size with the unit square.

In all cases, the corresponding convergence factors given by $\rho_h(\nu = 3)$ are very similar to those on the unit square.

As is usual, the multigrid convergence factors for V-cycles are somewhat worse than those for W-cycles. The convergence factors are nevertheless so good that the V-cycle efficiency (convergence factor per computational work) is better than that of the W-cycle.

### 2.8.4 Multigrid Components for Cell-centered Discretizations

Up to now the discussion in this book was oriented to the vertex-centered discretization of scalar PDEs. As already mentioned in Section 1.2, other discretization approaches exist. One of them, the cell-centered discretization, uses unknowns located at the centers of the grid cells. In the case of Poisson's equation, there is no difference in the order of accuracy of the solution of a vertex- or a cell-centered discretization. Often the choice of discretization depends on the applications, the boundary conditions, but also on personal taste or history. In particular, the treatment of boundary conditions is different in vertex- and cell-centered discretizations.

Efficient multigrid methods can also be developed for cell-centered discretizations. As smoothers, we can use the same schemes as for the vertex-centered discretizations considered so far, for example, $\omega$-JAC, GS-RB or GS-LEX.

The main difference in the multigrid algorithm is that the coarse grid points do not form a subset of the fine grid points (see Fig. 2.15 showing a fine and a coarse grid in the case of a cell-centered location of unknowns). Therefore, we will briefly discuss the transfer operators, which, of course, depend on the arrangement of the unknowns.

A frequently used *restriction operator* $I_h^{2h}$ for cell-centered discretizations is the four-point average

$$
\begin{aligned}
I_h^{2h} d_h(x, y) &= \frac{1}{4} \begin{bmatrix} 1 & & 1 \\ & \cdot & \\ 1 & & 1 \end{bmatrix}_h^{2h} d_h \\
&= \frac{1}{4} \left[ d_h \left( x - \frac{h}{2}, y - \frac{h}{2} \right) + d_h \left( x - \frac{h}{2}, y + \frac{h}{2} \right) \right. \\
&\quad \left. + d_h \left( x + \frac{h}{2}, y - \frac{h}{2} \right) + d_h \left( x + \frac{h}{2}, y + \frac{h}{2} \right) \right].
\end{aligned}
\tag{2.8.14}
$$

Figure 2.16 presents a fine grid with the symbols for the fine and coarse grid points corresponding to the interpolation formula (2.8.15).

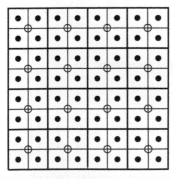

**Figure 2.15.** Arrangement of unknowns on the $h$- and $2h$-grids for a cell-centered discretization with fine grid points (•) and coarse grid points (○).

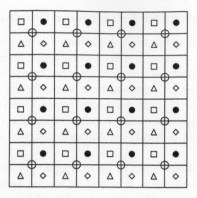

**Figure 2.16.** A fine grid with symbols explaining the bilinear interpolation (2.8.15) used for the transfer from the coarse grid (○) to the fine grid.

The *bilinear interpolation operator* for cell-centered discretizations is given by

$$
I_{2h}^h \hat{v}_{2h}(x, y) =
\begin{cases}
\dfrac{1}{16}\Big[9\hat{v}_{2h}\Big(x - \dfrac{h}{2}, y - \dfrac{h}{2}\Big) + 3\hat{v}_{2h}\Big(x - \dfrac{h}{2}, y + \dfrac{3h}{2}\Big) \\
\qquad + 3\hat{v}_{2h}\Big(x + \dfrac{3h}{2}, y - \dfrac{h}{2}\Big) + \hat{v}_{2h}\Big(x + \dfrac{3h}{2}, y + \dfrac{3h}{2}\Big)\Big] & \text{for } \bullet \\[6pt]
\dfrac{1}{16}\Big[3\hat{v}_{2h}\Big(x - \dfrac{3h}{2}, y - \dfrac{h}{2}\Big) + \hat{v}_{2h}\Big(x - \dfrac{3h}{2}, y + \dfrac{3h}{2}\Big) \\
\qquad + 9\hat{v}_{2h}\Big(x + \dfrac{h}{2}, y - \dfrac{h}{2}\Big) + 3\hat{v}_{2h}\Big(x + \dfrac{h}{2}, y + \dfrac{3h}{2}\Big)\Big] & \text{for } \square \\[6pt]
\dfrac{1}{16}\Big[3\hat{v}_{2h}\Big(x - \dfrac{h}{2}, y - \dfrac{3h}{2}\Big) + 9\hat{v}_{2h}\Big(x - \dfrac{h}{2}, y + \dfrac{h}{2}\Big) \\
\qquad + \hat{v}_{2h}\Big(x + \dfrac{3h}{2}, y - \dfrac{3h}{2}\Big) + 3\hat{v}_{2h}\Big(x + \dfrac{3h}{2}, y + \dfrac{h}{2}\Big)\Big] & \text{for } \diamond \\[6pt]
\dfrac{1}{16}\Big[\hat{v}_{2h}\Big(x - \dfrac{3h}{2}, y - \dfrac{3h}{2}\Big) + 3\hat{v}_{2h}\Big(x - \dfrac{3h}{2}, y + \dfrac{h}{2}\Big) \\
\qquad + 3\hat{v}_{2h}\Big(x + \dfrac{h}{2}, y - \dfrac{3h}{2}\Big) + 9\hat{v}_{2h}\Big(x + \dfrac{h}{2}, y + \dfrac{h}{2}\Big)\Big] & \text{for } \triangle,
\end{cases}
$$

$$(2.8.15)$$

which in stencil notation simply corresponds to

$$
I_{2h}^h \hat{v}_{2h}(x, y) = \frac{1}{16}
\begin{bmatrix}
1 & 3 & 3 & 1 \\
3 & 9 & 9 & 3 \\
3 & 9 & 9 & 3 \\
1 & 3 & 3 & 1
\end{bmatrix}_{2h}^{h}
\hat{v}_{2h}(x, y). \tag{2.8.16}
$$

**Remark 2.8.1**  The above transfer operators for cell-centered locations of unknowns and discretizations are $O(h^2)$ accurate and thus well-suited for Poisson's equation (see Remark 2.7.1).                                                                      ≫

## 2.9 MULTIGRID IN 3D

### 2.9.1 The 3D Poisson Problem

Multigrid has essentially the same properties (complexity) for 3D as for 2D problems. We start with the discussion of a *standard 3D multigrid* method for the 3D Poisson equation, the 3D analog of Model Problem 1 (1.4.1):

**Model Problem 2**

$$-\Delta_h u_h(x, y, z) = f_h^\Omega(x, y, z) \quad ((x, y, z) \in \Omega_h)$$
$$u_h(x, y, z) = f_h^\Gamma(x, y, z) \quad ((x, y, z) \in \Gamma_h)$$

(2.9.1)

*in the unit cube* $\Omega = (0, 1)^3 \subset \mathbb{R}^3$ *with* $h = 1/n, \ n \in \mathbb{N}$.

In 3D the infinite grid $\mathbf{G}_h$ is defined by

$$\mathbf{G}_h := \{(x, y, z): x = ih_x, \ y = jh_y, \ z = kh_z; \ i, j, k \in \mathbf{Z}\}\ ,$$

(2.9.2)

where $\boldsymbol{h} = (h_x, h_y, h_z)$ is a vector of fixed mesh sizes. In the special case of cubic Cartesian grids, we simply identify $h = h_x = h_y = h_z$. $\Delta_h$ denotes the standard seven-point $O(h^2)$-approximation of the 3D Laplace operator $\Delta$ and can be written as

$$L_h u_h(x, y, z) = -\Delta_h u_h(x, y, z)$$

$$= \frac{1}{h^2} \begin{bmatrix} 0 & 0 & 0 \\ 0 & -1 & 0 \\ 0 & 0 & 0 \end{bmatrix}_h u_h(x, y, z - h)$$

$$+ \frac{1}{h^2} \begin{bmatrix} 0 & -1 & 0 \\ -1 & 6 & -1 \\ 0 & -1 & 0 \end{bmatrix}_h u_h(x, y, z)$$

$$+ \frac{1}{h^2} \begin{bmatrix} 0 & 0 & 0 \\ 0 & -1 & 0 \\ 0 & 0 & 0 \end{bmatrix}_h u_h(x, y, z + h),$$

(2.9.3)

where the 2D stencils are applied to the $x$- and $y$-coordinates as in 2D. Introducing the *3D stencil notation*, we write in short

$$-\Delta_h u_h = \frac{1}{h^2} \left[ \begin{bmatrix} 0 & 0 & 0 \\ 0 & -1 & 0 \\ 0 & 0 & 0 \end{bmatrix}_h \begin{bmatrix} 0 & -1 & 0 \\ -1 & 6 & -1 \\ 0 & -1 & 0 \end{bmatrix}_h \begin{bmatrix} 0 & 0 & 0 \\ 0 & -1 & 0 \\ 0 & 0 & 0 \end{bmatrix}_h \right] u_h.$$

### 2.9.2 3D Multigrid Components

*Standard coarsening* is as in 2D ($H = 2h, h = 1/n, n$ even). Figure 2.17 shows part of a fine grid with the coarse grid points.

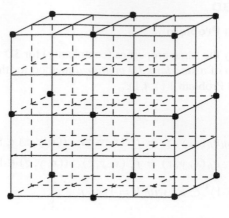

● : coarse grid point

**Figure 2.17.** A (vertex-centered) 3D fine grid with a coarse grid (standard coarsening).

For the transfer operators, we obtain the following generalizations of the 2D operators. The 3D FW operator is given by

$$\frac{1}{64}\left[\begin{bmatrix} 1 & 2 & 1 \\ 2 & 4 & 2 \\ 1 & 2 & 1 \end{bmatrix}_h^{2h} \begin{bmatrix} 2 & 4 & 2 \\ 4 & 8 & 4 \\ 2 & 4 & 2 \end{bmatrix}_h^{2h} \begin{bmatrix} 1 & 2 & 1 \\ 2 & 4 & 2 \\ 1 & 2 & 1 \end{bmatrix}_h^{2h}\right]. \tag{2.9.4}$$

HW is represented by the stencil

$$\frac{1}{12}\left[\begin{bmatrix} 0 & 0 & 0 \\ 0 & 1 & 0 \\ 0 & 0 & 0 \end{bmatrix}_h^{2h} \begin{bmatrix} 0 & 1 & 0 \\ 1 & 6 & 1 \\ 0 & 1 & 0 \end{bmatrix}_h^{2h} \begin{bmatrix} 0 & 0 & 0 \\ 0 & 1 & 0 \\ 0 & 0 & 0 \end{bmatrix}_h^{2h}\right]. \tag{2.9.5}$$

The generalization of the 2D *bilinear interpolation operator* introduced in (2.3.7) to 3D is *trilinear interpolation*. Along planes that contain coarse grid points the formulas from (2.3.7) are still valid (with an additional index for the respective third dimension). One extra formula is necessary for the trilinear interpolation to the fine grid points not belonging to any plane of the coarse grid (those with eight coarse grid neighbors).

$$\begin{aligned}
\hat{v}_h(x, y, z) &= I_{2h}^h \hat{v}_{2h}(x, y, z) \\
&= \tfrac{1}{8}[\hat{v}_{2h}(x + h, y + h, z + h) + \hat{v}_{2h}(x + h, y + h, z - h) \\
&\quad + \hat{v}_{2h}(x + h, y - h, z + h) + \hat{v}_{2h}(x + h, y - h, z - h) \\
&\quad + \hat{v}_{2h}(x - h, y + h, z + h) + \hat{v}_{2h}(x - h, y + h, z - h) \\
&\quad + \hat{v}_{2h}(x - h, y - h, z + h) + \hat{v}_{2h}(x - h, y - h, z - h)].
\end{aligned} \tag{2.9.6}$$

Formula (2.9.6) is illustrated in Fig. 2.18, in which such a fine grid point and the eight coarse grid neighbor points are shown.

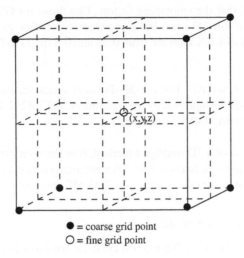

● = coarse grid point
○ = fine grid point

**Figure 2.18.** The 3D interpolation (2.9.6).

The *smoothers* $\omega$-JAC, GS-LEX and GS-RB as described for Model Problem 1 can also be generalized immediately to 3D. It is also straightforward to determine the *smoothing factor* (2.1.8) of $\omega$-JAC for Model Problem 2. First, the discrete eigenfunctions of the discrete 3D operator $\Delta_h$ are

$$\varphi_h^{k,\ell,m}(x, y) = \sin k\pi x \, \sin l\pi y \, \sin m\pi z$$
$$((x, y, z) \in \Omega_h \ (k, \ell, m = 1, \dots, n - 1)). \tag{2.9.7}$$

The corresponding eigenvalues of the $\omega$-JAC relaxation operator are

$$\chi_h^{k,\ell} = \chi_h^{k,\ell}(\omega) = 1 - \frac{\omega}{3}(3 - \cos k\pi h - \cos \ell\pi h - \cos m\pi h). \tag{2.9.8}$$

For standard coarsening, the low frequency components are the $\varphi_h^{k,\ell,m}$ with $\max(k, \ell, m) < n/2$, the high frequency components are those with $n/2 \leq \max(k, \ell, m) < n$. Thus, the smoothing factor $\mu(h; \omega)$ of $S_h$ (representing the worst factor by which high frequency error components are reduced per relaxation step) and its supremum $\mu^*$ over $h$ are

$$\mu(h; \omega) := \max\{|\chi_h^{k,\ell,m}(\omega)| : n/2 \leq \max(k, \ell, m) \leq n - 1\}$$
$$\mu^*(\omega) := \sup_{h \in \mathcal{H}} \mu(h; \omega). \tag{2.9.9}$$

It is straightforward to find the smoothing factor of $\omega$-JAC for Model Problem 2 as

$$\mu^*(\omega) = \max\{|1 - \omega/3|, |1 - 2\omega|\}.$$

Optimal smoothing is obtained for $\omega = 6/7$ ($\approx 0.857$) for which we find the smoothing factor $\mu^* = 5/7$ ($\approx 0.714$).

It should be noted, that the smoothing factors, like those for GS-LEX and GS-RB, in 3D are not identical to the 2D case, but somewhat worse. For example, the smoothing factors of GS-RB and GS-LEX increase from 0.25 and 0.5 in 2D to 0.445 and 0.567 in 3D, respectively.

**Remark 2.9.1 (3D $\omega$-GS-RB)**   For the 3D Poisson equation, the smoothing factor of GS-RB can be substantially improved by *overrelaxation* [427, 428]. For example, choosing $\omega = 1.15$ gives a smoothing factor $\mu \approx 0.23$.                                          $\gg$

For the FMG version of a 3D multigrid method, it is necessary to apply a 3D analog of the 2D FMG interpolation (see Section 2.6). Tricubic interpolation is a direct generalization of the bicubic interpolation. It can be used as a 3D FMG interpolation procedure.

### 2.9.3 Computational Work in 3D

The *computational work $W_\ell$* per 3D multigrid cycle $\Omega_\ell$ for Model Problem 2 is recursively given by (2.4.8) as in 2D. In case of *3D standard coarsening* with fixed cycle index $\gamma$ we have

$$N_k \doteq 8 N_{k-1} \quad (k = 1, 2, \ldots, \ell) \tag{2.9.10}$$

where $N_k = \#\Omega_k$ (number of grid points on $\Omega_k$) and "$\doteq$" again means equality up to lower order terms (boundary effects). Similarly as in the 2D case (see Section 2.4.3), one immediately obtains the following estimate for the *total computational work $W_\ell$ of one complete multigrid cycle in 3D* from (2.4.14). The computational work for the F-cycle is found similarly as in Remark 2.4.8:

$$W_\ell \dot{\leq} \begin{cases} \frac{8}{7} C N_\ell & \text{for } \gamma = 1 \\ \frac{4}{3} C N_\ell & \text{for } \gamma = 2 \\ \frac{8}{5} C N_\ell & \text{for } \gamma = 3 \\ \frac{64}{49} C N_\ell & \text{for F-cycle.} \end{cases} \tag{2.9.11}$$

This estimate of $W_\ell$ together with the $h$-independent convergence shows the asymptotic optimality of iterative 3D multigrid methods. For standard coarsening, multigrid cycles remain asymptotically optimal up to $\gamma \leq 7$ (in 2D they were asymptotically optimal only up to $\gamma \leq 3$).

The computational work $W_\ell^{\text{FMG}}$ needed for the FMG method is again easily estimated. By arguments similar to those in Section 2.6.2, one obtains, for example

$$W_\ell^{\text{FMG}} \dot{\leq} \frac{8}{7} r W_\ell + \frac{8}{7} W_{\ell-1}^{\text{INT}} \quad \text{for standard coarsening in 3D} \tag{2.9.12}$$

(neglecting lower order terms). Here $W_{\ell-1}^{\text{INT}}$ denotes the work needed for the FMG interpolation process from grid $\Omega_{\ell-1}$ to the finest grid $\Omega_\ell$ and $W_\ell$ is the work required for one multigrid cycle on the finest level $\Omega_\ell$.

# 3

# ELEMENTARY MULTIGRID THEORY

In this chapter, we will introduce some elementary multigrid theory. The main purpose of this theory is to show the *h-independent fast convergence* of multigrid iterations. This means that we will derive estimates for the multigrid convergence factor $\rho(M_h)$ of the form

$$\rho(M_h) \leq \text{const} \ll 1. \tag{3.0.1}$$

Together with the fact that we can estimate the computational work per multigrid cycle by $W_\ell \leq \text{const } N$, (we can actually count the number of operations, see Section 2.4.3), this estimate gives what is often called the "optimality" of multigrid methods. We have already seen from practical measurements (see Section 2.5.1) that (3.0.1) indeed holds for the RBMPS.

Our approach is *quantitative*. This means that we are interested in *h*-independent *realistic bounds* of $\rho(M_h)$ which are close to those that can be measured numerically. The theoretical results can then be used to compare different multigrid algorithms and to find suitable, or even optimal, components.

In addition to the multigrid convergence factor $\rho(M_h)$ and the error reduction factor $||M_h||$ (in some appropriate norm $|| \cdot ||$), we are also interested in realistic quantitative bounds of

- smoothing factors $\mu(S_h)$,
- two-grid convergence factors $\rho(M_h^{2h})$ and corresponding norms $||M_h^{2h}||$,
- FMG error estimates.

All these quantities will be derived in this chapter and Chapter 4. The tools that we will use are all elementary: some linear algebra, Fourier analysis and simple estimates.

The elementary multigrid theory in this chapter is only one approach for analyzing multigrid methods theoretically. We will distinguish three other theoretical approaches:

- the *local Fourier analysis* (LFA),
- the *classical* (qualitative) multigrid theory,
- the (qualitative) multigrid theory based on subspace splitting.

LFA is closely related to the elementary multigrid theory and is also quantitative. It is the kernel of Brandt's theoretical understanding of multigrid [58]. In Chapter 4, we will present LFA not as a theory [69, 373], but as a tool. *With respect to multigrid practice, we regard LFA as the most important tool for the design of multigrid algorithms.*

The "classical" qualitative multigrid theory goes back to Hackbusch [173]. We will outline its idea in Section 3.4.5. The other qualitative multigrid theories rely on the idea of space splitting and subspace correction methods, see, for example [54, 55, 298, 425]. Appendix B of this book gives an introduction into this theory. The qualitative multigrid theories are more general than the elementary theory presented in this chapter. However, they do not reflect quantitatively the convergence factors of particular multigrid methods that are practically observed and measured.

Before we present the results, we will survey the elementary multigrid theory. We will guide the reader through the theoretical fields and recommend where multigrid beginners and practitioners may start reading in this book.

## 3.1 SURVEY

The multigrid theory which we will present here consists of two parts.

(1) The *two*-grid operator $M_h^{2h}$ is analyzed. Concretely, estimates for $\rho(M_h^{2h})$ and $||M_h^{2h}||$ in an appropriate norm are derived (see Section 3.3).

(2) Based on these results, the *multi*grid operator $M_h$ is analyzed. Estimates for $||M_h||$ and FMG error estimates are derived (see Section 3.2).

Here, the first part turns out to be the hard one. For the second part, elementary estimations give the desired results, under rather general assumptions for W-cycles.

Therefore, we will start with the second part, which answers the question: *Why is it sufficient to derive realistic two-grid convergence factors?* Here we derive bounds for $||M_h||$ (see Section 3.2.1) and FMG error estimates (see Section 3.2.2).

---

**Norms for** $||M_h||$   If $h$-independent small bounds for $||M_h^{2h}||$ are known, a simple and general recursive estimate yields small bounds for $||M_h||$ for W-cycles. This estimate is directly based on the recursive definition of $M_h$ in Theorem 2.4.1.

**FMG error estimate**   On the basis of bounds for $||M_h||$, a similarly simple and general estimate yields bounds for

$$|| u_h - u_h^{\text{FMG}} || \quad (\text{or } || u - u_h^{\text{FMG}} ||, \text{ respectively})$$

where $u_h^{\text{FMG}}$ is the FMG solution of the given problem. In particular, we will show that an approximate solution up to discretization accuracy can be obtained by the FMG process in $O(N)$ operations, with a small constant of proportionality.

---

**Remark 3.1.1 (convergence for V-cycles)** The estimates fail for V-cycles, i.e. for $\gamma = 1$ and $h \to 0$ (with a fixed coarsest grid). For V-cycles, a more sophisticated theoretical analysis is needed [48, 49, 52, 176]. A V-cycle multigrid convergence proof is also given in Appendix B. $\gg$

The remaining part of this chapter (and also parts of Chapter 4) deals with the harder question: *How to derive realistic two-grid convergence factors $\rho(M_h^{2h})$ and $||M_h^{2h}||$?* In Section 3.3, we will present corresponding results for the RBMPS introduced in Section 2.5.1. The results are based on what we call *rigorous Fourier analysis*. The essential (six) steps of the two-grid rigorous Fourier analysis presented in Section 3.3.4 are characteristic. We do not urge those readers who are more interested in practical questions to follow all the details of that elementary, but somewhat lengthy, section. Those readers are referred to the LFA in Chapter 4, which is more important for the practical design of multigrid algorithms. The connections between rigorous Fourier analysis and LFA are briefly addressed in Section 3.4.4.

The range of rigorous Fourier analysis is, however, rather limited. Section 3.4 discusses its range of applicability and sets pointers to the field of qualitative analysis, where many results have been obtained, starting with Hackbusch's distinction between *smoothing property* and *approximation property* (see Section 3.4.5, [173, 176] and the references therein). We will not discuss these qualitative approaches in detail.

## 3.2 WHY IT IS SUFFICIENT TO DERIVE TWO-GRID CONVERGENCE FACTORS

### 3.2.1 $h$-Independent Convergence of Multigrid

In this section we will prove the following fundamental result on the multigrid convergence:

> If a given two-grid method converges sufficiently well, i.e.
>
> $$||M_h^{2h}|| \leq \sigma^*,$$
>
> with $\sigma^*$ small enough and independent of $h$, then the corresponding multigrid method with $\gamma \geq 2$ will have similar convergence properties, under natural assumptions.

For the simple proof of this statement, it is natural to make use of Theorem 2.4.1 and Remark 2.4.3. Remark 2.4.3 shows that a multigrid method can be regarded as a perturbed two-grid method. The perturbation is formally represented by (2.4.7). We summarize this observation as follows:

**Corollary 3.2.1** *For $k = 1, \ldots, \ell - 1$, the equations*

$$M_{k+1} = M_{k+1}^k + A_k^{k+1}(M_k)^\gamma A_{k+1}^k \tag{3.2.1}$$

*hold, where*

$$A_k^{k+1} := (S_{k+1})^{\nu_2} I_k^{k+1} : \mathcal{G}(\Omega_k) \to \mathcal{G}(\Omega_{k+1}),$$
$$A_{k+1}^k := (L_k)^{-1} I_{k+1}^k L_{k+1} (S_{k+1})^{\nu_1} : \mathcal{G}(\Omega_{k+1}) \to \mathcal{G}(\Omega_k) \tag{3.2.2}$$

*and $M_{k+1}^k$ ($\mathcal{G}(\Omega_{k+1}) \to \mathcal{G}(\Omega_{k+1})$) is as in (2.4.6) with $k+1$ instead of $k$.*

For the following central theorem, we have to assume that estimates for $||M_{k+1}^k||$ and for $||A_k^{k+1}||$ and $||A_{k+1}^k||$ are known. Here, the assumption on $||M_{k+1}^k||$ is the crucial one. Bounds for $||A_k^{k+1}||$ and $||A_{k+1}^k||$ are usually easy to calculate. We do not make specific assumptions with respect to the norms of the operators under consideration. The spectral norm $|| \cdot ||_S$ is a natural choice here.

**Theorem 3.2.1** *Let the following estimates hold uniformly with respect to $k$ ($\leq \ell - 1$):*

$$|| M_{k+1}^k || \leq \sigma^*, \qquad || A_k^{k+1} || \cdot || A_{k+1}^k || \leq C. \tag{3.2.3}$$

*Then we have $|| M_\ell || \leq \eta_\ell$ where $\eta_\ell$ is recursively defined by*

$$\eta_1 := \sigma^*, \qquad \eta_{k+1} := \sigma^* + C\eta_k^\gamma \quad (k = 1, \ldots, \ell - 1). \tag{3.2.4}$$

*If we additionally assume that*

$$4C\sigma^* \leq 1 \quad and \quad \gamma = 2, \tag{3.2.5}$$

*we obtain the following uniform estimate for $M_h = M_\ell$: ($h = h_\ell$):*

$$\boxed{|| M_h || \leq \eta := (1 - \sqrt{1 - 4C\sigma^*})/2C \leq 2\sigma^* \quad (\ell \geq 1)}. \tag{3.2.6}$$

*Proof.* (3.2.4) immediately follows from (3.2.1) and (3.2.3). If $\gamma = 2$ and $4C\sigma^* \leq 1$, we obtain (3.2.6) from (3.2.4) using the limit equation $\eta = \sigma^* + C\eta^2$.  $\square$

**Remark 3.2.1 (optimality of the multigrid cycle)** The existence of this $h$-independent upper bound for the convergence of the multigrid cycle (3.2.6) *and* the fact that the number of operations per cycle is $O(N)$ (see (2.4.12) or (2.4.14)), together imply the optimality of the multigrid cycle. In order to achieve a (fixed) error (or defect) reduction by a factor of $\varepsilon$, $O(N \log \varepsilon)$ operations are sufficient.  $\gg$

This optimality result is particularly impressive if we calculate the constants which are characteristic in the $O$-statement in Remark 3.2.1. We will see in Section 3.3.2 that we can indeed derive small realistic bounds of $\sigma^*$ for typical two-grid methods. In such cases, we will see that $\eta \approx \sigma^*$ in (3.2.6). For example, if $C = 1$, we obtain from (3.2.6)

$$\eta \leq 0.113 \quad \text{if } \sigma^* \leq 0.1,$$

i.e., if the two-grid reduction factor $||M_h^{2h}|| \leq 0.1$, then the multigrid reduction factor $||M_h|| \leq 0.113$. Typically, the constant $C$ is $\geq 1$, but not very large. For Model Problem 1 and $|| \cdot || = || \cdot ||_S$, for example, we have $C \leq \sqrt{2}$ for all choices of $\nu_1$ and $\nu_2$ and $C \searrow 1$ if $\nu_1 \to \infty$.

> In this respect, for the construction of multigrid methods, it is *usually sufficient to analyze only the corresponding two-grid method*. Furthermore, it is usually not necessary to work with $\gamma > 2$ if $\sigma^*$ is sufficiently small.

**Remark 3.2.2 (fixed coarsest grid size)**  For the theoretical investigations of multigrid methods ($h$-independence of convergence factors $\rho$), one needs to consider the case that $h = h_\ell$ tends to 0. For such asymptotic investigations, *we always consider the coarsest grid to be fixed (mesh size $h_0$) and let the number $\ell$ of grids tend to infinity*. If the coarsest mesh size $h_0$ is not fixed, but the *number of levels* is fixed, with $h$ tending to 0, the coarsest mesh size also tends to 0. Theory becomes easier then, but *the algorithm can no longer be guaranteed to be an $O(N)$ algorithm* (because the coarsest grid problem becomes larger and larger). We do not discuss such methods systematically in this book.                    $\gg$

**Remark 3.2.3 (modified V-cycles)**  If $\gamma = 1$, i.e. if V-cycles are used, Theorem 3.2.1 does not give an $h$-independent (i.e. $\ell$-independent) upper bound for $|| M_\ell ||$. However, instead of $\gamma = 1$ one can use, for example, a combination of a V-cycle (on the other levels) and of a W-cycle (on the other levels), i.e., $\gamma = \gamma_k$ with $\gamma_k = 1$ if $\ell \geq k \geq \ell - \ell_0$ (for some $\ell_0$) and $\gamma_k = 2$ otherwise. For larger values of $\ell_0$ (many levels of V-cycle type), this would result in only a slight increase of the computational work compared to the V-cycle. For a cycle of this type, Theorem 3.2.1 could, in principle, be used to derive $\ell$-independent bounds for $|| M_\ell ||$. For large $\ell_0$, we would then, however, have to assume $\sigma^*$ to be extremely small and the estimate would become unrealistic from a practical point of view.                    $\gg$

### 3.2.2 A Theoretical Estimate for Full Multigrid

In this section, we will derive an elementary, but general estimate for FMG. The proof is similarly simple as the proof of Theorem 3.2.1. The main result of this section will be the following statement.

> Under natural assumptions, FMG provides approximations with discretization accuracy (e.g. $O(h^2)$ for Model Problem 1). Together with the fact that the number of operations in FMG is $O(N)$ (see Section 2.6), this is the reason for the optimality of FMG. Discretization accuracy is achieved in $O(N)$ operations.

We recall the procedures and the notation introduced in Section 2.6.

For simplicity, we again consider only the case of standard coarsening. We do not make specific assumptions about the norms $|| \cdot ||$ on $\mathcal{G}(\Omega_k)$ and the corresponding operator norms.

(A standard choice is $|| \cdot ||_2$ and, correspondingly, $|| \cdot ||_S$, see, however, Remark 3.2.5.) We make the following natural assumptions:

(1) Let the norm of the multigrid iteration operator $M_k$ be uniformly bounded

$$||M_k|| \leq \eta < 1 \quad (k = 1, 2, \ldots).$$

(2) Let the norm of the FMG interpolation operator $\Pi_{k-1}^k (k = 1, 2, \ldots)$ be uniformly bounded

$$||\Pi_{k-1}^k|| \leq P \quad (k = 1, 2, \ldots). \tag{3.2.7}$$

(3) The discretization error and the FMG interpolation error are assumed to be of order $\kappa$ and $\kappa_{FMG}$, respectively:

$$||u - u_h|| \leq K h^\kappa \quad (h = h_k; \ k = 1, 2, \ldots), \tag{3.2.8}$$

$$||u - \Pi_{k-1}^k u|| \leq \bar{K} h^{\kappa_{FMG}} \quad (h = h_k; \ k = 1, 2, \ldots). \tag{3.2.9}$$

Assumption (1) is exactly what we have established in the previous section. Assumption (2) can usually be easily verified in typical cases. $P$ turns out to be a (small) constant ($\geq 1$). We will first additionally assume that the FMG-interpolation accuracy is higher than the discretization accuracy:

$$\kappa_{FMG} > \kappa.$$

The case $\kappa_{FMG} = \kappa$ will be discussed in Remark 3.2.4.

Under these assumptions, one or a few multigrid cycles ($r$ = number of cycles) are sufficient to achieve the discretization accuracy for $u_\ell^{FMG}$ provided $\eta$ is small enough.

**Theorem 3.2.2** *Let the assumptions (1), (2) and (3) be fulfilled and assume additionally that*

$$\eta^r < \frac{1}{2^\kappa P}.$$

*Then the following estimate holds (for any $\ell \geq 1$):*

$$||u_\ell - u_\ell^{FMG}|| \leq \delta h^\kappa \quad (h = h_\ell), \tag{3.2.10}$$

*where*

$$\delta = \eta^r \frac{B}{1 - \eta^r A},$$

*with $A = 2^\kappa P$ and any bound $B$ such that $K(1 + A) + \bar{K} h^{\kappa_{FMG} - \kappa} \leq B$.*

*Proof.* By definition of the FMG method, we have for all $\ell \geq 1$

$$u_\ell^{FMG} - u_\ell = (M_\ell)^r (u_\ell^0 - u_\ell), \quad u_\ell^0 = \Pi_{\ell-1}^\ell u_{\ell-1}^{FMG}.$$

Using the identity

$$u_\ell^0 - u_\ell = \Pi_{\ell-1}^\ell (u_{\ell-1}^{FMG} - u_{\ell-1}) + \Pi_{\ell-1}^\ell (u_{\ell-1} - u) + (\Pi_{\ell-1}^\ell u - u) + (u - u_\ell),$$

we obtain the recursive estimate

$$\delta_\ell \leq \eta^r (A\delta_{\ell-1} + KA + \bar{K}h_\ell^{\kappa_{FMG} - \kappa} + K),$$

where $\delta_k := \| u_k^{FMG} - u_k \| / h_k^\kappa$ $(k = \ell, \ell-1, \ldots, 1)$. From this, (3.2.10) follows by a simple calculation. $\qquad\square$

## Supplement

(1) If we additionally have a lower bound for the discretization error

$$\| u - u_\ell \| \geq \hat{K}(h_\ell)^\kappa \quad (\hat{K} > 0, \text{ independent of } \ell),$$

we obtain from (3.2.10),

$$\| u_\ell - u_\ell^{FMG} \| \leq \beta \| u - u_\ell \| \quad \text{with } \beta := \frac{\delta}{\hat{K}}. \qquad (3.2.11)$$

and, trivially,

$$\| u - u_\ell^{FMG} \| \leq (1 + \beta)\| u - u_\ell \|.$$

(2) If we also assume that an asymptotic expansion

$$u_\ell = u + (h_\ell)^\kappa e + o(h_\ell^\kappa)$$

exists, we can replace $K/\hat{K}$ by 1, so that $\beta$ becomes asymptotically

$$\beta^* = \eta^r \frac{1 + A}{1 - \eta^r A}. \qquad (3.2.12)$$

**Remark 3.2.4 (order of FMG interpolation)** The above proof is also valid if $\kappa_{FMG} = \kappa$. In that case, the contribution of the FMG interpolation constant $\bar{K}$ to the bounds $\delta$ and $\beta$ may become arbitrarily large (depending on $u$). (For details see Section 6.4 in [378].)

*Therefore, based on practical experience, we recommend readers to choose an FMG interpolation of an order which is higher than that of the discretization.* In this case the contribution of the FMG interpolation constant $\bar{K}$ to the bounds $\delta^*$ and $\beta^*$ vanishes asymptotically.

**Remark 3.2.5** We have made no explicit assumption about the norms that are used in this section. The interpretation of the results in this section, however, may be different for different norms. For instance, the order $\kappa_{FMG}$ (and also the order $\kappa$) may depend on the choice of norms. (For example, norms including discrete derivatives lead to lower orders.)

## 3.3 HOW TO DERIVE TWO-GRID CONVERGENCE FACTORS BY RIGOROUS FOURIER ANALYSIS

In this section, we show how Fourier analysis can be used to derive sharp bounds for $\rho(M_h^H)$ and $\| M_h^H \|$. The purpose of this section is two-fold: first, we want to introduce the idea of rigorous Fourier analysis and to explain its structure. Secondly, we want to derive concrete results for the RBMPS and for some variants (see Section 3.3.4). In Sections 3.3.1 and 3.3.2, we present the results of this analysis. The bounds for $\rho(M_h^H)$ and $\| M_h^H \|$ are small ($\approx 0.1$) and $h$-independent.

### 3.3.1 Asymptotic Two-grid Convergence

When being applied to Model Problem 1, the rigorous Fourier analysis gives the following convergence theorem (the essential steps of the proof will be given in Section 3.3.4):

**Theorem 3.3.1** *For the two-grid version of the RBMPS as described in Section 2.5, we obtain the asymptotic convergence factors*

$$\rho^* = \sup_{h \in \mathcal{H}} \rho(M_h^{2h}), \tag{3.3.1}$$

*listed in Table 3.1 (where $\mathcal{H}$ represents the set of admissible mesh sizes). Here, the number $\nu$ of smoothing steps is varied and the FW and HW restriction operators are compared.*

**Remark 3.3.1** According to a detailed efficiency analysis (Section 8.2 in [378]), which takes into account the convergence behavior and the computational effort per cycle, the two algorithms

$$\text{GS-RB}, \quad \nu_1 = \nu_2 = 1, \quad \text{FW}$$

and

$$\text{GS-RB}, \quad \nu_1 = 2, \quad \nu_2 = 1, \quad \text{HW}$$

turn out to be the most efficient multigrid Poisson solvers (within this class of algorithms), which is in agreement with the results in Section 2.5.3.

In fact, the two-grid convergence factors in Table 3.1 are quite close to the *multigrid W-cycle convergence factors* observed in Fig. 2.10 and Table 2.2. ≫

**Table 3.1.** Two-grid convergence factors $\rho^* = \rho^*(\nu)$ for Model Problem 1 using GS-RB.

|    | $\nu = 1$ | $\nu = 2$ | $\nu = 3$ | $\nu = 4$ |
|----|-----------|-----------|-----------|-----------|
| FW | 0.250 | 0.074 | 0.053 | 0.041 |
| HW | 0.500 | 0.125 | 0.033 | 0.025 |

**Table 3.2.** Two-grid convergence factors $\rho^* = \rho^*(\nu)$ obtained with $\omega$-JAC and FW for Model Problem 1.

|  | $\nu = 1$ | $\nu = 2$ | $\nu = 3$ | $\nu = 4$ |
|---|---|---|---|---|
| $\omega = 4/5$ | 0.600 | 0.360 | 0.216 | 0.137 |
| $\omega = 1/2$ | 0.750 | 0.563 | 0.422 | 0.316 |

If we replace the GS-RB smoother in the above algorithm by $\omega$-JAC relaxation, we obtain significantly worse convergence factors. For FW and $\omega = 4/5$ (optimal $\omega$) or $\omega = 1/2$, we obtain the two-grid convergence factors $\rho^*(\nu)$ in Table 3.2 (for improvements with varying $\omega$, see Section 7.4.2).

From Tables 3.1 and 3.2 we see that (asymptotically) an error reduction of $10^{-6}$ is obtained by four two-grid cycles with GS-RB smoothing (HW, $\nu = 3$) (since $0.033^4 \approx 10^{-6}$), whereas with $\omega$-JAC ($\nu = 3$), nine and 16 cycles are required (for $\omega = 4/5$ and $\omega = 1/2$, respectively).

**Remark 3.3.2** Based on the proof given in Section 3.3.4, also a general analytical formula for $\rho^*(\nu)$ of GS-RB with FW and for Model Problem 1 has been derived [378]:

$$\rho^*(\nu) = \begin{cases} \dfrac{1}{4} & \text{for } \nu = 1 \\ \dfrac{1}{2\nu}\left(\dfrac{\nu}{\nu+1}\right)^{\nu+1} & \text{for } \nu \geq 2. \end{cases} \tag{3.3.2}$$

For the optimal $\nu$ with respect to efficiency ($\nu = 2$), we obtain $\rho^*(2) = 2/27$. $\gg$

**Remark 3.3.3 (number of smoothing steps per cycle)** Formula (3.3.2) implies that

$$\rho^*(\nu) \sim \frac{1}{2e\nu} = O\left(\frac{1}{\nu}\right) \quad \text{for } \nu \to \infty.$$

As pointed out in Section 2.5.3, the general conclusion is that it makes no sense to perform many smoothing steps within one multigrid cycle as the convergence improvement is too small compared to the amount of additional work. $\gg$

### 3.3.2 Norms of the Two-grid Operator

Although the spectral radius $\rho(M_h^{2h})$ gives insights into the *asymptotic convergence* behavior of a two-grid method, norms are needed to estimate the error (or defect) reduction *of one iteration step*. In particular, norms of $M_h^{2h}$ are needed in the theoretical investigations of complete multigrid iterations and of FMG (as seen in Sections 3.2.1 and 3.2.2).

As we will see, the norms of the two-grid operator depend strongly on $\nu_1$ and $\nu_2$, whereas the spectral radius depends only on the sum $\nu = \nu_1 + \nu_2$. This corresponds to the general observation that the spectral radius is less sensitive to algorithmic details than the norms are.

There are several reasonable norms. Of course, different choices of norms will, in general, lead to different results.

We consider the operator norm $|| \cdot ||_S$, corresponding to the Euclidean inner product on $\mathcal{G}(\Omega_h)$, i.e. the spectral norm

$$||B_h||_S = \sqrt{\rho(B_h B_h^*)},\qquad(3.3.3)$$

where $B_h$ denotes any linear operator $B_h : \mathcal{G}(\Omega_h) \to \mathcal{G}(\Omega_h)$.

Here, we are interested in the quantity

$$\sigma_S = \sigma_S(h) = ||M_h^{2h}||_S.\qquad(3.3.4)$$

This quantity depends in particular on $h$, $\nu_1$, $\nu_2$. By $\sigma_S^*$ we denote the supremum of $\sigma_S$ with respect to $h$

$$\sigma_S^* := \sup_{h \in \mathcal{H}} \{\sigma_S(h)\}.\qquad(3.3.5)$$

**Supplement to Theorem 3.3.1**   *For the two-grid method in Theorem 3.3.1 (GS-RB, FW or HW), we obtain the values of $\sigma_S^*$ listed in Table 3.3.*

For comparison, the corresponding spectral radii are also included in this table. Obviously, the norms can be larger than one (and even tend to $\infty$ in the HW case). This means that if only one multigrid cycle is performed, it may lead to an enlarged error in this norm although, asymptotically, the convergence factor (represented by $\rho^*(\nu)$) is much smaller than one. For the multigrid components considered here, norms greater than one appear for the two-grid cycles *without presmoothing* ($\nu_1 = 0$). As we have seen in our description of the FMG method, there are situations where one would like to perform only *one* multigrid cycle ($r = 1$ in (2.6.4)) and still guarantee a sufficiently good reduction of the error (see also Section 3.2.2). In those cases, it is necessary to apply presmoothing steps.

**Table 3.3.**  $\sigma_S^*(\nu_1, \nu_2)$ and $\rho^*(\nu)$ for Model Problem I using GS-RB and FW or HW.

| $(\nu_1, \nu_2)$ | FW | | HW | |
|---|---|---|---|---|
|  | $\rho^*(\nu)$ | $\sigma_S^*$ | $\rho^*(\nu)$ | $\sigma_S^*$ |
| (1,0) | 0.250 | 0.559 | 0.500 | 0.707 |
| (0,1) |  | 1.414 |  | $\infty$ |
| (2,0) | 0.074 | 0.200 | 0.125 | 0.191 |
| (1,1) |  | 0.141 |  | 0.707 |
| (0,2) |  | 1.414 |  | $\infty$ |
| (3,0) | 0.053 | 0.137 | 0.033 | 0.115 |
| (2,1) |  | 0.081 |  | 0.070 |
| (1,2) |  | 0.081 |  | 0.707 |
| (0,3) |  | 1.414 |  | $\infty$ |

**Table 3.4.** FMG estimates for Model Problem I and RBMPS: asymptotic values of $\beta^*$ according to (3.2.12) with $A = 4P = 4$.

|         | FW, W(1,1) | HW, W(2,1) |
|---------|------------|------------|
| $r = 1$ | 2.65       | 0.75       |
| $r = 2$ | 0.12       | 0.04       |

### 3.3.3 Results for Multigrid

We now combine the results of Section 3.2 with the results for $||M_h^{2h}||$ as given in Table 3.3.

**Result 3.3.1** For the W-cycle of the RBMPS and its HW variant in Table 3.1, we obtain the norm estimates

$$||M_h||_S \leq \begin{cases} \eta = 0.17 & \text{(FW, W(1,1))} \\ \eta = 0.08 & \text{(HW, W(2,1))}. \end{cases}$$

Furthermore, the FMG estimate (3.2.11) holds with the asymptotic values of $\beta^*$ in (3.2.12) as given in Table 3.4. Here, we have assumed that cubic interpolation is used for FMG interpolation.

For the proof, simply use the results of Theorems 3.2.1 ($||M_h|| \leq \eta$) and (3.2.12) with $\sigma_S^*$ in Table 3.3 ($\sigma_S^* \leq 0.141$ for FW, $(\nu_1, \nu_2) = (1, 1)$ and $\sigma_S^* \leq 0.070$ for HW, $(\nu_1, \nu_2) = (2, 1)$). Note that $\kappa_{\text{FMG}} = 4$ and $P = 1$ for the cubic FMG interpolation, and, of course, $\kappa = 2$ here.

> This result provides a full quantitative analysis for an efficient multigrid Poisson solver.

### 3.3.4 Essential Steps and Details of the Two-grid Analysis

In the introducion of the multigrid idea in Chapter 1 and in the study of $\omega$-Jacobi relaxation in Section 2.1, we have already used some facts, which are characteristic for the *rigorous Fourier two-grid analysis*. We again start with $\varphi_h^{k,\ell}$ ($k, \ell = 1, \ldots, n - 1$) from (2.1.4), the discrete eigenfunctions of $L_h$. If we apply $\omega$-JAC for smoothing, according to Section 2.1, the $\varphi_h^{k,\ell}$ are also eigenfunctions of the corresponding smoothing operator. This is no longer true for GS-RB and the $\varphi_h^{k,\ell}$ are not eigenfunctions of the two-grid iteration operator $M_h^{2h}$, either. However, simple (at most) four-dimensional spaces

$$E_h^{k,\ell} \quad \left( k, \ell = 1, \ldots, \frac{n}{2} \right),$$

spanned by the $\varphi_h^{k,\ell}$, will turn out to be invariant under $K_h^{2h}$ and the GS-RB smoothing operator. As a consequence, the determination of $\rho(M_h^{2h})$ and $\|M_h^{2h}\|$ can be reduced to the determination of the largest eigenvalues of $4 \times 4$ matrices.

We distinguish six steps in the analysis and use the representation

$$M_h^{2h} = S_h^{\nu_2} K_h^{2h} S_h^{\nu_1} = S_h^{\nu_2} \left( I_h - I_{2h}^h (L_{2h})^{-1} I_h^{2h} L_h \right) S_h^{\nu_1}.$$

We separately consider the coarse grid operator $K_h^{2h}$ and the smoothing operator $S_h$.

**Step (1)** *Discrete eigenfunctions of $L_h$ and $L_{2h}$*
The Fourier analysis is based on the fact that the functions

$$\boxed{\varphi_h^{k,\ell}(x, y) = \sin k\pi x \, \sin \ell\pi y \quad (x, y) \in \Omega_h \ (k, \ell = 1, \dots, n-1)} \tag{3.3.6}$$

are the (discrete) eigenfunctions of $L_h$ (and of $(L_h)^{-1}$). These eigenfunctions are orthogonal with respect to the discrete inner product (1.3.6).

Correspondingly, the functions

$$\boxed{\varphi_{2h}^{k,\ell}(x, y) = \sin k\pi x \, \sin \ell\pi y \quad (x, y) \in \Omega_{2h} \ \left(k, \ell = 1, \dots, \frac{n}{2} - 1\right)} \tag{3.3.7}$$

are the discrete eigenfunctions of $L_{2h}$ (and of $(L_{2h})^{-1}$).

**Step (2)** *Harmonics, spaces of harmonics*
*For any $k$ and $\ell$ ($k, \ell = 1, \dots, n/2 - 1$), the $\Omega_h$-functions*

$$\varphi_h^{k,\ell}, \quad \varphi_h^{n-k,n-\ell}, \quad \varphi_h^{n-k,\ell}, \quad \varphi_h^{k,n-\ell}$$

*coincide (up to their sign) on $\Omega_{2h}$*. This means that

$$\varphi_{2h}^{k,\ell}(x, y) = \varphi_h^{k,\ell}(x, y) = -\varphi_h^{n-k,\ell}(x, y)$$
$$= -\varphi_h^{k,n-\ell}(x, y) = \varphi_h^{n-k,n-\ell}(x, y) \quad (\text{for } (x, y) \in \Omega_{2h}).$$

We call these four linearly independent $\Omega_h$-functions "*harmonics*" and define the four-dimensional spaces of harmonics

$$\boxed{E_h^{k,\ell} = \text{span}\left[\varphi_h^{k,\ell}, \varphi_h^{n-k,n-\ell}, -\varphi_h^{n-k,\ell}, -\varphi_h^{k,n-\ell}\right] \quad \text{for } k, \ell = 1, \dots, \frac{n}{2}.} \tag{3.3.8}$$

For $k$ or $\ell = n/2$ and for $k = \ell = n/2$, the corresponding spaces $E_h^{k,\ell}$ degenerate to two- and one-dimensional spaces, respectively.

**Step (3)** *Transfer operators*

One can show that the transfer operators $I_h^{2h}$ and $I_{2h}^h$ employed in the RBMPS (*FW* and *bilinear interpolation*, respectively) have the following properties.

*Restriction* (FW):

$$\boxed{I_h^{2h} : E_h^{k,\ell} \longrightarrow \text{span}[\varphi_{2h}^{k,\ell}] \quad \left(k, \ell = 1, \ldots, \frac{n}{2} - 1\right)}. \tag{3.3.9}$$

More concretely:

$$I_h^{2h} \begin{Bmatrix} \varphi_h^{k,\ell} \\ \varphi_h^{n-k,n-\ell} \\ -\varphi_h^{n-k,\ell} \\ -\varphi_h^{k,n-\ell} \end{Bmatrix} = \begin{Bmatrix} (1-\xi)(1-\eta) \\ \xi\eta \\ \xi(1-\eta) \\ (1-\xi)\eta \end{Bmatrix} \varphi_{2h}^{k,\ell} \tag{3.3.10}$$

with

$$\xi = \sin^2\left(\frac{k\pi h}{2}\right), \qquad \eta = \sin^2\left(\frac{\ell\pi h}{2}\right). \tag{3.3.11}$$

Also $I_h^{2h}\varphi_h^{k,\ell} = 0$ for $k = n/2$ or $\ell = n/2$. This representation can be verified using trigonometric identities.

*Prolongation* (bilinear interpolation):

$$\boxed{I_{2h}^h : \text{span}[\varphi_{2h}^{k,\ell}] \longrightarrow E_h^{k,\ell} \quad \left(k, \ell = 1, \ldots, \frac{n}{2} - 1\right)}. \tag{3.3.12}$$

In fact,

$$I_{2h}^h\varphi_{2h}^{k,\ell} = (1-\xi)(1-\eta)\varphi_h^{k,\ell} + \xi\eta\varphi_h^{n-k,n-\ell} \\ - \xi(1-\eta)\varphi_h^{n-k,\ell} - (1-\xi)\eta\varphi_h^{k,n-\ell} \quad \text{on } \Omega_h. \tag{3.3.13}$$

To prove (3.3.13), we use the definition of bilinear interpolation and trigonometric identities and obtain

$$I_{2h}^h\varphi_{2h}^{k,\ell}(x,y) = \begin{cases} \delta_1\varphi_h^{k,\ell}(x,y) & \text{if } x/h \text{ and } y/h \text{ even} \\ \delta_2\varphi_h^{k,\ell}(x,y) & \text{if } x/h \text{ and } y/h \text{ odd} \\ \delta_3\varphi_h^{k,\ell}(x,y) & \text{if } x/h \text{ odd, } y/h \text{ even} \\ \delta_4\varphi_h^{k,\ell}(x,y) & \text{if } x/h \text{ even, } y/h \text{ odd} \end{cases}, \tag{3.3.14}$$

where $\delta_1 = 1$, $\delta_2 = \cos k\pi h \cos \ell\pi h$, $\delta_3 = \cos k\pi h$ and $\delta_4 = \cos \ell\pi h$. With this representation, we obtain (3.3.13) from the following more general remark.

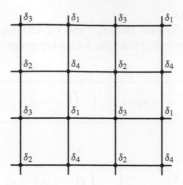

**Figure 3.1.** Coefficients at grid points according to (3.3.15).

**Remark 3.3.4**  Any grid function $c(x, y)\varphi_h^{k,\ell}(x, y)$ with

$$c(x, y) = \begin{Bmatrix} \delta_1 & \text{if } x/h \text{ and } y/h \text{ even} \\ \delta_2 & \text{if } x/h \text{ and } y/h \text{ odd} \\ \delta_3 & \text{if } x/h \text{ odd, } y/h \text{ even} \\ \delta_4 & \text{if } x/h \text{ even, } y/h \text{ odd} \end{Bmatrix} \qquad (3.3.15)$$

(see Fig. 3.1) can be represented on $\Omega_h$ as a *linear combination*

$$a^{00}\varphi_h^{k,\ell} + a^{11}\varphi_h^{n-k,n-\ell} - a^{10}\varphi_h^{n-k,\ell} - a^{01}\varphi_h^{k,n-\ell},$$

where

$$\begin{pmatrix} a^{00} \\ a^{11} \\ a^{10} \\ a^{01} \end{pmatrix} = \frac{1}{4} \begin{pmatrix} 1 & 1 & 1 & 1 \\ 1 & 1 & -1 & -1 \\ 1 & -1 & -1 & 1 \\ 1 & -1 & 1 & -1 \end{pmatrix} \begin{pmatrix} \delta_1 \\ \delta_2 \\ \delta_3 \\ \delta_4 \end{pmatrix}. \qquad (3.3.16)$$

$\gg$

To prove this remark one again has to use trigonometric identities.

**Remark 3.3.5**  The fact that FW and bilinear interpolation are *adjoint* to each other is reflected by (3.3.10) and (3.3.13). The coefficients of the harmonics coincide.        $\gg$

Representations of the above type are also valid for other restriction and interpolation operators such as, for instance, HW (2.3.6) or injection (2.2.22) and cubic interpolation [379].

**Step (4)**   *Invariance of $E_h^{k,\ell}$ under the coarse grid operator $K_h^{2h}$*
For the coarse grid operator

$$K_h^{2h} = I_h - I_{2h}^h (L_{2h})^{-1} I_h^{2h} L_h,$$

we have found

$$L_h : \text{span}[\varphi_h^{k,\ell}] \rightarrow \text{span}[\varphi_h^{k,\ell}]$$
$$I_h^{2h} : E_h^{k,\ell} \rightarrow \text{span}[\varphi_{2h}^{k,\ell}]$$
$$(L_{2h})^{-1} : \text{span}[\varphi_{2h}^{k,\ell}] \rightarrow \text{span}[\varphi_{2h}^{k,\ell}]$$
$$I_{2h}^{h} : \text{span}[\varphi_{2h}^{k,\ell}] \rightarrow E_h^{k,\ell}$$

and therefore

$$\boxed{K_h^{2h} : E_h^{k,\ell} \rightarrow E_h^{k,\ell} \quad \left( k,\ell = 1,\ldots,\frac{n}{2} \right)} .$$

Because of this invariance property, the representation of $K_h^{2h}$ with respect to the spaces $E_h^{k,\ell}$ leads to a block-diagonal matrix

$$K_h^{2h} \triangleq \left[ \hat{K}_h^{2h}(k,\ell) \right]_{k,\ell=1,\ldots,n/2}, \tag{3.3.17}$$

in which the blocks $\hat{K}_h^{2h}(k,\ell)$ are $4 \times 4$ (respectively, $2 \times 2$ or $1 \times 1$) matrices. From the representations above, we find (for FW and bilinear interpolation)

$$\hat{K}_h^{2h}(k,\ell) = \begin{cases} I - [b_i c_j]_{4,4}/\Lambda & \text{(if } k,\ell < n/2) \\ 2 \times 2 - \text{identity matrix} & \text{(if } k = n/2 \text{ or } \ell = n/2) \\ 1 \times 1 - \text{identity matrix} & \text{(if } k = \ell = n/2) \end{cases} \tag{3.3.18}$$

with $\Lambda = \xi(1-\xi) + \eta(1-\eta)$ and

$$\begin{aligned} b_1 &= (1-\xi)(1-\eta) & c_1 &= (1-\xi)(1-\eta)(\xi+\eta) \\ b_2 &= \xi\eta & c_2 &= \xi\eta(2-\xi-\eta) \\ b_3 &= \xi(1-\eta) & c_3 &= \xi(1-\eta)(1-\xi+\eta) \\ b_4 &= (1-\xi)\eta & c_4 &= (1-\xi)\eta(1+\xi-\eta), \end{aligned}$$

with $\xi, \eta$ as in (3.3.11).

**Step (5)** *Smoothing operators*
The spaces $E_h^{k,\ell}$ are also invariant under the smoothing operators $S_h$ considered here. This is trivial for the $\omega$-JAC smoothing operator, because already the $\varphi_h^{k,\ell}$ themselves are eigenfunctions of $S_h$ in that case. For the GS-RB operator, the two-dimensional subspaces of $E_h^{k,\ell}$

$$\text{span}[\varphi_h^{k,\ell}, \varphi_h^{n-k,n-\ell}] \quad \text{for } k+\ell < n$$

(and the degenerate one-dimensional space span $[\varphi_h^{n/2,n/2}]$) are invariant under $S_h$. To see this, we split $S_h$ into a product

$$S_h = S_h^{\mathrm{BLACK}} \cdot S_h^{\mathrm{RED}} \tag{3.3.19}$$

where $S_h^{\mathrm{RED}}$ and $S_h^{\mathrm{BLACK}}$ are "partial step operators". They represent the two half-steps of one full smoothing step. Each of these partial steps is of Jacobi type, applied, however, only to the red or the black grid points.

Let $w_h$ be any approximation of $u_h$ and

$$v_h = u_h - w_h.$$

If $\bar{w}_h$ is the approximation after one of the partial step operators has been applied to $w_h$, we obtain for the corresponding error $\bar{v}_h$

$$\bar{v}_h(x, y) = \begin{cases} (I_h - (L_h^0)^{-1} L_h) v_h(x, y) & (x, y) \in \tilde{\Omega}_h \\ v_h(x, y) & (x, y) \in \Omega_h \setminus \tilde{\Omega}_h, \end{cases} \tag{3.3.20}$$

where the stencil $L_h^0 = (1/h^2)[4]_h$ is the "diagonal" part of $L_h$. $\tilde{\Omega}_h$ is the subset of $\Omega_h$ consisting of those grid points of $\Omega_h$ which have been relaxed by the respective partial step operator: the red points for $S_h^{\mathrm{RED}}$ and the black points for $S_h^{\mathrm{BLACK}}$.

From the representation (3.3.20) we can conclude, that $\varphi_h^{k,\ell}$ is mapped into

$$\bar{\varphi}_h^{k,\ell}(x, y) = \begin{cases} (1 - (h^2/4)\lambda_{k,\ell}) \varphi_h^{k,\ell}(x, y) & (x, y) \in \tilde{\Omega}_h \\ \varphi_h^{k,\ell}(x, y) & (x, y) \in \Omega_h \setminus \tilde{\Omega}_h, \end{cases}$$

where, the $\lambda_{k,\ell}$ are the eigenvalues of $L_h$. This is not yet a Fourier representation of $S_h^{\mathrm{RED}}$ and $S_h^{\mathrm{BLACK}}$, respectively, but the $\bar{\varphi}_h^{k,\ell}$ can easily be written as a linear combination of $\varphi_h^{k,\ell}$ and $\varphi_h^{n-k,n-\ell}$ (see Remark 3.3.4).

**Step (6)**   $M_h^{2h}$ *representation with respect to the* $E_h^{k,\ell}$
In the final step, we combine the representations of $S_h$ and $K_h^{2h}$ with respect to the $E_h^{k,\ell}$ spaces. Altogether we obtain a representation of $M_h^{2h}$ that is characterized by a block-diagonal matrix

$$M_h^{2h} \triangleq \left[ \hat{M}_h^{2h}(k, \ell) \right]_{k,\ell=1,\dots,n/2}. \tag{3.3.21}$$

Here the blocks $\hat{M}_h^{2h}(k, \ell)$ are $4 \times 4$ matrices if $k, \ell < n/2$, $2 \times 2$ ($1 \times 1$) matrices if either $k = n/2$ or $\ell = n/2$ ($k = \ell = n/2$). Note that the coefficients of these matrices depend on the smoothing operator $S_h$ ($\omega$-JAC or GS-RB) and on $\nu_1$, $\nu_2$.

This representation is finally used to prove Theorem 3.3.1 and Equation (3.3.2) as well as to calculate the values in Table 3.1: we have

$$\rho\big(M_h^{2h}\big) = \max_{1 \le k, \ell \le n/2} \rho\big(\hat{M}_h^{2h}(k, \ell)\big). \qquad (3.3.22)$$

Thus we have to determine the spectral radii (eigenvalues) of the (at most) $4 \times 4$ matrices $\hat{M}_h^{2h}(k, \ell)$, find their maximum with respect to $k$ and $\ell$ and finally consider the limit $\rho^*$ as $h \to 0$.

The norms considered in Section 3.3.2 can also be determined from the block representation of $M_h^{2h}$ (3.3.21). Here, we can make use of the fact that the similarity transformation $M_h^{2h} \leftrightarrow [\hat{M}_h^{2h}(k, l)]_{k, \ell = 1, \dots, n/2}$ is orthogonal and obtain

$$\sigma_S(h) = \max\{\|\hat{M}_h^{2h}(k, \ell)\|_S : \max(k, l) \le n/2\}. \qquad (3.3.23)$$

The computation of $\sigma_S^*$ can be performed analogously to the computation of $\rho^*$.

**Remark 3.3.6** In practice, the coefficients of the corresponding blocks $\hat{M}_h^{2h}(k, \ell)$ and the eigenvalues are computed numerically. This means that the analysis is the subject of a simple *Fourier analysis computer program*. ≫

## 3.4 RANGE OF APPLICABILITY OF THE RIGOROUS FOURIER ANALYSIS, OTHER APPROACHES

So far, we have used the analysis only for Model Problem 1 and a specific algorithm. The rigorous Fourier analysis can be applied to a number of other problems. We indicate the range of application of the rigorous Fourier analysis (in particular in Section 3.4.3). In Section 3.4.4, we make some remarks on the relationship between the rigorous Fourier analysis and the local Fourier analysis, which is treated systematically in Chapter 4. Other theoretical approaches are sketched in Section 3.4.5.

Before we start with these general remarks, we briefly discuss how the rigorous Fourier analysis has to be modified in order to treat 3D problems (in Section 3.4.1) and other boundary conditions (in Section 3.4.2).

### 3.4.1 The 3D Case

The Fourier analysis, introduced in Sections 3.3.2 and 3.3.4 for the 2D case, can also be applied in 3D with some natural adaptations:

- As mentioned in Section 2.9, the discrete eigenfunctions $\varphi_h^{k, \ell, m}(x, y)$ of the discrete 3D operator $\Delta_h$ are given by (2.9.7) and the low frequencies in the case of standard coarsening are represented by $\max(k, \ell, m) < n/2$, high frequencies by $n/2 \le \max(k, \ell, m) < n$.
- Similarly as in 2D, we have spaces $E_h^{k, \ell, m}$

$$
\boxed{
\begin{aligned}
E_h^{k,\ell,m} &= \mathrm{span}\Big[\varphi_h^{k,\ell,m},\ -\varphi_h^{n-k,n-\ell,n-m},\ -\varphi_h^{n-k,\ell,m},\ \varphi_h^{k,n-\ell,n-m}, \\
&\qquad -\varphi_h^{k,n-\ell,m},\ \varphi_h^{n-k,\ell,n-m},\ -\varphi_h^{k,\ell,n-m},\ \varphi_h^{n-k,n-\ell,m}\Big] \\
&\text{for } k,\ell,m = 1,\dots,\frac{n}{2}.
\end{aligned}
}
\tag{3.4.1}
$$

which turn out to be invariant under the 3D GS-RB smoothing operator. Most of these spaces of harmonics are eight-dimensional. If one (two, three) of the indices $k, \ell, m$ equal $n/2$, the $E_h^{k,\ell,m}$ degenerate to four- (two-, one-) dimensional spaces.

- Correspondingly, the invariance of $E_h^{k,\ell,m}$ under the coarse grid operator $K_h^{2h}$ is maintained. The representation of $K_h^{2h}$ with respect to the spaces $E_h^{k,\ell,m}$ leads to a block-diagonal matrix

$$
K_h^{2h} \overset{\triangle}{=} \left[\hat{K}_h^{2h}(k,\ell,m)\right]_{k,\ell,m=1,\dots,n/2},
\tag{3.4.2}
$$

where the blocks $\hat{K}_h^{2h}(k,\ell,m)$ are $8 \times 8$ matrices (if $k, \ell, m < n/2$), or $4 \times 4, 2 \times 2$ or $1 \times 1$ matrices in the degenerate cases.

- A similar block-diagonal representation is obtained for the two-grid operator $M_h^{2h}$.
- The particular two-grid convergence results and smoothing factors in 3D are somewhat worse than in 2D. For example, the smoothing factor of GS-RB increases from 0.25 in 2D to 0.44 in 3D.

**Remark 3.4.1 (3D $\omega$-GS-RB)** In [427, 428] it has been shown that an overrelaxation parameter, $\omega > 1$, improves the convergence of the RBMPS analog (with FW) for $d$-dimensional Poisson-type problems. The extra computational work for performing the overrelaxation is particularly worthwhile for $d \geq 3$ (see Remark 2.9.1). In 2D, the two-grid convergence factor improves from 0.25 to 0.16 for $\nu = 1$ and from 0.074 to 0.052 for $\nu = 2$, with an optimal $\omega$. By the 3D rigorous Fourier two-grid analysis, we are able to predict the (more impressive) convergence improvement accurately, see Table 3.5. Average numerical convergence factors are compared to $\rho(M_h^{2h})$. The number of multigrid levels used is 5, 6 and 6, respectively, for the $32^3$, $64^3$ and $96^3$ problems.                    ≫

**Table 3.5.** W(1,1)-cycle measured multigrid and two-grid convergence factors ($\rho(M_h^{2h})$) for the 3D Poisson equation.

| $1/h$ | $\omega = 1$ | | $\omega = 1.1$ | | $\omega = 1.15$ | |
|---|---|---|---|---|---|---|
| | $\hat{q}^{(100)}$ | $\rho(M_h^{2h})$ | $\hat{q}^{(100)}$ | $\rho(M_h^{2h})$ | $\hat{q}^{(100)}$ | $\rho(M_h^{2h})$ |
| 32 | 0.192 | 0.194 | 0.089 | 0.091 | 0.070 | 0.072 |
| 64 | 0.196 | 0.197 | 0.091 | 0.092 | 0.074 | 0.074 |
| 96 | 0.196 | 0.198 | 0.091 | 0.093 | 0.074 | 0.075 |

### 3.4.2 Boundary Conditions

The discrete sine-functions $\varphi_h^{k,\ell}$ in (3.3.6), which we have used for the rigorous Fourier analysis above, correspond to the Dirichlet boundary conditions considered so far. If we consider the same problem *with Neumann boundary conditions* (discretized as described in Section 5.6.2), we are able to obtain very similar results as those in Section 3.3. However, the analysis in this case has to be based on the discrete eigenfunctions

$$\varphi_h^{k,\ell}(x, y) = \cos k\pi x \cos \ell\pi y \quad (k, \ell = 0, 1, \ldots, n) \tag{3.4.3}$$

instead of the sine-functions in (3.3.6). The $\varphi_h^{k,\ell}$ in (3.4.3) also form four-dimensional invariant subspaces of the form (3.3.8).

If we consider the discrete Poisson equation *with periodic boundary conditions* (see Fig. 5.21 for the corresponding grid $\Omega_h$ and Section 5.6.3 for details of the discretization), the discrete eigenfunctions are

$$\varphi_h^{k,\ell}(x, y) = e^{i2\pi kx} e^{i2\pi \ell y} \quad (k, \ell = 0, 1, \ldots, n - 1). \tag{3.4.4}$$

Because of periodicity, we have

$$e^{i2\pi k'x} \equiv e^{i2\pi kx} \quad \text{on } \Omega_h \text{ if } k' = k(\text{mod } n). \tag{3.4.5}$$

Shifting $k$ and $\ell$ and assuming $n$ to be even, we can thus renumber the eigenfunctions (3.4.4) as follows

$$0 \leq k, \ell < n - 1 \longleftrightarrow -\frac{n}{2} \leq k, \ell < \frac{n}{2}. \tag{3.4.6}$$

This numeration is convenient and is customary in the local Fourier analysis context described in Chapter 4.

**Remark 3.4.2** For both Neumann and periodic boundary conditions, the case $k = \ell = 0$ characterizes the discrete eigenfunction which is identical to 1 (corresponding to eigenvalue 0). ≫

If we have Dirichlet, Neumann and periodic boundary conditions at different parts of the boundary, combinations of the above eigenfunctions can be used accordingly.

### 3.4.3 List of Applications and Limitations

Based on rigorous Fourier analysis many results have been obtained for different model problems in 2D and 3D, for different discretizations, coarsening strategies, transfer operators and smoothing procedures, e.g. [378, 379, 389].

Rigorous Fourier analysis has been applied in the following situations (see, for example [378], for most of these cases):

- *Differential equations:* Poisson, Helmholtz equation, anisotropic equations, biharmonic equation.
- *Boundary conditions:* Dirichlet, Neumann, periodic, certain combinations of these.
- *Discretizations:* standard central differences, Mehrstellen method, symmetric higher order methods.
- *Coarsenings:* standard (2D, 3D, . . .), semi (2D, 3D, . . .), alternating, red–black.
- *Smoothers:* Jacobi-type, GS-RB, multicolor, zebra-line, zebra-plane.
- *Fine-to-coarse transfer:* injection, HW, FW, larger stencil weighting.
- *Coarse-to-fine transfer:* bi-, trilinear interpolation, cubic interpolation, higher-order interpolation.
- *Coarse grid operators:* natural $L_H$ discretization, Galerkin coarse grid discretization.

Nevertheless, the rigorous Fourier analysis is restricted to the above types of model problems and model algorithms. As soon as we deal with some nonconstant coefficients in the differential equation, nonsymmetric operators or nonrectangular domains, just to mention three conditions, we will, in general, not be able to apply the rigorous Fourier analysis.

---

**Remark 3.4.3 (failure of rigorous two-grid analysis for lexicographic relaxation)** One important smoothing procedure, namely Gauss–Seidel relaxation with *lexicographic* ordering of the grid points (GS-LEX), cannot be analyzed by the *rigorous two-grid Fourier analysis*. If we consider, for example, Model Problem 1, then no simple low-dimensional spaces can be formed which are invariant under GS-LEX with the eigenfunctions $\varphi_h^{k,\ell}$ (3.3.6). In particular, the $E_h^{k,\ell}$ (3.3.8) are not invariant under GS-LEX; instead, the $\varphi_h^{k,\ell}$ are all intermixed by the GS-LEX operator $S_h$.

The case of GS-LEX relaxation can, however, be easily analyzed by the local Fourier analysis in Chapter 4.                                                                  ≫

---

## 3.4.4 Towards Local Fourier Analysis

As seen in the previous section, the rigorous Fourier analysis can be used only for a limited class of problems. So, what can the practitioner do, when he has to solve a new problem and wants to design an efficient multigrid algorithm? Qualitative multigrid theories, which we will survey briefly in the next section, are usually not suited for the design and comparison of concrete algorithms for complex problems. This is different for the *local Fourier analysis* (LFA). We will introduce the basic ideas and formalism of LFA in Chapter 4.

In this section, we will point out that the rigorous Fourier analysis and the LFA are closely related. The mathematical substance of both is very similar. In fact, the rigorous Fourier analysis results can also be derived within the framework of LFA.

Formally, the relationship is most easily explained by reconsidering the discrete 2D Laplacian $\Delta_h$ with *periodic boundary conditions* in $\Omega_h = (0, 1)^2 \cap \mathbf{G}_h$. In this case, the discrete eigenfunctions are given by (3.4.4). With the notation

$$\boldsymbol{\theta} = (\theta_1, \theta_2) := \left(2\pi \frac{k}{n}, 2\pi \frac{\ell}{n}\right) \tag{3.4.7}$$

and

$$T_h = \left\{\boldsymbol{\theta} = \left(2\pi \frac{k}{n}, 2\pi \frac{\ell}{n}\right): -\frac{n}{2} \le k, \ell < \frac{n}{2}; \ k, \ell \in \mathbb{Z}\right\}, \tag{3.4.8}$$

we can write the discrete eigenfunctions (3.4.4) in the form

$$\varphi_h(\boldsymbol{\theta}, x, y) = e^{i\theta_1 x/h} e^{i\theta_2 y/h} \quad (\boldsymbol{\theta} \in T_h). \tag{3.4.9}$$

Obviously,

$$T_h \subset [-\pi, \pi)^2, \quad \#T_h = n^2. \tag{3.4.10}$$

In case of standard coarsening we can split $T_h$ into subsets corresponding to low and high frequencies

$$T_h^{\text{low}} := \left\{\boldsymbol{\theta} = \left(2\pi \frac{k}{n}, 2\pi \frac{\ell}{n}\right): -\frac{n}{4} \le k, \ell < \frac{n}{4}; \ k, \ell \in \mathbb{Z}\right\} \subset \left[-\frac{\pi}{2}, \frac{\pi}{2}\right)^2$$

$$T_h^{\text{high}} := T_h \setminus T_h^{\text{low}} \subset [-\pi, \pi)^2 \setminus \left[-\frac{\pi}{2}, \frac{\pi}{2}\right)^2. \tag{3.4.11}$$

> **Remark 3.4.4** Assuming periodic boundary conditions, the grid functions (3.4.9) form a basis of eigenfunctions for any discrete operator $L_h$ with constant coefficients. $\gg$

The LFA formalism is based on the grid functions (3.4.9). However, instead of considering only a finite number of these functions (varying $\boldsymbol{\theta}$ in the *finite* set $T_h$), we will allow $\boldsymbol{\theta}$ to vary continuously in $[-\pi, \pi)^2$. This makes the formulation easier and more convenient. Correspondingly, low frequencies $\boldsymbol{\theta}$ vary in $[-\pi/2, \pi/2)^2$ and high frequencies $\boldsymbol{\theta}$ in $[-\pi, \pi)^2 \setminus [-\pi/2, \pi/2)^2$.

**Remark 3.4.5** For clarification, if all $\boldsymbol{\theta}$ vary continuously in $[-\pi, \pi)^2$, the $\varphi_h(\boldsymbol{\theta}, x, y)$ no longer fulfill the discrete periodic boundary conditions in $[0, 1]^2$ (only those $\varphi_h$ with $\boldsymbol{\theta} \in T_h$ fulfil them). Consequently, we will consider the $\varphi(\boldsymbol{\theta}, \cdot)$ on the infinite grid $\mathbf{G}_h$.

If one wants to provide this approach with a rigorous formal framework, as in [378], one has to be aware that the $\varphi(\theta, \cdot)$ form a nondenumerable basis of a (consequently nonseparable) Hilbert space.

An ambitious theoretical framework for LFA has been given in [373] and in [69] (*rigorous* LFA). $\gg$

### 3.4.5 Smoothing and Approximation Property; a Theoretical Overview

Another approach to analyzing multigrid is to develop a general multigrid convergence theory. Although, the convergence estimates of such approaches do not, in general, give satisfactorily sharp predictions of actual multigrid convergence, they guarantee $h$-independent multigrid convergence in general situations. These theories discuss general requirements on regularities of solutions, on discretizations, grids, smoothers, coarse grid corrections etc. for multigrid convergence.

The classical multigrid theory is based on the study of the *approximation and smoothing property* of a multigrid method as introduced by Hackbusch [171–173]. The basis of this theory is a splitting of the two-grid operator $M_h^{2h}$, which leads, for $\nu_1 = \nu$, $\nu_2 = 0$, to the estimate

$$||M_h^{2h}|| \leq ||K_h^{2h}(L_h)^{-1}|| \cdot ||L_h S_h^{\nu}|| \qquad (3.4.12)$$

(with suitably chosen norms). Based on this splitting, the *approximation property*

$$||K_h^{2h}(L_h)^{-1}|| = ||(L_h)^{-1} - I_{2h}^h(L_{2h})^{-1}I_h^{2h}|| \leq Ch^{\delta} \ (\delta > 0), \qquad (3.4.13)$$

and the *smoothing property*

$$||L_h S_h^{\nu}|| \leq \eta(\nu)h^{-\delta} \quad \text{with } \eta(\nu) \to 0 \ (\nu \to \infty) \qquad (3.4.14)$$

are fundamental. The smoothing property states, in principle, that the smoother reduces the high frequency components of the error (without amplifying the low frequency components). The approximation property requires the coarse grid correction to be reasonable. Both properties are connected via the choice of the norms.

In particular, for symmetric second-order problems with sufficient regularity, the approximation and smoothing property can be fulfilled, by a suitable choice of multigrid components, with, for instance, $|| \cdot ||_S$, $\delta = 2$ and $\eta(\nu) \sim 1/\nu$. Assuming these properties, the $h$-independent boundedness of $||M_h^{2h}||$ ($\leq$ const $< 1$) follows immediately for $\nu$ large enough.

Hackbusch was able to verify the smoothing and approximation property for many finite difference and finite element discretizations with sufficient regularity on uniform grids and for Jacobi- and Gauss–Seidel-type smoothing methods, including anisotropic problems and corresponding block smoothers. A detailed explanation of this theoretical approach with many results is given in his classical multigrid book [176]. Based on the smoothing property, the robustness of a smoother of ILU type (see Section 7.5) for anisotropic problems has been shown in [420].

The convergence for the V-cycle was first proved by Braess [48, 49]. In his proofs he also reduced the number of smoothing iterations needed to a fixed number ($\nu = 2$, for instance). Braess together with Hackbusch [52] proved the convergence of the V-cycle for general symmetric problems with sufficient regularity.

Whereas the classical theory allows one to prove $h$-independent W-cycle convergence without full elliptic regularity (which is missing, for example, for problems on L-shaped domains, see Section 5.5), a corresponding proof for the V-cycle was finally given in [88].

A different qualitative multigrid theory aiming to solve some of the open questions started with the study of hierarchical, i.e., multi*level preconditioners* for CG-type methods, introduced by Yserentant [433]. In particular, the hierarchical bases of finite element spaces can be used to prove $O(N \log N)$ complexity on nonuniform meshes. $O(N)$ complexity is obtained with the BPX preconditioners [55] for V-cycles on nonuniform grids. The corresponding theory for multilevel, multigrid and domain-decomposition methods based on subspace decompositions is due to Bramble, Pasciak, Oswald and Xu. This is the basis of the *multigrid subspace correction theory* (see for example [55, 295, 425]). Sharper convergence estimates based on this theoretical approach, have been presented [45, 114, 295] and new insights have been gained in classical iteration methods [163, 164].

The theory based on subspace decompositions is valid for *symmetric* problems without regularity assumptions on the PDE and on nonuniform grids. Furthermore, it is based on Galerkin coarse grid operators. An introduction in this theoretical approach is presented in Appendix B.

For *nonsymmetric problems*, which are very common in practice, this theory is not yet well-established. LFA (see Chapter 4) and its variants can help in constructing efficient multigrid solution methods.

**Remark 3.4.6 (Cascadic multigrid)**   An approach with some interesting theoretical properties is the so-called cascadic multigrid method [46, 123]. In cascadic multigrid, the algorithm processes only from coarse to fine grids, with smoothing iterations and/or Krylov subspace acceleration employed on each grid. For model problems, cascadic multigrid can be shown to lead to $O(N)$ algorithms *if the convergence is measured in the energy norm* $\| \cdot \|_E$. The convergence cannot be proved and is not valid [47] in the practically relevant $\| \cdot \|_2$ or $\| \cdot \|_\infty$ norms.                                                              $\gg$

# 4

# LOCAL FOURIER ANALYSIS

The local Fourier analysis (LFA) can be introduced and used in different ways. In this book, we present it as a formal tool. The emphasis is laid on explaining and demonstrating how it is applied in practice.

> In our view, LFA is the most powerful tool for the quantitative analysis and the design of efficient multigrid methods for general problems. We strongly recommend this approach to multigrid practitioners.

LFA (and also the idea of rigorous Fourier analysis) was introduced by Brandt [58] and extended and refined in [63, 69]. Contributions have been made by many others [378, 415]. Brandt prefers the term *local mode analysis* instead of LFA. Both terms denote the same approach.

In Section 4.1 we will discuss the LFA philosophy, in particular its local nature and its objective. LFA terminology will be introduced in Section 4.2. This section may appear somewhat formal at first sight, but the terminology is all elementary, transparent and useful. Some basic examples will be treated in Section 4.3, where the *smoothing factor* $\mu_{\text{loc}}$ will be defined. The smoothing factor is a very important quantity for the design of (efficient) multigrid methods. Using LFA $\mu_{\text{loc}}$ can easily be calculated for many smoothers as seen in Section 4.3. For red–black and other "pattern" smoothers, however, an extension of the definition of the smoothing factor is necessary, which we will introduce in Section 4.5.

Section 4.4 will describe the two-grid LFA. Two-grid LFA is needed if one wants more insight into the design and structure of a multigrid algorithm than smoothing analysis can give. Mathematically, the content of this section is closely related to the rigorous two-grid Fourier analysis (in Section 3.3.4).

In Section 4.6, we will list LFA results for basic multigrid algorithms. In Section 4.7, the notation and concept of $h$-ellipticity will be introduced, based on [63, 66]. The $h$-ellipticity establishes a qualitative criterion for the existence of local smoothers for a given discrete operator $L_h$. $h$-ellipticity is directly connected to the definition of the smoothing factor.

Some practical guidelines on the use of LFA for the design of efficient multigrid algorithms are given in Section 4.6.5. A computer program for performing LFA in general situations is available from `http://www.gmd.de/SCAI/multigrid/book.html`. This program was written by R. Wienands.

## 4.1 BACKGROUND

We first want to explain the *local* nature of LFA:

---

**Local nature of LFA**  Under general assumptions, any general discrete operator, nonlinear, with nonconstant coefficients, can be linearized *locally* and can be replaced locally (by freezing the coefficients) by an operator with constant coefficients. Thus, general *linear discrete operators with constant coefficients* are considered in LFA. Formally, they are defined on an infinite grid.

---

All considerations in the context of LFA are based on grid functions of the form

$$\varphi(\theta, x) = e^{i\theta x/h} \tag{4.1.1}$$

(actually on multidimensional analogs of them). Here, $x$ varies in the given infinite grid $\mathbf{G}_h$ and $\theta$ is a continuous parameter that characterizes the frequency of the grid function under consideration.

---

In the context of LFA, all operators which we will deal with (including $L_h$, the smoothing and intergrid transfer operators $S_h$, $I_h^H$, $I_H^h$, the coarse grid operator $L_H$ as well as the coarse grid correction and two-grid operators $K_h^H$ and $M_h^H$, respectively) will operate on these grid functions $\varphi(\theta, \cdot)$ and are considered on the infinite grids $\mathbf{G}_h$ and $\mathbf{G}_H$, respectively.

The goal of LFA is to determine smoothing factors for $S_h$ and two-grid convergence factors as well as error reduction factors for $M_h^H$. We denote these quantities by

$$\mu_{\text{loc}}(S_h), \quad \rho_{\text{loc}}(M_h^H), \quad \sigma_{\text{loc}}(M_h^H).$$

---

The spectral radius $\rho_{\text{loc}}$ gives insight into the *asymptotic convergence* behavior, whereas $\sigma_{\text{loc}}(M_h^H)$ refers to the error reduction *in one iteration step* measured in an appropriate norm.

Since the $\varphi(\theta, \cdot)$ are defined on the infinite grid $\mathbf{G}_h$, the *influence of boundaries and of boundary conditions is not taken into account*. At first sight, this may be regarded as a deficiency of the LFA approach. However, the idea of LFA is compatible with neglecting the effects of boundaries in the following sense.

> The objective of LFA is to determine the quantitative convergence behavior and effi-
> ciency an appropriate multigrid algorithm *can* attain *if a proper boundary treatment
> is included.*

A proper boundary treatment typically requires only additional suitable *relaxations near the
respective boundaries*, an amount of work which is negligible for $h \to 0$. In the following
chapters, we will see examples of such boundary relaxation.


## 4.2 TERMINOLOGY

In describing the formalism of LFA, we confine ourselves to the 2D case and to standard
coarsening in this section, for simplicity and in order to avoid formal complications (like
many indices). But we now use a vector terminology that immediately carries over to $d$
dimensions. We write $x = (x_1, x_2)$ instead of $(x, y)$ etc.

In the following,

$$h = (h_1, h_2),$$

is a fixed mesh size (i.e., a vector). With $h$ we associate the infinite grid $\mathbf{G}_h$ (1.3.3), which
we now write in the form

$$\mathbf{G}_h = \{x = kh := (k_1 h_1, k_2 h_2), \ k \in \mathbb{Z}^2\}.$$

On $\mathbf{G}_h$, we consider a discrete operator $L_h$ corresponding to a difference stencil

$$L_h \stackrel{\triangle}{=} [s_\kappa]_h \quad (\kappa = (\kappa_1, \kappa_2) \in \mathbb{Z}^2) \tag{4.2.1}$$

$$\text{i.e.} \quad L_h w_h(x) = \sum_{\kappa \in V} s_\kappa w_h(x + \kappa h) \tag{4.2.2}$$

with *constant coefficients* $s_\kappa \in \mathbb{R}$ (or $\mathbb{C}$), which, of course, will usually depend on $h$. Here,
$V$ is again a finite index set.

The fundamental quantities in the LFA are the grid functions

$$\boxed{\varphi(\boldsymbol{\theta}, x) = e^{i\boldsymbol{\theta} \cdot x / h} := e^{i\theta_1 x_1 / h_1} e^{i\theta_2 x_2 / h_2} \quad \text{for } x \in \mathbf{G}_h} \ . \tag{4.2.3}$$

We assume that $\boldsymbol{\theta}$ varies continuously in $\mathbb{R}^2$. One recognizes that

$$\varphi(\boldsymbol{\theta}, x) \equiv \varphi(\boldsymbol{\theta}', x) \quad \text{for } x \in \mathbf{G}_h$$

if and only if

$$\theta_1 = \theta_1' (\text{mod } 2\pi) \quad \text{and} \quad \theta_2 = \theta_2' (\text{mod } 2\pi)$$

(i.e. if the difference between $\theta_1$ and $\theta_1'$ and the difference between $\theta_2$ and $\theta_2'$ are both
multiples of $2\pi$). Therefore, it is sufficient to consider

$$\varphi(\boldsymbol{\theta}, x) \quad \text{with } \boldsymbol{\theta} \in [-\pi, \pi)^2.$$

For $\theta \in [-\pi, \pi)^2$, we also use the notation $-\pi \leq \theta < \pi$, where the inequalities refer to both components $\theta_1$ and $\theta_2$.

The grid functions $\varphi$ for $-\pi \leq \theta < \pi$ are linearly independent on $\mathbf{G}_h$. Since the $\varphi(\theta, x)$ are defined on $\mathbf{G}_h$ and in that respect depend on $h$, we sometimes write $\varphi_h(\theta, x)$ instead of $\varphi(\theta, x)$, for clarity.

**Lemma 4.2.1** *For $-\pi \leq \theta < \pi$, all grid functions $\varphi(\theta, x)$ are (formal) eigenfunctions of any discrete operator which can be described by a difference stencil as in (4.2.1). The relation*

$$L_h \varphi(\theta, x) = \tilde{L}_h(\theta) \varphi(\theta, x) \quad (x \in \mathbf{G}_h)$$

*holds, with*

$$\boxed{\tilde{L}_h(\theta) = \sum_\kappa s_\kappa e^{i\theta \cdot \kappa}} \tag{4.2.4}$$

*We call $\tilde{L}_h(\theta)$ the* formal eigenvalue *or the* symbol of $L_h$.

The proof of Lemma 4.2.1 is straightforward.

**Example 4.2.1**  The symbol of the standard discrete Laplace operator $L_h = -\Delta_h$ is

$$\tilde{L}_h(\theta) = \frac{1}{h^2}(4 - (e^{i\theta_1} + e^{i\theta_2} + e^{-i\theta_1} + e^{-i\theta_2})) = \frac{2}{h^2}(2 - (\cos\theta_1 + \cos\theta_2)). \quad \triangle$$

In addition to $\mathbf{G}_h$, we assume an (infinite) coarse grid

$$\mathbf{G}_H = \{x = \kappa H : \kappa \in \mathbb{Z}^2\}$$

is obtained by standard coarsening of $\mathbf{G}_h$ ($H = (2h_1, 2h_2)$). Other coarsenings can be treated similarly by LFA (see Section 4.6.3).

For the smoothing and two-grid analysis, we again have to distinguish *high and low frequency components* on $\mathbf{G}_h$ with respect to $\mathbf{G}_H$. The definition is based on the trivial but fundamental phenomenon that only those frequency components

$$\varphi(\theta, \cdot) \quad \text{with} -\frac{\pi}{2} \leq \theta < \frac{\pi}{2}$$

are distinguishable on $\mathbf{G}_H$. For each $\theta' \in [-\pi/2, \pi/2)^2$, three other frequency components $\varphi(\theta, \cdot)$ with $\theta \in [-\pi, \pi)^2$ coincide on $\mathbf{G}_H$ with $\varphi(\theta', \cdot)$ and are not distinguishable (not "visible") on $\mathbf{G}_H$. Actually, we have

$$\boxed{\varphi(\theta, x) = \varphi(\theta', x) \quad \text{for } x \in \mathbf{G}_H \text{ if and only if } \theta = \theta' (\text{mod } \pi)} \tag{4.2.5}$$

(see also Lemma 4.4.1 below). This leads to the following definition.

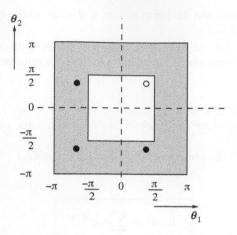

**Figure 4.1.** Low frequencies (interior white region) and high frequencies (shaded region); for a given low frequency $\theta(\circ)$, the three frequencies $\theta$ for which the corresponding $\varphi(\theta, x)$ coincide on $\mathbf{G}_H$ $(H = (2h_1, 2h_2))$ are marked by $\bullet$.

### Definition 4.2.1 (high and low frequencies for standard coarsening)

$$\varphi \text{ low frequency component} \iff \theta \in T^{\text{low}} := \left[ -\frac{\pi}{2}, \frac{\pi}{2} \right)^2$$

$$\varphi \text{ high frequency component} \iff \theta \in T^{\text{high}} := [-\pi, \pi)^2 \Big\backslash \left[ -\frac{\pi}{2}, \frac{\pi}{2} \right)^2$$

(see also Fig. 4.1).

This definition is analogous to (2.1.8) and (3.4.11), respectively. In addition to speaking of high and low frequency components $\varphi(\theta, \cdot)$, we will, for simplicity, sometimes call the corresponding $\theta$s high frequency or low frequency.

## 4.3 SMOOTHING ANALYSIS I

We will start our discussion and definition of the smoothing factor $\mu$ by looking at some very simple but fundamental examples. Considering the discrete Laplace operator $L_h = -\Delta_h$ $(h = h_1 = h_2)$, we want to show how LFA is used to analyze the smoothing behavior of GS-LEX and similar relaxation schemes and how the smoothing factor is actually calculated. These examples are fundamental for two reasons. First, GS-LEX is the classical relaxation method which has been used, generalized and analyzed since the beginning of numerical analysis. Secondly, GS-LEX has natural smoothing properties, but the rigorous Fourier analysis presented in Chapter 3 cannot be applied directly to it. LFA, however, is applicable and shows the smoothing properties of GS-LEX very clearly.

Let us consider a discretized PDE, $L_h u_h = f_h$, and assume that a relaxation method can be written *locally* as

$$L_h^+ \bar{w}_h + L_h^- w_h = f_h, \qquad (4.3.1)$$

where $w_h$ corresponds to the old approximation of $u_h$ (approximation before the relaxation step) and $\bar{w}_h$ to the new approximation (after the step). In this sense, the relaxation is characterized by a splitting

$$L_h = L_h^+ + L_h^-. \qquad (4.3.2)$$

**Example 4.3.1 (GS-LEX)**  For GS-LEX applied to the 2D Laplace operator $L_h = -\Delta_h$, the splitting reads

$$L_h^+ = \frac{1}{h^2} \begin{bmatrix} & 0 & \\ -1 & 4 & 0 \\ & -1 & \end{bmatrix}, \qquad L_h^- = \frac{1}{h^2} \begin{bmatrix} & -1 & \\ 0 & 0 & -1 \\ & 0 & \end{bmatrix}. \qquad (4.3.3)$$

$\triangle$

**Example 4.3.2 (GS-LEX in 3D)**  In 3D, GS-LEX for $L_h = -\Delta_h$ is characterized by

$$L_h^+ = \frac{1}{h^2} \left[ \begin{bmatrix} 0 & 0 & 0 \\ 0 & -1 & 0 \\ 0 & 0 & 0 \end{bmatrix}_h \begin{bmatrix} 0 & 0 & 0 \\ -1 & 6 & 0 \\ 0 & -1 & 0 \end{bmatrix}_h \begin{bmatrix} 0 & 0 & 0 \\ 0 & 0 & 0 \\ 0 & 0 & 0 \end{bmatrix}_h \right]$$

and

$$L_h^- = \frac{1}{h^2} \left[ \begin{bmatrix} 0 & 0 & 0 \\ 0 & 0 & 0 \\ 0 & 0 & 0 \end{bmatrix}_h \begin{bmatrix} 0 & -1 & 0 \\ 0 & 0 & -1 \\ 0 & 0 & 0 \end{bmatrix}_h \begin{bmatrix} 0 & 0 & 0 \\ 0 & -1 & 0 \\ 0 & 0 & 0 \end{bmatrix}_h \right].$$

$\triangle$

**Example 4.3.3 ($\omega$-JAC)**  In 2D, $\omega$-JAC for $L_h = -\Delta_h$ is characterized by

$$L_h^+ = \frac{1}{h^2} \begin{bmatrix} & 0 & \\ 0 & 4/\omega & 0 \\ & 0 & \end{bmatrix}, \qquad L_h^- = \frac{1}{h^2} \begin{bmatrix} & -1 & \\ -1 & 4(1 - 1/\omega) & -1 \\ & -1 & \end{bmatrix}. \qquad (4.3.4)$$

$\triangle$

Subtracting (4.3.1) from the discrete equation $L_h u_h = f_h$, we obtain the local relation

$$L_h^+ \bar{v}_h + L_h^- v_h = 0$$

or

$$\bar{v}_h = S_h v_h$$

for the errors $\bar{v}_h = u_h - \bar{w}_h$, $v_h = u_h - w_h$, where $S_h$ is the resulting smoothing operator. From this we immediately get the (infinite grid) Fourier representation of $L_h^-$, $L_h^+$, $S_h$.

Applying $L_h^-$ and $L_h^+$ to the formal eigenfunctions $\varphi(\theta, x)$, we obtain

$$L_h^- e^{i\theta \cdot x/h} = \tilde{L}_h^-(\theta)e^{i\theta \cdot x/h}$$
$$L_h^+ e^{i\theta \cdot x/h} = \tilde{L}_h^+(\theta)e^{i\theta \cdot x/h},$$

where the $\tilde{L}_h^-$ and $\tilde{L}_h^+$ are the *symbols* (formal eigenvalues) of the operators $L_h^-$ and $L_h^+$, respectively.

---

**Lemma 4.3.1** *Under the assumptions (4.3.1) and (4.3.2), all $\varphi(\theta, \cdot)$ with $\tilde{L}_h^+(\theta) \neq 0$ are eigenfunctions of $S_h$:*

$$S_h\varphi(\theta, x) = \tilde{S}_h(\theta)\varphi(\theta, x) \quad (-\pi \le \theta < \pi) \tag{4.3.5}$$

*with the amplification factor*

$$\tilde{S}_h(\theta) := -\frac{\tilde{L}_h^-(\theta)}{\tilde{L}_h^+(\theta)}. \tag{4.3.6}$$

---

**Example 4.3.4** For GS-LEX, we have, for example,

$$L_h^+ e^{i\theta \cdot x/h} = \frac{1}{h^2}\begin{bmatrix} & 0 & \\ -1 & 4 & 0 \\ & -1 & \end{bmatrix} e^{i\theta \cdot x/h}$$
$$= \frac{1}{h^2}(4 - e^{-i\theta_1} - e^{-i\theta_2})e^{i\theta \cdot x/h}.$$

The symbols $\tilde{L}_h^-, \tilde{L}_h^+, \tilde{S}_h$ of the splitting in Example 4.3.1 are thus

$$\tilde{L}_h^+(\theta) = \frac{1}{h^2}(4 - e^{-i\theta_1} - e^{-i\theta_2})$$
$$\tilde{L}_h^-(\theta) = -\frac{1}{h^2}(e^{i\theta_1} + e^{i\theta_2})$$
$$\tilde{S}_h(\theta) = -\frac{\tilde{L}_h^-(\theta)}{\tilde{L}_h^+(\theta)} = \frac{e^{i\theta_1} + e^{i\theta_2}}{4 - e^{-i\theta_1} - e^{-i\theta_2}}. \qquad \triangle$$

Based on Lemma 4.3.1, we can immediately define the smoothing factor. According to Definition 4.2.1, we distinguish high and low frequency components $\varphi(\theta, \cdot)$ and define the smoothing factor $\mu_{\text{loc}}(S_h)$ by

**Definition 4.3.1**

$$\mu_{\text{loc}} = \mu_{\text{loc}}(S_h) := \sup\{|\tilde{S}_h(\theta)| : \theta \in T^{\text{high}}\}. \tag{4.3.7}$$

For GS-LEX and standard coarsening, for example, this definition results in

$$\mu_{loc}(S_h) = \sup \left\{ \left| \frac{e^{i\theta_1} + e^{i\theta_2}}{e^{-i\theta_1} + e^{-i\theta_2} - 4} \right| : \boldsymbol{\theta} \in T^{high} \right\}.$$

Since $\tilde{S}_h(\boldsymbol{\theta})$ attains its maximum 0.5 for $(\theta_1, \theta_2) = (\pi/2, \arccos 4/5)$, we obtain

$$\boxed{\mu_{loc}(S_h) = 0.5 \quad \text{for GS-LEX}}.$$

The above definition of a smoothing factor and its value of 0.5 for GS-LEX were first given by Brandt [57].

**Example 4.3.5 (smoothing factor for $\omega$-JAC)** For $L_h = -\Delta_h$, the application to $\omega$-JAC is straightforward. According to (4.3.4) and Definition 4.3.1, the symbol $\tilde{S}_h(\boldsymbol{\theta})$ is

$$\tilde{S}_h(\omega, \boldsymbol{\theta}) = 1 - \frac{\omega}{2}(2 - \cos\theta_1 - \cos\theta_2).$$

For the smoothing factor, we obtain

$$\mu_{loc}(S_h(\omega)) = \max \left\{ \left| 1 - \frac{\omega}{2} \right|, |1 - 2\omega| \right\}. \qquad \triangle$$

This is the same result that we have obtained in Section 2.1.2 for $\mu^*(\omega)$. So, $\omega$-JAC for symmetric operators can be treated by both rigorous Fourier analysis and LFA. The results are the same, apart from the following slight formal difference: in comparison with (2.1.9), we recognize that the $h$-dependence of $\mu(S_h(\omega))$ has disappeared here. This is due to the fact that $\boldsymbol{\theta}$ varies continuously in $T^{high}$.

**Remark 4.3.1** If $L_h$ corresponds to a differential operator which contains derivatives of different orders, $\mu_{loc}(S_h)$ will, in general, depend explicitly on $h$. $\qquad \gg$

**Example 4.3.6 (SOR or $\omega$-GS-LEX)** We again consider the discrete Laplacian $L_h = -\Delta_h$. If we apply GS-LEX with an additional relaxation parameter $\omega$, we speak of $\omega$-GS-LEX relaxation corresponding to the classical SOR method [431].

We can apply LFA (smoothing analysis) to investigate the influence of the parameter $\omega$ on the smoothing properties of SOR. The splitting of $L_h$ for SOR is given by

$$L_h^+ = \frac{1}{h^2} \begin{bmatrix} & 0 & \\ -1 & 4/\omega & 0 \\ & -1 & \end{bmatrix}, \qquad L_h^- = \frac{1}{h^2} \begin{bmatrix} & -1 & \\ 0 & 4(1 - 1/\omega) & -1 \\ & 0 & \end{bmatrix}. \qquad (4.3.8)$$

Figure 4.2 shows the behavior of the smoothing factor $\mu_{loc}(S_h(\omega))$ with respect to $\omega$. Using a relaxation parameter $\omega \neq 1$ in $\omega$-GS-LEX hardly improves its smoothing property for this problem in 2D. The optimal value of $\omega$ is not exactly 1 but very close to 1. The gain by using the optimal $\omega$ would be very small and does not pay if the additional work (two operations per point per smoothing step) is taken into account. $\qquad \triangle$

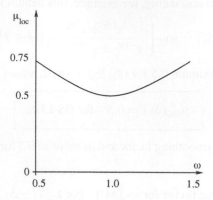

**Figure 4.2.** $\mu_{\text{loc}}$ ($\omega$-GS-LEX) as a function of $\omega$.

Many practically important smoothing procedures fulfil the assumptions of Lemma 4.3.1 and the corresponding Definition 4.3.1 of a smoothing factor can be applied to them. As we have seen, pointwise GS-LEX and $\omega$-JAC type relaxations are among them, but also *block versions* of GS and $\omega$-JAC (line and plane relaxation), which we will discuss in Sections 5.1 and 5.2, can be treated this way. Furthermore, ILU-type smoothers, which will be treated in Section 7.5, belong to this class of smoothers.

On the other hand, relaxation methods like *red–black* Gauss–Seidel do *not* fulfil the above assumptions. In particular, they cannot be represented directly by a splitting as in (4.3.1). However, the above assumptions can be generalized to cover red–black-type relaxations (see Section 4.5).

## 4.4 TWO-GRID ANALYSIS

Let us now apply LFA to the *two-grid operator*

$$M_h^H = S_h^{\nu_2} K_h^H S_h^{\nu_1} \quad \text{with the coarse grid correction operator } K_h^H. \tag{4.4.1}$$

In order to calculate convergence factors and other relevant quantities of $M_h^H$, we will analyze how the operators $L_h$, $I_h^H$, $L_H$, $I_H^h$ and $S_h$ act on the Fourier components $\varphi(\boldsymbol{\theta}, \cdot)$. The analysis uses the fact that quadruples of the $\varphi(\boldsymbol{\theta}, \cdot)$ coincide on $\mathbf{G}_H$. For any low frequency $\boldsymbol{\theta} = (\theta_1, \theta_2) \in T^{\text{low}} = [-\pi/2, \pi/2)^2$, we consider the frequencies

$$\boldsymbol{\theta}^{(0,0)} := (\theta_1, \theta_2), \qquad \boldsymbol{\theta}^{(1,1)} := (\bar{\theta}_1, \bar{\theta}_2),$$
$$\boldsymbol{\theta}^{(1,0)} := (\bar{\theta}_1, \theta_2), \qquad \boldsymbol{\theta}^{(0,1)} := (\theta_1, \bar{\theta}_2),$$

where

$$\bar{\theta}_i := \begin{cases} \theta_i + \pi & \text{if } \theta_i < 0 \\ \theta_i - \pi & \text{if } \theta_i \geq 0 \end{cases} \tag{4.4.2}$$

(see also Fig. 4.1).

**Lemma 4.4.1**

*(1) For any low frequency $\theta^{(0,0)} \in T^{\text{low}}$, we have*

$$\varphi(\theta^{(0,0)}, x) \equiv \varphi(\theta^{(1,1)}, x) \equiv \varphi(\theta^{(1,0)}, x) \equiv \varphi(\theta^{(0,1)}, x) \quad (x \in \mathbf{G}_{2h}). \quad (4.4.3)$$

*(2) Each of these four Fourier components $\varphi(\theta^\alpha, \cdot) = \varphi_h(\theta^\alpha, \cdot)$ with $\alpha \in \{(0, 0), (1, 1), (1, 0), (0, 1)\}$ coincides on $\mathbf{G}_{2h}$ with the respective grid function $\varphi_{2h}(2\theta^{(0,0)}, \cdot)$:*

$$\varphi_h(\theta^\alpha, x) \equiv \varphi_{2h}(2\theta^{(0,0)}, x) \quad (x \in \mathbf{G}_{2h}). \quad (4.4.4)$$

The proof of this lemma is straightforward.

For $\theta^{(1,1)}$, for example, we find

$$\begin{aligned}
\varphi_h(\theta^{(1,1)}, x) &= e^{i\bar{\theta}_1 x_1/h_1} e^{i\bar{\theta}_2 x_2/h_2} = e^{i(\theta_1 \pm \pi) x_1/h_1} e^{i(\theta_2 \pm \pi) x_2/h_2} \\
&= e^{i\theta_1 x_1/h_1} e^{\pm i\pi 2k_1} e^{i\theta_2 x_2/h_2} e^{\pm i\pi 2k_2} \quad (x_i = 2k_i h \in \mathbf{G}_{2h}, \; k_i \in \mathbb{Z}) \\
&= e^{i\theta_1 2x_1/(2h_1)} e^{i\theta_2 2x_2/(2h_2)} = \varphi_{2h}(2\theta^{(0,0)}).
\end{aligned}$$

**Definition 4.4.1 (harmonics for standard coarsening)**   The corresponding four $\varphi(\theta^\alpha, \cdot)$ (and sometimes also the corresponding frequencies $\theta^\alpha$) are called *harmonics* (of each other). For a given $\theta = \theta^{(0,0)} \in T^{\text{low}}$, we define its *four-dimensional space of harmonics* by

$$E_h^\theta := \text{span}[\varphi(\theta^\alpha, \cdot): \alpha = (\alpha_1, \alpha_2) \in \{(0, 0), (1, 1), (0, 1), (1, 0)\}]. \quad (4.4.5)$$

> *The significance of these spaces $E_h^\theta$ is that* they turn out to be invariant *under the two-grid operator $M_h^H$ under general assumptions.*

All $\varphi_h(\theta, \cdot)$ are formal eigenfunctions of $L_h$, and under the assumptions made in the previous section also of $S_h$. The operator $K_h^H$ intermixes Fourier components with each other. This is a consequence of the fact that the two different grids, $\mathbf{G}_h$ and $\mathbf{G}_H$, are involved. In the following, we will discuss this behavior in detail. For that purpose, we consider an arbitrary $\psi \in E_h^\theta$. We can represent $\psi$ in the form

$$\psi = A^{(0,0)}\varphi(\theta^{(0,0)}, \cdot) + A^{(1,1)}\varphi(\theta^{(1,1)}, \cdot) + A^{(1,0)}\varphi(\theta^{(1,0)}, \cdot) + A^{(0,1)}\varphi(\theta^{(0,1)}, \cdot) \quad (4.4.6)$$

with uniquely defined coefficients or amplitudes $A^\alpha$. We will analyze how the coefficients $A^\alpha$ are transformed if the multigrid components in (4.4.1) are applied to $\psi$. We will treat each of the multigrid components separately in the following, but first summarize the results with respect to $K_h^H$ and $M_h^H$. For clarity, we now write $2h$ instead of $H$.

For the following theorem, we make the general assumption that $L_h$, $I_h^{2h}$, $L_{2h}$ and $I_{2h}^h$ are represented by stencils on $\mathbf{G}_h$ and $\mathbf{G}_{2h}$. Furthermore, in forming $K_h^{2h}$ and $M_h^{2h}$

according to (4.4.1), we have implicitly assumed that $(L_{2h})^{-1}$ exists. This assumption will be discussed below. All other assumptions are made explicitly in the theorem itself.

---

**Theorem 4.4.1**

(1) *Under the above assumptions, the coarse grid correction operator $K_h^{2h}$ is represented on $E_h^\theta$ by the $(4 \times 4)$-matrix $\hat{K}_h^{2h}(\theta)$*

$$\hat{K}_h^{2h}(\theta) = \hat{I}_h - \hat{I}_{2h}^h(\theta)(\hat{L}_{2h}(2\theta))^{-1}\hat{I}_h^{2h}(\theta)\hat{L}_h(\theta) \qquad (4.4.7)$$

*for each $\theta \in T^{\text{low}}$. Here, $\hat{I}_h$, $\hat{L}_h(\theta)$ are $(4 \times 4)$-matrices, $\hat{I}_h^{2h}(\theta)$ is a $(4 \times 1)$-matrix, $(\hat{L}_{2h}(2\theta))^{-1}$ is a $(1 \times 1)$-matrix, and $\hat{I}_{2h}^h(\theta)$ is a $(1 \times 4)$-matrix.*

*In other words: If we apply $K_h^{2h}$ to any $\psi \in E_h^\theta$, the corresponding coefficients $A^\alpha$ in (4.4.6) are transformed according to*

$$\begin{pmatrix} A^{(0,0)} \\ A^{(1,1)} \\ A^{(1,0)} \\ A^{(0,1)} \end{pmatrix} \Leftarrow \hat{K}_h^{2h}(\theta) \begin{pmatrix} A^{(0,0)} \\ A^{(1,1)} \\ A^{(1,0)} \\ A^{(0,1)} \end{pmatrix}. \qquad (4.4.8)$$

(2) *If the spaces $E_h^\theta$ are invariant under the smoothing operator $S_h$, i.e.*

$$S_h : E_h^\theta \to E_h^\theta \quad \text{for all } \theta \in T^{\text{low}}, \qquad (4.4.9)$$

*we also obtain a representation of $M_h^{2h}$ on $E_h^\theta$ by a $(4 \times 4)$-matrix $\hat{M}_h^{2h}(\theta)$ with respect to $E_h^\theta$. Here,*

$$\hat{M}_h^{2h}(\theta) = \hat{S}_h(\theta)^{\nu_2} \hat{K}_h^{2h}(\theta) \, \hat{S}_h(\theta)^{\nu_1} \qquad (4.4.10)$$

*with $\hat{K}_h^{2h}(\theta)$ from (4.4.7) and the $(4 \times 4)$-matrix $\hat{S}_h(\theta)$ which represents $S_h$. This means that $M_h^{2h}\psi$ can be written as*

$$\begin{aligned} M_h^{2h}\psi = {} & B^{(0,0)}\varphi(\theta^{(0,0)}, \cdot) + B^{(1,1)}\varphi(\theta^{(1,1)}, \cdot) \\ & + B^{(1,0)}\varphi(\theta^{(1,0)}, \cdot) + B^{(0,1)}\varphi(\theta^{(0,1)}, \cdot) \end{aligned} \qquad (4.4.11)$$

*where*

$$\begin{pmatrix} B^{(0,0)} \\ B^{(1,1)} \\ B^{(1,0)} \\ B^{(0,1)} \end{pmatrix} = \hat{M}_h^{2h}(\theta) \begin{pmatrix} A^{(0,0)} \\ A^{(1,1)} \\ A^{(1,0)} \\ A^{(0,1)} \end{pmatrix}. \qquad (4.4.12)$$

We will prove the above theorem and specify the $(4 \times 4)$-matrices $\hat{K}_h^{2h}(\boldsymbol{\theta})$ and $\hat{M}_h^{2h}(\boldsymbol{\theta})$ in more detail below. But first, we discuss the assumption that $(L_{2h})^{-1}$ exists. Apart from the existence of $(L_{2h})^{-1}$, the existence of $(L_h)^{-1}$ is also implicitly assumed. If $(L_h)^{-1}$ does not exist, we can, in general, not expect that any iterative method will be convergent. As the following example shows, these assumptions are *not formally fulfilled* even in standard situations.

**Example 4.4.1** Let $L_h = -\Delta_h$ and $L_{2h} = -\Delta_{2h}$ be discretizations of the Laplacian on the infinite grid. Then, obviously, for $\boldsymbol{\theta} = 0$ (i.e. $\varphi(\boldsymbol{\theta}, \boldsymbol{x}) \equiv 1$), the corresponding symbols (formal eigenvalues) are zero:

$$\tilde{L}_h(\boldsymbol{\theta}) = \tilde{L}_{2h}(\boldsymbol{\theta}) = 0.$$

In addition, $\tilde{L}_{2h}(\boldsymbol{\theta}) = 0$ for all four harmonics which coincide on $\mathbf{G}_{2h}$ with $\varphi \equiv 1$, i.e. $\boldsymbol{\theta} = (\theta_1, \theta_2) \in \{(0, 0), (-\pi, 0), (0, -\pi), (-\pi, -\pi)\}$. We will find the same behavior for those $L_h (= [s_{\kappa,h}])$ and $L_{2h} (= [s_{\kappa,2h}])$, for which

$$\sum_{\kappa \in V} s_{\kappa,h} = 0 \quad \text{and} \quad \sum_{\kappa \in V} s_{\kappa,2h} = 0, \text{ respectively.}$$

$\triangle$

**Remark 4.4.1** These sums are 0 for consistent discretizations of those differential operators $L$ on an infinite grid which contain only derivatives of $u$. $\gg$

The complications illustrated in Example 4.4.1 are *only formal ones* (due to the infinite grid assumption). If we have, for example, an application with Dirichlet boundary conditions, we do not have to bother about these symbols and the corresponding eigenfunctions. If we have a singular discrete operator (due to, for example, periodic or pure Neumann boundary conditions), this singular behavior has to be taken into account separately by any iterative solver. In order to make sure that $M_h^{2h}$ can be formed and gives a reasonable iterative operator, we remove all $\boldsymbol{\theta}$ from our analysis for which $L_h$ or $L_{2h}$ have the symbol 0. That is, we will exclude the set

$$\Lambda = \left\{ \boldsymbol{\theta} \in \left[ -\frac{\pi}{2}, \frac{\pi}{2} \right)^2 : \tilde{L}_h(\boldsymbol{\theta}) = 0 \text{ or } \tilde{L}_{2h}(\boldsymbol{\theta}) = 0 \right\}. \tag{4.4.13}$$

From the above theorem, we can draw conclusions for the convergence behavior of the two-grid method under consideration: *we have reduced the problem to the investigation (of spectral radii and norms) of $(4 \times 4)$-matrices.*

**Supplement**   Based on the representation of $M_h^{2h}$ by the $(4 \times 4)$-matrices $\hat{M}_h^{2h}(\boldsymbol{\theta})$, we can calculate the *asymptotic convergence factor*

$$\rho_{\text{loc}}(M_h^{2h}) = \sup\{\rho_{\text{loc}}(\hat{M}_h^{2h}(\boldsymbol{\theta})): \boldsymbol{\theta} \in T^{\text{low}}, \ \boldsymbol{\theta} \notin \Lambda\}. \tag{4.4.14}$$

Here, $\rho_{\text{loc}}(\hat{M}_h^{2h}(\boldsymbol{\theta}))$ is the spectral radius of the $(4 \times 4)$-matrix $\hat{M}_h^{2h}(\boldsymbol{\theta})$.

Similarly, we can introduce *an error reduction factor* $\sigma_{\text{loc}}(M_h^{2h})$ with respect to an appropriate norm. In particular, we consider

$$\sigma_{\text{loc},S}(M_h^{2h}) = \sup\{\|\hat{M}_h^{2h}(\boldsymbol{\theta})\|: \boldsymbol{\theta} \in T^{\text{low}}, \ \boldsymbol{\theta} \notin \Lambda\}. \tag{4.4.15}$$

Here $\|\cdot\|$ denotes the spectral norm associated with the Euclidean vector norm in $\mathbb{C}^4$.

We will now derive the representation of $M_h^{2h}$ by $(4 \times 4)$-matrices in Theorem 4.4.1 by considering each multigrid component separately.

**Operators $L_h$, $I_h$**   Under the general assumption that $L_h$ is characterized by the difference stencil $L_h \overset{\wedge}{=} [s_\kappa]_h$, Lemma 4.2.1 immediately implies that the $A^\alpha$ are transformed by $L_h$ as follows:

$$\begin{pmatrix} A^{(0,0)} \\ A^{(1,1)} \\ A^{(1,0)} \\ A^{(0,1)} \end{pmatrix} \Leftarrow \begin{pmatrix} \tilde{L}_h(\boldsymbol{\theta}^{(0,0)}) & & & \\ & \tilde{L}_h(\boldsymbol{\theta}^{(1,1)}) & & \\ & & \tilde{L}_h(\boldsymbol{\theta}^{(1,0)}) & \\ & & & \tilde{L}_h(\boldsymbol{\theta}^{(0,1)}) \end{pmatrix} \begin{pmatrix} A^{(0,0)} \\ A^{(1,1)} \\ A^{(1,0)} \\ A^{(0,1)} \end{pmatrix}. \tag{4.4.16}$$

We denote this $(4 \times 4)$-matrix by $\hat{L}_h(\boldsymbol{\theta})$. Trivially, the identity operator $I_h$ is represented by the $(4 \times 4)$-identity matrix on $E_h^\theta$.

**Restriction operator $I_h^{2h}$**   We also assume that the transfer operator $I_h^{2h}$ is characterized by a stencil

$$I_h^{2h} \overset{\wedge}{=} [\hat{t}_\kappa]_h^{2h}, \qquad \text{i.e. } I_h^{2h} w_h(\boldsymbol{x}) = \sum_{\kappa \in V} \hat{t}_\kappa w_h(\boldsymbol{x} + \kappa \boldsymbol{h}), \quad (\boldsymbol{x} \in \mathbf{G}_{2h}) \tag{4.4.17}$$

with a finite index set $V$. Then we obtain

$$I_h^{2h} \varphi_h(\boldsymbol{\theta}^\alpha, \cdot) = \tilde{I}_h^{2h}(\boldsymbol{\theta}^\alpha) \varphi_{2h}(2\boldsymbol{\theta}^{(0,0)}, \cdot) \tag{4.4.18}$$

(see (4.4.4)) with

$$\tilde{I}_h^{2h}(\boldsymbol{\theta}^\alpha) := \sum_{\kappa \in V} \hat{t}_\kappa e^{i\boldsymbol{\theta}^\alpha \cdot \kappa}. \tag{4.4.19}$$

For the coefficients $A^\alpha$, this results in the transformation

$$A_{2h} = \hat{I}_h^{2h}(\boldsymbol{\theta}) \begin{pmatrix} A^{(0,0)} \\ A^{(1,1)} \\ A^{(1,0)} \\ A^{(0,1)} \end{pmatrix}, \tag{4.4.20}$$

where $A_{2h}$ is the resulting coefficient of the $2h$ Fourier component $\varphi_{2h}(2\boldsymbol{\theta}^{(0,0)}, \cdot)$ and $\hat{I}_h^{2h}(\boldsymbol{\theta})$ is the $(1 \times 4)$-matrix

$$[\tilde{I}_h^{2h}(\boldsymbol{\theta}^{(0,0)}), \ \tilde{I}_h^{2h}(\boldsymbol{\theta}^{(1,1)}), \ \tilde{I}_h^{2h}(\boldsymbol{\theta}^{(1,0)}), \ \tilde{I}_h^{2h}(\boldsymbol{\theta}^{(0,1)})].$$

**Example 4.4.2 (injection and FW)**   For injection, the Fourier symbol is 1 for all frequencies:

$$\tilde{I}_h^{2h}(\boldsymbol{\theta}^\alpha) = 1$$

For FW, we obtain

$$\tilde{I}_h^{2h}(\boldsymbol{\theta}^\alpha) = \tfrac{1}{4}(1 + \cos \bar{\theta}_1)(1 + \cos \bar{\theta}_2)$$

from (4.4.19) with $\bar{\theta}_i$ as in (4.4.2). $\qquad\qquad\qquad\qquad\qquad\qquad\qquad\qquad\qquad\triangle$

**Solution on $2h$-grid**   According to the general assumption, $L_{2h}$ is also given by a stencil:

$$L_{2h} \stackrel{\wedge}{=} [s_{\kappa,2h}]_{2h}.$$

For $\boldsymbol{\theta} = \boldsymbol{\theta}^{(0,0)}$ and $\boldsymbol{\theta} \in T^{\text{low}}$, we thus have

$$L_{2h}\varphi_{2h}(2\boldsymbol{\theta}, \cdot) = \tilde{L}_{2h}(2\boldsymbol{\theta})\varphi_{2h}(2\boldsymbol{\theta}, \cdot) \tag{4.4.21}$$

with the symbol

$$\boxed{\tilde{L}_{2h}(2\boldsymbol{\theta}) = \sum_{\kappa \in V} s_{\kappa,2h} e^{i2\boldsymbol{\theta}\cdot\kappa}} \ . \tag{4.4.22}$$

The solution on the $2h$-grid induces (for $2\boldsymbol{\theta} \notin \Lambda$) for the respective coefficient

$$A_{2h} \Longleftarrow \frac{1}{(\tilde{L}_{2h}(2\boldsymbol{\theta}))} A_{2h}. \tag{4.4.23}$$

**Interpolation operator** $I_{2h}^h$    For the interpolation operator, we use the stencil nota-tion from Section 2.3.4. We can prove the following representation for $I_{2h}^h$ if applied to $\varphi_{2h}(2\theta^{(0,0)}, \cdot)$:

$$I_{2h}^h\varphi_{2h}(2\theta^{(0,0)}, \cdot) = \sum_{\alpha} \tilde{I}_{2h}^h(\theta^{\alpha})\varphi(\theta^{\alpha}, \cdot) \quad \text{with } \tilde{I}_{2h}^h(\theta^{\alpha}) = \frac{1}{4}\sum_{\kappa \in V} t_{\kappa}e^{i\theta^{\alpha}\cdot\kappa}.$$

Interpreting this representation for the coefficients $A^{\alpha}$ and $A_{2h}$ of $\varphi_h(\theta^{\alpha}, \cdot)$ and $\varphi_{2h}(2\theta^{(0,0)}, \cdot)$, we obtain

$$\begin{pmatrix} A^{(0,0)} \\ A^{(1,1)} \\ A^{(1,0)} \\ A^{(0,1)} \end{pmatrix} \Leftarrow \hat{I}_{2h}^h(\theta^{(0,0)})A_{2h} \quad \text{with } \hat{I}_{2h}^h(\theta^{(0,0)}) = \begin{pmatrix} \tilde{I}_{2h}^h(\theta^{(0,0)}) \\ \tilde{I}_{2h}^h(\theta^{(1,1)}) \\ \tilde{I}_{2h}^h(\theta^{(1,0)}) \\ \tilde{I}_{2h}^h(\theta^{(0,1)}) \end{pmatrix}. \tag{4.4.24}$$

For the proof of the above representation, we can apply an argument like that already used in Section 3.3.4 (step 3) in the context of rigorous Fourier analysis. Starting with the stencil representation of $I_{2h}^h$, we can find a representation of $I_{2h}^h\varphi_{2h}(2\theta^{(0,0)}, \cdot)$ in terms of $\varphi_h(\theta^{(0,0)}, \cdot)$ similar to the one in (3.3.14). Here the coefficients $\delta_1, \delta_2, \delta_3, \delta_4$ correspond to the four types of grid points characterized by the indices $(i, j)$ with $x_1 = ih$, $x_2 = jh$: even–even, odd–odd, odd–even, even–odd. We use the following remark, which corresponds directly to Remark 3.3.4.

**Remark 4.4.2**  Any grid function $c(x_1, x_2)e^{i\theta\cdot x/h}$ with $c(x_1, x_2)$ as in (3.3.15) can be represented on $\mathbf{G}_h$ as a linear combination

$$a^{(0,0)}e^{i\theta^{(0,0)}\cdot x/h} + a^{(1,1)}e^{i\theta^{(1,1)}\cdot x/h} + a^{(1,0)}e^{i\theta^{(1,0)}\cdot x/h} + a^{(0,1)}e^{i\theta^{(0,1)}\cdot x/h},$$

where the coefficients $a^{(0,0)}, a^{(1,1)}, a^{(1,0)}, a^{(0,1)}$ are given by (3.3.16).                                    $\gg$

**Example 4.4.3**  For the bilinear interpolation operator (2.3.8), we obtain (with $\theta = \theta^{(0,0)}$)

$$I_{2h}^h\varphi_{2h}(2\theta, x) = \varphi_{2h}(2\theta, x) \cdot \begin{cases} 1 & \text{if } x_1/h_1 \text{ and } x_2/h_2 \text{ even} \\ \cos\theta_1\cos\theta_2 & \text{if } x_1/h_1 \text{ and } x_2/h_2 \text{ odd} \\ \cos\theta_1 & \text{if } x_1/h_1 \text{ odd, } x_2/h_2 \text{ even} \\ \cos\theta_2 & \text{if } x_1/h_1 \text{ even, } x_2/h_2 \text{ odd.} \end{cases} \tag{4.4.25}$$

The above remark gives

$$\begin{pmatrix} \tilde{I}_{2h}^h(\theta^{(0,0)}) \\ \tilde{I}_{2h}^h(\theta^{(1,1)}) \\ \tilde{I}_{2h}^h(\theta^{(1,0)}) \\ \tilde{I}_{2h}^h(\theta^{(0,1)}) \end{pmatrix} = \frac{1}{4}\begin{pmatrix} (1+\cos\theta_1)(1+\cos\theta_2) \\ (1-\cos\theta_1)(1-\cos\theta_2) \\ (1+\cos\theta_1)(1-\cos\theta_2) \\ (1-\cos\theta_1)(1+\cos\theta_2) \end{pmatrix} \tag{4.4.26}$$

or simply

$$\tilde{I}_{2h}^h(\boldsymbol{\theta}^\alpha) = \tfrac{1}{4}(1 + \cos\bar\theta_1)(1 + \cos\bar\theta_2),$$

which is the same as for the FW operator (up to a constant scaling factor). This also reflects the fact that bilinear interpolation and FW are transpose to each other.                                              $\triangle$

**Remark 4.4.3**  It can be shown that the symbol of linear interpolation in $d$ dimensions can be written as the product

$$\tilde{I}_{2h}^h(\boldsymbol{\theta}^\alpha) = \prod_{j=1}^d \frac{1 + \cos\bar\theta_j}{2}.$$

If in $d$ dimensions, $2^d I_h^{2h}$ is the transpose of some interpolation $I_{2h}^h$, we have $\hat{I}_h^{2h}(\boldsymbol{\theta}) = 2^{-d}\hat{I}_{2h}^h(\boldsymbol{\theta})^T$.                                              $\gg$

From the representations of $L_h$, $I_h^{2h}$, $I_{2h}^h$, $I_h$ and the solution on the coarse grid, we immediately obtain the representation of $K_h^{2h}$. In order to extend it to $M_h^{2h}$, we have to include the representations of $S_h$ in the analysis.

**Smoothing operator $S_h$**    This is particularly simple if the assumptions of the last section are fulfilled. In that case all Fourier components $\varphi(\boldsymbol{\theta}, .)$ are formal eigenfunctions of $S_h$, and $S_h$ can be represented by the diagonal $(4 \times 4)$-matrix

$$
\begin{pmatrix} A^{(0,0)} \\ A^{(1,1)} \\ A^{(1,0)} \\ A^{(0,1)} \end{pmatrix}
\Leftarrow
\begin{pmatrix}
\tilde{S}_h(\boldsymbol{\theta}^{(0,0)}) & & & \\
& \tilde{S}_h(\boldsymbol{\theta}^{(1,1)}) & & \\
& & \tilde{S}_h(\boldsymbol{\theta}^{(1,0)}) & \\
& & & \tilde{S}_h(\boldsymbol{\theta}^{(0,1)})
\end{pmatrix}
\begin{pmatrix} A^{(0,0)} \\ A^{(1,1)} \\ A^{(1,0)} \\ A^{(0,1)} \end{pmatrix}
\tag{4.4.27}
$$

for each $\boldsymbol{\theta} \in T^{\text{low}}$. Under the more general assumption (4.4.9), $S_h$ is represented on $E_h^\theta$ by a $(4 \times 4)$-matrix $\hat{S}_h(\boldsymbol{\theta})$.

## 4.5 SMOOTHING ANALYSIS II

In Section 4.3, we considered smoothing procedures which have the property that all $\varphi(\boldsymbol{\theta}, \cdot)$ are eigenfunctions of the smoothing operators $S_h$. For such smoothers, the definition of a smoothing factor $\mu_{\text{loc}}(S_h)$ is straightforward. In the two-grid analysis in the last section, we made the more general assumption that only the *invariance property*

$$\boxed{\; S_h : E_h^\theta \to E_h^\theta \quad \text{for all } \boldsymbol{\theta} \in T^{\text{low}} \;}
\tag{4.5.1}$$

is valid for the smoothing operator under consideration. This assumption is fulfilled for a larger class of smoothers, including GS-RB relaxation as well as "multicolor" and "zebra"-type relaxations which will be described in the following chapters.

In the following, we will generalize the definition of the smoothing factor for all smoothers which have the invariance property (4.5.1).

We will describe a concrete class of smoothing procedures including red–black type smoothers, based on a straightforward generalization of the splitting formalism in Section 4.3.

If a smoother has the invariance property (4.5.1), high and low frequencies may be intermixed by $S_h$. In order to be able to measure the smoothing properties of $S_h$, the idea is to assume *an "ideal" coarse grid operator* $Q_h^{2h}$ (instead of the real coarse grid operator $K_h^{2h}$), *which annihilates the low frequency error components and leaves the high frequency components unchanged*. More precisely, $Q_h^{2h}$ is a projection operator, defined on $E_h^\theta$ by

$$
Q_h^{2h} \varphi(\theta, \cdot) = \begin{cases} 0 & \text{if } \theta = \theta^{(0,0)} \in T^{\text{low}} \\ \varphi(\theta, \cdot) & \text{if } \theta \in \{\theta^{(1,0)}, \theta^{(0,1)}, \theta^{(1,1)}\}. \end{cases} \tag{4.5.2}
$$

As a consequence, the $(4 \times 4)$-matrix $\hat{K}_h^{2h}(\theta)$ in (4.4.7) is replaced by the $(4 \times 4)$-projection matrix

$$
\hat{Q}_h^{2h}(\theta) = \hat{Q}_h^{2h} = \begin{pmatrix} 0 & & & \\ & 1 & & \\ & & 1 & \\ & & & 1 \end{pmatrix} \quad \text{for} \quad \theta \in T^{\text{low}} \tag{4.5.3}
$$

for the smoothing analysis. Furthermore, we replace

$$
M_h^{2h} = S_h^{\nu_2} K_h^{2h} S_h^{\nu_1} \quad \text{by} \quad S_h^{\nu_2} Q_h^{2h} S_h^{\nu_1}
$$

and $\rho_{\text{loc}}(M_h^{2h})$ in (4.4.14) by

$$
\rho_{\text{loc}}(S_h^{\nu_2} Q_h^{2h} S_h^{\nu_1}) := \sup \left\{ \rho(\hat{S}_h(\theta)^{\nu_2} \hat{Q}_h^{2h} \hat{S}_h(\theta)^{\nu_1}) : \theta \in T^{\text{low}} \right\}. \tag{4.5.4}
$$

This quantity $\rho_{\text{loc}}(S_h^{\nu_2} Q_h^{2h} S_h^{\nu_1})$ is a good measure of the total smoothing effect resulting from $\nu = \nu_1 + \nu_2$ smoothing steps. In addition, one can expect that this quantity gives a realistic prediction of the convergence factor $\rho_{\text{loc}}(M_h^{2h})$ as long as the replacement of $K_h^{2h}$ by $Q_h^{2h}$ is acceptable. Practically this means that the transfer and the coarse grid operator should be sufficiently good and that not too many smoothing steps are carried out per two-grid cycle. Since

$$
\rho_{\text{loc}}(\hat{S}_h(\theta)^{\nu_2} \hat{Q}_h^{2h} \hat{S}_h(\theta)^{\nu_1}) = \rho_{\text{loc}}(\hat{Q}_h^{2h} \hat{S}_h(\theta)^{\nu}) \quad (\text{with } \nu = \nu_1 + \nu_2),
$$

we arrive at the following definition of a general LFA smoothing factor.

**Definition 4.5.1** Under the assumption that $S_h$ has the invariance property (4.5.1), we define the *smoothing factor* $\mu_{\text{loc}}(S_h, \nu)$ of $S_h$ by

$$
\mu_{\text{loc}}(S_h, \nu) := \sup \left\{ \sqrt[\nu]{\rho_{\text{loc}}(\hat{Q}_h^{2h} \hat{S}_h(\theta)^{\nu})}, \ \theta \in T^{\text{low}} \right\}. \tag{4.5.5}
$$

Here, $\mu$ depends on $\nu$!

**Remark 4.5.1** For the practical calculation of $\mu_{\text{loc}}(S_h, \nu)$, note that the first row of the $(4 \times 4)$-matrix $Q_h^{2h} S_h^{\nu}$ is zero. $\gg$

**Remark 4.5.2** For the special case that all $\varphi(\boldsymbol{\theta}, \cdot)$ are eigenfunctions of $S_h$, the above definition coincides with Definition 4.3.1 given in Section 4.3. In that case $\hat{S}_h(\boldsymbol{\theta})^{\nu}$ is the diagonal matrix in (4.4.27) and $\hat{Q}_h^{2h} \hat{S}_h(\boldsymbol{\theta})^{\nu}$ contains only the high frequency symbols $\tilde{S}_h(\boldsymbol{\theta}^{(1,0)})^{\nu}, \tilde{S}_h(\boldsymbol{\theta}^{(0,1)})^{\nu}, \tilde{S}_h(\boldsymbol{\theta}^{(1,1)})^{\nu}$. $\gg$

### 4.5.1 Local Fourier Analysis for GS-RB

We now apply the above general definition of a smoothing factor. For that purpose, we first generalize the assumption on smoothing procedures in Section 4.3 in such a way that red–black type smoothers and similar "pattern" smoothers, like four-color relaxation, are included.

A corresponding smoothing operator $S_h^{\text{partial}}$ is then given by

$$S_h^{\text{partial}} v_h(\boldsymbol{x}) = \begin{cases} -(L_h^+)^{-1} L_h^- v_h(\boldsymbol{x}) & \text{for } \boldsymbol{x} \in \tilde{\mathbf{G}}_h \\ v_h(\boldsymbol{x}) & \text{for } \boldsymbol{x} \in \mathbf{G}_h \setminus \tilde{\mathbf{G}}_h \end{cases} \tag{4.5.6}$$

(assuming that $(L_h^+)^{-1}$ exists). Here, $\tilde{\mathbf{G}}_h$ is a subset of $\mathbf{G}_h$, typically defined by a pattern that is characteristic for the smoothing procedure under consideration. Obviously, only the grid points of $\tilde{\mathbf{G}}_h$ are processed in the above partial relaxation step; the remaining points $(\mathbf{G}_h \setminus \tilde{\mathbf{G}}_h)$ are not treated.

A typical example for such a smoothing operator is the GS-RB smoother for $L_h = -\Delta_h$. Using the above notation, GS-RB consists of two partial steps, each of which is of Jacobi type. In the first partial step, $\tilde{\mathbf{G}}_h$ consists of the "red" and in the second step $\tilde{\mathbf{G}}_h$ consists of the "black" points of $\mathbf{G}_h$. For both steps, we now use (4.3.4) with a relaxation parameter $\omega$.

The partial step smoothing operators, denoted by $S_h^{\text{RED}}$ and $S_h^{\text{BLACK}}$, respectively, then both have the form (4.5.6), with $\tilde{\mathbf{G}}_h$ being the red/black points of $\mathbf{G}_h$, respectively. The complete GS-RB smoothing operator is, of course, the product

$$S_h^{\text{RB}} = S_h^{\text{BLACK}} S_h^{\text{RED}}. \tag{4.5.7}$$

In order to calculate the smoothing factor (4.5.5) of $S_h^{\text{RB}}$, we must find the $(4 \times 4)$-matrix representation $\hat{S}_h^{\text{RB}}(\omega, \boldsymbol{\theta})$ for $S_h^{\text{RB}}(\omega)$. Since we have

$$S_h(\omega) = -(L_h^+)^{-1} L_h^- = I_h - \frac{\omega h^2}{4} L_h$$

for $\omega$-JAC (4.3.4), we find from (4.5.6) that

$$S_h^{\text{partial}} \varphi(\boldsymbol{x}) = \begin{cases} (1 - (\omega h^2/4) \tilde{L}_h(\boldsymbol{\theta})) \varphi(\boldsymbol{\theta}, \boldsymbol{x}) & (\boldsymbol{x} \in \tilde{\mathbf{G}}_h) \\ \varphi(\boldsymbol{\theta}, \boldsymbol{x}) & (\boldsymbol{x} \in \mathbf{G}_h \setminus \tilde{\mathbf{G}}_h). \end{cases} \tag{4.5.8}$$

This is not yet a Fourier representation of $S_h^{\text{partial}}$. Note, however, that any $\psi \in E_h^\theta$

$$\psi(\theta, x) = \begin{cases} \beta_1 \varphi(\theta, x) & (x \in \tilde{G}_h) \\ \beta_2 \varphi(\theta, x) & (x \in G_h \backslash \tilde{G}_h) \end{cases} \tag{4.5.9}$$

(with a fixed Fourier component $\varphi$ and constants $\beta_1$ and $\beta_2$) can easily be written as a linear combination of $\varphi(\theta^\alpha, \cdot)$ with $\theta^\alpha = \theta$ and $\varphi(\theta^{\bar{\alpha}}, \cdot)$, where $\bar{\alpha} = (1, 1) - \alpha$. This is actually a special case of Remark 4.4.2. We obtain the following $(4 \times 4)$-matrix representations for $S_h^{\text{RED}}$ and $S_h^{\text{BLACK}}$ with respect to $E_h^\theta (-\pi/2 \leq \theta < \pi/2)$:

$$\hat{S}_h^{\text{RED}}(\theta, \omega) = \frac{1}{2} \begin{pmatrix} A^{(0,0)} + 1 & A^{(1,1)} - 1 & 0 & 0 \\ A^{(0,0)} - 1 & A^{(1,1)} + 1 & 0 & 0 \\ 0 & 0 & A^{(1,0)} + 1 & A^{(0,1)} - 1 \\ 0 & 0 & A^{(1,0)} - 1 & A^{(0,1)} + 1 \end{pmatrix}$$

$$\hat{S}_h^{\text{BLACK}}(\theta, \omega) = \frac{1}{2} \begin{pmatrix} A^{(0,0)} + 1 & -A^{(1,1)} + 1 & 0 & 0 \\ -A^{(0,0)} + 1 & A^{(1,1)} + 1 & 0 & 0 \\ 0 & 0 & A^{(1,0)} + 1 & -A^{(0,1)} + 1 \\ 0 & 0 & -A^{(1,0)} + 1 & A^{(0,1)} + 1 \end{pmatrix}$$

with

$$A^\alpha := 1 - \frac{\omega h^2}{4} \tilde{L}_h(\theta^\alpha).$$

From this representation, one can compute the smoothing factor $\mu_{\text{loc}}(S_h^{\text{RB}}(\omega), \nu)$. For the standard value $\omega = 1$, we obtain

$$\mu_{\text{loc}}(\nu) = \max \left\{ \tfrac{1}{4}, \chi(\nu) \right\} \tag{4.5.10}$$

(see [378]), where

$$\chi(\nu) = \left( \frac{2\nu - 1}{2\nu} \right)^2 \Big/ \sqrt[\nu]{2(2\nu - 1)}. \tag{4.5.11}$$

Some particular values of $\chi(\nu)$ are $\chi(1) = 0.125$, $\chi(2) = 0.229$ and $\chi(3) = 0.322$. $\chi(\nu) \sim \sqrt[\nu]{1/(4e\nu)}$ for $\nu \to \infty$.

**Remark 4.5.3** The smoothing analysis of $\omega$-GS-RB can also be carried out in the context of the rigorous Fourier analysis (see Section 7.5 of [378]). The representation of the $(4 \times 4)$-block matrices is the same as in the LFA case. $\gg$

## 4.6 SOME RESULTS, REMARKS AND EXTENSIONS

In this section we will give some additional LFA results. More results can be found in [58, 378]. We also outline straightforward extensions of LFA with respect to other coarsening strategies. Furthermore, a simplified two-grid analysis is presented, which gives a basic insight into the quality of the coarse grid correction.

### 4.6.1 Some Local Fourier Analysis Results for Model Problem 1

We consider $L_h = -\Delta_h$ and a multigrid method based on GS-LEX smoothing, FW and bilinear interpolation. LFA two-grid analysis gives the two-grid convergence factors $\rho_{\mathrm{loc}}$ as listed in Table 4.1. For comparison, experimentally measured convergence factors for Model Problem 1 are included (W-cycle, $h = 1/128$). The correspondence of theoretical and practical values is excellent and typical for many applications.

We also recognize a good prediction of the two-grid convergence factors by the smoothing analysis. This good agreement can be expected only for small numbers $\nu$ of smoothing steps. If too many smoothing steps are performed per multigrid cycle, the coarse grid approximation can no longer cope with the smoothing effect. Therefore, the smoothing factors $\mu_{\mathrm{loc}}^{\nu}$ are too optimistic for $\nu \geq 4$ in this case.

Table 4.2 compares the influence of the restriction operator on the convergence factors $\rho_{\mathrm{loc}}$ for GS-LEX, comparing full weighting (FW) and injection (INJ). The results show that INJ is a satisfactory restriction operator *if combined with GS-LEX* for Model Problem 1. (This is particularly true if the computational effort is taken into account.) INJ is cheaper than FW. Intuitively, it seems to be clear that INJ is the better, the smoother the defects are. This is reflected by the convergence factors for $\nu \geq 3$, where $\rho_{\mathrm{loc}}(\mathrm{INJ})$ is even smaller than $\rho_{\mathrm{loc}}(\mathrm{FW})$.

The application of the INJ operator has, however, the disadvantage that the spectral norm of the corresponding two-grid operators is not bounded, see Table 4.3, where also the influence of the pre- and postsmoothing on the norm $\sigma_{\mathrm{loc},S}$ is presented.

**Table 4.1.** LFA smoothing factors, LFA two-grid convergence factors and measured W-cycle convergence factors (Model Problem 1).

| $\nu_1, \nu_2$ | $\mu_{\mathrm{loc}}^{\nu_1+\nu_2}$ | $\rho_{\mathrm{loc}}$ | W-cycle ($h = 1/128$) |
|---|---|---|---|
| 1,0 | 0.500 | 0.400 | 0.40 |
| 1,1 | 0.250 | 0.193 | 0.19 |
| 2,1 | 0.125 | 0.119 | 0.12 |
| 2,2 | 0.063 | 0.084 | 0.08 |

**Table 4.2.** $\rho_{\mathrm{loc}}$ for GS-LEX relaxation (for $L_h = -\Delta_h$).

| $\nu$ | $\rho_{\mathrm{loc}}(\mathrm{FW})$ | $\rho_{\mathrm{loc}}(\mathrm{INJ})$ |
|---|---|---|
| 1 | 0.400 | 0.447 |
| 2 | 0.193 | 0.200 |
| 3 | 0.119 | 0.089 |
| 4 | 0.084 | 0.042 |

**Table 4.3.** $\sigma_{\text{loc},S}$ corresponding to the methods in Table 4.2.

| $(\nu_1, \nu_2)$ | $I_h^{2h}$ (FW) | $I_h^{2h}$ (INJ) |
|---|---|---|
| (1,0) | 0.447 | $\infty$ |
| (0,1) | 1.000 | $\infty$ |
| (2,0) | 0.208 | $\infty$ |
| (1,1) | 0.203 | $\infty$ |
| (0,2) | 1.000 | $\infty$ |
| (3,0) | 0.128 | $\infty$ |
| (2,1) | 0.119 | $\infty$ |
| (1,2) | 0.131 | $\infty$ |
| (0,3) | 1.000 | $\infty$ |

One thus needs to be careful with this operator, in particular if it is used in FMG or, more generally, if only one or a small number of cycles is used. Remember that injection is a restriction of order 0 (see Section 2.7). Again, FW is more robust and reliable.

*All the values* given for $\rho_{\text{loc}}$ above can be confirmed by numerical measurements. In all cases, the measured convergence factors are close to $\rho_{\text{loc}}$ if $h$ is chosen to be small enough (here for instance $h = 1/128$).

In the following chapters, we will use LFA for many different operators and multigrid approaches, 3D cases, singularly perturbed problems, operators with mixed derivatives, systems of PDEs and so on. In most cases, the LFA results will be very helpful in designing multigrid algorithms and understanding difficulties. Of course, LFA also has its limitations. If the local view which is characteristic for LFA is not sufficient for the description of certain global phenomena (e.g. singularly perturbed or hyperbolic situations), other considerations have to be added.

Brandt has proposed and developed extensions and generalizations of LFA. One extension is the so-called half-space analysis [63], which allows the inclusion of boundary effects into the analysis.

### 4.6.2 Additional Remarks

In this section we add some remarks about LFA approaches, modifications of LFA and about the relationship between LFA and rigorous Fourier analysis.

LFA smoothing analysis is usually the first step in analyzing a given problem and its multigrid solution. Smoothing analysis is typically much easier than two-grid analysis. This is particularly true if the $e^{i\theta \cdot x/h}$ functions are eigenfunctions of $S_h$ as assumed in Section 4.3. But even if only the spaces $E_h^\theta$ are invariant under $S_h$, the smoothing analysis is simpler since it uses the "ideal" coarse grid operator given in Section 4.5.

**Remark 4.6.1 (coarse grid correction and simplified two-grid analysis)** In order to avoid the complete two-grid analysis which may become rather involved, in particular in 3D cases and for systems of PDEs, one can analyze the smoothing procedure and the coarse grid correction separately. Therefore, in addition to smoothing analysis, "coarse grid correction analysis" approaches have been proposed. One idea in this context is the so-called *simplified two-grid analysis*, which is based on the *first differential approximation* (FDA) [426]. The goal here is to obtain some insight into the quality of the approximation of the operator $L_h$ by $L_{2h}$ for very low frequencies. The analysis neglects high frequencies and the coupling of harmonics. Furthermore, for very low frequencies, the transfer operators act nearly like identities. Therefore, the behavior of $K_h^{2h}$ can be approximately described by the quantity

$$\tilde{I}_h - (\tilde{L}_{2h})^{-1} \tilde{L}_h \quad \text{with} \quad \tilde{I}_h = 1$$

(for very low frequencies). The analysis of this term gives some insight into the quality of the coarse grid correction especially for problems with *characteristic directions* (like convection–diffusion equations, see Chapter 7). If we consider a low frequency $\boldsymbol{\theta} = (\theta_1, \theta_2)$ along a characteristic direction with $\theta_2 = c\theta_1$, we can compute

$$\lim_{\theta_1 \to 0} \left( 1 - \frac{\tilde{L}_h(\boldsymbol{\theta})}{\tilde{L}_{2h}(2\boldsymbol{\theta})} \right).$$

This estimate of the coarse grid approximation for the lowest frequencies should be a very small number. If it is not, the multigrid performance will be negatively influenced by a bad coarse grid correction due to very low frequencies (see Section 7.2.3 for an example).    $\gg$

**Remark 4.6.2**   Some authors use the LFA approach without allowing $\boldsymbol{\theta}$ to vary continuously in $-\pi \leq \boldsymbol{\theta} < \pi$. In this case, $\boldsymbol{\theta}$ varies in the finite set $T_h$ as introduced in (3.4.8). Such approaches can be useful if the real convergence factors are substantially dependent on $h$.    $\gg$

**Remark 4.6.3**   Table 4.4 illustrates which of the analysis approaches can be used to analyze the three smoothers $\omega$-JAC, GS-LEX and GS-RB.    $\gg$

**Remark 4.6.4**   As we have pointed out before, GS-LEX is a smoothing procedure which can be analyzed by LFA, not, however, by the rigorous Fourier analysis presented in Section 3.3. The reason is that the GS-LEX smoothing operator $S_h$ for the infinite grid is different from the GS-LEX smoothing operator for periodic boundary conditions on a

**Table 4.4.** Which analysis can be applied to which smoothing scheme?

|  | Rigorous Fourier analysis | LFA |
|---|---|---|
| $\varphi(\boldsymbol{\theta}, \cdot)$ eigenfunctions of $S_h$ | $\omega$-JAC | $\omega$-JAC<br>GS-LEX |
| $E_{h,\boldsymbol{\theta}}$ invariant under $S_h$ | $\omega$-JAC | $\omega$-JAC<br>GS-LEX |
|  | GS-RB | GS-RB |

finite domain. On an infinite grid there is no natural starting point for GS-LEX, contrary to the finite grid situation. If one tried to mimic the "infinite grid GS-LEX" in a finite domain situation with periodic boundary conditions, an *implicit* smoothing scheme would result.                                                                                          ≫

**Remark 4.6.5 (how to use LFA in practice)**   The two-grid LFA is rather complicated at the first glance. It is usually carried out with *a computer program* (see http://www.gmd.de/SCAI/multigrid/book.html for such a program by R. Wienands). An even simpler program can be used for the LFA smoothing analysis. Here, we present some general guidelines for the use of LFA:

If one wants to design (or analyze) a multigrid algorithm for a concrete problem $Lu = f$, we recommend using the following LFA ladder:

(1) Find an appropriate discretization for the problem, e.g. with good "$h$-ellipticity" $E_h(L_h)$ (to be discussed in Section 4.7).
(2) Find a smoothing scheme with a satisfactory smoothing factor.
(3) Choose appropriate transfer operators and check whether the two-grid LFA convergence factors are close to the smoothing factors.
(4) Check whether the convergence of multigrid cycles in your program approximates the LFA prediction.
(5) Use FMG and check whether you obtain discretization accuracy.

Of course, one could often skip some of these steps, e.g. check directly whether the multigrid program (Step 4) produces the efficiency anticipated by the smoothing factors (Step 2). If not, *then* perform the intermediate steps.                                                                                 ≫

**Remark 4.6.6 (debugging)**   In the debugging phase, it is advisable to use as coarse grids as possible.

If a multigrid code does not show a good convergence for a particular problem, a useful debugging approach is to start with an example, for which the exact discrete solution is known. This is easy if the example can be chosen such that the discretization error is 0, e.g. $u(x, y) = x + 2y$ in the case of the standard Laplacian.

A discrete approximation from the algorithm can then be compared with the exact solution of the discrete problem. One can check, whether the error between the approximation and the reference solution is smooth after the relaxation. Maybe it is not smooth near a boundary (see Section 5.6 for the boundary treatment). Also the defect can give some insight. It might be large in only a few grid points, for example. This debugging can be carried out graphically, e.g. by plotting the defects or the errors with available tools such as "gnuplot".

If a problem without a discretization error is not known, one can solve the problem first (possibly with a single grid solver), store the obtained solution and compare the multigrid approximations with this discrete solution. Additional insight is gained by printing norms of the defect on all grid levels, for example, after presmoothing and after the coarse grid correction and postsmoothing. These defects should also decrease on coarser grids.    ≫

### 4.6.3 Other Coarsening Strategies

In introducing LFA in this chapter, we have assumed standard coarsening, $\boldsymbol{H} = (2h_1, 2h_2)$. LFA can be extended to other coarsening strategies in a straightforward manner. *What has to be changed is the definition of low and high frequencies.* As a consequence, the definitions of spaces of harmonics $E_h^\theta$ and of the central quantities $\mu_{\mathrm{loc}}(S_h)$, $\rho_{\mathrm{loc}}(M_h^H)$ and $\sigma_{\mathrm{loc}}(M_h^H)$ have to be adapted.

In Fig. 4.3, we consider the following coarsening strategies:

(1) $x$-semicoarsening ($\boldsymbol{H} = (2h_1, h_2)$, upper left picture in Fig. 4.3),
(2) $y$-semicoarsening ($\boldsymbol{H} = (h_1, 2h_2)$, upper right picture in Fig. 4.3),
(3) red–black-coarsening (lower left picture in Fig. 4.3),
(4) $(h, 4h)$-coarsening (lower right picture in Fig. 4.3).

Instead of formally defining the respective low and high frequency terms, we simply show in Fig. 4.3 how Fig. 4.1 is to be changed for these four cases.

Note that the spaces of harmonics are two-dimensional in the first three coarsening strategies listed above, but 16-dimensional for $(h, 4h)$-coarsening. These facts have obvious and straightforward implications for the definition of $\mu_{\mathrm{loc}}(S_h)$, $\rho_{\mathrm{loc}}(M_h^H)$ and $\sigma_{\mathrm{loc}}(M_h^H)$.

**Remark 4.6.7** If GS-RB is used in connection with red–black-coarsening, the two-dimensional spaces of harmonics (characterized in Fig. 4.3) are invariant under $S_h$ and $K_h^H$ and thus $M_h^H$.

If GS-RB is used, however, in connection with $x$- or $y$-semicoarsening, the corresponding two-dimensional spaces that are invariant under $S_h$ and under $K_h^H$ are not the same. In this case, the four-dimensional "standard" spaces $E_h^\theta$ in (4.4.5) have to be used again for the two-grid LFA. They are again invariant under $S_h$ and $K_h^H$.                    $\gg$

### 4.7 $h$-ELLIPTICITY

In Section 1.1, we gave a first impression of the properties a PDE should fulfill for "standard" multigrid to work properly. The *ellipticity* of a PDE was explicitly addressed in this context. In describing basic multigrid algorithms in Chapter 2, we considered discrete elliptic differential operators $L_h$, or more concretely, Poisson-like equations in 2D or 3D. Finally, all the concrete theoretical results in Chapter 3 were related to the discrete Poisson equation.

Ellipticity of a differential operator $L$ is, however, neither a necessary nor a sufficient condition for the applicability of multigrid. In the following chapters we will also consider differential operators which may lack ellipticity or become nonelliptic depending on some characteristic parameters: anisotropic operators, convection-dominated operators, the full potential equation, Euler equations and Navier–Stokes equations.

In this section, we will use the LFA terminology to (qualitatively) characterize the properties a discrete operator $L_h$ should have to be suitable for multigrid. For that purpose, we will introduce the concept of $h$-*ellipticity* proposed by Brandt [61]. This property is

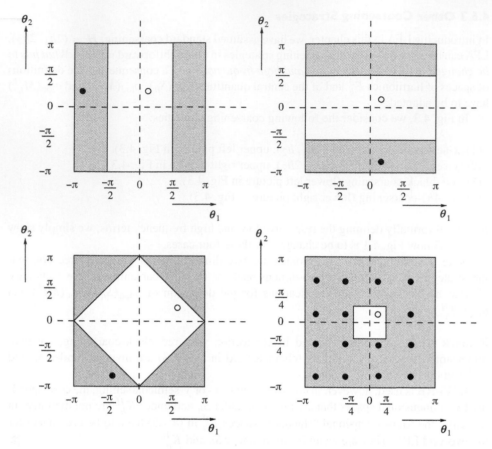

**Figure 4.3.** Low frequencies (interior white regions) and high frequencies (shaded regions) for various coarsening strategies; for a given low frequency $\theta(\circ)$, the other frequencies $\theta$ for which the corresponding $\varphi(\theta, x)$ coincide on $\mathbf{G}_H$ are marked by $\bullet$.

defined for operators $L_h$ given on an infinite grid $\mathbf{G}_h$. Throughout this section, we make the same assumptions on $L_h$ as for the LFA.

> The central result of this section is that, in a certain sense, the $h$-ellipticity of $L_h$ is a necessary and sufficient condition for the existence of pointwise smoothing procedures for $L_h$.

If a discrete operator lacks $h$-ellipticity, the addition of some "artificial ellipticity" (or "discrete ellipticity") is one way to make $L_h$ sufficiently $h$-elliptic. The idea of introducing *artificial viscosity* is a classical approach in the case of convection-dominated problems.

In Section 4.7.1 we will introduce the concept of $h$-ellipticity and define the *h-ellipticity measure* of a discrete operator $L_h$ and give some examples. In Section 4.7.2, we will discuss the relation of smoothing and $h$-ellipticity in detail.

The concept of $h$-ellipticity is of particular importance for systems of equations. We will give a corresponding generalization in Section 8.3.2.

### 4.7.1 The Concept of $h$-Ellipticity

The $h$-ellipticity concept is based on the LFA and also represents the LFA philosophy. When speaking of $h$-ellipticity in the following, we implicitly assume standard coarsening $H = 2h$ unless mentioned otherwise explicitly. This means again that $\boldsymbol{\theta}$ is a high frequency if $\boldsymbol{\theta} \in T^{\text{high}} := [-\pi, \pi)^2 \setminus [-\pi/2, \pi/2)^2$.

**Definition 4.7.1** The $h$-ellipticity measure $E_h$ of $L_h$ can be defined as

$$E_h(L_h) := \frac{\min\{|\tilde{L}_h(\boldsymbol{\theta})|: \boldsymbol{\theta} \in T^{\text{high}}\}}{\max\{|\tilde{L}_h(\boldsymbol{\theta})|: -\pi \le \boldsymbol{\theta} < \pi\}} \tag{4.7.1}$$

where $\tilde{L}_h(\boldsymbol{\theta})$ represents the Fourier symbol of $L_h$.

**Remark 4.7.1** The denominator in this definition is a scaling factor, which guarantees that

$$0 \le E_h(L_h) \le 1.$$

Other scalings are also used, for example,

$$\bar{E}_h(L_h) := \min\left\{ \frac{|\tilde{L}_h(\boldsymbol{\theta})|}{|L_h|} : \boldsymbol{\theta} \in T^{\text{high}} \right\} \tag{4.7.2}$$

where $|L_h|$ is the sum of the absolute values of the coefficients in the stencil. In this section, we will work with $E_h$, later on we will also use $\bar{E}_h$ if convenient.  ≫

Here, and in the following chapters, we will use the notation that an operator $L_h$ is "$h$-elliptic" or has a "good measure of $h$-ellipticity". This is a loose way of saying that

$$E_h(L_h) \ge \text{ const } > 0,$$

i.e., $E_h$ is sufficiently far from 0 for certain ranges of parameters on which $L_h$ depends. Heuristically, an $h$-ellipticity measure $E_h(L_h)$, which is close to zero, indicates that very small eigenvalues exist for high frequencies.

We will discuss the $h$-ellipticity for various examples. They illustrate that the lack of $h$-ellipticity of $L_h$ may have different reasons: It may be due to the fact that $L$ is *not elliptic* itself (and artificial ellipticity is not introduced in $L_h$). It may, however, also be caused by an *instability of the $L_h$ discretization*, even if $L$ is uniformly elliptic.

**Example 4.7.1** The standard five-point discrete Laplacian on a square mesh is a stable discretization of $\Delta$. For its $h$-ellipticity measure, we find

$$E_h(\Delta_h) = 0.25. \tag{4.7.3}$$

The minimum in the numerator ($= 2/(h^2)$) is attained, for example, at $(0, \pi/2)$, the maximum in the denominator ($= 8/(h^2)$), for example, at $(\pi, \pi)$. $\triangle$

**Example 4.7.2** The discrete operator

$$L_h = \frac{1}{2h}[-1 \quad 0 \quad 1]_h, \tag{4.7.4}$$

which is a central $O(h^2)$-discretization of the first derivative, $Lu = u_x$ is not $h$-elliptic: $E_h(L_h) = 0$.

The one-sided first-order operator

$$L_h = \frac{1}{h}[-1 \quad 1 \quad 0]_h,$$

however, which also approximates the first derivative $u_x$ (with $O(h)$-accuracy), is $h$-elliptic, $E_h(L_h) = 1/\sqrt{2}$. The above operator (4.7.4) is a (typical) example for non-$h$-ellipticity due to central differencing. $\triangle$

**Example 4.7.3** The discrete operator

$$L_h = \frac{1}{4h^2} \begin{bmatrix} -1 & 0 & 1 \\ 0 & 0 & 0 \\ 1 & 0 & -1 \end{bmatrix}_h, \tag{4.7.5}$$

which is a discretization of the hyperbolic operator $Lu = u_{xy}$, is not $h$-elliptic: $E_h(L_h) = 0$. The discretization

$$L_h = \frac{1}{h^2} \begin{bmatrix} 0 & 0 & 0 \\ -1 & 1 & 0 \\ 1 & -1 & 0 \end{bmatrix}_h,$$

is not $h$-elliptic either: $E_h(L_h) = 0$. In these two cases, the non-$h$-ellipticity is basically due to the underlying differential operator (mixed derivatives). $\triangle$

**Example 4.7.4** The discretization

$$L_h = \frac{1}{h^2} \begin{bmatrix} 0 & -1 & 0 \\ 0 & 2 & 0 \\ 0 & -1 & 0 \end{bmatrix}_h$$

of the parabolic operator $Lu = -u_{yy}$ (in 2D) is not $h$-elliptic: $E_h(L_h) = 0$. $\triangle$

**Example 4.7.5** A simple example of a discretization of the elliptic operator $L = -\Delta$ which is not $h$-elliptic is (in stencil notation)

$$L_h = -\Delta_{\sqrt{2}h} = \frac{1}{2h^2} \begin{bmatrix} -1 & 0 & -1 \\ 0 & 4 & 0 \\ -1 & 0 & -1 \end{bmatrix}_h \quad \text{on a square grid with mesh size } h$$

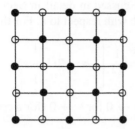

**Figure 4.4.** A $4 \times 4$ grid with red (i.e. white) and black points indicated.

(the skew five-point stencil for $-\Delta$). We obtain $E_h(L_h) = 0$ here since $\tilde{L}_h(\boldsymbol{\theta}) = 0$ for $\boldsymbol{\theta} = (\pi, \pi)$. Indeed, on the infinite grid $\mathbf{G}_h$, the red and the black points of $\mathbf{G}_h$ are separated by $L_h$. Therefore, any function which has constant values at the red points and constant values of a different size at the black points lies in the kernel of the discrete operator. This is an example of a uniformly elliptic operator $\Delta$, a discretization of which lacks $h$-ellipticity. Because of the decoupling of red and black points in this discretization, this phenomenon is often referred to as the *checkerboard instability*.

If we use $L_h$ with Dirichlet boundary conditions on a finite domain (see Fig. 4.4, where $h = 1/4$), the "Dirichlet binding" of the discrete solution makes the discretization asymptotically stable in the classical sense in some appropriate norm. In the case of periodic boundary conditions, however, the discretization is unstable in the classical sense (if $h = 1/n$, $n$ even). $\triangle$

**Example 4.7.6** An example with positive and negative operator elements is the $O(h^4)$ discretization of $-\Delta$

$$\frac{1}{12h^2} \begin{bmatrix} & & 1 & & \\ & & -16 & & \\ 1 & -16 & 60 & -16 & 1 \\ & & -16 & & \\ & & 1 & & \end{bmatrix}_h. \qquad (4.7.6)$$

The fourth-order accuracy of $L_h$ is easily verified by a Taylor expansion. The resulting matrix is not an $M$-matrix. Nevertheless, the $h$-ellipticity measure $E_h \approx 0.22 \ (= 7/32)$ is satisfactory. Multigrid methods for the Poisson equation discretized by this stencil are described in Example 5.4.1 and, in particular, in Remark 5.4.3. $\triangle$

**Example 4.7.7** The parabolic operator $Lu = -u_{yy}$ above is the limit case of the anisotropic operator

$$L(\varepsilon)u = -\varepsilon u_{xx} - u_{yy} \quad (0 < \varepsilon \ll 1). \qquad (4.7.7)$$

Anisotropic problems will be discussed in detail in Section 5.1. For any $\varepsilon > 0$, $L(\varepsilon)$ is elliptic, but not uniformly elliptic with respect to $\varepsilon$. For the discrete analog of $L(\varepsilon)$

$$L_h(\varepsilon) = \frac{1}{h^2} \begin{bmatrix} & -1 & \\ -\varepsilon & 2(1+\varepsilon) & -\varepsilon \\ & -1 & \end{bmatrix}_h, \qquad (4.7.8)$$

we obtain $E_h(L_h) = \varepsilon/(2+2\varepsilon) \to 0$ for $\varepsilon \to 0$ (for a coarsening different from $H = 2h$ see Remark 4.7.5). $\triangle$

## 4.7.2 Smoothing and $h$-Ellipticity

In this section, we will show that $h$-ellipticity is a necessary and sufficient condition for the existence of pointwise smoothers. Again we consider pointwise smoothers based on a splitting

$$L_h = L_h^+ + L_h^-, \qquad (4.7.9)$$

as in Section 4.3. Obviously, $E_h(L_h) = 0$ means that $E_h(\theta^*) = 0$ for at least one high frequency $\theta^*$, and therefore $\tilde{L}_h^+(\theta^*) = -\tilde{L}^-(\theta^*)$. If we additionally assume

$$\tilde{L}_h^+(\theta) \neq 0, \qquad (4.7.10)$$

for all $\theta \in T^{\text{high}}$, we obtain

$$|\tilde{L}_h^-(\theta^*)|/|\tilde{L}_h^+(\theta^*)| = 1$$

and therefore, according to (4.3.6) and (4.3.7),

$$\mu_{\text{loc}}(S_h) \geq 1.$$

In other words: no splitting of the form (4.7.9) can lead to a reasonable smoother if $E_h(L_h) = 0$ and (4.7.10) is excluded.

This simple result is summarized as Statement (1) in the following theorem.

---

**Theorem 4.7.1**

*(1) If $E_h(L_h) = 0$, we have*

$$\mu_{\text{loc}}(S_h) \geq 1$$

*for any smoother $S_h$ satisfying (4.7.9) and (4.7.10).*

*(2) If, on the other hand, $E_h(L_h)$ is bounded away from 0 by some constant $c > 0$: $E_h(L_h) \geq c > 0$ (for $h \to 0$), we can always construct a pointwise smoothing procedure $S_h$ with a smoothing factor $\mu_{\text{loc}} < 1$ that is bounded away from 1 by some constant that depends only on c.*

---

(3) *If (4.7.9) holds and we additionally assume that $L_h$ is characterized by a symmetric stencil*

$$s_{i,j} = s_{-i,-j} \quad (i, j = 0, 1, 2, \ldots), \text{ with } s_{0,0} > 0,$$

*and that $\tilde{L}_h(\boldsymbol{\theta}) > 0$ for $\boldsymbol{\theta} \neq 0$, then a pointwise $\omega$-JAC smoother $S_h$ exists for which*

$$\mu_{\text{loc}}(S_h(\omega\text{-}JAC)) \leq \text{const} < 1.$$

*Proof.* We first consider the symmetric case (3). $\tilde{L}_h(\boldsymbol{\theta})$ and $s_{0,0}$ are thus real numbers. For the $\omega$-JAC smoother, we have

$$S_h(\omega) = I_h - \frac{\omega}{s_{0,0}} L_h \quad \text{and} \quad \tilde{S}_h(\omega, \boldsymbol{\theta}) = 1 - \frac{\omega}{s_{0,0}} \tilde{L}_h(\boldsymbol{\theta})$$

and therefore

$$\mu_{\text{loc}} = \mu_{\text{loc}}(S_h(\omega)) = \sup \left\{ |1 - \frac{\omega}{s_{0,0}} \tilde{L}_h(\boldsymbol{\theta})| : \boldsymbol{\theta} \in T^{\text{high}} \right\}.$$

With $m := \min_{\boldsymbol{\theta} \in T^{\text{high}}} \tilde{L}_h(\boldsymbol{\theta})$ and $M := \max_{-\pi \leq \theta < \pi} \tilde{L}_h(\boldsymbol{\theta})$, (4.7.1) reads

$$E_h(L_h) = \frac{m}{M}.$$

We now choose

$$\omega = \omega^* = \frac{2s_{0,0}}{m + M} \tag{4.7.11}$$

and obtain

$$\mu_{\text{loc}}(S_h(\omega^*)) = \frac{M - m}{M + m} = \frac{1 - E_h(L_h)}{1 + E_h(L_h)}. \tag{4.7.12}$$

Obviously, we have $\mu_{\text{loc}} \leq \text{const} < 1$ if $E_h(L_h) \geq c > 0$ with const $= (1 - c)/(1 + c)$.

The proof of Statement (2) is similar, but technically more involved. For nonsymmetric $L_h$, instead of $\omega$-JAC for $L_h$, an $\omega$-JAC relaxation for the operator $L_h L_h^T$ is used for smoothing. This means that smoothing is applied to the system

$$L_h L_h^T w_h = f_h,$$

where $u_h = L_h^T w_h$ and $w_h$ is a new grid function. $L_h L_h^T$ is symmetric and has the symbol

$$\widetilde{L_h L_h^T}(\boldsymbol{\theta}) = |\tilde{L}_h(\boldsymbol{\theta})|^2.$$

With similar arguments as above, we obtain

$$\mu_{\text{loc}} = \mu_{\text{loc}}(\omega^*) = \frac{1 - E_h^2(L_h)}{1 + E_h^2(L_h)}$$

for

$$\omega^* = \frac{2s_{0,0}(L_h L_h^T)}{m^2 + M^2} s_{0,0}(L_h L_h^T) = \sum_{i,j} (s_{i,j})^2 > 0.$$

For details, see [63, 240].                                                                                 □

**Remark 4.7.2 (Kaczmarz relaxation)**  The relaxation used for the proof of Statement (2) is also known as an $\omega$-Jacobi–Kaczmarz relaxation [207] or Cimmino method [108]. It is a special type of "distributive" relaxation (see Section 8.2.5). According to the above theorem, this relaxation always gives smoothing (for $E_h > 0$) but it is often far from being an *efficient* smoother.

This can easily be seen for $L_h = -\Delta_h$ (Example 4.7.1). Since $E_h(L_h) = 0.25$, the above relaxation has the smoothing factor $\mu_{\mathrm{loc}} = 15/17 = 0.882$ which is far worse than that of the cheaper standard relaxation schemes introduced in Chapter 2.        ≫

**Example 4.7.8**  In Example 4.7.1, we have seen that $m = 2/h^2$ and $M = 8/h^2$. With $s_{0,0}$ being $4/h^2$, we immediately obtain from (4.7.11) and (4.7.12),

$$\omega^* = 8/(2+8) = 4/5 \quad \text{and} \quad \mu_{\mathrm{loc}} = 8 - 2/(8+2) = 3/5,$$

for $\omega$-JAC in the case of Poisson's equation.                                                      △

**Example 4.7.9**  In Example 4.7.2, we have seen that $E_h(L_h) = 0$ for the central discretization of a first derivative. If we consider the 1D *convection–diffusion operator* $L$,

$$Lu = -\varepsilon u_{xx} + u_x,$$

with the central discretization

$$L_h = -\frac{\varepsilon}{h^2}[1 \quad -2 \quad 1]_h + \frac{1}{2h}[-1 \quad 0 \quad 1]_h,$$

we find the Fourier symbol

$$\tilde{L}_h(\theta) = \frac{2\varepsilon}{h^2}(1 - \cos\theta) + i\frac{\sin\theta}{h},$$

which is $O(\varepsilon)$ for $\theta = \pi$. Correspondingly,

$$\tilde{S}_h(\theta) = \left| \frac{(\varepsilon - h/2)e^{i\theta}}{2\varepsilon - (\varepsilon + h/2)e^{-i\theta}} \right| \quad \text{for GS-LEX,}$$

and we obtain $\mu_{\mathrm{loc}} \to 1$ as $\varepsilon \to 0$. The convection–diffusion equation will be discussed in detail in Section 7.1.                                                                        △

---

**Remark 4.7.3**  Statement (1) of the theorem refers to the case $E_h(L_h) = 0$. If $E_h(L_h) > 0$, but is very small, we will be unable to construct *efficient* pointwise smoothers. In that case, small highly oscillating variations in the discrete right-hand side result in large oscillations of the discrete solution.        ≫

**Remark 4.7.4** Statements (2) and (3) guarantee the existence of *pointwise* smoothers, whereas Statement (1) at first sight seems to say that *no* splitting-based smoother can exist if $E_h(L_h) = 0$. This is, however, only true for pointwise smoothers. In Chapter 5, we will consider anisotropic equations like $\varepsilon u_{xx} + u_{yy} = f$. Although $E_h(L_h) = O(\varepsilon)$ for such equations (see Example 4.7.4), we will find smoothers with a smoothing factor $\mu_{\mathrm{loc}}(S_h) \leq \mathrm{const} < 1$, where the constant is independent of $\varepsilon$. Such smoothers are also based on a splitting of the form (4.7.9). They are, however, not pointwise but linewise smoothers.

The above theorems can be generalized to cover line smoothers [240].            ≫

Since the $h$-ellipticity measure depends on the set of high frequencies, a different coarsening also has an impact on the $h$-ellipticity measure. The semicoarsening strategies

$$(H_x, H_y) = (2h_x, h_y), \qquad (H_x, H_y) = (h_x, 2h_y),$$

for anisotropic operators will be discussed in Chapter 5.

**Remark 4.7.5 (semi-$h$-ellipticity)** We consider the anisotropic operator (4.7.7) in 2D. As we will see in Chapter 5, semicoarsening in the $y$-direction is a natural coarsening for this operator. We define

$$E_h^S(L_h) := \frac{\min\{|\tilde{L}_h(\boldsymbol{\theta})|\colon \theta_2 \in [-\pi, \pi) \setminus [-\pi/2, \pi/2)\}}{\max\{|\tilde{L}_h(\boldsymbol{\theta})|\colon -\pi \leq \boldsymbol{\theta} < \pi\}}. \tag{4.7.13}$$

For the operator (4.7.8), we find $E_h^S(L_h) = 1/(2 + 2\varepsilon) \to 1/2$ for $\varepsilon \to 0$ instead of $E_h(L_h) = \varepsilon/(2+2\varepsilon) \to 0$ for $\varepsilon \to 0$ in case of standard coarsening. Thus, this coarsening strategy can be combined with pointwise smoothers for the anisotropic problem under consideration. (4.7.13) is an example for the so-called *semi-h-ellipticity* [66] which gives guidelines for the design of multigrid solvers based on semicoarsening or line relaxation.    ≫

# 5

# BASIC MULTIGRID II

In Chapter 2, we introduced basic multigrid. Although the formal description has been quite general, the examples we have discussed all refer to Model Problem 1, Poisson's equation, or similarly simple problems. In this chapter we will discuss straightforward modifications of the previously introduced multigrid components so that more general second-order problems can be handled well.

An emphasis is laid on efficient multigrid methods for 2D and 3D anisotropic equations (see Sections 5.1 and 5.2) and for nonlinear equations (see Section 5.3). In particular, we will introduce some other smoothing methods: line smoothers in Section 5.1, plane smoothers in Section 5.2 and nonlinear smoothing variants in Section 5.3.

Furthermore, in Section 5.4, we focus on the application of multigrid methods to high order discretizations. In Section 5.5, we consider problems with reentrant corners, which may lead to discrete solutions with an accuracy of less than second order.

The multigrid treatment of problems whose boundary conditions are more general than Dirichlet boundary conditions is discussed in Section 5.6. A problem that sometimes occurs, for example due to certain boundary conditions, is that the resulting system of discrete equations is singular. The Poisson equation with periodic boundary conditions is presented as a specific example. The discussion of an appropriate multigrid treatment of such systems in more general cases is also part of Section 5.6.

In Section 5.7, we discuss the idea of finite volume discretization in the context of curvilinear grids. In Section 5.8, we resume the discussion on grid structures and comment on the multigrid treatment on unstructured grids.

> For all examples in this chapter, we will obtain the typical multigrid efficiency as for Model Problem 1.

**Remark 5.0.1 (LFA and rigorous Fourier analysis)**   In this chapter, we can often apply both LFA (as described in Chapter 4) *and* rigorous Fourier analysis (see Chapter 3). The resulting smoothing and convergence factors are the same for sufficiently small $h$. We will thus mainly use LFA.                                                                          ≫

## 5.1 ANISOTROPIC EQUATIONS IN 2D

As a first step away from Poisson-like elliptic equations, we consider anisotropic elliptic problems.

**Model Problem 3** *(2D anisotropic model problem)*

$$-\varepsilon u_{xx} - u_{yy} = f^{\Omega}(x, y) \quad (\Omega = (0, 1)^2)$$
$$u = f^{\Gamma}(x, y) \quad (\partial\Omega) \tag{5.1.1}$$

*with $0 < \varepsilon \ll 1$. (The case $\varepsilon \gg 1$ is similar, with interchanged roles of x and y.)*

If we discretize Model Problem 3 by the standard five-point difference operator, we obtain the discrete problem

$$L_h(\varepsilon)u_h(x, y) = f_h^{\Omega}(x, y) \quad (\Omega_h)$$
$$u_h(x, y) = f_h^{\Gamma}(x, y) \quad (\Gamma_h), \tag{5.1.2}$$

where $\Omega_h = \mathbf{G}_h \cap \Omega$ is the square grid (1.3.3) with $h = h_x = h_y$, $\Gamma_h$ is again the set of discrete intersection points of the grid lines with the boundary $\Gamma$ and

$$L_h(\varepsilon) = \frac{1}{h^2} \begin{bmatrix} & -1 & \\ -\varepsilon & 2(1+\varepsilon) & -\varepsilon \\ & -1 & \end{bmatrix}_h . \tag{5.1.3}$$

Here, we restrict ourselves to the case where the discrete anisotropy is "aligned" with the grid. In 2D such problems are characterized by coefficients in front of the $u_{xx}$ and $u_{yy}$ terms, which may differ by orders of magnitude. Anisotropic problems play an important role in practice. The discretization may also introduce (discrete) anisotropies, in the form of stretched grids.

**Remark 5.1.1 (stretched grids)** The same discrete operator (5.1.3) is obtained, if we discretize the Laplace operator (in Model Problem 1) by the standard five-point difference operator on a *stretched* grid with mesh sizes $h_x = h_y/\sqrt{\varepsilon}$.                    ≫

### 5.1.1 Failure of Pointwise Relaxation and Standard Coarsening

In Example 4.7.7, we have seen that the $h$-ellipticity measure of an anisotropic operator tends to 0 for $\varepsilon \to 0$:

$$E_h(L_h(\varepsilon)) = O(\varepsilon) \longrightarrow 0 \quad \text{for } \varepsilon \to 0.$$

According to the discussion in Section 4.7.2, it is expected that the smoothing properties of standard pointwise smoothing schemes will deteriorate for $\varepsilon \to 0$.

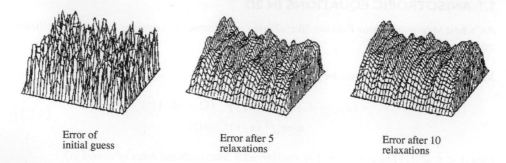

| Error of initial guess | Error after 5 relaxations | Error after 10 relaxations |

**Figure 5.1.** Influence of (pointwise) GS-LEX on the error for the 2D anisotropic model problem with $\varepsilon \ll 1$.

In particular, highly anisotropic problems cannot be treated efficiently with standard coarsening and the multigrid components introduced so far. Therefore, they are a first and a very illustrative example of the fact that multigrid, in general, has to be adapted and tuned to the problem at hand. However, the proper ways to treat anisotropic problems by multigrid are fully understood and can easily be explained.

If we apply a standard pointwise relaxation such as $\omega$-JAC, GS-LEX or GS-RB to the above discrete system, we find that the smoothing effect of this relaxation is very poor with respect to the $x$-direction. The reason is that pointwise relaxation has a smoothing effect only with respect to the "strong coupling" in the operator, i.e. in the $y$-direction. This effect is illustrated in Fig. 5.1 for $\varepsilon = 10^{-2}$.

If we consider GS-LEX, for example, we find the error relation

$$\bar{v}_h(x_i, y_j) = \frac{1}{2(\varepsilon + 1)} [\varepsilon \bar{v}_h(x_{i-1}, y_j) + \varepsilon v_h(x_{i+1}, y_j) + \bar{v}_h(x_i, y_{j-1}) + v_h(x_i, y_{j+1})], \tag{5.1.4}$$

which for $\varepsilon \to 0$ becomes

$$\bar{v}_h(x_i, y_j) = \tfrac{1}{2} [\bar{v}_h(x_i, y_{j-1}) + v_h(x_i, y_{j+1})]. \tag{5.1.5}$$

Obviously, there is no averaging effect with respect to the $x$-direction and, hence, no smoothing with respect to this direction is achieved. (For $\varepsilon \gg 1$, it is the other way round.) Such nonsmooth errors can no longer be efficiently reduced by means of a coarser grid which is obtained by standard coarsening, i.e. by doubling the mesh size in *both* directions.

This failure can be directly explained by applying LFA smoothing analysis as in Section 4.3 to the GS-LEX pointwise smoother for Model Problem 3. By adapting the

**Table 5.1.** Two-grid convergence factors $\rho_{\text{loc}}$ for the anisotropic model problem using GS-RB ($\nu$ = three smoothing steps), FW and bilinear interpolation for different values of $\varepsilon$.

| $\varepsilon$ | 0.001 | 0.01 | 0.1 | 0.5 | 1 | 2 | 10 | 100 | 1000 |
|---|---|---|---|---|---|---|---|---|---|
| $\rho_{\text{loc}}$ | 0.99 | 0.94 | 0.56 | 0.088 | 0.053 | 0.088 | 0.56 | 0.94 | 0.99 |

splitting from (4.3.3) to (5.1.3) we obtain

$$L_h^- = \frac{1}{h^2} \begin{bmatrix} & -1 & \\ 0 & 0 & \varepsilon \\ & 0 & \end{bmatrix}, \qquad L_h^+ = \frac{1}{h^2} \begin{bmatrix} & 0 & \\ -\varepsilon & 2+2\varepsilon & 0 \\ & -1 & \end{bmatrix}. \qquad (5.1.6)$$

Following the procedure in Section 4.3, the smoothing factor $\mu_{\text{loc}}$ of GS-LEX for (5.1.3) is found to be

$$\mu_{\text{loc}} = \sup_{\theta \in T^{\text{high}}} \left| \tilde{S}_h(\theta) \right| = \sup_{\theta \in T^{\text{high}}} \left| \frac{\varepsilon e^{i\theta_1} + e^{i\theta_2}}{\varepsilon e^{-i\theta_1} + e^{-i\theta_2} - 2 - 2\varepsilon} \right|. \qquad (5.1.7)$$

For $\varepsilon \to 0$ (and for $\varepsilon \to \infty$), we obtain

$$\lim_{\varepsilon \to 0} \mu_{\text{loc}} = \lim_{\varepsilon \to 0} \tilde{S}_h(\pi, 0) = 1.$$

Table 5.1 presents corresponding two-grid convergence factors $\rho_{\text{loc}} = \rho_{\text{loc}}(\nu)$ for GS-RB pointwise smoothing, standard coarsening, FW and linear interpolation. For large or small values of $\varepsilon$, standard pointwise smoothers fail to achieve satisfactory two-grid (and thus also multigrid) convergence.

> *Pointwise relaxation and standard coarsening is not a reasonable combination for highly anisotropic problems. The multigrid convergence factor will increase towards* 1 *for $\varepsilon \to 0$ or $\varepsilon \to \infty$.*

**Remark 5.1.2** It is to some extent possible to keep the GS-RB pointwise smoother for moderate anisotropies and improve the two-grid factors from Table 5.1 *by overrelaxation.* This has been shown in [427], where analytic formulas are presented for optimal relaxation parameters $\omega$. For $\varepsilon = 0.1$, for example, $\omega_{\text{opt}} = 1.41$ leads to $\rho_{\text{loc}} = 0.12$. For $\varepsilon = 0.01$, $\omega_{\text{opt}} = 1.76$ results in $\rho_{\text{loc}} = 0.45$, which is a major improvement compared to the results presented in Table 5.1. For $\varepsilon = 0.001$, $\omega_{\text{opt}} = 1.92$ and $\rho_{\text{loc}} = 0.78$. More robust multigrid remedies to master the difficulty of highly anisotropic problems are presented in the subsequent subsections.                                                    $\gg$

### 5.1.2 Semicoarsening

The first possibility is to keep pointwise relaxation for smoothing, but to change the grid coarsening according to the one-dimensional smoothness of errors. The coarse grid is defined by doubling the mesh size only in that direction in which the errors are smooth. Semicoarsening in the $y$-direction, as introduced in Section 2.3.1, is appropriate if $\varepsilon \ll 1$. (The 1D restriction operator for semicoarsening has been introduced, for example, in Remark 2.3.2.)

In the case of semicoarsening, LFA can also be applied. For example, we can use (5.1.7) for GS-LEX and $y$-semicoarsening. But the range of high frequencies is changed now to

$$\left\{ (\theta_1, \theta_2) : \theta_2 \in [-\pi, \pi) \setminus \left[ -\frac{\pi}{2}, \frac{\pi}{2} \right) \right\}$$

(see Section 4.6.3). Maximizing over this range, we find

$$\mu_{\text{loc}} = \frac{1 + \varepsilon}{\sqrt{5 + \varepsilon}}.$$

For $\varepsilon \to 0$, we obtain the satisfactory value $\mu_{\text{loc}} \to 1/\sqrt{5} \approx 0.45$. This shows that the quality of a smoother depends on the range of high frequencies and thus on the choice of the coarse grid. (For $\varepsilon \gg 1$, $\mu_{\text{loc}}$ rapidly increases towards 1. In this case GS-LEX provides good smoothing if $x$-*semicoarsening* is employed.)

The operator $L_H$ on the coarse grid $\Omega_H$ is

$$L_H(\varepsilon) = \frac{1}{H^2} \begin{bmatrix} & -1 & \\ -4\varepsilon & 2(1 + 4\varepsilon) & -4\varepsilon \\ & -1 & \end{bmatrix}_H \tag{5.1.8}$$

in the case of $y$-semicoarsening. Compared to the fine grid operator (5.1.4), the anisotropy has decreased. If we continue the $y$-semicoarsening process, the anisotropy will decrease further.

**Remark 5.1.3** For *isotropic* problems such as Model Problem 1 on a standard grid with $h = h_x = h_y$, repeated semicoarsening in the same direction will not give satisfactory convergence results if pointwise smoothers are used. This occurs because a discrete anisotropy that arises from cells that have been stretched too much will be built up by the semicoarsening process (see also Remark 5.1.1). ≫

### 5.1.3 Line Smoothers

An alternative multigrid approach for highly anisotropic problems is to keep the standard multigrid coarsening, but to change the relaxation procedure from pointwise relaxation to *linewise* relaxation (called line relaxation in short). That is, *all unknowns on a line are updated collectively* (simultaneously). Line relaxations are thus block iterations in which each block of unknowns corresponds to a line. For Model Problem 3, the collective solution of all equations corresponding to a line means the solution of a *tridiagonal system of equations*.

Figure 5.2 shows the order in which the unknowns are relaxed for *lexicographic x-* and *y*-line Gauss–Seidel relaxation.

| 4 | 4 | 4 | 4 |  | 1 | 2 | 3 | 4 |
|---|---|---|---|---|---|---|---|---|
| 3 | 3 | 3 | 3 |  | 1 | 2 | 3 | 4 |
| 2 | 2 | 2 | 2 |  | 1 | 2 | 3 | 4 |
| 1 | 1 | 1 | 1 |  | 1 | 2 | 3 | 4 |

(a)                                                         (b)

**Figure 5.2.** Order in which unknowns are solved for collectively by lexicographic line Gauss–Seidel relaxation; (a) $x$-line Gauss–Seidel, (b) $y$-line Gauss–Seidel relaxation.

Gauss–Seidel-type line relaxations are particularly efficient smoothers for anisotropic problems (if the anisotropy is aligned with the grid). This is due to the general observation that *errors become smooth if strongly connected unknowns are updated collectively.*

The good smoothing properties of (lexicographic) line Gauss–Seidel can also be seen by LFA. Using the same notation as in Section 4.3, the lexicographic $y$-line Gauss–Seidel for (5.1.3) can be written as

$$\frac{1}{h^2}\begin{bmatrix} -1 \\ 2(1+\varepsilon) \\ -1 \end{bmatrix}_h \bar{w}_h = f_h + \frac{1}{h^2}([\varepsilon \quad 0 \quad 0]_h \bar{w}_h + [0 \quad 0 \quad \varepsilon]_h w_h). \qquad (5.1.9)$$

It is thus characterized by the splitting

$$L_h^- = \frac{1}{h^2}\begin{bmatrix} 0 \\ 0 \quad 0 \quad -\varepsilon \\ 0 \end{bmatrix}, \qquad L_h^+ = \frac{1}{h^2}\begin{bmatrix} -1 \\ -\varepsilon \quad 2+2\varepsilon \quad 0 \\ -1 \end{bmatrix}, \qquad (5.1.10)$$

and the smoothing factor $\mu_{\text{loc}}$ for (5.1.3) can be calculated as in Section 4.3. We find

$$\mu_{\text{loc}} = \sup_{\boldsymbol{\theta} \in T^{\text{high}}} |\tilde{S}_h(\boldsymbol{\theta})| = \sup_{\boldsymbol{\theta} \in T^{\text{high}}} \left| \frac{\varepsilon e^{i\theta_1}}{\varepsilon e^{-i\theta_1} + e^{-i\theta_2} - 2 - 2\varepsilon + e^{i\theta_2}} \right|. \qquad (5.1.11)$$

$\mu_{\text{loc}}$ of lexicographic $y$-line Gauss–Seidel for arbitrary $\varepsilon > 0$ turns out to be

$$\mu_{\text{loc}} = \max\left( \frac{1}{\sqrt{5}}, \frac{\varepsilon}{2+\varepsilon} \right), \qquad (5.1.12)$$

the first value being obtained for $(\theta_1, \theta_2) = (\pi/2, 0)$ and the latter one for $(\theta_1, \theta_2) = (0, \pi/2)$. Equation (5.1.12) implies that

$$\mu_{\text{loc}} = 1/\sqrt{5} \quad \text{for } 0 < \varepsilon \le 1.$$

The influence of the $y$-line smoother on the error for Model Problem 3 with $\varepsilon = 10^{-2}$ is presented in Fig. 5.3.

For $\varepsilon \gg 1$ the smoothing factor of $y$-line relaxation tends to 1. In such cases, $x$-line relaxation is to be used instead: line relaxation parallel to the $y$-axis (*y-line relaxation*) is suitable if $\varepsilon < 1$, and $x$-*line relaxation* if $\varepsilon > 1$.

Line smoothers other than lexicographic line Gauss–Seidel are, for example, *line $\omega$-Jacobi* (with underrelaxation), or *zebra line Gauss–Seidel smoothing*. Zebra line Gauss–Seidel relaxation is the line analog to pointwise RB Gauss–Seidel relaxation. Smoothing again consists of two half-steps. First all odd lines are processed, then all even ones. In the second half-step, the updated approximations on the odd lines are used. Figure 5.4 presents the points that are updated collectively in an $x$- and in a $y$-line zebra smoothing procedure, by the solution of tridiagonal systems.

Table 5.2 contains two-grid convergence factors $\rho_{\text{loc}}$ for zebra line GS obtained by two-grid LFA (see Section 4.4). It can be seen that the line smoother in the direction of the

| Error of initial guess | Error after 5 relaxations | Error after 10 relaxations |

**Figure 5.3.** Influence of $y$-line Gauss–Seidel relaxation on the error for the 2D anisotropic model problem with $\varepsilon = 10^{-2}$.

| 1 | 1 | 1 | 1 | 1 |   | 1 | 2 | 1 | 2 | 1 |
| 2 | 2 | 2 | 2 | 2 |   | 1 | 2 | 1 | 2 | 1 |
| 1 | 1 | 1 | 1 | 1 |   | 1 | 2 | 1 | 2 | 1 |
| 2 | 2 | 2 | 2 | 2 |   | 1 | 2 | 1 | 2 | 1 |
| 1 | 1 | 1 | 1 | 1 |   | 1 | 2 | 1 | 2 | 1 |

|                (a)                |                (b)                |

**Figure 5.4.** Zebra line Gauss–Seidel relaxation: approximations in points marked by 1 are updated in the first, those marked by 2 in the second half-step of the relaxation. (a) $x$-zebra line Gauss–Seidel (xZGS), (b) $y$-zebra line Gauss–Seidel relaxation (yZGS).

**Table 5.2.** Two-grid convergence factors $\rho_{loc}$ for the anisotropic model problem using $x$- or $y$-zebra line Gauss–Seidel ($\nu = 2$ smoothing steps), FW and bilinear interpolation for different values of $\varepsilon$.

| $\varepsilon$ | 0.001 | 0.01 | 0.1 | 0.5 | 1 | 2 | 10 | 100 | 1000 |
|---|---|---|---|---|---|---|---|---|---|
| xZGS: | 0.996 | 0.96 | 0.68 | 0.20 | 0.063 | 0.028 | 0.047 | 0.052 | 0.053 |
| yZGS: | 0.053 | 0.052 | 0.047 | 0.028 | 0.063 | 0.20 | 0.68 | 0.96 | 0.996 |

"weak coupling" fails to achieve satisfactory convergence. Line smoothing in the direction of the strong coupling works perfectly.

The cost of the zebra line smoother is the same as that of lexicographic line Gauss–Seidel.

**Remark 5.1.4** One advantage of using zebra line Gauss–Seidel is that its smoothing factors and thus also the multigrid convergence factors are better than those of lexicographic line Gauss–Seidel relaxations (e.g. $\mu_{loc}$ ($\nu = 1$) $= 0.125$ for yZGS and $\varepsilon \leq 0.5$ [378] versus $1/\sqrt{5}$ in the lexicographic case). $\gg$

**Remark 5.1.5** The degree of parallelism of zebra line Gauss–Seidel is higher than that of lexicographic line Gauss–Seidel relaxation. For five (or compact nine-point stencils (1.3.11)), the degree of parallelism is

$$\text{par-deg(zebra GS)} \geq \tfrac{1}{2}\sqrt{\#\Omega_h}.$$

Half of the lines can be processed in parallel. It depends on the concrete choice of the tridiagonal solver whether or not there is a further degree of parallelism inherent and how it can be exploited (see Section 6.4.2). Of course, the degree of parallelism is even better with line Jacobi relaxation, but, similarly to the point Jacobi for Model Problem 1, its smoothing factor turns out to be worse than that of zebra line Gauss–Seidel. For example, we find that the smoothing factor $\mu_{loc}$ of $x$-line Jacobi relaxation is $\mu_{loc} = 1/3$ for $\varepsilon = 1000$ and the optimal $\omega = 2/3$. $\gg$

**Remark 5.1.6 ($h$-dependence of convergence factors)** Generally, line relaxation methods are "less local" than pointwise relaxation; often, long range effects cannot be neglected. This is the reason why the measured multigrid convergence factors can differ considerably for different values of $h$ (and can become much better than predicted by LFA). For example, for Model Problem 3 $\rho_{loc}$ is first observed on a grid with $h = 1/2048$; on coarser grids a much better convergence is observed. This can be confirmed by rigorous Fourier analysis (for a fixed $h$). $\gg$

### 5.1.4 Strong Coupling of Unknowns in Two Directions

Up to now we have discussed the multigrid remedies for anisotropic problems with constant coefficients. In practice, $\varepsilon$ is often not constant but varying: $\varepsilon = \varepsilon(x, y)$. Then, $\varepsilon$ may be much larger than 1 in some parts of the domain, whereas it may be much smaller than 1 in other parts. As an example we can think of Poisson's equation on stretched grids with stretching in two directions towards the domain boundaries. In this case stretched cells that are thin and long *and* cells that are wide and short occur (see the examples in Fig. 5.6).

The two approaches introduced above to regain optimal multigrid convergence can be adapted to this more general situation.

### (1) Alternating line relaxation

An alternating linewise relaxation method is a combination of an $x$-line and a $y$-line smoother. After the $x$-line smoothing step a $y$-line smoothing step is performed. The cost of an alternating line smoother is twice that of a pure $x$- or $y$-line relaxation.

In contrast to pure $x$-line or $y$-line relaxations, alternating line relaxations are suitable smoothers for problems with strong coupling of unknowns in varying directions (see Section 5.1.5 for an example). As we see in Table 5.3, the alternating zebra line smoother (aZGS) provides excellent two-grid (and multigrid) convergence for Model Problem 3, independent of the size and the direction of the anisotropy. The aZGS smoother considered here consists of four partial steps in the following ordering: first the odd $x$-lines are updated, then the even $x$-lines. After that the even $y$-lines are treated and finally the odd $y$-lines. This smoother leads to results that are symmetric in $\varepsilon$.

**Remark 5.1.7 (robustness)**   An advantage of alternating line smoothers is their robustness: we can obtain excellent smoothing properties for a large class of problems. Not only the discrete isotropic or anisotropic model problems are solved satisfactorily by the corresponding multigrid method but, as we will see in Chapters 7 and 8, other types of equations can also be solved very well with multigrid based on alternating line smoothers.      ≫

**Remark 5.1.8 (segment relaxation)**   A more sophisticated approach than alternating line relaxation is to use block relaxation on segments (parts) of lines instead of on whole lines [339]. For example, if the strong coupling in one direction is restricted to some part of the domain only, it is sufficient to perform the block relaxation only in that part. In the remaining part another appropriate smoothing procedure can be applied locally (see Section 10.1.3 for an example of segment relaxation).      ≫

**Table 5.3.** Two-grid convergence factors $\rho_{\text{loc}}$ for the anisotropic model problem using alternating zebra-line GS ($\nu = 2$ smoothing steps), FW and bilinear interpolation for different values of $\varepsilon$.

| $\varepsilon$ | 0.001 | 0.01 | 0.1 | 0.5 | 1 | 2 | 10 | 100 | 1000 |
|---|---|---|---|---|---|---|---|---|---|
| aZGS: | 0.053 | 0.051 | 0.038 | 0.013 | 0.009 | 0.013 | 0.038 | 0.051 | 0.053 |

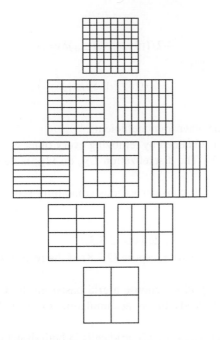

**Figure 5.5.** Multiple semicoarsened grids in 2D with a finest grid consisting of $9 \times 9$ grid points.

## (2) Multiple semicoarsening

We have seen that using only $x$-semicoarsening or only $y$-semicoarsening (in connection with pointwise smoothers) for the general case of varying anisotropic coefficients is not appropriate since in both cases errors are not smoothed in some part of the domain. In this general case, advanced and robust approaches based on *point smoothers and multiple semicoarsening* can be employed. Multiple semicoarsening is based on using more coarse grids than in standard coarsening or in semicoarsening (see the diamond-like structure of coarse grids in Fig. 5.5). There are various possibilities for processing the coarse grids and defining proper weighted interpolations of the corrections from these coarse grids (see [275, 277, 289] for different choices). V- or F-cycles still require $O(N)$ operations, whereas the W-cycle is more expensive ($O(N \log N)$).

In advanced multiple semicoarsening approaches, only a subset of the coarse grids is processed [411].

### 5.1.5 An Example with Varying Coefficients

We will apply multigrid to the discrete Poisson equation on a *stretched grid* $\Omega_h$ with variable stretching, which leads to anisotropies in different directions. Instead of a discretization with $h = h_x = h_y$, we have to take into account the variation of the mesh sizes $\delta x_i = x_i - x_{i-1}$ etc. (We still use the index $h$ in the following in order to distinguish the fine stretched grid from the next coarser one.) The difference operator $L_h$ for Poisson's equation then reads

$$
L_h = \left[
\begin{array}{ccc}
 & -2/\big((\delta y_j + \delta y_{j+1})\delta y_{j+1}\big) & \\
-2/\big((\delta x_i + \delta x_{i+1})\delta x_i\big) & -\sum & -2/\big((\delta x_i + \delta x_{i+1})\delta x_{i+1}\big) \\
 & -2/\big((\delta y_j + \delta y_{j+1})\delta y_j\big) &
\end{array}
\right]_h ,
$$

$$(5.1.13)$$

where $\sum$ is the sum of the other four coefficients.

The coarse grid $\Omega_{2h}$ is obtained by removing every other grid point in both directions from the stretched grid $\Omega_h$. The discretization $L_{2h}$ on $\Omega_{2h}$ is defined in the same way as $L_h$ on $\Omega_h$.

For this problem, we will consider various smoothing schemes for a multigrid algorithm with

– restriction $I_h^{2h}$: *standard* FW of defects (2.3.3),
– prolongation $I_{2h}^h$: the (scaled) transpose (2.3.8) of the restriction operator.

As a special stretched grid we choose a grid based on the Gauss–Lobatto–Legendre (GLL) points [37]. GLL leads to an accumulation of grid points near the domain boundaries.

The grid we consider consists of 128 grid cells in both directions. An indication of the grid stretching is that for the $128^2$ grid on the unit square there are five points between 0.99 and 1. This GLL grid is shown in the left picture of Fig. 5.6.

In addition to the GLL grid, which has a particular theoretical background, we consider an extremely (continuously) stretched $128^2$ grid (see right picture in Fig. 5.6). The extreme

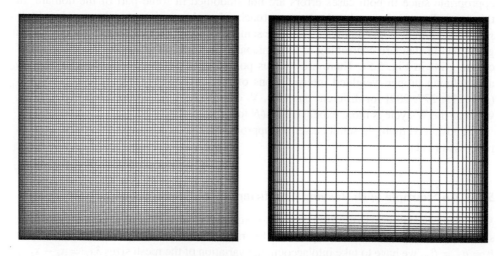

**Figure 5.6.** The stretched GLL grid (left) and an extremely stretched grid (right).

**Table 5.4.** Measured multigrid convergence factors on the stretched grids shown in Fig. 5.6 (using alternating zebra line smoothing).

| Cycle | V(0,1) | F(0,1) | V(1,1) | F(1,1) |
|---|---|---|---|---|
| GLL grid | 0.15 | 0.06 | 0.04 | 0.02 |
| Extremely stretched grid | 0.18 | 0.11 | 0.10 | 0.04 |

stretching may be somewhat artificial here, but is needed for singularly perturbed problems, where strong gradients in the solution occur near boundaries of a domain, for example in typical CFD applications. In such cases a high resolution of grid points perpendicular to the boundary is required.

Table 5.4 compares the measured convergence of the multigrid algorithm using the alternating zebra line smoother for both stretched grids shown in Fig. 5.6. Excellent convergence factors are obtained for the GLL grid as well as for the extremely stretched grid. Results for W-cycles are very similar to those for F-cycles. They are therefore omitted.

If the GS-RB pointwise smoother is used instead of the alternating line smoother, we obtain, e.g. for the F(1,1)-cycle, the convergence factors 0.70 on the GLL grid and 0.96 on the extremely stretched grid. The convergence when using $\omega$-GS-RB, is also unsatisfactory. Comparing these numbers with those in Table 5.4, the efficiency of alternating line smoothers is obvious.

Figure 5.7 compares the convergence history for multigrid using the pointwise smoother GS-RB, the $y$-line zebra smoother (yZGS) and the alternating zebra line smoother (aZGS) on the extremely stretched grid. For aZGS we consider the convergence of the F(1,1)- and the V(1,1)-cycle. For the first two smoothers the convergence is presented for cycles with $\nu_1 = \nu_2 = 2$ to make their cost comparable to that of the alternating smoothers.

The pointwise smoother and the $y$-line smoother do not show satisfactory multigrid convergence results. Their poor convergence behavior is almost identical. However, the convergence is excellent if the alternating line smoother is used.

## 5.2 ANISOTROPIC EQUATIONS IN 3D

In this section, we introduce multigrid methods for 3D anisotropic elliptic problems (with anisotropies aligned with the grid). The principal phenomena in the 3D anisotropic case are similar to those in 2D and the theoretical background for the development of fast multigrid algorithms is fully understood. However, with respect to optimal algorithms, the situation is somewhat more involved than in 2D. There are more possibilities for choosing different multigrid components, especially for coarsening and smoothing operators.

In Section 5.1, we have seen that basically two strategies can be applied for 2D anisotropic problems. The first is to maintain the standard multigrid coarsening and change the smoother to the problem at hand. The second is to keep the smoothing procedure, but to adapt the coarsening according to the problem. These two approaches can be generalized to

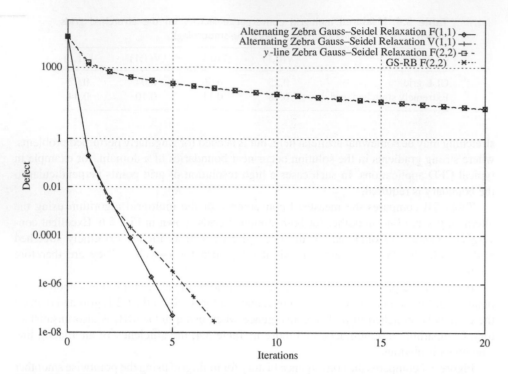

**Figure 5.7.** The convergence history of different multigrid cycles and smoothers for the Poisson equation on an extremely stretched grid. The discrete $L_2$ norm of the defect is plotted in a logarithmic scale along the $y$-axis.

the 3D anisotropic situation. One can either adapt the smoother or the coarsening or both, as will be discussed in more detail in the following subsections.

We start our discussion with the introduction of the 3D anisotropic model problem:

**Model Problem 4**

$$-au_{xx} - bu_{yy} - cu_{zz} = f^{\Omega}(x, y, z) \quad (\Omega = (0, 1)^3)$$
$$u = f^{\Gamma}(x, y, z) \quad (\Gamma = \partial\Omega) \tag{5.2.1}$$

We assume that the operator is elliptic, that all coefficients $a$, $b$ and $c$ have the same sign and that (5.2.1) is discretized on $\mathbf{G}_h$ (2.9.2) by the well-known seven-point discretization. In the 3D stencil notation, the discrete operator is

$$\frac{1}{h^2} \left[ \begin{bmatrix} 0 & 0 & 0 \\ 0 & -c & 0 \\ 0 & 0 & 0 \end{bmatrix}_h \begin{bmatrix} 0 & -b & 0 \\ -a & 2(a+b+c) & -a \\ 0 & -b & 0 \end{bmatrix}_h \begin{bmatrix} 0 & 0 & 0 \\ 0 & -c & 0 \\ 0 & 0 & 0 \end{bmatrix}_h \right]. \tag{5.2.2}$$

For constant coefficients, four representative parameter sets can be distinguished (without loss of generality):

$$
\begin{aligned}
&\text{Case 1:} \quad a \approx b \approx c \\
&\text{Case 2:} \quad a \gg b \approx c \\
&\text{Case 3:} \quad a \approx b \gg c \\
&\text{Case 4:} \quad a \gg b \gg c.
\end{aligned}
\tag{5.2.3}
$$

Different multigrid components are needed for good multigrid convergence in these four cases. Case 1 has already been discussed in Section 2.9, where it has been found that standard coarsening combined with GS-RB results in an efficient multigrid solver.

### 5.2.1 Standard Coarsening for 3D Anisotropic Problems

In this section suitable multigrid solution methods based on standard coarsening are introduced for the cases (5.2.3). Compared with the 3D multigrid solver in Section 2.9, only the smoother is changed in order to obtain the typical multigrid efficiency. This means that the computational work estimates presented in Section 2.9.3 are still valid, except for a possible change in the cost of the smoother.

A general rule in the case of standard coarsening, which carries over from the 2D situation, is that we obtain good smoothing of errors in all coordinate directions if we relax all strongly coupled unknowns collectively.

We will see in the following that all the above cases (5.2.3) can be handled with standard coarsening if *plane relaxation* is employed, in which all unknowns lying in the plane of strongly coupled unknowns are relaxed simultaneously.

Although plane relaxation is robust, other more efficient methods exist for some of the cases which we will detail in this section.

In Case 2, for example, we have a strong coupling of unknowns only in the $x$-direction. Correspondingly, we relax the unknowns of each single line in that direction collectively. $x$-line relaxation as presented in Section 5.1.3 can be safely used and will result in excellent smoothing.

One might think that one would obtain a robust 3D smoothing method by a straightforward generalization of the 2D case, i.e. by using alternating line relaxation with respect to three directions. This, however, is not true.

For Cases 3 and 4 in (5.2.3), the situation is more involved. In Case 3, all unknowns lying in the same $(x, y)$-plane are strongly coupled. This means that all these unknowns should be relaxed collectively (plane relaxation).

Plane relaxation schemes (similar to line relaxation) are block iterative methods. However, in contrast to line relaxation, which leads to tridiagonal matrices that are easily solved, plane relaxation is basically different. *For each plane we have to solve a discrete 2D problem similar to (2.8.2).*

A natural way to perform such a plane relaxation for Model Problem 4 is to use 2D multigrid in a plane as is analyzed in [388] and performed in [151].

For Case 4 the situation is similar: proper smoothing is guaranteed by $(x, y)$-plane relaxation. If the plane relaxation is performed by 2D multigrid, one has to take into account that the corresponding discrete 2D problems are now anisotropic themselves (in each plane). *Thus $x$-line smoothing should be used within the 2D multigrid plane solver.*

All these considerations can be verified by LFA and by rigorous Fourier analysis, taking into account the remarks on the generalization to 3D in Section 3.4. Table 5.5 presents some results for the four cases in (5.2.3). Standard coarsening ($H = 2h$) is employed and $L_H$ is the seven-point discretization of $L$ on $\Omega_H$. The transfer operators $I_h^H$ and $I_H^h$ are FW and trilinear interpolation. As smoothing schemes, we consider point, line or plane relaxation according to the cases considered. More specifically, we choose 3D *red–black point relaxation*, $x$-line relaxation *in a 2D red–black ordering of lines* and $(x, y)$-plane relaxation *in a zebra ordering of planes.*

The correct choice of smoothers for the four cases considered is emphasized in Table 5.5. The corresponding convergence factors are excellent. Table 5.5 confirms that alternating $x$-/$y$-line smoothing is not suited for Case 3.

---

**Remark 5.2.1 (plane relaxation in practice)**   In general, it is not necessary to solve the 2D problems arising in plane relaxation exactly. In [388] it has been shown by a rigorous two-grid Fourier analysis and by numerical experiments that the use of *one* V(1,1)- or *one* F(1,1)-cycle within each plane is sufficient to obtain the convergence factors predicted by the analysis in Table 5.5. The cost of this approach, employing plane smoothers based on point relaxation (for Case 3) and on line relaxation (for Case 4) is then about twice the cost of the solvers based on standard multigrid with a point smoother (for Case 1), or with a line smoother (for Case 2). In fact, it turns out that the use of one V(1,0)- or F(1,0)-cycle per plane relaxation is even *more efficient*. A small deterioration of the 3D convergence factors is compensated for by the reduced cost (one relaxation step instead of two) of these plane relaxation schemes.                                              ≫

---

**Remark 5.2.2 (optimality of 3D multigrid)**   An important result of this analysis is the asymptotic optimality of 3D multigrid for Cases 1–4 because a fixed (and small) number of appropriate smoothing steps is sufficient to guarantee excellent $h$-independent convergence.

In particular, using FMG, discretization accuracy can be obtained in $O(N)$ operations for all four cases, with a small constant of proportionality.                ≫

---

**Remark 5.2.3**   In principle, it is also possible to use other fast methods for the solution of the corresponding 2D problems per plane (e.g. preconditioned Krylov methods [291]).

**Table 5.5.** Two-grid convergence factors $\rho_{\text{loc}}$ for Model Problem 4 and standard coarsening ($\nu = 2$ smoothing steps being performed on the fine grid).

| Case | $a$ | $b$ | $c$ | Relaxation | $\rho_{\text{loc}}$ |
|------|-----|-----|-----|------------|---------------------|
| Case 1 | 1 | 1 | 1 | Point | **0.198** |
| Case 2 | 100 | 1 | 1 | Point | 0.961 |
|        |     |   |   | $x$-line | **0.074** |
| Case 3 | 100 | 100 | 1 | Point | 0.980 |
|        |     |     |   | Alt. $x$-/$y$-line | 0.938 |
|        |     |     |   | $(x, y)$-plane | **0.052** |
| Case 4 | 10 000 | 100 | 1 | Point | 1.0 |
|        |        |     |   | $x$-line | 0.961 |
|        |        |     |   | $(x, y)$-plane | **0.052** |

But the numerical cost and convergence of the method applied is crucial for the overall performance of the 3D multigrid solver. The solution of these 2D problems per plane should be as economic as possible, consistent with obtaining sufficient smoothing within the overall 3D multigrid algorithm (see Remark 5.2.1).                                    ≫

**Remark 5.2.4 (alternating plane relaxation)**   For 3D anisotropic problems with (smoothly varying) variable coefficients $a(x, y)$, $b(x, y)$ and $c(x, y)$ in (5.2.1), a *robust* multigrid method exists which has good convergence properties, independent of the size of $a(x, y)$, $b(x, y)$ and $c(x, y)$. This robust 3D method is based on the combination of standard coarsening and *alternating plane relaxation* for smoothing (with, for example FW and trilinear interpolation as transfer operators). Each smoothing step consists of applying three plane relaxations (an $(x, y)$-, a $(y, z)$- and an $(x, z)$-plane relaxation). If 2D multigrid is used as the plane solver, alternating line relaxation within each plane solver is required in order to guarantee good smoothing properties for all choices of $a, b$ and $c$.                                                                                    ≫

### 5.2.2 Point Relaxation for 3D Anisotropic Problems

In the 3D case, we also have the option of using point relaxation and semicoarsening strategies instead of block (line or plane) relaxations and standard coarsening. However, as already mentioned, in 3D there are more cases and possible combinations than in 2D.

For Case 2, the 1D semicoarsening strategy from Section 5.1.2 still holds. The strong coupling is in one direction and *x-semicoarsening*, i.e. coarsening only in the $x$-direction, will result in an efficient solver if combined with point smoothing. (Similarly, *y-semicoarsening* and *z-semicoarsening* will result in satisfactory solution methods if the strong coupling is in only one of these directions.) Figure 5.8(a) shows a coarse grid in the

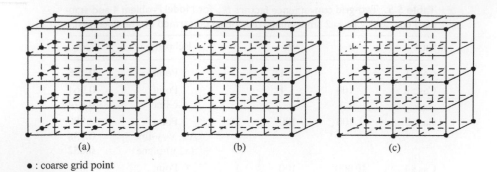

● : coarse grid point

**Figure 5.8.** A 3D fine grid with different coarse grids coming from (a) 1D semicoarsening in $x$-direction, (b) 2D semicoarsening in $(x, y)$-directions, (c) standard coarsening.

**Table 5.6.** Two-grid factors $\rho_{\text{loc}}$ for Model Problem 4 ($\nu = 2$ GS-RB steps being performed on the fine grid).

| Case | $a$ | $b$ | $c$ | Coarsening | $\rho_{\text{loc}}$ |
|---|---|---|---|---|---|
| Case 2 | 100 | 1 | 1 | $(x, y)$-semi | 0.961 |
| | | | | $x$-semi | **0.017** |
| Case 3 | 100 | 100 | 1 | $(x, y)$-semi | **0.074** |
| Case 4 | 10 000 | 100 | 1 | $(x, y)$-semi | 0.961 |
| | | | | $x$-semi | **0.009** |

case of 1D semicoarsening. As in the case of 1D semicoarsening in 2D problems, the number of grid points is reduced by a factor of 2 on coarse grids in the case of 1D semicoarsening for 3D problems. Based on the same consideration as in Remark 2.4.9, the computational work of an F-cycle is still $O(N)$, but that of a W-cycle is not.

A straightforward generalization to Case 3 means that coarsening should take place *in two directions*. Indeed, 2D semicoarsening in $(x, y)$-direction (see Fig. 5.8(b)) together with red–black or lexicographic point GS relaxation results in an efficient solver. Coarsening in two directions means that $\#\Omega_H \approx \#\Omega_h/4$. According to (2.4.14), the computational work for W-cycles is still $O(N)$.

An efficient semicoarsening strategy for Case 4 is the $x$-semicoarsening (as in Fig. 5.8(a)).

Some Fourier two-grid analysis results are presented in Table 5.6. $L_H$ is the seven-point discretization of $L$ on $\Omega_H$ and the transfer operators $I_h^H$ and $I_H^h$ are FW and trilinear interpolation. We use 3D GS-RB for smoothing in all cases.

### 5.2.3 Further Approaches, Robust Variants

So far, we have discussed two variants to handle the cases presented in (5.2.3). The first one is standard coarsening with different smoothing methods. A robust variant is then based on alternating plane smoothing. The second approach is based on pointwise relaxation and different semicoarsening strategies.

Efficient 3D multigrid solution methods for the different cases can also be based on "intermediate" variants. They are based on the combination of semicoarsening and block (line, plane) relaxation. Block relaxation is then applied along a subset of the coordinates and the coarsening takes place only in the other coordinate directions. Although these techniques seem to be rather complicated at first sight, they can result in very efficient and robust methods.

An alternative for Case 4 is, for example, $(x, y)$-semicoarsening (see Fig. 5.8(b)), but then combined with $x$-(zebra)-line relaxation instead of point relaxation; its asymptotic two-grid convergence factor $\rho_{\mathrm{loc}}$ is then 0.052 for $a = 10\,000$, $b = 100$ and $c = 1$ [388].

The smoothing factor of such an approach is the same as that of the lower dimensional problem obtained by neglecting the directions along which the grid is not coarsened [427]. For example, if in (5.2.1) we know that $a \approx b$, but have no information on $c$, we can use $z$-line (zebra) relaxation and coarsen only in $x$ and $y$. The resulting smoothing factor is the same as that of GS-RB for the 2D equation

$$-au_{xx} - bu_{yy} = f^{\Omega}.$$

If there is no information at all about the relative size of the coefficients, then a robust variant based on semicoarsening is to coarsen only along one coordinate, say $z$, while employing plane relaxation for $x$ and $y$. This plane relaxation itself may be carried out by a multigrid cycle, which employs line relaxation in one direction and semicoarsening in the other [122]. This (complicated) method is more efficient than alternating plane relaxations (but less straightforward) in the general case.

**Remark 5.2.5 (3D multiple semicoarsening)** Another alternative is to use multiple semicoarsening strategies as sketched in Section 5.1.4 for the 2D case. In general, too many coarse grids are processed in situations with general anisotropies if a point smoother is used. An efficient compromise is to use a 2D multiple *semicoarsening* strategy with line relaxation in the third dimension. In order to obtain a *robust* solver for general anisotropies, this third dimension should not be coarsened [411]. ≫

## 5.3 NONLINEAR PROBLEMS, THE FULL APPROXIMATION SCHEME

So far, we have considered linear problems. We will now discuss how multigrid methods can be used to solve nonlinear problems. In principle, there are two approaches.

The first is to apply some *global linearization method* like Newton's iteration to the nonlinear problem. In each iteration step, we then have to solve a linear problem. Under suitable conditions, multigrid can be used to solve each of these linear problems. We will describe this approach in detail in Section 5.3.3.

In the second approach, multigrid is applied directly to the nonlinear problem. The two multigrid ingredients, the error smoothing and the coarse grid correction ideas are not restricted to linear situations but can immediately be used for the nonlinear problem itself. This leads to the so-called full approximation scheme (FAS) [58]. We will describe this method in Section 5.3.4. In Section 5.3.6, we will apply FAS to a quasilinear problem from fluid dynamics (the full potential equation). For linear problems, FAS and linear multigrid as described in Chapter 2 are equivalent.

However, the FAS is also the basis of a number of advanced numerical techniques (see Section 5.3.7). In this respect, the FAS is also of interest for linear problems. A simple example of such a technique is the so-called $\tau$-extrapolation which provides solutions of higher accuracy.

Before we describe these multigrid methods, we will review some classical approaches for nonlinear problems in Sections 5.3.1 and 5.3.2.

### 5.3.1 Classical Numerical Methods for Nonlinear PDEs: an Example

For the numerical treatment of nonlinear PDEs, one has, in principle, a variety of choices of how to proceed. We give a rough survey of possible approaches here, emphasizing those aspects which are relevant in the multigrid context.

We will use the general notation

$$Nu = f^\Omega \quad (\Omega) \tag{5.3.1}$$

$$Bu = f^\Gamma \quad (\Gamma) \tag{5.3.2}$$

in the following. Here $N$ is assumed to be a nonlinear elliptic differential operator and $B$ is a boundary operator.

We assume that (5.3.1) and (5.3.2) have been discretized on a given grid $\Omega_h$. We denote a nonlinear system of discrete equations by

$$N_h u_h = f_h \quad (\Omega_h) \tag{5.3.3}$$

with a nonlinear operator $N_h : \mathcal{G}(\Omega_h) \to \mathcal{G}(\Omega_h)$. For ease of presentation, we assume that the boundary conditions have been eliminated and that they are implicitly contained in the discrete right-hand side $f_h$.

In order to be concrete and for illustration, we will again consider a model problem.

### Model Problem 5
*We consider the semilinear equation*

$$-\Delta u(x, y) + g(x, y, u) = f^\Omega(x, y) \quad ((x, y) \in \Omega) \tag{5.3.4}$$

*with Dirichlet boundary conditions*

$$u(x, y) = f^{\Gamma}(x, y) \quad ((x, y) \in \Gamma = \partial\Omega)$$

*in a bounded domain* $\Omega \subset \mathbb{R}^2$. *Here, g,* $f^{\Omega}$ *and* $f^{\Gamma}$ *as well as* $\Gamma$ *are assumed to be sufficiently smooth. Furthermore, we assume that the problem has a uniquely defined (classical) solution u.*

*Discretizing the operator* $\Delta$ *in (5.3.4) with the standard five-point stencil on* $\Omega_h$, *the system (5.3.3) is given by*

$$-\Delta_h u_h(x, y) + g(x, y, u_h(x, y)) = f_h(x, y) \quad ((x, y) \in \Omega_h). \tag{5.3.5}$$

$$\triangle$$

In the description of classical approaches for solving (5.3.3), we will distinguish between *global linearization methods* and *(local) relaxation-type methods*. This distinction is not rigid in all cases since some methods can be interpreted in either way and certain approaches are very similar.

From *global linearization*, we obtain a sequence of linear discretized PDEs. A classical method of this type is Newton's iteration

$$u_h^{m+1} = u_h^m - (K_h^m)^{-1}(N_h u_h^m - f_h) \quad (m = 0, 1, 2, \dots) \tag{5.3.6}$$

with some initial approximation $u_h^0$. Here,

$$K_h^m = N_h{}'[u_h^m] \tag{5.3.7}$$

is the Jacobian of $N_h$, i.e. in matrix terminology, the matrix of first partial derivatives. For the practical use of Newton's iteration, instead of (5.3.6) the equivalent defect correction notation

$$u_h^m \longrightarrow d_h^m := f_h - N_h u_h^m \longrightarrow \boxed{K_h^m \hat{v}_h^m = d_h^m} \longrightarrow u_h^{m+1} = u_h^m + \hat{v}_h^m$$

is suitable. Here, the box highlights the linear equation which has to be solved at the $m$th step of Newton's iteration.

**Example 5.3.1 (Newton's method)** In Model Problem 5, $K_h^m \hat{v}_h^m = d_h^m$ is represented by

$$-\Delta_h \hat{v}_h^m + c_h^m(x, y)\hat{v}_h^m = d_h^m(x, y) \quad ((x, y) \in \Omega_h)$$

with

$$c_h^m(x, y) = \frac{\partial g}{\partial u}(x, y, u_h^m(x, y)). \tag{5.3.8}$$

Thus, in each Newton step, a discrete Helmholtz-like equation (with nonconstant Helmholtz coefficient $c_h^m(x, y)$) has to be solved in this case.                                                    △

It is well known that Newton's iteration converges *quadratically* under suitable conditions, i.e.

$$\|v_h^{m+1}\| \leq \text{const } \|v_h^m\|^2 \quad (m = 0, 1, 2, \ldots)$$

holds for the errors $v_h^m = u - u_h^m$ and $v_h^{m+1} = u - u_h^{m+1}$. However, a drawback of Newton's method is that its domain of attraction is usually small. Therefore, Newton's method is often preceded by a slower but more robust algorithm in order to enlarge the domain of attraction. If the approximations are close enough to the solution, the method is switched to Newton's iteration in order to obtain the quadratic convergence.

Next to Newton's iteration in its original form (5.3.6), many variants and simplifications are used in practice. We list some of them in the following. They all belong to the class of global linearization methods. The *simplified Newton iteration* is obtained if

$$K_h^m = K_h = N_h'[u_h^0]$$

is used instead of (5.3.7). In this case, the Jacobian needs to be calculated only once and the discrete linear differential operator $K_h$ is the same for all iteration steps. The simplified Newton iteration converges only linearly (under suitable conditions).

**Example 5.3.2 (simplified Newton's method)**   For Model Problem 5, the operator $K_h^m$ is then as in Example 5.3.1, but $c_h^m$ is replaced by

$$c_h^0(x, y) = \frac{\partial g}{\partial u}(x, y, u_h^0(x, y)).$$                                       △

**Remark 5.3.1**   A different simplification, which also leads to linear convergence, is to replace $c_h^m(x, y)$ in Example 5.3.1 by a constant, e.g.

$$\hat{c}_h^m := \left( \min_{x,y} c_h^m(x, y) + \max_{x,y} c_h^m(x, y) \right)/2.$$

In this case, $K_h^m$ is the discrete Helmholtz operator

$$K_h^m = -\Delta_h + \hat{c}_h^m I_h,$$

where $I_h$ denotes the identity operator. This choice is of interest if one of the special *fast elliptic solvers* from Table 1.1 are to be applied (in simple domains) for the solution of $K_h^m v_h^m = d_h^m$. These solvers are usually only applicable to elliptic problems *with constant coefficients*. Such restrictions are irrelevant when using multigrid.                            ≫

**Remark 5.3.2 (Picard's method)**   An even simpler choice of $K_h^m$ is to use

$$K_h^m = K_h = -\Delta_h$$

for Model Problem 5, which gives *Picard's iteration*.                                              ≫

**Remark 5.3.3 (Newton-SOR)**   For the linear problems which have to be solved in each step of global linearization methods, in principle, any linear solver can be applied. Iterative methods like SOR have been discussed in the traditional numerical literature [286]. If, for example, Newton's iteration is used for global linearization and SOR for the resulting linear systems, this combination is usually called Newton-SOR.                                    ≫

### 5.3.2 Local Linearization

Apart from global linearization, the second principal approach for the solution of nonlinear systems is to employ methods using only *local* linearization. We use a somewhat different notation here. For instance, let

$$N_i(u_1, \ldots, u_r) = 0 \quad (i = 1, \ldots, r)$$

denote a nonlinear system of algebraic equations for the unknowns $u_1, \ldots, u_r$. A nonlinear Jacobi iteration to solve for the $i$th unknown from the $i$th equation then reads

$$N_i(u_1^m, \ldots, u_{i-1}^m, u_i^{m+1}, u_{i+1}^m, \ldots, u_r^m) = 0$$

(here, the indices $m$ and $m+1$ denote the current and the new approximations).

Correspondingly, a nonlinear Gauss–Seidel iteration reads

$$N_i(u_1^{m+1}, \ldots, u_{i-1}^{m+1}, u_i^{m+1}, u_{i+1}^m, \ldots, u_r^m) = 0 \quad (i = 1, 2, \ldots, r).$$

In both cases, a *single* nonlinear equation for a (single) unknown $u_i^{m+1}$ has to be solved. One option is to use Newton's method again, which, of course, is an iteration for a single equation then.

In the following, we will consider such relaxation-type methods in the context of the nonlinear discrete problems (5.3.3). For simplicity, we confine our description to Model Problem 5.

**Example 5.3.3**   For Model Problem 5, the nonlinear Jacobi iteration is

$$\frac{1}{h^2}[4]u_h^{m+1} + g(x, y, u_h^{m+1}) = f_h - \frac{1}{h^2}\begin{bmatrix} & -1 & \\ -1 & 0 & -1 \\ & -1 & \end{bmatrix}u_h^m, \qquad (5.3.9)$$

while the nonlinear Gauss–Seidel relaxation with lexicographic ordering is

$$\frac{1}{h^2}\begin{bmatrix} & 0 & \\ -1 & 4 & 0 \\ & -1 & \end{bmatrix}u_h^{m+1} + g(x, y, u_h^{m+1}) = f_h - \frac{1}{h^2}\begin{bmatrix} & -1 & \\ 0 & 0 & -1 \\ & 0 & \end{bmatrix}u_h^m.$$

If the nonlinearity (in $g$) is treated by one step of Newton's iteration, we obtain the "Jacobi–Newton" relaxation

$$\frac{1}{h^2}[4]u_h^{m+1} + c_h^m(x, y)u_h^{m+1}$$

$$= c_h^m(x, y)u_h^m - g(x, y, u_h^m) + f_h - \frac{1}{h^2}\begin{bmatrix} & -1 & \\ -1 & 0 & -1 \\ & -1 & \end{bmatrix}u_h^m$$

with $c_h^m(x, y)$ as in (5.3.8). A corresponding formula is obtained for "Gauss–Seidel–Newton". $\triangle$

With respect to *relaxation parameters* $\omega \neq 1$, similar results as in the linear case can be derived. In particular, the convergence properties of Gauss–Seidel relaxations can usually be improved by a parameter $\omega > 1$.

> **Remark 5.3.4** As in the linear case, relaxation-type methods like Jacobi–Newton and Gauss–Seidel–Newton turn out to be inefficient *solvers* for typical nonlinear elliptic problems. However, if applied in the context of (nonlinear) multigrid methods, which we will describe in Section 5.3.4, these relaxations are natural as *smoothers*.   ≫

**Remark 5.3.5** Instead of Jacobi–Newton or Gauss–Seidel–Newton, sometimes $g(x, y, u_h^{m+1})$ is simply replaced by $g(x, y, u_h^m)$ in Example 5.3.3. We call the corresponding relaxations *Jacobi–Picard* and *Gauss–Seidel–Picard*, respectively. Typically, these relaxations are less expensive since the derivatives of $g$ need not to be calculated in the relaxation. For *smoothing* purposes, this simplification often works well (see Section 5.3.5).   ≫

**Remark 5.3.6** We have made a major distinction between global linearization and local linearization in this section. In that respect, the two algorithms *Newton–Gauss–Seidel* (i.e. global linearization by Newton's iteration and Gauss–Seidel iteration for the resulting linear systems) and *Gauss–Seidel–Newton* (i.e. pointwise nonlinear Gauss–Seidel relaxation and Newton's iteration for the corresponding nonlinear scalar equations in each grid point) are basically different. However, if applied to the semilinear Model Problem 5, the algorithms are very similar. For a systematic comparison see, for example, [286].   ≫

**Remark 5.3.7 (exchangeability of discretization and linearization processes)** In our description of numerical methods for nonlinear PDEs, we have assumed that we have discretized the equation $Nu = f$ first and then applied a global linearization of the form (5.3.7) to the discrete problem $N_h u_h = f_h$. However, we can also apply a linearization approach like Newton's iteration to the nonlinear PDEs first and discretize the linearized PDEs afterwards. This approach can have advantages (like more flexibility with respect to the discretization) and may be regarded as more elegant. In many cases both ways lead to the same algorithm.

Formally, the linearization of the PDE is more ambitious since appropriate derivatives in adequate spaces have to be defined.                                                                    ≫

### 5.3.3 Linear Multigrid in Connection with Global Linearization

The linear problems which arise in the context of global linearization can efficiently be solved by linear multigrid. Here, not only *one* linear problem has to be solved but the global linearization leads to a *sequence of linear problems*, which typically are closely related. If multigrid cycles are used, an outer iteration (global linearization) is combined with an inner iteration (linear multigrid). In this section, we discuss how to adapt the convergence properties of the outer and the inner iteration so that the overall efficiency is satisfactory.

One way to combine Newton's method with an iterative linear multigrid method for $K_h^m v_h^m = d_h^m$ is to choose the number of multigrid iterations in each Newton step such that the convergence speed of Newton's method is fully exploited. This means that the number of multigrid iterations should roughly be doubled from one Newton step to the next as soon as Newton's method converges quadratically. We will refer to this method as *Method I*.

The main problem in this method is that one has to establish an appropriate control mechanism in order to obtain the required information on the convergence of Newton's method. If, for instance, too many multigrid cycles are carried out per Newton step, the overall efficiency of this approach will be reduced.

Another possibility is to *fix* the number of multigrid iterations per Newton step. For example, one may perform only *one* multigrid iteration per Newton step. A control mechanism is not needed in this case. As a consequence, Newton's method is, of course, truncated to a linearly convergent method. This method has the disadvantage that the Jacobian (5.3.7) needs to be calculated more often. We will refer to this as *Method II*. Some results for Methods I and II are shown in the following example.

**Example 5.3.4**   Consider the problem

$$
\begin{aligned}
Nu &= -\Delta u + e^u = f^\Omega(x, y) & ((x, y) \in \Omega), \\
u &= f^\Gamma(x, y) & ((x, y) \in \Gamma),
\end{aligned}
\tag{5.3.10}
$$

where the domain $\Omega$ sketched in Fig. 5.9 has a boundary composed of semicircles and straight lines and where $f^\Omega$ and $f^\Gamma$ are chosen such that the solution $u$ is $u(x, y) = \sin 3(x + y)$. The Laplace operator is discretized by the standard five-point formula (with $\boldsymbol{h} = (h_x, h_y) = (h, h)$) except for grid points near the boundary, where the Shortley–Weller approximation (2.8.13) is used.

Table 5.7 compares the methods discussed above for this problem. In Method II, one multigrid cycle is performed per (global) Newton step, in Method I the number of multigrid cycles is doubled from one Newton step to the next. In both cases the multigrid cycle for the linear problems uses GS-RB, FW and bilinear interpolation of the corrections. For all methods, the zero grid function $u_h^0 = 0$ is used as the initial multigrid approximation. Further parameters are $\nu_1 = 2$, $\nu_2 = 1$ and $\gamma = 2$ (W-cycles).

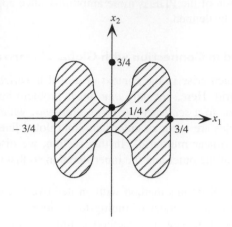

**Figure 5.9.** Domain $\Omega$ in (5.3.10).

**Table 5.7.** Behavior of $\| u_h - u_h^m \|_2$ ($h = 1/32$) for (5.3.10); for the Methods I and II and FAS, horizontal lines indicate that a new (global) Newton step is performed.

| Number of cycles ($m$) | Method I | Method II | FAS |
|---|---|---|---|
| 1 | 0.18 (+2) | 0.18 (+2) | 0.14 (+2) |
| 2 | 0.20 (0) | 0.20 (0) | 0.20 (0) |
| 3 | 0.86 (−2) | 0.55 (−2) | 0.54 (−2) |
| 4 | 0.14 (−3) | 0.14 (−3) | 0.14 (−3) |
| 5 | 0.43 (−5) | 0.42 (−5) | 0.42 (−5) |
| 6 | 0.13 (−6) | 0.13 (−6) | 0.13 (−6) |
| 7 | 0.47 (−8) | 0.39 (−8) | 0.38 (−8) |
| 8 | 0.13 (−9) | 0.12 (−9) | 0.12 (−9) |
| 9 | 0.42 (−11) | 0.40 (−11) | 0.39 (−11) |

If we compare the results of Methods I and II, we observe that the quadratic convergence speed of Newton's method is fully exploited in Method I. An accuracy of $10^{-8}$ is reached after three Newton steps. The linearly convergent Method II needs seven steps of the "multigrid truncated" Newton iteration to reach the same accuracy. However, if we compare the errors after the same number of *multigrid iterations*, the accuracy is nearly the same in both cases.

Furthermore, the work and the storage requirements for Methods I and II are similar. In that respect, the efficiency of both methods is essentially the same. The fact that the Newton iteration in Method I converges quadratically, whereas that of Method II does not, does not say anything about the overall efficiency of the methods. In other words, both methods give quadratic convergence if we define one algorithmical unit to consist of $1, 2, 4, \ldots$ multigrid steps.

The third column in Table 5.7 corresponds to the nonlinear multigrid method (FAS) which will be discussed in detail in Section 5.3.4. In principle, it is a nonlinear analog of a linear red black multigrid solver described in Section 2.5, where GS-RB relaxation is essentially replaced by a corresponding nonlinear GS-RB relaxation method.                    △

### 5.3.4 Nonlinear Multigrid: the Full Approximation Scheme

Similar to the linear case, the nonlinear FAS *multigrid* method can be recursively defined on the basis of a *two-grid* method. Thus we start with the description of one iteration cycle of the nonlinear $(h, H)$ two-grid method for solving (5.3.3), computing $u_h^{m+1}$ from $u_h^m$. The fundamental idea of nonlinear multigrid is the same as in the linear case. First, the errors to the solution have to be smoothed such that they can be approximated on a coarser grid. An analog of the linear defect equation is transferred to the coarse grid. The *coarse grid corrections* are interpolated back to the fine grid, where the errors are finally smoothed. However, formally we do not work with the errors, but with full approximations to the discrete solution on the coarse grid.

In the nonlinear case the (exact) defect equation on $\Omega_h$ is given by

$$N_h(\bar{u}_h^m + v_h^m) - N_h \bar{u}_h^m = \bar{d}_h^m \tag{5.3.11}$$

and this equation is approximated on $\Omega_H$ by

$$N_H(\bar{u}_H^m + \hat{v}_H^m) - N_H \bar{u}_H^m = \bar{d}_H^m, \tag{5.3.12}$$

where $N_H$ is an appropriate discrete operator on $\Omega_H$. An illustration of the corresponding two-grid cycle, which is similar to the one given in Fig. 2.4 for the linear correction scheme is given in Fig. 5.10.

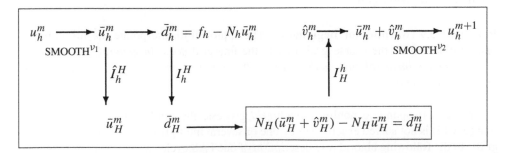

**Figure 5.10.** FAS $(h, H)$ two-grid method.

In this description, SMOOTH stands for a *nonlinear* relaxation procedure (for example, one of those described in Section 5.3.2) which has suitable error smoothing properties. As in the linear case, $\nu_1$ smoothing steps are performed before and $\nu_2$ smoothing steps after the coarse grid correction.

---

In contrast to the linear case, not only the defect $d_h^m$ is transferred to the coarse grid (by some restriction operator $I_h^H$) in the FAS two-grid method, but also the relaxed approximation $\bar{u}_h^m$ itself (by some restriction operator $\hat{I}_h^H$, which may be different from $I_h^H$).

---

On the coarse grid $\Omega_H$, we deal with the problem

$$N_H w_H = f_H, \qquad (5.3.13)$$

where $w_H = \bar{u}_H^m + \hat{v}_H^m$ and where the right-hand side $f_H$ is defined by

$$f_H := I_h^H(f_h - N_h \bar{u}_h^m) + N_H \hat{I}_h^H \bar{u}_h^m. \qquad (5.3.14)$$

The transfer of the current approximation to the coarse grid is used to obtain $\bar{u}_H^m = \hat{I}_h^H \bar{u}_h^m$. The most common choice for $\hat{I}_h^H$ is injection (for vertex-centered grids).

**Remark 5.3.8**  If $N_h$ and $N_H$ are linear operators, the FAS two-grid method is equivalent to the (linear) correction scheme introduced in Section 2.4.2. This can be seen immediately from (5.3.12).                                                                ≫

As mentioned above, in the FAS method, the correction $\hat{v}_H^m$ is transferred back to the fine grid $\Omega_h$ as in the linear case. This is important since only correction grid functions (i.e. errors) are smoothed by relaxation processes and can therefore be approximated well on coarser grids (see the explanations in Sections 2.1 and 2.2, which, in principle, also apply to the nonlinear case). $\hat{v}_H^m$ is computed as the difference of $\hat{I}_h^H \bar{u}_h^m$ and $w_H = \bar{u}_H^m + \hat{v}_H^m$ after solution on the coarse grid.

**Remark 5.3.9 (warning for beginners)**  The approach to interpolate the full approximation obtained on the coarse grid back to the fine grid *does, in general, not lead to a converging nonlinear solution method*. This is the source of an error that is often made by multigrid beginners.                                                        ≫

In the corresponding nonlinear *multigrid* process, the nonlinear coarse grid equation (5.3.12) is not solved exactly, but by one or several multigrid cycles using still coarser grids. In the following algorithmic description of one FAS cycle, we use a similar notation as in Section 2.4. In particular, we assume a sequence of grids $\Omega_k$ and grid operators $N_k$,

$I_k^{k-1}$, $\hat{I}_k^{k-1}$, $I_{k-1}^k$ etc. to be given. One *FAS multigrid cycle* starting on the finest level $k = \ell$ (more precisely: FAS $(\ell + 1)$-*grid cycle*) for the solution of

$$N_\ell u_\ell = f_\ell \quad (\ell \geq 1, \text{fixed}) \tag{5.3.15}$$

proceeds as follows. If $k = 1$, we just have the two-grid method described above with $\Omega_0$ and $\Omega_1$ instead of $\Omega_H$ and $\Omega_h$, respectively. For general $k = 1, \ldots, \ell$, we have:

---

**FAS multigrid cycle** $u_k^{m+1} = \text{FASCYC}(k, \gamma, u_k^m, N_k, f_k, \nu_1, \nu_2)$

(1) Presmoothing

 – Compute $\bar{u}_k^m$ by applying $\nu_1 (\geq 0)$ smoothing steps to $u_k^m$

$$\bar{u}_k^m = \text{SMOOTH}^{\nu_1}(u_k^m, N_k, f_k).$$

(2) Coarse-grid correction
 – Compute the defect           $\bar{d}_k^m = f_k - N_k \bar{u}_k^m.$
 – Restrict the defect          $\bar{d}_{k-1}^m = I_k^{k-1} \bar{d}_k^m.$
 – Restrict $\bar{u}_k^m$        $\bar{u}_{k-1}^m = \hat{I}_k^{k-1} \bar{u}_k^m.$
 – Compute the right-hand side   $f_{k-1} = \bar{d}_{k-1}^m + N_{k-1} \bar{u}_{k-1}^m.$
 – Compute an approximate solution
   $\hat{w}_{k-1}^m$ of the coarse grid equation on $\Omega_{k-1}$

$$N_{k-1} w_{k-1}^m = f_{k-1}. \tag{5.3.16}$$

 If $k = 1$ employ a fast solver for this purpose.
 If $k > 1$ solve (5.3.16) by performing $\gamma (\geq 1)$ FAS $k$-grid cycles
  using $\bar{u}_{k-1}^m$ as initial approximation

$$\hat{w}_{k-1}^m = \text{FASCYC}^\gamma (k - 1, \gamma, \bar{u}_{k-1}^m, N_{k-1}, f_{k-1}, \nu_1, \nu_2).$$

 – Compute the correction         $\hat{v}_{k-1}^m = \hat{w}_{k-1}^m - \bar{u}_{k-1}^m.$
 – Interpolate the correction      $\hat{v}_k^m = I_{k-1}^k \hat{v}_{k-1}^m.$
 – Compute the corrected
   approximation on $\Omega_k$      $u_k^{m,\text{after CGC}} = \bar{u}_k^m + \hat{v}_k^m.$
(3) Postsmoothing

 – Compute $u_k^{m+1}$ by applying $\nu_2 (\geq 0)$ smoothing steps to $u_k^{m,\text{after CGC}}$

$$u_k^{m+1} = \text{SMOOTH}^{\nu_2}(u_k^{m,\text{after CGC}}, N_k, f_k).$$

---

One can observe from this description that *no global linearization is needed in the FAS multigrid process, except on the coarsest grid*. Only nonlinear relaxation methods are required as well as (linear) fine-to-coarse and coarse-to-fine transfer operators. These transfer operators are thus often chosen as in the linear case: FW, HW and linear interpolation. As discussed in Section 5.3.1, there are usually several nonlinear analogs to a given linear relaxation method [286] which correspond to a (locally) linearized problem. (See also Section 5.3.5 for some remarks about simple nonlinear relaxation methods and their smoothing properties.)

> The combination of FAS with full multigrid (FMG), i.e, starting FAS on the coarsest grid, is also easily possible and a natural choice in many cases.
>
> In the nonlinear case, again bicubic interpolation is typically used as the FMG interpolation.

**Remark 5.3.10 (continuation)**   In order to obtain a reasonable initial approximation on a fine grid, i.e. an approximation that lies in the domain of attraction, a *continuation* process can be incorporated within an FMG iteration combined with FAS. The idea of this approach is to start with a weakly nonlinear or even linear problem on the coarsest grid and increase the strength of the nonlinearity step by step, when going to finer levels in FMG.                                                                                                      ≫

**Remark 5.3.11 (FAS versus global linearization)**   In Example 5.3.4, the results of the FAS nearly coincide with those of Method II (see the errors in Table 5.7). This is a general observation. Although the FAS and the indirect multigrid Methods I and II may look quite different at first sight, they often show similar convergence.

In fact, if we consider one iteration step of Method II (see Section 5.3.3, one linear multigrid cycle per linearization step) and one FAS cycle, the main differences lie in the solution process on the coarsest grid and in the relaxation process (which in the one case refers to $N_h$ and in the other case refers to its current linearization $N_h^m$).

An advantage of FAS compared to Methods I and II is the memory requirement of FAS. It is not necessary to compute and store the (fine grid) Jacobian in the FAS process, as is necessary in the Newton-based solution methods.                                                           ≫

**Remark 5.3.12**   A more general proposal for a nonlinear multigrid method is Hackbusch's NLMG method [176]. The main difference between NLMG and FAS is the choice of the initial approximation on the coarse grid ($\bar{u}_H^m$ in the FAS). In principle, any grid function can be used as initial guess on a coarse grid. NLMG uses, for example, a coarse grid approximation from the FMG process as a first approximation on the coarse grid since this is a solution of the nonlinear coarse grid equation, whereas in the FAS the restriction of a current fine grid solution $u_h^m$ is employed.

In addition, NLMG allows a scaling factor $s$ in front of the restricted defect $\bar{d}_H^m$ in the right-hand side of the coarse grid equation (and the factor $1/s$ in front of the correction $\hat{v}_h^m$

back to the fine grid). This parameter can be used to ensure the solvability of the coarse grid equation [176].                                                           ≫

### 5.3.5 Smoothing Analysis: a Simple Example

In the nonlinear multigrid context, we are interested in the *smoothing properties* of nonlinear relaxation methods. We want to discuss this question briefly for (5.3.5). For simplicity, we consider only Jacobi-type iteration. A corresponding analysis can, of course, also be carried out for other relaxation methods.

We compare $\omega$-Jacobi–Newton and $\omega$-Jacobi–Picard relaxation (see Remark 5.3.5). They differ in the way in which the nonlinear function $g$ is treated in the relaxation. Newton linearization uses

$$g(x, y, u_h^m) + \frac{\partial g}{\partial u}(x, y, u_h^m)(u_h^{m+1}(x, y) - u_h^m(x, y)), \qquad (5.3.17)$$

whereas Picard uses

$$g(x, y, u_h^m) \qquad (5.3.18)$$

as an approximation for $g(x, y, u_h^{m+1})$ while performing the relaxation at any fixed grid point.

**Remark 5.3.13**  We consider only the *linear* case $g(x, u) = cu$, with constant $c > 0$. The relaxation operators of the $\omega$-Jacobi–Newton and the $\omega$-Jacobi–Picard methods are given by

$$S_h^N = \left(1 - \frac{\omega^N ch^2}{4 + ch^2}\right) I_h - \frac{\omega^N h^2}{4 + ch^2} \Delta_h, \qquad (5.3.19)$$

$$S_h^P = \left(1 - \frac{\omega^P ch^2}{4}\right) I_h - \frac{\omega^P h^2}{4} \Delta_h. \qquad (5.3.20)$$

Obviously, both operators coincide if

$$\omega^N = \frac{4 + ch^2}{4}\omega^P. \qquad (5.3.21)$$

It is therefore sufficient to analyze $S_h^N = S_h^N(\omega)$.

By considerations similar to those in Section 2.1, we obtain the eigenvalues $\chi_h^{N,k,\ell}$ and the smoothing factor $\mu(h; \omega)$ for $S_h^N(\omega)$:

$$\chi_h^{N,k,\ell} = \frac{4}{4 + ch^2}\chi_h^{k,\ell} + \frac{(1 - \omega)ch^2}{4 + ch^2}$$

$$= 1 - \frac{\omega}{4 + ch^2}(4 + ch^2 - 2\cos k\pi h - 2\cos \ell\pi h)$$

$$\mu(h; \omega)^N = \max\left\{\left|1 - \omega\left(1 - \frac{2\cos \pi h}{4 + ch^2}\right)\right|, \left|1 - \omega\left(1 + \frac{4\cos \pi h}{4 + ch^2}\right)\right|\right\},$$

with $\chi_h^{k,\ell} = 1 - (\omega/2)(2 - \cos k\pi h - \cos \ell\pi h)$ from (2.1.5). Of course, the $\omega$-Jacobi–Newton relaxation coincides with $\omega$-JAC in this linear case. From (5.3.22), we see that, for any fixed $0 < \omega < 1$ (and any fixed $h$), the smoothing properties of $S_h^N$ improve for increasing $c$:

$$\mu(h; \omega) \to 1 - \omega \quad (0 \le c \to \infty).$$

The Jacobi–Picard $\omega$-relaxation method should, however, be used with some care [378]. From (5.3.21), we can see that for any fixed $\omega = \omega^P$, $0 < \omega^P < 1$, the $\omega$-Jacobi–Picard method has no smoothing properties, if $ch^2$ is sufficiently large. $\gg$

More generally, this means that for the nonlinear case the $\omega$-Jacobi–Picard relaxation with fixed $\omega$ cannot be used for smoothing purposes whenever

$$h^2 \frac{\partial g}{\partial u}(x, y, u_h(x, y))$$

is large compared to 1 for certain $(x, y) \in \Omega$. This is quite likely to occur, at least on coarse levels of the FAS algorithm. If, however,

$$0 \le h^2 \frac{\partial g}{\partial u}(x, y, u_h(x, y)) < 1 \quad ((x, y) \in \Omega_h)$$

and if $u_h^m$ is sufficiently close to $u_h$, the Jacobi–Picard method should give results similar to those of the Jacobi–Newton method.

Even in the general case, typically one Newton step in relaxing each single equation turns out to be sufficient for many applications. We finally remark that all other smoothing methods for linear problems (point, line, plane relaxations) have natural nonlinear analogs.

### 5.3.6 FAS for the Full Potential Equation

The full potential equation describes a steady, inviscid, isentropic and irrotational flow and is a nonlinear differential equation for the potential $\Phi$. For convenience, we use the scaled potential $\phi = \Phi/u_\infty$, where $u_\infty$ is the flow velocity of the free undisturbed flow, which we assume to be in $x$-direction ($v_\infty = 0$). In 2D, the full potential equation can then be written as

$$-\Delta\phi + \frac{1}{c^2}\left((\phi_x)^2\phi_{xx} + 2\phi_x\phi_y\phi_{xy} + (\phi_y)^2\phi_{yy}\right) = 0, \qquad (5.3.22)$$

where $c = c(\phi_x, \phi_y)$ is the (local) speed of sound defined by

$$c^2 = \frac{1}{(M_\infty)^2} + \frac{\gamma - 1}{2}(1 - ((\phi_x)^2 + (\phi_y)^2))$$

and $M_\infty = u_\infty/c_\infty$ is the Mach number ($c_\infty$ the speed of sound) of the undisturbed flow. The parameter $\gamma$ is the ratio of the specific heats at constant pressure and volume ($\approx 1.4$ for air). The flow velocities (also scaled by $u_\infty$) in the $x$- and $y$-directions are

$$u = \phi_x \quad \text{and} \quad v = \phi_y.$$

In the limit $c \to \infty$ (incompressible flow), (5.3.22) reduces to the Laplace equation for the potential $\phi$.

It can easily be seen that the full potential equation is *elliptic* if the velocity is smaller than the speed of sound, i.e.

$$u^2 + v^2 < c^2 \quad \text{(subsonic flow)},$$

and *hyperbolic* if

$$u^2 + v^2 > c^2 \quad \text{(supersonic flow)}.$$

*For the range of subsonic potential flow, (5.3.22) is elliptic and standard nonlinear multigrid works very efficiently.*

As an example, we consider the full potential flow around the unit circle. This flow is subsonic up to $M_\infty \approx 0.4$. (For larger Mach numbers $M_\infty$, the flow becomes transonic, i.e. there will be regions in which the flow is supersonic.) We assume that the undisturbed flow is in the $x$-direction. Due to symmetry, we can restrict ourselves to the flow around a quarter of the circle. In this example, it is convenient to use polar coordinates $(r, \theta)$. Figure 5.11 presents the geometry of this problem in Cartesian and in polar coordinates.

In polar coordinates, the corresponding boundary value problem is given by the PDE

$$-\Delta\phi + \frac{1}{c^2}\left(\phi_r{}^2\phi_{rr} + \frac{2}{r^2}\phi_r\phi_\theta\phi_{r\theta} + \frac{1}{r^4}\phi_\theta{}^2\phi_{\theta\theta} - \frac{1}{r^3}\phi_r\phi_\theta{}^2\right) = 0 \quad (\Omega) \qquad (5.3.23)$$

with

$$\Delta\phi = \phi_{rr} + \frac{1}{r^2}\phi_{\theta\theta} + \frac{1}{r}\phi_r \qquad (5.3.24)$$

and

$$c^2 = \frac{1}{(M_\infty)^2} + \frac{\gamma - 1}{2}\left(1 - \left(\phi_r{}^2 + \frac{1}{r^2}\phi_\theta{}^2\right)\right)$$

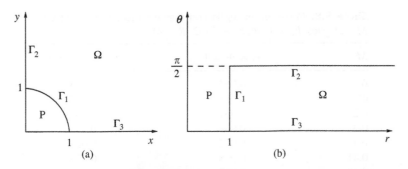

**Figure 5.11.** Relevant domains $\Omega$ for the computation of the full potential flow around the circle (indicated by P) in (a) Cartesian coordinates, (b) polar coordinates.

and the boundary conditions

$$\phi = 0 \quad (\Gamma_2) \tag{5.3.25}$$

$$\phi_n = 0 \quad (\Gamma_1, \Gamma_3) \tag{5.3.26}$$

$$\phi \longrightarrow r \cos \theta \quad (\text{for } r \longrightarrow \infty). \tag{5.3.27}$$

Instead of the far-field condition (5.3.27), we use the nonlinear boundary condition

$$\phi\phi_n = r \cos^2 \theta \quad (\text{for } r = R \text{ with } R \text{ sufficiently large}).$$

which is a better approximation than the Dirichlet boundary condition $\phi = r \cos \theta$ or the Neumann boundary condition $\phi_r = \cos \theta$ [345]. (The proper multigrid treatment of boundary conditions will be described in Section 5.6.)

To solve this system by nonlinear multigrid (FAS), we discretize all occurring derivatives by standard second-order central differences. It can be seen in the linear case, i.e. the Laplace operator in polar coordinates (5.3.24), that the resulting PDE becomes anisotropic for large values of $r$. Therefore, we use a zebra line relaxation in the radial direction. Within the (nonlinear) line relaxation, one Newton step is used for each line. FW and bilinear prolongation are used as transfer operators.

Table 5.8 shows measured multigrid convergence factors in dependence of the Mach number of the undisturbed flow $M_\infty$.

Obviously, the convergence rates hardly depend on $M_\infty$ as long as the flow remains subsonic. They are very close to those of the pure Laplacian ($M_\infty = 0$).

For $M_\infty$ larger than about 0.4, the flow is locally hypersonic and, as a consequence, the discrete equation is no longer $h$-elliptic as long as central differences are used. According to the discussion in Section 4.7.2, the divergence of the algorithm has thus to be expected for such Mach numbers. In such cases the discretization has to be modified [369].

Note that in the transonic case, the assumptions leading to the full potential equation are not fulfilled physically; the compressible Euler equations describe the physical flow much better.

**Table 5.8.** Measured multigrid convergence rates depending on $M_\infty$ (4 grids, $h_r = 1/16$, $h_\theta = \pi/32$, $R = 4$).

| $M_\infty$ | $\nu_1 = \nu_2 = 1$ | $\nu_1 = 2, \nu_2 = 1$ |
|------------|---------------------|------------------------|
| 0          | 0.064               | 0.035                  |
| 0.1        | 0.065               | 0.035                  |
| 0.2        | 0.067               | 0.037                  |
| 0.3        | 0.066               | 0.039                  |
| 0.4        | 0.062               | 0.037                  |
| 0.41       | 0.068               | 0.037                  |
| 0.42       | 0.067               | 0.039                  |
| 0.43       | div.                | div.                   |

### 5.3.7 The $(h, H)$-Relative Truncation Error and $\tau$-Extrapolation

A variety of more sophisticated multigrid techniques is based on the FAS method. In the following, we use the notation of a linear operator $L_h$ for convenience, although these techniques can also be applied in nonlinear applications.

The coarse grid equation (5.3.13), (5.3.14) in the FAS scheme can be written in the form

$$L_{2h} w_{2h}^m = I_h^{2h} f_h + \tau_h^{2h}(\bar{u}_h^m) \tag{5.3.28}$$

where

$$\tau_h^{2h}(u_h) := L_{2h} \hat{I}_h^{2h} u_h - I_h^{2h} L_h u_h. \tag{5.3.29}$$

Clearly, the identity

$$L_{2h}(\hat{I}_h^{2h} u_h) = I_h^{2h} f_h + \tau_h^{2h}(u_h) \tag{5.3.30}$$

holds for the discrete solution $u_h$. $\tau_h^{2h}(u_h)$ is called the $(h, 2h)$-relative truncation error (with respect to $I_h^{2h}$, $\hat{I}_h^{2h}$). With respect to the grids $\Omega_h$ and $\Omega_{2h}$, $\tau_h^{2h}$ plays a role similar to the truncation error (local discretization error)

$$\tau_h(u) := L_h \hat{I}_h u - \hat{I}_h L u \tag{5.3.31}$$

of the continuous solution $u$ with respect to $\Omega$ and $\Omega_h$. (Here $\hat{I}_h$ denotes the injection operator from $\Omega$ to $\Omega_h$.)

---

If, in particular, $\Omega_{2h} \subset \Omega_h$ and $\hat{I}_h^{2h}$ is the injection operator from $\Omega_h$ to $\Omega_{2h}$, we see from (5.3.30) that $\tau_h^{2h}(u_h)$ is that quantity which has to be added to the right-hand side $I_h^{2h} f_h$ to obtain the values of the *fine grid* solution $u_h$ on $\Omega_{2h}$. By solving (5.3.30), we obtain the *fine grid solution* represented by $\hat{I}_h^{2h} u_h$ on the coarse grid.

---

The quantity $\tau_h^{2h}$ is the starting point for several techniques, all of which are based on a somewhat different interpretation of multigrid. This is the *dual point of view* of multigrid [66], in which the coarse grid is regarded as the primary grid. In this view, the fine grid is used to provide fine grid accuracy to the problem on the coarse grid, instead of regarding the coarse grid as a means of providing a correction for the solution of the problem on the fine grid. For this purpose, of course, a suitable approximation for $\tau_h^{2h}(u_h)$ has to be provided. Some of the advanced techniques are

- $\tau$-estimation, where $\tau_h^{2h}$ is used to estimate the discretization error,
- *adaptive mesh refinement techniques*, where the utilization of locally refined grids can be based on criteria using $\tau_h^{2h}$ (see Chapter 9)
- $\tau$-extrapolation, which we discuss in some more detail in the following.

For details and further techniques based on $\tau_h^{2h}$ (e.g. so-called *frozen-$\tau$ techniques* which can be useful for problems in which a sequence of similar systems of equations needs to be solved, like in time-dependent problems) see [66, 447].

Employing $\tau$-extrapolation is one way to obtain more accurate approximations. However, in order to obtain high order approximations in general situations, we refer to the defect correction approach (see Section 5.4.1).

Let us consider a discrete linear problem

$$L_h u_h = f_h, \tag{5.3.32}$$

where $L_h$ is a $p$th order discretization. If $\tau_h^{2h}(u_h)$ is added to the right-hand side of

$$L_{2h} u_{2h} = I_h^{2h} f_h,$$

we will obtain fine grid accuracy. A simple modification of $\tau_h^{2h}$, however, provides an approximation $u_{2h}^\tau$, the accuracy of which will be better than that of $u_h$. A formula for the modification is easily derived if we assume asymptotic expansions of the form

$$L_h u - Lu = e_1 h^p + O(h^q) \tag{5.3.33}$$
$$L_{2h} u - Lu = e_1 2^p h^p + O(h^q) \tag{5.3.34}$$
$$u_h - u = e_2 h^p + O(h^q) \tag{5.3.35}$$

($q > p$) and further assume that all functions appearing are sufficiently smooth. We consider the $2h$-grid problem

$$L_{2h} u_{2h}^\tau = I_h^{2h} f_h + \frac{2^p}{2^p - 1} \tau_h^{2h}(u_h). \tag{5.3.36}$$

Choosing injection as the restriction operator $I_h^{2h}$ and omitting it for ease of presentation, we find from (5.3.33–5.3.35) that

$$
\begin{aligned}
L_{2h} u_{2h}^\tau &= f_h + \frac{2^p}{2^p - 1}(L_{2h} u - L_h u) + \frac{2^p}{2^p - 1} h^p (L_{2h} e_2 - L_h e_2) + O(h^q) \\
&= f_h + \frac{2^p}{2^p - 1}(L_{2h} u - L_h u) + O(h^{\tilde{q}}) \quad \text{(where } \tilde{q} = \min\{2p, q\}) \\
&= Lu + 2^p h^p e_1 + O(h^{\tilde{q}}) \\
&= L_{2h} u + O(h^{\tilde{q}}).
\end{aligned}
$$

Obviously, the order of accuracy of $u_{2h}^\tau$ is higher than that of $u_h$.

**Remark 5.3.14** In practice, (5.3.36) can also be used if the asymptotic expansion (5.3.35) does not hold or cannot be proved. From (5.3.33) and (5.3.34) we see that

$$L_{2h} u = Lu + \frac{2^p}{2^p - 1}(L_{2h} u - L_h u) + O(h^q). \tag{5.3.37}$$

≫

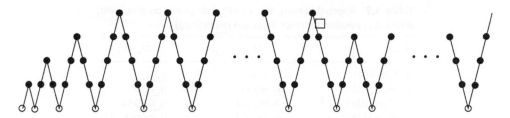

**Figure 5.12.** An example of how to apply $\tau$-extrapolation. $\circ$, solution on coarsest grid; $\bullet$, smoothing; $\square$, $\tau$-extrapolation, i.e. setting up the right-hand side of the coarse grid equation (5.3.36).

$\tau$-extrapolation means that the transfer from the finest grid is modified. Here, (5.3.36) is used as the problem on the $2h$-grid. The additional amount of work compared to standard FAS is only the multiplication of $\tau_h^{2h}$ with the constant factor $(2^p)/(2^p - 1)$ per coarse grid point. Note that only $u_{2h}^\tau$ is higher order accurate, $u_h$ is the original low order approximation. Therefore, relaxation should not be employed anymore on the finest grid if $\tau$-extrapolation has been applied.

Figure 5.12 shows a simple, although not asymptotically optimal method, for which high order accuracy is obtained. The basic idea is to avoid any relaxation on the fine grid, once the $\tau$-restriction has been applied. The approximation before the $\tau$-restriction has to be sufficiently accurate, and the number of cycles afterwards (on the coarse grid) has to be large enough to really reach high order accuracy.

For optimal approaches and a more detailed description we refer to [40, 346].

**Remark 5.3.15** *The use of injection in (5.3.36) provides good approximations of $L_h u_h$ and $f_h$ at the coarse grid points.*

FW, however, is not the appropriate scheme in (5.3.36). This is because FW itself provides only second-order accurate approximations. In the usual FAS iteration, FW is applied *to the defect* which tends to zero during the multigrid iterations and therefore the second-order does not damage the accuracy of the discrete solution. In the computation of the right-hand side on the coarse grid (5.3.36), however, $f_h$ and $L_h u_h$ have been restricted with different weights (1 and $2^p/(2^p - 1)$). $\gg$

**Remark 5.3.16** If we have non-Dirichlet boundary conditions, $\tau$-extrapolation should be applied separately to the discrete boundary conditions and to the discrete equations of the PDEs, i.e. not to the eliminated equations at the boundary (see [346] for details). $\gg$

Finally, we apply the $\tau$-extrapolation to the subsonic full potential equation (see Section 5.3.6).

**Example 5.3.5** In order to show that the $\tau$-extrapolation can also be used for nonlinear problems, we reconsider the full potential flow around the unit circle in 2D (see Fig. 5.11

**Table 5.9.** Approximations $\tilde{M}^*_\infty$ of $M^*_\infty$ computed on grids with different resolution (correct digits are underlined).

| $(h_z, h_\theta)$ | $\tilde{M}^*_\infty$ | $\tilde{M}^*_\infty(\tau)$ |
|---|---|---|
| $(1/8, \pi/16)$ | 0.3956 | 0.3956 |
| $(1/16, \pi/32)$ | 0.39759 | 0.39767 |
| $(1/32, \pi/64)$ | 0.397908 | 0.397954 |
| $(1/64, \pi/128)$ | 0.397953 | 0.397968728 |
| $(1/128, \pi/256)$ | 0.3979646 | 0.3979688565 |

and [375]). Here, we want to compute the lowest Mach number $M^*_\infty$ for which supersonic regions in the flow exist.

We transform the full potential equation to polar coordinates and perform a further transformation to $(1/r, \theta)$-coordinates which corresponds to a refinement of the grid near the profile. Table 5.9 shows approximations $\tilde{M}^*_\infty$ and $\tilde{M}^*_\infty(\tau)$ of $M^*_\infty$ computed with and without $\tau$-extrapolation for different grid sizes. For this purpose, the critical Mach number has been determined from solutions $\phi_h(M_\infty)$ and $\phi^\tau_h(M_\infty)$. For each Mach number $M_\infty$, we have checked whether or not supersonic flow appears at the top of the circle.

The number of significant digits of $\tilde{M}^*_\infty(\tau)$ is increased by at least one if the mesh size is reduced by a factor of 2. The accuracy obtained without $\tau$-extrapolation is worse.     △

## 5.4 HIGHER ORDER DISCRETIZATIONS

So far, we have mainly considered second-order accurate discretizations of PDEs in detail. In many situations, however, higher order discretizations have advantages. One benefit of higher order discretizations is that the same accuracy can be achieved on a much coarser grid provided the solution is sufficiently smooth. If an efficient solution method for a higher order discretization is available, a large gain in computing time (and in computer memory) can be achieved.

Uniformly elliptic problems tend to have smooth solutions. However, the shape of a domain or the type of the boundary conditions involved may cause complications. In the case of uniformly elliptic problems, "low order" discretization usually means $O(h^2)$ and "high order" usually $O(h^4)$ accuracy.

We will present two ways of combining multigrid with high order discretizations:

- the direct multigrid solution (long stencil or "Mehrstellen" discretization in Section 5.4.2),
- the solution via defect correction (see Section 5.4.1).

Here, we will restrict our discussion to uniformly elliptic problems. The defect correction approach, however, is more general. In Section 7.1, for example, we will use it for (singularly perturbed) convection–diffusion problems. For such problems, even efficient solvers for second-order accurate discretizations cannot be obtained easily.

**Example 5.4.1** We want to solve the Poisson equation $-\Delta u = f$ with Dirichlet boundary conditions $u = g$ on the domain $\Omega = (0, 1)^2$ with fourth-order accuracy. For $L_h$, we choose the $O(h^4)$-accurate discretization

$$L_h = -\frac{4}{3}\Delta_h + \frac{1}{3}\Delta_{2h} = \frac{1}{12h^2}\begin{bmatrix} & & 1 & & \\ & & -16 & & \\ 1 & -16 & 60 & -16 & 1 \\ & & -16 & & \\ & & 1 & & \end{bmatrix}_h . \qquad (5.4.1)$$

The fourth-order accuracy of $L_h$ is easily verified by a Taylor expansion.

At grid points adjacent to boundary points, the nine-point stencil (5.4.1) cannot be applied since it has entries which are outside $\Omega_h$. Here, we use a modified stencil, e.g. near the left (west) boundary:

$$L_h = \frac{1}{12h^2}\begin{bmatrix} & & 1 & & \\ & & -16 & & \\ 0 & -12 & 54 & -12 & 0 \\ & & -16 & & \\ & & 1 & & \end{bmatrix}_h . \qquad (5.4.2)$$

The stencil (5.4.2) is obtained by the use of fourth-order extrapolation in order to eliminate the entries of (5.4.1), which are outside of $\Omega_h$.

We set the boundary conditions $g$ and the right-hand side $f$ such that the analytical solution is $u(x, y) = e^{xy}$. In order to see the benefits of a higher order discretization, we compare the errors $\|u - u_h\|_\infty$ of the fourth order and the standard $O(h^2)$ five-point stencil. Table 5.10 presents the difference between the analytical and numerical solutions and the measured order $p$ of accuracy $h^p$ for several grid sizes ($p$ is obtained using the asymptotic relation $2^p \approx \|u_{2h} - u\|_\infty / \|u_h - u\|_\infty$). Obviously, the accuracy of the $O(h^2)$-discretization on the $512^2$ grid is the same as that of the $O(h^4)$-discretization on the $64^2$ grid. Solving the $64^2$ problem with fourth-order accuracy is much cheaper than solving the $512^2$ problem with second-order accuracy.                                    $\triangle$

For simple high order operators such as (5.4.1), multigrid can be applied directly. We have already seen in Example 4.7.6 that the $h$-ellipticity measure of this operator is satisfactory. For (5.4.1), the smoothing factor $\mu_{\text{loc}}$ of GS-LEX and the two grid factor $\rho_{\text{loc}}(\nu)$ (for the components GS-LEX, FW and linear interpolation) are $\mu_{\text{loc}} = 0.53$, $\rho_{\text{loc}}(\nu = 1) = 0.43$ and $\rho_{\text{loc}}(\nu = 2) = 0.24$.

This is an example for the direct multigrid treatment of higher order discretizations. In general, the efficient solution of problems discretized by higher order schemes is more difficult than that of lower order schemes since it becomes more difficult to find efficient smoothing schemes.

**Remark 5.4.1 (a trivial hint)** Even if a solver for a higher order discretization is not asymptotically optimal, it may be much more efficient than an asymptotically optimal

**Table 5.10.** Higher order accuracy for Poisson's equation.

| Grid | (Second order) $-\Delta_h$ | | (Fourth order ) $L_h$ (5.4.1) | |
|------|------------------|-----|------------------|-----|
|      | $\|u_h - u\|_\infty$ | $p$ | $\|u_h - u\|_\infty$ | $p$ |
| $8^2$   | $4.6 \times 10^{-5}$ | –   | $1.9 \times 10^{-5}$  | –   |
| $16^2$  | $1.2 \times 10^{-5}$ | 1.9 | $2.0 \times 10^{-6}$  | 3.2 |
| $32^2$  | $3.1 \times 10^{-6}$ | 2.0 | $1.7 \times 10^{-7}$  | 3.6 |
| $64^2$  | $7.7 \times 10^{-7}$ | 2.0 | $1.2 \times 10^{-8}$  | 3.8 |
| $128^2$ | $1.9 \times 10^{-7}$ | 2.0 | $8.5 \times 10^{-10}$ | 3.8 |
| $256^2$ | $4.8 \times 10^{-8}$ | 2.0 | $5.7 \times 10^{-11}$ | 3.9 |
| $512^2$ | $1.2 \times 10^{-8}$ | 2.0 | $3.6 \times 10^{-12}$ | 4.0 |

procedure for a lower order discretization. Let us assume, for example, that a second-order accurate solution for a 2D problem is obtained by an optimal FMG algorithm and that a fourth-order accurate approximation is calculated in $O(N \log N)$ operations. In order to achieve a similar discretization accuracy, FMG based on the low order discretization would require $O(N^2)$ grid points and thus $O(N^2)$ operations, which is obviously much less efficient than the high order approach.                                                    ≫

### 5.4.1 Defect Correction

The *high order defect correction* iteration offers a general possibility to employ low order schemes and obtain high order accuracy. The basic idea of defect correction is simple. Consider the problem

$$L_h u_h = f_h \quad (\Omega_h),$$

where $L_h$ is a high order discretization of $L$. A general defect correction iteration can be written as

$$\hat{L}_h u_h^m = \hat{f}_h \quad \text{with } \hat{f}_h := f_h - L_h u_h^{m-1} + \hat{L}_h u_h^{m-1}, \tag{5.4.3}$$

where $\hat{L}_h$ is a low(er) order discretization of $L$.

If $\rho(I - (\hat{L})_h^{-1} L_h) < 1$, the iterated defect correction procedure converges to the solution of the high order discrete problem. Defect correction can be applied to linear and nonlinear problems. For linear problems, it can even be shown that, under suitable assumptions, one defect correction step gains at least one order of accuracy (e.g. from $O(h^2)$ to $O(h^3)$) if the low order problem is solved sufficiently accurately [176].

There are various possibilities for combining defect correction with multigrid. Some of them are outlined in [7, 185]. The simplest approach to combining the defect correction iteration with multigrid is to solve the $\hat{L}_h$ equation by an efficient multigrid method in each defect correction step such that the defect correction is the "outer iteration" given by (5.4.3) and multigrid is the "inner iteration" used to obtain an approximation of $u_h^m$ in (5.4.3). In this approach it is not necessary to develop smoothing schemes for higher order discretizations.

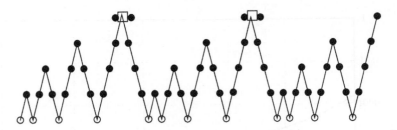

**Figure 5.13.** A defect correction strategy with one F-cycle per defect correction step; ○, solution on coarsest grid; •, smoothing; □, correction of the right-hand side.

The low order equation (5.4.3) does not have to be solved exactly by multigrid. A simple example for multigrid with defect correction as an outer iteration is depicted in Fig. 5.13. We demonstrate the result of this procedure in the Example 5.4.2.

**Example 5.4.2 (Poisson's equation and defect correction)**   We solve the Poisson equation $-\Delta u = f$ with Dirichlet boundary conditions on $\Omega = (0, 1)^2$ with the $O(h^4)$ discretization (5.4.1), (5.4.2) by defect correction. For $\hat{L}_h$ we choose the five-point Laplace operator $-\Delta_h$.

We use only one F(1,1)-cycle of the RBMPS from Section 2.5 per defect correction step. Nested iteration (FMG) is used to obtain an initial approximation on the finest grid. Figure 5.14 shows the convergence of the errors $\|u_h^m - u\|_\infty$ and of the defects $\|f_h - L_h u_h^m\|_\infty$ on a $256^2$-grid with boundary conditions and right-hand side such that $u = e^{xy}$ is the analytical solution.                    △

In this example, we observe that, although the defect for the higher order discretization is still relatively large after three defect correction steps, the discretization accuracy is already reached. *This is a more generally observed behavior.*

Figure 5.15 presents the Fourier symbols of the defect correction iteration matrix $I - (\hat{L}_h)^{-1} L_h$ in this case. The spectral radius $\rho_{\text{loc}}(I - (\hat{L}_h)^{-1} L_h)$ is found to be 0.33. This convergence is also seen for the defects in the above experiment. However, Fig. 5.15 also shows that the convergence of the *low frequency components* of the solution is faster than that of the high frequency components. This is important since, typically, the low frequency parts of a solution determine the accuracy. This is the reason why, in general, the defect-corrected approximation is better than indicated by the reduction of the defect, which is governed by the worse convergence of high frequency error components.

In order to illuminate this statement, we consider the corresponding 1D case. We compare the Fourier symbols of $L = -d^2/dx^2$ and of its discretizations

$$\hat{L}_h = -\frac{1}{h^2}[1 \quad -2 \quad 1], \qquad L_h = -\frac{1}{12h^2}[-1 \quad 16 \quad -30 \quad 16 \quad -1].$$

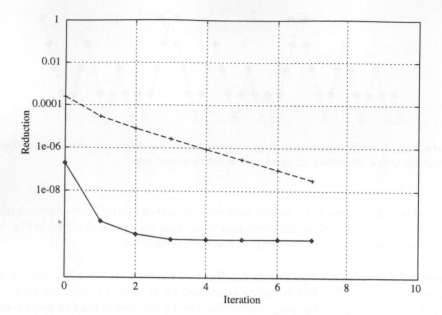

**Figure 5.14.** Behavior of differential error and defects for an example of a defect correction iteration on a $256^2$ grid. $\diamondsuit$, $||u_h^m - u||_\infty$ (lower curve); $+$, $||f_h - L_h u_h^m||_\infty$ (upper curve).

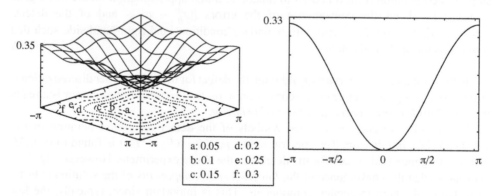

**Figure 5.15.** The Fourier symbol of $I - (\hat{L}_h)^{-1} L_h$ for Poisson's equation. Left, the 2D case with isolines; right, the 1D case.

The corresponding symbols (formal eigenvalues of the operators with respect to $\varphi = e^{i\theta x/h}$) are

$$\tilde{L} = \frac{\theta^2}{h^2}, \qquad \tilde{\tilde{L}}_h = \frac{2 - 2\cos\theta}{h^2}, \qquad \tilde{L}_h = \frac{30 - 32\cos\theta + 2\cos(2\theta)}{12h^2}.$$

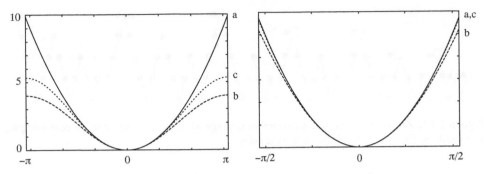

**Figure 5.16.** Fourier symbols $\tilde{L}$, $\tilde{L}_h$, $\tilde{\tilde{L}}_h$ for Poisson's equation. (a) $\tilde{L}$; (b) $\tilde{\tilde{L}}_h$; (c) $\tilde{L}_h$ (all multiplied by $h^2$). Left, the region $[-\pi, \pi]$; right, zooming the region $[-\pi/2, \pi/2]$.

They are presented in Fig. 5.16. Indeed, the high frequency components of the discrete operators do not approximate the symbols of $L$ well, whereas the low frequency part of $L$ is actually better approximated by the fourth-order operator. This, together with the fast convergence of the low frequencies in defect correction, as found in Fig. 5.15, is an indication for the fast convergence of the solution to higher order accuracy with defect correction.

Often, however, there is no other convergence measure available than the defect reduction.

**Remark 5.4.2** When replacing $\theta$ by $k\pi h$, i.e. considering a discrete spectrum for Dirichlet boundary conditions as in the rigorous Fourier analysis, one can apply a Taylor expansion to the cosine terms with respect to $h$. In this case, one finds that the two discretizations approximate the eigenvalues $\pi^2 k^2$ of the continuous problem by second- and fourth-order, respectively. Again, we see that small values of $k$ (low frequencies) are much better approximated than high frequencies. $\gg$

**Remark 5.4.3** As we have seen above, it is also possible to use multigrid directly for the operator (5.4.1). From a programmer's point of view, however, even in this case it may be convenient to work with the $O(h^2)$ discretization of the Poisson equation and obtain $O(h^4)$ accuracy "easily" by only changing the right-hand side on the finest grid, i.e. via defect correction. $\gg$

**Remark 5.4.4** The computational work when starting defect correction on the finest grid is at least $O(N \log N)$. The amount of work is not only governed by the convergence speed of defect correction, but also by the number of required multigrid cycles. If the inner iteration consists only of one multigrid cycle, the convergence of the defect correction is also limited by the convergence factor of one multigrid cycle. $O(N)$-methods can be developed by integrating the defect correction process into FMG (see, for example, Fig. 5.17) [346]. $\gg$

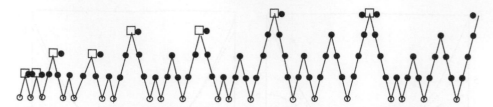

**Figure 5.17.** An F-cycle FMG defect correction strategy with $r = 2$ (∘: solution on coarsest grid, •: smoothing, □: correction of the right-hand side).

### 5.4.2 The Mehrstellen Discretization for Poisson's Equation

Fourth-order accuracy can also be achieved with a 2D Mehrstellen discretization which is based on a compact nine-point stencil,

$$\begin{bmatrix} s_{-1,1} & s_{0,1} & s_{1,1} \\ s_{-1,0} & s_{0,0} & s_{1,0} \\ s_{-1,-1} & s_{0,-1} & s_{1,-1} \end{bmatrix}_h . \tag{5.4.4}$$

We consider again the 2D Poisson equation on the unit square. On a square Cartesian grid $\Omega_h$, a 2D Mehrstellen discretization [111] is

$$-\Delta_h^M u_h = R_h^M f_h, \tag{5.4.5}$$

where $\Delta_h^M$ is the compact nine-point stencil

$$-\Delta_h^M = \frac{1}{6h^2} \begin{bmatrix} -1 & -4 & -1 \\ -4 & 20 & -4 \\ -1 & -4 & -1 \end{bmatrix}_h \quad \text{and} \quad R_h^M = \frac{1}{12} \begin{bmatrix} & 1 & \\ 1 & 8 & 1 \\ & 1 & \end{bmatrix}_h .$$

It can be easily verified by Taylor's expansion of $u$ and $f$ that this discretization is $O(h^4)$ accurate. The $h$-ellipticity measure of $-\Delta_h^M$ is $E_h = 0.375$.

The transfer operators such as injection, FW and linear interpolation need not be changed for this compact nine-point stencil. Furthermore, the smoothing schemes $\omega$-JAC, GS-LEX and GS-RB can be applied immediately and have similar features to the five-point stencils.

**Remark 5.4.5 (parallel smoothers for compact nine-point stencils)** For GS-RB, however, the situation is somewhat different with respect to parallelization. Each partial step of GS-RB when applied to five-point stencils can be carried out fully in parallel since there are no data dependencies. However, GS-RB is not directly parallelizable for nine-point stencils since it contains data dependencies with respect to the *diagonal* stencil elements. For nine-point stencils, there are basically two different parallel generalizations of GS-RB.

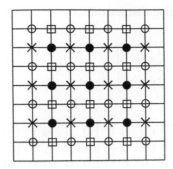

**Figure 5.18.** Four-color distribution of grid points in $\Omega_h$.

**Table 5.11.** Results for the nine-point Mehrstellen operator ($\nu_1 = \nu_2 = 1$).

|  | $\omega$-JAC ($\omega = 10/11$) | GS-RB | GS-FC | GS-LEX | JAC-RB |
|---|---|---|---|---|---|
| $(\mu_{\text{loc}})^2$ | 0.21 | 0.060 | 0.048 | 0.22 | 0.040 |
| $\rho_{\text{loc}}$ | 0.21 | 0.044 | 0.039 | 0.13 | 0.066 |

The first extension of the GS-RB idea is the multicolor Gauss–Seidel relaxation. Multicoloring allows the parallel execution of Gauss–Seidel relaxation if larger stencils are used for discretization. A standard example is *four-color pointwise Gauss–Seidel relaxation* (GS-FC) in 2D, illustrated in Fig. 5.18. Here, a full relaxation step consists of four "quarter steps", corresponding to the four colors of the grid points ($\bullet$, $\circ$, $\square$, $\times$). All grid points of one type can be treated simultaneously and independently if the corresponding difference stencil is a compact nine-point stencil.

A second generalization of GS-RB is the two-color compromise JAC-RB, i.e. performing a Jacobi sweep over the red points, followed by a Jacobi sweep over the black points using the updated values at the red points. The degree of parallelism is twice as large as that of GS-FC, and the convergence is often very satisfactory. This smoother often turns out to be better than either $\omega$-JAC or GS-LEX for smoothing compact nine-point operators.                                                                                  $\gg$

**Example 5.4.3**  Table 5.11 presents LFA smoothing and two-grid factors for various smoothers, i.e., for $\omega$-JAC, GS-RB, GS-FC, GS-LEX and JAC-RB. The transfer operators are FW and bilinear interpolation. The optimal parameter for the smoothing properties of $\omega$-JAC is found to be $\omega = 10/11$ [58].

From Table 5.11, we see that GS-RB, GS-FC and JAC-RB have better smoothing and two-grid convergence factors than GS-LEX and $\omega$-JAC. We obtain the typical multigrid efficiency in a straightforward way.                                                                $\triangle$

**Remark 5.4.6 (3D Mehrstellen discretization)** A corresponding $O(h^4)$ discretization also exists in 3D. $\Delta_h^M$ is then

$$-\Delta_h^M = \frac{1}{6h^2} \left[ \begin{bmatrix} 0 & -1 & 0 \\ -1 & -2 & -1 \\ 0 & -1 & 0 \end{bmatrix}_h \begin{bmatrix} -1 & -2 & -1 \\ -2 & 24 & -2 \\ -1 & -2 & -1 \end{bmatrix}_h \begin{bmatrix} 0 & -1 & 0 \\ -1 & -2 & -1 \\ 0 & -1 & 0 \end{bmatrix}_h \right]$$

and $R_h^M$ is

$$R_h^M = \frac{1}{12} \left[ \begin{bmatrix} 0 & 0 & 0 \\ 0 & 1 & 0 \\ 0 & 0 & 0 \end{bmatrix}_h \begin{bmatrix} 0 & 1 & 0 \\ 1 & 6 & 1 \\ 0 & 1 & 0 \end{bmatrix}_h \begin{bmatrix} 0 & 0 & 0 \\ 0 & 1 & 0 \\ 0 & 0 & 0 \end{bmatrix}_h \right].$$

The typical multigrid efficiency is also easily obtained in the 3D case. In fact, for this nineteen-point stencil it is sufficient to use four colors for a parallel GS variant.                    $\gg$

## 5.5 DOMAINS WITH GEOMETRIC SINGULARITIES

In this subsection we consider a class of domains with geometric singularities, as sketched in Fig. 5.19.

All of these domains are nonconvex and have a reentrant corner or a cut, leaving an (interior) angle $\alpha\pi$, $1 < \alpha \le 2$. We consider Poisson's equation with Dirichlet boundary conditions. For the discretization we use a Cartesian grid in these domains. (For a general angle $\alpha$, one can employ the Shortley–Weller discretization (2.8.13) near the boundary.)

In general, for smooth right-hand sides $f^\Omega$ and smooth boundary conditions the solution $u$ can be represented as

$$u = \tilde{u} + \text{const } u_s,$$

where $\tilde{u}$ is smooth and $u_s$ reads

$$u_s(r, \phi) = \sqrt[\alpha]{r} \sin\left(\frac{\phi}{\alpha}\right). \tag{5.5.1}$$

Here $(r, \phi)$ are polar coordinates with respect to the singular point $(0, 0)$. The function $u_s$ has the characteristic singular behavior. Obviously, the larger $\alpha$, the stronger is the

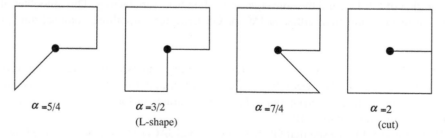

$\alpha = 5/4$       $\alpha = 3/2$       $\alpha = 7/4$       $\alpha = 2$

(L-shape)       (cut)

**Figure 5.19.** Some domains with geometric singularities with an angle $\alpha\pi$ at the reentrant corner $(0, 0)$ marked by ●.

singularity. The four problems sketched in Fig. 5.19 are mathematically well understood, in particular with regard to the influence of the singular behavior on the discretization accuracy. More precisely, the following estimate is valid under natural assumptions [228]. For any $1 \leq \alpha \leq 2$ and any $\varepsilon > 0$, there exists a constant $C$ such that

$$|u(r, \phi) - u_h(r, \phi)| \leq Ch^{2/\alpha - 2\varepsilon} r^{-1/\alpha + \varepsilon} = \begin{cases} O\left(h^{2/\alpha - 2\varepsilon}\right) & \text{if } r \text{ fixed} \\ O\left(h^{1/\alpha - \varepsilon}\right) & \text{if } r = O(h). \end{cases} \quad (5.5.2)$$

Therefore, these examples are good candidates for *qualitative and quantitative* case studies, for example, on how standard multigrid methods are influenced by such singularities, which complications occur and how these complications can be mastered. Here, we will discuss the following questions.

(1) How is the convergence speed affected by the singularities if we use uniform global grids for discretization and standard multigrid iterations (V-cycles, W-cycles)?
(2) How can a deterioration of the multigrid convergence speed (that we will observe) be overcome?
(3) What are the effects in the context of FMG (on uniform global grids)?

Concerning Question (1), we recall that on a rectangular domain with $h_x = h_y = h$ the corresponding two-grid convergence factor (with HW and $\nu = 3$) is given by $\rho^* \approx 0.033$ (see Section 3.3.1). In Section 2.8.3, we have seen that the measured W-cycle convergence factors are very close to this value for more general domains if there are no reentrant corners. For two domains with reentrant corners ($\alpha = 3/2$ and $\alpha = 2$, see Fig. 5.19), the observed multigrid convergence is presented in Table 5.12.

The worst convergence factors can be observed for the domain with a cut. Nevertheless, the F- and the W-cycle convergence are satisfactory even for this problem. The convergence of the V-cycle is worse (and $h$-dependent).

We point out that the given convergence factors remain essentially unchanged, if the cycle index $\gamma$ is further increased.

Concerning Question (2), it is generally observed that standard error smoothing is somewhat less efficient near the singularity [14, 320]. Figure 5.20 shows that the error between the current approximation after smoothing and the exact discrete solution for the L-shaped

**Table 5.12.** Measured convergence factors for problems with reentrant corners, $h = 1/128$, varying restriction operator and number of smoothing steps.

| | L-shape ($\alpha = 3/2$) | | Cut ($\alpha = 2$) | |
| | FW | HW | FW | HW |
| Cycle | ($\nu_1 = \nu_2 = 1$) | ($\nu_1 = 2, \nu_2 = 1$) | ($\nu_1 = \nu_2 = 1$) | ($\nu_1 = 2, \nu_2 = 1$) |
|---|---|---|---|---|
| V | 0.26 | 0.14 | 0.37 | 0.26 |
| F | 0.11 | 0.051 | 0.16 | 0.10 |
| W | 0.11 | 0.050 | 0.15 | 0.10 |

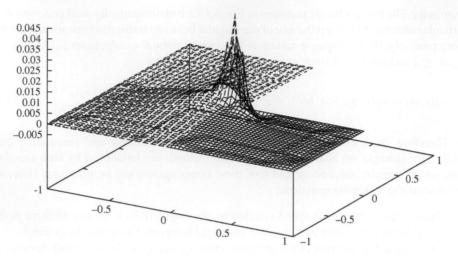

**Figure 5.20.** Typical error after smoothing for Poisson's equation in an L-shaped domain.

**Table 5.13.** $||u - u_h^{FMG}||_\infty$ for the *L-shaped problem*; grid size, cycle type and number of relaxations are varied (the exponent $E$ in $10^E$ is in brackets).

| $h^{-1}$ | $||u - u_h||_\infty$ | V(1,1) | V(2,1) | F(0,1) | F(1,1) | F(2,1) |
|---|---|---|---|---|---|---|
| 32 | 0.83 (−2) | 0.98 (−2) | 0.92 (−2) | 0.11 (−1) | 0.92 (−2) | 0.89 (−2) |
| 64 | 0.52 (−2) | 0.63 (−2) | 0.58 (−2) | 0.72 (−2) | 0.58 (−2) | 0.56 (−2) |
| 128 | 0.33 (−2) | 0.40 (−2) | 0.37 (−2) | 0.45 (−2) | 0.37 (−2) | 0.35 (−2) |
| 256 | 0.21 (−2) | 0.25 (−2) | 0.23 (−2) | 0.45 (−2) | 0.23 (−2) | 0.22 (−2) |

problem is not smooth in the neighborhood of (0,0). The addition of *local smoothing sweeps* in a neighborhood of the singularity is one possibility to improve the smoother in a cheap way and to overcome this convergence degradation. It has been shown [14] that the region where these local smoothing sweeps are needed grows slightly for $h \to 0$. Also the number of local smoothing sweeps should be increased in order to achieve the convergence factor 0.033. The additional work, however, is still negligible.

The feasibility of this approach has been investigated in detail for Poisson's equation on domains with reentrant corners in [320].

Regarding Question (3), for the *full multigrid* version, we use the cycle structure $r = 1$ (see Section 2.6.1), V- or F-cycles and cubic FMG interpolation. Tables 5.13–5.15 present the FMG results for the two domains with reentrant corners. For some values of $h$, both the error $||u_h^{FMG} - u_h||_\infty$ and the discretization error $||u_h - u||_\infty$ are shown in Table 5.13. The maximum norm measures, in particular, the behavior of the error near the singularity. The corresponding errors in the $L_2$ norm are presented for the L-shaped problem in Table 5.14

**Table 5.14.** $||u - u_h^{FMG}||_2$ for the *L-shaped problem*; grid size, cycle type and number of relaxations are varied.

| $h^{-1}$ | $||u - u_h||_2$ | V(1,1) | V(2,1) | F(0,1) | F(1,1) | F(2,1) |
|------|------------|---------|---------|---------|---------|---------|
| 32  | 0.11 (−2) | 0.18 (−2) | 0.15 (−2) | 0.20 (−2) | 0.13 (−2) | 0.12 (−2) |
| 64  | 0.45 (−3) | 0.83 (−3) | 0.65 (−3) | 0.87 (−3) | 0.53 (−3) | 0.49 (−3) |
| 128 | 0.18 (−3) | 0.38 (−3) | 0.28 (−3) | 0.37 (−3) | 0.21 (−3) | 0.20 (−3) |
| 256 | 0.72 (−4) | 0.17 (−3) | 0.12 (−3) | 0.16 (−3) | 0.85 (−4) | 0.79 (−4) |

**Table 5.15.** $||u - u_h^{FMG}||_2$ for the problem with a *cut*; grid size, cycle type and number of relaxations are varied.

| $h^{-1}$ | $||u - u_h||_2$ | V(1,1) | V(2,1) | F(0,1) | F(1,1) | F(2,1) |
|------|------------|---------|---------|---------|---------|---------|
| 32  | 0.58 (−2) | 0.10 (−1) | 0.82 (−2) | 0.10 (−1) | 0.71 (−2) | 0.66 (−2) |
| 64  | 0.29 (−2) | 0.59 (−2) | 0.46 (−2) | 0.57 (−2) | 0.36 (−2) | 0.33 (−2) |
| 128 | 0.15 (−2) | 0.35 (−2) | 0.26 (−2) | 0.31 (−2) | 0.19 (−2) | 0.17 (−2) |
| 256 | 0.73 (−3) | 0.21 (−2) | 0.15 (−2) | 0.17 (−2) | 0.95 (−3) | 0.86 (−3) |

and for the problem with the cut in Table 5.15. In all examples, the continuous solution is $u = u_s$ as in (5.5.1).

As we have seen in Table 5.12, the multigrid convergence factors are somewhat worse for the domains with reentrant corners for increasing $\alpha$ $(1 < \alpha \leq 2)$. One then asks, whether or not the corresponding convergence speed is still sufficient for satisfactory performance of FMG (see Theorem 3.2.2).

On the other hand, the discretization error is also worse for increasing $\alpha$ (see (5.5.2)). The value of $\kappa$ in (3.2.10) is smaller than 2 in these cases.

Tables 5.14 and 5.15 show indeed that the main objective of FMG (see Section 2.6), namely to obtain approximate solutions $u_h^{FMG}$ with

$$||u_h^{FMG} - u||_2 \leq (1 + \beta)||u - u_h||_2$$

can be achieved for the examples considered (for example, when using the F(1,1)-cycle): The loss of multigrid convergence speed is, so to speak, compensated for by a loss of discretization accuracy.

We will return to this class of problems in Chapter 9. There, it will be shown that *local grid refinement near the singularity* can help to improve the overall accuracy of the solution.

## 5.6 BOUNDARY CONDITIONS AND SINGULAR SYSTEMS

For ease of presentation of the basic ideas of multigrid, we have assumed eliminated Dirichlet boundary conditions. In practice, however, it is often necessary and/or convenient to use the separated, i.e. *noneliminated form* of boundary conditions.

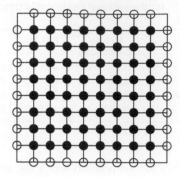

**Figure 5.21.** The grid $\Omega_h$ with interior and boundary points.

### 5.6.1 General Treatment of Boundary Conditions in Multigrid

For Dirichlet boundary conditions, noneliminated boundary conditions have the advantage that one can apply the five-point difference stencil at all interior points and that one need not use a different eliminated stencil near boundary points in a computer program. Of course, in the case of Dirichlet boundary conditions, the discrete boundary unknowns are usually initialized with the correct values (in the first relaxation) so that all boundary conditions are fulfilled from the very beginning. Figure 5.21 indicates at which points of a grid the interior equations (the grid points characterized by •) and at which points the discrete Dirichlet boundary equations (those grid points marked by ∘) are located.

We here consider the 2D problem

$$L^{\Omega} u = f^{\Omega} \quad (\Omega)$$
$$L^{\Gamma} u = f^{\Gamma} \quad (\Gamma = \partial\Omega)$$

with a discretization

$$L_h^{\Omega} u_h = f_h^{\Omega} \quad (\Omega_h) \qquad\qquad (5.6.1)$$
$$L_h^{\Gamma} u_h = f_h^{\Gamma} \quad (\Gamma_h). \qquad\qquad (5.6.2)$$

In the multigrid *correction scheme*, the general idea of how to treat the boundary conditions can be summarized in the following way. (The generalization to the FAS is straightforward.)

---

• The smoothing procedure consists of a relaxation of the interior difference equations (5.6.1) and of *a relaxation of the discrete boundary conditions* (5.6.2). In the simplest case, this means that we have a loop over the boundary points and relax the discrete boundary conditions (see, however, Remark 5.6.1).

- The transfer of the defects from the fine grid to the coarse grid is performed *separately* for the boundary conditions (5.6.2) and for the interior equations (5.6.1). This separation is important since the discrete boundary conditions and the discrete inner equations have different operators, typically with different powers of $h$.
- The defects $d_h^\Gamma = f_h^\Gamma - L_h^\Gamma \bar{u}_h^m$ of the discrete boundary conditions are transferred to the coarse grid via injection or via 1D restriction operators: $d_H^\Gamma = I_h^H d_h^\Gamma$ (see the next section for an example).
- *The complete discrete problem (5.6.1)–(5.6.2) is thus represented on the coarse grid* by

$$L_H^\Omega v_H = d_H^\Omega \tag{5.6.3}$$

$$L_H^\Gamma v_H = d_H^\Gamma, \tag{5.6.4}$$

where $v_H$ denotes the coarse grid correction and $d_H^\Omega$ and $d_H^\Gamma$ denote the defects at interior and boundary points, respectively.
- The coarse grid correction $v_H$ is interpolated to the fine grid including its boundary, $\Omega_h \cup \Gamma_h$.

If all discrete boundary conditions are satisfied before the coarse grid correction, the defects of the boundary conditions are zero and, consequently, we have homogeneous boundary conditions on the coarse grid.

**Remark 5.6.1** Often, a more specific and more involved boundary relaxation is required than that described above. One reason is that a boundary relaxation may spoil the smoothness of the errors near the boundary. In order to overcome this problem, two practical approaches are to add local relaxations near the boundary (as in Section 5.5) or to relax the boundary conditions collectively with equations at adjacent interior points. We will discuss these approaches in the context of systems of PDEs in Chapter 8.                              ≫

In the following sections, we will give some concrete examples for various types of boundary conditions and discuss a proper multigrid treatment in each case.

For Poisson's equation with pure periodic or pure Neumann boundary conditions, the resulting boundary value problem is *singular*, i.e. it depends on the right-hand side whether or not a solution exists. If a solution exists, then it is not unique. Any constant function is a solution of the homogeneous system. In order to separate the two topics (treatment of boundary conditions and treatment of singular systems), we start with the discussion of a nonsingular system with Neumann and Dirichlet boundary conditions at different parts of the boundary in Section 5.6.2. In Section 5.6.3, we will then discuss multigrid for periodic boundary conditions which is a first example of a singular system. In Section 5.6.4, we will consider a general singular case.

### 5.6.2 Neumann Boundary Conditions

We consider Poisson's equation in the unit square, with a Neumann boundary condition at one of the sides, for instance, at the left boundary (see Fig. 5.22) and Dirichlet boundary

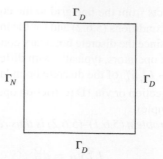

**Figure 5.22.** Domain with Neumann and Dirichlet boundary conditions.

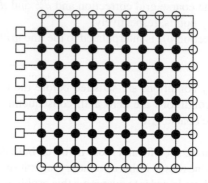

**Figure 5.23.** The grid with auxiliary points (□).

conditions at the others:

$$L^\Omega u = -\Delta u = f^\Omega \qquad \Omega = (0,1)^2$$
$$L^{\Gamma_N} u = u_n = f^{\Gamma_N} \qquad \Gamma_N = \{(x,y): x = 0, 0 < y < 1\} \qquad (5.6.5)$$
$$L^{\Gamma_D} u = u = f^{\Gamma_D} \qquad \Gamma_D = \partial\Omega \setminus \Gamma_N$$

where $u_n$ denotes the normal derivative of $u$ in the direction of the outward normal.

For the discretization of the Neumann boundary conditions, we use the central second-order approximation

$$L_h^{\Gamma_N} u_h(x,y) = u_{n,h}(x,y) = \frac{1}{2h}(u_h(x-h,y) - u_h(x+h,y)).$$

Here, we make use of an extended grid with auxiliary points □ outside $\bar{\Omega}$ (see Fig. 5.23). Such auxiliary points are also called *ghost points*.

With this discretization, we introduce unknowns at the ghost points. In order to close the discrete system, we assume that $f^\Omega$ is extended to $\Gamma_N$ and discretize the Laplace operator $\Delta$ by the five-point approximation $\Delta_h$ not only in the interior of $\Omega$ but also on the Neumann boundary $\Gamma_N$. Figure 5.23 shows all the points at which the standard five-point discretization $\Delta_h$ is applied (i.e., all those marked by •). The discrete system then is

$$-\Delta_h u_h = f_h^{\Omega} \quad (\Omega_h) \tag{5.6.6}$$

$$L_h^{\Gamma_N} u_h = f_h^{\Gamma_N} \quad (\Gamma_{N,h}) \tag{5.6.7}$$

$$-\Delta_h u_h = f_h^{\Omega} \quad (\Gamma_{N,h}) \tag{5.6.8}$$

$$u_h = f_h^{\Gamma_D} \quad (\Gamma_{D,h}). \tag{5.6.9}$$

The discrete equations for the unknowns at $\Gamma_N$ and at the corresponding ghost points are the discrete Poisson equation at boundary points (5.6.8) and the discrete Neumann boundary condition (5.6.7). Consequently, the discrete Poisson equation is used to provide a new approximation at the boundary whereas the discrete Neumann boundary condition is used to update the unknowns at the ghost points. In principle, the relaxation of these two equations can be done *one after the other* or *simultaneously*, the latter meaning that a $2 \times 2$-system is solved per pair of grid points.

We restrict the defects of the discrete Neumann boundary condition and of the discrete Poisson equation at Neumann boundary points separately. For the restriction of the discrete Neumann boundary condition (5.6.7) to the coarse grid, the 1D FW operator (see Remark 2.3.2) is an appropriate choice. For the restriction of the discrete Poisson equation at Neumann boundary points, Condition (2.3.5) in Remark 2.3.1 leads to the so-called *modified FW operator*, which, for example at vertical boundaries is

$$\frac{1}{16} \begin{bmatrix} 2 & 2 \\ 4 & 4 \\ 2 & 2 \end{bmatrix}_h^{2h} . \tag{5.6.10}$$

Figure 5.24 illustrates this modified FW restriction operator.

**Example 5.6.1** If we apply the 1D FW for the discrete boundary conditions and the modified FW operator for the discrete Poisson equation at the Neumann boundary points, we obtain measured multigrid convergence factors of 0.14 and 0.09 for the V(1,1)- and W(1,1)-cycle, respectively (using GS-RB, linear interpolation and FW at interior points). $\triangle$

**Remark 5.6.2** If Neumann boundary conditions are present in two adjacent edges of the boundary (corner points), the condition in Remark 2.3.1 gives the restriction operator

$$\frac{1}{16} \begin{bmatrix} 4 & 4 \\ 4 & 4 \end{bmatrix}_h^{2h} . \tag{5.6.11}$$

$\gg$

**Remark 5.6.3 (elimination of Neumann boundary conditions)** In principle, it is also possible to eliminate the unknowns at ghost points. In the case of Poisson's equation, we

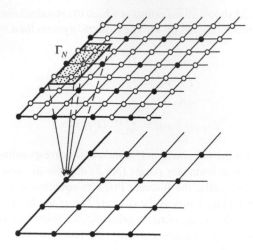

**Figure 5.24.** Modified FW operator (ghost points are not shown).

obtain, e.g. for a point $(x, y)$ on the left boundary $\Gamma_N$

$$f_h(x, y) = f_h^\Omega(x, y) + \frac{2}{h} f_h^\Gamma(x, y) \qquad (5.6.12)$$

$$L_h u_h(x, y) = \frac{1}{h^2} \begin{bmatrix} & -1 & \\ 0 & 4 & -2 \\ & -1 & \end{bmatrix}_h u_h(x, y). \qquad (5.6.13)$$

The relaxation of the eliminated boundary condition is then equivalent to the *collective* relaxation of both equations located on the boundary in the noneliminated case described above.

At boundary points, the use of the modified FW operator (5.6.10) for the eliminated Neumann boundary conditions is equivalent to the noneliminated approach in Example 5.6.1 (use of 1D FW for the discrete Neumann boundary condition in combination with modified FW for the discrete Poisson equation at boundary points).

**Warning for beginners**   We point out that, especially for eliminated Neumann boundary conditions, the use of injection (or HW modified similarly to FW) at boundary points leads to an incorrect scaling in the coarse grid right-hand side at boundary points. This incorrect scaling leads to a substantial deterioration of multigrid convergence. With injection, for example, the restriction of the right-hand side of (5.6.12) gives

$$f_H^\Omega + \frac{2}{h} f_H^\Gamma \qquad (5.6.14)$$

instead of

$$f_H^\Omega + \frac{2}{H} f_H^\Gamma \qquad (5.6.15)$$

which would be the coarse grid analog to the right-hand side in (5.6.12) (and which is also obtained if the original noneliminated boundary conditions are transferred to the coarse grid using injection and are eliminated afterwards). The modified FW operators (5.6.10) and (5.6.11), however, weight the crucial term $2f_h^\Gamma/h$ automatically by a factor of 1/2 in the restriction since this term appears only in the right-hand sides of the discrete equations *at boundary points.*                                                                 ≫

### 5.6.3 Periodic Boundary Conditions and Global Constraints

As indicated above, we obtain *a singular system of equations* if we have periodic or pure Neumann boundary conditions for the Poisson equation at all boundaries. In the following, we will discuss how such singular systems can be treated efficiently by multigrid.

We consider the problem

$$-\Delta u = f \quad (\Omega = (0, 1)^2)$$

with periodic boundary conditions

$$\begin{aligned}
u(0, y) &= u(1, y) \\
u(x, 0) &= u(x, 1) \\
u_x(0, y) &= u_x(1, y) \\
u_y(x, 0) &= u_y(x, 1).
\end{aligned} \tag{5.6.16}$$

An alternative formulation of the periodicity condition is

$$u(x, y) = u(x, y + 1) = u(x + 1, y) \tag{5.6.17}$$

at and near boundary points.

Solutions of the continuous boundary value problem exist if (and only if) the *compatibility condition*

$$\int_{[0,1]^2} f(\tilde{x}, \tilde{y}) \, d\Omega = 0 \tag{5.6.18}$$

is satisfied. If a solution exists, it is determined only up to a constant since the constant functions are solutions of the homogeneous problem. Of course, the singularity has to be taken into account in the discretization *and in the multigrid solution of the problem.* We consider the standard $O(h^2)$ discretization of the Laplacian

$$-\Delta_h u_h(x_i, y_j) = f_h(x_i, y_j) \quad ((x_i, y_j), i, j = 1, 2, \ldots, n),$$

where $x_i = ih, y_j = jh$ and $h = 1/n$, with the discrete boundary conditions

$$\begin{aligned}
u_h(x_0, y_j) &= u_h(x_n, y_j) & j &= 1, 2, \ldots, n \\
u_h(x_i, y_0) &= u_h(x_i, y_n) & i &= 1, 2, \ldots, n \\
u_h(x_1, y_j) &= u_h(x_{n+1}, y_j) & j &= 1, 2, \ldots, n \\
u_h(x_i, y_1) &= u_h(x_i, y_{n+1}) & i &= 1, 2, \ldots, n
\end{aligned} \tag{5.6.19}$$

(see Fig. 5.25). In the discrete problem, the ∘ points at the approximations at the left (lower) boundary can be identified with those at the corresponding points marked by • at the right

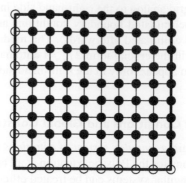

**Figure 5.25.** The grid $\Omega_h$ with interior and boundary points in the case of periodic boundary conditions.

(upper) boundary in Fig. 5.25. Note that we have neither a boundary nor explicit boundary conditions on this logical torus structure. We only have interior equations in all points marked by $\bullet$.

This *discrete* system has a solution (which is unique up to a constant) if and only if the *discrete compatibility condition*

$$\sum_{k,l=1}^{n} f_h(x_k, y_l) = 0 \qquad (5.6.20)$$

is satisfied. Obviously, the discrete and the continuous compatibility conditions are not equivalent; the discrete one can be interpreted as an approximation to the continuous one. An easy modification, which guarantees that discrete solutions exist, is to replace $f_h$ by

$$\tilde{f}_h = f_h - h^2 \sum_{k,l=1}^{n} f_h(x_k, y_l).$$

Since $\tilde{f}_h = f_h + O(h^2)$, the consistency order of the discretization, $O(h^2)$, is maintained by this replacement.

For the unique determination of the discrete solution, we have to fix the constant. This is often done by a *global constraint*, for example, by setting the average of the discrete solution to 0,

$$\sum_{k,l=1}^{n} u_h(x_k, y_l) = 0. \qquad (5.6.21)$$

In applying multigrid to this problem with the global constraint (5.6.21), it is, in general, not necessary to *relax* or fulfill global constraints on all grid levels (this may be expensive and may influence the smoothness of errors). It is sufficient to compute its defect and to transfer it to the next coarser grid. Only on the coarsest grid (or on some of the coarse grids), is the global constraint to be fulfilled.

**Remark 5.6.4** If we use the FAS and handle the global constraint in the case of periodic boundary conditions only on the coarsest grid by (5.6.21), without using any information on the defects of the global constraint on the fine grid, we already obtain a solution in the typical multigrid efficiency. The numerically observed convergence factor for a W(2,1)-cycle consisting of GS-RB, FW and linear interpolation on a $256^2$ grid is 0.025 (and 0.040 for the corresponding V-cycle). FMG again provides an $O(h^2)$-accurate approximative solution.

This procedure fixes the constant and gives fast convergence, the condition (5.6.21) is, however, not fulfilled on the finest grid. At the end of the multigrid algorithm, the average of the solution can easily be set to zero.

(Of course, even if we do not determine the constant at all during the multigrid algorithm, we will, in general, observe fast convergence of *the defects*.)                                      $\gg$

## 5.6.4 General Treatment of Singular Systems

Poisson's equation with periodic boundary conditions is a first example of a singular system. A similar problem occurs, for example, if the periodic boundary conditions are replaced by Neumann boundary conditions. The compatibility condition for this differential problem is

$$\int_\Omega f^\Omega(\tilde{x}, \tilde{y})\, d\Omega = \int_{\partial\Omega} f^\Gamma(\tilde{x}, \tilde{y})\, ds. \tag{5.6.22}$$

We will now discuss a multigrid treatment of such problems, in which the compatibility conditions do not have to be known. This treatment can be applied to more general situations, e.g. to problems with variable coefficients, nonsymmetric problems and to systems of PDEs.

For convenience, we switch to matrix notation. A reasonable multigrid treatment is based on the following result.

**Lemma 5.6.1** *Consider the linear system*

$$Au = f \tag{5.6.23}$$

*with a singular $N \times N$ matrix A, the range of which is $N - 1$ dimensional. We assume that $\lambda = 0$ is a simple eigenvalue of A with eigenvector $\varphi$ and that $\varphi^*$ is an eigenvector with eigenvalue 0 of the adjoint matrix $A^*$. Finally, assume two vectors v, w with $\langle v, \varphi \rangle \neq 0$ and $\langle w, \varphi^* \rangle \neq 0$ (where $\langle \cdot, \cdot \rangle$ denotes the usual Euclidean inner product).*

*Then, the augmented system*

$$\hat{A}\hat{u} := \begin{pmatrix} A & w \\ v^T & 0 \end{pmatrix} \begin{pmatrix} u \\ \xi \end{pmatrix} = \begin{pmatrix} f \\ 0 \end{pmatrix} =: \hat{f} \tag{5.6.24}$$

*has a unique solution, i.e. $\hat{A}$ is a regular matrix.*

*Proof.* We have to show that $\hat{A}$ is a regular matrix, i.e. that $\hat{A}\hat{u} = 0$ implies $\hat{u} = 0$. $\hat{A}\hat{u} = 0$ means that

$$Au + \xi w = 0 \quad \text{and} \quad \langle u, v \rangle = 0. \tag{5.6.25}$$

From the first equality, it follows that

$$\langle Au, \varphi^* \rangle + \xi \langle w, \varphi^* \rangle = \langle u, A^* \varphi^* \rangle + \xi \langle w, \varphi^* \rangle = 0.$$

Since $A^* \varphi^* = 0$ and $\langle w, \varphi^* \rangle \neq 0$, it follows that $\xi = 0$. Hence, (5.6.25) implies $Au = 0$, which means that $u = \alpha \varphi$. The additional equation in the augmented system, $\langle u, v \rangle = 0$, gives $\langle u, v \rangle = \alpha \langle \varphi, v \rangle = 0$. Since the latter inner product is not 0, we have $\alpha = 0$ which means that $u = 0$.                                                                    □

The following important conclusions can be drawn from Lemma 5.6.1.

(1) If $\langle f, \varphi^* \rangle = 0$, we find $\xi = 0$ and $u$ is a solution of $Au = f$ due to the uniqueness of the solution of (5.6.24).

(2) For a solution $\begin{pmatrix} u \\ \xi \end{pmatrix}$ of (5.6.24), we have $\langle f - \xi w, \varphi^* \rangle = 0$. This means that $\xi w$ is a correction of the right-hand side of $Au = f$, which makes this system *compatible*. As a consequence, by solving (5.6.24), a correction of the right-hand side is determined *automatically*. In particular, we need not formulate or handle the compatibility condition explicitly.

**Remark 5.6.5** In practical applications, one often deals with problems whose matrix $A$ has only zero row sums (like in the case of the discretization of $\Delta$ with pure periodic or pure Neumann boundary conditions). In such cases, $\varphi^T = (1, 1, \ldots, 1)$ is an eigenvector of $A$ and we can choose $v^T = (1, 1, 1, \ldots, 1)$. Usually, one can also choose $w^T = (1, 1, 1, \ldots, 1)$. Although $\varphi^*$ is not known in general, it is improbable that $\langle w, \varphi^* \rangle = 0$.                    ≫

Based on these considerations, we outline a proper *multigrid treatment of the augmented system*. We assume that $v^T = w^T = (1, 1, 1, \ldots, 1)$. The augmentation of the system then consists of two parts. The first is the addition of the variable $\xi$ to each discrete equation. The variable $\xi$ can be interpreted as the average defect of the discrete (nonaugmented) system. An update of $\xi$ ("normalization" of the defect) can thus be easily calculated. Again, it is typically sufficient to update $\xi$ only on the very coarse grids. This means that the normalization of the defect is transferred to the coarse grids and determined on the coarsest grid (or the very coarse grids). On coarse grids, the system is thus augmented in the same way as on the finest grid.

The second part of the augmentation is the normalization of $u$, i.e. the condition $\langle v, u \rangle = 0$. The normalization can be dealt with similarly as described for periodic boundary conditions in Section 5.6.3.

**Example 5.6.2** We now consider the multigrid convergence obtained by a RBMPS adapted to Poisson's equation with *pure Neumann boundary conditions*. All multigrid components in the interior are the same as those defined for the RBMPS. At the boundary, we apply a collective relaxation at each boundary point. The transfer of the defects at the boundary to the coarse grid is as described in Section 5.6.2. Treating the augmented system in the way described above, we determine $\xi$ only on the coarsest grid. Table 5.16 shows that the multigrid convergence factors for the V-, F- and W-cycles are again very good.                    △

**Table 5.16.** Measured multigrid convergence factors for Poisson's equation with Neumann boundary conditions.

| V(1,1) | F(1,1) | W(1,1) |
|--------|--------|--------|
| 0.13   | 0.09   | 0.09   |

## 5.7 FINITE VOLUME DISCRETIZATION AND CURVILINEAR GRIDS

Here, we will introduce a finite volume discretization and discuss the multigrid treatment of a Poisson-like equation on a curvilinear grid. The *finite volume approximation* is an important and natural discretization approach for general curvilinear and also for unstructured grids. Since *conservation properties* in PDEs can be preserved easily with finite volume discretizations, they are commonly used for problems from computational fluid dynamics [195]. The finite volume method can also be interpreted from a finite element point of view (see [41] for details).

We will describe the finite volume approach for a Poisson-like diffusion problem with diffusion coefficient $a(x, y)$:

$$-\nabla \cdot (a\nabla u) = -\frac{\partial}{\partial x}(a\,\frac{\partial u}{\partial x}) - \frac{\partial}{\partial y}(a\,\frac{\partial u}{\partial y}) = f^{\Omega} \quad (\Omega) \qquad (5.7.1)$$

$$u = f^{\Gamma} \quad (\partial\Omega). \qquad (5.7.2)$$

For the discretization, we assume that $\Omega$ is divided into small (nonoverlapping) finite volumes $\Omega_{i,j}$. In each of the finite volumes $\Omega_{i,j}$ (as depicted for two grids in Fig. 5.26), the integral formulation of (5.7.1) for the volume $\Omega_{i,j}$ has the form

$$-\int_{\Omega_{i,j}} \nabla \cdot (a\nabla u)\,d\Omega = \int_{\Omega_{i,j}} f\,d\Omega. \qquad (5.7.3)$$

A simple approximation of the right-hand side in (5.7.3) is given by

$$\int_{\Omega_{i,j}} f\,d\Omega \approx |\Omega_{i,j}|f_{i,j},$$

where $f_{i,j}$ denotes the value of the function $f$ at the center of the volume $\Omega_{i,j}$ and $|\Omega_{i,j}|$ is the size of the volume.

Using the Gaussian theorem, the left-hand side of (5.7.3) can be reformulated to an integral over the boundary $\partial\Omega_{i,j} = \bigcup_{\kappa=1}^{4} \partial\Omega_{i,j,\kappa}$

$$-\int_{\Omega_{i,j}} \nabla \cdot (a\nabla u)\,d\Omega = -\sum_{\kappa=1}^{4} \int_{\partial\Omega_{i,j,\kappa}} a\nabla u \cdot \mathbf{n}\,dS_{\kappa} \qquad (5.7.4)$$

if $a$ is continuous in $\Omega_{i,j}$. Here $\mathbf{n}$ denotes the unit outward normal vector. The integral $\int_{\partial\Omega_{i,j,\kappa}} \mathbf{w} \cdot \mathbf{n}\,dS_{\kappa}$ is also called *the flux of* $\mathbf{w}$ *through* $\partial\Omega_{i,j,\kappa}$.

In order to obtain a finite volume discretization of the left-hand side of (5.7.3) on general curvilinear grids, the outward normal vectors **n** and volumes $|\Omega_{i,j}|$ need to be defined appropriately [195, 405]. The resulting discrete equation is the so-called *flux balance* per finite volume which is typical for finite volume discretizations.

In the following, we will restrict ourselves to the *Cartesian case* (see the right domain in Fig. 5.26), for simplicity. In this case, we have

$$-\sum_{\kappa=1}^{4} \int_{\partial\Omega_{i,j,\kappa}} (a\nabla u) \cdot \mathbf{n} \, dS_\kappa \tag{5.7.5}$$

$$= \int_A^B a \frac{\partial u}{\partial y} \, dx - \int_B^C a \frac{\partial u}{\partial x} \, dy - \int_C^D a \frac{\partial u}{\partial y} \, dx + \int_D^A a \frac{\partial u}{\partial x} \, dy \tag{5.7.6}$$

(see Fig. 5.27).

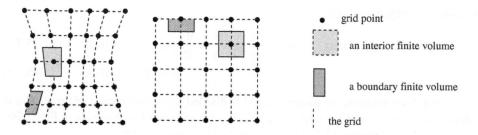

**Figure 5.26.** Definition of finite volumes $\Omega_{i,j}$ around the grid points on a curvilinear grid and on a Cartesian grid.

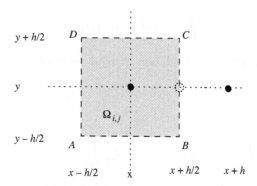

**Figure 5.27.** A volume $\Omega_{i,j}$.

We approximate the second integral along $BC$ in (5.7.6) (the other ones are approximated similarly) by

$$
\int_B^C a \frac{\partial u}{\partial x} dy \approx ha\left(x + \frac{h}{2}, y\right)(u(x+h, y) - u(x, y))/h
$$
$$
= a\left(x + \frac{h}{2}, y\right)(u(x+h, y) - u(x, y)). \tag{5.7.7}
$$

Summarizing, the discrete stencil obtained by the finite volume discretization is

$$
\begin{bmatrix} & s_n & \\ s_w & s_c & s_e \\ & s_s & \end{bmatrix}_h
$$

with

$$
\begin{aligned}
s_w &= -a(x - h/2, y), \\
s_e &= -a(x + h/2, y), \\
s_n &= -a(x, y + h/2), \\
s_s &= -a(x, y - h/2), \\
s_c &= -(s_n + s_w + s_e + s_s).
\end{aligned}
$$

**Remark 5.7.1** In this case, obviously, the finite volume discretization results in the same discretization as a finite difference approximation, apart from the fact that the equation is multiplied by $|\Omega_{i,j}|$. Such a scaling has to be taken into account in multigrid (see Remark 2.7.5). $\gg$

**Remark 5.7.2** Finite volume discretizations require special attention near boundaries. There, half volumes $\Omega_{i,j}$ are used in the discretization, as illustrated in Fig. 5.26. $\gg$

**Example 5.7.1** We consider the Poisson equation in a nonrectangular domain $\Omega$, presented in Fig. 5.28, and discretize it by the finite volume technique on a curvilinear grid $\Omega_h$. The resulting stencil for the Poisson equation depends on the details of the grid, i.e. on the aspect ratios and angles of the grid cells. Three different domains $\Omega = \Omega(\delta)$ are considered. The height and the width of $\Omega$ are set to one. The parameter $\delta$ determines the deviation from the straight line as indicated in Fig. 5.28 and thus the amount of curvature. We use $\delta = 0.05$, with almost rectangular grid cells, $\delta = 0.15$ and $\delta = 0.25$ with a strong curvature of the cells.

Table 5.17 shows the measured multigrid V(1,1) convergence factors with the point smoothers GS-RB, $\omega$-GS-RB and GS-LEX, with the line smoothers lexicographical $x$-line GS, $y$-line GS and with alternating line GS. The operator (2.3.3) is used for the restriction and (2.3.8) for interpolation. The results for the W(1,1)-cycle are very similar. The influence of the strong curvature on the convergence is obvious.

We have seen in Section 5.1 that grid cells with significant stretching lead to large anisotropies in the discrete equations. The same is true for grids with curvature. The point

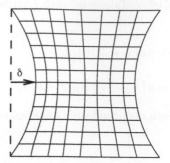

**Figure 5.28.** A curvilinear grid in a curved domain.

**Table 5.17.** Measured V(1,1)-cycle convergence factors on a $128^2$ curvilinear grid.

| $\delta$ | GS-RB | $\omega(=1.15)$-GS-RB | GS-LEX | $x$-line GS | $y$-line GS | alt. line GS |
|------|-------|-----------------------|--------|-------------|-------------|--------------|
| 0.0  | 0.10  | 0.04                  | 0.17   | 0.12        | 0.12        | 0.05         |
| 0.05 | 0.12  | 0.06                  | 0.19   | 0.12        | 0.14        | 0.05         |
| 0.15 | 0.27  | 0.16                  | 0.31   | 0.31        | 0.16        | 0.09         |
| 0.25 | 0.51  | 0.38                  | 0.48   | 0.53        | 0.30        | 0.18         |

smoothers combined with standard coarsening are efficient on curvilinear grids without severe stretching. For Poisson's equation on $\Omega(\delta)$ with $\delta = 0.05$, the analog of the RBMPS still has good multigrid efficiency. On general curvilinear grids, alternating line smoothers (or multiple semi-coarsening strategies) have better convergence factors. Another option is to use the ILU type smoothers (to be discussed in Sections 7.5 and 7.6).                    $\triangle$

## 5.8 GENERAL GRID STRUCTURES

In this section, we return to the discussion of multigrid methods for discrete problems on general grid structures, which we have already touched upon in Section 1.2. The introduction of multigrid has been based on simple Cartesian grid structures. In general, Cartesian grids can also be used for complicated (3D) geometries, e.g. for the *Euler* equations [1, 2, 441], based on Shortley–Weller type discretizations (2.8.13) and adaptive refinement (see Chapter 9).

Mostly, however, non-Cartesian grids are used for problems on geometrically complicated domains. As we have already seen for one example of Poisson's equation in Section 5.7, curvilinear grids can be treated by multigrid as regular Cartesian grids with similar convergence factors.

For complicated geometries, a global curvilinear grid is often unavailable or too difficult to construct. In such cases, *block-structured grids* are an option. The given domain is divided into subdomains and for each subdomain a (local) curvilinear grid is constructed such that the local grids (blocks) fit "nicely" together, i.e. continuously or even smoothly. The block-structured grid approach is quite general and is used for many applications even in complicated geometries. Multigrid is compatible with block-structured grids. In programs working on block-structured grids, line or plane smoothers are usually restricted to the blocks [248].

Of course, *finite element discretizations* can also be used on structured grids. For simple finite elements, the corresponding discrete equations can then be treated efficiently by multigrid without any principal difficulties. For example, the standard five-point stencil for the Laplacian is also obtained by linear finite elements on a standard regularly structured grid and the stencil in Example 7.7.1 in Section 7.7.4 refers to the corresponding situation with bilinear finite elements. We will not describe multigrid for finite element discretizations in this book but refer to the literature [50, 89, 176, 351].

In many commercial codes, *unstructured* irregular grids and corresponding finite volume or finite element discretizations are used. Unstructured grids can be generated easily and automatically by commercial *grid generators*. These grids are more flexible and can easily be adapted to the boundary of a general domain.

However, unstructured grids cannot be handled by multigrid as easily as structured ones. We will mention two technical complications. First, *smoothers* which correspond to line or plane relaxation for anisotropic problems may be difficult to realize on unstructured grids. An alternative is to use more general block relaxations like ILU (to be discussed in Section 7.5) or box-type relaxations (see Section 8.7). Secondly, the construction of a *hierarchy of coarse grids* is not trivial for a general unstructured grid. (This is, of course, not a problem if the fine grid itself is constructed by using some hierarchical refinement starting from an unstructured coarsest grid.) One way to bypass this coarsening difficulty is to use a different coarsening instead, as referred to in the following remark.

**Remark 5.8.1 (Multilevel $p$ solvers)**   If higher order finite elements (so-called $p$ elements [13]) are used on unstructured grids, a multigrid related approach is to use *lower order finite element spaces* for the "coarse grid correction" instead of coarse grids. All other multigrid components like smoothing, transfer operators etc. can be essentially maintained. This approach is used, for instance in [253, 449]. The corresponding *multilevel p solver* is usually applied as a preconditioner. This type of coarsening and conventional grid coarsening can, if applicable, also be combined ($h$-$p$ elements).                                    $\gg$

An elegant multigrid approach for unstructured grids is the algebraic multigrid method (AMG). AMG allows us, for a certain class of problems, to define a coarsening strategy so that problems on completely unstructured grids can also be solved efficiently. It combines an algebraically defined problem-dependent coarsening with *point smoothers*.

> We propose using AMG for *problems on unstructured grids*. It generates the coarse grids *automatically* (see Appendix A for a detailed description of AMG).

Typically, the application of AMG for low order finite elements, for example, for Poisson-like equations is straightforward and efficient.

An overview of multilevel methods for problems on unstructured grids is given in [103]. Finally, we would like to mention that several general multigrid-based software packages for unstructured grids and finite elements exist, like FEATFLOW [393], PLTMG [20], UG [30] or the DLR-$\tau$-code [155].

# 6

# PARALLEL MULTIGRID IN PRACTICE

In this chapter, we will discuss the parallel aspects of multigrid in more detail. In the previous chapters, we have already made several remarks about parallelism and multigrid. Although suitable multigrid *components* may be highly parallel, the *overall structure* of standard multigrid is intrinsically not fully parallel for two reasons. The first reason is that the grid levels are run through sequentially in standard multigrid. The second reason is that the *degree of parallelism* of multigrid is different on different grid levels (i.e. small on coarse grids). A basic theoretical complexity discussion is presented in Section 6.1. The problem of *communication*, which is an important aspect in practical parallel computing, is not addressed in this discussion.

In Sections 6.2–6.3, we deal with the question of how grid and multigrid algorithms are implemented on parallel computers *in practice*. We will start with some general remarks on parallel architectures and discuss two basic rules for an efficient parallelization. Here, the question of communication comes into play.

In Section 6.2 the fundamental concept of *grid partitioning* will be introduced. Grid partitioning is a genuinely geometric parallelization approach. We will introduce terms such as *speed-up*, *parallel efficiency*, *scalability* and discuss the *boundary-volume effect*. Furthermore, we will make some remarks and assumptions about parallel computer architectures. We will mainly use the terminology of parallel systems with *distributed memory* in this chapter. For example, data is assumed to be *communicated* between processors by *sending* and *receiving messages*. Our discussion covers also other parallel architectures, for example shared memory computers, where other data transfer (and/or synchronization) techniques are common.

In Section 6.3, we will discuss the extension of *grid partitioning* to *multigrid*. Whereas (point) relaxation methods are *local* methods (i.e. all operations at a grid point involve only values at points in a local neighborhood), multigrid has nonlocal features. On coarse grids the relation of computation and communication becomes worse than on fine grids. Thus, the definition of the coarsest grid and the corresponding solution process may have to be reconsidered. We will also discuss the boundary-volume effect and scalability in the context of multigrid and we will give some concrete hints on programming parallel systems.

Among the multigrid components, certain types of *smoothing* methods such as line smoothers are not inherently parallel. In Section 6.4, we will discuss parallel versions of line and plane smoothers.

Generally speaking, there are two different types of approaches for the parallel treatment of PDEs. Both of them are based on a geometric decomposition of the domain, on which the PDE is to be solved. *The first approach is to use a fast sequential solver for the given problem and to parallelize this solver as efficiently as possible.* In practice, this means that a fast multigrid method is used for the solution of the global problem and *grid partitioning* is used for parallelization. Typically, the parallel versions of multigrid *are then equivalent to their sequential counterparts*.

The second approach is to start with *a decomposition of the given problem into a number of subproblems* on subdomains with or without an overlap. This is the basic principle of the domain decomposition methods (DD). Although, in principle, the philosophy of global multigrid partitioning and domain decomposition are quite different, in practice certain variants of both approaches are very similar or may even lead to the same algorithm. We will briefly discuss DD methods in Section 6.5.

Since the communication requirements in parallel multigrid may be relatively high, certain modifications have been proposed to avoid high communication overhead (see Sections 6.2.2 and 6.5). "New" parallel multigrid related algorithms have been considered by several authors: in particular, additive variants of multigrid [55, 161, 425] and the parallel superconvergent multigrid method [143] belong to this class. In our opinion, these extensions may be useful in theory and in special situations, but so far they have not been proven to lead to general efficient parallel methods in practice. We will briefly describe some of them in Section 6.5.

All considerations in this chapter refer to "*nonadaptive*" multigrid algorithms where the grid structures are known in advance. Adaptive multigrid algorithms and their parallel versions will be treated in Chapter 9.

In general, the presentation in this chapter is elementary. Many details are explained for simple cases like Model Problem 1. Most of the ideas and results, however, carry over to more general cases. Only Section 6.4, in which we deal with parallel versions of smoothers for anisotropic problems, is technically somewhat more involved.

## 6.1 PARALLELISM OF MULTIGRID COMPONENTS

The *degree of parallelism* reflects the number of processors that can contribute to the computation of an algorithm or of algorithmic components. It can be interpreted as the number of operations that can be carried out in parallel. A parallel system that would exploit the degree of parallelism of a fully parallel relaxation method would need to have at least as many processors $P$ as grid points $N$. (In practice, only a limited number of processors is usually available.)

As we have seen in Chapters 2 and 5, a standard multigrid algorithm generally is characterized by its overall hierarchical structure (MGI or FMG; V-, F- or W-cycle) and the specification of its components. The considerations on "parallel multigrid" in the previous chapters refer to the *parallelism of the multigrid components*. This, of course, does not mean that the choice of the overall MG-structure is unimportant for the parallel efficiency of the algorithm. In fact, certain choices of the overall structure turn out to be advantageous and others to be disadvantageous.

### 6.1.1 Parallel Components for Poisson's Equation

We reconsider Model Problem 1 and a corresponding multigrid method with standard coarsening. The crucial multigrid component with respect to parallelism is usually the *smoothing procedure*. We first recall the parallel properties of $\omega$-JAC, GS-LEX and GS-RB on $\Omega_h$ from Section 2.1: $\omega$-Jacobi relaxation is *fully $\Omega_h$ parallel*. We say that its degree of parallelism is #$\Omega_h$. Correspondingly, for GS-LEX the degree of parallelism is less than or equal to $(\#\Omega_h)^{1/2}$ and for GS-RB the degree of parallelism is $(1/2)\#\Omega_h$. For Model Problem 1, GS-RB has the best smoothing properties of these relaxations *and* is highly parallel. For discretizations with larger stencils, *multicolor Gauss–Seidel* relaxation or JAC-RB as introduced in Section 5.4.2 have good parallelization properties.

The three other multigrid components can be applied in parallel for all (interior) grid points:

- *Calculation of defects* The defect computations at different grid points are independent of each other and can be performed in parallel for all points of $\Omega_h$.
- *Fine-to-coarse transfer* The computations and transfer operations at different grid points are again independent of each other and can be performed in parallel for all points of the coarse grid $\Omega_{2h}$. This applies to all of the restriction operators (FW, HW and, trivially, injection) discussed so far.
- *Coarse-to-fine transfer* Interpolation from coarse to fine refers to the $\Omega_h$ grid points. Typically, the operations to be performed are different for different types of grid points (see, for example, (2.3.7) and Fig. 2.6, for the case of bilinear interpolation), but the operations can be performed in parallel.

So far, the discussion of the parallelism in multigrid components has been oriented to the levels $\Omega_h$ and $\Omega_{2h}$.

> Clearly, all considerations carry over to any other (coarser) pair of levels $h_k$, $h_{k-1}$, with the corresponding (smaller) degree of parallelism.

**Remark 6.1.1 (parallelization below grid level)** In the above considerations, we have looked for parallelism only *on the grid level*, i.e. we have regarded an operation such as the evaluation of $L_h$ or $S_h$ etc. as *one* unit, to be applied for each single grid point of $\Omega_h$. In fact, such an evaluation typically consists of several arithmetic operations and therefore has an additional potential for parallel execution. We will, however, not discuss this topic here. $\gg$

### 6.1.2 Parallel Complexity

On the coarse grids, the degree of parallelism decreases substantially since

$$\#\Omega_k \ll \#\Omega_\ell \quad \text{for } k < \ell,$$

until finally $\#\Omega_0 = 1$ if the coarsest grid consists of only one grid point. The problem of the very coarse grids leads to multigrid specific parallel complications which do not occur in classical single-grid algorithms.

> This crucial impact of the coarse grids increases, the more often the coarse grids are processed in each cycle. A parallel W-cycle, for example, has a *substantially* different parallel complexity from that of a parallel V-cycle.

This can be seen by the following results of a simple analysis of parallel multigrid complexity. Here, the parallel complexity is the number of parallel steps/operations that a processor has to carry out, assuming that the degree of parallelism is fully exploited.

**Result 6.1.1** *Let $N$ denote the overall number of fine grid points (unknowns) in case of Model Problem 1:*

$$N = \#\Omega_h = (n-1)^2$$

*and let us consider the RBMPS. Then we obtain the sequential and parallel complexities (the sequential complexity being simply the number of required floating point operations) listed in Table 6.1. For MGI (i.e. multigrid iteration) we assume an accuracy of $\varepsilon$ (error reduction) as stopping criterion.*

*Proof.* The sequential complexities (total computational work) in Table 6.1 have already been derived in Sections 2.4.3 and 2.6.2. The parallel complexities are obtained by summing the number of times the multigrid levels are processed in the corresponding algorithms. In the case of the W-cycle, for example, the number of visits of the level $\Omega_{\ell-k}$ per cycle is $O(2^k)$ which sums to $O(2^\ell) = O(\sqrt{N})$. The other results can be obtained similarly.

**Table 6.1.** Sequential and parallel complexities of 2D-multigrid.

|  | Cycle type | Sequential | Parallel |
|---|---|---|---|
| MGI | V | $O(N \log \varepsilon)$ | $O(\log N \log \varepsilon)$ |
|  | F | $O(N \log \varepsilon)$ | $O(\log^2 N \log \varepsilon)$ |
|  | W | $O(N \log \varepsilon)$ | $O(\sqrt{N} \log \varepsilon)$ |
| FMG | V | $O(N)$ | $O(\log^2 N)$ |
|  | F | $O(N)$ | $O(\log^3 N)$ |
|  | W | $O(N)$ | $O(\sqrt{N} \log N)$ |
| Lower bound (for any solver) |  | $O(N)$ | $O(\log N)$ |

That at least $O(\log N)$ parallel steps are required for any solver follows from the fact that the solution of a discrete elliptic problem in general depends on all discrete entries. Therefore, full (weighted) sums of all entries have to be evaluated which requires at least $O(\log N)$ parallel steps.                                                                                    $\square$

The two main observations from Table 6.1 are:

---

- The step from V- to W-cycles gives a substantial increase of parallel complexity, from

$$O(\log N) \quad \text{to} \quad O(\sqrt{N})$$

(whereas in the sequential case the number of operations is only multiplied by a factor of $3/2$).
- The parallel FMG approach (based on a V-cycle) needs

$$O(\log^2 N)$$

parallel steps instead of the theoretical bound of $O(\log N)$ (whereas in the sequential case FMG is optimal in the sense that it requires $O(N)$ operations).

---

The increase of complexity for W-cycles means that V-cycles are preferable from a parallel point of view. This is relevant if we want to work with a highly parallel system which consists of nearly as many processors as grid points.

**Remark 6.1.2** The results of Table 6.1 carry over to other multigrid algorithms and other 2D problems provided that:

- we have $h$-independent error reduction per multigrid cycle,
- all multigrid components employed are $\#\Omega_h$-parallel (or $O(\#\Omega_h)$-parallel),
- FMG produces an approximation with discretization accuracy,
- the treatment of boundary conditions can also be performed in parallel.

Similar results are also obtained in 3D.                                                          $\gg$

## 6.2 GRID PARTITIONING

The existence of a sufficiently high degree of parallelism of an algorithm, as discussed in the previous section, is a prerequisite for the utilization of parallel computers. If a multigrid algorithm is to be implemented on a parallel system, many additional (mathematical and technical) aspects have to be taken into account. Here, we are interested in the practical questions such as the objective of minimizing the corresponding parallelization overhead.

Typically, practical questions are related to the architecture of the parallel computer at hand and to the programming model employed.

### 6.2.1 Parallel Systems, Processes and Basic Rules for Parallelization

Parallel computer architectures have been developing and changing very rapidly. The answer to the question, how to design an ideal parallel multigrid algorithm, clearly depends on the concrete parallel architecture to be employed: whether we use a parallel (multiprocessor) system with shared, distributed or some hierarchical memory, whether it consists of vector, cache or scalar processors and which type of interconnection network is used (a static or a dynamic one, the type of topology etc.). We also regard workstation and PC clusters as parallel systems.

Figure 6.1 shows such a parallel system, a cluster consisting of 16 workstations. We assume that the workstations are connected by some network in order to exchange data and to communicate with each other.

For such (and other) architectures, the memory/cache organization may have an essential impact on the overall (parallel) efficiency of an algorithm (with the phenomenon of "superlinear speed-ups" etc.). We regard this as a more technical question, which will not be discussed in detail in this chapter.

The real performance of a parallel algorithm on a concrete parallel system is also influenced by other details like whether the hardware and the software allow an overlap of computation and communication etc. and, of course, the operating system and the compiler may also be important.

Overall, we have tried to find a compromise: discussing practical questions but not confining ourselves too narrowly to a specific parallel computer model.

**Remark 6.2.1 (parallel processes)**  Generally, we consider an abstract model of a parallel computer that consists of a (possibly large) number of processors, each of which has, *logically*, its own memory and can work (and be programmed) independently of all other processors. The processors can communicate and exchange data over some suitable interconnection, which is not further specified. We assume that a parallel application consists of a number of logical *processes* with their own address spaces. A process cannot directly access data of another process. If such *remote* data are required by a process, the process possessing them has to make them available for the requesting process.

In the case of distributed memory architectures, communication between processes is based on *message passing*. However, other parallel systems can also be represented

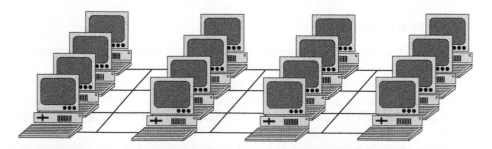

**Figure 6.1.** 16 workstations as an example of 16 connected processors.

by this concept. On shared memory architectures, for instance, "communication between processes" corresponds to "data copying" and/or synchronization.

For simplicity, we will *always assume one process per processor when discussing parallel features of an algorithm* in the following. In that respect, we will not distinguish between processes and processors. ≫

There are two obvious reasons why an algorithm and/or a parallel system may perform unsatisfactorily: load imbalance, and communication overhead. *Load imbalance* means that some processors have to do much more work than most of the others. In this case, most of the processors have to wait for others to finish their computation before a data exchange can be carried out. A purely sequential algorithm or a sequential phase in a parallel algorithm produces extreme load imbalance as only one processor is busy in that case.

---

**Remark 6.2.2 (basic rule for load balance)**   For architectures with many processors, in general, it does no harm if one (or a few) processors have much less computational work than the average, but it is crucial for the performance of the parallel application if one (or a few) processors have much more computational work than the average. Then, most of the processors will be idle and have to wait for the overloaded ones to finish their parts of the computations. ≫

---

*Communication overhead* means that the communication and data transfer between the processors takes too much time compared to the effective computing time. This overhead may even lead to slow-down instead of speed-up when more and more processors are used. Summarizing:

---

Avoiding load imbalance and limiting the communication overhead are the two most important principles in the parallelization of a given sequential algorithm.

---

### 6.2.2 Grid Partitioning for Jacobi and Red–Black Relaxation

If grid applications are to be implemented on parallel computers, *grid partitioning* is a natural approach. In this approach, the original grid $\Omega_h$ is split into $P$ parts (subdomains, subgrids), such that $P$ available processors can jointly solve the underlying discrete problem.

Each subgrid (and the corresponding "subproblem", i.e. the equations and the unknowns located in the subgrid) is assigned to a different process such that each process is responsible for the computations in its part of the domain.

The grid partitioning idea is widely independent of the particular boundary value problem to be solved and of the particular parallel architecture to be used. It is applicable to general $d$-dimensional domains, structured and unstructured grids, linear and nonlinear equations and systems of partial differential equations. Here and in the following sections, we focus on this approach.

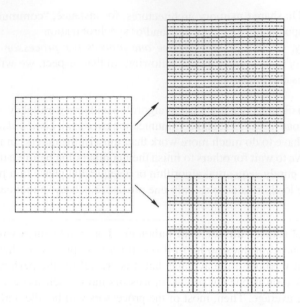

**Figure 6.2.** Partitioning of a $16 \times 16$ grid into $4 \times 4$ or $1 \times 16$ (rectangular) subgrids.

**Example 6.2.1**  If we consider a parallel system consisting of 16 processors as in Fig. 6.1, we are interested in a grid partitioning into 16 subgrids. Obviously, a grid can be partitioned in various ways. Two examples, a $4 \times 4$ (2D) and a $1 \times 16$ (1D) partitioning, are shown in Fig. 6.2. The partitionings generate certain artificial boundaries within the original domain.

When applying a $4 \times 4$ partitioning in the case of 16 processors, each process is responsible for the computations in about $n/4 \times n/4$ of the computational grid ($n$ being the number of grid points in each direction).                                                                    △

In order to illustrate the basic ideas of grid partitioning as simply as possible, we start our discussion with the parallel treatment of $\omega$-JAC as an iterative *method* for Model Problem 1 on the square Cartesian grid.

Remember that $\omega$-JAC is fully $\Omega_h$ parallel. When distributing the work performed during $\omega$-JAC iterations to the available processors (all of the same performance), it is crucial that each processor obtains roughly the same amount of work at any stage of the solution procedure. Since the work of $\omega$-JAC is the same at each interior grid point of $\Omega_h$, a good load balance can easily be obtained. $\Omega_h$ is split into as many (rectangular) subgrids as processors are available so that each subgrid contains *approximately the same number of grid points*. (If the processors have different performances, the distribution of grid points has to be adjusted accordingly.)

**Remark 6.2.3**  Even for a regular grid consisting of $2^\ell \times 2^\ell$ grid cells, often, a certain amount of load imbalance cannot be avoided. For instance, in the $4 \times 4$ partitioning of

Figure 6.2, nine of the 16 subgrids have $2^{\ell-2} \times 2^{\ell-2}$ grid points, six have $(2^{\ell-2} - 1) \times 2^{\ell-2}$ points and one has $(2^{\ell-2} - 1) \times (2^{\ell-2} - 1)$ points. This imbalance is a harmless case of load imbalance. Parallel systems are typically used for the solution of large problems (large $\ell$), for which such imbalances are negligible. $\gg$

Starting with some approximation $u_h^m$ on the partitioned grid, each process can compute a new $\omega$-JAC iterate $u_h^{m+1}$ for each point in the interior of its subgrid. Near the subgrid boundaries, each process needs the old approximations $u_h^m$ located at those points which are direct *neighbors* of its subgrid (see upper picture in Fig. 6.3). In principle, each process can obtain this data from the neighbor process by communicating it in a pointwise fashion, whenever such data is needed for the computations. But this approach would require a very large number of messages to be sent and received during the computations which would result in a large communication overhead due to the corresponding *large start-up time of sending many small messages* (see Section 6.2.4).

An efficient and elegant approach is obtained if each process not only stores the data belonging to its subgrid but also a copy of the data located in neighbor subgrids in a *thin overlap area of a certain width* $w$, for example an overlap of one grid point ($w = 1$) (see lower picture in Fig. 6.3).

Then, each process can perform a full $\omega$-JAC iteration without any communication in between. After an iteration, the copies in the overlap areas have to be updated by communication so that the next $\omega$-JAC iteration can be carried out. In our example, this communication can easily be realized. Each process sends all its data belonging to one side of the overlap area of a neighbor subgrid collectively (i.e. in one long message) to the corresponding "neighbor" process and receives the data corresponding to its own overlap area from that neighbor. This communication via the interconnection network of the processors is a typical example of *message passing* in a parallel system.

The situation is very similar if we replace $\omega$-JAC by GS-RB. Starting with an approximation $u_h^m$, we can perform the first half-step of GS-RB fully in parallel. Before the relaxation of the black points, we have to update the approximations $u_h^{m+1}$ at the *red points of the overlap regions*. After the second half-step of GS-RB, we have to update the approximation $u_h^{m+1}$ at the *black* points of the overlap regions. We thus need two communication phases per GS-RB iteration instead of one for $\omega$-JAC, but with only half of the points being exchanged in each step.

In this respect, computing phases and communication phases alternate during the execution of the parallel program for both $\omega$-JAC and GS-RB. This alternation of computing and communication phases is natural for parallel grid and multigrid programs based on the grid partitioning idea.

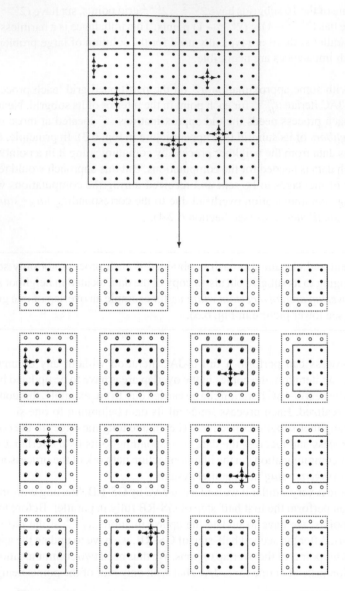

**Figure 6.3.** Introduction of an overlap area around the subgrids.

**Remark 6.2.4 (algorithmic equivalence)**    We point out that the parallel algorithms $\omega$-JAC and GS-RB as described above are *algorithmically equivalent* to their sequential versions. The results of both algorithms are the same. This algorithmical equivalence is, however, not naturally achieved if sequential multigrid components such as GS-LEX are modified to improve their parallel properties.          ≫

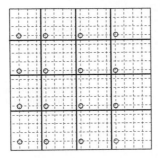

**Figure 6.4.** Modified GS-LEX: the subgrids are relaxed in parallel. A GS-LEX iteration is used in each block starting, for example, at the grid points (○).

**Remark 6.2.5 (parallel modification of GS-LEX)**   As we have seen before, the degree of parallelism of GS-LEX is not satisfactory. An update of unknowns in GS-LEX depends on previously calculated values.

A modification of GS-LEX, which better suits the grid partitioning concept is to apply the GS-LEX smoother *only within a subgrid of a grid-partitioned application*. As a consequence, the resulting relaxation procedure is no longer a classical GS relaxation, but a combination of Jacobi-type relaxation and GS-LEX: All the subgrids are treated simultaneously (block-Jacobi) and within each subgrid (block) GS-LEX is used (see Fig. 6.4).

$\gg$

In the case of five-point stencils, approximations at corner points are not needed when applying GS-RB or $\omega$-JAC so that these points need not be included in the update of the overlap region. The situation is different for discretizations with compact nine-point stencils (5.4.4).

**Remark 6.2.6 (compact nine-point stencils)**   In this case, the corner points in the overlap area are also important. They correspond to *diagonal* neighbor subgrids (and the corresponding processors). Nevertheless, explicit communication between these processors can be avoided: One can separate the communication in the $x$-direction and in the $y$-direction such that, for example, the exchange of data in the $x$-direction is completed before the communication in the $y$-direction starts. The correct approximations at the corner points are obtained automatically if the exchange in the $y$-direction includes the approximations at the corner points of subgrids (see Fig. 6.5 for a schematic representation).     $\gg$

**Remark 6.2.7 (overlap width $w = 2$)**   Using an overlap of width $w = 2$ grid points can be beneficial for parallel GS-RB on certain parallel systems. In this case, we can perform both relaxation half-steps without an intermediate communication if we *additionally relax* the points in that part of the overlap region which corresponds to an overlap of width $w = 1$. Then only one exchange per GS-RB step is required. However, these additional

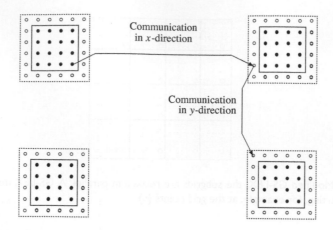

**Figure 6.5.** Avoiding diagonal sends by separating communication in $x$- and $y$-direction

computations during the first relaxation half-step at some of the overlap points represent an "overhead of computation" (as opposed to an overhead of communication). On parallel systems with slow communication (more accurately: very high start-up times, see Section 6.2.4), this approach may still be beneficial.

For larger stencils, for example, the higher order discretization for Poisson's equation (5.4.1), it is natural to choose the overlap width $w = 2$.                                    ≫

### 6.2.3 Speed-up and Parallel Efficiency

In the following, we assume that we have a homogeneous parallel system with at least $P$ processors (all of the same performance). We consider an algorithm for the solution of a given problem which runs on $P \geq 1$ processors in a computing time $T(P)$. Then the *speed-up* $S(P)$ and the *parallel efficiency* $E(P)$ are defined as

$$S(P) = \frac{T(1)}{T(P)}, \qquad E(P) = \frac{S(P)}{P}. \qquad (6.2.1)$$

We are interested in the behavior of $S$ and $E$ as a function of $P$. Ideally, one would like to achieve

$$S(P) \approx P, \text{ or equivalently } E(P) \approx 1.$$

This would mean that we are able to accelerate the computation by a factor close to $P$ if we use $P$ processors instead of one. In many cases, however, an efficiency $E$ close to 1 cannot be achieved. Usually, *the cost of interprocessor communication is not negligible*, in particular not, if for a given problem the number of processors $P$ is increased. In other words, the efficiency $E$ will be low if a large parallel system ($P$ large) is used for a small problem.

> Increasing the number of processors is usually reasonable if the size of the problem (characterized, for example, by the number of grid points or unknowns $N$) is also "*scaled up*". Therefore, we are interested in $S = S(P, N)$, $E = E(P, N)$ as a function of $P$ *and* $N$.

The above definitions and assumptions are somewhat problematic in certain situations. *S and E as defined above usually do not merely represent the parallel properties of a given algorithm but also include other effects due to special processor characteristics.* For example, it may be unrealistic to assume that the entire application can be executed on one processor of the system considered. The memory of only *one* processor may not be large enough for the whole application. Similar phenomena and complications occur when considering cache-based processors or vector processors. Here, the computing time may strongly depend on the arrangement of the unknowns in the memory and on the order and range of the loops in the computations. In such cases, the above definition would not be applicable, or, if used one would perhaps observe a *superlinear* speed-up ($E(P) > 1$).

With respect to these complications, one always has to interpret the meaning of $S$ and $E$ according to the situation considered. One may use the above definitions, but should not interpret them naively.

It may also be appropriate to modify the above definitions. One suitable modification is the following definition of parallel efficiency $\tilde{E}(P)$:

$$\tilde{E}(P) = \frac{\sum_i a_i}{P \max_i (a_i + c_i)}.$$

Here $a_i$ denotes the CPU time for arithmetic computations in processor $i$ and $c_i$ the corresponding total communication time (including idle time), assuming that computation and communication are not overlapping. On one processor, only the CPU time is taken into account in this definition. Here, we assume that the times $a_i$ and $c_i$ can be measured during the execution of the parallel algorithm.

Most importantly, the above quantities $S$, $E$ and $\tilde{E}$ have a limited numerical significance. The fact that an algorithm has a "good" efficiency on a certain parallel machine, does not at all mean that it is a *numerically* efficient algorithm.

> Often numerically inefficient algorithms give (much) better parallel efficiencies $E$ than more sophisticated and numerically efficient ones since those are much easier to parallelize. A typical example is a single grid iterative method like classical Jacobi-type iteration compared to multigrid. As the Jacobi-type iteration is fully parallelizable and *local*, one will find a very good parallel efficiency. Multigrid, however, is more involved so that its parallel efficiency will be worse than that of Jacobi. Nevertheless, the overall efficiency (i.e. the total computing time to solve a given problem) will typically be *better by far* for the multigrid approach.

**Remark 6.2.8**   In order to avoid this *paradox* (high parallel efficiency for numerically inefficient methods and low parallel efficiency for numerically efficient ones), some authors have introduced a different "speed-up" quantity

$$S^*(P) = \frac{T^*(1)}{T(P)}.$$

Here a concrete problem is considered and $T(P)$ and $T^*(1)$ are computing times needed to solve this problem. $T(P)$ is the computing time needed for the parallel algorithm under consideration on $P$ processors (same quantity as above in (6.2.1)). $T^*(1)$, however, is the computing time needed for the *fastest sequential* algorithm (on one processor of the same system). In general, two different algorithms are compared. Although the above paradox disappears with this definition, we do not use it in this book since we regard it as not being very practical. On different computers different algorithms or different implementations of the same algorithm may be the fastest.                                                                        $\gg$

### 6.2.4 A Simple Communication Model

Typically, concrete results on the parallel efficiency of an algorithm are obtained by measurements on a particular parallel computer. Often not only an algorithm is evaluated, but different parallel computer architectures (and computer products from different vendors) are also implicitly included in a comparison.

Some quantitative and qualitative results can also be derived theoretically using computer and communication models ("performance prediction"). A substantial number of performance models have been developed, some of which are very sophisticated. Some of them have been used for the evaluation of parallel multigrid methods, for example [197, 219, 261, 268]. (Parallel adaptive multigrid methods have also been studied in [197, 268, 446].)

One of the simplest communication models is already useful in practice. Here, the time needed for sending a message of length $L$ is modeled by the formula

$$t_{\text{comm}} = \alpha + \beta L$$

with parameters $\alpha$ and $\beta$: $\alpha$ is the so-called *start-up time* which has to be spent whenever a message is sent, and $1/\beta$ is the *bandwidth* of the respective communication channel. For a realistic evaluation of the performance of a solution method on a particular parallel system, $t_{\text{comm}}$ has to be compared with the computing time $t_{\text{comp}}$ needed, e.g. for an arithmetic operation.

For a concrete parallel computer, it may be useful or even necessary to take the size of $\alpha$ and $\beta$ into account when a specific algorithm is parallelized.

> If $\alpha$ is large, the *number* of messages should be minimized, and if $\beta$ is large, the communication *volume* is the issue.

In any concrete case, it depends on the hardware and on the application which of these parameters is crucial. On distributed memory systems, often $\alpha$ is the important parameter.

As an example, reconsider the two partitioning options in Fig. 6.2: in the square (2D) partitioning the communication volume is smaller (less than half) than in the strip (1D) partitioning, but the number of messages is larger (four versus two per subgrid). The amount of data to be communicated is proportional to the total length of the interior boundaries, the number of messages equals the total number of edges of the subgrids.

If all the communication features (number of messages, volume of data, distance of communication, etc.) are taken into account in a concrete algorithm, this *communication complexity* is a new and valuable criterion, in addition to time and memory space. We will not, however, discuss this field in this book.

### 6.2.5 Scalability and the Boundary-volume Effect

In this section, we will show that the boundary-volume effect is the reason that grid partitioning leads, in general, to satisfactory efficiencies for sufficiently large problems. We first study the behavior of an algorithm if the number of processors $P$ is fixed, but the size of the problem $N$ is increased. We make the following assumptions:

- the given problem is characterized by local dependencies, as in finite difference and finite volume discretizations,
- the solution method has a sufficiently high degree of parallelism (for example, proportional to the number of grid points) and is sufficiently local,
- the number of grid points and the number of arithmetic operations per grid point are (asymptotically, for $N \to \infty$) equal for all subgrids.

Under these assumptions, we obtain

$$\boxed{E(P, N) \to 1 \quad \text{for } P \text{ fixed, } N \to \infty} \tag{6.2.2}$$

for a large class of applications and algorithms. The result is known as the *boundary-volume effect*. The reason for this is that the ratio $T_{\text{comm}}/T_{\text{comp}}$ (i.e. the overall time for communication versus the overall time for computation) behaves like the number of *boundary* grid points of the subgrids versus the number of *interior* grid points of the subgrids. For $N \to \infty$ and $P$ fixed this means that $T_{\text{comm}}/T_{\text{comp}} \to 0$ and this, together with the other assumptions, implies $E(P, N) \to 1$.

For local methods, like $\omega$-JAC and GS-RB, the boundary-volume effect holds trivially. But as we will see in Section 6.3.3, standard multigrid methods with sufficiently parallel smoothers also exhibit the boundary-volume effect.

The situation is less trivial if both the grid size $N$ and the number of processors $P$ are increased. In the grid partitioning applications considered here, it is reasonable to assume that the number of grid points per processor is constant if $P$ is increased:

$$N/P = \text{const for } P \to \infty.$$

The term *"scalability"*refers to this situation and assumption. We call a parallel algorithm and a corresponding application *"E-scalable"*, if

$$E(P, N) \geq \text{const} > 0 \quad \text{for } P \to \infty, \quad N/P = \text{const}.$$

Typically, local parallel algorithms like $\omega$-JAC and GS-RB relaxation methods turn out to be $E$-scalable. However, *numerically* highly efficient algorithms, like multigrid methods are not $E$-scalable (see Section 6.3.3).

As mentioned before, the terms "parallel efficiency" and "$E$-scalability" are questionable because they do not take the numerical efficiency into account.

**Remark 6.2.9** The term "scalability" was originally used as a feature of parallel hardware. It addresses the fact that parallel systems can be built according to the same architectural concept in a scalable way: small systems with a few, medium systems with a medium number and large systems with a large number of processors. But the software and the parallel algorithms should also be "scalable" in order to fully exploit the hardware scalability. This led to the above definition of algorithmic scalability.                                $\gg$

## 6.3 GRID PARTITIONING AND MULTIGRID

So far, we have described the idea of grid partitioning and discussed some details of its realization only for "local" iterative methods such as $\omega$-JAC and GS-RB. Grid partitioning is also the natural parallelization approach for multigrid. *Its extension of the single grid case to parallel multigrid is straightforward.* As a typical example, we consider Model Problem 1 in this section.

The mapping of the subgrids on different multigrid levels to processors will be discussed in Section 6.3.1. We also deal with some details of the communication on coarse grids and in the intergrid transfer. The *very* coarse grids have to be treated properly on a parallel system, particularly if the number of processors to be employed is large (see Section 6.3.2). Typically, the boundary-volume effect is maintained for multigrid methods, not, however, $E$-scalability in its *strict* form (see Section 6.3.3). Section 6.3.4 contains some remarks on programming parallel multigrid.

### 6.3.1 Two-grid and Basic Multigrid Considerations

We consider two grids in a specific multigrid algorithm (correction scheme) in order to discuss the impact of the parallelization on the multigrid components. We choose the components GS-RB for smoothing, HW for restriction (we will see that it requires less communication than FW) and bilinear interpolation and assume that the number of grid points on the coarse grid $\Omega_H$ is (much) larger than the number of available processors.

Let us further assume that we perform the parallel computations during the smoothing procedure on the fine grid as described in Section 6.2.2, with an overlap width of $w = 1$. The computations on the coarse grid can be assumed to be of the same type as those on the fine grid. In particular, the arithmetic relations are *local* with respect to the grid level on both grids. Since the coarse grid problem is a direct analog of the fine grid problem, we can perform grid partitioning on the coarse grid accordingly. In general, there is no reason to change the partitioning of the subdomains and the mapping to the processors. On the contrary, if the information on the same *geometric* points on different grids $\Omega_h$ and $\Omega_H$ were allocated to different processors, additional communication among these processors would be required during the intergrid transfers. Therefore, the grid partitioning of the coarse grid is carried out as for the fine grid.

> If extended to multigrid, this means that one processor is responsible for the computations on a sequence of grid levels on the same subdomain.

Of course, the overlap idea has to be adapted according to the different grid levels. On each grid, we need an overlap region of at least $w = 1$ in order to be able to perform the parallel smoothing steps and defect calculations. Since the distance between two adjacent grid points increases on coarse grids, the "geometric size of overlap regions" will be different on different grid levels (see Example 6.3.1).

> As on the fine grid, communication on the coarser grids is "local" with respect to the corresponding grid level.

**Example 6.3.1** Figures 6.6 and 6.7 show a typical grid partitioning on the coarse and on the fine grid. Here, we show possible effects on partitionings with nonsquare subdomains. In this example, we have $11 \times 9$ grid points on the fine grid. Accordingly, the coarse grid consists of $6 \times 5$ grid points. The coarse and fine grids are mapped to $2 \times 2$ processors. Obviously, the geometric regions of the overlap are not uniform. In some parts of the domain and in the corresponding processes the overlap on the coarse grid corresponds to a larger geometric region than on the fine grid (simply because $H > h$). $\triangle$

Let us now assume that we have performed one (or several) GS-RB smoothing steps on the fine grid and that the approximations in the overlap regions have been updated afterwards. Then we can apply the HW restriction. There is no need for any kind of communication since it coincides with half injection if preceded by GS-RB smoothing (see Remark 2.7.3). All data required for the computation of defects at coarse grid points is available (see Fig. 6.7).

For the prolongation back to the fine grid, we have a similar situation. After performing one or several smoothing steps on the coarse grid and updating the overlap regions afterwards, we can immediately perform bilinear interpolation because all coarse grid data required for the interpolation is available in the same process (as can be seen from Fig. 6.7).

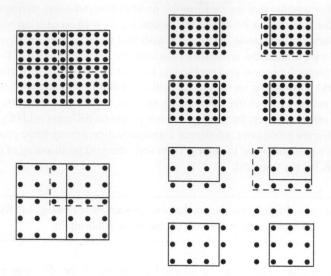

**Figure 6.6.** Distribution of subdomains and grid points to four processes on a fine and a coarse grid (overlap $w = 1$).

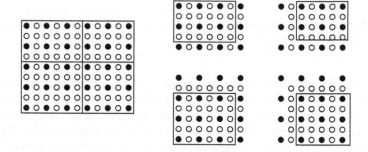

**Figure 6.7.** Overlap regions of the four processes: $\bullet$, coarse grid points; $\circ$, fine grid points.

It is thus sufficient in this parallel multigrid algorithm to perform communication only after each smoothing half-step. With an overlap width of $w = 2$, an update of the overlap regions after each full smoothing step is sufficient as discussed in Remark 6.2.7.

**Remark 6.3.1 (full weighting)** If we replace HW in the above parallel multigrid algorithm by FW, there is an increase of the communication requirements because the defects in the overlap areas are then needed. They can either be obtained by an additional communication step or be computed by each process itself by the use of an overlap width $w = 2$. ≫

**Remark 6.3.2 (cubic interpolation)**  Performing cubic interpolation (e.g. in FMG) with symmetric interpolation formulas near the internal subgrid boundaries is not directly possible if $w = 1$. The two possibilities discussed for FW, i.e. to use a larger overlap or additional communication to provide the data for the cubic interpolation, are also applicable here. Alternatively, one can apply *nonsymmetric* cubic interpolation formulas in such points. Usually, the results of FMG are hardly affected by this modification. With such a modification, the sequential and the parallel algorithms are, however, no longer equivalent. Moreover, the algorithm is different (not equivalent) for any two different partitionings. $\gg$

**Remark 6.3.3 (FAS)**  An additional communication step is needed if we use the FAS instead of the correction scheme. After the restriction to the coarse grid, we do not have any current approximations in the overlap regions. We thus need an overlap update *before the smoothing procedure starts* on the coarse grid. In the correction scheme, this communication is not necessary because we usually start with 0 as the initial approximation on the coarse grid. $\gg$

### 6.3.2 Multigrid and the Very Coarse Grids

If we process coarser and coarser grids during a multigrid cycle, first smaller and smaller numbers of grid points are mapped to each process. Then more and more processes no longer have grid points on these very coarse grids, and finally only one or a few processes have one (or a few) grid points. At the same time, the relative communication overhead on the grids (as compared to the time for arithmetic computations) increases and may finally dominate the arithmetic work on the very coarse grids. This can result in a significant loss of efficiency for the overall parallel application. In particular, in W-cycles these coarse grids are frequently processed.

Whereas idling processes on very coarse grids seem to be the main problem at first sight, experience and theoretical considerations show that *the large communication overhead on the very coarse grids is usually more annoying than the idling processes*. Algorithmically, the computations continue as usual in the processes which still have grid points. The other ones are idle until the multigrid algorithm returns to a level, on which these processes have grid points again. We summarize the important aspects:

- On coarse grids, the ratio between communication and computation becomes worse than on fine grids, up to a (possibly) large communication overhead on very coarse grids.
- The time spent on very coarse grids in W-cycles may become unacceptable.
- On very coarse grids we may have (many) idle processes.
- On coarse grids, the communication is no longer local (in the sense of finest grid locality).

---

**Remark 6.3.4**  In practice, it depends on the particular application under consideration to which extent the coarse grids reduce the parallel efficiency. For Poisson's equation on a Cartesian grid (a scalar equation with few arithmetic operations per grid point), for example, the communication overhead on the coarse grids may have a strong impact on the parallel efficiency. In that respect Poisson's equation is a hard test case for multigrid on a parallel system.

For more complex applications such as nonlinear systems of equations on general curvilinear grids (applications with many arithmetic operations per grid point), the effect is much less severe. Really large-scale parallel applications (which are the ones that need parallel processing) are dominated by computational work on fine grids, at least in the case of V- or F-cycles.  ≫

There are various approaches to overcome the *problem of the communication overhead*.

(1) First, if the parallel efficiency on a particular parallel system is not strongly affected by the communication on coarse grids, we can stay with this approach. This variant may often be the best one.

(2) A possibility to reduce frequent communication on very coarse grids is known as the *agglomeration technique* [192]. The idea of this approach is to "agglomerate" grid points to new "process units" and to redistribute these new units to a subset of the active processes. For example, instead of using 64 processors for 64 coarse grid points and performing communication between them, it can be more efficient to group these 64 grid points to sets of four or 16 points (corresponding to 16 or four process units, see Fig. 6.8) or even to one group of 64 points (to one process unit). This means that only some of the processors are responsible for the computations on these very coarse grids, the majority of processors being idle. *Communication is avoided by this grouping or agglomeration*, and the fact that most of the processors become idle turns out to be acceptable for many computers (although, of course, the agglomeration itself requires communication).

Agglomeration can also be applied within FMG. On the very coarse grids only a subset of the available processes is then responsible for the computations, whereas the full number of processes is employed on finer grids.

(3) A third approach to treating the coarse grid problem is to redefine what the coarsest grid is, i.e. reduce the number of levels. One way is to define the coarsest grid such that *each process has at least one grid point*. Since the coarsest grid then consists of $O(P)$ points, the parallel algorithm is, in general, different from the sequential one: the parallel algorithm does not process the *coarsest possible grid* (truncated cycle, see Fig. 6.9). The efficiency of this strategy then depends on the solution procedure on the coarsest grid. One will usually employ an iterative method on the coarsest grid. Then it is important to use parallel iteration schemes with good convergence properties. Remember that, for example, $\omega$-GS-RB (GS-RB with an optimal overrelaxation parameter $\omega$) has much better *convergence* properties than GS-RB for Model Problem 1.

In the following, we will briefly show that *it is often not necessary to solve the coarsest grid problem exactly*, which is particularly interesting for the third approach. Of course, this is also valid in the sequential case, but a direct solution of the coarsest grid problem is often not problematic in that case.

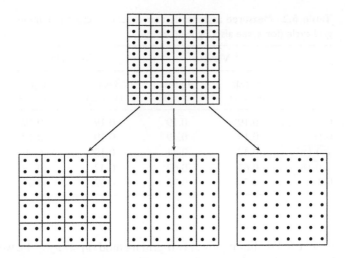

**Figure 6.8.** Agglomeration of 64 grid points to 16, 4 or 1 processes.

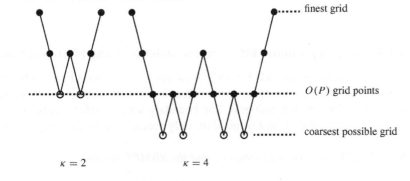

**Figure 6.9.** Truncated and nontruncated W-cycle.

In multigrid cycles the coarsest grid may be processed more than once, depending on the cycle type. For example, while a V-cycle processes the coarsest grid once, a W-cycle over $\ell$ levels involves $\kappa = 2^{\ell-2}$ visits of the coarsest grid. This offers an opportunity to reduce the accuracy criterion for these visits. For each visit of the coarsest grid, one may expect that a defect reduction of roughly $\rho^{1/\kappa}$ will be sufficient, where $\rho$ denotes the expected multigrid convergence factor. In this way, truncated W- or F-cycles become an interesting alternative in the parallel case [241].

**Example 6.3.2** In order to illustrate the feasibility of this approach, Table 6.2 presents convergence factors for Model Problem 1 with $128^2$ fine grid cells. Here, grid coarsening is *truncated* so that only three or four levels are used. The relaxation method used is GS-LEX. On the coarsest grid, GS-LEX iterations are performed until a defect reduction by a factor

**Table 6.2.** Measured asymptotic convergence factors per multi-grid cycle (for $\kappa$ see also Figure 6.9).

| $\varepsilon$ | V-cycle | | W-cycle | |
|---|---|---|---|---|
| | 3 grids ($\kappa = 1$) | 4 grids ($\kappa = 1$) | 3 grids ($\kappa = 2$) | 4 grids ($\kappa = 4$) |
| 0.1 | 0.19 | 0.19 | 0.19 | 0.19 |
| 0.18 | 0.19 | 0.19 | 0.19 | 0.19 |
| $\sqrt{0.18}$ | 0.42 | 0.41 | 0.19 | 0.19 |
| $\sqrt[4]{0.18}$ | 0.65 | 0.62 | 0.42 | 0.19 |
| 0.8 | 0.79 | 0.78 | 0.62 | 0.37 |

of $\varepsilon$ is achieved *per visit*. For the employed two-grid method ($\nu_1 = \nu_2 = 1$), we have $\rho_{\text{loc}} = 0.19$. Table 6.2, shows which choices of $\varepsilon$ are sufficient to obtain the expected multigrid convergence factors. Choosing $\varepsilon \leq 0.18^{1/\kappa}$ does indeed give the expected convergence for this example, whereas a larger value of $\varepsilon$ leads to a significant deterioration.                    △

### 6.3.3 Boundary-volume Effect and Scalability in the Multigrid Context

As discussed in Section 6.2.5, simple local iterative methods like $\omega$-JAC and GS-RB are $E$-scalable, and the boundary-volume effect is trivially fulfilled for them. The situation is somewhat different for multigrid. The boundary-volume effect is also valid for multigrid methods, but instead of the $E$-scalability, only a somewhat weaker property can be achieved.

**Result 6.3.1**    *For Model Problem 1 and the RBMPS, we have*

$$E(P, N) \longrightarrow 1 \quad for\ N \longrightarrow \infty,\ P\ fixed.$$

*With respect to E-scalability, we obtain the results in Table 6.3 for $N \to \infty$, $N/P = const$.*

*Proof:* On the fine grids, the ratio between communication and computation tends to 0 if $N \to \infty$ and $P$ is fixed. For $N \to \infty$, the relative influence of the fine grids will thus finally dominate the influence of the coarse grids, so that the boundary-volume effect is maintained.

The statements on $E$-scalability are obtained by similar arguments to those used in Section 6.1.2 on parallel multigrid complexity. From

$$E(P) = \frac{S(P)}{P} = \frac{T(1)}{T(P) \cdot P}, \qquad \frac{N}{P} = \text{const} \Rightarrow P \sim N,$$

**Table 6.3.** Asymptotic parallel efficiencies for $N \to \infty$, $N/P = $ const.

| Cycle | $E(P, N)$ |
|-------|-----------|
| V | $O(1/\log P)$ |
| F | $O(1/\log^2 P)$ |
| W | $O(1/\sqrt{P})$ |
| FMG (V-cycle) | $O(1/\log^2 P)$ |

we find

$$\text{V-cycle:} \quad T(P) = O(\log P \log \varepsilon) \Rightarrow E(P) = O(1/\log P)$$

$$\text{F-cycle:} \quad T(P) = O(\log^2 P \log \varepsilon) \Rightarrow E(P) = O(1/\log^2 P)$$

$$\text{W-cycle:} \quad T(P) = O(\sqrt{P} \log \varepsilon) \Rightarrow E(P) = O(1/\sqrt{P})$$

$$\text{FMG (based on V-cycle):} \quad T(P) = O(\log^2 P) \Rightarrow E(P) = O(1/\log^2 P) \qquad \square$$

Compared to $E$-scalability, we loose a logarithmic term for V-cycles and a square root term for W-cycles.

**Remark 6.3.5 (vector computers)** Many of our parallel considerations carry over to the situation of vector computing. The parallelism of multigrid components can also be exploited on vector computers, and the complications on coarse grids (lower degree of parallelism) are also seen there (shorter vectors on coarse grids). $\gg$

### 6.3.4 Programming Parallel Systems

In general, writing programs for parallel computers with distributed memory can still be a difficult job for a programmer. It is therefore advisable to first check whether or not appropriate software packages are available. Several communication libraries (including freely available ones) exist, which can be used, for example, for all mapping and communications tasks. These libraries allow a strict separation of computation and communication in a program, which leads to easier and safer programming and to portability from one parallel system to another (see [448] for an example). In Section 10.4, we will describe the parallel multigrid software package $L_i SS$.

A typical parallel multigrid program for solving a PDE on a parallel computer with distributed memory proceeds as follows.

- First, the input data has to be made available to all processes. This can happen in two ways: each process reads the input file(s), or only one process (or a subset of the processes) performs this task and then distributes the information to the others. The initial phase includes also the identification of the process neighbors, the set up of the overlap regions and a distribution of parameters for the algorithm.

- After that, the processes start their calculations. After certain computational steps, for example, after a relaxation step and after the defect restriction to coarser grids, data is exchanged between process neighbors in order to update the overlap areas.
- The grid is coarsened as long as all processes are kept busy, i.e. until the last grid is reached on which each process contains at least one point (Strategy (3) in Section 6.3.2). On the coarsest grid, a fast (iterative) parallel solver is used. As alternatives, agglomeration techniques (Strategy (2) in Section 6.3.2) can be employed to continue to even coarser grids or the idling of processes (Strategy (1)) is allowed.
- During the computation, global results (i.e. results whose computation requires data from all grid points like residual norms) are assembled treewise and written to an output file. This allows controlling the convergence of the algorithm.
- After the computation, the distributed solution is collected and the results are written to the result file(s).

## 6.4 PARALLEL LINE SMOOTHERS

So far, we have discussed parallel multigrid and given some results for multigrid algorithms with point smoothers. In this section we will discuss parallel multigrid strategies for anisotropic problems.

**Remark 6.4.1 (semicoarsening)**   If the anisotropic operator $-\varepsilon u_{xx} - u_{yy}$ is treated with point smoothers and *semicoarsening techniques*, all the considerations in the previous sections still apply accordingly. In particular, the grid partitioning does not lead to any additional complications.                                                                                      $\gg$

We restrict our considerations to parallel line and plane smoothers for compact stencils (nine-point stencils in 2D and 27-point stencils in 3D). In Section 6.4.1, we will present a particular tridiagonal solver for line relaxation, which provides an additional potential for parallelization. In Section 6.4.2, we will show how this line solver can be parallelized in practice. Parallel line solvers are also important for parallel multigrid plane relaxation if the plane smoother uses line relaxation. Some considerations concerning parallel plane relaxation are summarized in Section 6.4.3.

In the following remark, we first discuss a straightforward combination of grid partitioning and zebra line and plane relaxations.

**Remark 6.4.2**   Line and plane smoothers of Gauss–Seidel type with an obvious degree of parallelism are the *zebra* line and *zebra* plane smoothers (see Sections 5.1 and 5.2). These smoothing schemes have a natural independence between every other line/plane, similar to the case of GS-RB.

Zebra line smoothing is easily parallelized if the grid partitioning is parallel to the lines to be solved. A simple example for this situation is shown in the $1 \times 16$ partitioning in Fig. 6.2. Since a whole line is located in one process, grid partitioning can be applied without modification. After every relaxation half-step, the overlap areas have to be updated. This holds, of course, for both 2D and 3D applications.

The situation is similar for plane relaxation in 3D applications. If the grid partitioning is parallel to the plane to be solved, i.e. if a plane always belongs to a single process, the parallelization is again straightforward. ≫

In order to exploit this natural parallelism of the zebra smoothers, the grid partitioning needs to be chosen appropriately. If the grid partitioning is orthogonal to the lines to be solved, the lines are no longer fully available in one process. Similarly, this approach is no longer suitable if *alternating* line or alternating plane relaxation is to be employed, for example, for a highly anisotropic problem with varying coefficients. In such cases, there are lines or planes, which do not belong to one process and we need parallel variants of them. We will discuss a parallel tridiagonal solver in the next section.

### 6.4.1  1D Reduction (or Cyclic Reduction) Methods

Line smoothing requires efficient solvers for tridiagonal matrices (or more generally, band matrices). One efficient way to solve these tridiagonal systems is a *1D reduction approach* originally introduced in [340] (better known as cyclic reduction). Here, we present the basic idea of this algorithm. In the following section, we will describe its parallelization.

Let $Au = f$ be an $(n - 1) \times (n - 1)$ linear system ($n = 2^\ell$) with a tridiagonal matrix:

$$A = \begin{bmatrix} b_1 & -c_1 & & & & \\ -a_2 & b_2 & -c_2 & & 0 & \\ & \ddots & \ddots & \ddots & & \\ & & \ddots & \ddots & \ddots & \\ 0 & & & \ddots & \ddots & -c_{n-2} \\ & & & & -a_{n-1} & b_{n-1} \end{bmatrix}, \quad b_i \neq 0 \text{ for all } i. \quad (6.4.1)$$

The idea behind the 1D reduction is to eliminate, in a first reduction step, $u_1$ and $u_3$ from the second equation, $u_3$ and $u_5$ from the fourth and so on, i.e. to decouple the odd and the even unknowns. The first step of the reduction method thus leads to the following system

$$\begin{bmatrix} & & & & \\ & P & & 0 & \\ & & & & \\ \hline & & b_1 & & \\ & & & b_3 & \\ & Q & & \ddots & \\ & & & & b_{n-1} \end{bmatrix} \begin{bmatrix} u_2 \\ u_4 \\ \cdot \\ u_{n-2} \\ u_1 \\ u_3 \\ \cdot \\ u_{n-1} \end{bmatrix} = \begin{bmatrix} R_2 \\ R_4 \\ \cdot \\ R_{n-2} \\ f_1 \\ f_3 \\ \cdot \\ f_{n-1} \end{bmatrix}. \quad (6.4.2)$$

Here, the upper part $P$ of this matrix is

$$
\begin{bmatrix}
B_2 & -C_2 & & & \\
-A_4 & B_4 & -C_4 & & \\
& \cdots & \cdots & \cdots & \\
& & \cdots & \cdots & \cdots \\
& & & -A_{n-2} & B_{n-2}
\end{bmatrix}
\begin{bmatrix}
u_2 \\
u_4 \\
\cdot \\
\cdot \\
u_{n-2}
\end{bmatrix}
=
\begin{bmatrix}
R_2 \\
R_4 \\
\cdot \\
\cdot \\
R_{n-2}
\end{bmatrix}
\qquad (6.4.3)
$$

where, the values of $A_i$, $B_i$, $C_i$, $R_i$ $(i = 2, 4, \ldots, n - 2)$ are obtained from the values of $a_j$, $b_j$, $c_j$, $f_j$ by

$$
\alpha_i = a_i / b_{i-1}, \qquad \gamma_i = c_i / b_{i+1},
$$
$$
A_i = \alpha_i a_{i-1}, \qquad C_i = \gamma_i c_{i+1},
$$
$$
B_i = -\alpha_i c_{i-1} + b_i - \gamma_i a_{i+1},
$$
$$
R_i = \alpha_i f_{i-1} + f_i + \gamma_i f_{i+1}.
$$

On this remainder matrix $P$ the same procedure can be applied as on the original system (second reduction step). This can be continued until after $\ell - 1$ reduction steps a completely reduced system with a lower triangular matrix is obtained [342]. The computation of the solution is then trivially achieved by a back substitution.

**Remark 6.4.3**   In stencil notation, the original equations are written as

$$
\begin{bmatrix} -a_i & b_i & -c_i \end{bmatrix} u = f_i \quad (i = 1, 2, \ldots, n - 1) \qquad (6.4.4)
$$

and the equations of the remainder matrix $P$ are

$$
\begin{bmatrix} -A_i & B_i & -C_i \end{bmatrix} u = R_i \quad (i = 2, 4, \ldots, n) \qquad (6.4.5)
$$

$\gg$

The reduction method for tridiagonal systems can be regarded as a special case of a 1D multigrid method. Compared to other tridiagonal solvers, the cyclic reduction method is distinguished by its particular numerical stability [343, 450].

### 6.4.2 Cyclic Reduction and Grid Partitioning

The method of cyclic reduction is well suited for parallelization. Whereas the sequential version needs $O(n)$ operations, the parallel complexity (see Section 6.1.2) is $O(\log n)$. The number of reduction steps needed is $O(\log n)$ and each reduction step and each substitution step can be performed in parallel.

**Remark 6.4.4**   Tridiagonal cyclic reduction needs 17 operations per grid point, the usual tridiagonal Gaussian elimination needs eight operations. In the parallel case, however, the operation count of cyclic reduction remains unchanged, whereas the parallel Gaussian algorithm loses its advantage since fill-in (outside of the three diagonals) occurs during the elimination process. If, on the other hand, many lines are solved simultaneously, the parallel efficiency of standard Gauss algorithms improves compared to cyclic reduction. A detailed overview of parallel tridiagonal solvers including comparisons is presented, for example in [221, 267].                                                                                      ≫

In the following, we will describe cyclic reduction in a grid partitioning context. Let us assume that the grid lines to which the tridiagonal systems correspond, are distributed over several processes (see Fig. 6.10).

Consequently, we concentrate on the following problem. Given a number $J \geq 1$ of tridiagonal systems with $n - 1$ unknowns each, find a parallel algorithm for solving these systems on $P$ processes, assuming that process $k$ ($1 \leq k \leq P$) knows only a certain number of the equations of each of the $J$ tridiagonal systems.

**Remark 6.4.5**   First, one should note that a rearrangement so that each processor can work with full tridiagonal systems, i.e. a global redistribution of the data (such that each process receives about $J/P$ complete systems and a standard sequential solver can be applied) is, in general, rather expensive with respect to communication.                              ≫

Thus, we assume that the systems need to remain distributed during the solution procedure. Concretely, we assume a partitioning as indicated in Fig. 6.10 (where $n - 1 = 15$ and $P = 4$).

We now outline how the first reduction step can be carried out in parallel: For $k = 2, \ldots, P$, the first equation of process $k$ coincides with the last equation of process $k - 1$. We call the equations shared by two neighbor processes *interface equations* and the remaining ones the inner equations, corresponding to (and being known by) only one particular process. The inner unknowns of different processes are not directly coupled to each other. Thus, elimination operations only involving inner equations may be performed fully in parallel by the different processes. At the end of the reduction step, the interface equations are handled. The setup of the new equations at the interfaces requires communication.

Each process can apply standard reduction steps recursively until only interface equations are left. The handling of this remaining system is explained in detail in [153]. For

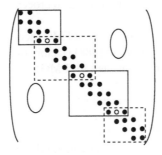

**Figure 6.10.** A tridiagonal system distributed over four processors. ○, elements shared by two processors.

sufficiently large $n$ (and $P$ fixed), the size of the interface system is negligible. Assuming that this system has been solved and each processor knows the values of its interface unknowns, the inner unknowns can be solved for by a parallel back substitution.

**Remark 6.4.6**   When applying a parallel tridiagonal solver like cyclic reduction in a 2D or 3D multigrid algorithm, an efficient implementation requires that the data which have to be communicated between different processes are *not* sent separately for each line, but *collectively*, i.e. in one single message for all lines in a process (see the discussion in Section 6.2.4).                                                                                   $\gg$

### 6.4.3 Parallel Plane Relaxation

For the implementation of plane relaxation in a 3D multigrid code, it is reasonable to use parallel 2D-multigrid algorithms. In order to illustrate an efficient incorporation of 2D multigrid into the smoothing components of a 3D multigrid program, we consider the implementation of $(x, y)$-plane relaxation.

Grid partitioning with respect to only one direction has already been discussed. If we now consider a general grid partitioning, the 2D multigrid algorithm also has to be parallel. Again, data to be exchanged between processes in each of the parallel 2D solvers should be collected (as for parallel line relaxations in Remark 6.4.6), according to the following rule.

*Keep the number of messages to a minimum.* This means that data belonging to different planes are collected and sent as a single message. After each 2D relaxation (or restriction or interpolation) applied simultaneously to all incorporated planes, the data to be sent to a specific neighbor process is collected (from all planes) and sent in a single message.

This leads to a considerably better parallel efficiency of the 3D multigrid method [153], especially on parallel systems with relatively large start-up times $\alpha$.

**Remark 6.4.7**   The degree of parallelism is lower for multigrid plane relaxation than for point relaxation. Plane relaxation by 2D multigrid requires more coarse grid processing. In particular, each multigrid plane relaxation step also processes coarser grids, independent of the grid level one starts on.

Correspondingly, the parallel complexity of FMG incorporating alternating plane relaxation is $O(\log^4 N)$ as has been shown in [388] (factors of $\log N$ for the V-cycle, the plane smoother, the line solver within a plane and for FMG).                                              $\gg$

**Remark 6.4.8 (parallel 3D semicoarsening)**   Also in 3D, methods based on pointwise smoothing and semicoarsening can easily be parallelized with grid partitioning. More robust variants, like *multiple semicoarsening* methods (see Remark 5.2.5) use many coarse grids. Parallelization with grid partitioning for these methods has larger communication requirements, since communication takes place on all coarse grids. A parallel variant of multiple semicoarsening based on line smoothing has been presented in [411].                              $\gg$

## 6.5 MODIFICATIONS OF MULTIGRID AND RELATED APPROACHES

Most of the parallel approaches described so far were oriented to, and essentially equivalent to, sequential multigrid. In the parallel context, however, other variants exist, where the equivalence to sequential multigrid is no longer maintained.

For example, in our description of parallel multigrid with grid partitioning, we have assumed that the data needed by neighbor processes is communicated *on all levels*. For parallel computers with low communication bandwith and/or high latency (like loosely coupled PC clusters), one may try to avoid communication on some grid levels and thus make the calculations on the subgrids (subdomains) more independent. Careful analysis on the algorithm is necessary if the communication on certain grid levels is completely removed. Some variants have been discussed in [73, 192]. If one tries to reduce communication in standard parallel multigrid, the resulting algorithms are more similar to versions of the domain decomposition (DD) methods.

### 6.5.1 Domain Decomposition Methods: a Brief Survey

Since there are two books [306, 362] on DD methods available, we will only make some remarks on connections to multigrid. Algorithmic relations between DD and multigrid have also been discussed in [362]. From a theoretical point of view, connections between multigrid and DD are also summarized in Appendix B.

One root of the domain decomposition development is the classical alternating Schwarz method. For simplicity, we consider the problem

$$
\begin{aligned}
-\Delta u &= f^\Omega(x, y) \quad (\Omega) \\
u &= f^\Gamma(x, y) \quad (\Gamma = \partial\Omega)
\end{aligned}
\tag{6.5.1}
$$

in the rectangular domain $\Omega = (0, 2) \times (0, 1)$. In order to illustrate the DD idea, we use a decomposition of $\Omega$ into two overlapping domains

$$
\Omega_1 = (0, 1 + \delta) \times (0, 1)
$$
$$
\Omega_2 = (1 - \delta, 2) \times (0, 1)
$$

(see Figure 6.11). The parameter $\delta$ controls the overlap $\Omega_1 \cap \Omega_2$. By $\Gamma_1$ and $\Gamma_2$, we denote the interior boundary lines

$$
\Gamma_1 = \{(1 + \delta, y): 0 \le y \le 1\}, \qquad \Gamma_2 = \{(1 - \delta, y): 0 \le y \le 1\}.
$$

In the classical alternating Schwarz method, the subproblems in $\Omega_1$ and in $\Omega_2$ are solved alternatingly, according to the iteration

$$
\begin{array}{llll}
-\Delta u_1^{m+1/2} = f^\Omega & (\Omega_1) & -\Delta u_2^{m+1} = f^\Omega & (\Omega_2) \\
u_1^{m+1/2} = f^\Gamma & (\partial\Omega_1 \setminus \Gamma_1) & u_2^{m+1} = f^\Gamma & (\partial\Omega_2 \setminus \Gamma_2) \\
u_1^{m+1/2} = u_2^m & (\Gamma_1) & u_2^{m+1} = u_1^{m+1/2} & (\Gamma_2)
\end{array}
\tag{6.5.2}
$$

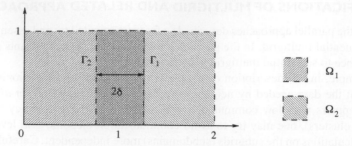

**Figure 6.11.** A domain $\Omega$ divided into two overlapping parts $\Omega_1$ and $\Omega_2$.

with, for example, $u^0 = 0$. The convergence speed of the above iteration depends on the overlap parameter $\delta$. For the above example, the convergence factor $\rho$ behaves like

$$\rho \approx 1 - \alpha\delta + O(\delta^2)$$

with some constant $\alpha$: *the smaller $\delta$, the slower the convergence; for $\delta \to 0$ the convergence factor tends to 1.*

From a practical point of view, a solver for each of the subproblems has to be applied in each iteration step. For that purpose, we consider discrete analogs of the above subproblems and thus obtain a discrete version of the alternating Schwarz method on $\Omega_{1,h}$ and $\Omega_{2,h}$. *A trivial combination of DD and multigrid is to use multigrid for the solution of the subproblems.* Of course, the fact that multigrid allows a fast solution of the subproblem does not help with respect to the limited overall convergence which is determined by the size of the overlap.

Many extensions and modifications of the classical method have been proposed: extension to many subdomains, so-called additive versions, DD with one or several coarse levels etc. For a rough survey on DD methods, we will use matrix terminology to avoid formalizing the discrete versions of different DD approaches and specifying the corresponding grids and spaces, which is not needed for our purposes. In this formulation, $A$ is the matrix corresponding to the discrete version of the original problem (6.5.1) and $u^m, u^{m+1/2}, d^m, d^{m+1/2}$ ($m = 0, 1, \dots$) are "full" vectors corresponding to $\Omega_h$.

We start with the alternating Schwarz method. We denote by $A_1$ and $A_2$ the matrices belonging to the discrete analogs of the problems on $\Omega_1$ and $\Omega_2$ respectively. Using a defect formulation for both half-steps, one complete step of the Schwarz iteration reads

$$u^{m+1/2} = u^m + P_1 A_1^{-1} R_1 d^m, \quad \text{where } d^m = f - Au^m$$
$$u^{m+1} = u^{m+1/2} + P_2 A_2^{-1} R_2 d^{m+1/2}, \quad \text{where } d^{m+1/2} = f - Au^{m+1/2}.$$

Here $R_1$, $R_2$ denote matrices which restrict the full $\Omega_h$ vectors to $\Omega_{h,1}$ and $\Omega_{h,2}$, respectively, whereas $P_1$, $P_2$ extend the vectors defined on $\Omega_{h,1}$, $\Omega_{h,2}$ to full $\Omega_h$ vectors (extension by 0). That only the $\Omega_{1,h}$ and the $\Omega_{2,h}$ parts of $u^{m+1/2}$ and $u^{m+1}$ are updated at each half-step of the iteration is reflected by the terms $P_1 A_1^{-1} R_1$ and $P_2 A_2^{-1} R_2$. From the above representation we find that the iteration matrix for a complete iteration of the Schwarz method is given by

$$M = I - P_1 A_1^{-1} R_1 A - P_2 A_2^{-1} R_2 A + P_2 A_2^{-1} R_2 A P_1 A_1^{-1} R_1 A. \qquad (6.5.3)$$

Here the last term is characteristic for the *alternating* character of the Schwarz method, in which the $\Omega_1$ problem is solved in the first half-step and the result is used (on $\Gamma_2$) to solve the $\Omega_2$ problem. If we neglect this term, we obtain the *additive* variant of Schwarz's method characterized by

$$M_{\text{add}} = I - P_1 A_1^{-1} R_1 A - P_2 A_2^{-1} R_2 A. \qquad (6.5.4)$$

In this setting, the $\Omega_1$ and the $\Omega_2$ problem can be solved simultaneously. In that respect, the additive variant is the natural parallel version of the Schwarz method. For distinction, the original alternating approach (6.5.3) is called *multiplicative* ($M = M_{\text{mult}}$). The additive variant corresponds to a block Jacobi-type method and the multiplicative variant to a block Gauss–Seidel-type method where the $\Omega_1$ and the $\Omega_2$ problems characterize the blocks. If we generalize the additive and the multiplicative version of Schwarz's method to $p$ domains, the corresponding iteration matrices become

$$M_{\text{add}} = I - \sum_{j=1}^{p} P_j A_j^{-1} R_j A \qquad (6.5.5)$$

and

$$M_{\text{mult}} = \prod_{j=1}^{p} (I - P_j A_j^{-1} R_j A). \qquad (6.5.6)$$

The fact that the pure Schwarz methods, whether multiplicative or additive, are only slowly convergent for small overlap is worse for the many domains case. Therefore *acceleration techniques* have been introduced. We mention three of them.

**Remark 6.5.1 (DD as a preconditioner)**   In the first approach, the Schwarz methods are not used as iterative *solvers*, but as *preconditioners*. The use of Schwarz methods as preconditioners is included in the above formalism by writing $M$ in the form $M = I - CA$ where

$$C_{\text{add}} = \sum_{j=1}^{p} P_j A_j^{-1} R_j$$

in the additive case and

$$C_{\text{mult}} = \left[ I - \prod_{j=1}^{p} (I - P_j A_j^{-1} R_j A) \right] A^{-1}$$

in the multiplicative case. (The original additive and multiplicative Schwarz methods are preconditioned Richardson iterations, with $\tau = 1, (1.6.2)$, in this interpretation.) In practice, the above preconditioners are used with Krylov subspace methods (see Section 7.8). Under certain assumptions, the additive preconditioner $C_{add}$ can be arranged in a symmetric way (for example, by choosing $P_j = R_j^T$) so that the conjugate gradient method can be used for acceleration. In the multiplicative case, GMRES (see Section 7.8.3) is usually chosen for acceleration.                                                                        ≫

**Remark 6.5.2 (DD with a coarse grid)**   The second acceleration possibility which is usually combined with the first one, is characterized by a *coarse grid* $\Omega_H$ (in addition to $\Omega_h$). This turns out to be necessary for satisfactory convergence, in particular if $p$ is large.

Formally, this "two-level approach" can be included in the above formalism by replacing

$$\sum_{j=1}^{p} \text{ by } \sum_{j=0}^{p} \quad \text{and} \quad \prod_{j=1}^{p} \text{ by } \prod_{j=0}^{p}.$$

Here, the term corresponding to index $j = 0$ characterizes the coarse grid part; it has the same form as the other summands (or factors).

Let us denote the typical diameter of a subdomain by $H$. It has been shown [132] that in this case the condition number $\kappa(C_{add}A)$ fulfills

$$\kappa(C_{add}A) \leq \text{const} \left(1 + \frac{H}{\delta}\right).$$

So, the condition number of the preconditioned system for the two-level overlapping Schwarz method (with exact solution of the subdomain problems) is bounded independently of $h$ and $H$, if the overlap is uniformly of width $O(H)$. Similar bounds also hold for the convergence factors of the multiplicative version (see the discussion in [132] and Chapter 5 of [362]). The proof for this statement is outlined in Appendix B.          ≫

**Remark 6.5.3 (DD as a smoother for multigrid)**   The third idea to improve the efficiency of Schwarz's method is not to use multigrid as a solver *in an overall DD context*, but to use *the DD idea for smoothing in an overall multigrid context*. In this sense, the roles of DD and MG are exchanged and the efficiency of multigrid is maintained. The convergence factors of multigrid with DD-type smoothing turn out to be small and independent of the overlap width $\delta$, under general assumptions. This approach has been studied for model problems [176, 339, 378]. Whereas in [378] the idea of the alternating Schwarz method (more precisely, of one Gauss–Seidel type iteration for each subdomain) is used for DD-smoothing, a DD-smoother based on a Jacobi-type iteration for each subdomain, which can thus be executed in parallel, is presented in [176]. Some results are shown there, which compare the influence of these two DD-smoothers on the multigrid convergence for the above example. The multigrid convergence factors are approximately the same for both approaches ($\rho \approx 0.06$ for $h = 1/128$). These methods are very close to grid partitioning.          ≫

Finally, we would like to mention that a lot of theoretical and practical work is devoted to *nonoverlapping DD approaches* (also known as *iterative substructuring* methods). In the nonoverlapping approaches, the idea of *parallel* treatment of the subproblems is more strongly addressed than in the overlapping case. For a general description and survey articles, see, e.g., [131, 133, 138, 237, 306, 362, 418] and the references therein.

Among the best known and most tested nonoverlapping domain decomposition methods for elliptic PDEs are the FETI and Neumann–Neumann families of iterative substructuring algorithms. The former is a domain decomposition method with Lagrange multipliers which are introduced to enforce the "intersubdomain continuity". The algorithm iterates on the Lagrange multipliers (the dual variables) and is often called a dual domain decomposition method (see [138, 217, 255, 324] and the references therein). The Neumann–Neumann method is a "primal" domain decomposition algorithm which iterates on the original set of unknowns (the primal variables). For an introduction to the Neumann–Neumann algorithm and references to earlier work on it and its predecessors, see [237]. Some further references are [131, 133, 254]. A unified analysis for both, FETI and Neumann–Neumann algorithms, has been given [217].

**Remark 6.5.4** In our view, DD is the more natural, the more weakly the respective subproblems are coupled. In particular, *multidisciplinary applications*, with different PDEs on different parts of a domain, for which "coupling algorithms" are needed, may benefit from DD theory and practice. $\gg$

### 6.5.2 Multigrid Related Parallel Approaches

In this section, we will list some multigrid related approaches, which are of interest in the context of parallel computing, in the form of remarks.

**Remark 6.5.5 (additive multigrid)** In the additive multigrid approach [55, 161, 425], smoothing is performed on all levels simultaneously, i.e. the grids are *not* processed sequentially.

The *additive multigrid V-cycle* on level $k$, $u_k^{m+1} = \mathrm{ADMG}(k, u_k^m, L_k, f_k, \nu)$ consists of the following steps:

- Compute the defect on level $k$ (if $k > 1$) and restrict it:  $d_{k-1}^m = I_k^{k-1}(f_k - L_k u_k^m)$.
- Perform $\nu$ smoothing steps on level $k$:  $\overline{u}_k^m = \mathrm{SMOOTH}^\nu(u_k^m, L_k, f_k)$.
- If $k = 1$, use a direct or fast iterative solver for $L_{k-1}\hat{v}_{k-1}^m = d_{k-1}^m$.
  If $k > 1$, perform one $(k-1)$-grid additive cycle to $L_{k-1}\hat{v}_{k-1}^m = d_{k-1}^m$ using the zero grid function as the first approximation

$$\hat{v}_{k-1}^m = \mathrm{ADMG}(k-1, 0, L_{k-1}, d_{k-1}^m, \nu).$$

- Interpolate the correction:  $u_k^{m+1} = \overline{u}_k^m + \omega_I I_{k-1}^k \hat{v}_{k-1}^m$.

Here, $\omega_I$ is an interpolation weight. In the additive multigrid V-cycle, the smoothing steps on different grid levels are independent of each other and can be performed in parallel. Restriction and interpolation are performed sequentially.

An additive multigrid method is usually employed as a preconditioner (see Section 7.8) rather than as an iterative solver. (The convergence is, in general, not guaranteed if used as a solver.) A well-known example of an additive multigrid preconditioner is the BPX method [55].

Standard multigrid is also called *multiplicative* multigrid if the difference to the additive version is to be emphasized. A detailed study of additive multigrid versus standard (multiplicative) multigrid has been made, for example, in [32]. Additive multigrid has some interesting theoretical features (see Appendix B), but does not compete with standard multigrid in most practical applications.                                              ≫

**Remark 6.5.6 (parallel superconvergent multigrid method)**   The parallel superconvergent multigrid method (PSMG) [143, 144] and related ideas essentially refer to massively parallel computing. To keep all processors of a massively parallel system busy especially on coarse grids, PSMG works simultaneously on many different grids, instead of working only on the standard coarse grid hierarchy. In this way, one can try to accelerate the multigrid convergence (see [144] for a detailed discussion and for results on Poisson's equation). ≫

**Remark 6.5.7 (sparse grids, combination method)**   In the context of the so-called sparse grid discretizations [443], the "combination method" [165] is a variant, that is naturally parallel. Mathematically, the combination method can be interpreted as a specific extrapolation technique, in which the sparse grid solution can be computed directly from the solutions of several small problems. A particular feature of this method is that no communication is needed during long phases of the computation, i.e. the small problems are fully independent of each other, so that this method may be interesting for clusters of workstations or PCs. After the computation, the different parts must be collected to obtain the approximation of the complete solution.

In [152], for example, it has been demonstrated for Model Problem 2 how the combination method can be combined with 3D anisotropic multigrid. The sparse grid approach itself already leads to anisotropic grid structures.                                              ≫

# 7

# MORE ADVANCED MULTIGRID

In this chapter, we will introduce more advanced multigrid methods. Such methods are needed for important classes of more involved problems like the convection–diffusion equation and problems characterized by mixed derivatives $u_{xy}$. Such problems can lead to bad $h$-ellipticity measures, as we have seen in Section 4.7.

> As we have discussed already for anisotropic diffusion problems, there are basically two strategies to deal with problems that cannot be solved efficiently with the basic multigrid components. One strategy is to adapt the smoother to the problem at hand but to keep the "standard" coarse grid correction components. In most parts of this chapter, we will apply this strategy to deal with more difficult problems and thus consider standard coarsening.
>
> In Appendix A, in the context of algebraic multigrid, the second strategy, i.e. to maintain pointwise smoothing and to change the coarse grid correction, is applied to some of the problem classes discussed in this chapter.

In Sections 7.1 and 7.3, we discuss first- and higher order accurate upwind-type discretizations of the convection–diffusion equation in detail. In Sections 7.2 and 7.4, we focus on the corresponding multigrid treatment. We consider various smoothers and point out certain problems arising from the coarse grid correction.

In Section 7.5, incomplete factorization (ILU type) smoothers are introduced. They are an interesting alternative to line smoothers. These smoothers can handle problems with mixed derivatives as discussed in Section 7.6.

In Section 7.7, we will introduce a proper multigrid treatment for problems characterized by jumping coefficients in front of derivatives. Here, the standard coarse grid correction is modified.

Section 7.8 discusses the use of multigrid as a preconditioner, which can also be interpreted as a recombination of multigrid iterants. With this approach the convergence of difficult problems can often be improved by an outer iteration ("accelerated") while standard multigrid components are maintained.

## 7.1 THE CONVECTION–DIFFUSION EQUATION: DISCRETIZATION I

In studying anisotropic problems, we have discussed the multigrid treatment of an important class of problems with "vanishing ellipticity" for $\varepsilon \to 0$. By using line/plane relaxations and/or semicoarsening techniques, we are able to solve these problems with similar multigrid efficiency as observed for Poisson's equation.

Another, equally important, class of problems is the class of convection–diffusion problems. We again consider a 2D model problem.

**Model Problem 6**

$$Lu = -\varepsilon \Delta u + au_x + bu_y = f^{\Omega}(x, y) \quad (\Omega = (0, 1)^2) \tag{7.1.1}$$

with, for example, Dirichlet boundary conditions $u = f^{\Gamma}(x, y)$. Here, the parameter $\varepsilon > 0$ is constant, and $a, b$ are constants $\in \mathbb{R}$ or (smooth) functions $a(x, y), b(x, y)$.

The convection–diffusion equation (7.1.1) is an important problem in computational fluid dynamics. The combination of diffusion, modeled by $\Delta u$, and convection, modeled by $au_x + bu_y$, is often found in nature. The parameter $\varepsilon$ determines the ratio between diffusion and convection. Obviously, the convection–diffusion equation is a *singularly perturbed* equation: in the limit case $\varepsilon \to 0$ it is no longer elliptic, but hyperbolic (with the convection direction $(a, b)^T$). Furthermore, the convection–diffusion equation is a linear model for any of the momentum equations of the incompressible Navier–Stokes equations (to be discussed in Sections 8.6–8.8). This is another reason for our detailed considerations.

Specific complications already arise with the *discretization* of convection–diffusion equations. Whereas the Laplace operator in (7.1.1) can be discretized by the standard five-point approximation, the convective terms $au_x$ and $bu_y$ have to be discretized with care.

In the following, we will discuss several discretization schemes for the convective terms and the corresponding multigrid approaches. We will describe the difficulties in terms of finite differences on the Cartesian grid $\mathbf{G}_h$; the description also applies, however, to finite volume discretizations. We start with some simple considerations in 1D.

### 7.1.1 The 1D Case

Some of the discretization issues that occur in the context of convection–diffusion equations can be illustrated by the simple 1D problem

$$-\varepsilon u_{xx}(x) + au_x(x) = f^{\Omega}(x) \quad (0 < x < 1)$$
$$u(0) = f_0^{\Gamma}, \qquad u(1) = f_1^{\Gamma}. \tag{7.1.2}$$

The standard second-order discretization operator with central differences is

$$L_h = \frac{1}{h^2}[-ah/2 - \varepsilon \quad 2\varepsilon \quad ah/2 - \varepsilon]_h. \tag{7.1.3}$$

We first notice that this operator and the corresponding matrix $A_h$ are *not symmetric*.

**Example 7.1.1**  In this example we will see that central differencing leads to difficulties for *small $\varepsilon > 0$*.

For $a = 1$ and $f^\Omega = 0$, we obtain

$$-\varepsilon u_{xx} + u_x = 0 \quad (0 < x < 1).$$

If we prescribe the boundary conditions

$$u(0) = 0, \qquad u(1) = 1,$$

the solution of this problem, $u(x) = (1 - e^{x/\varepsilon})/(1 - e^{1/\varepsilon})$, has a sharp gradient (*boundary layer*) near the boundary $x = 1$. The width of the layer depends on the parameter $\varepsilon$. For $\varepsilon = 1/100$, this solution is presented graphically in Fig. 7.1.

The discrete solution of this problem discretized by central differencing is

$$u_h(x_k) = \frac{1 - q^k}{1 - q^n} \quad \text{with } q = \frac{2\varepsilon + h}{2\varepsilon - h},$$

where $h = 1/n$ is the mesh size and $x_k = kh$ $(k = 0, \ldots, n)$. This discrete solution is highly oscillating for $q < -1$, or, equivalently for $h > 2\varepsilon$. Such oscillations do not correspond to the solution of the differential problem. The central difference solution does, however, approximate the differential solution reasonably if $h \le 2\varepsilon$.                                        △

For (7.1.3), the condition for the suitability of central discretization is

$$\frac{h}{\varepsilon}|a| \le 2, \tag{7.1.4}$$

which is also called the *Péclet* condition.

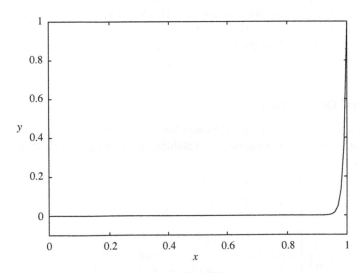

**Figure 7.1.** Analytical solution $(1 - e^{x/\varepsilon})/(1 - e^{1/\varepsilon})$ of the 1D convection–diffusion problem for $\varepsilon = 1/100$.

If the Péclet condition is not fulfilled, central differencing of the convective terms is not appropriate.

A discretization that overcomes the instability complications for small $\varepsilon$, is an *upwind* discretization. In a *first-order upwind* scheme, first derivatives are approximated by *one-sided* differences such that only upstream grid points are used in the discretization of the convection term. Depending on the sign of $a$, $u_x$ is discretized by

$$u_{x,h,\text{upwind}} = \begin{cases} 1/h[0 \quad -1 \quad 1]_h u_h & \text{if } a < 0 \\ 1/h[-1 \quad 1 \quad 0]_h u_h & \text{if } a > 0. \end{cases} \tag{7.1.5}$$

For $a > 0$, the first-order upwind discretization for $u_x$ leads to

$$L_h = -\frac{\varepsilon}{h^2}[1 \quad -2 \quad 1]_h + \frac{a}{h}[-1 \quad 1 \quad 0]_h. \tag{7.1.6}$$

A major drawback of this discretization is that it is only $O(h)$ accurate. For Example 7.1.1 this results in a "smearing" of the boundary layer in the discrete solution.

An important observation can, however, immediately be made here.

---

From (7.1.6), one can see that the stencil degenerates to

$$\frac{a}{h}[-1 \quad 1 \quad 0]_h$$

for $\varepsilon \to 0$. In that case, the discrete solution $u_h$ at $x_k = kh$ can be calculated immediately from $u_h(x_{k-1})$. Therefore, a *pointwise Gauss–Seidel relaxation* with "downsteam" numbering becomes an exact solver for $\varepsilon \to 0$. For small $\varepsilon$, only a few iterations are needed for convergence. This holds also in higher dimensions if first-order upwind discretization is used.

Such an iteration is called *downstream relaxation* or also downstream marching.

---

### 7.1.2 Central Differencing

We now consider standard central differences for the discretization of the convective terms in the 2D case. In stencil notation, the resulting (nonsymmetric) discrete operator for (7.1.1) reads

$$L_h = \frac{\varepsilon}{h^2} \begin{bmatrix} & -1 & \\ -1 & 4 & -1 \\ & -1 & \end{bmatrix}_h + \frac{a}{2h}[-1 \quad 0 \quad 1]_h + \frac{b}{2h} \begin{bmatrix} 1 \\ 0 \\ -1 \end{bmatrix}_h$$

$$= \frac{1}{h^2} \begin{bmatrix} & bh/2 - \varepsilon & \\ -ah/2 - \varepsilon & 4\varepsilon & ah/2 - \varepsilon \\ & -bh/2 - \varepsilon & \end{bmatrix}_h. \tag{7.1.7}$$

As in the 1D case, difficulties with central differences arise for *small* $\varepsilon > 0$.

**Example 7.1.2** Choosing $a = b = 1$ and $f^\Omega = 0$, the 2D convection–diffusion model problem corresponds to

$$-\varepsilon \Delta u + u_x + u_y = 0 \quad (\Omega = (0, 1)^2). \tag{7.1.8}$$

We prescribe Dirichlet boundary conditions

$$
\begin{aligned}
u &= 0 \quad &&\text{for} \quad x = 0, \ 0 \le y < 1, \ \text{and} \ y = 0, \ 0 \le x < 1, \\
u &= 1 \quad &&\text{for} \quad x = 1, \ 0 \le y < 1, \ \text{and} \ y = 1, \ 0 \le x \le 1.
\end{aligned}
$$

Using the central discretization (7.1.7), the discrete solution $u_h$ is presented in Fig. 7.2(a) along the line $y = 1/2$ for $\varepsilon = 10^{-2}$ and $h = 1/16$. As in the 1D case, the solution has wiggles, i.e. nonphysical oscillations, near the boundary layer. Even more oscillations appear for smaller $\varepsilon / h$ as is seen in Fig. 7.2(b) with $\varepsilon = 10^{-5}$ and $h = 1/64$. Obviously, $L_h$ obtained from central differencing becomes *unstable* for both cases. $\triangle$

In 2D the *Péclet* condition is

$$Pe := \frac{h}{\varepsilon} \max(|a|, |b|) \le 2. \tag{7.1.9}$$

The left-hand side of (7.1.9) is called the *mesh-Péclet number Pe*.

> *If the Péclet condition is fulfilled, $L_h$ based on central differencing gives a reasonable and stable $O(h^2)$-approximation for L. If it is not fulfilled, we have to expect stability complications (oscillations).*

Already in Fig. 7.2(a) ($\varepsilon = 1/100$, $h = 1/16$), we have $Pe = 6.25$ and the Péclet condition (7.1.9) is not satisfied.

With respect to the matrix corresponding to $L_h$ from (7.1.7), we see that some off-diagonal elements of the matrix become positive, if the Péclet condition is not satisfied. As a consequence, the matrix is no longer an *M-matrix* [403].

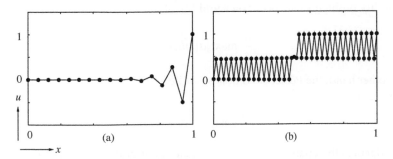

**Figure 7.2.** Wiggles in the discrete solutions obtained by the central discretization. (a) $\varepsilon = 10^{-2}$, $h = 1/16$; (b) $\varepsilon = 10^{-5}$, $h = 1/64$.

Another indication for the instability of the discrete operator $L_h$ is its *h-ellipticity measure*, for example,

$$\bar{E}_h(L_h) := \min \left\{ \frac{|\tilde{L}_h(\theta)|}{|L_h|} : \theta \in T^{\text{high}} \right\} \tag{7.1.10}$$

(see the discussion in Section 4.7). Here,

$$\tilde{L}_h(\theta, h) = \frac{\varepsilon}{h^2}(4 - 2\cos\theta_1 - 2\cos\theta_2) + \frac{i}{h}(a\sin\theta_1 + b\sin\theta_2)$$

represents the symbol of $L_h$ and

$$|L_h| = \frac{4\varepsilon}{h^2} + \left| \frac{a}{2h} - \frac{\varepsilon}{h^2} \right| + \left| -\frac{a}{2h} - \frac{\varepsilon}{h^2} \right| + \left| \frac{b}{2h} - \frac{\varepsilon}{h^2} \right| + \left| -\frac{b}{2h} - \frac{\varepsilon}{h^2} \right|$$

is the sum of the absolute values of the coefficients in the stencil. (For nonsymmetric operators $\bar{E}_h$ is easier to compute than $E_h$.) For small $\varepsilon$ and $h$ fixed, we find

$$\bar{E}_h(L_h) \le \frac{4\varepsilon}{4\varepsilon + h(|a| + |b|)},$$

the upper bound being obtained for $\theta = (0, \pi)$. For $\varepsilon \to 0$ and $h$ and $(a, b)$ $(\ne (0, 0))$ fixed, this leads to

$$\bar{E}_h(L_h) \longrightarrow 0.$$

**Remark 7.1.1**   For $\varepsilon = 0$, we have a *checkerboard-type instability* of $L_h$ (see Section 4.7): Any grid function which has constant values at the red (odd) grid points and constant values of a different size at the black (even) points lies in the kernel of the discrete operator.   $\gg$

**Remark 7.1.2 (multigrid behavior on coarse grids)**   Even if the $h$-ellipticity measure of central differencing is satisfactory on a fine grid, it decreases on coarse grids, which indicates corresponding problems for smoothers on coarse grids. If we use central differencing on all grids, stability problems may occur on a grid $\Omega_H$ if

$$\frac{H}{\varepsilon} \max(|a|, |b|) \ge 2.$$

If, on the other hand, the Péclet condition is fulfilled *on all grids* occurring in the multigrid hierarchy, i.e. if

$$\frac{h_0}{\varepsilon} \max(|a|, |b|) < 2,$$

where $h_0$ denotes the coarsest mesh size, standard multigrid algorithms will solve (7.1.1) without complications with a very high efficiency, since we have a "Poisson-like situation".   $\gg$

### 7.1.3 First-order Upwind Discretizations and Artificial Viscosity

Many flow phenomena are characterized by very small values of $\varepsilon$, realistic values being $\varepsilon = 10^{-5}$ or even smaller for typical flow problems (in water or air). For such problems, central differencing is only reasonable on extremely fine grids. However, this would require too much CPU time and memory and, hence, is not a realistic possibility even with state-of-the-art (parallel) supercomputers.

The instability complications for small $\varepsilon$ can be avoided by the use of first-order *upwind* discretizations. With $u_x$ discretized as in (7.1.5) and $u_y$ discretized correspondingly, we obtain the following first-order upwind stencil for (7.1.1) and general $a$ and $b$:

$$
L_h = \frac{\varepsilon}{h^2} \begin{bmatrix} & -1 & \\ -1 & 4 & -1 \\ & -1 & \end{bmatrix}_h + \frac{1}{2h} \begin{bmatrix} & b - |b| & \\ -a - |a| & 2(|a| + |b|) & a - |a| \\ & -b - |b| & \end{bmatrix}_h
$$

$$
= \frac{1}{h^2} \begin{bmatrix} & \frac{1}{2}h(b - |b|) - \varepsilon & \\ -\frac{1}{2}h(a + |a|) - \varepsilon & h(|a| + |b|) + 4\varepsilon & \frac{1}{2}h(a - |a|) - \varepsilon \\ & -\frac{1}{2}h(b + |b|) - \varepsilon & \end{bmatrix}_h . \quad (7.1.11)
$$

This discretization does not lead to stability problems. The corresponding matrix is an $M$-matrix. Also the $h$-ellipticity measure $E_h$ is good. We find, for example, $E_h \approx 0.3$, for $a = b = 1, h = 1/64$ and $\varepsilon = 10^{-5}$. However, as mentioned in the 1D case, the $O(h)$ accuracy is often not sufficient for engineering applications.

**Example 7.1.3** Here, we consider the width of the boundary layer for the problem from Example 7.1.2 with $\varepsilon = 1/100$ and $h = 1/64$. In this case, the Péclet condition is satisfied, so that we can compare the first-order upwind solution with the one obtained by central differencing. Figure 7.3 indicates that the width of the "discrete boundary layer" from the first-order upwind discretization is too "thick" (it is $O(h)$ instead of $O(\varepsilon)$). $\triangle$

A discretization similar to first-order upwind is obtained by the introduction of *artificial viscosity* in central difference discretizations. This means that $\varepsilon$ is replaced by some $\varepsilon_h \geq \varepsilon$ in (7.1.7). Here, $\varepsilon_h$ is chosen such that the Péclet condition (7.1.9) is fulfilled, for instance,

$$
\varepsilon_h = \max\left(\varepsilon, \frac{h}{2} \max(|a|, |b|)\right).
$$

This approach also leads to a stable discretization, but the drawback of only an $O(h)$ approximation remains (for $\varepsilon < h/2 \max(|a|, |b|)$).

In fact, the upwind discretization can be regarded as a special case of the artificial viscosity approach since, e.g. for $a > 0$,

$$
au_{x,h,\text{upwind}} = \frac{a}{h}[-1 \quad 1 \quad 0]_h u_h = \frac{a}{2h}[-1 \quad 0 \quad 1]_h u_h + \frac{a}{2h}[-1 \quad 2 \quad -1]_h u_h
$$

$$
(7.1.12)
$$

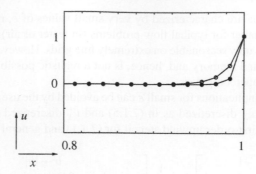

**Figure 7.3.** The boundary layer width for the central discretization (•) and first-order upwind discretization (○); $\varepsilon = 1/100$, $h = 1/64$.

(correspondingly for the $y$-direction). That is, the first-order upwind discretization of $au_x$ is equivalent to a combination of the central difference term and an extra viscous (or "dissipative") term

$$-\frac{ah}{2}u_{xx,h} = \frac{a}{2h}[-1 \quad 2 \quad -1]_h u_h.$$

Implicitly, this latter term is added to the diffusion term of the discrete convection–diffusion equation, so that the factor $\varepsilon_h := ah/2$ can be interpreted as "artificial viscosity".

## 7.2 THE CONVECTION–DIFFUSION EQUATION: MULTIGRID I

In this section, we will discuss multigrid features for the first-order upwind discretization of the convection–diffusion equation.

We first recall that downstream GS-LEX is close to an exact solver in the 1D case if $\varepsilon$ is small (see Section 7.1.1). The same is true in 2D and 3D for first-order upwind discretizations of convection-dominated problems. The *convection equation*, obtained for $\varepsilon = 0$, is hyperbolic. In that case, information propagates from the boundary along the characteristics governed by $(a, b)$.

Anyway, we will discuss the multigrid treatment of the convection–diffusion equation with a first-order upwind discretization in detail in the following subsections. This is because usually the convection–diffusion operator is part of more complicated operators in systems of equations, for which the specific behavior mentioned above cannot be easily exploited. Furthermore, moderate values of $\varepsilon$ are also of interest. The basic insights are also useful for other discretizations of first derivatives (see Section 7.3).

### 7.2.1 Smoothers for First-order Upwind Discretizations

To find a proper *smoother*, we use the LFA as introduced in Section 4.3. We consider Model Problem 6 with the *first-order upwind* discretization (7.1.11). If we apply $\omega$-JAC and assume that $a \geq 0$ and $b \geq 0$ , the smoothing analysis gives

$$\tilde{S}_h(\theta_1, \theta_2) = 1 - \omega + \omega \frac{2\varepsilon \cos \theta_1 + 2\varepsilon \cos \theta_2 + ahe^{-i\theta_1} + bhe^{-i\theta_2}}{4\varepsilon + ah + bh}. \tag{7.2.1}$$

For the special high frequency $(\theta_1, \theta_2) = (\pi, 0)$, for example, we find

$$\tilde{S}_h(\pi, 0) = 1 - \omega + \omega \frac{(-ah + bh)}{4\varepsilon + ah + bh}, \tag{7.2.2}$$

which tends to 1 for $\varepsilon \to 0$ and $a = 0$, independent of the parameter $\omega$. Obviously, $\omega$-JAC *is not a reasonable smoother in this situation.*

**Remark 7.2.1**   This result is interesting in the context of Theorem 4.7.1. Statement (3) of that theorem says that a pointwise $\omega$-JAC smoother exists with a smoothing factor $\mu_{\text{loc}} \leq$ const $< 1$ if the $h$-ellipticity measure $E_h$ is good and if the stencil of $L_h$ is symmetric. Since $E_h \approx 0.3$ (see Section 7.1.3) the above result shows that the symmetry of the stencil is indeed a necessary condition here. $\gg$

Now, we consider the smoothing properties of other Gauss–Seidel- and Jacobi-type smoothers. Table 7.1 presents smoothing factors $\mu_{\text{loc}}$ of several smoothers for three different combinations of $(a, b)$ and two values of $\varepsilon$, $10^{-2}$ and $10^{-6}$. Here, we have used underrelaxation ($\omega < 1$) for the Jacobi-type smoothers and, in the case of $\varepsilon = 10^{-6}$, also for GS-RB and zebra-type relaxations. According to the LFA, $\omega = 0.8$ is suitable in all these cases and has therefore been used.

Several conclusions can be drawn from this table. The smoothing factor of most smoothers depends sensitively on the parameters $(a, b)$ and $\varepsilon$. For $\varepsilon = 10^{-2}$, all smoothers are acceptable. For $\varepsilon = 10^{-6}$, however, one has to be careful with the choice of the smoother. The Jacobi-type smoothers considered are not competitive in many of the cases listed.

Smoothing analysis for GS-LEX (with the standard lexicographical ordering) results in satisfactory smoothing factors for positive constants $a$ and $b$. For $a < 0$ and $b \leq 0$, however, the smoothing factor tends to 1 as $\varepsilon \to 0$. This can easily be seen from the symbol of GS-LEX

$$\tilde{S}_h(\theta_1, \theta_2) = \frac{e^{i\theta_1}(\varepsilon - ah) + \varepsilon e^{i\theta_2}}{4\varepsilon - ah + bh - \varepsilon e^{-i\theta_1} - (\varepsilon + bh)e^{-i\theta_2}}. \tag{7.2.3}$$

For example, for $b = 0$, inserting the high frequency $(\theta_1, \theta_2) = (0, \pi)$ results in

$$\tilde{S}_h(0, \pi) = \frac{-ah}{4\varepsilon - ah}, \tag{7.2.4}$$

which tends to 1 for $\varepsilon \to 0$.

On the other hand, GS-BackLEX, point Gauss–Seidel relaxation with *backward lexicographical ordering* (see Fig. 7.4), results in good smoothing for this case.

> More generally, it can be shown that pointwise Gauss–Seidel relaxation is a good smoother if the ordering of the grid points is according to the direction of the convection vector $(a, b)$. We speak of *downstream* relaxation in that case. As mentioned above, the downstream relaxation is even a good *solver* if $\varepsilon$ is very small.

Examples for the efficiency of downstream relaxation are presented, for example, in [41, 86, 430].

Table 7.1 shows a similar behavior for *line relaxations*. Line Gauss–Seidel relaxation against the convection direction has bad smoothing properties. The alternating line Gauss–Seidel relaxation shows good smoothing factors for all the cases considered in Table 7.1.

**Remark 7.2.2** From Table 7.1 one might think that a combination of forward and backward lexicographical point Gauss–Seidel relaxations would be satisfactory smoothers for any

$$
\begin{array}{cccc}
4 & 3 & 2 & 1 \\
8 & 7 & 6 & 5 \\
12 & 11 & 10 & 9 \\
16 & 15 & 14 & 13
\end{array}
$$

**Figure 7.4.** Backward (lexicographical) ordering of grid points.

**Table 7.1.** Smoothing factors $\mu_{\text{loc}} = \mu_{\text{loc}}(\varepsilon, h)$ of different smoothers for the convection–diffusion equation with $h = 1/256$.

| Smoother | $(a, b)$ | $(1, 0)$ | $(-1, 0)$ | $(1, 1)$ | $(1, 0)$ | $(-1, 0)$ | $(1, 1)$ |
|---|---|---|---|---|---|---|---|
| $\varepsilon$ | | | $10^{-2}$ | | | $10^{-6}$ | |
| $\omega$-JAC | | 0.64 | 0.64 | 0.60 | 1.0 | 1.0 | 0.85 |
| $\omega$ $x$-line JAC | | 0.60 | 0.60 | 0.60 | 0.60 | 0.60 | 0.84 |
| $\omega$ $y$-line JAC | | 0.64 | 0.64 | 0.60 | 1.0 | 1.0 | 0.84 |
| $\omega$ alt. line JAC | | 0.30 | 0.30 | 0.28 | 0.60 | 0.60 | 0.46 |
| $\omega$ GS-RB | | 0.30 | 0.30 | 0.26 | 1.0 | 1.0 | 0.52 |
| $\omega$ $x$-zebra line GS | | 0.21 | 0.21 | 0.25 | 0.20 | 0.20 | 0.57 |
| $\omega$ $y$-zebra line GS | | 0.30 | 0.30 | 0.25 | 1.0 | 1.0 | 0.57 |
| $\omega$ alt. zebra line GS | | 0.05 | 0.05 | 0.05 | 0.20 | 0.20 | 0.27 |
| GS-LEX | | 0.48 | 0.55 | 0.42 | 0.45 | 1.0 | $10^{-4}$ |
| GS-BackLEX | | 0.55 | 0.48 | 0.58 | 1.0 | 0.45 | 1.0 |
| $x$-line GS | | 0.45 | 0.45 | 0.36 | 0.45 | 0.45 | $10^{-4}$ |
| $y$-line GS | | 0.36 | 0.54 | 0.36 | 0.33 | 1.0 | $10^{-4}$ |
| Alt. line GS | | 0.15 | 0.18 | 0.11 | 0.15 | 0.45 | $10^{-8}$ |

| 16 | 15 | 14 | 13 | | 1 | 2 | 3 | 4 |
|----|----|----|----|---|---|---|---|---|
| 12 | 11 | 10 | 9 | | 5 | 6 | 7 | 8 |
| 8 | 7 | 6 | 5 | | 9 | 10 | 11 | 12 |
| 4 | 3 | 2 | 1 | | 13 | 14 | 15 | 16 |

**Figure 7.5.** Point Gauss–Seidel ordering starting in two different corners of the grid.

| 1 | 1 | 1 | 1 | | 4 | 3 | 2 | 1 |
|---|---|---|---|---|---|---|---|---|
| 2 | 2 | 2 | 2 | | 4 | 3 | 2 | 1 |
| 3 | 3 | 3 | 3 | | 4 | 3 | 2 | 1 |
| 4 | 4 | 4 | 4 | | 4 | 3 | 2 | 1 |

**Figure 7.6.** Backward $x$- and $y$-line Gauss–Seidel ordering of grid points.

values $a$ and $b$. This is, however, not true. For certain combinations of $a$ and $b$, point Gauss–Seidel smoothers that correspond to the orderings in Fig. 7.5 are needed. For similar reasons, one may need backward line Gauss–Seidel relaxations (see Fig. 7.6). Smoothing analysis, but also the heuristics to adapt the marching direction to the direction of convection, shows which ordering is appropriate. For example, if $a < 0, b > 0$, the grid point ordering in the left picture of Fig. 7.5 corresponds to the convection direction; for $a > 0, b < 0$, it is the ordering in the right picture. $\gg$

**Example 7.2.1** We consider Example 7.1.2 with $a = b = 1$ and Dirichlet boundary conditions. Standard multigrid works well for first-order upwind discretizations of this particular example with $\varepsilon = 10^{-2}$ and $\varepsilon = 10^{-6}$ if a suitable smoother is employed.

We compare the F(1,1)-cycle multigrid convergence history with GS-RB and GS-LEX on two grids with $h = 1/64$ and $h = 1/256$. Only for $\varepsilon = 10^{-6}$, do we choose underrelaxation $\omega = 0.8$ in GS-RB as in Table 7.1. The other multigrid components are FW and bilinear interpolation. Figure 7.7 shows that the convergence for $\varepsilon = 10^{-2}$ is very good in all cases.

For $\varepsilon = 10^{-6}$, the convergence with GS-LEX is extremely fast. GS-LEX itself is a fast solver in this case. The convergence with GS-RB is satisfactory, as predicted by the smoothing analysis. $\triangle$

### 7.2.2 Variable Coefficients

In the following, we consider the general case $a = a(x, y)$ and $b = b(x, y)$. From the above results and considerations, one can conclude that the ordering of the grid points is important for the smoothing properties in the context of convection-dominated problems. In particular, downstream-type relaxation schemes with an ordering that corresponds to the characteristic direction $(a, b)$ can be regarded as candidates for good smoothing in the

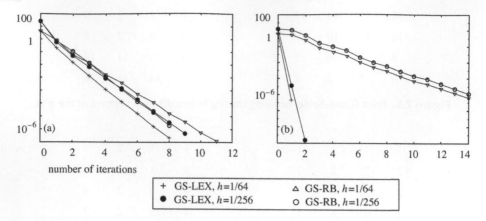

number of iterations

| + GS-LEX, $h=1/64$ | △ GS-RB, $h=1/64$ |
|---|---|
| ● GS-LEX, $h=1/256$ | ○ GS-RB, $h=1/256$ |

**Figure 7.7.** Measured F(1,1) multigrid convergence with GS-LEX and GS-RB on grids with $h = 1/64$ and $h = 1/256$. (a) $\varepsilon = 10^{-2}$; (b) $\varepsilon = 10^{-6}$.

general case. However, downstream relaxation may become complicated and difficult to implement in general situations, e.g. in cases with *variable coefficients* or for nonlinear applications with $a$ and $b$ depending on $u$ or derivatives of $u$. This is true in 2D, but even more in 3D.

In addition, the implementation of *parallel* downstream-type relaxations is even more difficult. The concept of a static grid partitioning does not match well with such relaxations. A sophisticated smoothing variant based on downstream relaxation, which partly avoids some of the complications is based on the segment relaxation mentioned in Remark 5.1.8.

An approach to become more independent of a specific convection direction is to use *four-direction GS-LEX point smoothing*. This smoother consists of four relaxation sweeps, each of which starts in a different corner of the domain so that all grid points are processed four times per smoothing step. The parallelization may be performed as discussed in Remark 6.2.5.

Smoothers that also show very satisfactory smoothing factors for all values of $a$ and $b$ are *alternating symmetric line smoothers*. In particular, the alternating *symmetric* line Gauss–Seidel relaxation should be mentioned. This relaxation consists of four steps: a forward and a backward line relaxation step in each direction. Alternating symmetric line smoothers are robust, but quite expensive.

Furthermore, the $\omega$ alternating line Jacobi and the $\omega$ alternating zebra Gauss–Seidel relaxation are satisfactory smoothers in 2D for all $a$ and $b$. Contrary to the lexicographical smoothers, these smoothers are "direction-free", which makes them less expensive in the general case. The parallelization of line smoothers has already been discussed in Section 6.4.

**Remark 7.2.3** An important advantage of the line smoothers is that excellent smoothing factors can also be obtained for anisotropic diffusion problems, so that flow problems involving a combination of anisotropy and dominating convection are handled well by these smoothers.                                                                                   ≫

Finally, it should be noted that variants of incomplete factorization smoothers such as ILU (discussed in Section 7.5) are also good smoothers for the 2D convection–diffusion problem. An overview of smoothing analysis results for different smoothers for (7.1.1) with $a = \cos \beta$ and $b = \sin \beta$ with $\beta$ varying can be found in [415].

> We can summarize: a variety of reasonable smoothing procedures exists for convection–diffusion problems and first-order upwind discretization. Pure downstream-type relaxations are preferable, but tend to be complicated in general situations and are limited with respect to parallelism.

### 7.2.3 The Coarse Grid Correction

Although the multigrid convergence in Example 7.2.1 is convincing, we will see in this section that difficulties may arise for certain convection–diffusion problems with very small $\varepsilon$ (i.e. "dominating convection"). The main problem in constructing robust and efficient multigrid solvers for convection-dominated problems is not the choice of a good smoother, but the coarse grid correction. With standard coarsening and the first-order upwind discretization, we face a fundamental problem.

> From a two-grid LFA, we find that the two-grid convergence factor is limited by 0.5 for $\varepsilon \to 0$, independent of the number of smoothing steps [43, 63, 87].

This factor can already be seen by the simplified two-grid analysis as introduced in Remark 4.6.1. Applying this analysis, we find for the first-order upwind discretization $L_{2h}$ of Model Problem 6:

$$1 - \frac{\tilde{L}_h(\boldsymbol{\theta})}{\tilde{L}_{2h}(2\boldsymbol{\theta})} = 1 - 4\frac{\varepsilon_h(4 - 2\cos\theta_1 - 2\cos\theta_2) + ih(a\sin\theta_1 + b\sin\theta_2)}{\varepsilon_{2h}(4 - 2\cos 2\theta_1 - 2\cos 2\theta_2) + i2h(a\sin 2\theta_1 + b\sin 2\theta_2)}.$$
$$(7.2.5)$$

In order to understand the two-grid convergence for the lowest frequencies, we consider $\theta_2 = c\theta_1$ and obtain

$$\lim_{\theta_1 \to 0} \left(1 - \frac{\tilde{L}_h(\theta_1, \theta_2)}{\tilde{L}_{2h}(2\theta_1, 2\theta_2)}\right) = \begin{cases} 0 & \text{if } c \neq -a/b \ (b \neq 0) \\ 1 - (\varepsilon_h/\varepsilon_{2h}) & \text{if } c = -a/b \end{cases}$$

from (7.2.5). Here, $c = -a/b$ means that $a\theta_1 + b\theta_2 = 0$. These $\theta_1$ and $\theta_2$ correspond to the so-called *characteristic Fourier components*, which are constant along the characteristics of the convection operator. The reason for the two-grid factor of 0.5 is that first-order upwind differencing for $L_{2h}$ produces an *artificial viscosity* on the coarse grid $\Omega_{2h}$, which is twice as large as on the fine grid $\Omega_h$. Thus, we find $\varepsilon_{2h} \approx 2\varepsilon_h$ for $\varepsilon \ll 1$, resulting in an error

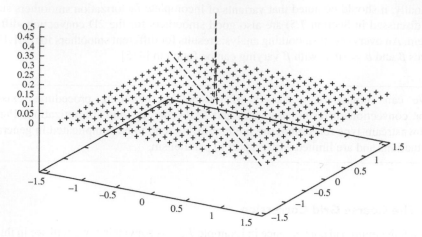

**Figure 7.8.** A discrete representation of the spectrum of the two-grid operator obtained by two-grid LFA.

reduction factor $1 - \varepsilon_h/\varepsilon_{2h} \approx 0.5$. In other words, the low frequency characteristic Fourier components of the error are not "correctly" approximated on the coarse grid: "They lose a factor of two in amplitude."

**Example 7.2.2** Figure 7.8 shows the result of a two-grid LFA for a symmetric alternating GS line smoother, $a = b = 1$, $\varepsilon = 10^{-6}$. Indeed, the components, which are responsible for the problematic factor of 0.5, are the characteristic Fourier components. △

**Remark 7.2.4** A two-level factor of 0.5 caused by the coarse grid correction leads to even worse *multigrid* convergence. The convergence factor of a multigrid cycle can be roughly estimated from the two-grid factor $\rho_{\text{loc}}$. If we assume that we have a good smoothing of the high frequencies, that the low frequencies remain unchanged under smoothing, and that the two-level factor $\rho_{\text{loc}}$ is determined by the coarse grid approximation of (some) low frequencies, we can give a rough prediction for the convergence factor of multigrid cycles [346].

$$\hat{\rho}_2 = \rho_{\text{loc}}, \qquad \hat{\rho}_\ell = 1 - (1 - \rho_{\text{loc}})(1 - (\hat{\rho}_{\ell-1})^\gamma), \quad \ell = 3, 4, 5, \ldots \qquad (7.2.6)$$

Here, $\gamma$ is the cycle index and $\hat{\rho}_\ell$ is the predicted multigrid convergence factor when using $\ell$ grids. Table 7.2 presents results obtained by the recursion (7.2.6) for $\rho_{\text{loc}} = 0.5$.

Obviously, the estimates $\hat{\rho}_\ell$ increase towards 1 for V- and W-cycles ($\gamma = 1$ and $\gamma = 2$) if the number of grids $\ell$ grows. However, for $\gamma = 3$, convergence factors independent of $h$ and the number of levels may be expected. In 2D the choice $\gamma \geq 4$ does not result in an $O(N)$ method; the computational complexity for $\gamma = 4$ is $O(N \log N)$. ≫

**Table 7.2.** Asymptotic estimates $\hat{\rho}_\ell$ for the convergence factors of different multigrid cycles if the two-level convergence factor $\rho_{loc} = 0.5$.

| $\ell$ | $\hat{\rho}_\ell$ | | | |
|---|---|---|---|---|
| | $\gamma = 1$ | $\gamma = 2$ | $\gamma = 3$ | $\gamma = 4$ |
| 2 | 0.5 | 0.5 | 0.5 | 0.5 |
| 3 | 0.75 | 0.63 | 0.56 | 0.53 |
| 4 | 0.88 | 0.70 | 0.59 | 0.54 |
| 5 | 0.94 | 0.74 | 0.60 | 0.54 |
| ⇓ | ⇓ | ⇓ | ⇓ | ⇓ |
| ∞ | 1.0 | 1.0 | 0.62 | 0.54 |

**Table 7.3.** Measured W(0,1)-cycle convergence factors for the recirculating convection–diffusion equation with $\varepsilon = 10^{-6}$.

| $1/h$ | 32 | 64 | 128 | 256 |
|---|---|---|---|---|
| 4 dir. point GS | 0.49 | 0.58 | 0.66 | 0.71 |
| Alt. symmetric line GS | 0.40 | 0.51 | 0.59 | 0.66 |
| Zebra line GS | 0.59 | 0.66 | 0.73 | 0.76 |

In principle, one has to be aware of the above reduction factor of 0.5, when dealing with singularly perturbed problems with upwind or artificial viscosity schemes. An example, for which the 0.5 convergence factor is observed, is the following *convection-dominated recirculating flow problem*.

**Example 7.2.3 (the recirculating convection–diffusion problem)** We consider the problem

$$-\varepsilon\Delta u + a(x, y)\frac{\partial u}{\partial x} + b(x, y)\frac{\partial u}{\partial y} = 0 \quad (\Omega = (0, 1)^2) \qquad (7.2.7)$$

with $\varepsilon = 10^{-6}, a(x, y) = -\sin(\pi x)\cdot\cos(\pi y), b(x, y) = \sin(\pi y)\cdot\cos(\pi x)$ (recirculating convection) and Dirichlet boundary conditions

$$u|_\Gamma = \sin(\pi x) + \sin(13\pi x) + \sin(\pi y) + \sin(13\pi y).$$

For the first-order upwind discretization from Section 7.1.3, we obtain the measured multigrid convergence factors in Table 7.3. Here, the four direction point GS smoother, the alternating symmetric line Gauss–Seidel and the zebra line Gauss–Seidel smoother are compared. Furthermore, the transfer operators FW and bilinear interpolation have been used.

The multigrid convergence factors increase due to the coarse grid problem described above. The increase in the convergence factors is not as drastic as predicted in Table 7.2, since the smoothers also reduce low frequency errors to some extent. △

*In many practical applications, however, the bad two-grid convergence is not observed* (see Example 7.2.1). We clarify this behavior in the following remark.

**Remark 7.2.5 (two-grid factor 0.5—what is observed in practice?)**
- The LFA does not take into account any special kind of boundary conditions. For convection-dominated problems, however, convergence often depends on the type of boundary conditions (Dirichlet, Neumann, periodic etc.). Of course, this cannot be observed by LFA. In this sense, the LFA smoothing and two-grid factors can be considered as *pessimistic values* for which positive influences of boundary conditions are not taken into account.
- In particular, downstream relaxations have good *convergence* properties for problems with Dirichlet boundary conditions at the upstream (inflow) boundary (they become direct solvers for $\varepsilon \to 0$). The two-grid factor of 0.5 is thus not observed in such cases. Concrete flow situations with this property are so-called *entering flows*, i.e. flows that enter the domain $\Omega$ through a boundary. If the characteristic component responsible for the bad two-grid factor meets upstream Dirichlet boundaries (with zero errors at these boundaries), such error components are automatically reduced by the downstream smoothers.
- If boundary conditions influence the convergence properties, the *half-space mode analysis* [64] (an extension of the LFA) can be used to obtain a quantitative prediction of the dependence of the multigrid convergence on the distance from the boundary and its relation to the discretization error.                                                      $\gg$

If the 0.5 difficulty occurs, there are different remedies to overcome it. We will mention some of them in the following remark.

**Remark 7.2.6 (remedies)**
**Coarsening and discretization**   The choice of the coarse grid and the definition of the coarse grid operator have significant influence on the multigrid convergence for convection–diffusion problems. In the discussion above, we have assumed standard coarsening and the first-order upwind discretization on coarse grids. In AMG the factor 0.5 is *not* observed (see Appendix A). AMG defines the coarse grid in a problem-dependent way and *Galerkin operators* (see Sections 7.7.4 and 7.7.5) are used on the coarse grids.

**Defect overweighting**   Another remedy is defect overweighting [87, 445]. Here, the defects that are transferred to the coarse grid are multiplied by some constant factor $\eta$ between 1 and 2. Thus, the approximation of the crucial low frequency error components on the coarse grid is improved. On the other hand, $\eta$ should not be too large because those components which normally would receive a good correction are now overcorrected. By choosing an optimal $\eta$, the two-grid convergence factor can be improved to $1/3$, but this factor is, in general, not maintained for *multi*grid. In advanced approaches, the overweighting should be chosen in a level-dependent way.

**Krylov subspace acceleration**   It is possible to use multigrid as a preconditioner in connection with other solution methods such as the Krylov subspace acceleration to improve the convergence. We will discuss these approaches in some detail in Section 7.8.      ≫

In summary, multigrid methods can be applied to convection–diffusion problems with first-order upwind differencing, even for small $\varepsilon$. The coarse grid correction, if based on standard coarsening and the first-order upwind discretization $L_H$, leads to the LFA two-grid factor of 0.5. For typical inflow–outflow problems, the factor of 0.5 can be avoided, for instance, by downstream-type relaxation. For harder problems (like recirculating flows) more sophisticated approaches may have to be used, as described in Remark 7.2.3, since the implementation of downstream relaxation may be too complicated.

## 7.3 THE CONVECTION–DIFFUSION EQUATION: DISCRETIZATION II

We have seen that central differencing is of $O(h^2)$ consistency, but leads to unstable schemes for fixed $h$ and $\varepsilon \to 0$. The stability problems can be overcome by introducing artificial viscosity or by using first-order upwind schemes. However, we then achieve only $O(h)$ accuracy which is insufficient in practice.

*What one really wants to achieve in practice is (at least) $O(h^2)$ accuracy and sufficiently stable behavior of the discretization.*

### 7.3.1 Combining Central and Upwind Differencing

Here, we mention three ideas on how to combine the stable $O(h)$ approximation with the unstable central $O(h^2)$ approximation. These ideas are historically interesting, but have been superseded by the higher order upwind discretizations that we will discuss together with their multigrid treatment in the next subsections.

**Remark 7.3.1 (hybrid upwind schemes)**   An idea, which is particularly useful if the advection direction depends on the grid coordinates $(a, b) = (a(x, y), b(x, y))$ is to use first-order upwind discretizations only at those grid points where the Péclet condition is not satisfied. At all other grid points, it is possible and useful to use central differencing. For satisfactory multigrid convergence it is important (based on experience) to choose a "smooth" switch between both discretizations. For example, for the mesh-Péclet number (7.1.9) $Pe < 1.8$, one chooses the central discretization; for $Pe \geq 2$ the first-order upwind discretization, whereas for $1.8 \leq Pe < 2$ a linear combination of both discretizations is used.

Although the resulting discretization is still usually $O(h)$-accurate globally, hybrid upwind schemes may be beneficial, for example, for the convergence of defect correction as described in the next remark.      ≫

**Remark 7.3.2 (defect correction)**   The second idea is to use the defect correction principle for a combination of a first-order stable discretization $\hat{L}_h$ (e.g. first-order upwind) with a second-order, unstable, $L_h$ (central discretization). Of course, if one allows the number of defect correction steps to tend to infinity, this would not help as the limit (if it exists) would be the unstable solution corresponding to $L_h$. The idea is therefore to apply only very few defect correction steps assuming that $O(h^2)$ accuracy is approximately obtained after a few steps while the instability is not yet relevant.                                    $\gg$

**Remark 7.3.3 (double discretization)**   Another idea of combining $\hat{L}_h$ (first-order upwind) with $L_h$ (central) in a more sophisticated way in the multigrid context is to use $\hat{L}_h$ in all smoothing processes and $L_h$ in calculating the defects (residuals). This means that two different discrete operators are used simultaneously and therefore this approach is called "double discretization" [66]. One disadvantage of this approach is that the respective defects will not tend to 0 in multigrid cycling, i.e. we will not observe algebraic convergence.   $\gg$

### 7.3.2 Higher Order Upwind Discretizations

In this section, we address the problem of achieving *both* $O(h^2)$ consistency *and* stability of the discretization by using *higher order upwind-biased schemes*. For such schemes the stability concept becomes somewhat more involved. Such upwind techniques are also important in the context of compressible and incompressible flow equations (discussed in Sections 8.8 and 8.9). Many of these discretizations have been studied, for example in [183, 196, 233, 287].

Here, we will discuss only one important class of upwind-biased higher order schemes, the so-called $\kappa$-schemes proposed by van Leer [233]. For the convective term $au_x$, the $\kappa$-scheme is defined as

$$
(au_x)_h :=
\begin{cases}
(a/2h)[-1 \quad \underline{0} \quad 1]_h u_h - (a/h)(1-\kappa)/4[-1 \quad 3 \quad \underline{-3} \quad 1]_h u_h & \text{for } a > 0 \\
(a/2h)[-1 \quad \underline{0} \quad 1]_h u_h - (a/h)(1-\kappa)/4[-1 \quad \underline{3} \quad -3 \quad 1]_h u_h & \text{for } a < 0.
\end{cases}
$$
$$(7.3.1)$$

The central elements in the stencils are underlined. The discretization of $bu_y$ is analogous.

The right-hand side of (7.3.1) is the sum of the central difference scheme and a second-order dissipation term, which can be interpreted as an (upwind) approximation of $ah^2(\kappa - 1) u_{xxx}/4$. From a Taylor expansion, we see that (7.3.1) is at least $O(h^2)$ consistent for $-1 \le \kappa \le 1$. For specific values of $\kappa$, we obtain well-known schemes:

- for $\kappa = 1$ the central differencing scheme,
- for $\kappa = 0$ Fromm's scheme [146],
- for $\kappa = 1/2$ the QUICK scheme ("quadratic upwind interpolation for convective kinematics", see [236]),
- for $\kappa = 1/3$ the CUI ("cubic upwind interpolation") scheme, which is even $O(h^3)$ consistent,
- for $\kappa = -1$ the "usual" second-order upwind scheme.

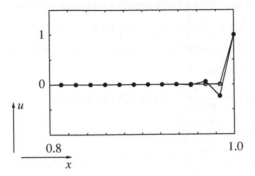

**Figure 7.9.** Zooming in on the solution along the line $y = 1/2$ of Example 7.1.2: •, discretization by Fromm's scheme; ○, with van Leer limiter.

With respect to stability, a discretization based on $\kappa$-schemes can handle a wider range of problems than the central discretization schemes. Nevertheless, a problem remains. If we evaluate Example 7.1.2, we observe that the $\kappa$-schemes still produce (local) unphysical oscillations near sharp gradients or discontinuities in a solution, shown in Fig. 7.9 for the example of $\kappa = 0$. In some sense, the dissipation term mentioned above is too small to avoid the wiggles completely.

In order to suppress such spurious oscillations in the solution, so-called *limiters* have been introduced in the $\kappa$-schemes. The background for these limiters are *total variation diminishing* (TVD) [183] concepts. We here present only some basic ideas, for a detailed discussion we refer to overviews [196, 444].

TVD schemes are constructed in such a way that the coefficients, which multiply differences of unknowns at neighbor grid points, e.g. $u_{i,j} - u_{i-1,j}$ and $u_{i-1,j} - u_{i-2,j}$, are guaranteed to have the same sign. We explain this in some more detail for the usual second-order upwind scheme, i.e. (7.3.1) with $\kappa = -1$. For $a > 0$, the TVD scheme can be written as

$$(au_x)_h = L_1 + \hat{L}_\alpha + \hat{L}_\beta \tag{7.3.2}$$

where

$$L_1 = \frac{a}{h}(u_{i,j} - u_{i-1,j})$$

is the first-order upwind scheme. The terms $\hat{L}_\alpha$ and $\hat{L}_\beta$ are obtained by multiplying

$$L_\alpha := \frac{a}{2h}(u_{i,j} - u_{i-1,j}), \qquad L_\beta := -\frac{a}{2h}(u_{i-1,j} - u_{i-2,j})$$

by functions $\Psi = \Psi(R)$ called *limiters*:

$$\hat{L}_\alpha = \Psi(R_{i-1/2})L_\alpha, \qquad \hat{L}_\beta = \Psi(R_{i-3/2})L_\beta.$$

Here, the arguments $R$ are ratios of differences (variations) of $u$ in neighbor points,

$$R_{i-1/2} = (u_{i+1,j} - u_{i,j})/(u_{i,j} - u_{i-1,j}),$$
$$R_{i-3/2} = (u_{i,j} - u_{i-1,j})/(u_{i-1,j} - u_{i-2,j}).$$

Some well-known limiters are

$$\Psi(R) = \frac{R^2 + R}{R^2 + 1} \quad \textit{(van Albada limiter)} \tag{7.3.3}$$

$$\Psi(R) = \frac{|R| + R}{R + 1} \quad \textit{(van Leer limiter)} \tag{7.3.4}$$

$$\Psi(R) = \frac{(|R| + R)(3R + 1)}{2(R + 1)^2} \quad \textit{(ISNAS limiter [444]).} \tag{7.3.5}$$

The goal behind these limiters is to guarantee both stability (the "TVD stability") *and* the $O(h^2)$ accuracy, at least in most parts of the computational domain (*locally*, at extrema, the schemes reduce to first-order accuracy). All of the above limiters are *nonlinear* and the resulting discretizations become *nonlinear, even for linear problems*. However, this cannot be avoided: any *linear* TVD scheme can be shown to be only first-order accurate, as stated in Godunov's order barrier theorem [158].

The above limiters lie in the so-called TVD region:

$$R \leq \Psi(R) \leq 2R \quad (0 \leq R \leq 1)$$

(see Fig. 7.10 where the TVD region is indicated and the limiters are shown). With limiters in this region, unphysical oscillations near strong gradients are avoided. If $R = 1$ (see (7.3.3)–(7.3.5)), the $\kappa = -1$ upwind scheme is obtained for all the limiters presented (which ensures $O(h^2)$ accuracy in regions without sharp gradients).

Many more limiters have been proposed for model problems and real applications, each of which has certain advantages for specific problems under consideration.

**Example 7.3.1 (the Smith–Hutton problem; discretization)**   We illustrate the above considerations on discretizations of the convection–diffusion equation for a problem with discontinuous boundary conditions. For such a so-called *contact discontinuity* that often

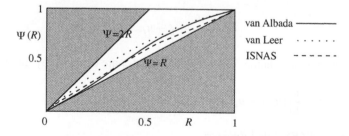

**Figure 7.10.** Three limiters and the monotonicity region (indicated by the white area) in an $(R, \Psi(R))$-diagram.

results in a strong gradient in the solution, similar problems as for boundary layer solutions can be observed. We compare different discrete solutions of the *Smith–Hutton problem* [361]

$$-\varepsilon \Delta u + a u_x + b u_y = f^{\Omega}(x, y) \quad \text{on } \Omega = (-1, 1) \times (0, 1),$$

with

$$a = 2y(1 - x^2), \qquad b = -2x(1 - y^2), \qquad \varepsilon = 10^{-5} \quad \text{and} \quad f^{\Omega} = 0$$

and discontinuous boundary conditions:

$$u = 2 \quad \text{on } -\tfrac{1}{2} \le x \le 0, \ y = 0$$
$$\frac{\partial u}{\partial y} = 0 \quad \text{on } 0 < x \le 1, \ y = 0$$
$$u = 0 \quad \text{elsewhere.}$$

The solution contains a step-like discontinuity that is convected along the characteristics of the convection operator. Figure 7.11 presents the problem geometry and also the curve $C$ along which the discontinuity is convected.

We discretize this problem on a $128^2$ Cartesian grid by: (a) central differencing; (b) first-order upwind discretization; (c) Fromm's ($\kappa = 0$) discretization; and (d) the limited $\kappa = 0$ upwind-biased discretization with the van Albada limiter (7.3.3). Figure 7.12 presents the corresponding discrete solutions. The left pictures show isolines of the solution on the whole domain, whereas the right pictures zoom in on the solution along the line $y = 1/4$. By central differencing, an unstable solution is obtained (see Fig. 7.12(a)). With first-order upwind the solution smears out; too much numerical diffusion leads to an unphysical first-order upwind solution (see Fig. 7.12(b)). Some unphysical oscillations are still visible for the second-order ($\kappa = 0$) upwind-biased discretization at the discontinuity (see Fig. 7.12(c)). The solution with the limiter is the most reasonable one: sharp gradients are obtained without oscillations (see Fig. 7.12(d)).

This is a general observation. Especially for the *Euler equations*, discussed in Section 8.9, with shock discontinuities in transonic and supersonic flow examples,

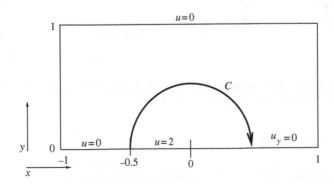

**Figure 7.11.** Smith–Hutton problem with discontinuous boundary conditions.

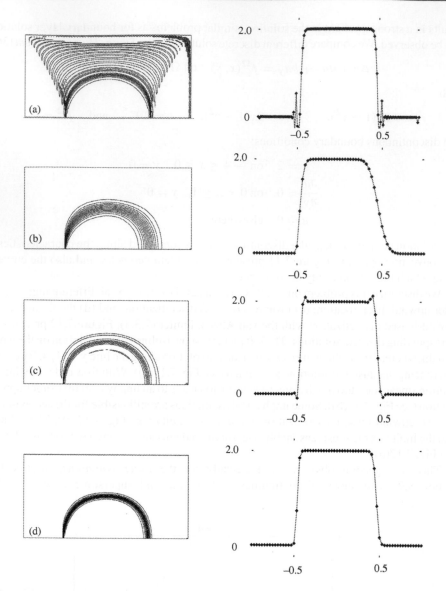

**Figure 7.12.** The solution of the Smith–Hutton problem with discontinuous boundary conditions. (a) central differencing; (b) first-order upwind; (c) Fromm's ($\kappa = 0$) upwind; (d) $\kappa = 0$ upwind with van Albada limiter.

limited upwind discretizations are commonly used to obtain sharp shocks without wiggles.

The multigrid convergence for the upwind discretizations with limiter will be discussed in Example 7.4.3 below.                                                                                    $\triangle$

## 7.4 THE CONVECTION–DIFFUSION EQUATION: MULTIGRID II

In this section, we will discuss multigrid related issues for the higher order discretization schemes introduced in the previous section. Whereas the discretization question (based on the TVD stability) is answered satisfactorily by the schemes, the efficient solution of the corresponding discrete problems is not easy.

Complications arise from the following facts.

- If limiters are used, the discrete problems are nonlinear.
- In defining efficient smoothing procedures, we have similar phenomena as discussed in the first-order upwind case (directional dependence etc.). In addition, the difference stencils are more complicated, with large coefficients of different signs.
- Finally, a coarse grid correction problem (an even more severe analog of the 0.5 two-grid factor difficulty as discussed in Section 7.2.3) occurs.

These difficulties have been discussed in detail in [293]. We are not going to repeat all results, but rather summarize how the difficulties can be mastered and what can be achieved with appropriate multigrid components.

Before, however, we return to the defect correction approach in the following remark.

**Remark 7.4.1 (defect correction)** Because of the complications listed above, one may prefer not to construct multigrid components for the second-order upwind-type discretizations directly, but use the defect correction approach instead: Proceed according to the general description in Section 5.4.1, where $\hat{L}_h$ now corresponds to the (stable) first-order upwind discretization and where $L_h$ corresponds to the (stable) higher order upwind-type discretization. Note that, in general, $L_h$ is a nonlinear operator (due to the nonlinear limiters). However, $L_h$ is used only for calculating the right-hand side in the outer defect correction iteration, whereas the multigrid solver itself is based only on the linear $\hat{L}_h$.

The defect correction approach is simple and works satisfactorily in many cases [186, 220]. However, the algebraic convergence of the defect correction iteration is often quite slow (see, for example, Example 7.4.3 or [293]), although the "differential" error typically behaves better.                                                                     ≫

Nevertheless, if efficient smoothers and coarse grid correction components are designed appropriately, the multigrid method, applied directly to the higher order discretizations, will usually be faster than the defect correction approach. Furthermore, fast *algebraic convergence* can often be obtained, which is regarded as an important criterion for many engineers.

### 7.4.1 Line Smoothers for Higher Order Upwind Discretizations

The straightforward application of smoothers discussed so far is not suitable for the $\kappa$-schemes (and their nonlinear modifications). For instance, the smoothing factors $\mu_{\text{loc}}$ of standard relaxation methods such as GS point or line smoothers for the $\kappa$-schemes of a convection-dominated problem often turn out to be larger than 1, which can be shown by

LFA. The corresponding multigrid method will then diverge. Consequently, one has to look for more sophisticated smoothing processes. One possibility is to *use smoothers, which are motivated by the defect correction approach.*

Let $L_h$ be a $\kappa$-scheme-based discrete operator (or its nonlinear analog). For lexicographical $x$-line relaxation, for instance, we consider the following splitting of $L_h$:

$$L_h = L_h^+ + L_h^- \tag{7.4.1}$$

with

$$L_h^+ = \begin{bmatrix} & & 0 & & \\ & & 0 & & \\ 0 & a_{-10}^{(1)} & a_{00}^{(1)} & a_{10}^{(1)} & 0 \\ & & a_{0-1}^{(2)} & & \\ & & a_{0-2}^{(2)} & & \end{bmatrix}. \tag{7.4.2}$$

Here, the coefficients $a_{**}^{(1)}$ correspond to the first-order upwind operator $\hat{L}_h$, the coefficients $a_{**}^{(2)}$ to the second-order upwind-type scheme. The line relaxation step is written as in (5.1.9), with a tridiagonal solver for $[a_{-10}^{(1)} \quad a_{00}^{(1)} \quad a_{10}^{(1)}]$. The approximation after one smoothing step, $\bar{w}_h$, is then obtained by

$$L_h^+ \bar{w}_h = f_h - L_h^- w_h.$$

**Example 7.4.1**   For an $x$-line solver, the above splitting for Fromm's scheme ($\kappa = 0$) without limiters is given by

$$L_h^+ = \begin{bmatrix} & & 0 & & \\ & & 0 & & \\ 0 & -a/h - \varepsilon/h^2 & a/h + 4\varepsilon/h^2 + b/h & -\varepsilon/h^2 & 0 \\ & & -5b/4h - \varepsilon/h^2 & & \\ & & b/4h & & \end{bmatrix},$$

$$L_h^- = \begin{bmatrix} & & 0 & & \\ & & b/4h - \varepsilon/h^2 & & \\ a/4h & -a/4h & -(a+b)/4h & a/4h & 0 \\ & & 0 & & \\ & & 0 & & \end{bmatrix},$$

for $a, b > 0$.                                                                                              $\triangle$

In addition, an underrelaxation parameter $\omega$ can be introduced in the usual way. Of course, we can use similar splittings for $y$-line relaxation and for the backward line relaxations. In this way, we can form an *alternating symmetric line smoother*, which is called the

KAPPA smoother in [293]. This smoother consists of four line relaxation steps, two $x$-line and two $y$-line partial steps with forward and backward numbering of lines. Note that a *robust* smoother for convection-dominated problems has to process all directions in order to handle convection in all directions. For special problems it may, of course, be possible to choose the direction of line smoothing downstream ("with the flow").

In the pure diffusion case, i.e. $a = b = 0$, the KAPPA smoother, is the symmetric alternating $\omega$ line GS smoother.

**Example 7.4.2** We consider Model Problem 6 with constant coefficients $a = \cos \beta$, $b = \sin \beta$ and varying parameters $\beta$ and $\varepsilon$, discretized by Fromm's scheme. Table 7.4 compares LFA smoothing factors with numerical measurements for W(0,1)-cycles for the KAPPA smoother with $\omega = 1$ for three representative values of $\beta$ and for two values of $\varepsilon$. Other angles, $\beta = 120°$ or $\beta = 135°$ for example, lead to identical results as for $\beta = 60°$ and $\beta = 45°$, respectively, with the symmetric smoother. In the numerical calculations, Dirichlet boundary conditions have been set.

According to Table 7.4, $\mu_{\text{loc}}$ gives a good indication of the actual asymptotic multigrid convergence.

Furthermore, LFA indicates that the splitting (7.4.1), (7.4.2) is robust for the $\kappa$-range $-0.3 \le \kappa \le 0.5$. There are also other splittings, which can handle a larger $\kappa$-range (see Remark 7.4.4 at the end of this section). The smoother presented above, however, is easier to implement for complicated systems of equations. $\triangle$

**Example 7.4.3 (the Smith–Hutton problem; multigrid treatment)** The Smith–Hutton problem with discontinuous boundary conditions, discussed in Example 7.3.1 is reconsidered and the multigrid convergence is checked for this problem. We have seen in Example 7.3.1 that limiters are necessary to assure an accurate solution without oscillations. The van Albada limiter (7.3.3) is employed here for $\kappa = 0$ leading to a nonlinear discrete problem.

The problem is solved by the FAS multigrid method with FW and bilinear interpolation. We employ the alternating symmetric KAPPA smoother and choose $\omega = 0.7$ for fast convergence. (The choice $\omega = 1$ leads to convergence difficulties here.) Figure 7.13 shows

**Table 7.4.** A comparison of LFA smoothing factors with measured multigrid convergence factors for the convection–diffusion equation for $\varepsilon = 10^{-3}$ and $\varepsilon = 10^{-6}$, $h = 1/256$ and $\kappa = 0$.

| $\beta$ | | $\varepsilon = 10^{-3}$ | $\varepsilon = 10^{-6}$ |
|---|---|---|---|
| 0 | $\mu_{\text{loc}}$ | 0.048 | 0.079 |
| | $W(0, 1)$ | 0.041 | 0.080 |
| 45 | $\mu_{\text{loc}}$ | 0.043 | 0.177 |
| | $W(0, 1)$ | 0.093 | 0.180 |
| 60 | $\mu_{\text{loc}}$ | 0.046 | 0.220 |
| | $W(0, 1)$ | 0.094 | 0.140 |

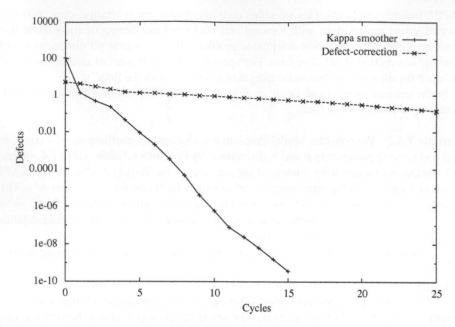

**Figure 7.13.** The (algebraic) convergence for the Smith–Hutton problem with defect correction and with the symmetric alternating KAPPA smoother on a $256 \times 128$ grid.

that the multigrid convergence on the $128^2$ grid obtained by V(1,1)-cycles is satisfactory. The algebraic convergence of the defect correction iteration is much worse. (The differential convergence, i.e. the convergence of the discrete solution is, however, better than indicated by the algebraic convergence of the defect correction.)                                    △

In the above examples, we do not observe any problems with the coarse grid correction. The measured multigrid convergence factors are about 0.2, even for the strongly convection-dominated situation ($\varepsilon = 10^{-6}$). In this specific case, even a *single grid KAPPA iteration* already performs well [293].

**Remark 7.4.2**   More generally, for the convection-dominated problems with entering flow, the KAPPA smoother also *converges* satisfactorily, i.e. it can be used as a solver.      ≫

Nevertheless, the coarse grid problem that we have described for the first-order upwind discretization (reflected by the LFA two-grid factor of 0.5) is, in principle, also present for second-order upwind-type discretizations.

**Remark 7.4.3**   Based on two-grid LFA, we observe, as in the first-order upwind case considered in Section 7.2.1, that with Fromm's discretization characteristic components, which are constant along the characteristics of the convection operator, are not correctly

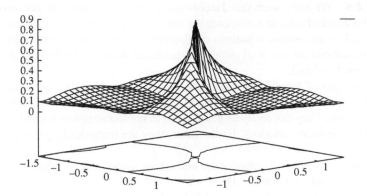

**Figure 7.14.** A representation of the spectrum of the two-grid operator obtained by two-grid LFA if Fromm's scheme is employed for the convection-dominated operator (with $(a, b) = (1/\sqrt{2}, 1/\sqrt{2})$, $\varepsilon = 10^{-6}, h = 1/64$).

reduced by the coarse grid correction operator. This phenomenon can be seen from the representation of the eigenvalues from the two-grid LFA. For $\beta = 45°$ and $\varepsilon = 10^{-6}$ these eigenvalues are shown in Fig. 7.14, where a maximum of about 0.9 ($\approx \rho_{loc}$) is observed along the characteristic direction.

Such unsatisfactory multigrid convergence, in fact, occurs in the context of convection-dominated *recirculating flow problems*. One way to overcome this problem to some extent is described in Section 7.8, where Krylov subspace acceleration of the multigrid is discussed. ≫

**Remark 7.4.4 (a modification of the KAPPA smoother)** As mentioned above, it is possible to enlarge the $\kappa$-range for which the KAPPA smoother is effective by modifying the smoother, more precisely, by modifying the coefficients $a_{**}^{(1)}$ in the splitting (7.4.1), (7.4.2) in such a way that appropriate ("positive") parts of the higher order operator are included (see [293] for details and examples). ≫

### 7.4.2 Multistage Smoothers

The KAPPA smoothers described in the previous section lead to fast convergence for various higher order upwind discretizations of the convection–diffusion equation. They are, however, somewhat involved and expensive per iteration. Therefore, the KAPPA smoothers are not in general use. Other smoothing methods that are simpler to program and that also work for the type of equations and the discretizations discussed above are the so-called "multiparameter" or "multistage" smoothers. Typically, they are of point Jacobi-type, which already indicates some of their advantages and disadvantages as summarized in the following remark.

**Remark 7.4.5**   We have seen that Jacobi-type point relaxations are not very efficient smoothers for Model Problem 1 (compare the smoothing properties of $\omega$-JAC with GS-RB, for example). For anisotropic equations or discrete problems on severely stretched grids, pointwise relaxations are not at all suitable if combined with standard coarsening (without additional modifications).

The advantages of Jacobi-type smoothers are, however, that they are simple and that they have at least some smoothing properties for a relatively large class of problems, as we will discuss below. They can easily be generalized to nonlinear equations and to systems of equations, like the Euler equations. Furthermore, they are intrinsically parallel.         $\gg$

In the following, we will consider the *multistage* smoothers and the connection with multiparameter Jacobi relaxation. There are several variants of such smoothers in use, which differ, in particular, for systems of PDEs. For scalar PDEs, they coincide or differ only slightly.

Considering a discrete scalar problem, $L_h u_h = f_h$, a *multistage* (more precisely, $p$-stage) smoother is defined by

$$
\begin{aligned}
u_h^{(0)} &= u_h^m \\
u_h^{(j)} &= u_h^{(0)} + \beta_j\big(f_h - L_h u_h^{(j-1)}\big) \quad (j = 1, 2, \ldots, p) \\
\bar{u}_h^m &= u_h^{(p)},
\end{aligned}
\tag{7.4.3}
$$

with parameters $\beta_j$ to be determined. (These smoothers are also sometimes called multistage "Runge–Kutta" smoothers [202]. The connection of these schemes with the Runge–Kutta time stepping schemes for ODEs will be explained in Section 8.9.4.)

A generalization is obtained if (7.4.3) is replaced by

$$
u_h^{(j)} = u_h^{(0)} + \alpha_j (D_h)^{-1}\big(f_h - L_h u_h^{(j-1)}\big) \quad (j = 1, 2, \ldots, p).
\tag{7.4.4}
$$

(usually called multistage Jacobi smoothers [126, 318]). Here, $D_h$ denotes the diagonal part of $L_h$. It is possible to represent $\bar{u}_h^m$ in the polynomial form

$$
\bar{u}_h^m = P_p((D_h)^{-1} L_h) u_h^m + Q_{p-1}((D_h)^{-1} L_h) f_h \quad (j = 1, \ldots, p),
\tag{7.4.5}
$$

with polynomials $P_p$ and $Q_{p-1}$ of degree $p$ and $p - 1$, respectively. The so-called amplification polynomial $P_p$ represents the smoothing operator $S_h$, which, of course, depends on $\alpha_1, \ldots, \alpha_p$:

$$
S_h = S_h(\alpha_1, \ldots, \alpha_p) = P_p((D_h)^{-1} L_h).
$$

Using LFA, optimal multistage parameters $\alpha_j$ (or $\beta_j$) can be determined by minimizing the smoothing factors:

$$
\mu_{\text{loc}}^{\text{opt}} = \min_{\alpha_j}\{\mu_{\text{loc}}(\alpha_1, \ldots, \alpha_p)\}.
$$

**Remark 7.4.6** The above multistage smoother is closely related to the *multiparameter Jacobi relaxation*

$$
\begin{aligned}
u_h^{(0)} &= u_h^m \\
u_h^{(j)} &= u_h^{(j-1)} + \omega_j D_h^{-1}(f_h - L_h u_h^{(j-1)}) \quad (j = 1, 2, \ldots, p) \\
\bar{u}_h^m &= u_h^{(p)}.
\end{aligned}
\tag{7.4.6}
$$

Here, we perform $p$ steps of standard $\omega_j$-JAC relaxation, using different parameters $\omega_1, \ldots, \omega_p$ in each step. The relation between the two formulas (7.4.4) and (7.4.6) is given by the theorem of Vieta. In particular, the two-stage and two-parameter methods can be identified via

$$
\alpha_2 = \omega_1 + \omega_2, \qquad \alpha_1\alpha_2 = \omega_1\omega_2.
$$

Similarly, the three-stage and three-parameter methods coincide if

$$
\begin{aligned}
\alpha_3 &= \omega_1 + \omega_2 + \omega_3 \\
\alpha_2\alpha_3 &= \omega_1\omega_2 + \omega_2\omega_3 + \omega_3\omega_1 \\
\alpha_1\alpha_2\alpha_3 &= \omega_1\omega_2\omega_3.
\end{aligned}
$$

$\gg$

**Example 7.4.4** For Model Problem 1 (Poisson's equation), a straightforward extension of the definition of the smoothing factor $\mu^*$ to the multiparameter Jacobi relaxation is obtained from the corresponding definition in Section 2.1:

$$
\begin{aligned}
\mu(h; \omega_i) &:= \max\{|\chi_h^{k,\ell}(\omega_1)| \cdot \ldots \cdot |\chi_h^{k,\ell}(\omega_p)|: n/2 \le \max(k, \ell) \le n - 1\}, \\
\mu^*(\omega) &= \max\{|(1 - \omega_1 t)| \cdot \ldots \cdot |(1 - \omega_p t)|: 1/2 \le t \le 2\}.
\end{aligned}
\tag{7.4.7}
$$

Minimizing $\mu^*$ with respect to $\omega_1, \ldots, \omega_p$ gives the optimal parameters

$$
\omega_j = \left(\frac{5}{4} + \frac{3}{4}\cos\left(\frac{2j-1}{2p}\pi\right)\right)^{-1} \quad (j = 1, \ldots, p).
\tag{7.4.8}
$$

(The cosine terms in (7.4.8) are the zeros of the Chebychev polynomials.)

The corresponding optimal parameters $\alpha_j$ for the two-stage Jacobi relaxation (7.4.4) are $\alpha_1 = 0.400$ and $\alpha_2 = 1.95$, leading to the smoothing factor $\mu_{\text{loc}}^{\text{opt}} = 0.22$. The optimal three-stage Jacobi method is obtained for $\alpha_1 = 0.27$, $\alpha_2 = 0.88$, $\alpha_3 = 2.99$ with $\mu_{\text{loc}}^{\text{opt}} = 0.074$.

Using the underrelaxation parameters (7.4.8), Table 7.5 presents some values for $\mu^*$ and $\rho^*$ obtained by the rigorous Fourier analysis. The use of different relaxation parameters improves smoothing for the case considered to some extent (compared to $p$ steps of standard $\omega$-JAC). $\triangle$

As mentioned in the beginning of this subsection, multistage Jacobi smoothers can also be applied to more general problems, for example, to the convection–diffusion equation discretized by higher order upwind-type schemes discussed in Section 7.3.2.

**Table 7.5.** Smoothing factors of multistage Jacobi relaxation with optimal parameters (7.4.8).

| $p$ | 1 | 2 | 3 | 4 |
|-----|-----|-----|-----|-----|
| $\mu^*$ | 0.60 | 0.22 | 0.074 | 0.025 |
| $\rho^*$ | 0.60 | 0.22 | 0.13 | 0.11 |

**Table 7.6.** Three-stage Jacobi smoothing with optimal parameters for standard upwind and Fromm's discretizations of the convection terms with $(a, b) = (1, 1)$, $h = 1/128$.

| Discretization | $\varepsilon$ | $\alpha_1$ | $\alpha_2$ | $\alpha_3$ | $\mu_{\text{loc}}$ |
|-----|-----|-----|-----|-----|-----|
| $O(h)$ upwind | $10^{-2}$ | 0.25 | 0.85 | 2.82 | 0.16 |
| $O(h^2) \, \kappa = 0$ | $10^{-2}$ | 0.27 | 0.85 | 2.66 | 0.25 |
| $O(h)$ upwind | $10^{-3}$ | 0.28 | 0.82 | 2.16 | 0.31 |
| $O(h^2) \, \kappa = 0$ | $10^{-3}$ | 0.22 | 0.62 | 1.31 | 0.58 |
| $O(h)$ upwind | $10^{-4}$ | 0.27 | 0.77 | 1.93 | 0.35 |
| $O(h^2) \, \kappa = 0$ | $10^{-4}$ | 0.22 | 0.54 | 1.03 | 0.67 |

**Example 7.4.5** For Problem 7.1.1 with fixed coefficients $(a, b) = (1, 1)$, the optimal parameters for a three-stage Jacobi smoother are given in Table 7.6 together with some LFA smoothing factors (for $h = 1/128$). It can be seen that the multistage Jacobi smoother gives reasonable smoothing factors for this convection–diffusion example. They are, however, much worse than the smoothing factors of the KAPPA smoother found in Table 7.4 for the same problem. △

For other values of $a$ and $b$, for instance for $(a, b) = (1, 0)$ and $\varepsilon = 10^{-4}$ or smaller, the best smoothing factor obtained with a three-stage Jacobi smoother and standard coarsening is $\mu_{\text{loc}} = 0.96$. This indicates that it is not easy to obtain a satisfactory smoothing factor with a three-stage Jacobi method for Fromm's upwind-type discretization if the convection direction is *aligned with the grid*. A remedy is to use semicoarsening in the appropriate direction or to employ multistage linewise smoothing methods.

The coarse grid correction problem, however, remains. It will be further discussed in Section 7.8.1.

## 7.5 ILU SMOOTHING METHODS

In this section, we leave the convection–diffusion equation behind and discuss a different relaxation approach, the "ILU-type" (incomplete LU matrix decomposition) smoothers. ILU smoothers represent a class of smoothing procedures all of which are based on an incomplete LU-type factorization of the matrix $A_h$ corresponding to $L_h$. Actually, a variety of such smoothing procedures has been proposed [211, 212, 414, 442] and investigated [415,

420–422, 439, 440]. Certain versions of ILU type smoothing, for example the one discussed in Section 7.5.4, are regarded as particularly robust for anisotropic 2D problems with the anisotropy not necessarily aligned with the grid. For 3D problems, however, robust ILU smoothers for general anisotropic equations are more complicated and expensive.

**Remark 7.5.1** We will not give a detailed formal description of an ILU algorithm here (see, for example, [9] for algorithmic details). Furthermore, we assume here that an ILU decomposition exists. Existence is proved, for example, for $M$-matrices in [264].     ≫

## 7.5.1 Idea of ILU Smoothing

Since ILU has originally been developed in the framework of *matrices*, we first describe the ILU decomposition in this form. Later, we will also use a corresponding stencil notation.

We consider a problem $A_h u_h = f_h$ with a structured sparse $N \times N$ matrix $A_h$. For simplicity, we assume here a regular structure consisting of $2 + 3 + 2$ diagonals, as sketched in Fig. 7.15. This structure corresponds to a seven-point discretization (in 2D) which is somewhat more general than the five-point stencils considered before. Two examples of seven-point discretizations will be given in Section 7.6.1.

If the usual (complete) LU decomposition is applied to $A_h$, i.e. $A_h = \mathcal{L}_h \mathcal{U}_h$, then the matrices $\mathcal{L}_h$ and $\mathcal{U}_h$ are also band matrices, but the bands are filled with nonzero elements (see Fig. 7.16). The number of interior "zero diagonals" will increase like $h^{-1}$ if $h \to 0$. After the decomposition, the triangular systems $\mathcal{L}_h v_h = f_h$ and $\mathcal{U}_h u_h = v_h$ are solved successively to compute $u_h$.

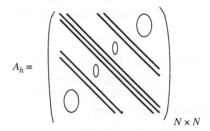

**Figure 7.15.** A structured sparse matrix.

**Figure 7.16.** An LU-decomposition of a structured sparse matrix with fill-in (indicated in grey).

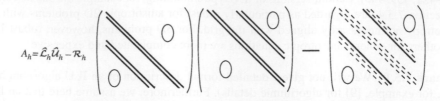

$$A_h = \hat{\mathcal{L}}_h \hat{\mathcal{U}}_h - \mathcal{R}_h$$

**Figure 7.17.** An ILU-decomposition of a banded matrix $A_h$.

In an *incomplete* LU (ILU) decomposition, $A_h = \hat{\mathcal{L}}_h \hat{\mathcal{U}}_h - \mathcal{R}_h$, one forces the matrices $\hat{\mathcal{L}}_h$ and $\hat{\mathcal{U}}_h$ to be sparse also: *a nonzero sparsity pattern is prescribed for the matrices $\hat{\mathcal{L}}_h$ and $\hat{\mathcal{U}}_h$.* In the simplest case, $\hat{\mathcal{L}}_h$ and $\hat{\mathcal{U}}_h$ have the same sparsity pattern as $A_h$ restricted to the respective triangular parts.

Roughly speaking, the ILU decomposition (with the same sparsity pattern as the original matrix $A_h$) is computed as follows: for the calculation of the coefficients $\hat{l}_{ij}$ (of $\hat{\mathcal{L}}_h$) and of the coefficients $\hat{u}_{ij}$ (of $\hat{\mathcal{U}}_h$) proceed as for the complete LU decomposition, but replace in the algorithm (step by step) those $\hat{l}_{\rho\sigma}$ and $\hat{u}_{\rho\sigma}$ by 0 for which $(\rho, \sigma)$ corresponds to one of the zero diagonals of $A_h$.

Note that ILU depends on the ordering of the grid points. In particular, the entries in ILU are different for row and column ordering. This allows the definition of *alternating ILU decompositions* used later in this section, where two ILU decompositions, corresponding to different orderings (e.g. row and column) are combined.

Since the ILU decomposition is not exact, an error matrix $\mathcal{R}_h$ remains. In the case of a matrix $A_h$ according to Fig. 7.15, the error matrix $\mathcal{R}_h$ contains two additional diagonals as indicated in Fig. 7.17. (The *dashed* lines in $\mathcal{R}_h$ are zeros and correspond to the nonzero entries in $A_h$. They are shown here to illustrate the location of the additional diagonals of $\mathcal{R}_h$.)

**Remark 7.5.2**    ILU was introduced in [264] in its incomplete Cholesky (IC) variant and proposed as a preconditioner for the conjugate gradient method (ICCG). Various modifications have been proposed since then. In the incomplete Cholesky decomposition of a symmetric positive definite matrix $A_h$, we have

$$A_h = \mathcal{L}_h \mathcal{L}_h^T - \mathcal{R}_h \quad \text{with } \mathcal{R}_h = \mathcal{R}_h^T$$

(see [264] for details).                                                                $\gg$

In principle, the ILU decomposition can be used iteratively (see also (1.6.7)) to solve $A_h u_h = f_h$:

$$u_h^0 = 0, \qquad u_h^{m+1} = u_h^m + \hat{v}_h^m \quad (m = 0, 1, 2, \dots),$$

where

$$\hat{\mathcal{L}}_h \hat{\mathcal{U}}_h \hat{v}_h^m = d_h^m = f_h - A_h u_h^m. \tag{7.5.1}$$

This iteration can also be written in the form

$$\hat{\mathcal{L}}_h \hat{\mathcal{U}}_h u_h^{m+1} = \mathcal{R}_h u_h^m + f_h, \quad m = 0, 1, \dots. \qquad (7.5.2)$$

$\gg$

> However, the ILU iteration has, in general, poor convergence properties. Nevertheless, its *smoothing properties* are good for a wide range of problems. ILU can therefore be used as a smoother in multigrid methods.

### 7.5.2 Stencil Notation

In order to investigate the smoothing properties of the above ILU iteration by LFA, we now switch to the stencil notation and extend all occurring operators to the infinite grid $\mathbf{G}_h$.

For a given discrete seven-point difference operator $L_h$,

$$L_h = \begin{bmatrix} & \star & \star & \\ \star & \star & \star \\ & \star & \star & \end{bmatrix}_h,$$

a *seven-point ILU decomposition* can be represented in stencil notation by

$$L_h = \hat{L}_h \hat{U}_h - R_h, \qquad (7.5.3)$$

where the stencils $\hat{L}_h$, $\hat{U}_h$ and $R_h$ correspond to the matrices $\hat{\mathcal{L}}_h$, $\hat{\mathcal{U}}_h$ and $\mathcal{R}_h$, respectively. The multiplication of stencils corresponds directly to the multiplication of operators (for the formulas, see [342]). Depending on the ordering of grid points, in principle, eight different ILU decompositions can be defined. If the ordering of grid points is first in east direction (E), and then in north direction (N), this EN decomposition is of the form

$$\hat{L}_h : \begin{bmatrix} 0 & 0 & \\ \star & \star & 0 \\ & \star & \star \end{bmatrix}_h, \quad \hat{U}_h : \begin{bmatrix} \star & \star & \\ 0 & \star & \star \\ & 0 & 0 \end{bmatrix}_h, \quad R_h : \begin{bmatrix} \star & 0 & 0 & \\ & 0 & 0 & 0 \\ & & 0 & 0 & \star \end{bmatrix}_h.$$

$$(7.5.4)$$

Here, $\star$ stands for possible nonzero elements. This decomposition corresponds to the one in Fig. 7.17. In the following, however, we will analyze the "NE decomposition" with ordering first in the north direction then eastwards. The corresponding stencils read

$$\hat{L}_h : \begin{bmatrix} \star & 0 & \\ \star & \star & 0 \\ & \star & 0 \end{bmatrix}_h, \quad \hat{U}_h : \begin{bmatrix} 0 & \star & \\ 0 & \star & \star \\ & 0 & \star \end{bmatrix}_h, \quad R_h : \begin{bmatrix} \star & & \\ 0 & 0 & \\ 0 & 0 & 0 \\ & 0 & 0 \\ & & \star \end{bmatrix}_h. \qquad (7.5.5)$$

**Remark 7.5.3** Another decomposition is obtained by a north-west ordering (NW),

$$\hat{L}_h : \begin{bmatrix} 0 & 0 & \\ 0 & \star & \star \\ & \star & \star \end{bmatrix}_h, \quad \hat{U}_h : \begin{bmatrix} \star & \star & \\ \star & \star & 0 \\ & 0 & 0 \end{bmatrix}_h, \quad R_h : \begin{bmatrix} 0 & 0 & \star \\ 0 & 0 & 0 \\ \star & 0 & 0 \end{bmatrix}_h. \quad (7.5.6)$$

≫

**Remark 7.5.4** Other ILU decompositions can be obtained by using larger stencils for $\hat{L}_h$ and $\hat{U}_h$. Such decompositions do no longer reflect the sparsity pattern of $L_h$ (and are somewhat more expensive). For example, *nine-point ILU decompositions* have been used in [415] with good smoothing properties. ≫

In stencil notation, the ILU iteration (7.5.2) reads

$$\hat{L}_h \hat{U}_h u_h^{m+1} = R_h u_h^m + f_h \quad \text{or} \quad (L_h + R_h) u_h^{m+1} = R_h u_h^m + f_h. \quad (7.5.7)$$

We denote the symbols of $L_h$, $\hat{L}_h$, $\hat{U}_h$ and $R_h$ here by

$$\lambda_h(\boldsymbol{\theta}), \quad \lambda_h^L(\boldsymbol{\theta}), \quad \lambda_h^U(\boldsymbol{\theta}), \quad \lambda_h^R(\boldsymbol{\theta}) \quad (-\pi \le \boldsymbol{\theta} < \pi),$$

respectively. Correspondingly, the smoothing operator $S_h$, defined by (7.5.7), is represented by

$$S_h \varphi(\boldsymbol{\theta}, x) = \tilde{S}_h(\boldsymbol{\theta}) \varphi(\boldsymbol{\theta}, x) \quad (-\pi \le \boldsymbol{\theta} < \pi) \quad (7.5.8)$$

with

$$\tilde{S}_h(\boldsymbol{\theta}) := \frac{\lambda_h^L(\boldsymbol{\theta}) \lambda_h^U(\boldsymbol{\theta}) - \lambda_h(\boldsymbol{\theta})}{\lambda_h^L(\boldsymbol{\theta}) \lambda_h^U(\boldsymbol{\theta})} = \frac{\lambda_h^R(\boldsymbol{\theta})}{\lambda_h(\boldsymbol{\theta}) + \lambda_h^R(\boldsymbol{\theta})},$$

(assuming that the denominator $\ne 0$).

LFA smoothing results for Model Problem 1 (Poisson's equation) are given in the next subsection together with results for anisotropic operators (see also [285, 378, 415] for further results and details). Before we discuss smoothing properties, we point out in the following two remarks that the LFA results for ILU smoothing must be interpreted with some care.

**Remark 7.5.5** For similar reasons as explained for GS-LEX, ILU smoothing cannot be analyzed by the *rigorous* Fourier analysis. However, LFA can be used.

In this context, note that: for an operator $L_h$ with constant coefficients on an infinite grid, the coefficients of $\hat{L}_h$, $\hat{U}_h$ and $R_h$ are also constant. Due to boundary conditions, for example, the nonzero entries in the matrices corresponding to a finite grid problem are generally *not constant*. ≫

**Remark 7.5.6** As also discussed for linewise relaxation in Remark 5.1.6, ILU is "less local" than typical pointwise relaxation methods; often, long range effects can only be neglected asymptotically (for $h \to 0$). We will see in the following section that the ILU decomposition may even closely resemble an LU decomposition for certain problems. The

**Table 7.7.** LFA results ($\mu_{\text{loc}}$, $\rho_{\text{loc}}$) and measured V(1,0) convergence factors $\rho_h(h = 1/128)$ for the anisotropic model operator using *seven-point ILU* (7.5.5) for smoothing.

| $\varepsilon$ | $10^{-4}$ | $10^{-3}$ | $10^{-2}$ | 0.1 | 1 | 10 | $10^2$ | $10^3$ | $10^4$ |
|---|---|---|---|---|---|---|---|---|---|
| $\mu_{\text{loc}}$ | 0.17 | 0.17 | 0.17 | 0.17 | 0.13 | 0.27 | 0.61 | 0.84 | 0.95 |
| $\rho_{\text{loc}}$ | 0.17 | 0.17 | 0.17 | 0.17 | 0.13 | 0.27 | 0.61 | 0.84 | 0.94 |
| $\rho_h$ | 0.011 | 0.10 | 0.16 | 0.16 | 0.12 | 0.27 | 0.58 | 0.65 | 0.23 |

error operator $R_h$ then contains very small elements. This is the reason why the measured multigrid convergence factors can differ considerably for different values of $h$ and are often much better than predicted by LFA.                                           ≫

### 7.5.3 ILU Smoothing for the Anisotropic Diffusion Equation

ILU smoothers can be used for 2D anisotropic problems. We consider (5.1.1) with the five-point stencil $L_h(\varepsilon)$ (5.1.3).

**Remark 7.5.7**   For five-point stencils, in principle, a five-point ILU decomposition can be used. However, already for the anisotropic diffusion equation $L(\varepsilon)u = -\varepsilon u_{xx} - u_{yy}$, the five-point ILU decomposition turns out to be a bad smoother for (very) small and for (very) large values of $\varepsilon$ [415]. The seven-point "north-east" ILU iteration (7.5.5), for example, has much better smoothing properties for small values of $\varepsilon$. The seven-point north-west ILU decomposition (7.5.6), on the other hand, is identical to the five-point decomposition, because of zero entries in $\hat{L}_h$ and $\hat{U}_h$. We therefore consider the seven-point north-east ILU (7.5.5) for five-point stencils in the following. In this case, the sparsity pattern of the decomposition is not identical to that of $L_h$.                                         ≫

Table 7.7 shows LFA smoothing and two-grid factors $\mu_{\text{loc}}$ and $\rho_{\text{loc}}$ for the 2D anisotropic model problem together with measured numerical V(1,0)-cycle convergence obtained on a $128^2$ grid. On the coarse grids, the direct discretization of the PDE is used to define $L_H$. Furthermore, FW and bilinear interpolation are employed as the transfer operators.

Obviously, the convergence properties of the above method are not symmetric with respect to $\varepsilon$. We have good convergence for $\varepsilon \le 1$ but $\mu_{\text{loc}}$ and $\rho_{\text{loc}}$ tend to 1 for $\varepsilon \to \infty$. This behavior is similar to that observed in connection with $y$-line ZEBRA relaxation in Table 5.2. For large $\varepsilon$ the measured convergence is nevertheless satisfactory for $h$ not very small. The prediction of LFA is too pessimistic for a certain range of $\varepsilon$ ($\varepsilon = 10^{\pm 4}$ in Table 7.7). This is due to the fact that, for a fixed mesh size $h$ and $\varepsilon \to \infty$ or $\varepsilon \to 0$, ILU becomes almost a direct solver (see Remark 7.5.6). Consequently, the resulting fast ILU-convergence "supersedes" the multigrid convergence expected (see also [211, 285]). This is not reflected by the LFA because this effect vanishes for fixed $\varepsilon$ and $h \to 0$: LFA predicts the asymptotic case $h \to 0$.

### 7.5.4 A Particularly Robust ILU Smoother

The numerical results discussed above show that multigrid methods with the seven-point ILU smoother (7.5.5) do not always lead to good convergence. The asymmetric behavior of ILU smoothing has to be overcome in order to obtain a robust 2D solution method. This can be achieved by two modifications of the original decomposition.

- **Alternating ILU**    As discussed above, the ILU decomposition can be used for smoothing in an alternating manner (like the alternating zebra-line relaxation method in Section 5.1.4), giving a fast method for 2D anisotropic equations, independent of $\varepsilon$. That is, a standard row-ordered ILU is followed by a standard column-ordered ILU or vice versa. In this way, the above nonsymmetric behavior disappears. The smoothing properties are good for both small *and* large $\varepsilon$ (see [285]). Here, we combine the northeast (NE) ILU (7.5.5) with the west-south (WS) ILU. The alternating ILU decomposition is, however, not yet an efficient smoother for more general problems which involve, for example, mixed derivatives (see Section 7.6). For that purpose, the following additional modification is also important [210].

- **Modified ILU**    In the case of the usual ILU decomposition, the matrix $\mathcal{R}_h$ contains zeros at all positions which correspond to the nonzero structure of $\hat{\mathcal{L}}_h$ and $\hat{\mathcal{U}}_h$, especially along the main diagonal, i.e. $r_{ii} = 0$ (as in Fig. 7.17). With a parameter $\delta$ a *modified ILU decomposition*

$$A_h = \hat{\mathcal{L}}_h^\delta \hat{\mathcal{U}}_h^\delta - \mathcal{R}_h^\delta \tag{7.5.9}$$

can be defined, for which the diagonal of $\mathcal{R}_h^\delta$ is no longer zero. Assuming, for example, that the incomplete factorization of $A_h$ is computed row by row, the main diagonal of $\mathcal{R}_h$ and the $u_{ii}$ of $\hat{\mathcal{U}}_h$ can be modified in such a way that

$$r_{ii} \leftarrow \delta \sum_{j \neq i} |r_{ij}|, \qquad u_{ii} \leftarrow u_{ii} + r_{ii}$$

immediately before the coefficients of row number $i + 1$ are computed. Reasonable choices for $\delta$ are in the range $0 < \delta \leq 1$, where $\delta = 0$ corresponds to the nonmodified ILU decomposition. The modified ILU decomposition can also be analyzed by LFA. Based on the LFA and confirmed by measured convergence factors, it has been shown [285] that the modification of ILU with $\delta = 1$ results in better convergence and in improved robustness. This modification is useful, for instance, in the case that a main diagonal element in the $\hat{\mathcal{U}}_h$-part of the decomposition becomes zero or less than zero numerically. It keeps this main diagonal at a certain positive value and adapts the error matrix $\mathcal{R}_h$ accordingly.

The combination of both modifications presented above leads to *alternating modified ILU* as a smoother, which makes 2D multigrid impressively robust. Here, we show that it handles the anisotropic model problem very well for the whole $\varepsilon$-range. More results will follow in the next section. Table 7.8 gives numerical $V(1, 0)$ multigrid convergence results for the

**Table 7.8.** Measured V(1,0) multigrid convergence factors $\rho_h$ for the anisotropic model operator $L_h(\varepsilon)$ using seven-point modified ($\delta = 1$) alternating NE–WS ILU for smoothing.

| $\varepsilon$ | $10^{-4}$ | $10^{-3}$ | $10^{-2}$ | 0.1 | 1 | 10 | $10^2$ | $10^3$ | $10^4$ |
|---|---|---|---|---|---|---|---|---|---|
| $\rho_h$ | 0.001 | 0.027 | 0.048 | 0.048 | 0.040 | 0.047 | 0.049 | 0.027 | 0.001 |

anisotropic Model Problem 3 discretized on a grid with $h = 1/128$. Alternating modified ILU with $\delta = 1$ is the smoother.

**Remark 7.5.8**   The alternating modified ILU decomposition (and also block ILU decompositions) also have good smoothing properties for the convection–diffusion model problem [211, 415].

**Remark 7.5.9 (ILU and parallelism)**   It is not at all easy to efficiently parallelize an ILU decomposition and maintain exactly the same method as in the sequential situation. In principle, parallel variants of ILU can be obtained by employing a "parallel ordering" of grid points (like red–black ordering). However, this approach can reduce the smoothing properties of ILU and the corresponding multigrid convergence significantly (for sophisticated modifications leading to satisfactory parallel behavior, see [153]).

≫

**Remark 7.5.10 (3D ILU smoothers)**   For general 3D problems, it is not straightforward to define a robust ILU-based smoother in combination with standard coarsening as is indicated by LFA, for example, in [212]. Only expensive ILU-based smoothers have been proposed so far. An efficient 3D ILU smoother can, however, be developed in the case of one dominant direction or in the case of two dominant directions [212, 213].   ≫

## 7.6 PROBLEMS WITH MIXED DERIVATIVES

We will now consider a 2D differential operator $L^\tau u$ with a mixed derivative $u_{xy}$. In particular, we will treat the model equation

$$L^\tau u = -\Delta u - \tau u_{xy} = f \quad (\Omega). \tag{7.6.1}$$

For this equation, any reasonable discretization leads to at least seven- or nine-point stencils with *diagonal entries*. Mixed derivatives are no longer aligned with the $x$- and the $y$-axes, but can lead to *diagonal* anisotropies.

Equation (7.6.1) is interesting for two reasons. First, it is a very simple equation, where we can study the multigrid treatment of mixed derivatives systematically. In principle, we have already dealt with mixed derivatives in the context of the full potential equation in Section 5.3.6. From the good convergence results there, we may conclude that their occurrence presents no major problems.

The second interesting feature of this equation is that it becomes hyperbolic for $|\tau| > 2$. Therefore, we may expect difficulties for $|\tau| \to 2$. Indeed, we will see that the combination of standard coarsening and Gauss–Seidel-type smoothers will become problematic. In principle, there are (again) two possible ways to overcome this problem. One is to use more robust smoothers, like modified ILU, which also reduces certain *low* frequency error components on the fine grid. We will present this approach here. The second possibility is to change the grid coarsening and adapt it to the problem at hand, while keeping a (cheap) point smoother. The latter approach is presented in Appendix A, where this test problem is also solved by AMG. Efficient solution methods based on semicoarsening for similar problems are also presented in [74, 75].

**Remark 7.6.1** Equation (7.6.1) is a special case of a more general problem which occurs, for example, if the anisotropic operator $L(\varepsilon)u = -\varepsilon u_{xx} - u_{yy}$ is transformed by a rotation. The rotated operator then has the form

$$L(\varepsilon; \beta)u = -(s^2 + \varepsilon c^2)u_{xx} + 2(1 - \varepsilon)csu_{xy} - (c^2 + \varepsilon s^2)u_{yy}, \qquad (7.6.2)$$

where $s = \sin\beta$, $c = \cos\beta$, $\varepsilon > 0$ and $\beta$ is the angle of rotation.

This is a common test problem to study the quality of smoothers [415] and of coarse grid correction. The direction of strong coupling in (7.6.2) depends on $\beta$, which means that it is *not aligned* with the coordinate axes, except for $\beta = 0°, 90°$ and $180°$. This causes unsatisfactory multigrid convergence for many smoothers if combined with standard coarsening.

For $\beta = 45°$, we obtain the operator in (7.6.1) with $\tau = 2(1 - \varepsilon)/(1 + \varepsilon)$, on which we will concentrate here.                                                                        »

### 7.6.1 Standard Smoothing and Coarse Grid Correction

Basically, three different second-order discretizations are used for (7.6.1), represented by a nine-point and two seven-point stencils:

$$L_h^{\tau,9} \triangleq \frac{1}{h^2} \begin{bmatrix} \tau/4 & -1 & -\tau/4 \\ -1 & 4 & -1 \\ -\tau/4 & -1 & \tau/4 \end{bmatrix}_h, \qquad (7.6.3)$$

$$L_h^{\tau,7} \triangleq \frac{1}{h^2} \begin{bmatrix} \tau/2 & -1 - (\tau/2) & 0 \\ -1 - (\tau/2) & 4 + \tau & -1 - (\tau/2) \\ 0 & -1 - (\tau/2) & \tau/2 \end{bmatrix}_h, \qquad (7.6.4)$$

$$L_h^{\tau,7*} \triangleq \frac{1}{h^2} \begin{bmatrix} 0 & -1 + (\tau/2) & -\tau/2 \\ -1 + (\tau/2) & 4 - \tau & -1 + (\tau/2) \\ -\tau/2 & -1 + (\tau/2) & 0 \end{bmatrix}_h. \qquad (7.6.5)$$

For $\tau = 0$, we have the discrete five-point Laplace operator in all cases. Note that the description of ILU in the previous section was based on a seven-point discretization with the same structure as (7.6.4).

**Table 7.9.** Smoothing and two-grid LFA results for the nine-point operator (7.6.3) using GS-RB and GS-LEX relaxation, $\nu = 2$, $I_h^{2h}$: FW restriction; $I_{2h}^h$: linear interpolation.

| | GS-RB | | GS-LEX | |
| --- | --- | --- | --- | --- |
| $\tau$ | $\mu_{\mathrm{loc}}$ | $\rho_{\mathrm{loc}}$ | $\mu_{\mathrm{loc}}$ | $\rho_{\mathrm{loc}}$ |
| $-2.0$ | 0.41 | 0.76 | 0.52 | 0.76 |
| $-1.9$ | 0.38 | 0.66 | 0.50 | 0.66 |
| $-1.7$ | 0.32 | 0.54 | 0.46 | 0.54 |
| $-1.5$ | 0.27 | 0.44 | 0.42 | 0.46 |
| $-1.0$ | 0.16 | 0.27 | 0.35 | 0.33 |
| 0.0 | 0.063 | 0.074 | 0.25 | 0.19 |
| 1.0 | 0.16 | 0.27 | 0.29 | 0.26 |
| 1.5 | 0.27 | 0.44 | 0.34 | 0.42 |
| 1.7 | 0.32 | 0.54 | 0.37 | 0.52 |
| 1.9 | 0.38 | 0.66 | 0.41 | 0.65 |
| 2.0 | 0.41 | 0.76 | 0.43 | 0.76 |

Since these stencils are not (axially) symmetric for $\tau \neq 0$, the *rigorous* Fourier analysis of Chapter 3 cannot be applied in a straightforward way. We will apply LFA to analyze the behavior of various smoothers and transfer operators in order to develop efficient multigrid algorithms for (7.6.1).

In the case of the *nine-point stencil* (7.6.3), Table 7.9 presents LFA results for $|\tau| \leq 2$ with GS-LEX and GS-RB as the smoothers, FW and bilinear interpolation as the transfer operators. We observe that GS-RB and GS-LEX behave qualitatively similarly. The smoothing and two-grid convergence factors of both methods increase for $|\tau| \to 2$. This is not so much a problem of smoothing but of the coarse grid correction (see Table 7.9).

**Remark 7.6.2 (simplified two-grid analysis)** The situation can also be analyzed by the simplified two-grid analysis introduced in Remark 4.6.1, where we assume that the transfer operators act as identities and neglect the coupling of harmonics. We consider again

$$1 - \frac{\tilde{L}_h(\boldsymbol{\theta})}{\tilde{L}_{2h}(2\boldsymbol{\theta})}.$$

Applying a Taylor expansion to

$$\tilde{L}_h(\boldsymbol{\theta}) = \frac{1}{h^2}\left(4 - 2\cos\theta_1 - 2\cos\theta_2 + \frac{\tau}{2}\cos(\theta_1 - \theta_2) - \frac{\tau}{2}\cos(\theta_1 + \theta_2)\right)$$

and to the corresponding $\tilde{L}_{2h}(\boldsymbol{\theta})$, we obtain

$$1 - \frac{\tilde{L}_h(\boldsymbol{\theta})}{\tilde{L}_{2h}(2\boldsymbol{\theta})} \to 0.75$$

for the frequencies given by $\theta_2 = \mp\theta_1$ for $\tau \to \pm 2$ and $|\boldsymbol{\theta}| \to 0$.

**Table 7.10.** Analog to Table 7.9 with the seven-point discretization (7.6.4) and *FW and bilinear interpolation as transfer operators.*

| | GS-RB | | GS-LEX | |
|---|---|---|---|---|
| $\tau$ | $\mu_{\text{loc}}$ | $\rho_{\text{loc}}$ | $\mu_{\text{loc}}$ | $\rho_{\text{loc}}$ |
| $-2.0$ | 1.0 | $> 1$ | 1.0 | $> 1$ |
| $-1.95$ | 0.91 | $> 1$ | 0.82 | $> 1$ |
| $-1.9$ | 0.82 | $> 1$ | 0.69 | 0.68 |
| $-1.7$ | 0.56 | 0.70 | 0.44 | 0.41 |
| $-1.5$ | 0.38 | 0.53 | 0.36 | 0.32 |
| $-1.0$ | 0.15 | 0.25 | 0.29 | 0.23 |
| 0.0 | 0.063 | 0.074 | 0.25 | 0.19 |
| 1.0 | 0.20 | 0.30 | 0.38 | 0.33 |
| 1.5 | 0.29 | 0.47 | 0.46 | 0.50 |
| 1.7 | 0.33 | 0.56 | 0.49 | 0.58 |
| 1.9 | 0.37 | 0.67 | 0.52 | 0.68 |
| 2.0 | 0.39 | 0.76 | 0.53 | 0.76 |

Obviously the bad two-grid convergence is caused by a bad coarse grid approximation of these low frequencies. ≫

Similar observations can be made if the *seven-point stencil* (7.6.4) is used for the discretization of (7.6.1). Again, we use the transfer operators FW and bilinear interpolation. Table 7.10 shows that the corresponding multigrid algorithm results in similar convergence, if $\tau$ is not too close to $-2$. For $\tau \to -2$, the stencil becomes diagonal. The $h$-ellipticity measure $E_h(L_h)$ becomes 0. Consequently, the smoothing factors $\mu_{\text{loc}}$ of the point smoothers GS-LEX and GS-RB tend to 1. For $\tau = 2$, however, this stencil is not diagonal and the $h$-ellipticity measure is satisfactory. (For the seven-point stencil (7.6.5) it is the other way around.) In Table 7.10, we observe that the corresponding two-grid factors for $\tau \to -2$ are even larger than 1. Divergence of the corresponding multigrid algorithm has to be expected.

**Remark 7.6.3 (seven-point transfer operators)** For seven-point discretization stencils, it is an option to choose seven-point transfer operators. (They can, of course, also be employed for five-point or other discretizations). For standard coarsening $H = 2h$, a seven-point restriction operator is

$$I_h^{2h} := \frac{1}{8} \begin{bmatrix} 1 & 1 & 0 \\ 1 & 2 & 1 \\ 0 & 1 & 1 \end{bmatrix}_h^{2h} \tag{7.6.6}$$

and a corresponding scaled prolongation operator, with scaling $s = 2^d$, is

$$I_{2h}^h = s(I_h^{2h})^T. \tag{≫}$$

**Table 7.11.** Two-grid factors $\rho_{\text{loc}}$ for GS-RB and GS-LEX, $\nu = 2$, with the seven-point discretization (7.6.4) and *the seven-point transfer operators.*

| | GS-RB | GS-LEX |
|---|---|---|
| $\tau$ | $\rho_{\text{loc}}$ | $\rho_{\text{loc}}$ |
| −2.0 | 1.0 | 1.0 |
| −1.95 | 0.90 | 0.82 |
| −1.9 | 0.82 | 0.68 |
| −1.7 | 0.55 | 0.36 |
| −1.5 | 0.36 | 0.20 |
| −1.0 | 0.12 | 0.12 |
| 0.0 | 0.15 | 0.18 |
| 1.0 | 0.38 | 0.40 |
| 1.5 | 0.54 | 0.56 |
| 1.7 | 0.61 | 0.63 |
| 1.9 | 0.71 | 0.72 |
| 2.0 | 0.78 | 0.78 |

Table 7.11 presents the LFA two-grid factors, $\rho_{\text{loc}}$, for GS-RB and GS-LEX and the seven-point transfer operators in the case of the seven-point discretization (7.6.4). For $\tau$ not too close to −2, they are somewhat better for the seven-point transfer operators compared to FW and bilinear interpolation (see Table 7.10). We see from Table 7.11, of course, that the two-grid factors still deteriorate for $|\tau| \to 2$ for both smoothers.

The same problematic convergence behavior is found with linewise smoothers. In order to overcome the coarse grid problem, we look for smoothers which have good smoothing properties and additionally reduce the low frequency error components that are responsible for the bad two-grid convergence. Certain versions of *ILU smoothing* (Section 7.5), like the modified alternating ILU smoother discussed in Section 7.5.4, have this property for the problem under consideration. In the following, we discuss the seven-point discretization and seven-point ILU smoothers.

### 7.6.2 ILU Smoothing

In this section, we will concentrate on the seven-point stencil (7.6.4), first for $-2 \leq \tau \leq 0$.

Table 7.12 shows LFA results for (7.6.1) discretized by (7.6.4) with one smoothing iteration of the seven-point NE ILU smoothing method (7.5.5). Furthermore, the seven-point transfer operators are employed.

Obviously, the seven-point NE ILU smoother behaves somewhat better than the point smoothers above. Here, only one smoothing iteration is applied compared to two in the case of GS-RB and GS-LEX. The measured convergence factors in Table 7.12 show that multigrid methods with a seven-point NE ILU smoother do not lead to good convergence for $\tau \to -2$.

We also mention that this ILU smoother is not a good smoother for $\tau \to 2$ using the stencil (7.6.4). In that case even divergence is observed.

**Table 7.12.** $\mu_{\text{loc}}$, $\rho_{\text{loc}}$ and measured V(1,0) multigrid convergence factors $\rho_h(h = 1/256)$ for (7.6.4) using seven-point NE ILU smoothing and seven-point transfer operators.

| $\tau$ | $\mu_{\text{loc}}$ | $\rho_{\text{loc}}$ | $\rho_h$ |
|---|---|---|---|
| −1.99 | 0.68 | 0.68 | 0.67 |
| −1.95 | 0.46 | 0.46 | 0.45 |
| −1.9 | 0.36 | 0.36 | 0.35 |
| −1.7 | 0.22 | 0.21 | 0.20 |
| −1.5 | 0.21 | 0.15 | 0.15 |
| −1.0 | 0.19 | 0.17 | 0.17 |
| 0.0 | 0.13 | 0.13 | 0.13 |

**Table 7.13.** $\mu_{\text{loc}}$, $\rho_{\text{loc}}$ and measured V(1, 0) multigrid convergence factors $\rho_h(h = 1/128)$ for $L_h^{\tau}$ from (7.6.4) with seven-point modified ($\delta = 1$) alternating ILU for smoothing (and seven-point restriction and interpolation).

| $\tau$ | −1.99 | −1.9 | −1.7 | −1.0 | 0 | 1.0 | 1.7 | 1.9 | 1.99 |
|---|---|---|---|---|---|---|---|---|---|
| $\mu_{\text{loc}}$ | 0.19 | 0.11 | 0.059 | 0.017 | 0.004 | $< 10^{-3}$ | $< 10^{-5}$ | $< 10^{-6}$ | $< 10^{-8}$ |
| $\rho_{\text{loc}}$ | 0.19 | 0.10 | 0.062 | 0.028 | 0.044 | 0.067 | 0.089 | 0.099 | 0.10 |
| $\rho_h$ | 0.19 | 0.13 | 0.090 | 0.040 | 0.058 | 0.086 | 0.092 | 0.047 | 0.091 |

The *alternating modified ILU* smoother (Section 7.5.4) with $\delta = 1$, however, improves the multigrid convergence considerably for the whole range $-2 \le \tau \le 2$. This is confirmed by LFA results. Table 7.13 presents measured multigrid convergence factors for (7.6.4) on a $128^2$ grid using this smoother.

Similar convergence is obtained if the transfer operators FW and bilinear interpolation are employed.

For the nine-point discretization (7.6.3), discussed in Table 7.9, an appropriate *nine-point ILU* relaxation method can be defined that handles this case well [285, 378].

## 7.7 PROBLEMS WITH JUMPING COEFFICIENTS AND GALERKIN COARSE GRID OPERATORS

Here, we will discuss the proper multigrid treatment of problems with "jumping" coefficients. We will use a finite volume discretization which provides an accurate discretization for such applications in a natural way.

We also have to modify the multigrid components for such problems. First, we need "operator-dependent" transfer operators. In addition, we will make use of the Galerkin coarse grid operator. Although problems with jumping coefficients are important in practice, the methodological significance of the Galerkin coarse grid discretization is more general than this class of problems indicates.

> The *Galerkin coarse grid operator* approach is also employed in the finite element context, where these operators naturally arise, and in algebraic multigrid.

### 7.7.1 Jumping Coefficients

Let us reconsider problem (5.7.1),

$$-\nabla \cdot (a\nabla u) = f^{\Omega} \quad (\Omega = (0, 1)^2)$$
$$u = f^{\Gamma} \quad (\partial\Omega), \tag{7.7.1}$$

but now with a discontinuous (jumping) coefficient $a(x, y)$. Problems with jumping coefficients often occur in nature. When modeling diffusion in some inhomogeneous medium, it might happen that certain parts in the medium are so dense, that there is very little diffusion possible through this region (an impermeable region). A typical occurrence of such a phenomenon is oil reservoir simulation. Another example is the modeling of diffusive groundwater flow, where the main medium is sand and where rock (with very low permeability) is present in certain parts of the domain. In such a case, one has to deal with a coefficient $a$ that is discontinuous and strongly varying by several orders of magnitude in the domain of interest.

In the following considerations we assume, for simplicity, that $a$ jumps at an interface $\Gamma$.

It is well known that the function $\nabla u$ will not be continuous at the interface $\Gamma$ if the coefficient $a$ is discontinuous. However, *the flux can be assumed to be continuous at the interface*. For problem (7.7.1) the flux is given by $a\nabla u$. In our example, the *jumping condition* is

$$\lim_{x\uparrow\Gamma} a \frac{\partial u}{\partial x}(\Omega_l) = \lim_{x\downarrow\Gamma} a \frac{\partial u}{\partial x}(\Omega_r) \quad \text{on } \Gamma. \tag{7.7.2}$$

**Remark 7.7.1 (conservation of fluxes)** *Conservation of fluxes* is naturally achieved in finite volume discretizations if the flux between adjacent volumes is independent of the finite volume in which it is calculated. For example, the flux through the west face of a volume has to equal the flux through the east face of the west neighbor volume. If a discretization is not conservative, the flux-balance per finite volume is not equal to zero. Physically, this nonzero term can be interpreted as an artificial source term (in the right-hand side of the PDE).

In many fluid dynamics problems and in other applications, a conservative discretization of a PDE or a PDE system is important. Under general assumptions, Lax and Wendroff [230] have shown that, if a discretization is conservative and if the discrete solution is convergent for $h \to 0$, then the limit is a "weak" solution of the equations under consideration. $\gg$

The idea of the finite volume discretization discussed in Section 5.7 is based on Gauss's theorem (5.7.4) and on the approximation of the integrals over the boundaries of control

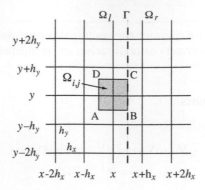

**Figure 7.18.** An example of a grid with an interface $\Gamma$ at which the coefficient $a$ jumps; two subdomains $\Omega_l$ and $\Omega_r$ and a finite volume $\Omega_{i,j}$.

volumes. These approximations of the boundary integrals now have to take into account the possible discontinuities in the coefficient $a$. As in Section 5.7, we need a discrete approximation of the right-hand side (5.7.6) (see Fig. 7.18 for the indices ABCD).

We assume a *vertex-centered* location of unknowns and consider here explicitly the case that the diffusion coefficient $a$ is continuous in $\Omega_{i,j}$, but may be discontinuous at the interface $\Gamma$, corresponding to $\partial\Omega_{i,j}$ (see Fig. 7.18). For other cases, see the remarks at the end of this section. The crucial integral is thus that over BC. We approximate this integral by

$$\int_B^C a\,\frac{\partial u}{\partial x}\,dy \approx h_y a(x, y)\frac{u(x + h_x/2, y) - u(x, y)}{h_x/2}. \tag{7.7.3}$$

Here, the derivative is approximated using one-sided differences such that only grid function values at one side of the discontinuity are used. At $(x + h_x/2, y)$, however, there is no function value available. An interpolation between the unknowns at the neighbor grid points is needed to define an approximation. In order to substitute the unknown $u(x + h_x/2, y)$ in (7.7.3), we use the discrete analog of the jumping condition (7.7.2)

$$\frac{2a(x, y)}{h_x}(u(x + h_x/2, y) - u(x, y))$$
$$= \frac{2a(x + h_x, y)}{h_x}(u(x + h_x, y) - u(x + h_x/2, y)), \tag{7.7.4}$$

We obtain

$$u(x + h_x/2, y) = \frac{a(x, y)u(x, y) + a(x + h_x, y)u(x + h_x, y)}{a(x, y) + a(x + h_x, y)}.$$

We can thus approximate

$$a\frac{\partial u}{\partial x}\bigg|_{(x+h_x/2,y)} \approx \frac{2a(x,y)a(x+h_x,y)}{a(x,y)+a(x+h_x,y)}(u(x+h_x,y)-u(x,y))/h_x.$$

Obviously, a *harmonic averaging* of the coefficients $a$ results from the finite volume discretization at an interface $\Gamma$, if the interface is not on a grid line but on the boundary of the finite volume. It can be shown that this discretization is of $O(h^2)$ consistency [415].

Summarizing, the resulting discrete equations in the case $h = h_x = h_y$ are

$$\begin{bmatrix} & s_n & \\ s_w & s_c & s_e \\ & s_s & \end{bmatrix}_h u_h = h^2 f_h$$

with

$$s_w = -2a(x-h,y)a(x,y)/(a(x-h,y)+a(x,y)),$$
$$s_e = -2a(x+h,y)a(x,y)/(a(x+h,y)+a(x,y)),$$
$$s_n = -2a(x,y+h)a(x,y)/(a(x,y+h)+a(x,y)),$$
$$s_s = -2a(x,y-h)a(x,y)/(a(x,y-h)+a(x,y)),$$
$$s_c = -(s_n + s_s + s_e + s_w).$$

**Remark 7.7.2** If the discontinuity occurs at a grid line, e.g. at a constant $x_i = ih$, a different discretization has to be used. The finite volume discretization must take into account that the discontinuity is within a volume [403]. In this case, the *arithmetic average* of coefficients replaces the above harmonic average. For the integral along BC, for example, this leads to the stencil entry $s_e$,

$$s_e = -\tfrac{1}{2}(a(x+h_x/2, y+h_y/2) + a(x+h_x/2, y-h_y/2)). \qquad \gg$$

**Remark 7.7.3** In the case that a discontinuity interface is neither a grid line nor coincides with the boundary of a control volume, the distance from the interface to grid lines and control volume boundaries must be taken into account for an accurate discretization. The combination of weighted harmonic averaging and weighted arithmetic averaging, with the weighting according to these distances, leads to a general stencil that combines all the cases discussed above [311]. $\qquad \gg$

### 7.7.2 Multigrid for Problems with Jumping Coefficients

Based on this finite volume discretization, we will now discuss the multigrid treatment of (7.7.1) with jumping coefficients. The convergence, for example, of the analog of the

Red–Black Multigrid Poisson Solver from Section 2.5 is very different for strongly jumping and for smoothly varying coefficients. For strongly jumping coefficients, the convergence depends on the size of the jump and on its location relative to grid lines. Even divergence may occur.

These multigrid convergence problems arise from the fact that $\nabla u$ is not continuous (although $u$ and $a\nabla u$ are). If the coefficient $a$ jumps by several orders of magnitude, then so does $\nabla u$. With respect to multigrid, this means that, close to the location of a jump, the error in $(a\nabla)_h u_h$ can be smoothed by a smoothing procedure, not, however the error in $u_h$ or $\nabla_h u_h$. As we have discussed in Chapter 2, only smooth functions are well approximated on coarse grids. Therefore, the quantity that is approximated on the coarse grid and should be transferred back to the fine grid is a correction for $(a\nabla)_h u_h$. Standard multigrid methods and, in particular, the bilinear interpolation operator implicitly rely on the continuity of $\nabla u$.

> If $a$ jumps significantly (e.g., by several orders of magnitude), a more appropriate prolongation operator is an interpolation which exploits the continuity of $a\nabla u$.

### 7.7.3 Operator-dependent Interpolation

So, for the improvement of the interpolation in the case of jumping coefficients, one makes use of the "continuity" of the error in $a\nabla u$ in the definition of the prolongation operator. We start with a 1D example, where $(a\nabla)_h$ is represented by a stencil $L_h = [s_w \quad s_c \quad s_e]_h$, $(s_c = -s_e - s_w)$. The assumption $(a\nabla)_h v_h = 0$ for a correction $v_h$ leads to

$$s_w(v_h(x) - v_h(x - h)) = s_e(v_h(x + h) - v_h(x)) \quad \text{or}$$

$$v_h(x) = \frac{s_w v_h(x - h) + s_e v_h(x + h)}{s_w + s_e}.$$

The corresponding interpolation for the corrections, $\hat{v}_{2h}$, reads (see Fig. 7.19)

$$I_{2h}^h \hat{v}_{2h}(x) = \begin{cases} \hat{v}_{2h}(x) & \text{for } \bullet \\ -(1/s_c)[s_e\hat{v}_{2h}(x + h) + s_w\hat{v}_{2h}(x - h)] & \text{for } \square \end{cases} \tag{7.7.5}$$

and is characterized by the "operator-dependent" stencil

$$I_{2h}^h = \left] -\frac{s_e}{s_c} \quad 1 \quad -\frac{s_w}{s_c} \right[_h.$$

**Figure 7.19.** A 1D fine grid with symbols indicating the operator-dependent interpolation (7.7.5) used for the transfer from the coarse grid ($\bullet$).

In 2D, the definition of operator-dependent interpolation weights is somewhat more involved. We will consider the general nine-point stencil, which includes five-point stencils. In contrast to the 1D case, it is no longer possible to fulfil the condition $(a\nabla)_h v_h = 0$ at every interpolation point for standard coarsening. There are various possibilities to define the interpolation weights in the 2D case [3, 120, 439].

One common approach is to use *sums* of off-diagonal stencil elements and to work with these sums in the definition of the operator-dependent interpolation $I_{2h}^h$ [3, 120], yielding

$$
I_{2h}^h \hat{v}_{2h}(x, y) = \begin{cases}
\hat{v}_{2h}(x, y) & \text{for } \bullet \\
(1/\tilde{s}_{cy})[\tilde{s}_e \hat{v}_{2h}(x + h, y) + \tilde{s}_w \hat{v}_{2h}(x - h, y)] & \text{for } \square \\
(1/\tilde{s}_{cx})[\tilde{s}_n \hat{v}_{2h}(x, y + h) + \tilde{s}_s \hat{v}_{2h}(x, y - h)] & \text{for } \diamond \\
\text{choose } I_{2h}^h \text{ such that: } L_h I_{2h}^h \hat{v}_{2h}(x, y) = 0 & \text{for } \circ
\end{cases}
\tag{7.7.6}
$$

where

$$
\tilde{s}_w = s_{sw} + s_w + s_{nw} \qquad \tilde{s}_e = s_{se} + s_e + s_{ne}
$$

$$
\tilde{s}_{cy} = -s_n - s_s - s_c \qquad \tilde{s}_s = s_{sw} + s_s + s_{se}
\tag{7.7.7}
$$

$$
\tilde{s}_n = s_{nw} + s_n + s_{ne} \qquad \tilde{s}_{cx} = -s_e - s_w - s_c.
$$

(see Fig. 2.6 for the symbols used above). These interpolation weights coincide with (bi)linear interpolation if the coefficient $a$ is constant.

Often, the *restriction* operator is chosen as the transpose of the (operator-dependent) interpolation. This is especially useful in the context of the *Galerkin coarse grid operator*.

### 7.7.4 The Galerkin Coarse Grid Operator

The introduction of operator-dependent interpolation is not the only modification of standard multigrid components that is necessary in order to achieve a convergence which is comparable to the case of smoothly varying coefficients and *independent of the size of the jumps in a*. It is also important that the coarse grid problem represents the fine grid problem well enough. This property can get lost if the lines on which the coefficients jump are no longer coarse grid lines. Then, the direct discretization $L_{2h}$ on the coarse grid is not necessarily a good approximation of the fine grid problem. This is reflected by unsatisfactory multigrid convergence [3].

The most common approach to obtain a satisfactory coarse grid discretization is to use *Galerkin operators on the coarse grids* (mentioned already in Section 2.3.2).

For a given fine grid operator $L_\ell$, the Galerkin coarse grid operators $L_k$ are recursively defined by

$$L_k := I_{k+1}^k L_{k+1} I_k^{k+1} \quad (k = \ell - 1, \ldots, 0). \tag{7.7.8}$$

Here, $I_k^{k+1}$ and $I_{k+1}^k$ are the intergrid transfer operators. In this context, the restriction operator is usually chosen as the transpose of the interpolation operator. In that case, many important properties of $L_\ell$, for example, symmetry and positive definiteness, are transmitted automatically to $L_k$.

The Galerkin operator has several interesting theoretical properties. We refer here to Section A.2.4 of Appendix A, where these properties are discussed in more detail.

From the recursion (7.7.8), it is clear that all $L_k$ depend on the finest grid operator $L_\ell$. The general description of multigrid given in Chapter 2 is also immediately applicable to these $L_k$. The main practical difference is that *the $L_k$ are, in general, no longer known in advance but have to be calculated from the recursion formula (7.7.8)*.

**Example 7.7.1** If $I_k^{k-1}$ and $I_{k-1}^k$ ($k = 1, 2, \ldots$) are FW and bilinear interpolation, respectively, we obtain the following coarse grid Galerkin discretization for the *standard five-point* discrete Laplace operator $L_\ell$:

$$L_{\ell-1} = I_\ell^{\ell-1} L_\ell I_{\ell-1}^\ell = \frac{1}{h_{\ell-1}^2} \begin{bmatrix} -1/4 & -1/2 & -1/4 \\ -1/2 & 3 & -1/2 \\ -1/4 & -1/2 & -1/4 \end{bmatrix}_{h_{\ell-1}}. \tag{7.7.9}$$

For $\ell \to \infty$, $L_k$ ($k$ fixed) converges to a difference operator which is characterized by the stencil

$$\frac{1}{h_k^2} \begin{bmatrix} -1/3 & -1/3 & -1/3 \\ -1/3 & 8/3 & -1/3 \\ -1/3 & -1/3 & -1/3 \end{bmatrix}_{h_k}. \tag{7.7.10}$$

$\triangle$

This is a well-known approximation for the Laplace operator, which occurs in connection with *bilinear finite elements*. If (7.7.10) with $k = \ell$ is used as the difference operator $L_\ell$ on the finest grid $\Omega_\ell$ (and the transfer operators are chosen as above), then this $L_\ell$ is reproduced by the Galerkin recursion.

**Remark 7.7.4** The five-point Laplace operator $\triangle_h$ is reproduced by the Galerkin recursion, if we employ the seven-point transfer operators (see Remark 7.6.3). $\gg$

For the problem with jumping coefficients, it is important to choose the operator-dependent interpolation discussed in Section 7.7.2 (and the corresponding transpose as restriction) in the definition of the Galerkin coarse grid operators and in the multigrid algorithm.

**Example 7.7.2** Consider a 1D diffusion equation

$$-\frac{d}{dx}\left(a\,\frac{du}{dx}\right) = f,$$

where the coefficient $a$ jumps at an interior fine grid point $x_{i*}$, which is not a coarse grid point. Following the discussion in Remark 7.7.2, the fine grid discretization reads at the grid point $x_{i*}$:

$$L_h = [-a_{i*-1} \quad a_{i*-1} + a_{i*+1} \quad -a_{i*+1}]_h.$$

The 1D Galerkin operator with $I_H^h$ as defined in (7.7.5) and its transpose $I_h^H$ leads to a coarse grid operator $L_H$ with the following properties. The east stencil element $s_e$ of the operator $L_H$ at the coarse grid point $x_{i*-1}$ and the west element $s_w$ at the grid point $x_{i*+1}$ are given by

$$\frac{a_{i*-1}a_{i*+1}}{a_{i*-1} + a_{i*+1}}.$$

These are the same entries as one would find by a direct finite volume discretization with harmonic averaging on the coarse grid, if the jump occurs between two grid points (lines). In this sense, the coarse grid discretization is automatically "physically correct".    △

There are two further possibilities to find a satisfactory coarse grid discretization which are not based on the Galerkin approach.

**Remark 7.7.5** For applications with jumping coefficients the *direct discretization* of the PDE on coarse grids is an option. Note that the *relative position* of the interface with respect to grid lines can change, for example, if the interface corresponds to a fine grid line that is not in the coarse grid anymore. On the finest grid, one can then use the arithmetic averaging of coefficients $a$, as outlined in Remark 7.7.2, whereas on the coarse grids a (weighted) harmonic averaging should be used (see Remark 7.7.3). Without such adaptations, the quality of the coarse grid discretization $L_H$ and the coarse grid correction is not satisfactory and the corresponding multigrid convergence will be slow, even divergence may occur.

In combination with the use of a direct discretization, it is advisable to choose a "powerful" smoother, like ILU type smoothing.    ≫

**Remark 7.7.6** For PDEs with irregularly localized strongly varying coefficients, "homogenization techniques" are sometimes used in the definition of a grid discretization. These techniques are particularly interesting if, instead of the detailed fine scale behavior of a solution, a global (averaged) solution is of interest. The ideas of these techniques can also be used to define the coarse grid discretizations in multigrid. We will not discuss this in detail here. Examples are presented, for instance, in [3, 218, 273].    ≫

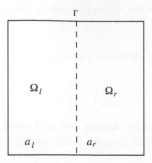

**Figure 7.20.** An example of a domain with an interface $\Gamma$ at which the coefficient $a$ jumps and two subdomains $\Omega_l$ and $\Omega_r$.

**Example 7.7.3** We consider the diffusion problem (7.7.1) with a jumping coefficient and Neumann boundary conditions. We assume that the discontinuity of $a$ is at the line $x = 1/2 + h$, as shown in Fig. 7.20. This grid line exists only on the finest grid. We choose $a_l = 1$ in $\Omega_l$ and $a_r = 10^p$ in $\Omega_r$. The parameter $p$ is varied between 0 and 5. Due to the constant coefficients, $a_l$ and $a_r$, a direct finite volume discretization (as in Remark 7.7.3) can easily be applied on all grids. Concretely, on the finest grid a straightforward discretization can be applied, whereas on the coarse grids the position of the interface relative to coarse grid lines is taken into account. We also have to employ the techniques described in Section 5.6.3 to properly deal with the singular system (due to the Neumann boundary conditions) and to obtain a unique solution.

Table 7.14 compares the direct coarse grid discretization (according to Remark 7.7.5) with the Galerkin coarse grid discretization on a $64^2$ grid. A multigrid $V(1, 1)$ cycle is employed with GS-RB as the smoother. The interpolation operator used is as defined in (7.7.6), (7.7.7), and the restriction operator is its transpose. For this simple example, both coarse grid discretizations lead to efficient multigrid algorithms, almost independent of the size of the jump $10^p$. △

**Example 7.7.4** As an additional example, amongst many others in the literature, we consider problem (7.7.1), discretized with $h = 1/64$. The discontinuous coefficient $a$ is defined as follows:

$$a = 1 \qquad \text{if } 0 < x < \tfrac{1}{2}, \ 0 < y < \tfrac{1}{2}$$
$$a = 1000 \quad \text{if } \tfrac{1}{2} \le x < 1, \ 0 < y < \tfrac{1}{2}$$
$$a = 10 \qquad \text{if } 0 < x < \tfrac{1}{2}, \ \tfrac{1}{2} \le y < 1$$
$$a = 100 \quad \text{if } \tfrac{1}{2} \le x < 1, \ \tfrac{1}{2} \le y < 1.$$

The measured asymptotic convergence factor for the multigrid $V(1, 1)$-cycle based on operator-dependent transfer operators, the Galerkin coarse grid operator and alternating lexocigraphic line relaxation is $\rho_h = 0.044$. △

**Table 7.14.** Measured $V(1, 1)$ multigrid convergence factors for a problem with a jumping coefficient of the order $10^p$.

| $p$ | 0 | 1 | 2 | 3 | 4 | 5 |
|---|---|---|---|---|---|---|
| $L_H$: Galerkin | 0.12 | 0.14 | 0.17 | 0.18 | 0.19 | 0.19 |
| $L_H$: direct | 0.09 | 0.13 | 0.19 | 0.20 | 0.20 | 0.21 |

### 7.7.5 Further Remarks on Galerkin-based Coarsening

We conclude this section with some more remarks on Galerkin coarse grid operators. It is, for example, also possible to use the Galerkin coarse grid approach for *nonsymmetric problems* like the convection–diffusion equation. Here, we discuss the use of a first-order upwind discretization on the finest grid.

If we define the Galerkin operators by FW as the restriction and by bilinear interpolation, the coarse grid operators will lead to unstable coarse grid discretizations [176, 438]. In fact, the resulting discretization on coarse grids converges towards a *central* differencing discretization (see Remark 7.1.2). An appropriate Galerkin approach takes the nonsymmetric "upwind" character of the discretization into account.

**Remark 7.7.7 (Galerkin for the convection–diffusion equation)** It has been shown [439] that the Galerkin coarse grid discretization can indeed lead to efficient multigrid algorithms for the first-order upwind discretization if operator-dependent interpolation and restriction are employed, in which the symmetric and nonsymmetric parts of the discrete operator $L_h$ are considered separately. The corresponding software code, MGD9V, developed by de Zeeuw, solves scalar PDEs discretized on structured grids and is available, for example, from the MGNET web page (http://www.mgnet.org).

In principle, the stencil elements of the Galerkin coarse grid discretizations for nonsymmetric problems can have large positive off-diagonal elements. It is even possible that they are not well suited for (point) smoothing. In MGD9V, for example, a linewise ILU-based smoother (ILLU [439]) is employed, which is more robust with respect to unpleasant stencil elements.

Another coarse grid discretization for nonsymmetric problems has been presented by Dendy [121]. Here, the prolongation operator is the same as in (7.7.7) and the restriction is its transpose. However, in the definition of the Galerkin operator a different restriction operator is used, namely an operator that is based on the transpose operator $(L_h)^T$ of $L_h$. The typical two-grid factor 0.5 which arises from the upwind discretization of the PDE on coarse grids (see Section 7.2.3) is not observed, with this operator. This is confirmed by LFA. This feature is also discussed in [429], where another coarse grid discretization with the same feature is introduced.

In order to deal with *positive* coarse grid stencil elements, a Kaczmarz line relaxation method has been proposed [121]. $\gg$

On the other hand, the Galerkin coarse grid operator also has the following general *disadvantages*.

- The resulting coarse grid stencils enlarge in general. In 2D, nine-point coarse grid stencils are commonly obtained from five-point stencils by the Galerkin approach. Even worse, in 3D 27-point coarse grid stencils are usually obtained from standard seven-point discretizations on the fine grid.
- The Galerkin coarse grid discretization can only be applied directly to linear discrete operators. This means, that it is not easily possible to combine it with the nonlinear FAS version of the multigrid method. If a global (Newton) linearization is applied, the Galerkin coarse grid discretizations may need to be redefined after every linearization step.
- Furthermore, it is not so easy to implement the Galerkin discretization. A point-by-point multiplication of the restriction operator "times" the discretization operator "times" the prolongation operator is, in general, more complicated than a direct discretization on $\Omega_H$. Finally, it is time and memory consuming to set up these coarse grid difference stencils.

## 7.8 MULTIGRID AS A PRECONDITIONER (ACCELERATION OF MULTIGRID BY ITERANT RECOMBINATION)

In this section, we discuss the acceleration of multigrid by Krylov subspace approaches like the conjugate gradient (CG) method [194] or the generalized minimal residual method (GMRES) [335]. This combination, which means that *multigrid is used as a preconditioner*, is particularly interesting with respect to robustness and efficiency in complex applications.

**Remark 7.8.1** With respect to terminology, we point out that "using multigrid as a *preconditioner* in connection with Krylov subspace methods" is identical to "*accelerating multigrid by a Krylov subspace method*" (briefly: Krylov subspace acceleration of multigrid).     ≫

We will motivate the use of multigrid as a preconditioner in Section 7.8.1 by revisiting the recirculating convection–diffusion problem (discussed in Section 7.2.3). From the multigrid point of view, multigrid as a preconditioner can also be interpreted as an *acceleration of multigrid by iterant recombination*. This interpretation, which will be described in Section 7.8.2, allows certain important generalizations, for example, to nonlinear problems and additional acceleration options on coarse grids. Details of Krylov subspace acceleration of multigrid will be discussed in Section 7.8.3. In Section 7.8.4, we will summarize our view of multigrid as a preconditioner.

### 7.8.1 The Recirculating Convection–Diffusion Problem Revisited

As discussed in Section 7.2.3, convection–diffusion problems with a dominating recirculating convection term are not easy to handle with multigrid based on standard coarsening.

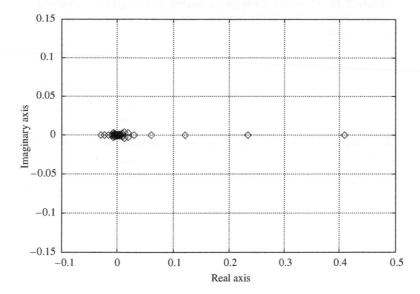

**Figure 7.21.** The spectrum of the multigrid iteration operator $M_h$ for the recirculating convection–diffusion problem with first-order upwind discretization ($h = 1/32$).

We have seen that the multigrid convergence factors increase for the first-order upwind discretization due to the coarse grid problem described in Section 7.2.3.

For the recirculating convection–diffusion problem (7.2.7), discretized on a grid with $h = 1/32$, the *spectrum of the corresponding W(1,1) multigrid operator*, $M_h$, with a symmetric alternating line Gauss–Seidel smoother is presented in Fig. 7.21. Most of the eigenvalues of $M_h$ are clustered around 0. There are only some isolated eigenvalues away from the cluster. The (isolated) largest eigenvalue (the spectral radius) near 0.4 limits the multigrid convergence on this relatively coarse grid (as seen in Table 7.15).

As is known from theory and experience [98, 292], such a situation is well-suited for Krylov subspace acceleration. *The eigenvectors belonging to the few isolated eigenvalues can be expected to be "captured" after only a few Krylov subspace iterations.* Indeed, Table 7.15 shows that multigrid can be substantially accelerated in this way (the results from Table 7.3 are included for convenience). Although we do not observe $h$-independent convergence in this example, the convergence is much better than the convergence without the acceleration technique. The *multigrid preconditioned* GMRES(15) [335] method used here will be presented in Section 7.8.3.

Qualitatively similar results are obtained if we use, for example, Fromm's second-order upwind discretization scheme for the convective terms. In that case, multigrid (with the KAPPA smoother) gives convergence factors of about 0.9 for this problem and the multigrid preconditioned GMRES(15) method reduces these factors to about 0.7.

**Table 7.15.** Measured convergence factors of multigrid and multigrid preconditioned GMRES(15) for the recirculating convection–diffusion equation with $\varepsilon = 10^{-5}$.

| $1/h$ | 32 | 64 | 128 | 256 |
|---|---|---|---|---|
| MG | 0.40 | 0.51 | 0.59 | 0.66 |
| MG with GMRES(15) | 0.0001 | 0.007 | 0.05 | 0.18 |

**Remark 7.8.2 (how to compute the spectrum of the multigrid operator)** In applying multigrid algorithms, it is, of course, not necessary to determine the multigrid iteration operator explicitly. If one needs it, for instance to determine its spectrum using a numerical software package, one straightforward possibility is to perform one multigrid cycle for each unit vector $e_i$ as initial approximation (with zero right-hand side and zero boundary conditions). Using $e_1 = (1, 0, \ldots, 0)$ as $u_h^0$, one obtains the approximation $u_h^1$ *which is the first column $m_1$ of the iteration matrix.* Then, one iteration with the unit vector $e_2 = (0, 1, \ldots, 0)$ is performed to obtain $m_2$ and so on. In this way, the whole multigrid iteration matrix is found. $\gg$

### 7.8.2 Multigrid Acceleration by Iterant Recombination

Multigrid acceleration by iterant recombination and multigrid preconditioning can be identified and lead to similar algorithms in practice. The iterant recombination fits naturally into the multigrid philosophy and can easily be described. Therefore, we discuss this approach first.

The acceleration of multigrid by iterant recombination starts from successive approximations $u_h^1, u_h^2, \ldots, u_h^m$, from previous multigrid cycles. In order to find an improved approximation $u_{h,\mathrm{acc}}$, we consider a linear combination of the $\tilde{m} + 1$ latest approximations $u_h^{m-i}, i = 0, \ldots, \tilde{m}$,

$$u_{h,\mathrm{acc}} = u_h^m + \sum_{i=1}^{\tilde{m}} \alpha_i (u_h^{m-i} - u_h^m), \tag{7.8.1}$$

(assuming $m \geq \tilde{m}$). For linear equations, the corresponding defect, $d_{h,\mathrm{acc}} = f_h - L_h u_{h,\mathrm{acc}}$, is given by

$$d_{h,\mathrm{acc}} = d_h^m + \sum_{i=1}^{\tilde{m}} \alpha_i (d_h^{m-i} - d_h^m), \tag{7.8.2}$$

where $d_h^{m-i} = f_h - L_h u_h^{m-i}$. In order to obtain an improved approximation $u_{h,\mathrm{acc}}$, the parameters $\alpha_i$ are determined in such a way that the defect $d_{h,\mathrm{acc}}$ is minimized. We will minimize $d_{h,\mathrm{acc}}$, i.e.

$$\left\| d_h^m + \sum_{i=1}^{\tilde{m}} \alpha_i (d_h^{m-i} - d_h^m) \right\|, \tag{7.8.3}$$

with respect to the $L_2$-norm $\| \cdot \|_2$.

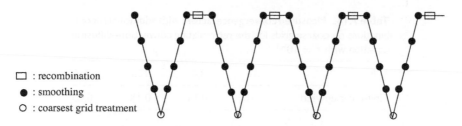

**Figure 7.22.** Recombination of multigrid iterants.

□ : recombination
● : smoothing
○ : coarsest grid treatment

This is a classical defect minimization problem. In principle, the optimal coefficients $\alpha_i$ can be determined by a (Gram–Schmidt) orthonormalization process. Here, however, we solve the system of linear (normal) equations

$$H \begin{pmatrix} \alpha_1 \\ \alpha_2 \\ \vdots \\ \alpha_{\tilde{m}} \end{pmatrix} = \begin{pmatrix} \beta_1 \\ \beta_2 \\ \vdots \\ \beta_{\tilde{m}} \end{pmatrix}, \tag{7.8.4}$$

where the matrix $H = (h_{ik})$ is defined by

$$h_{ik} = \langle d_h^{m-i}, d_h^{m-k} \rangle - \langle d_h^m, d_h^{m-i} \rangle - \langle d_h^m, d_h^{m-k} \rangle + \langle d_h^m, d_h^m \rangle \tag{7.8.5}$$
$$i = 1, \ldots, \tilde{m}, \quad k = 1, \ldots, \tilde{m},$$

with the standard Euclidean inner product $\langle ., . \rangle$ and

$$\beta_i = \langle d_h^m, d_h^m \rangle - \langle d_h^m, d_h^{m-i} \rangle. \tag{7.8.6}$$

Now, the current approximation $u_h^m$ is replaced by $u_{h,\mathrm{acc}}$. With this replaced approximation, the next multigrid cycle is performed leading to a new iterant $u_h^{m+1}$. The recombination (7.8.1) is again carried out with the latest iterants $u_h^{m+1-i}$, $i = 0, \ldots, \tilde{m}$ and so on.

In practice, the iterant recombination is already carried out with the very first multigrid iterants. Here, one can perform a recombination with the iterants already available. The resulting iterative method is sketched in Fig. 7.22.

**Remark 7.8.3** In general, for the minimization problem (7.8.3), it may happen that $H$ is an *ill-conditioned matrix*. In practice, however, $\tilde{m}$ is chosen small, for example 5 or 10. Such small matrices $H$ are usually still satisfactorily conditioned, so that the system (7.8.4) can be solved directly. For cases in which $H$ is a singular matrix, more details are given in [410].

A heuristic explanation of why small values of $\tilde{m}$ are sufficient is as follows. Often when multigrid is applied to complicated problems, there are only a few isolated large eigenvalues in the multigrid iteration matrix. The number $\tilde{m}$ of iterants that are to be recombined to obtain a considerable convergence improvement is related to the number of large (isolated) eigenvalues of the iteration matrix. $\gg$

**Table 7.16.** Measured convergence factors with additional recombinations on coarse grids for the recirculating convection–diffusion equation with $\varepsilon = 10^{-5}$.

| $1/h$ | 32 | 64 | 128 | 256 |
|---|---|---|---|---|
| First-order upwind | 0.10 | 0.14 | 0.18 | 0.23 |

**Remark 7.8.4 (recombination and nonlinear multigrid)**   It is possible to generalize the idea of iterant recombination to nonlinear situations, where a nonlinear multigrid method is used. In this case, the defect relation (7.8.2) does not hold exactly, due to the nonlinearity. The substitution of $u_h^m$ by $u_{h,\mathrm{acc}}$ is only carried out if the defect $d_{h,\mathrm{acc}}$ of $u_{h,\mathrm{acc}}$ is not much larger than the defect $d_h^m$. For details on the nonlinear variant, see [294, 410].      $\gg$

**Remark 7.8.5 (recombination on coarse grids)**   Typically, $\tilde{m} = 10$ is often sufficient for a considerable convergence acceleration. However, for "real life" problems, it may still be a problem to store these additional grid functions. If the reason for multigrid convergence difficulties lies in an insufficient coarse grid correction, $\tilde{m}$ can often be significantly reduced by applying iterant recombinations on coarse grids. In [294], one way to apply the acceleration on the coarse grids is presented using the FAS version of multigrid. Here, we only present one convergence result with this additional coarse grid acceleration for the recirculating convection–diffusion equation. Table 7.16 presents convergence factors, using this strategy with $\tilde{m} = 2$ on the finest grid and $\tilde{m}_c = 5$ on coarse grids, for the same recirculating convection–diffusion problem as in Table 7.15.

Compared to the convergence in Table 7.15 (second row), the convergence is somewhat worse (mainly because $\tilde{m}$ is now much smaller on the finest grid). The storage is, however, significantly reduced. For Fromm's discretization, the coarse grid iterant recombination ($\tilde{m} = 2, \tilde{m}_c = 5$) improves the convergence. In that case the convergence factors are about 0.65 [294].      $\gg$

### 7.8.3 Krylov Subspace Iteration and Multigrid Preconditioning

Here, we will briefly discuss the preconditioned Krylov subspace iteration methods. Since these methods are usually introduced in the framework of matrices, we adopt the matrix notation $Au = f$ in this section.

We will start with a Krylov subspace iteration and improve its convergence properties by multigrid preconditioning. In a certain sense, this description represents the Krylov subspace view of the approach. We make use of the following relations.

Let $u^0$ be an initial approximation and $d^0 = f - Au^0$ its defect. The *Krylov subspace* $K^m$ is then defined by

$$K^m := \mathrm{span}[d^0, Ad^0, \ldots, A^{m-1}d^0].    \tag{7.8.7}$$

This subspace can also be represented by

$$K^m = \text{span}[u^1 - u^0, u^2 - u^1, \ldots, u^m - u^{m-1}]$$
$$= \text{span}[u^0 - u^m, u^1 - u^m, \ldots, u^{m-1} - u^m],$$

where

$$u^i = (I - A)u^{i-1} + f$$

are the iterants from Richardson's iteration (with $\tau = 1$, see Section 1.6).

These representations are easily obtained by induction using

$$u^1 - u^0 = d^0, \qquad u^{i+1} - u^i = (I - A)(u^i - u^{i-1}).$$

The Krylov subspace approximation

$$u_{\text{acc}}^m \in u^0 + K^m = u^0 + \text{span}[d^0, Ad^0, \ldots, A^{m-1}d^0]$$

is then characterized as the approximation *with minimal defect* in a suitable norm (for $m = 1, 2, \ldots$).

The different Krylov subspace methods differ in the way, the minimization is carried out. The classical *conjugate gradient* (CG) method for s.p.d. matrices $A$ is characterized by minimizing the defect in the norm

$$||d|| = \langle d, A^{-1}d \rangle,$$

where $\langle . , . \rangle$ is the Euclidean inner product. Krylov subspace methods that are not restricted to s.p.d. matrices are, for example, the GMRES method [335] and the BiCGSTAB method [397]. GMRES is obtained by using the $|| \cdot ||_2$ norm for minimization. There are many different Krylov subspace iteration methods, which we will not discuss here in detail. Books giving an overview of these methods include [26, 159, 162, 337].

**Remark 7.8.6 (GMRES($\bar{m}$))**   In GMRES, all $m$ vectors are kept in the Krylov subspace for the calculation of the next iterant in order to guarantee the minimal norm of the defect. This leads to storage complications for large problems. A remedy for this disadvantage is given by the *restarted* GMRES($\bar{m}$) method [335], that uses $\bar{m}$ previous vectors after which the iteration is restarted with a new Krylov subspace. Other approaches are based on *truncation*, where the Krylov subspace is spanned by the latest $\bar{m}$ approximations. That is, the oldest iterant is removed from the subspace when a new iterant is added, whereas in GMRES($\bar{m}$) the subspace is completely removed and restarted after $\bar{m}$ iterants.      ≫

In general, stand-alone Krylov subspace iterations are slow. Their effectiveness depends strongly on the *condition number of the matrix $A$* and *on the distribution of its eigenvalues*. If the original system $Au = f$ is *preconditioned* with a suitable left or a right preconditioner, the condition number can often be significantly reduced (see the remarks in Section 1.6). Correspondingly, the performance of these solution methods is substantially improved [264].

In the case of a *right preconditioner $C$*, we solve $ACz = f$, where $z = C^{-1}u$.

**Remark 7.8.7 (preconditioned GMRES($\bar{m}$))**   In a C-type metalanguage, the GMRES($\bar{m}$) algorithm with a right preconditioner $C$ is given by:

**GMRES** ($\bar{m}$,$A$, $C$, $f$, $u^0$){
        Set the matrix $\tilde{\mathbf{H}} = \mathbf{0}$ with dim: $(\bar{m} + 1) \times \bar{m}$
        $d^0 = f - Au^0$; $\beta = ||d^0||_2$; $b_1 = d^0/\beta$;
        for $j = 1, \ldots, \bar{m}$ {
                $r_j := C^{-1}b_j$;
                $w := Ar_j$;
                for $i = 1, \ldots, j${
                        $h_{i,j} := \langle w, b_i \rangle$;
                        $w := w - h_{i,j}b_j$;
                        $h_{j+1,j} := ||w||_2, b_{j+1} = w/h_{j+1,j}$;
                }
        }
        Define $\mathbf{B}_{\bar{m}} := [b_1, \ldots, b_{\bar{m}}]$;
        $u^{\bar{m}} := u^0 + C^{-1}\mathbf{B}_{\bar{m}}y_{\bar{m}}$; with $y_{\bar{m}} = \min_y ||\beta e_1 - \tilde{\mathbf{H}}y||_2$, $(e_1 = [1, 0, \ldots, 0]^T)$;
        Compute $d^{\bar{m}} = f - Au^{\bar{m}}$;
        If satisfied **stop**, else restart $u^0 \leftarrow u^{\bar{m}}$;
}

The preconditioning step is formally denoted by $C^{-1}$. This means that $r_j$ is obtained from the right-hand side $b_j$ by one iteration with a preconditioner $C$. For GMRES($\bar{m}$) without preconditioning, set $C = I$.                                                                                    ≫

Many basic one-level iterative methods, like $\omega$-JAC, GS-LEX and ILU methods have been used as preconditioners. Although the resulting solvers became very popular in the 1980s, an $O(N)$ solver was not obtained for the Poisson equation. For example, with the incomplete Cholesky factorization (see Section 7.5.2) as a preconditioner, the overall complexity of preconditioned conjugate gradients (called ICCG [264]) for solving Poisson's equation is $O(N^{5/4})$ in 2D and $O(N^{9/8})$ in 3D.

---

**Remark 7.8.8 (multigrid as a preconditioner)**   In the same way as the classical single grid iterative methods can be used as preconditioners, it is also possible to use multigrid as a preconditioner. We choose $C = (I - M)A^{-1}$ where $M$ is the multigrid iteration operator.                                                                                    ≫

---

For s.p.d. problems, the robustness of multigrid as a preconditioner for CG has been demonstrated in [211]. By using multigrid as a preconditioner for GMRES or BiCGSTAB [397], nonsymmetric problems like the convection–diffusion equation can also be handled. This has been demonstrated, for example, in [292].

Our description of Krylov subspace iteration with multigrid preconditioning already indicates the close connection between the recombination of multigrid iterants and multigrid as a preconditioner. In particular, the multigrid acceleration by iterant recombination with the minimization of (7.8.3) in the $\| \cdot \|_2$-norm and restarting after $\bar{m}$ iterants is equivalent to the preconditioned GMRES($\bar{m}$) method with $C = (I - M)A^{-1}$, where $M$ denotes the multigrid iteration matrix. The multigrid iterant recombination as described in Section 7.8.2 corresponds to a truncated Krylov subspace variant as mentioned in Remark 7.8.6.

In spite of this theoretical equivalence, in practice the recombination and preconditioning differ algorithmically. A typical difference is the following.

**Remark 7.8.9** Whereas in the recombination of multigrid iterants, the system (7.8.4) is used for minimization, in the GMRES method the vectors used in the subspace are orthogonalized by a Gram–Schmidt process. The basis used in the Krylov subspace methods is thus numerically of a better condition. $\gg$

**Remark 7.8.10** With a *varying* preconditioner, like multigrid with a different cycle in each iteration, a Krylov subspace method in which the preconditioner can change from iteration to iteration is needed. The flexible GMRES method (FGMRES) [336] allows such a varying preconditioner. FGMRES stores defects and approximations of the solution. In this sense, it corresponds to the method of iterant recombination. $\gg$

**Example 7.8.1 (rotated anisotropic diffusion revisited)** Consider again the rotated anisotropic diffusion equation (7.6.2) with Dirichlet boundary conditions. The mixed derivative is approximated by the four-point stencil (4.7.5). A nine-point stencil results from the standard second-order discretization of all terms in (7.6.2). As we have discussed in Section 7.6, the multigrid convergence with point and line smoothers and standard coarsening is slow, for example, for parameters $\varepsilon = 10^{-5}$ and $\beta = 135°$. For these parameters, we compute the spectra of the F(0,2)-cycle multigrid iteration operator *with an alternating GS line smoother* on a $32^2$ and a $64^2$ grid. The spectra for the F(0,2)-cycle are shown in Fig. 7.23. One observes the mesh dependence of the multigrid convergence: the spectral radius increases as the grid gets finer. It is already larger than 0.6 for these coarse grid problems. However, many eigenvalues are clustered around 0 and only a limited number of eigenvalues are larger than 0.4 for both grid sizes. This indicates that Krylov methods will improve the convergence considerably.

The convergence of multigrid as a solver and as a preconditioner for GMRES is presented for the $32^2$ grid in Fig. 7.24. On finer grids the multigrid convergence factors further increase, whereas the convergence with the W-cycle as a preconditioner is hardly level dependent. Table 7.17 presents the number of iterations after which the initial defect is reduced by eight orders of magnitude with BiCGSTAB and GMRES(20) preconditioned by F(0,2) and W(0,2) multigrid cycles on three very fine grid sizes. Note that BiCGSTAB is twice as expensive per iteration as GMRES. The fastest method in CPU time is here the F-cycle as the preconditioner (for all three grid sizes). $\triangle$

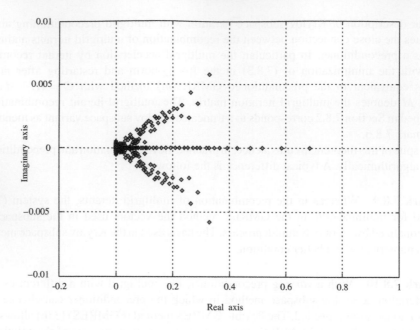

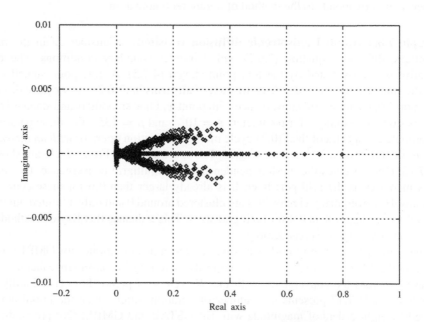

**Figure 7.23.** The eigenvalue spectra of the F(0,2)-cycle for the rotated anisotropic diffusion problem, $\varepsilon = 10^{-5}$, $\beta = 135°$ on a $32^2$ (upper picture) and a $64^2$ (lower picture) grid.

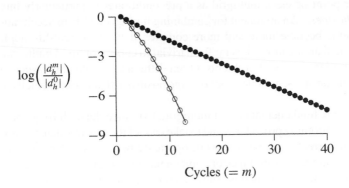

$$\log\left(\frac{|d_h^m|}{|d_h^0|}\right)$$

Cycles $(= m)$

**Figure 7.24.** The convergence of multigrid as a solver ($\bullet$) and as a preconditioner, with GMRES acceleration ($\circ$), for the rotated anisotropic diffusion equation ($h = 1/32$).

**Table 7.17.** Number of BiCGSTAB and GMRES iterations for a defect reduction of $10^{-8}$ for the rotated anisotropic diffusion equation ($\varepsilon = 10^{-5}$, $\beta = 135°$).

| | Grid | | | | | |
|---|---|---|---|---|---|---|
| | $256^2$ | | $512^2$ | | $768^2$ | |
| Cycle | BiCGSTAB | GMRES | BiCGSTAB | GMRES | BiCGSTAB | GMRES |
| F(0,2) | 17 | 31 | 21 | 43 | 25 | 48 |
| W(0,2) | 10 | 19 | 12 | 20 | 13 | 22 |

**Remark 7.8.11 (parallelism in multigrid preconditioned Krylov methods)** As discussed in Chapter 6, it is natural to use the grid partitioning technique for an efficient parallelization of multigrid. If we use multigrid as a preconditioner, it is also necessary that the Krylov acceleration method can be parallelized efficiently in the framework of grid partitioning. Krylov subspace methods, however, consist mainly of the computation of inner products and matrix–vector multiplications and do not introduce particular difficulties in parallelization. Matrix–vector and inner products need only simple communication strategies among the processors. $\gg$

### 7.8.4 Multigrid: Solver versus Preconditioner

We return to the question of whether multigrid should be used as a solver or as a preconditioner. In particular, the question to be answered is which approach should be used when.

*First, it is not useful to accelerate a highly efficient multigrid algorithm (with an ideal convergence factor) by Krylov subspace acceleration. The extra effort does not pay off.*

From our point of view, multigrid as a preconditioner is particularly interesting with respect to *robustness*. An argument for combining multigrid with an acceleration technique is that problems become more and more complex if we treat real-life applications. For such complicated applications, it is far from trivial to choose optimal multigrid components uniformly for a large class of problems. Often different complications such as convection dominance, anisotropies, nonlinearities or large positive off-diagonal stencil elements occur simultaneously.

Therefore, the fundamental idea of multigrid, to reduce the high frequency components of the error by smoothing procedures and to take care of the low frequency error components by coarse grid corrections, does not work optimally in all cases if straightforward multigrid approaches are used. Certain error components may remain large since they cannot be reduced by *standard* smoothing procedures combined with standard coarse grid approximations. These specific error components (and the corresponding eigenvectors/eigenvalues) are then responsible for the poor multigrid convergence. In such situations, the combination with Krylov subspace methods may have the potential of a substantial acceleration.

The recirculating convection–diffusion problem can be regarded as a simple example of this type of problem. In this case, more sophisticated multigrid components (e.g., flow-oriented downstream relaxation or Galerkin-type coarse grid correction, see Section 7.7.4) will also lead to very satisfactory convergence factors. But they are more involved to realize and implement. In other cases, the efficiency of the multigrid components may strongly depend on problem parameters. Then, the combination of multigrid with Krylov subspace may have advantages.

Multigrid as a preconditioner is also interesting for several other problems, for example, on unstructured grids, for problems with small geometric details, which are not visible on coarse grids [82], and for the problems with geometric singularities discussed in Section 5.5.

**Remark 7.8.12**   In Appendix A, we will see examples, for which AMG is used as a preconditioner. AMG is particularly suited as a preconditioner for sparse matrices and unstructured grids.                                                                          ≫

# 8

# MULTIGRID FOR SYSTEMS OF EQUATIONS

In describing the basic multigrid techniques, we have, so far, confined ourselves to scalar PDEs. *Systems of PDEs* can also be treated by multigrid, usually with efficiency similar to that of scalar equations.

The main messages we want to give in this chapter are:

- In principle, the extension of multigrid methods from scalar PDEs to systems of PDEs is straightforward.
- There are various ways to generalize scalar smoothing schemes to systems. A natural extension of smoothing by relaxation (in the scalar case) is smoothing by *collective* relaxation (in the systems case). That is, all unknowns at each single grid point are relaxed simultaneously. Sometimes, however, collective point or even collective versions of line smoothing are not sufficient and more complex relaxation schemes have to be employed. On the other hand, in simple cases, even *decoupled* relaxation already works fine.
- Special care has to be taken with the multigrid treatment of boundary conditions.
- For more involved problems, e.g. from fluid dynamics, we have to take care of several complications in the multigrid treatment: questions of stable discretization, singular perturbation behavior, nonellipticity etc. have to be considered.

We will treat all these topics in some detail.

The basic multigrid idea is the same as in the scalar case. Based on a suitable discretization of the PDE system, appropriate relaxation schemes are used to smooth the errors of the unknown grid functions. All other multigrid components can immediately be extended to systems of PDEs. Again, smoothing turns out to be the most crucial multigrid component.

In Section 8.2 we describe these multigrid components for systems of equations. Section 8.3 discusses the generalization of the scalar LFA smoothing analysis to systems, including the generalization of the $h$-ellipticity measure.

In the remaining sections of this chapter, we treat several specific systems of particular relevance, each of which is representative for a class of systems. The biharmonic system, discussed in Section 8.4, is a reformulation of the biharmonic equation. As a system, it is very simple and consists of two Poisson-type equations. Nevertheless, it is a first example of the rather common problem that the boundary conditions and the unknown functions do not necessarily match, e.g. that one has two boundary conditions for one function and none for another.

The linear shell problem in Section 8.5 is an example for which the difference of *collective* and of *decoupled* smoothing becomes clear. In the Stokes and incompressible Navier–Stokes equations, discussed in Sections 8.6–8.8 the natural equation for the pressure $p$ does not contain $p$ at all. The development of stable discretizations, stable smoothing schemes and other multigrid components is more involved then and will be discussed in detail. We treat both staggered and nonstaggered discretizations.

Compressible flow equations, which will be considered in Section 8.9, contain even more difficulties. Formally, they are no longer elliptic. Their solutions may have shocks.

Most of our presentation in this chapter is oriented to the 2D case for convenience. Typically, the generalization to 3D is straightforward. Throughout this chapter, we also assume standard coarsening unless we explicitly state otherwise.

## 8.1 NOTATION AND INTRODUCTORY REMARKS

In this section, we first consider a (linear) elliptic $q \times q$ system of PDEs in two dimensions,

$$L^\Omega u(x, y) = Lu(x, y) = f^\Omega(x, y) \tag{8.1.1}$$

with

$$u = (u^1, \ldots, u^q)^T, \qquad f = (f^1, \ldots, f^q)^T$$

on a domain $\Omega \subset \mathbb{R}^2$, together with a set of appropriate boundary conditions

$$L^\Gamma u(x, y) = Bu(x, y) = f^\Gamma(x, y) \tag{8.1.2}$$

at the boundary $\Gamma = \partial\Omega$. In the following we will use the notation $L$ and $B$ instead of $L^\Omega$ and $L^\Gamma$ for convenience. The system (8.1.1) can also be written in the form

$$\underbrace{\begin{pmatrix} L^{1,1} & \cdots & \cdots & L^{1,q} \\ \vdots & & & \vdots \\ \vdots & & & \vdots \\ L^{q,1} & \cdots & \cdots & L^{q,q} \end{pmatrix}}_{L} \begin{pmatrix} u^1 \\ \vdots \\ \vdots \\ u^q \end{pmatrix} = \begin{pmatrix} f^{\Omega,1} \\ \vdots \\ \vdots \\ f^{\Omega,q} \end{pmatrix}, \tag{8.1.3}$$

where the $L^{k,l}$ are scalar differential operators. Of course, the boundary conditions can be written correspondingly:

$$\underbrace{\begin{pmatrix} B^{1,1} & \cdots & \cdots & B^{1,q} \\ \vdots & & & \vdots \\ \vdots & & & \vdots \\ B^{\tilde{q},1} & \cdots & \cdots & B^{\tilde{q},q} \end{pmatrix}}_{B} \begin{pmatrix} u^1 \\ \vdots \\ \vdots \\ u^q \end{pmatrix} = \begin{pmatrix} f^{\Gamma,1} \\ \vdots \\ \vdots \\ f^{\Gamma,\tilde{q}} \end{pmatrix}. \tag{8.1.4}$$

Note that in general $q \neq \tilde{q}$, i.e. the number of boundary conditions does not necessarily coincide with the number of equations. Moreover, in general, there exists neither a natural relationship between an unknown function $u^s$ and a specific equation in the system nor a natural relationship between $u^s$ and one of the boundary conditions.

The number $\tilde{q}$ of required boundary conditions can be seen by considering the *principal part*, i.e. the terms including the highest derivative, of the operator determinant

$$\det L = \begin{vmatrix} L^{1,1} & \cdots & \cdots & L^{1,q} \\ \vdots & & & \vdots \\ \vdots & & & \vdots \\ L^{q,1} & \cdots & \cdots & L^{q,q} \end{vmatrix}. \tag{8.1.5}$$

The number of boundary conditions is typically determined by the order of the highest derivative, which is present in this determinant. For example, if the principal part of $\det L$ is of the form $(\Delta)^{\tilde{q}}$, $\tilde{q}$ boundary conditions are required.

Two examples of linear PDE systems are the biharmonic system and the Stokes equations, discussed in more detail in Sections 8.4 and 8.6, respectively. Each of these systems is a model problem representing a large and important class of PDEs (elasticity, CFD).

**Example 8.1.1** The biharmonic system is a reformulation of the biharmonic equation $\Delta\Delta u = f$. Introducing the function

$$v = \Delta u,$$

the biharmonic equation can be written as the $2 \times 2$ system

$$\Delta v(x, y) = f(x, y) \tag{8.1.6}$$
$$\Delta u(x, y) - v(x, y) = 0 \tag{8.1.7}$$

or in the form

$$\begin{pmatrix} \Delta & 0 \\ -I & \Delta \end{pmatrix} \begin{pmatrix} v \\ u \end{pmatrix} = \begin{pmatrix} f \\ 0 \end{pmatrix},$$

where the scalar differential operators $L^{k,l}$ are

$$L^{1,1} = \Delta, \qquad L^{1,2} = 0, \qquad L^{2,1} = -I \quad \text{and} \quad L^{2,2} = \Delta.$$

Obviously, we have $\det L = \Delta^2$, which means that two boundary conditions are appropriate.

$\triangle$

**Example 8.1.2** The Stokes equations describe incompressible flow situations governed by viscosity. We consider the stationary 2D Stokes equations

$$\begin{bmatrix} -\Delta & 0 & \partial_x \\ 0 & -\Delta & \partial_y \\ \partial_x & \partial_y & 0 \end{bmatrix} \begin{bmatrix} u \\ v \\ p \end{bmatrix} = \begin{bmatrix} 0 \\ 0 \\ 0 \end{bmatrix} \quad \text{in a domain } \Omega \in \mathbb{R}^2. \tag{8.1.8}$$

Here, the unknown functions $u = u(x, y)$, $v = v(x, y)$, $p = p(x, y)$ denote the velocity components in $x$- and $y$-directions and the pressure, respectively. The first two equations (i.e. the momentum equations) correspond to conservation of momentum and the third one (the continuity equation) to conservation of mass. The momentum equations are diffusion-based transport equations. In contrast to the incompressible Navier–Stokes equations (see Section 8.6), convection does not occur.

As in the previous example, the operator determinant is $\det L = \Delta^2$. Hence, only two boundary conditions are required for this $3 \times 3$ system of PDEs. The actual choice of the boundary conditions in an application is usually motivated physically. At solid walls, for example, $u = v = 0$ is a natural choice.

The first equation of this system is naturally related to the velocity $u$, the second one to $v$. The third unknown function, $p$, however, does not appear in the third equation.    $\triangle$

The discrete system

$$\begin{aligned} L_h u_h &= f_h^\Omega \\ B_h u_h &= f_h^\Gamma \end{aligned} \tag{8.1.9}$$

denotes the discrete analog of the PDE system (8.1.1) with boundary conditions (8.1.2).

Analogously to (8.1.3), this system can be written as

$$\underbrace{\begin{pmatrix} L_h^{1,1} & \cdots & \cdots & L_h^{1,q} \\ \vdots & & & \vdots \\ \vdots & & & \vdots \\ L_h^{q,1} & \cdots & \cdots & L_h^{q,q} \end{pmatrix}}_{L_h} \underbrace{\begin{pmatrix} u_h^1 \\ \vdots \\ \vdots \\ u_h^q \end{pmatrix}}_{u_h} = \underbrace{\begin{pmatrix} f_h^{\Omega,1} \\ \vdots \\ \vdots \\ f_h^{\Omega,q} \end{pmatrix}}_{f_h^\Omega} \tag{8.1.10}$$

$$
\underbrace{\begin{pmatrix} B_h^{1,1} & \cdots & \cdots & B_h^{1,q} \\ \vdots & & & \vdots \\ \vdots & & & \vdots \\ B_h^{\tilde{q},1} & \cdots & \cdots & B_h^{\tilde{q},q} \end{pmatrix}}_{\boldsymbol{B}_h} \underbrace{\begin{pmatrix} u_h^1 \\ \vdots \\ \vdots \\ u_h^q \end{pmatrix}}_{\boldsymbol{u}_h} = \underbrace{\begin{pmatrix} f_h^{\Gamma,1} \\ \vdots \\ \vdots \\ f_h^{\Gamma,\tilde{q}} \end{pmatrix}}_{\boldsymbol{f}_h^\Gamma}.
\tag{8.1.11}
$$

Correspondingly, we also consider nonlinear PDE systems of the form

$$
Nu = f,
\tag{8.1.12}
$$

consisting of $q$ nonlinear scalar PDEs

$$
\begin{pmatrix} N^1(u^1, u^2, \cdots, u^q) \\ \vdots \\ \vdots \\ N^q(u^1, u^2, \cdots, u^q) \end{pmatrix} = \begin{pmatrix} f^1 \\ \vdots \\ \vdots \\ f^q \end{pmatrix},
\tag{8.1.13}
$$

and their discrete analogs

$$
N_h u_h = f_h
\tag{8.1.14}
$$

or

$$
\begin{pmatrix} N_h^1(u_h^1, u_h^2, \cdots, u_h^q) \\ \vdots \\ \vdots \\ N_h^q(u_h^1, u_h^2, \cdots, u_h^q) \end{pmatrix} = \begin{pmatrix} f_h^1 \\ \vdots \\ \vdots \\ f_h^q \end{pmatrix},
\tag{8.1.15}
$$

together with an appropriate set of boundary conditions as in the linear case. The boundary conditions may also be nonlinear.

## 8.2 MULTIGRID COMPONENTS

As for scalar applications, any multigrid algorithm for PDE systems is characterized by the components smoothing, restriction, interpolation, solution on the coarsest grid and cycle type. In the following subsections, we will discuss how these components are generalized to systems of equations.

We will first concentrate on the treatment of the interior equations. Some remarks regarding boundary conditions are contained in Section 8.2.6.

### 8.2.1 Restriction

Let us assume that we have found a suitable smoothing scheme for the discrete system (8.1.9) and that we have performed one or several smoothing steps giving a current approximation

$\bar{u}_h$. The next step in multigrid is to perform the restriction to the next coarser grid with mesh size $H$. Remember that the discrete system at each interior point consists of $q$ linear equations of the form

$$L_h^k \boldsymbol{u}_h = f_h^k \quad \text{with } L_h^k = (L_h^{k,1}, \ldots, L_h^{k,q}) \tag{8.2.1}$$

for $k = 1, \ldots, q$. The restriction for each of these equations is done separately, in a straightforward generalization of the scalar case (see Section 2.2.2). For the correction scheme (CS), this means that the $q$ coarse grid equations

$$L_H^k \hat{\boldsymbol{v}}_H = d_H^k \tag{8.2.2}$$

are obtained by restricting the defects $d_h^k$ to $\Omega_H$ using the current approximations $\bar{\boldsymbol{u}}_h = (\bar{u}_h^1, \ldots, \bar{u}_h^q)^T$:

$$d_H^k := I_h^H d_h^k, \qquad d_h^k = f_h^k - L_h^k \bar{u}_h. \tag{8.2.3}$$

Here, $I_h^H$ is the (scalar) restriction operator, $L_H^k$ denotes the coarse grid analog of $L_h^k$ and $\hat{v}_H$ represents the solution of the coarse grid defect equations. If the FAS is to be employed, e.g. for a nonlinear system of the form (8.1.14), the coarse grid equations are defined by

$$N_H^k \boldsymbol{w}_H = f_H^k \tag{8.2.4}$$

with

$$f_H^k := I_h^H (f_h^k - N_h^k \bar{u}_h) + N_H^k \hat{I}_h^H \bar{u}_h \tag{8.2.5}$$

and

$$\hat{I}_h^H \bar{\boldsymbol{u}}_h := (\hat{I}_h^H \bar{u}_h^1, \ldots, \hat{I}_h^H \bar{u}_h^q)^T. $$

For $H = 2h$, a typical standard choice for the restriction operator $I_h^H$ is the scalar FW operator. The standard choice for $\hat{I}_h^H$ is the (scalar) injection applied to each unknown.

In general, the restriction operators need not be the same for all equations (see Section 8.2.3).

## 8.2.2 Interpolation of Coarse Grid Corrections

The interpolation and addition of the corrections from the coarse grid is carried out separately for each of the grid functions and looks exactly like in the scalar case. For the CS (see Section 2.2.3), we have

$$u_h^{k,\,\text{after CGC}} = \bar{u}_h^k + I_H^h \hat{v}_H^k, \tag{8.2.6}$$

where $\hat{v}_H^k$ is the correction computed on the coarse grid and $I_H^h$ the (scalar) interpolation operator. For the FAS (see Section 5.3.4), the coarse grid correction gives

$$u_h^{k,\,\text{after CGC}} = \bar{u}_h^k + I_H^h(\hat{w}_H^k - \hat{I}_h^H \bar{u}_h^k),\tag{8.2.7}$$

where $\hat{w}_H^k$ is an approximate solution of the coarse grid equation (8.2.4), (8.2.5).

A typical choice for the coarse-to-fine transfer $I_H^h$ is *bilinear interpolation for each unknown grid function*. Again, different interpolation operators may be applied to different grid functions (see Section 8.2.3).

### 8.2.3 Orders of Restriction and Interpolation

In the following, we assume that the interpolation of the coarse grid corrections and the restriction of the defects are performed as described above. In general, the required orders of the restriction and interpolation operators depend on the orders of the derivatives occurring in the PDE system. Let $m_{ij}$ denote the highest order of differentiation of the $j$th unknown in the $i$th equation of the PDE system. In order to avoid large amplifications of high frequencies by the coarse grid correction process, one should choose

$$m_i + m^j > m_{ij},\tag{8.2.8}$$

where $m_i$ denotes the order of the restriction of the $i$th equation and $m^j$ denotes the order of the interpolation of the corrections of the $j$th unknown grid function. This basic rule can be found by LFA, analyzing how the coarse grid correction amplifies the high frequency harmonics of the lowest frequencies [66, 69, 187].

**Remark 8.2.1 (full multigrid interpolation)** The FMG interpolation can be performed independently for each current approximation of the functions $u_h^k$ and can be chosen as in the scalar case (see Sections 2.6.1 and 3.2.2). $\gg$

### 8.2.4 Solution on the Coarsest Grid

As in the scalar case, the solution on the coarsest grid can be obtained with any suitable solver. However, the discrete systems on the coarsest grid may be much larger than in the scalar case, in particular for complex applications. The efficiency of a numerical algorithm used for solving the coarsest grid problem may, therefore, be more important than for scalar equations.

### 8.2.5 Smoothers

The immediate generalization of the scalar lexicographic point Gauss–Seidel relaxation scheme is the pointwise *collective* Gauss–Seidel relaxation. Like its scalar counterpart, this relaxation sweeps over all grid points in a lexicographic order. At each grid point $(x, y)$, all difference equations located there are solved simultaneously, changing the values

$u_h^1(x, y), \ldots, u_h^q(x, y)$ and using current values at the neighbor grid points. This means that a linear $q \times q$ system of equations is to be solved at each grid point. Of course, similar generalizations are possible for GS-RB or $\omega$-JAC resulting in collective GS-RB and collective $\omega$-JAC. This collective approach is also in wide use with line relaxation, ILU etc. However, collective relaxations are not always necessary.

For simple problems, relaxation schemes which do not collectively solve the $q$ difference equations located at the grid point $(x, y)$, but solve them one after the other (*decoupled relaxation*) may also have sufficient smoothing properties.

In some more involved cases, e.g. in the context of the Navier–Stokes equations, discussed in Section 8.6, the standard collective relaxations may fail to show satisfactory smoothing and convergence properties. More advanced smoothing schemes such as *box relaxations* or *distributive relaxations* may have to be applied then.

- The idea of box relaxation is to solve not only all equations at *one* grid point collectively, but all equations at a set of grid points (box). These boxes may or may not be overlapping. One smoothing step consists of a sweep over all boxes. In this sense, line relaxations are a special type of box relaxation. Typically, however, box relaxation employs a more compact set of points than a line. We will give examples of box relaxation in Sections 8.7.2 and 8.8.2.

- The idea of distributive relaxation is as follows. To relax the set of equations $L_h u_h = f_h$, we introduce a new variable $\hat{u}_h$ by $u_h = M_h \hat{u}_h$ and consider the (transformed) system $L_h M_h \hat{u}_h = f_h$. For example, $M_h$ is chosen such that the resulting operator $L_h M_h$ is suited for decoupled (equation-wise) relaxation. An example for such a distributive relaxation has already been given in Section 4.7.2 where the *Kaczmarz relaxation* was introduced. We will return to distributive relaxation in Sections 8.7.3 and 8.8.2.

The smoothing properties of a particular relaxation method for a given problem can again be evaluated by smoothing analysis (see Section 8.3).

### 8.2.6 Treatment of Boundary Conditions

The general idea of the multigrid treatment of boundary conditions for systems of equations remains essentially the same as in the scalar case. In particular, the transfer to coarse grids is performed separately for the boundary condition and for the interior equations. For 2D problems, 1D restriction operators are employed for boundary conditions (see Section 5.6).

There are, however, several complications compared to the scalar case. As already mentioned, the number of boundary conditions will, in general, differ from the number of unknown grid functions and from the number of PDEs. In particular, there is not necessarily a one-to-one correspondence of boundary conditions and grid functions (or PDEs).

The relaxation at boundaries often needs to be modified. In general, it is no longer sufficient to relax the boundary conditions separately and decoupled from the interior equations. Instead, the relaxation at a boundary point may have to be coupled with that at adjacent interior points. Box schemes which collectively update the unknowns at several adjacent boundary points together with the unknowns at adjacent interior points are then appropriate. For certain types of problems, additional local relaxations near the boundary may also have to be added as has been seen in the scalar case for problems with geometric singularities (see Section 5.5). We will return to these approaches and give an example in Section 8.4.

## 8.3 LFA FOR SYSTEMS OF PDEs

In this section, we will generalize the LFA smoothing analysis to systems of PDEs (in Section 8.3.1). The two-level LFA can be generalized in the same way. However, we will not treat it in detail. In Section 8.3.2, we will extend the concept of $h$-ellipticity to systems of PDEs.

### 8.3.1 Smoothing Analysis

In analogy to the scalar case (see Section 4.3), we consider the discrete system (8.1.10), where the $L_h^{k,l}$ are assumed to be scalar difference operators with constant coefficients on the infinite grid $\mathbf{G}_h$, i.e. in 2D

$$L_h^{k,l} u_h^l(\boldsymbol{x}) = \sum_{\boldsymbol{\kappa} \in V} s_{\kappa_1 \kappa_2}^{k,l} u_h^l(x_1 + \kappa_1 h_1, x_2 + \kappa_2 h_2)$$

with $s_{\kappa_1 \kappa_2}^{k,l} \in \mathbb{R}$ and a finite index set $V$.

Consider components of the form

$$\varphi(\boldsymbol{\theta}, \boldsymbol{x}) = \boldsymbol{a} e^{i \boldsymbol{\theta} \cdot \boldsymbol{x} / h},$$

where $\boldsymbol{a} = (1, \ldots, 1)^T \in \mathbb{R}^q, \boldsymbol{\theta} = (\theta_1, \theta_2)^T, \boldsymbol{x} = (x_1, x_2)^T, \boldsymbol{h} = (h_1, h_2)^T, e^{i \boldsymbol{\theta} \cdot \boldsymbol{x} / h} :=$ $e^{i \theta_1 x_1 / h_1} e^{i \theta_2 x_2 / h_2}$. Obviously, we have

$$L_h \varphi(\boldsymbol{\theta}, \boldsymbol{x}) = \underbrace{\begin{pmatrix} \tilde{L}_h^{1,1}(\boldsymbol{\theta}) & \cdots & \cdots & \tilde{L}_h^{1,q}(\boldsymbol{\theta}) \\ \vdots & & & \vdots \\ \vdots & & & \vdots \\ \tilde{L}_h^{q,1}(\boldsymbol{\theta}) & \cdots & \cdots & \tilde{L}_h^{q,q}(\boldsymbol{\theta}) \end{pmatrix}}_{=: \, \tilde{\boldsymbol{L}}_h(\boldsymbol{\theta}) \, \in \, \mathbb{C}^{q \times q}} \varphi(\boldsymbol{\theta}, \boldsymbol{x}),$$

where the terms

$$\tilde{L}_h^{k,l}(\boldsymbol{\theta}) = \sum_{\boldsymbol{\kappa} \in V} s_{\kappa_1 \kappa_2}^{k,l} e^{i \theta_1 \kappa_1} e^{i \theta_2 \kappa_2}$$

are the *symbols* of the scalar discrete operators $L_h^{k,l}$. Correspondingly, the matrix $\tilde{\boldsymbol{L}}_h(\boldsymbol{\theta})$ is called the *symbol of $\boldsymbol{L}_h$*.

As in the scalar case, we can distinguish low and high frequency error components. For standard coarsening, we obtain the following definition.

**Definition 8.3.1**

$$\varphi \text{ low frequency component} \ :\Longleftrightarrow\ \theta \in T^{\text{low}} = \left[-\frac{\pi}{2}, \frac{\pi}{2}\right)^2$$

$$\varphi \text{ high frequency component} \ :\Longleftrightarrow\ \theta \in T^{\text{high}} = [-\pi, \pi)^2 \setminus [-\tfrac{\pi}{2}, \tfrac{\pi}{2})^2$$

We will sometimes refer to $\theta$ as a high or a low frequency.

The *smoothing analysis for systems of equations* can now be performed as in the scalar case. Let us assume a linear $q \times q$ system of difference equations

$$L_h u_h = f_h$$

and a smoothing operator $S_h$ corresponding to a *splitting* of $L_h$

$$L_h = L_h^+ + L_h^-$$

($L_h^+, L_h^-$ are again $q \times q$ systems) such that the smoothing procedure can be described by

$$L_h^+ \bar{w}_h + L_h^- w_h = f_h.$$

Here, $w_h$ and $\bar{w}_h$ denote the approximations to $u_h$ before and after the smoothing procedure, respectively. Subtracting the discrete equation $L_h u_h = f_h$, we obtain the error equation

$$L_h^+ \bar{v}_h + L_h^- v_h = 0$$

or

$$\bar{v}_h = S_h v_h$$

where $v_h = u_h - w_h$ and $\bar{v}_h = u_h - \bar{w}_h$ denote the errors before and after the relaxation and where $S_h$ is the resulting smoothing operator.

Applying $L_h^-, L_h^+, S_h$ to the formal eigenfunctions $\varphi(\theta, x)$, we obtain

$$L_h^- a e^{i\theta \cdot x/h} = \tilde{L}_h^-(\theta) a e^{i\theta \cdot x/h}$$

$$L_h^+ a e^{i\theta \cdot x/h} = \tilde{L}_h^+(\theta) a e^{i\theta \cdot x/h}$$

$$S_h a e^{i\theta \cdot x/h} = \tilde{S}_h(\theta) a e^{i\theta \cdot x/h} = -\tilde{L}_h^+(\theta)^{-1} \tilde{L}_h^-(\theta) a e^{i\theta \cdot x/h},$$

where the symbols $\tilde{L}_h^-(\theta), \tilde{L}_h^+(\theta), \tilde{S}_h(\theta)$ are complex $q \times q$ matrices and where we assume that $\tilde{L}_h^+(\theta)^{-1}$ exists.

The *smoothing factor* for systems of equations can thus be defined as

$$\boxed{\mu_{\text{loc}} = \mu_{\text{loc}}(S_h) := \sup\left\{\left|\rho\big(\tilde{L}_h^+(\theta)^{-1}\tilde{L}_h^-(\theta)\big)\right| : \theta \text{ high frequency}\right\}}$$

(where $\rho$ denotes the spectral radius) or equivalently

$$\boxed{\mu_{\text{loc}} = \sup\left\{|\lambda(\theta)| : \det\big(\lambda(\theta)\tilde{L}_h^+(\theta) + \tilde{L}_h^-(\theta)\big) = 0; \theta \text{ high frequency}\right\}}. \qquad (8.3.1)$$

Obviously, $\mu_{\text{loc}}$ is the worst (asymptotic) amplification factor of all high frequency error components. This definition is consistent with the corresponding one in the scalar case (see Definition 4.3.1 in Section 4.3).

**Remark 8.3.1** In practice one has to evaluate the symbols $\tilde{L}_h^+(\theta)$ and $\tilde{L}_h^-(\theta)$ of the particular relaxation under consideration. The smoothing factor can then be determined numerically (see `http://www.gmd.de/SCAI/multigrid/book.html` for a program). $\gg$

**Remark 8.3.2 (generalizations)** The generalization of LFA for systems with *nonconstant* coefficients or *nonlinear* systems is exactly as in the scalar case by linearization and freezing of coefficients.

The generalization of LFA to GS-RB (or zebra line-type smoothing schemes) for systems follows the same basic considerations as presented in Section 4.5 for the scalar case. The same holds for the two-grid LFA. We will not discuss these approaches here, but, nevertheless, present some results of the two-grid LFA in the next section. $\gg$

**Remark 8.3.3 (coarse grid correction and boundary conditions)** In practice, there are two major reasons why the measured convergence of a multigrid algorithm may differ from what is predicted by smoothing LFA. The first is that the coarse grid correction may cause problems. This is a typical phenomenon of singularly perturbed problems, which we have discussed in detail in the context of the convection–diffusion problem. Similar effects also occur for various systems of equations, e.g. the incompressible Navier–Stokes equations at high Reynolds numbers. A proper analysis of this kind of situation requires a two-grid LFA (often a simplified two-grid analysis, as introduced in Remark 4.6.1 is sufficient).

The second reason is an unsuitable treatment of boundary conditions. Boundary conditions and their treatment by multigrid do not enter the LFA. The general experience is that the multigrid convergence predicted by the LFA smoothing factor can only be observed in practice if sufficient work at and near the boundary is invested. $\gg$

**Remark 8.3.4 (smoothing factors and factorization of $L$)** For complicated PDE systems a heuristic guideline of the question, which smoothing factors can be expected, is the following (discussed in detail in [66]).

*The smoothing factor of a smoothing procedure for a given PDE operator $L$ can be as good as the smoothing factors obtained for the factors of $\det L$.*

It can be shown [66] that if $\det L = L_1 L_2$, where each $L_i$ is a scalar differential operator, then one can factorize the $q \times q$ operator $L$ into $L = L_1 L_2$, where the $L_i$ are $q \times q$ matrix operators such that $\det L_i = L_i$. Factors, that often occur are the Laplacian $\Delta$ and the convection–diffusion operator $\Delta + a \cdot \nabla$.

A general possibility to relax the factorized system

$$L_1 L_2 u = f$$

is to introduce the auxiliary vector of unknown functions $v = L_2 u$ and relax the two systems

$$L_{1,h} v_h = f_h \quad \text{and} \quad L_{2,h} u_h = v_h \tag{8.3.2}$$

alternatingly.

The combined smoothing factor is not worse than the worst of the two systems. A simple example for this approach is the biharmonic equation, which we will discuss in Section 8.4.

If the smoothing factors $\mu_1$ and $\mu_2$ of the operators $L_{1,h}$ and $L_{2,h}$ differ significantly, a more advanced smoothing strategy for the system (8.3.2) is to relax $L_{1,h}v_h = f_h$ $\nu_1$ times and the other system $\nu_2$ times. Then we obtain for the smoothing factor $\mu$ of the whole system

$$\mu \leq \max\left(\mu_1^{\nu_1}, \mu_2^{\nu_2}\right). \qquad \gg$$

### 8.3.2 Smoothing and $h$-Ellipticity

In Section 4.7, we have discussed the question, which properties a (scalar) operator $L_h$ must have, so that a pointwise relaxation exists with $h$-independent smoothing factors $< 1$. The answer was that $L_h$ must be $h$-elliptic. This result carries over to systems of PDEs.

Of course, we first have to extend the definition of the $h$-ellipticity measure $E_h$ to systems. For standard coarsening, a natural generalization of $E_h$ to systems is

$$E_h(L_h) := \frac{\min\left\{\left|\det \tilde{L}_h(\boldsymbol{\theta})\right| : \boldsymbol{\theta} \in T^{\text{high}}\right\}}{\max\left\{\left|\det \tilde{L}_h(\boldsymbol{\theta})\right| : -\pi \leq \boldsymbol{\theta} < \pi\right\}},$$

where $\tilde{L}_h(\boldsymbol{\theta})$ denotes the symbol of $L_h$.

**Remark 8.3.5** As in the scalar case, the denominator in this definition is only a scaling factor, which guarantees $0 \leq E_h(L_h) \leq 1$. Other scalings are often used, e.g.

$$\bar{E}_h(L_h) := \min\left\{\frac{|\det \tilde{L}_h(\boldsymbol{\theta})|}{|\,|L_h|\,|} : \boldsymbol{\theta} \in T^{\text{high}}\right\}, \qquad (8.3.3)$$

where $|L_h|$ is a $q \times q$ matrix formed by replacing each $L_h^{k,j}$ by its *size*. Here, the size of a scalar discrete operator is defined as the sum of the absolute values of its entries in the stencil.

$$\gg$$

There is a direct analog of Theorem 4.7.1 for systems of PDEs. The first (trivial) part is that $\mu_{\text{loc}} \geq 1$ for any point relaxation described by a splitting $L_h = L_h^+ + L_h^-$ if $E_h(L_h) = 0$ and $\tilde{L}_h^+(\boldsymbol{\theta}) \neq 0$.

This is easily seen since

$$E_h(L_h) = 0 \implies \det \tilde{L}_h(\boldsymbol{\theta}) = 0 \quad \text{for at least one high frequency } \boldsymbol{\theta}.$$

This implies that $\mu_{\text{loc}} \geq 1$, because of (8.3.1) and $\tilde{L}_h(\boldsymbol{\theta}) = \tilde{L}_h^+(\boldsymbol{\theta}) + \tilde{L}_h^-(\boldsymbol{\theta})$.

The nontrivial part is that if $E_h(L_h)$ is bounded away from 0 by some constant $c > 0$:

$$E_h(L_h) \geq c > 0 \quad (\text{for } h \to 0),$$

then there exists a pointwise relaxation with smoothing factor $\mu_{\text{loc}} \leq \text{const} < 1$. The proof is similar to that for the scalar case [240].

In particular, one can show that a damped Kaczmarz relaxation of Jacobi-type exists that has the smoothing factor

$$\mu_{\text{loc}} = \frac{1 - E_h(L_h)^2}{1 + E_h(L_h)^2}.$$

**Remark 8.3.6** As in the scalar case, small $h$-ellipticity measures indicate that pointwise smoothers may be problematic. There are high frequencies which correspond to very small defects. $\gg$

**Remark 8.3.7** Semi-$h$-ellipticity can be defined as in the scalar case and allows corresponding generalizations to line smoothers etc. (see Remark 4.7.5). $\gg$

## 8.4 THE BIHARMONIC SYSTEM

The biharmonic equation models deflections in 2D plates. If the biharmonic equation $\Delta \Delta u = f$ is treated as a scalar fourth-order problem, discretized by standard second-order differences, the $O(h^2)$ accurate 13-point stencil

$$\Delta_h \Delta_h = \frac{1}{h^4} \begin{bmatrix} & & 1 & & \\ & 2 & -8 & 2 & \\ 1 & -8 & 20 & -8 & 1 \\ & 2 & -8 & 2 & \\ & & 1 & & \end{bmatrix}_h \tag{8.4.1}$$

is obtained. The smoothing factor of GS-LEX on a Cartesian grid is $\mu_{\text{loc}} = 0.8$, which is not satisfactory and causes a rather poor multigrid efficiency. For $\omega$-JAC-RB relaxation (see Section 5.4.2), we obtain $\mu_{\text{loc}} (\omega = 1) = 0.64$ and $\mu_{\text{loc}} (\omega = 1.4) = 0.512$ (if the underrelaxation is applied after the JAC-RB iteration) [379].

Better results are easily obtained *if the biharmonic problem is treated as a system of the form (8.1.6)–(8.1.7)*. This is trivial if we have the boundary conditions

$$u = f^{\Gamma,1} \quad \text{and} \quad \Delta u = f^{\Gamma,2}, \tag{8.4.2}$$

which describe the case that the edges of the plate are simply supported. With these boundary conditions, the biharmonic system is fully decoupled. One can solve the two discrete Poisson problems

$$\begin{aligned} \Delta_h v_h &= f_h^{\Omega} \quad (\Omega_h) \\ v_h &= f_h^{\Gamma,2} \quad (\Gamma_h) \end{aligned} \quad \text{and} \quad \begin{aligned} \Delta_h u_h &= v_h \quad (\Omega_h) \\ u_h &= f_h^{\Gamma,1} \quad (\Gamma_h) \end{aligned} \tag{8.4.3}$$

one after the other. Since multigrid works very well for the Poisson equation, we obtain a solution of the biharmonic problem with excellent numerical efficiency. Furthermore, the relaxation for the two Poisson problems (8.4.3) is simpler and cheaper than GS-LEX for the 13-point stencil (8.4.1).

For the boundary conditions

$$u = f^{\Gamma,1} \quad \text{and} \quad u_n = f^{\Gamma,2}, \tag{8.4.4}$$

which describe the case of clamped edges of the plate, the situation is more involved since the PDEs are coupled via the boundary conditions. Moreover, we have two boundary conditions for the function $u$, but none for $v$.

Since such situations often occur for PDE *systems*, we will discuss an appropriate treatment of this problem in Section 8.4.2. Excellent multigrid performance can also be achieved in this case, as will be shown in Section 8.4.3. The idea is to introduce a *modified collective* relaxation at the boundaries which treats boundary points together with adjacent interior grid points.

### 8.4.1 A Simple Example: GS-LEX Smoothing

In this section, we analyze the smoothing properties of GS-LEX for the discrete biharmonic system (see also Example 8.1.1)

$$\begin{pmatrix} \Delta_h & 0 \\ -I & \Delta_h \end{pmatrix} \begin{pmatrix} v_h \\ u_h \end{pmatrix} = \begin{pmatrix} f_h^\Omega \\ 0 \end{pmatrix} \quad \text{with } \Delta_h = \frac{1}{h^2} \begin{bmatrix} & 1 & \\ 1 & -4 & 1 \\ & 1 & \end{bmatrix}$$

in order to illustrate how LFA is used for systems.

The *collective* GS-LEX relaxation corresponds to the splitting

$$L_h^+ = \begin{pmatrix} \Delta_h^+ & 0 \\ -I & \Delta_h^+ \end{pmatrix}, \qquad L_h^- = \begin{pmatrix} \Delta_h^- & 0 \\ 0 & \Delta_h^- \end{pmatrix} \tag{8.4.5}$$

with

$$\Delta_h^+ = \frac{1}{h^2} \begin{bmatrix} & 0 & \\ 1 & -4 & 0 \\ & 1 & \end{bmatrix}, \qquad \Delta_h^- = \frac{1}{h^2} \begin{bmatrix} & 1 & \\ 0 & 0 & 1 \\ & 0 & \end{bmatrix}.$$

The symbols of $L_h^-$ and $L_h^+$ are easily computed to be

$$\tilde{L}_h^-(\boldsymbol{\theta}) = \frac{1}{h^2}\left(e^{i\theta_1} + e^{i\theta_2}\right) \begin{pmatrix} 1 & 0 \\ 0 & 1 \end{pmatrix},$$

$$\tilde{L}_h^+(\boldsymbol{\theta}) = \frac{1}{h^2}\begin{pmatrix} \left(e^{-i\theta_1} + e^{-i\theta_2} - 4\right) & 0 \\ h^2 & \left(e^{-i\theta_1} + e^{-i\theta_2} - 4\right) \end{pmatrix}.$$

Thus

$$\tilde{S}_h(\boldsymbol{\theta}) = -(\tilde{L}_h^+(\boldsymbol{\theta}))^{-1}\tilde{L}_h^-(\boldsymbol{\theta}) = -\frac{1}{h^2}\left(e^{i\theta_1} + e^{i\theta_2}\right)(\tilde{L}_h^+(\boldsymbol{\theta}))^{-1}$$

$$= -\frac{e^{i\theta_1} + e^{i\theta_2}}{e^{-i\theta_1} + e^{-i\theta_2} - 4}\begin{pmatrix} 1 & 0 \\ \star & 1 \end{pmatrix},$$

where $\star$ denotes some nonzero matrix element whose size does not influence the smoothing factor, and we obtain

$$\mu_{\text{loc}} = \mu(S_h) = \sup\left\{ \left| \frac{e^{i\theta_1} + e^{i\theta_2}}{e^{-i\theta_1} + e^{-i\theta_2} - 4} \right| : \boldsymbol{\theta} \text{ high frequency} \right\}.$$

This is identical to the smoothing factor of pointwise GS-LEX for Poisson's equation; we thus obtain $\mu_{\text{loc}} = 0.5$.

**Remark 8.4.1 (decoupled relaxation)** In the case of decoupled (noncollective) Gauss–Seidel relaxations, we obtain exactly the same smoothing factor for the biharmonic system. This can very easily be seen. If we apply a decoupled relaxation we have to distinguish two cases. In the first case, we first relax the first equation of the system in a point $(x, y)$ and afterwards the second one; in the second case, we perform the (scalar) relaxations the other way round.

The first case is described by exactly the same splitting (8.4.5) as the collective relaxation. Correspondingly, we obtain the same smoothing factor.

The second case is described by the splitting

$$L_h^- = \begin{pmatrix} \Delta_h^- & 0 \\ -I & \Delta_h^- \end{pmatrix}, \qquad L_h^+ = \begin{pmatrix} \Delta_h^+ & 0 \\ 0 & \Delta_h^+ \end{pmatrix} \tag{8.4.6}$$

which leads to the same value for the smoothing factor, too.

*Such behavior is not at all typical for general systems of equations.* In the case under consideration, the coincidence is due to the fact that the partial differential equations are decoupled. $\gg$

### 8.4.2 Treatment of Boundary Conditions

If we want to develop a suitable multigrid method for the biharmonic system (8.1.6)–(8.1.7) with boundary conditions (8.4.4), we have to explicitly take into account the fact that there are two boundary conditions for the function $u$, but none for $v$.

We will discuss a proper multigrid treatment of boundary conditions in the specific situation of $\Omega = (0, 1)^2$ and give some results in Section 8.4.3. Here, the $u_n$-boundary condition can be discretized exactly as the Neumann boundary conditions for Poisson's equation (see Section 5.6.2). Using an extended grid with external points outside $\bar{\Omega}$ as shown in Fig. 5.23, standard central second-order finite differences can be used.

In order to close the discrete system (to have as many equations as unknowns), (8.1.7) is also discretized on the boundary $\Gamma_h$, resulting in the discrete system

$$\Delta_h v_h = f_h^{\Omega} \quad (\Omega_h) \tag{8.4.7}$$

$$\Delta_h u_h - v_h = 0 \quad (\bar{\Omega}_h = \Omega_h \cup \Gamma_h) \tag{8.4.8}$$

$$u_h = f_h^{\Gamma,1} \quad (\Gamma_h) \tag{8.4.9}$$

$$(u_n)_h = f_h^{\Gamma,2} \quad (\Gamma_h). \tag{8.4.10}$$

**Remark 8.4.2** For standard second-order central differences on a square grid, (8.4.7)–(8.4.10) are equivalent to the 13-point approximation (8.4.1) of the biharmonic equation with the discrete boundary conditions (8.4.9) and (8.4.10).                      ≫

In this example, we consider eliminated boundary conditions. Using (8.4.8) and (8.4.10) on $\Gamma_h$, external grid points outside $\bar{\Omega}$ can be eliminated, resulting in eliminated boundary conditions, which, for example, at the left boundary read

$$\frac{1}{h^2}\begin{bmatrix} & 1 & \\ 0 & -4 & 2 \\ & 1 & \end{bmatrix} u_h - v_h = -\frac{2}{h} f_h^{\Gamma,2}. \tag{8.4.11}$$

This equation can also be written as

$$\frac{1}{h^2}\begin{bmatrix} 0 & 0 & 2 \end{bmatrix} u_h - v_h = -\frac{2}{h} f_h^{\Gamma,2} - \frac{1}{h^2}\begin{bmatrix} 1 \\ -4 \\ 1 \end{bmatrix} f_h^{\Gamma,1} \tag{8.4.12}$$

because of (8.4.9). Obviously, the value of $v_h$ at a given boundary point is only coupled with the value of $u_h$ at the adjacent interior point. In particular, there is no direct coupling *along* the boundary.

Since (8.4.12) is not at all diagonally dominant, special care has to be taken in selecting *an appropriate relaxation scheme near boundary points*. We will use collective GS-LEX *in the interior of* $\bar{\Omega}_h$ with the following modifications near boundary points.

(1) Whenever a neighbor point $(x, y)$ of a boundary point (e.g. $(x - h, y)$) is relaxed, (8.4.12) at the boundary point is included in the collective relaxation of (8.4.7) and (8.4.8) at $(x, y)$. Figure 8.1(a) illustrates which points are treated collectively by this relaxation. This means that we solve a $3 \times 3$ system near boundary points in order to update the approximations for $u_h(x, y)$, $v_h(x, y)$ and $v_h(x - h, y)$ simultaneously.

(2) At interior points near corners of $\bar{\Omega}_h$, both eliminated boundary equations at the adjacent boundary points are included in the collective relaxation (see Fig. 8.1(a)) such that near a corner point a $4 \times 4$ system is solved. The treatment at the corner points is not essential for the multigrid process since they are completely decoupled from all other points.

The eliminated boundary conditions are transferred to the coarse grid by the same restriction operators (5.6.10), (5.6.11) as in the corresponding case of Poisson's equation with Neumann boundary conditions (see Section 5.6.2 and, in particular, Remarks 5.6.2 and 5.6.3).

### 8.4.3 Multigrid Convergence

Based on the treatment of the boundary conditions as described in the previous section, we now present results for the multigrid solution of the biharmonic system with boundary

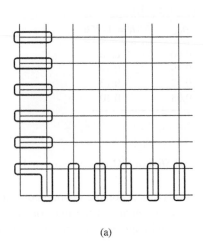 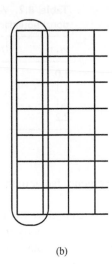

(a)                                    (b)

**Figure 8.1.** Simultaneous relaxation of grid points near boundaries. (a) Collective point relaxation near left and lower boundary; (b) collective line relaxation near left boundary.

**Table 8.1.** Asymptotic W-cycle convergence factors.

| $\nu_1, \nu_2$ | $(\mu_{loc})^{\nu_1+\nu_2}$ | $\rho_{loc}$ | $h = 1/16$ | $h = 1/32$ | $h = 1/64$ | $h = 1/128$ | $h = 1/256$ |
|---|---|---|---|---|---|---|---|
| 1,0 | 0.50 | 0.40 | 0.32 | 0.37 | 0.39 | 0.40 | 0.40 |
| 1,1 | 0.25 | 0.19 | 0.19 | 0.21 | 0.21 | 0.22 | 0.22 |
| 2,1 | 0.13 | 0.12 | 0.09 | 0.08 | 0.09 | 0.11 | 0.12 |
| 2,2 | 0.06 | 0.08 | 0.08 | 0.07 | 0.07 | 0.08 | 0.08 |

conditions (8.4.4). We choose standard multigrid components in the interior of $\Omega_h$ (standard coarsening, collective GS-LEX relaxation, $\nu = \nu_1 + \nu_2$ smoothing steps, FW, bilinear interpolation and direct solution on the coarsest grid always defined by a mesh size of $h_0 = 1/4$). Table 8.1 shows measured asymptotic convergence factors for *the W-cycle* (using $|| \cdot ||_\infty$). As we have seen in Section 8.4.1, the smoothing factor $\mu_{loc}$ of the decoupled and of the collective GS-LEX relaxation is 0.5 for the discrete biharmonic system. It can be seen from Table 8.1 (where collective GS-LEX has been used as an example) that the predictions obtained by LFA smoothing and LFA two-grid analysis ($(\mu_{loc})^{\nu_1+\nu_2}$ and $\rho_{loc}$) are excellent for the W-cycle.

**Remark 8.4.3 (boundary treatment and V-cycle convergence)** Unfortunately, the V-cycle does not show this behavior. In fact, the corresponding results are much worse than predicted by LFA. Moreover, the V-cycle multigrid convergence is level-dependent, i.e. it deteriorates when increasing the number of levels. For example, for $h = 1/256$, we

**Table 8.2.** V-cycle convergence factors with additional boundary relaxation.

| $\nu_1, \nu_2$ | $(\mu_{\text{loc}})^{\nu_1+\nu_2}$ | $\rho_{\text{loc}}$ | $h = 1/256$ |
|---|---|---|---|
| 1,0 | 0.50 | 0.40 | 0.40 |
| 1,1 | 0.25 | 0.19 | 0.19 |
| 2,1 | 0.13 | 0.12 | 0.13 |
| 2,2 | 0.06 | 0.08 | 0.10 |

observe a convergence factor of the V(1,1)-cycle of 0.56. For V(1,0)-cycles even divergence is observed.

This effect is caused by the boundary conditions since, for the decoupled boundary conditions (8.4.2) with exactly the same multigrid components in the interior of $\bar{\Omega}_h$, we obtain multigrid V-cycle convergence factors which are in agreement with LFA smoothing and two-grid analysis.

It has been shown [346] that the V-cycle convergence can be improved considerably if *additional collective line relaxations along all points which are adjacent to boundary points* (see Fig. 8.1(b)) are performed, where (8.4.12) at the boundary points is again included in the relaxation. If we add such a collective line relaxation along each side of the boundary before the restriction to a coarser grid and before the interpolation of corrections, we obtain V-cycle convergence factors which are again in good agreement with the results from LFA analysis. Table 8.2 shows that, for example, for $h = 1/256$ the measured V-cycle convergence factors agree well with the LFA two-grid factors $\rho_{\text{loc}}$. Obviously, the additional collective line relaxations (working on $v_h$ at boundary points and simultaneously on $u_h$ and $v_h$ at points adjacent to the boundary), lead to an impressive improvement of the multigrid convergence.

For a heuristic explanation, we focus again on the special structure of (8.4.12). This equation can be interpreted as the discrete boundary condition for $v_h$. In particular, we see that the approximation of $v_h$ at a boundary point $(x, y)$ depends strongly on the value of $u_h$ at the adjacent grid point in the interior of the domain (e.g. $u(x + h, y)$ at the left boundary). If $u_h$ has certain error components, $v_h$ obviously has much larger error components on fine grids since those of $u_h$ are amplified by the factor $1/h^2$. Such error components can grow by, for example, successive interpolations (without or with insufficient intermediate smoothing/damping of the errors). In the W-cycle, more smoothing iterations are applied on intermediate levels than in the V-cycle.                              $\gg$

**Remark 8.4.4 (local relaxation)**   Excellent multigrid convergence for the biharmonic problem with boundary conditions (8.4.4) can also be obtained by noneliminated $u_n$-boundary conditions and by performing additional relaxation sweeps near the boundary. According to [77], the number of the sweeps and their depth (i.e. the distance from the boundary up to which points have to be included in these extra relaxations) slightly increases, at least for V-cycles, with decreasing $h$ and increasing number of multigrid levels.        $\gg$

## 8.5 A LINEAR SHELL PROBLEM

In this section we discuss the multigrid treatment of a linear shell problem. With this example, we will focus on the difference between collective (coupled) and decoupled smoothing methods and make some general remarks on their ranges of applicability.

We consider thin elastic shells with weak curvature. The quantities to be computed are the stress $f(x, y)$ and the displacement $w(x, y)$ of the shell mean surface under a load $p$ (normal to the shell mean surface). The geometric form of the shell is described by the given function $z(x, y)$. The system of PDEs derived by linear shell theory, is

$$\Delta^2 f + \Lambda^2 \mathcal{K}(z, w) = 0$$
$$\Delta^2 w - \Lambda^2 \mathcal{K}(z, f) = p$$

$$(8.5.1)$$

with

$$\mathcal{K}(z, f) = z_{xx} f_{yy} - 2 z_{xy} f_{xy} + z_{yy} f_{xx}.$$

The positive parameter $\Lambda^2$ is proportional to the inverse of the thickness of the shell [363]. In the following, we restrict ourselves to the boundary conditions

$$f = 0, \quad w = 0, \quad \Delta f = 0 \quad \text{and} \quad \Delta w = 0,$$

which describe the case that the edges of the shell are simply supported.

**Remark 8.5.1** Boundary conditions for $f$, $w$, $f_n$ and $w_n$ describing clamped edges of the shell, are often of interest. Such boundary conditions can be treated, for example, in the same way as those for the corresponding biharmonic problem (see Section 8.4.2). Multigrid results for these boundary conditions can be found in [346]. $\gg$

The shell problem (8.5.1) consists of two biharmonic-like equations which are coupled via lower order terms. The strength of the coupling is proportional to the parameter $\Lambda^2$ and to the second derivatives of the given function $z(x, y)$. For $\Lambda^2 = 0$, for example, the system reduces to two (decoupled) biharmonic equations.

The system is elliptic. The type of the lower order terms, which are responsible for the coupling, depends on the shell geometry. Shells are called elliptic (hyperbolic, parabolic) if the linear lower order operator $\mathcal{K}(z, \cdot)$ is elliptic (hyperbolic, parabolic).

As for the biharmonic problem, we can split each of the equations into two Poisson-like equations if we introduce the functions $v = \Delta f$ and $u = \Delta w$ as additional unknowns. This leads to the system

$$\Delta v + \Lambda^2 \mathcal{K}(z, w) = 0 \quad (\Omega)$$
$$\Delta f - v = 0 \quad (\Omega)$$
$$\Delta u - \Lambda^2 \mathcal{K}(z, f) = p \quad (\Omega)$$
$$\Delta w - u = 0 \quad (\Omega)$$

$$(8.5.2)$$

for the four functions $u$, $v$, $w$ and $f$. Using second-order central differences, we obtain the discrete system

$$\Delta_h v_h + \Lambda^2 K_h(z, w_h) = 0 \quad (\Omega_h) \tag{8.5.3}$$

$$\Delta_h f_h - v_h = 0 \quad (\Omega_h) \tag{8.5.4}$$

$$\Delta_h u_h - \Lambda^2 K_h(z, f_h) = p_h \quad (\Omega_h) \tag{8.5.5}$$

$$\Delta_h w_h - u_h = 0 \quad (\Omega_h), \tag{8.5.6}$$

where

$$K_h(z, f_h) = z_{xx}(f_{yy})_h - 2z_{xy}(f_{xy})_h + z_{yy}(f_{xx})_h.$$

**Remark 8.5.2** For increasing $\Lambda^2$ and/or second derivatives of $z(x, y)$, the shell problem is *singularly perturbed*. As a consequence, standard central discretizations of the lower order operators cannot be expected to work for large $\Lambda^2$ if the diagonal dominance of the discrete lower order terms is lost. In particular, for $\Lambda^2 \to \infty$ and hyperbolic shells, it turns out that the $h$-ellipticity measure of the discrete system tends to zero. In such cases, different ($h$-elliptic) discretizations need to be used as discussed in detail for the convection–diffusion equation in Section 7.1. This is, however, not done here. ≫

## 8.5.1 Decoupled Smoothing

As indicated in Section 8.2.5, we can employ collective (coupled) or decoupled relaxation for smoothing. In this and the following sections, we will discuss the difference between these two approaches for *lexicographic y-line Gauss–Seidel* relaxation. Corresponding results are obtained for $x$-line relaxations because of symmetry.

A *decoupled* point (line) relaxation scheme consists of a sweep over the grid points (lines) and a sweep over the variables per point. In principle, we are free to choose the ordering of these sweeps and, in addition, the ordering within these sweeps (ordering of variables and ordering of points or lines).

As an example, we describe a decoupled line relaxation with the outer sweep being that over the lines. For each of the four equations of (8.5.2), we perform a separate scalar line relaxation, based only on the $\Delta_h$ operator. The relaxation of one line thus consists of four substeps, in each of which a tridiagonal system has to be solved as for a scalar Poisson-like operator. First, we update the approximations of $f_h$ in this line using (8.5.4), then $u_h$ using (8.5.5), then $w_h$ using (8.5.6) and, finally $v_h$ using (8.5.3) in such a way that new, updated approximations are already used whenever possible. We denote this smoothing procedure by $y$-line-decoupled relaxation (DEC). Changes of the ordering, in which the equations are relaxed, result in minor differences in the smoothing factors.

This relaxation corresponds to the splitting

$$L_h^+ = \begin{pmatrix} \Delta_h^+ & 0 & 0 & \Lambda^2 K_h^+ \\ 0 & \Delta_h^+ & 0 & 0 \\ 0 & -\Lambda^2 K_h^+ & \Delta_h^+ & 0 \\ 0 & 0 & -1 & \Delta_h^+ \end{pmatrix},$$

$$L_h^- = \begin{pmatrix} \Delta_h^- & 0 & 0 & \Lambda^2 K_h^- \\ -1 & \Delta_h^- & 0 & 0 \\ 0 & -\Lambda^2 K_h^- & \Delta_h^- & 0 \\ 0 & 0 & 0 & \Delta_h^- \end{pmatrix}$$

with

$$\Delta_h^+ = \frac{1}{h^2} \begin{bmatrix} & 1 & \\ 1 & -4 & 0 \\ & 1 & \end{bmatrix}, \qquad \Delta_h^- = \frac{1}{h^2} \begin{bmatrix} & 0 & \\ 0 & 0 & 1 \\ & 0 & \end{bmatrix}$$

and

$$K_h^+ = \frac{1}{h^2} \begin{bmatrix} -(1/2)z_{xy} & z_{yy} & 0 \\ z_{xx} & -2(z_{xx} + z_{yy}) & 0 \\ (1/2)z_{xy} & z_{yy} & 0 \end{bmatrix}, \qquad K_h^- = \frac{1}{h^2} \begin{bmatrix} 0 & 0 & (1/2)z_{xy} \\ 0 & 0 & z_{xx} \\ 0 & 0 & -(1/2)z_{xy} \end{bmatrix}$$

(8.5.7)

(using the standard central four-point second-order stencil for the mixed derivatives).

The smoothing factor $\mu_{loc}$ for this relaxation scheme depends on the shell geometry (which determines the character of $\mathcal{K}$) and on the factor

$$\kappa_h = \Lambda^2 h^2 \max\{z_{xx}, z_{xy}, z_{yy}\}. \tag{8.5.8}$$

Consequently, a change in the physical parameter $\Lambda^2$ has the same effect on the smoothing properties as a corresponding change in the mesh size $h$. In other words, effects caused by a strong coupling on fine grids will be similar to those caused by a moderate coupling on coarse grids (large $h$).

Table 8.3 shows smoothing factors of $y$-line-DEC for various values of $\kappa_h$ for an elliptic and for a hyperbolic shell. For $\Lambda^2 = 100$, which is a representative value for typical applications, $\kappa_h$ in Table 8.3 corresponds to mesh sizes $h$ between $1/4$ and $0$. The smoothing properties are satisfactory *on fine grids*. For a stronger coupling of the discrete equations, however, caused by larger $\Lambda^2$ or by coarser grids (larger $h$), the smoothing factors increase significantly until the smoothing properties are totally lost. This qualitative behavior is independent of the form of the shell and occurs for elliptic, parabolic and hyperbolic shells.

Obviously, the sole use of decoupled relaxations such as $y$-line-DEC as a smoother is not suitable here. In decoupled relaxations, the strength of the coupling of the equations is not taken into account *explicitly*. Here, the strength of the coupling of the equations, indicated by $\kappa_h$, is proportional to $h^2$ and thus becomes large on coarse grids.

**Table 8.3.** LFA smoothing factors depending on shell geometry ($z_{xx}$, $z_{xy}$, $z_{yy}$) and $\kappa_h$ for $y$-line-DEC.

| $z_{xx}$ | $z_{xy}$ | $z_{yy}$ | $\kappa_h = 0$ | $\kappa_h = 0.025$ | $\kappa_h = 0.1$ | $\kappa_h = 0.4$ | $\kappa_h = 1.6$ | $\kappa_h = 6.4$ |
|----|----|----|----|----|----|----|----|----|
| 1 | 0 | 1  | 0.45 | 0.45 | 0.47 | 0.56 | > 1 | > 1 |
| 1 | 0 | -1 | 0.45 | 0.45 | 0.47 | 0.56 | > 1 | > 1 |

As long as we have good smoothing properties on all grid levels (the coarsest grid may be omitted if, for example, a direct solver is applied here), multigrid can be applied with this relaxation. If the smoothing factors on coarse grids are worse than on fine grids but still less than 1, F- of W-cycles or level-dependent numbers of smoothing steps can be used in order to compensate for an insufficient smoothing on coarse grids.

In the next section, we will discuss a corresponding *collective* smoothing scheme, which turns out to be more robust for strong coupling.

### 8.5.2 Collective Versus Decoupled Smoothing

By collective Gauss–Seidel $y$-line relaxation with lexicographic ordering of lines, $y$-line-collective relaxation (COL), all four grid functions are updated collectively by the solution of the coupled system (8.5.3)–(8.5.6) in all points of a line. As a consequence, collective relaxation is more expensive than the corresponding decoupled smoother. For $y$-line-COL, for example, we have to solve a banded system with 12 nonzero entries in every row of the matrix whereas $y$-line-DEC requires the solution of four tridiagonal systems all with the same matrix resulting from the $\Delta_h$-operator. $y$-line-COL corresponds to the splitting

$$
L_h^+ = \begin{pmatrix}
\Delta_h^+ & 0 & 0 & \Lambda^2 K_h^+ \\
-1 & \Delta_h^+ & 0 & 0 \\
0 & -\Lambda^2 K_h^+ & \Delta_h^+ & 0 \\
0 & 0 & -1 & \Delta_h^+
\end{pmatrix},
$$

$$
L_h^- = \begin{pmatrix}
\Delta_h^- & 0 & 0 & \Lambda^2 K_h^- \\
0 & \Delta_h^- & 0 & 0 \\
0 & -\Lambda^2 K_h^- & \Delta_h^- & 0 \\
0 & 0 & 0 & \Delta_h^-
\end{pmatrix}
$$

with $\Delta_h^+$, $\Delta_h^-$, $K_h^+$ and $K_h^-$ as in (8.5.7).

As in the decoupled case, the smoothing factors depend only on the shell geometry and on $\kappa_h$. As indicated above, $\kappa_h$ is proportional to $h^2$.

The collective relaxation proves to be more robust than its decoupled counterpart (see Table 8.4). For the elliptic shell, we observe good smoothing for all values of $\kappa_h$. For the hyperbolic shell, the situation is different. For $\Lambda^2 = 100$, for example, we have satisfactory smoothing properties for $h \lesssim 1/8$ ($\kappa_h \lesssim 1.6$). For larger $\kappa_h$, the smoothing factors become larger than one. As discussed in Remark 8.5.2, this behavior has to be expected due to the vanishing $h$-ellipticity for the hyperbolic shell considered here.

**Table 8.4.** LFA smoothing factors depending on shell geometries and $\kappa_h$ for $y$-line-COL.

| $z_{xx}$ | $z_{xy}$ | $z_{yy}$ | $\kappa_h = 0$ | $\kappa_h = 0.4$ | $\kappa_h = 1.6$ | $\kappa_h = 6.4$ | $\kappa_h = 25.6$ | $\kappa_h = 6500$ |
|---|---|---|---|---|---|---|---|---|
| 1 | 0 | 1 | 0.45 | 0.48 | 0.48 | 0.45 | 0.45 | 0.45 |
| 1 | 0 | −1 | 0.45 | 0.48 | 0.65 | > 1 | > 1 | > 1 |

**Table 8.5.** Measured V(2,1)-cycle convergence factors in dependence of $\Lambda^2$.

| Smoother | $z_{xx}$ | $z_{xy}$ | $z_{yy}$ | $\Lambda^2 = 0$ | $\Lambda^2 = 40$ | $\Lambda^2 = 60$ | $\Lambda^2 = 80$ | $\Lambda^2 = 100$ |
|---|---|---|---|---|---|---|---|---|
| $y$-line-DEC | 1 | 0 | 1 | 0.06 | 0.32 | Div. | Div. | Div. |
| | 1 | 0 | −1 | 0.06 | 0.15 | 0.19 | 0.31 | Div. |
| $y$-line-COL | 1 | 0 | 1 | 0.06 | 0.06 | 0.06 | 0.06 | 0.06 |
| | 1 | 0 | −1 | 0.06 | 0.13 | 0.17 | 0.21 | 0.23 |

Table 8.5 compares V-cycle convergence factors for the coupled and the decoupled smoothing scheme for increasing values of the coupling parameter $\Lambda^2$. Here, the multigrid algorithm uses five grids with $h_0 = 1/2$.

For small $\Lambda^2$, both relaxations employed in a V-cycle are sufficient to obtain good convergence. The combination of V-cycle and decoupled smoothing is suitable up to $\Lambda^2 \approx 20$. For $\Lambda^2 \gtrsim 60$, the V-cycle using decoupled smoothing starts to diverge for one of the shells, for $\Lambda^2 = 100$ it diverges for both. The collective approach shows convergence in all cases considered here.

An interpretation of the multigrid convergence factors with the LFA results in Tables 8.3 and 8.4 is not trivial since the smoothing factors are level-dependent. The worst smoothing factors are obtained on the coarsest grids. On coarse grids, however, the LFA is only a very rough prediction since it neglects the influence of the boundary conditions, which is large on coarse grids (each point is close to a boundary). For $h = 1/4$, for example, there are more boundary points than interior points on the unit square. This is a heuristic explanation of the reasons why we will still observe convergence in some cases of Table 8.5 though the smoothing factor on the next-to-coarsest level is already larger than one.

### 8.5.3 Level-dependent Smoothing

The results in the previous subsection have demonstrated that the decoupled smoother $y$-line-DEC is not suitable for medium to large values of $\kappa_h$. On the other hand, the more robust collective relaxation $y$-line-COL is much more expensive than the decoupled one. Since, on fine grids, both relaxation schemes have comparable smoothing factors, it is reasonable to *combine* these two smoothers, namely *to employ the cheap decoupled relaxation on fine grids and the collective relaxation only on coarse grids, where its cost is not an issue*.

Table 8.6 shows measured convergence factors of V(2, 1)- and W(2, 1)-cycles which employ such a level-dependent smoothing strategy (called LDS here). We consider the

**Table 8.6.** Multigrid convergence factors for different shells, smoothing strategies and cycle types.

| $z_{xx}$ | $z_{xy}$ | $z_{yy}$ | V(LDS) | V(COL) | W(LDS) | W(COL) |
|---|---|---|---|---|---|---|
| 1 | 0 | 1 | 0.08 | 0.06 | 0.05 | 0.05 |
| 1 | 0 | −1 | 0.25 | 0.23 | 0.08 | 0.07 |

example $\Lambda^2 = 100$ and employ five grids with $h_0 = 1/2$. On the three coarsest meshes we employ $y$-line-COL and on the finer ones $y$-line-DEC.

We observe a similar convergence behavior for both the fully collective and the much cheaper level-dependent smoothing. The CPU times of multigrid algorithms using the collective, the decoupled and the level-dependent smoothing have been compared in [346]. A multigrid cycle based on collective line relaxation requires about ten times the computing time of a cycle based on $y$-line-DEC. The computing times of the level-dependent strategy, which has similar convergence properties to the collective variant, is comparable to that of multigrid employing the decoupled smoother. It is thus a *robust and efficient* alternative.

## 8.6 INTRODUCTION TO INCOMPRESSIBLE NAVIER–STOKES EQUATIONS

In this and the following sections we review both the discretization aspects and corresponding multigrid methods for the *stationary incompressible Navier–Stokes equations*. Due to the tremendous amount of related literature and the large variety of approaches, we have to restrict ourselves to a subset of available techniques. We focus on the discussion of some basic and fundamental concepts which are essential for a proper multigrid solution. A survey on the derivation of flow equations can be found in [33].

### 8.6.1 Equations and Boundary Conditions

We start with the 2D incompressible Navier–Stokes equations in conservative form in *primitive variable formulation* i.e. using the velocities $u$ and $v$ and the pressure $p$ as primary variables. The corresponding nonlinear PDE system consists of the momentum equations (8.6.1)–(8.6.2), which describe the momentum conservation, and the continuity equation (8.6.3),which can be deduced from the mass conservation law:

$$-\Delta u + Re((u^2)_x + (uv)_y + p_x) = 0 \qquad (8.6.1)$$

$$-\Delta v + Re((uv)_x + (v^2)_y + p_y) = 0 \qquad (8.6.2)$$

$$u_x + v_y = 0. \qquad (8.6.3)$$

Here, $Re$ is the so-called Reynolds number, which is proportional to a characteristic velocity (the unit in terms of which $u$ and $v$ are measured), to a characteristic length (the unit for $x$ and $y$) and to $1/\nu$, where $\nu$ is the kinematic viscosity of the flow. In this formulation, the variables $u$, $v$ and $p$ are dimensionless; the (only) relevant parameter is the Reynolds number. The nonconservative form of the incompressible Navier–Stokes equations

$$-\Delta u + Re(uu_x + vu_y + p_x) = 0 \qquad (8.6.4)$$

$$-\Delta v + Re(uv_x + vv_y + p_y) = 0 \qquad (8.6.5)$$

$$u_x + v_y = 0 \qquad (8.6.6)$$

is often the starting point for the solution of flow problems.

For $Re = 0$, we have the special case of the *Stokes equations* already introduced in Example 8.1.2, which describe highly viscous incompressible flows characterized by the diffusion terms in the momentum equations. In contrast to the incompressible Navier–Stokes equations, the Stokes equations are linear and simpler.

For high Reynolds numbers, the momentum equations become singularly perturbed. With $\varepsilon = 1/Re$, we have an analogy to the convection–diffusion equation, but with nonlinear convective parts of the operator.

**Remark 8.6.1** We here focus on the stationary equations and, correspondingly, on steady-state solutions. It is well known that, depending on the problem under consideration, the physical flow becomes unsteady for high Reynolds numbers. In such cases, of course, the time-dependent formulation of the equations is appropriate. Moreover, for high Reynolds numbers the flow may become turbulent, which causes additional complications. We will not consider turbulent and time-dependent flows in our discussion.          ≫

The momentum equations are naturally associated with $u$ and $v$, respectively. But there is no natural equation for $p$ in these systems. In particular, $p$ is not present in the continuity equation. This gives rise to complications in the discretization and in the numerical treatment.

The equations become *inhomogeneous* (i.e. they have nonzero right-hand sides) if external sources of momentum or mass are present in the application under consideration.

In the following, we will start with the discretization and numerical treatment of the nonconservative form of the incompressible Navier–Stokes equations. It is well-known that a discretization of a nonconservative formulation of a conservation law may introduce unacceptable errors in regions with strong gradients or discontinuities and the conservative form should then be preferred [196]. For the incompressible Navier–Stokes equations with smooth solutions, however, the nonconservative form can often be used safely.

**Remark 8.6.2 (boundary conditions)** The PDE systems (8.6.1)–(8.6.3) and (8.6.4)–(8.6.6) require only two boundary conditions. This has been demonstrated for the Stokes equations in Example 8.1.2 and carries over to the nonlinear case. If, however, boundary conditions for $u$ and $v$ are prescribed, the pressure is determined only up to a constant. One has a similar situation to that described in Section 5.6.4 for the Poisson equation with pure Neumann boundary conditions. A solution exists only if a compatibility condition is fulfilled. Similar techniques, as described in Section 5.6.4, can be used in such cases.     ≫

There are many possibilities to define proper boundary conditions for these systems depending on the particular application under consideration. At solid walls, it is appropriate to use $u = v = 0$ and no boundary conditions for $p$. At inflow and outflow boundaries, velocity profiles or Neumann boundary conditions for the velocities may be prescribed. If $p$ is prescribed at inflow and outflow boundaries (flows may be governed by pressure differences), only one further boundary condition at these boundaries is adequate. Other boundary conditions are, e.g. free stress, free slip or symmetric boundary conditions (see, for example, [33] for a physical motivation and formulation).

The actual choice of boundary conditions depends on the application under consideration. For the multigrid treatment of boundary conditions, we refer to the general discussion in Sections 5.6 and 8.2.6.

### 8.6.2 Survey

In this and the following sections, we will discuss the discretization and corresponding multigrid treatment of the stationary incompressible Navier–Stokes equations in detail.

One prerequisite of any multigrid method is the availability of a relaxation process with satisfactory error smoothing properties, the *stability of the underlying discretization scheme* being a necessary condition for this. For the incompressible Navier–Stokes equations, there are two sources of discrete instabilities.

The first is already present in the limit case of the Stokes equations and does not depend on the size of the Reynolds number. This *checkerboard instability* appears if central differencing of first-order derivatives in the pressure terms and in the continuity equation is applied and if all variables are located at the grid points. We have already seen scalar examples of this instability, e.g. in Example 4.7.5. In Section 8.6.3, we will discuss the checkerboard instability for the Stokes equations.

One approach to overcome the checkerboard instability is to use so-called *staggered* locations of unknowns (briefly: staggered grids or staggered discretization), for which the unknown grid functions are located at different places in a grid cell. In Section 8.7, we describe proper multigrid components such as box and distributive smoothers for staggered discretizations.

In Section 8.8, we consider nonstaggered (vertex-centered) discretizations. In that case, the checkerboard instability is overcome by the introduction of an artificial pressure term in the continuity equation. Also in this case, appropriate smoothing operators are still of box or distributive type. Straightforward collective relaxations can, however, be applied if suitable reformulations of the incompressible Navier–Stokes equations are used for the discretization (see Section 8.8.3). Most of the discussion up to Section 8.8.3 is on problems at low or moderate Reynolds numbers (up to $Re = 1000$).

The second source of instability is caused by the singular perturbation character of the momentum equations. For high Reynolds numbers, the $h$-ellipticity measure of standard (central) discretization schemes decreases and one has to introduce additional "artificial viscosity" to keep the discrete equations stable. This is a similar phenomenon as we have discussed in detail for the convection–diffusion equation for $\varepsilon \to 0$ (see Section 7.1), where upwind schemes for the convection terms were proposed. Of course, the situation is somewhat more involved for the *nonlinear system* of the Navier–Stokes equations compared to the scalar and linear convection–diffusion model problem. Nevertheless, straightforward (higher order) upwind-type schemes can be used for moderate Reynolds numbers.

In Section 8.8.4, we describe an example for an upwind-type scheme, which is particularly well-suited for *high* Reynolds numbers. This *flux splitting discretization scheme* is based on the conservative formulation of the incompressible Navier–Stokes equations and allows the use of collective point or line relaxations as smoothers.

**Remark 8.6.3** Another interesting discretization approach for the convective terms is to use so-called *narrow stencils* (see, for example, [86]). However, we do not discuss this approach here.       &#8811;

**Remark 8.6.4 (single grid solvers)** A well-known *classical solver* for the incompressible Navier–Stokes equations is the so-called SIMPLE algorithm ("semi-implicit method for pressure-linked equations") [301]. The SIMPLE and related algorithms are iterative solvers, which treat the momentum equations and a "pressure equation" separately in an outer iteration. Within this iteration, the pressure is updated using a Poisson-type equation. For such codes with an outer iteration, it is easy to replace the most time-consuming component of the solver, the solution of the Poisson-type equation for the pressure, by an efficient multigrid solver.

Although this "acceleration" approach reduces the computing times [247, 364], the overall convergence will, however, be unchanged since the outer SIMPLE iteration is not accelerated. The multigrid solution for the whole system will typically be much faster.

On the other hand, the SIMPLE approach allows a relatively straightforward numerical solution of more general and complicated PDE systems than the Navier–Stokes equations. It is, for example, relatively easy to add turbulence equations in the SIMPLE framework.       &#8811;

### 8.6.3 The Checkerboard Instability

If the Stokes or Navier–Stokes equations are discretized by means of standard central differencing with all unknowns at grid vertices (nonstaggered grid), pressure values are directly coupled only between grid points of distance $2h$. (The same is true if all variables are located at the cell centers.)

Therefore, the grid $\mathbf{G}_h$ can be subdivided into four subgrids (in a four-color fashion) among which the pressure values are decoupled. The pressure unknown at $(x, y)$, for instance, is only coupled with the grid points $(x + 2h, y)$, $(x, y + 2h)$, $(x + 2h, y + 2h)$, ... and we have similar couplings for the unknowns at the grid points $(x + h, y)$, $(x, y + h)$, $(x + h, y + h)$, .... We detail this phenomenon for the Stokes case:

$$
L_h \boldsymbol{u}_h = \begin{bmatrix} -\Delta_h & 0 & (\partial_x)_h \\ 0 & -\Delta_h & (\partial_y)_h \\ (\partial_x)_h & (\partial_y)_h & 0 \end{bmatrix} \begin{bmatrix} u_h \\ v_h \\ p_h \end{bmatrix} = 0
$$

with

$$-\Delta_h = \frac{1}{h^2}\begin{bmatrix} & -1 & \\ -1 & 4 & -1 \\ & -1 & \end{bmatrix}, \quad (\partial_x)_h = \frac{1}{2h}\begin{bmatrix} -1 & 0 & 1 \end{bmatrix}, \quad (\partial_y)_h = \frac{1}{2h}\begin{bmatrix} 1 \\ 0 \\ -1 \end{bmatrix}.$$

The *symbol* of $L_h$ is given by

$$\tilde{L}_h(\theta_1, \theta_2) = \frac{1}{h^2}\begin{pmatrix} 4 - 2\cos\theta_1 - 2\cos\theta_2 & 0 & ih\sin\theta_1 \\ 0 & 4 - 2\cos\theta_1 - 2\cos\theta_2 & ih\sin\theta_2 \\ ih\sin\theta_1 & ih\sin\theta_2 & 0 \end{pmatrix},$$

that is, for $v_h = \begin{pmatrix} 1 \\ 1 \\ 1 \end{pmatrix} e^{i\theta_1 x/h} e^{i\theta_2 y/h}$, we have

$$L_h v_h = \tilde{L}_h(\theta_1, \theta_2) v_h.$$

As in Example 4.7.5, we find that $L_h u_h = 0$ has *highly oscillating solutions* on the infinite grid, e.g.

$$u_h = 0, \qquad v_h = 0, \qquad p_h(x_i, y_j) = (-1)^{i+j},$$

the so-called *checkerboard mode*. This can also be seen from

$$\det \tilde{L}_h(\theta_1, \theta_2) = 0 \iff \begin{Bmatrix} \theta_1 = \pi, \ \theta_2 = \pi \\ \theta_1 = 0, \ \theta_2 = \pi \\ \theta_1 = \pi, \ \theta_2 = 0 \\ \theta_1 = 0, \ \theta_2 = 0 \end{Bmatrix},$$

which means that some high frequencies are annihilated by $L_h$. This is equivalent to the fact that there are high frequency error components which do not contribute to the defect. According to the discussion in Section 8.3.2, $L_h$ is not $h$-elliptic, $E_h(L_h) = 0$: Pointwise relaxation schemes do not have reasonable smoothing properties for such a discretization.

**Remark 8.6.5** This unstable behavior is also reflected by $\det L_h = \Delta_h \Delta_{2h}$. The checkerboard instability is implicitly present in the operator $\Delta_{2h}$ when applied on $\Omega_h$. This carries over to the system. $\gg$

## 8.7 INCOMPRESSIBLE NAVIER–STOKES EQUATIONS: STAGGERED DISCRETIZATIONS

One remedy for the checkerboard instability in the case of the incompressible Navier–Stokes equations is to use a *staggered* distribution of unknowns [182] instead of a nonstaggered one. This can be seen most easily for the Stokes equations and carries over to the incompressible Navier–Stokes case.

In a staggered arrangement, the discrete pressure unknowns $p_h$ are defined at cell centers (the ×-points), and the discrete values of $u_h$ and $v_h$ are located at the grid cell faces in the o- and •-points, respectively (see Fig. 8.2).

The discrete analog of the continuity equation

$$u_x + v_y = 0$$

is defined at the ×-points and the discrete momentum equations are located at the o- and the •-points, respectively. In the case of the Stokes equations, the discrete momentum equations read

$$-\Delta_h u_h + (\partial_x)_{h/2} p_h = 0, \qquad -\Delta_h v_h + (\partial_y)_{h/2} p_h = 0.$$

Here, we have used the standard five-point discretization for $\Delta_h$ (for $u_h$ on the o grid and for $v_h$ on the • grid) and the approximations

$$(\partial_x)_{h/2} p_h(x, y) := \frac{1}{h}\left(p_h\left(x + \frac{h}{2}, y\right) - p_h\left(x - \frac{h}{2}, y\right)\right)$$

$$(\partial_y)_{h/2} p_h(x, y) := \frac{1}{h}\left(p_h\left(x, y + \frac{h}{2}\right) - p_h\left(x, y - \frac{h}{2}\right)\right).$$

The staggered discretization of the Stokes equations leads to the system

$$\begin{bmatrix} -\Delta_h & 0 & (\partial_x)_{h/2} \\ 0 & -\Delta_h & (\partial_y)_{h/2} \\ (\partial_x)_{h/2} & (\partial_y)_{h/2} & 0 \end{bmatrix} \begin{bmatrix} u_h \\ v_h \\ p_h \end{bmatrix} = \begin{bmatrix} 0 \\ 0 \\ 0 \end{bmatrix}. \tag{8.7.1}$$

This $O(h^2)$-accurate discretization of the Stokes system is $h$-elliptic. The $h$-ellipticity can be seen from the determinant of the discrete operator, which is $\det L_h = (\Delta_h)^2$; therefore $E_h(L_h) = E_h((\Delta_h)^2) > 0$.

The stability of the staggered location of unknowns is also reflected by the fact that differencing of the first-order derivatives is now done with a distance of $h$ rather than $2h$ (see also Remark 8.6.5).

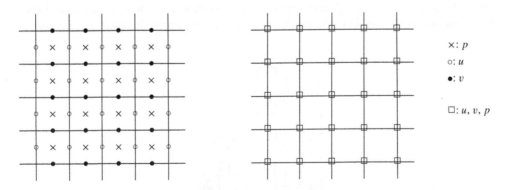

**Figure 8.2.** Staggered (left) and nonstaggered (right) location of unknowns.

Furthermore, the staggered approach leads to a natural discrete form of the continuity equation on a Cartesian grid: in accordance with the continuous problem, pressure values are not needed at physical boundaries.

> Although we now have a stable discretization, *a zero diagonal block* appears in the discrete system (8.7.1). For multigrid, this has the consequence that it is not possible to relax the discrete equations directly in a *decoupled* way.
>
> Moreover, the unknowns are not defined at the same locations. It is thus not immediately clear how to define standard *collective* relaxation.

We will present a generalization of collective relaxations in Section 8.7.2 ("box relaxation"). Another approach is to use *distributive relaxation* (see Section 8.7.3), which can be regarded as a generalization of decoupled relaxation.

All these considerations can be generalized from the Stokes equations to the incompressible Navier–Stokes equations in a straightforward manner. Clearly, when dealing with the (nonlinear) incompressible Navier–Stokes equations, the FAS version of the multigrid method (or global linearization) can be used. For high $Re$ numbers, the singular perturbation character of the momentum equations has to be taken into account additionally (*upwind*-type discretizations have to be used for the discretization of the convective terms) as pointed out in Section 8.6.2.

We will start the discussion on multigrid for staggered discretizations with a description of some appropriate transfer operators.

### 8.7.1 Transfer Operators

In staggered discretizations, the transfer operators depend on the relative locations of the unknowns with respect to the fine grid $\Omega_h$ and the coarse grid $\Omega_H$ (here, $\Omega_{2h}$) (see Fig. 8.3). Transfer operators for the different unknowns in staggered grids can easily be obtained. For the restriction $\hat{I}_h^H$ of the current approximation $\bar{u}_h$ in the FAS, the mean value of the unknowns at neighbor grid points is used to define approximations on the coarse grid:

$$\bar{u}_H = \frac{1}{2} \begin{bmatrix} 1 \\ \cdot \\ 1 \end{bmatrix} \bar{u}_h,$$

$$\bar{v}_H = \frac{1}{2} \begin{bmatrix} 1 & \cdot & 1 \end{bmatrix} \bar{v}_h, \qquad (8.7.2)$$

$$\bar{p}_H = \frac{1}{4} \begin{bmatrix} 1 & & 1 \\ & \cdot & \\ 1 & & 1 \end{bmatrix} \bar{p}_h,$$

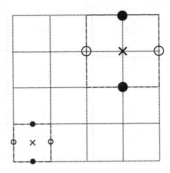

**Figure 8.3.** A fine and a coarse grid cell with corresponding staggered fine and coarse grid unknowns.

where the dot denotes the position on the coarse grid at which the restriction is applied. The defects can be transferred to the coarse grid in the following way:

$$d_H^u = \frac{1}{8} \begin{bmatrix} 1 & 2 & 1 \\ & \cdot & \\ 1 & 2 & 1 \end{bmatrix} d_h^u, \tag{8.7.3}$$

$$d_H^v = \frac{1}{8} \begin{bmatrix} 1 & & 1 \\ 2 & \cdot & 2 \\ 1 & & 1 \end{bmatrix} d_h^v, \tag{8.7.4}$$

$$d_H^p = \frac{1}{4} \begin{bmatrix} 1 & & 1 \\ & \cdot & \\ 1 & & 1 \end{bmatrix} d_h^p. \tag{8.7.5}$$

Two appropriate interpolation schemes for the coarse grid corrections are the (properly scaled) transpose of the restriction and *bilinear interpolation*. Let $\hat{w}_H$ denote the coarse grid correction for the fine grid function $u_h$. Linear interpolation of $\hat{w}_H$ means

$$\hat{w}_h(x, y) = \tfrac{1}{4}(\hat{w}_H(x, y + 3h/2) + 3\hat{w}_H(x, y - h/2) \qquad \text{(C)}$$

$$\hat{w}_h(x, y) = \tfrac{1}{8}(\hat{w}_H(x - h, y + 3h/2) + \hat{w}_H(x + h, y + 3h/2)$$
$$+3\hat{w}_H(x - h, y - h/2) + 3\hat{w}_H(x + h, y - h/2) \qquad \text{(D)}$$

$$\hat{w}_h(x, y) = \tfrac{1}{4}(\hat{w}_H(x, y - 3h/2) + 3\hat{w}_H(x, y + h/2) \qquad \text{(A)} \qquad \text{(8.7.6)}$$

$$\hat{w}_h(x, y) = \tfrac{1}{8}(\hat{w}_H(x - h, y - 3h/2) + \hat{w}_H(x + h, y - 3h/2)$$
$$+3\hat{w}_H(x - h, y + h/2) + 3\hat{w}_H(x + h, y - h/2) \qquad \text{(B)},$$

where the geometric position $(x, y)$ of the points A, B, C, D is shown in Fig. 8.4. The interpolation formulas for the correction of $v_h$ are similar. For the pressure, the interpolation formulas for cell-centered grids can be applied (see Section 2.8.4).

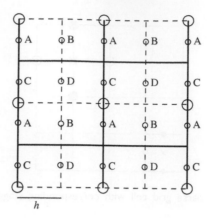

**Figure 8.4.** Bilinear interpolation of coarse grid corrections for $u_h$ on a staggered grid.

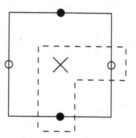

**Figure 8.5.** One possibility for collective relaxation on a staggered grid.

### 8.7.2 Box Smoothing

For the staggered discretization, one could define a collective local relaxation which is applied to the equations for $u_h$, $v_h$ and $p_h$ at adjacent locations (see Fig. 8.5 for an example). However, this approach is not a good smoother [401]. Better results are obtained with the so-called *box relaxation* [401, 402].

The basic idea of box relaxation [401] is to solve the discrete Navier–Stokes equations *locally* cell by cell involving *all* discrete equations which are located in the cell ("box"). This means that all five unknowns, sketched in Fig. 8.6, are updated *collectively*, using the respective four momentum equations at the cell boundaries and the continuity equation in the center of the box. Thus, for each box, one has to solve a $5 \times 5$ system of equations to obtain corrections for the unknowns. Using this scheme, each velocity component is updated *twice* and the pressure once per relaxation. Of course, the boxes are processed in a Gauss-Seidel manner (lexicographically or red–black like).

**Remark 8.7.1 (box-line relaxation)**   If significant anisotropies exist, e.g. in the case of strongly stretched grids, one has to update all strongly coupled unknowns collectively if

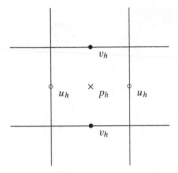

**Figure 8.6.** Unknowns updated collectively by box relaxation in the staggered case.

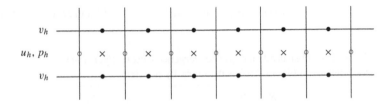

**Figure 8.7.** Unknowns updated collectively by box-line relaxation (staggered discretization).

standard coarsening is applied. In such cases, box relaxation can be replaced by a corresponding box-line version [390]. For instance, in case of box $x$-line relaxation, this means that all unknowns marked in Fig. 8.7 are updated collectively.                  $\gg$

Box relaxation turns out to have robust smoothing properties not only for small $Re$ numbers. However, underrelaxation is typically required, depending on $Re$. Box relaxation has been analyzed within the framework of LFA [242, 360]. The smoothing factor of (lexicographic) box relaxation for the Stokes problem on a staggered grid is $\mu_{\text{loc}} = 0.32$ using an underrelaxation parameter of $\omega = 0.7$ [360].

**Example 8.7.1** As an example for the 2D incompressible Navier–Stokes equations, we consider the driven cavity flow for $Re = 100$ and $Re = 1000$ in $\Omega = (0, 1)^2$ with boundary conditions $(u, v) = (1, 0)$ at $y = 1$ and homogeneous Dirichlet boundary conditions for $u$ and $v$ elsewhere (see Fig. 8.8). We use a hybrid discretization as sketched in Remark 7.3.1, where the switching is now governed by the *mesh Reynolds number*, i.e. by $hu_h Re$. We compare the multigrid convergence for box and box-line smoothing. For both Reynolds numbers, the smoothing steps are performed in alternating directions (although this is not really necessary for $Re = 100$).

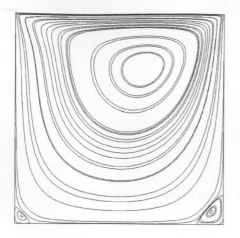

**Figure 8.8.** Streamlines for the driven cavity problem at $Re = 100$ (left) and at $Re = 1000$ (right).

**Table 8.7.** Measured multigrid convergence factors: W(1, 1)-cycle for solving the Navier–Stokes driven cavity problem.

|       | $Re = 100$ | | $Re = 1000$ | |
|-------|------|----------|------|----------|
| $1/h$ | Box  | Box-line | Box  | Box-line |
| 16    | 0.21 | 0.12     | 0.52 | 0.19     |
| 32    | 0.15 | 0.10     | 0.57 | 0.34     |
| 64    | 0.15 | 0.07     | 0.56 | 0.44     |
| 128   | 0.14 | 0.06     | 0.52 | 0.49     |

Here, for $Re = 100$ an underrelaxation parameter $\omega = 0.7$ and for $Re = 1000$, $\omega = 0.3$ has been employed. Other values of $\omega$ are, for example, used in [401]. LFA applied to the linearized incompressible Navier–Stokes equations indicates which values for $\omega$ are appropriate.

Table 8.7 presents measured average convergence factors of multigrid W(1, 1)-cycles. Here, the restriction operators (8.7.3)–(8.7.5) and the bilinear interpolation of corrections (8.7.6) have been applied. For $Re = 100$, the multigrid iteration converges very well. For $Re = 1000$, the convergence factors are worse. Note that this is an example of a *recirculating flow*. We have thus an analog to the convection-dominated convection–diffusion problem discussed in Section 7.2. The multigrid convergence is limited by the coarse grid correction, rather than by the smoothing properties. Accordingly, the somewhat worse convergence factors for the high Reynolds number $Re = 1000$ are not surprising.

Of course, the remedy proposed in Section 7.8, i.e. the recombination of iterants, can also be applied here to improve the convergence. △

### 8.7.3 Distributive Smoothing

Whereas the box relaxation schemes are clearly collective in type, other relaxation schemes have been proposed in the literature, some of which belong to the general class of so-called *distributive relaxation* schemes. Distributive relaxation was considered first in [72]. For a theoretical description and corresponding analysis, we refer to [419, 422, 423].

Before we give a general description of distributive relaxations, we would like to point out that various distributed relaxation schemes have been introduced as *predictor–corrector* type schemes. In the predictor step, a new velocity field $(u_h^*, v_h^*)$ is computed from $(u_h^m, v_h^m, p_h^m)$ by performing a few relaxations to the momentum equations. In the corrector step, both velocity and pressure are updated

$$\bar{u}_h^m = u_h^* + \delta u_h, \qquad \bar{v}_h^m = v_h^* + \delta v_h, \qquad \bar{p}_h^m = p_h^* + \delta p_h \qquad (8.7.7)$$

so that

(1) the continuity equation is satisfied and
(2) the momentum defect remains unchanged (or changes only slightly).

The resulting schemes differ in the way in which these updates are actually computed.

In the distributive Gauss–Seidel scheme (DGS) introduced in [72], the predictor step consists of a standard GS-LEX-type relaxation of each of the momentum equations to obtain $u_h^*$ and $v_h^*$. The corrector step is also carried out lexicographically over the points. Since we will describe only the basic idea of DGS, we restrict ourselves to the Stokes equations. Let $d_h^p(x, y)$ be the defect of the discrete continuity equation at a point $(x, y)$ just before the corrector step is applied at that point. The corrector step then consists of the nine updates

$$
\begin{aligned}
u_h(x + h/2, y) &\leftarrow u_h(x + h/2, y) + c \\
v_h(x, y + h/2) &\leftarrow v_h(x, y + h/2) + c \\
u_h(x - h/2, y) &\leftarrow u_h(x - h/2, y) - c \\
v_h(x, y - h/2) &\leftarrow v_h(x, y - h/2) - c \\
p_h(x, y) &\leftarrow p_h(x, y) + 4c/h \\
p_h(x, y + h) &\leftarrow p_h(x, y + h) - c/h \\
p_h(x, y - h) &\leftarrow p_h(x, y - h) - c/h \\
p_h(x + h, y) &\leftarrow p_h(x + h, y) - c/h \\
p_h(x - h, y) &\leftarrow p_h(x - h, y) - c/h,
\end{aligned}
\qquad (8.7.8)
$$

where $c = h d_h^p(x, y)/4$ (see Fig. 8.9 for the geometric position of the updates). After these updates, the defect of the discrete continuity equation at $(x, y)$ is zero. Moreover, the pressure changes are such that the defects of the momentum equations at all points remain unchanged.

An elegant and general way to describe distributive relaxations is to introduce a *right preconditioner* in the smoothing procedure [419, 422].

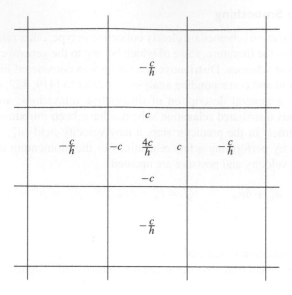

**Figure 8.9.** Geometric position of the updates in the corrector step of DGS.

The introduction of a right preconditioner means that we introduce new variables $\hat{u}_h$, where $u_h = C_h\hat{u}_h$, and consider the transformed system $L_hC_h\hat{u}_h = f_h$. $C_h$ is chosen in such a way that the resulting operator $L_hC_h$ is suited for *decoupled* (noncollective) relaxation. In particular, the *zero block in $L_h$ resulting from the continuity equation in (8.7.1) should disappear* in $L_hC_h$.

The distributive relaxation can then be described in the following way.

(1) Transform the system $L_hu_h = f_h$ to a "simpler" one by a suitable preconditioning with $C_h$ (the *distributor*).
(2) Choose a point- or linewise relaxation process, preferably for each of the equations of the transformed system separately, of the form

$$\bar{\hat{u}}_h^m = \hat{u}_h^m + B_h(f_h - L_hC_h\hat{u}_h^m) \qquad (8.7.9)$$

with $B_h$ being some approximation of the inverse of $L_hC_h$. Note that, depending on the choice of $B_h$, Jacobi- or Gauss–Seidel-type iterations are obtained.
(3) Reformulate this relaxation scheme in terms of the original operator and unknowns by using $\bar{u}_h^m = C_h\hat{u}_h$:

$$\bar{u}_h^m = u_h^m + C_hB_h(f_h - L_hu_h^m). \qquad (8.7.10)$$

This interpretation of distributive relaxation as a "standard" relaxation method for a properly preconditioned system may also serve as a general basis on which to construct smoothing schemes for the incompressible Navier–Stokes equations. Then, the operator $L_h$ is interpreted as the operator of the linearized equations (nonlinear terms frozen to current values by, for instance, a Picard-type linearization).

The above description leads to an elegant approach for *analyzing* the smoothing properties of distributive schemes by means of LFA.

**Example 8.7.2** For the Stokes equations, the DGS described above is represented by the preconditioner

$$C_h = \begin{pmatrix} I_h & 0 & -(\partial_x)_{h/2} \\ 0 & I_h & -(\partial_y)_{h/2} \\ 0 & 0 & -\Delta_h \end{pmatrix}. \tag{8.7.11}$$

In order to see the equivalence of the DGS approach described above and the use of this preconditioner, consider (8.7.10). Note that, according to (8.7.9), $B_h d_h^m = B_h(f_h - L_h u_h^m)$ determines the corrections $\hat{u}_h^{m+1} - \hat{u}_h^m$ obtained by a standard GS-LEX type relaxation with respect to the operator

$$L_h C_h = \begin{pmatrix} -\Delta_h & 0 & 0 \\ 0 & -\Delta_h & 0 \\ (\partial_x)_{h/2} & (\partial_y)_{h/2} & -\Delta_h \end{pmatrix}. \tag{8.7.12}$$

According to (8.7.10), $C_h$ then indicates how these corrections are to be *distributed* to $u_h$. With these considerations, the equivalence of the preconditioning approach to (8.7.8) can be derived.

For DGS, the LFA smoothing analysis can be performed in terms of the product operator $L_h C_h$. Here, the smoothing properties only depend on the diagonal terms $\Delta_h$. The DGS smoothing factor $\mu_{\text{loc}}$ is just that of pointwise GS-LEX for $L_h C_h$ and thus equal to $\mu_{\text{loc}}(-\Delta_h) = 0.5$ [72].

A detailed two-grid analysis can be found in [280], where stretched grids are also included in the discussion. $\triangle$

Box relaxation and DGS work well for low Reynolds numbers. Box relaxation is, however, somewhat more expensive than DGS. For higher Reynolds numbers, distributive relaxation (in its "classical form" [72]) is known to become worse than box relaxation. A variety of modified distributive schemes has been proposed (see for example [53, 66, 148, 149, 419]).

**Remark 8.7.2 (pressure correction-type smoothers)** The SIMPLE algorithm [301] mentioned in Remark 8.6.4 is another example of a distributive scheme. It belongs to the class of the so-called *pressure correction schemes*, which have been used as smoothers (for example, in [24, 247, 352]). They are obtained by using a distributor of the form

$$C_h = \begin{pmatrix} I_h & 0 & -\hat{Q}_h^{-1}(\partial_x)_{h/2} \\ 0 & I_h & -\hat{Q}_h^{-1}(\partial_y)_{h/2} \\ (\partial_x)_{h/2} & (\partial_y)_{h/2} & I_h \end{pmatrix}. \tag{8.7.13}$$

with $\hat{Q}_h$ being an approximation of $Q_h := -\Delta_h + \mathrm{Re}\, u_h(\partial_x)_{h/2} + \mathrm{Re}\, v_h(\partial_y)_{h/2}$. A suitably chosen relaxation scheme is then employed for the resulting product system $L_h C_h$ (e.g. line Gauss–Seidel successively for velocity and pressure).

Smoothing analysis results for pressure correction schemes are available in [352]. The resulting smoothing factor is $\mu_{\mathrm{loc}} = 0.6$ for the Stokes equations. In [53], the pressure correction smoothers are, however, not recommended as smoothers for the stationary incompressible Navier–Stokes equations.                                                     ≫

**Remark 8.7.3**  Distributive relaxation may not be suitable for the boundary conditions and should then be replaced by box relaxation at and near the boundary.                      ≫

**Remark 8.7.4**  When applied to the conservative form of the Navier–Stokes equations, usually not only a right preconditioner is introduced, but additionally a left one. The resulting schemes are sometimes called weighted distributive relaxations (see, for example, Appendix C).                                                                               ≫

---

> Summarizing our discussion, the incompressible Navier–Stokes equations can be solved efficiently by multigrid using staggered discretizations. Typical smoothers are of box or distributive type. The use of box relaxation may be considered as more straightforward and more convenient than that of distributive relaxation.

---

## 8.8 INCOMPRESSIBLE NAVIER–STOKES EQUATIONS: NONSTAGGERED DISCRETIZATIONS

In complex flow domains nonorthogonal curvilinear meshes (boundary fitted grids) are often used. The staggered discretization of the incompressible Navier–Stokes can be generalized to such grids. However, in order to obtain a stable discretization, one has to work with covariant velocities (i.e. velocities tangential to the cell faces) or contravariant velocities (velocities normal to the cell faces) as new unknowns. The numerical accuracy of the "straightforward" staggered discretization on boundary fitted grids depends sensitively on the nonorthogonality of a grid. This can be overcome by even more sophisticated staggered variants [330, 331, 417].

On curvilinear meshes, a nonstaggered discretization is much easier to implement since the equations can be discretized directly in terms of the original Cartesian velocity components [317], e.g. with a finite volume discretization. On the other hand, nonstaggered schemes have to overcome the checkerboard instability by artificial stabilization terms. We will discuss various possibilities of stabilizing the nonstaggered discretization in the following sections. A detailed discussion on the advantages and disadvantages of staggered and nonstaggered grids on curvilinear grids can, for example, be found in [327].

Flux splitting concepts as described in Section 8.8.4 are particularly suited for flows at high Reynolds numbers.

### 8.8.1 Artificial Pressure Terms

Stabilization for nonstaggered discretizations can be achieved by adding an artificial elliptic pressure term, e.g.

$$-\omega h^2 \Delta_h p_h \tag{8.8.1}$$

to the continuity equation $(u_x)_h + (v_y)_h = 0$. For $h \to 0$, the artificial pressure term tends to 0. Since this term is proportional to $h^2$, *second-order accuracy is maintained* if all other terms in the PDE system are discretized with second-order accuracy.

For the Stokes equations and for low Reynolds numbers, central second-order discretizations can be employed in a straightforward way when using the artificial pressure term.

The addition of an artificial pressure term is implicitly also used in other Stokes or Navier–Stokes discretizations. For instance, the SIMPLE method for nonstaggered discretizations as discussed in [302] can be seen to implicitly solve a discrete continuity equation augmented by an artificial pressure term of the form (8.8.1) with $\omega = 1/8$. Also, in the framework of (stabilized) finite element approximations, artificial pressure terms are quite common. Flux splitting discretizations such as the flux difference splitting described in Section 8.8.4 also implicitly introduce artificial pressure terms.

The parameter $\omega$ has to be chosen small enough to maintain a good accuracy but large enough so that the discrete Navier–Stokes system becomes sufficiently $h$-elliptic. The $h$-ellipticity measure $\bar{E}_h(L_h)$ indicates a proper choice of $\omega$. For simplicity, we restrict ourselves to the Stokes equations. The symbol of the discrete Stokes operator, $\tilde{L}_h(\boldsymbol{\theta})$ is given by

$$\tilde{L}_h(\boldsymbol{\theta}) = \begin{pmatrix} -\tilde{\Delta}_h(\boldsymbol{\theta}) & 0 & \tilde{\partial}_{x,h}(\boldsymbol{\theta}) \\ 0 & -\tilde{\Delta}_h(\boldsymbol{\theta}) & \tilde{\partial}_{y,h}(\boldsymbol{\theta}) \\ \tilde{\partial}_{x,h}(\boldsymbol{\theta}) & \tilde{\partial}_{y,h}(\boldsymbol{\theta}) & -\omega h^2 \tilde{\Delta}_h(\boldsymbol{\theta}) \end{pmatrix}, \tag{8.8.2}$$

where $\tilde{\Delta}_h, \tilde{\Delta}_{2h}, \tilde{\partial}_{x,h}$ and $\tilde{\partial}_{y,h}$ denote the symbols of the respective scalar difference operators $\Delta_h, \Delta_{2h}$,

$$(\partial_x)_h := 1/(2h) \begin{bmatrix} -1 & 0 & 1 \end{bmatrix}_h \quad \text{and} \quad (\partial_y)_h := 1/(2h) \begin{bmatrix} 1 \\ 0 \\ -1 \end{bmatrix}_h.$$

We obtain

$$\det \tilde{L}_h(\boldsymbol{\theta}) = -\tilde{\Delta}_h(\omega h^2 \tilde{\Delta}_h \tilde{\Delta}_h - \tilde{\Delta}_{2h}). \tag{8.8.3}$$

From this, we can compute $\bar{E}_h = \bar{E}_h(\omega)$. To simplify this computation, note that only the second term of $\det \tilde{L}_h$ (i.e., the term in parentheses in (8.8.3)) is crucial. Only this term may become small for high frequencies. Omitting the factor $\tilde{\Delta}_h$ in (8.8.3), we obtain

$$\bar{E}_h(\omega) \sim \begin{cases} 8\omega(1 - 8\omega) & (\omega \le 1/16) \\ 1/4 & (1/16 \le \omega \le 1/12) \\ (1 + 1/(4\omega))/16 & (\omega \ge 1/12) \end{cases}$$

which does not depend on $h$. This means that the $h$-ellipticity of the discrete Stokes equations, stabilized by (8.8.1), is $h$-independent.

As expected, the $h$-ellipticity measure is zero for $\omega = 0$. The maximum of $\bar{E}_h(\omega)$ $(= 1/4)$ is obtained for $1/16 \leq \omega \leq 1/12$. The smallest $\omega$ for which the maximum is attained is $\omega = 1/16$. This choice turns out to be a reasonable choice in practice, with respect to both accuracy and stability. This value of $\omega$ can also be used for the Navier–Stokes equations at low Reynolds numbers.

**Remark 8.8.1 (high Reynolds numbers)** For high Reynolds numbers, the difficulties of dominating advection operators have to be taken into account again (see Section 8.6.2). Moreover, the artificial pressure term also has to be suitably adapted. For high Reynolds numbers, when a significant amount of artificial viscosity (e.g. proportional to $\max\{Re|u_h|h, Re|v_h|h\}$) is introduced in the momentum equations, the choice of an appropriate $\omega$ is somewhat more involved and can be based on an analysis of the $h$-ellipticity of the linearized operator. $\gg$

**Remark 8.8.2** Central differencing of $\partial p/\partial x, \partial p/\partial y$ on nonstaggered grids formally requires pressure values at *physical boundaries*. They can most easily be obtained by means of *extrapolation* from interior pressure values. However, in order to obtain second-order accuracy, extrapolation has to be sufficiently accurate. In many cases, linear extrapolation is sufficient.

If the pressure is not prescribed at *any* boundary, the system is singular ($p$ is determined only up to a constant) and one has to take into account the considerations in Section 5.6.4 (see also Remark 8.6.2). $\gg$

**Remark 8.8.3** $h$-elliptic nonstaggered discretizations can also be obtained without introducing an artificial pressure term by simply using *noncentral* discretizations, e.g., forward differences for $p$ in the momentum equations and backward differences for $u$ and $v$ in the continuity equation. This is most easily seen by considering the determinant of the operator. If we replace, for example, the second-order central differences in (8.8.2) by the respective first-order one-sided schemes (and set $\omega = 0$), $h$-ellipticity follows immediately from $\det \tilde{L}_h = \tilde{\Delta}_h \tilde{\Delta}_h$ in the Stokes case. Some multigrid results are reported in [149]. In order to obtain an overall second-order accuracy, the nonsymmetric (forward and backward) upwind-type discretizations also have to be of second-order accuracy. $\gg$

### 8.8.2 Box Smoothing

When discretizing the incompressible Navier–Stokes equations with standard second-order discretizations and the artificial pressure term as discussed in Section 8.8.1, the development of smoothing schemes is not straightforward. LFA shows that the smoothing factors of standard *collective* point relaxations are not satisfactory. Although, according to the discussion on $h$-ellipticity in Section 8.3.2, there exist pointwise smoothing schemes with smoothing factors bounded below 1, the smoothing factor of such a scheme may be rather poor. One possibility to overcome this problem is to extend the idea of box relaxation, which has proved to be a suitable smoother in the case of staggered discretizations, to the nonstaggered case [243] (see Fig. 8.10).

In Cartesian coordinates, box relaxation is defined essentially in the same way as in the staggered case, except that the side lengths of the boxes are now twice as large. For each box, again a $5 \times 5$ system is solved, using the respective four momentum equations along the edges and the continuity equation in the center of the box.

**Remark 8.8.4 (Box-line relaxation)**  As in the staggered case, box relaxation should be replaced by a corresponding box-line version, as indicated in Fig. 8.11, if significant anisotropies occur.                                                                        ≫

The smoothing factor, $\mu_{\text{loc}}$, of lexicographic box relaxation for the Stokes operator on a nonstaggered grid with $\omega = 1/16$ is $\mu_{\text{loc}} = 0.63$ [242]. For box-line relaxation one obtains $\mu_{\text{loc}} = 0.56$. Using underrelaxation for the pressure, these smoothing factors can be improved to values below 0.5 [243].

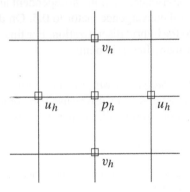

**Figure 8.10.** Unknowns, □, updated collectively by box relaxation in the nonstaggered case.

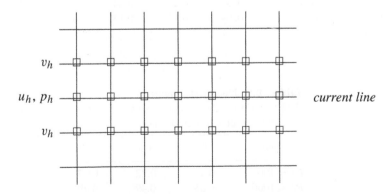

**Figure 8.11.** Unknowns, □, updated simultaneously by box-line relaxation.

**Example 8.8.1** For a demonstration of the multigrid convergence, we consider the nonstaggered discretization of the Stokes equations (in the unit square) with $\omega = 1/16$ in the artificial pressure term. The algorithm uses linear pressure extrapolation at physical boundaries, FW and bilinear interpolation. Table 8.8 gives convergence factors using W(2,1)-cycles and compares box- and box-line smoothing (without underrelaxation).

Ideally, according to the smoothing factors given above, one would expect a convergence factor per cycle of around 0.24 ($\approx 0.63^3$) and 0.18 ($\approx 0.56^3$) in the case of lexicographic box smoothing and box-line smoothing, respectively. This is in good agreement with the observed results in Table 8.8.                                    △

**Example 8.8.2** Table 8.9 shows the corresponding multigrid convergence for the *driven cavity flow problem* introduced in Example 8.7.1 both for $Re = 100$ and $Re = 1000$. We use a hybrid discretization (see Remark 7.3.1).

For $Re = 100$, the multigrid convergence factors are similar to those in the Stokes case. However, they become worse for $Re = 1000$ (unless the finest grid size is sufficiently small). Again, this effect is a consequence of the $h$-dependent amount of artificial viscosity used, which limits the two-grid convergence factor to 0.5. On the finest grid, however, the convergence improves due to the hybrid discretization. For this application, it is possible to apply central differencing at many fine grid points.                                    △

**Remark 8.8.5** Although *distributive relaxations* were originally developed for discretizations on staggered grids, these schemes can also be used on nonstaggered grids, with similar smoothing and convergence properties similar to the staggered case.                                    ≫

**Table 8.8.** Measured multigrid convergence factors: W(2,1)-cycle for the Stokes problem.

| $n = 1/h$ | 16 | 32 | 64 | 128 | 256 |
|---|---|---|---|---|---|
| Box | 0.24 | 0.22 | 0.24 | 0.24 | 0.24 |
| Box-line | 0.14 | 0.13 | 0.13 | 0.14 | 0.14 |

**Table 8.9.** Measured multigrid convergence factors: W(2,1)-cycle for the Navier–Stokes driven cavity problem, with lexicographic box smoothing.

| $n = 1/h$ | 32 | 64 | 128 | 256 |
|---|---|---|---|---|
| $Re = 100$ | 0.16 | 0.22 | 0.23 | 0.23 |
| $Re = 1000$ | 0.58 | 0.49 | 0.49 | 0.41 |

### 8.8.3 Alternative Formulations

As discussed above, difficulties in the numerical solution of the incompressible Navier–Stokes equations are caused by the special form of the continuity equation. As mentioned above standard collective point relaxations are not suited as smoothers for the nonstaggered discretization if stabilized by the artificial pressure term. It is, however, possible to *reformulate* the incompressible Navier–Stokes system in such a way that standard collective relaxations become appropriate for second-order accurate central discretizations. The applicability of these reformulations is, however, limited to certain classes of problems.

As an example, we consider a reformulation of the *stationary* incompressible Navier–Stokes system which is well suited to low Reynolds numbers. In this formulation, a Poisson-like equation for $p$ is obtained. To guarantee equivalence between the two systems of equations, the continuity equation has to be satisfied at the boundary of the domain. We will discuss this approach and a proper multigrid treatment in some detail in the following.

**Result 8.8.1** *Assume that $\Omega$ is a bounded domain in $\mathbb{R}^2$ and that $u(x, y)$, $v(x, y)$, $p(x, y)$ and $\partial\Omega$ are sufficiently smooth. Then the two systems*

$$-\Delta u + Re(uu_x + vu_y + p_x) = 0 \quad (\Omega) \tag{8.8.4}$$

$$-\Delta v + Re(uv_x + vv_y + p_y) = 0 \quad (\Omega) \tag{8.8.5}$$

$$u_x + v_y = 0 \quad (\bar{\Omega} = \Omega \cup \partial\Omega) \tag{8.8.6}$$

*and*

$$-\Delta u + Re(vu_y - uv_y + p_x) = 0 \quad (\Omega) \tag{8.8.7}$$

$$-\Delta v + Re(uv_x - vu_x + p_y) = 0 \quad (\Omega) \tag{8.8.8}$$

$$\Delta p + 2(v_x u_y - u_x v_y) = 0 \quad (\Omega) \tag{8.8.9}$$

$$u_x + v_y = 0 \quad (\partial\Omega) \tag{8.8.10}$$

*are equivalent.*

*Proof.* The momentum equations (8.8.4)–(8.8.5) can be written as (8.8.7)–(8.8.8) because of (8.8.6). Differentiating (8.8.7) with respect to $x$ and (8.8.8) with respect to $y$, adding these equations and using again (8.8.6), we obtain (8.8.9).

In the other direction, (8.8.6) can be regained from (8.8.7)–(8.8.9): the difference of (8.8.9) and the sum of the derivatives of (8.8.7) with respect to $x$ and (8.8.8) with respect to $y$ is $\Delta(u_x + v_y) = 0$. This PDE with the homogeneous Dirichlet boundary condition (8.8.10) interpreted as a boundary value problem for $u_x + v_y$ has the unique solution (8.8.6). $\square$

The problems related to the checkerboard instability in the original Navier–Stokes system disappear for the new boundary value problem. The "new continuity equation" (8.8.9) naturally contains the elliptic term $\Delta p$. In the Stokes case, we obtain the Laplace equation itself.

In the same way and under similar assumptions as in 2D, a corresponding 3D system can be derived [347].

For low Reynolds numbers, we can use standard second-order discretizations for all derivatives occurring. This discretization has a good $h$-ellipticity measure. A good smoothing algorithm for the resulting discrete Navier–Stokes problem can easily be found: the principal part of each difference equation is characterized by the $\Delta_h$-operator. The three equations are coupled by lower order terms. So, a standard collective point relaxation is suitable, which updates the three variables $u_h$, $v_h$ and $p_h$ simultaneously by the solution of a $3 \times 3$ system in every grid point. On a square grid the corresponding collective (pointwise) GS-LEX relaxation has a smoothing factor of $\mu_{loc} = 0.5$ for the Stokes problem.

**Remark 8.8.6 (treatment of boundary conditions)**  Two boundary conditions are necessary for the solution of the original Navier–Stokes equations (8.6.4)–(8.6.6). The new system (8.8.7)–(8.8.9) requires three, one of them being the continuity equation (8.8.10) on the boundary. We end up with three conditions for the velocity components but no condition for the pressure $p$ if, for example, $u$ and $v$ are prescribed by functions $u_0$ and $v_0$ on the boundary. This problem can be treated in the same way as the *biharmonic problem* with $u$ and $u_n$ boundary conditions (see Section 8.4.2), introducing auxiliary points along the boundary $\Gamma_h (\Gamma = \partial\Omega)$. The discrete equations at the boundary $\Gamma_h$ are (8.8.7)–(8.8.9), discretized by standard second-order central differences and

$$u_h = u_0, \qquad v_h = v_0, \qquad (\partial_x)_h u_h + (\partial_y)_h v_h = 0. \qquad (8.8.11)$$

These are six equations for six unknowns at the boundary ($u_h$, $v_h$ and $p_h$ at any boundary point and at the corresponding auxiliary point).

For a collective update of the unknowns located at a boundary point $P$, one has to include the three equations at that grid point of the interior of $\Omega_h$, which is the direct neighbor of $P$, in the same way as described in detail for the biharmonic system.               $\gg$

**Example 8.8.3**  Finally, we apply a corresponding multigrid algorithm to a *driven cavity Stokes flow problem* on $\Omega = (0, 1)^2$ here with

$$u_0(x, y) = \begin{cases} 0 & \text{if } y < 1 \\ 16x^2(1 - x)^2 & \text{if } y = 1. \end{cases}$$

$$v_0(x, y) = 0.$$

The FAS algorithm uses F(2,1)-cycles, FW and bilinear interpolation. The mesh size is $1/128$ on the finest grid and $1/4$ on the coarsest one.

The measured convergence factor is about 0.11, which is in good agreement with the smoothing factor $\mu_{loc} = 0.5$ of the collective GS-LEX relaxation ($0.5^3 = 0.125$).               $\triangle$

In the above example, the measured multigrid convergence factors are satisfactory for *Re* less than about 50.

**Remark 8.8.7 (stream function–vorticity formulation)** Another possibility to reformulate the incompressible Navier–Stokes equations, particularly suitable in 2D, is the so-called stream function–vorticity formulation. Here, the unknowns to be solved for are the vorticity (sometimes denoted by $\boldsymbol{\omega} = \nabla \times \boldsymbol{u}$) and the stream function, $\Psi$ ($u = \partial \Psi / \partial y$, $v = -\partial \Psi / \partial x$). In this formulation, the solution of the Navier–Stokes system reduces to the solution of two Poisson-like equations for the stream function and the vorticity, respectively. Multigrid methods have been developed which are appropriate even for high Reynolds numbers (see, for example, [157]).

In 3D, however, this approach becomes more complicated and loses its attractive features. In particular, the 3D vorticity–velocity formulation requires the determination of six unknown functions (three components of the vorticity and of the velocity). A corresponding multigrid algorithm for the incompressible Navier–Stokes equations has been proposed [201], however, on a staggered grid employing a *distributive relaxation*. $\gg$

### 8.8.4 Flux Splitting Concepts

In the following, we will present a discretization for the incompressible Navier–Stokes equations which is particularly suitable for flows at high Reynolds numbers. Up to now, we have motivated the discretizations for the incompressible Navier–Stokes equations by departing from the linear case of the Stokes equations. Here, the situation is different. The structure of the convective terms of the incompressible Navier–Stokes system resembles that of PDE systems describing *compressible* flow.

Many upwind discretizations for compressible flow problems make use of so-called *flux splitting concepts* in various forms [196]. Examples are the flux vector splitting method of van Leer [231], the flux difference splitting method of Roe [328] or of Osher *et al.* [97, 287, 288]. These discretization schemes lead naturally to an upwind-type discretization for systems of nonlinear equations. We will show how such a method can be applied to *incompressible* flow problems.

In flux formulation, the 2D incompressible Navier–Stokes equations in conservation form (8.6.1)–(8.6.3) can be written as

$$\frac{\partial \boldsymbol{f}}{\partial x} + \frac{\partial \boldsymbol{g}}{\partial y} = \frac{\partial \boldsymbol{f}_v}{\partial x} + \frac{\partial \boldsymbol{g}_v}{\partial y}, \tag{8.8.12}$$

where $\boldsymbol{f}$ and $\boldsymbol{g}$ are the *convective fluxes* and $\boldsymbol{f}_v$ and $\boldsymbol{g}_v$ are the *viscous fluxes*:

$$\boldsymbol{f} = \begin{pmatrix} u^2 + p \\ uv \\ c^2 u \end{pmatrix}, \qquad \boldsymbol{g} = \begin{pmatrix} uv \\ v^2 + p \\ c^2 v \end{pmatrix},$$

$$\boldsymbol{f}_v = \begin{pmatrix} (1/Re)(\partial u/\partial x) \\ (1/Re)(\partial v/\partial x) \\ 0 \end{pmatrix}, \qquad \boldsymbol{g}_v = \begin{pmatrix} (1/Re)(\partial u/\partial y) \\ (1/Re)(\partial v/\partial y) \\ 0 \end{pmatrix}. \tag{8.8.13}$$

Here, $c$ is a constant reference velocity introduced to homogenize the eigenvalues of the system matrices, as will be discussed below. The unknown functions (sometimes also called state variables, or state vector) are $\boldsymbol{u} = (u, v, p)^T$.

The starting point for a *finite volume discretization* is the integral formulation of (8.8.12) in a control volume $\Omega_{i,j}$ (see also Section 5.7)

$$\int_{\Omega_{i,j}} \left( \frac{\partial f}{\partial x} + \frac{\partial g}{\partial y} - \frac{\partial f_v}{\partial x} - \frac{\partial g_v}{\partial y} \right) d\Omega = 0.$$

According to Gauss's theorem, we obtain

$$\int_{\Omega_{i,j}} \left( \frac{\partial f}{\partial x} + \frac{\partial g}{\partial y} - \frac{\partial f_v}{\partial x} - \frac{\partial g_v}{\partial y} \right) d\Omega = \oint_{\partial \Omega_{i,j}} (n_x f + n_y g - n_x f_v - n_y g_v) \, dS,$$

where $n_x$ and $n_y$ denote the $x$- and the $y$-component, respectively, of the unit outward normal vector of the respective side.

A step towards a discrete formulation is to use an approximation

$$\oint_{\partial \Omega_{i,j}} (n_x f + n_y g - n_x f_v - n_y g_v) \, dS$$
$$\approx \sum_k (n_x f_h + n_y g_h)_k |S_k| - \sum_k (n_x f_{v,h} + n_y g_{v,h})_k |S_k|,$$

where $k$ is an appropriate index representing the sides $S_k$ of $\Omega_{i,j}$ and where $|S_k|$ denotes the length of a side of the control volume.

The discretization of the viscous fluxes $f_v$ and $g_v$ does not cause any problems and is performed as in the scalar case for the Laplacian (see Section 5.7).

For the convective fluxes $f$ and $g$, however, central differences do not yield stable discretizations.

The idea of many *upwind-type* discretizations starts from

$$\sum_k (n_x f_h + n_y g_h)_k |S_k| = F_{i+1/2,j} |S_{i+1/2}| + F_{i-1/2,j} |S_{i-1/2}|$$
$$+ F_{i,j+1/2} |S_{j+1/2}| + F_{i,j-1/2} |S_{j-1/2}|, \tag{8.8.14}$$

where each *flux vector* $F_{i+i_0, j+j_0}$ has to be defined in such a way that it is a sufficiently accurate approximation for $(n_x f_h + n_y g_h)_k$ at the corresponding side $S_k$ of $\Omega_{i,j}$. The indices $i + i_0$, $j + j_0$ with either $i_0$ or $j_0 \in \{-1/2, 1/2\}$ correspond naturally to a point on a control volume side $S_k$ (see Fig. 8.12). Note that the calculation of the flux vectors at grid points, denoted by $F_{i,j}$, is no problem since all grid functions are defined there.

The flux vector $F_{i+i_0, j+j_0}$ depends on the vector $u_h$ at the left-hand and right-hand side of a control volume boundary denoted by $u_h^L$ and $u_h^R$. An appropriate approximation for $F_{i+i_0, j+j_0} = F_R(u_h^L, u_h^R)$ has to be found. In a straightforward discretization, $u_h^L$ and $u_h^R$ are, for example, chosen as

$$u_h^L = u_{i,j} \quad \text{and} \quad u_h^R = u_{i+1,j}. \tag{8.8.15}$$

for the side $S_{i+1/2,j}$.

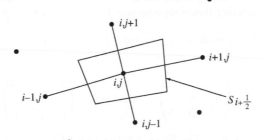

**Figure 8.12.** Control volume around point $i, j$.

**Remark 8.8.8**   The flux function $F_R(u_h^L, u_h^R)$ is usually based on the approximate solution of a so-called *Riemann problem* and is often called an *approximate Riemann solver* [196].

By 1987 [329], more than 100 different variants of such discretization schemes were listed. However, an "optimal" scheme, if it exists at all, has not yet been identified. We will not discuss this topic here in detail and refer to the literature for a more profound understanding (see [183, 196, 232] for a survey).

A general approach is

$$F_R(u_h^L, u_h^R) = \tfrac{1}{2}(F(u_h^L) + F(u_h^R) - d(u_h^L, u_h^R)) \tag{8.8.16}$$

which corresponds to a central approximation modified by a dissipative term $d$. Different approximative Riemann solvers use different definitions for the function $d$. Different choices of $u_h^L$ and $u_h^R$ lead to approximations of different accuracies (see the example below).  ≫

As one example, we describe a Roe-type flux difference splitting approach developed in [125] and discuss some of its features. Since the derivation of the scheme is technically rather complicated, we restrict ourselves first to (8.8.15), which will result in a *first-order discretization*.

Readers not interested in the (somewhat technical) details of the definition of the upwind flux function can move immediately to (8.8.25).

To define an upwind flux, we consider differences

$$\delta F_{i,i+1} = n_x \delta f_{i,i+1} + n_y \delta g_{i,i+1}, \tag{8.8.17}$$

where

$$\delta F_{i,i+1} = F_{i+1,j} - F_{i,j}, \qquad \delta f_{i,i+1} = f_{i+1,j} - f_{i,j},$$
$$\delta g_{i,i+1} = g_{i+1,j} - g_{i,j}.$$

The flux difference splitting approach makes use of the fact that the components of the flux vectors $f$ and $g$ (8.8.13) are *polynomials* in the primitive variables $u$, $v$ and $p$. For example, a difference $\delta(uv)$ can be written as

$$\delta(uv) = \bar{u}\delta v + \delta u\bar{v},$$

where the overbar denotes the algebraic mean of the differenced variables ("linearization"). Differences of the convective fluxes with respect to $\boldsymbol{u}$ can thus be expressed as

$$\delta f_{i,i+1} = A_1 \delta u_{i,i+1} \quad \text{and} \quad \delta g_{i,i+1} = A_2 \delta u_{i,i+1} \tag{8.8.18}$$

with

$$A_1 := \begin{pmatrix} 2\bar{u} & 0 & 1 \\ \bar{v} & \bar{u} & 0 \\ c^2 & 0 & 0 \end{pmatrix} \quad \text{and} \quad A_2 = \begin{pmatrix} \bar{v} & \bar{u} & 0 \\ 0 & 2\bar{v} & 1 \\ 0 & c^2 & 0 \end{pmatrix}, \tag{8.8.19}$$

where $\delta u_{i,i+1} = u_{i+1,j} - u_{i,j}$. The linear combination $\delta F_{i,i+1}$ in (8.8.17) is written as

$$\delta F_{i,i+1} = A \delta u_{i,i+1} \tag{8.8.20}$$

with

$$A = n_x A_1 + n_y A_2 = \begin{pmatrix} n_x \bar{u} + \bar{w} & n_y \bar{u} & n_x \\ n_x \bar{v} & n_y \bar{v} + \bar{w} & n_y \\ c^2 n_x & c^2 n_y & 0 \end{pmatrix} \tag{8.8.21}$$

and $\bar{w} = n_x \bar{u} + n_y \bar{v}$. For a unit normal vector $((n_x)^2 + (n_y)^2 = 1)$, $A$ has the eigenvalues

$$\lambda_1 = \bar{w}, \quad \lambda_2 = \bar{w} + a, \quad \lambda_3 = \bar{w} - a \quad \text{where } a = \sqrt{\bar{w}^2 + c^2} \tag{8.8.22}$$

and can thus be diagonalized:

$$A = R\Lambda L,$$

with the left and right eigenvector matrices $L$ and $R$ ($R = L^{-1}$) and the diagonal matrix $\Lambda$ which consists of the eigenvalues $\lambda_i$ of $A$.

*In its diagonalized form, the matrix $A$ can easily be split into positive and negative parts,* i.e. in a matrix $A^+$ with nonnegative and in a matrix $A^-$ with nonpositive eigenvalues,

$$A = A^+ + A^-, \quad A^+ = R\Lambda^+ L, \quad A^- = R\Lambda^- L, \tag{8.8.23}$$

where

$$\Lambda^+ = \text{diag}(\lambda_1^+, \lambda_2^+, \lambda_3^+), \quad \Lambda^- = \text{diag}(\lambda_1^-, \lambda_2^-, \lambda_3^-)$$

with

$$\lambda_i^+ = \max(\lambda_i, 0), \quad \lambda_i^- = \min(\lambda_i, 0).$$

Using (8.8.21) and (8.8.23), any linear combination of flux differences can now be written in terms of differences of the dependent variables $\boldsymbol{u}$ as

$$n_x \delta f + n_y \delta g = A^+ \delta u + A^- \delta u. \tag{8.8.24}$$

Based on (8.8.20) and using

$$|\delta F_{i,i+1}| := (A_{i,i+1}^+ - A_{i,i+1}^-)\delta u_{i,i+1}$$

we define an upwind flux (at the right boundary of the control volume) by

$$F_{i+1/2,j} := \tfrac{1}{2}(F_{i,j} + F_{i+1,j} - |\delta F_{i,i+1}|).\tag{8.8.25}$$

Note that we have obtained a special choice for the approximate Riemann solver introduced in (8.8.16) implicitly assuming $u_h^L = u_{i,j}$ and $u_h^R = u_{i+1,j}$ at the right side $S_{i+1/2}$ of the control volume, where the dissipative term $d$ is given by $|\delta F_{i,i+1}|$.

The upwind character of the definition (8.8.25) is clarified by the representations

$$F_{i+1/2,j} = F_{i,j} + A_{i,i+1}^{-}\delta u_{i,i+1}\tag{8.8.26}$$

$$= F_{i+1,j} - A_{i,i+1}^{+}\delta u_{i,i+1}.\tag{8.8.27}$$

If all eigenvalues of $A$ are positive, we have $F_{i+1/2} = F_i$, and if all eigenvalues of $A$ are negative, $F_{i+1/2} = F_{i+1}$.

The other sides of the control volume are treated in the same way. In this way, an overall upwind discretization for a system of PDEs can be defined. If we use (8.8.26) and sum up over all sides of the control volume, we obtain the flux balance

$$\oint_{\partial\Omega_{i,j}} (n_x f + n_y g)\,dS \approx |S_{i+1/2}|A_{i,i+1}^{-}\delta u_{i,i+1} + |S_{i-1/2}|A_{i,i-1}^{-}\delta u_{i,i-1}$$

$$+ |S_{j+1/2}|A_{j,j+1}^{-}\delta u_{j,j+1} + |S_{j-1/2}|A_{j,j-1}^{-}\delta u_{j,j-1}$$

(using $\sum_k |S_k| n_k = 0$).

By computing $A^-$ and $A^+$, the stencil of the discrete Navier–Stokes operator obtained by first-order flux difference splitting can be determined. On a Cartesian grid it has the form shown in Fig. 8.13.

$$\left(\begin{array}{ccc} -\frac{1}{Re}\Delta + 2u\partial_x + v\partial_y & u\partial_y & \partial_x \\[2pt] -\frac{h}{2}(|v|\partial_{yy} + \frac{2u^2+c^2}{\sqrt{u^2+c^2}}\partial_{xx}) & -\frac{h}{2}\frac{uv}{v^2+c^2}(2\sqrt{v^2+c^2}-|v|)\partial_{yy} & -\frac{h}{2}u(\frac{1}{\sqrt{u^2+c^2}}\partial_{xx} + \frac{\sqrt{v^2+c^2}-|v|}{v^2+c^2}\partial_{yy}) \\[20pt] v\partial_x & -\frac{1}{Re}\Delta + u\partial_x + 2v\partial_y & \partial_y \\[2pt] -\frac{h}{2}\frac{uv}{u^2+c^2}(2\sqrt{u^2+c^2}-|u|)\partial_{xx} & -\frac{h}{2}(|u|\partial_{xx} + \frac{2v^2+c^2}{\sqrt{v^2+c^2}}\partial_{yy}) & -\frac{h}{2}v(\frac{1}{\sqrt{v^2+c^2}}\partial_{yy} + \frac{\sqrt{u^2+c^2}-|u|}{u^2+c^2}\partial_{xx}) \\[20pt] c^2\partial_x & c^2\partial_y & \\[2pt] -\frac{h}{2}\frac{c^2 u}{\sqrt{u^2+c^2}}\partial_{xx} & -\frac{h}{2}\frac{c^2 v}{\sqrt{v^2+c^2}}\partial_{yy} & -\frac{h}{2}c^2(\frac{1}{\sqrt{u^2+c^2}}\partial_{xx} + \frac{1}{\sqrt{v^2+c^2}}\partial_{yy}) \end{array}\right)$$

**Figure 8.13.** Linearized discrete operator obtained by flux difference splitting on a Cartesian grid; the subscript $h$ and the overbars in $\bar{u}$ and $\bar{v}$ have been omitted.

In this figure, $\partial_x$, $\partial_y$, $\partial_{xx}$ and $\partial_{yy}$ have been used to indicate standard *central difference* operators, omitting the subscript $h$. Each element in this operator matrix consists of the operator representing the direct discretization of the linearized Navier–Stokes equations by central differences and additional stabilization terms. For clarity, we have written the central differences and the stabilization terms on separate lines. The stencil shows that, when using this first-order flux difference splitting, an artificial viscosity proportional to $h$ appears implicitly in the continuity equation. All other terms also contain artificial stabilization. This is significant for higher Reynolds numbers.

First-order upwind schemes are, in general, not accurate enough and second-order upwind discretizations are required. Second-order upwind-type discretizations can be obtained by defining the states $u_h^L$ and $u_h^R$ as

$$u_h^L = u_{i+1/2,j}^L := u_{i,j} + \frac{1+\kappa}{4}(u_{i+1,j} - u_{i,j}) + \frac{1-\kappa}{4}(u_{i,j} - u_{i-1,j}) \quad \text{and}$$

$$u_h^R = u_{i+1/2,j}^R := u_{i+1,j} + \frac{1+\kappa}{4}(u_{i,j} - u_{i+1,j}) + \frac{1-\kappa}{4}(u_{i+1,j} - u_{i+2,j}),$$

(8.8.28)

respectively, where $-1 \leq \kappa \leq 1$ (compare the $\kappa$-schemes introduced in Section 7.3.2). With this definition, we obtain

$$F_{i+1/2,j} := \tfrac{1}{2}(F(u_h^L) + F(u_h^R) - |\delta F_{L,R}|)$$

(8.8.29)

instead of (8.8.25).

The artificial terms will then be of higher order. A figure, corresponding to Fig. 8.13, for the second-order scheme can be found, for example, in [125]. For incompressible Navier–Stokes problems, it is, in general, not necessary to introduce a limiter in the discretization. Typically, even at high Reynolds numbers, spurious oscillations in the discrete solution do not appear.

### 8.8.5 Flux Difference Splitting and Multigrid: Examples

With respect to the multigrid convergence, LFA results [145] indicate that it is possible to use *collective pointwise and linewise relaxation* methods for the first-order accurate flux splitting discretization. Second-order accurate solutions can again be computed by the defect correction technique or directly using multigrid with the KAPPA smoother as described in Section 7.4.1.

In the following examples, we use the first-order discretization presented above and the second-order discretization based on Fromm's discretization (8.8.28) with $\kappa = 0$.

We will present some typical results of three possibilities (first-order directly, multigrid for the first-order discretization combined with defect correction of the second-order discretization and multigrid with the KAPPA smoother directly for the second-order discretization).

We choose the FAS version of multigrid, with nonlinear line relaxation. In order to keep the discussion as simple as possible, we use only W(1,1) cycles and fix the multigrid components for all problems considered here: we use FW and its transpose (2.3.8) as transfer operators. For the first-order discretization we always employ (collective) alternating symmetric line Gauss–Seidel smoothing. This smoother is also used within the defect correction procedure. The KAPPA smoother is also of alternating symmetric type. The generalization of this smoother to the case of the *system* of incompressible Navier–Stokes equations is straightforward. Only the coefficients of the first-order accurate discretization appear on the left-hand side of the linewise relaxation; the second-order discretization is used in the right-hand side. For all smoothers, we use an underrelaxation parameter of 0.9. In all cases, the calculation starts on the coarsest grid (nested iteration) to obtain a first approximation on the finest grid.

**Example 8.8.4 (analytic test problem)**   As a first example, we use the prescribed smooth solution

$$u = \sin \pi x \, \sin \pi y$$
$$v = \cos \pi x \, \cos \pi y \qquad\qquad (8.8.30)$$
$$p = \sin \pi x + \cos \pi y$$

on $\Omega = (0, 1)^2$, for which we can easily check the accuracy of the flux splitting discretization. The right-hand side of the incompressible system is set accordingly. Note that the right-hand side of the continuity equation is zero. This easy test can be used to evaluate the discretization scheme.

We prescribe Dirichlet boundary conditions for $u$ and $v$, but, not, however, for the pressure for which we use second-order extrapolation at the boundary. Table 8.10 shows the measured accuracy in the Euclidian norm $|| \cdot ||_2$ of the discrete solution with first- and second-order flux difference splitting discretization for $Re = 5000$. The $O(h)$ and $O(h^2)$ accuracy is obvious when comparing the discrete solutions for $h = 1/64$ and $h = 1/128$, respectively.

Table 8.11 presents the corresponding convergence factors. Convergence factors (defect reduction) for 20 iterations are presented in the maximum norm $|| \cdot ||_\infty$. In the defect correction iteration, one multigrid cycle for the $O(h)$ discretization is applied per defect correction step. The defect correction does not converge for $h = 1/16$. Multigrid using the KAPPA smoother is more robust and somewhat faster than the defect correction iteration for this problem.

**Table 8.10.** Measured accuracy for an analytic test problem with flux difference splitting, $Re = 5000$.

| $n = 1/h$ | | 16 | 32 | 64 | 128 |
|---|---|---|---|---|---|
| $O(h)$ discr. | $u$: | 5.7 (−2) | 3.0 (−2) | 1.7 (−2) | 9.3 (−3) |
| | $v$: | 5.5 (−2) | 3.4 (−2) | 2.0 (−2) | 1.0 (−2) |
| | $p$: | 6.7 (−2) | 3.8 (−2) | 2.1 (−2) | 1.1 (−2) |
| $O(h^2)$ discr. | $u$: | 1.9 (−2) | 5.4 (−3) | 1.4 (−3) | 3.6 (−4) |
| | $v$: | 1.1 (−2) | 3.1 (−3) | 8.2 (−4) | 2.0 (−4) |
| | $p$: | 8.7 (−3) | 3.1 (−3) | 9.4 (−4) | 2.6 (−4) |

**Table 8.11.** Measured convergence factors for Example 8.8.4.

| $n = 1/h$ | 16 | 32 | 64 | 128 |
|---|---|---|---|---|
| $O(h)$ discr. | 0.10 | 0.20 | 0.17 | 0.10 |
| $O(h^2)$ with defect correction | Div. | 0.58 | 0.64 | 0.66 |
| $O(h^2)$ with KAPPA smoother | 0.23 | 0.39 | 0.46 | 0.50 |

However, within only a few iterations of the defect correction iteration or of the multigrid cycle using the KAPPA smoother (less than five in both cases), the discretization accuracy from Table 8.10 is already achieved.

Qualitatively the same results with respect to accuracy are obtained at lower Reynolds numbers. The multigrid convergence depends, however, on the Reynolds number. For example, for $Re = 500$ the multigrid convergence factor with the KAPPA smoother on the $128^2$ grid is 0.20, for $Re = 50$ it is 0.46. For $Re = 5$, the convergence is only 0.81. This behavior reflects the fact that the flux difference discretization is oriented to compressible flow discretizations, which corresponds to high values of $Re$. (The unsatisfactory convergence for $Re = 5$ can be improved by a recombination of iterants (see Section 7.8) to 0.57 when using two iterants and to 0.53 when using five iterants.) △

**Example 8.8.5 (driven cavity flow)** As the next test problem, we consider the driven cavity flow problem introduced in Example 8.7.1 at $Re = 5000$, which has a boundary layer. Fig. 8.14 presents some streamlines obtained for the second-order discretization of this test problem.

Fig. 8.15 shows centerline velocity profiles, i.e. velocity $u$ is shown at the line $x = 1/2$, obtained with first- and second-order discretizations and compares them with a reference profile [157]. The need for second-order discretizations is obvious for this problem. Actually, only the second-order discretization approximates the reference profile well.

The fact that we have a boundary layer here can also be seen in Fig. 8.15. We find $u = 1$ at the top boundary $y = 1$, whereas $u \approx 0.4$ at a short distance from the boundary. A similar observation can be made for $y = 0$.

With respect to the multigrid convergence, note that we are dealing with a recirculating flow problem at a relatively high Reynolds number. For such flow problems, we cannot expect good multigrid convergence factors. This is confirmed by the results in Table 8.12. In this case, even if the first-order problems are solved exactly, the defect correction iteration does not converge. The recombination of multigrid iterants leads to a significant improvement of the multigrid convergence with the KAPPA smoother. Using only $\tilde{m} = 2$ improves the convergence factors from 0.74 to 0.51. △

**Example 8.8.6 (block-structured grid)** As an example of a flow problem in a non-Cartesian block-structured grid, we consider the domain sketched in Fig. 8.16. Here, two obstacles (plates) are placed inside the domain (indicated by the bold lines). The domain consists of five blocks (see Fig. 8.16). The grid is shown in the right picture of the same

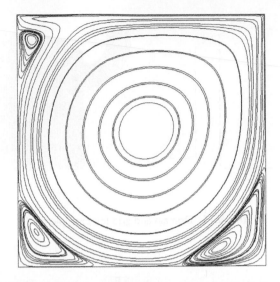

**Figure 8.14.** Streamlines for the driven cavity problem at $Re = 5000$.

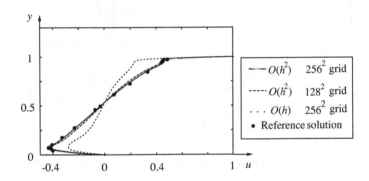

**Figure 8.15.** Centerline velocity profile for the driven cavity problem at $Re = 5000$.

**Table 8.12.** Measured multigrid W(1,1) cycle convergence for the cavity problem at $Re = 5000$, $h = 1/128$.

| | |
|---|---|
| $O(h)$ discr. | 0.59 |
| $O(h^2)$ with KAPPA smoother | 0.74 |
| $O(h^2)$ with KAPPA smoother and iterant recombination $\tilde{m} = 2$ | 0.51 |

figure. Block 1 and Block 5 consist of $16 \times 32$ cells, Block 2 of $8 \times 88$ cells, Block 3 of $32 \times 88$ cells and Block 4 of $16 \times 88$ cells. The inflow boundary is at the upper part of Block 1. A parabolic velocity profile is prescribed there. At the outflow, in Block 5, Neumann boundary-type conditions are set. At all other boundaries and at the obstacles, zero velocities are prescribed.

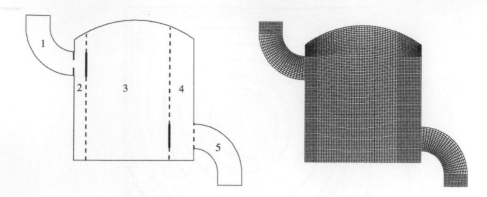

**Figure 8.16.** Domain and grid for Example 8.8.6.

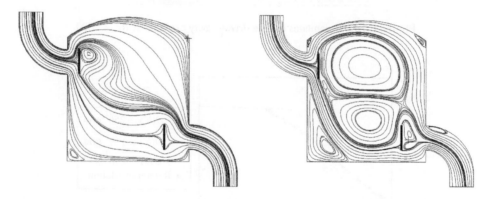

**Figure 8.17.** Streamlines of the flow in Example 8.8.6 at $Re = 100$ (left) and $Re = 1000$ (right).

In this block-structured application, collective line relaxation is performed "blockwise", i.e. lines in the context of the relaxation end at the boundaries of the blocks.

Figure 8.17 shows the streamlines for the laminar flow at $Re = 100$ (left picture) and $Re = 1000$ (right picture) with a second-order accurate discretization. The influence of the Reynolds number can be clearly seen. For $Re = 1000$ larger recirculation zones occur.

The corresponding multigrid convergence with the KAPPA smoother is 0.22 for $Re = 100$ and 0.63 for $Re = 1000$. △

**Example 8.8.7 (a 3D example)**   A well-known channel flow, which is often studied as a test case for 2D discretizations (for example in [215, 390]) is the laminar flow over a backward-facing step. This channel flow is solved in 3D here. The flux splitting discretization can be generalized to 3D in a straightforward way [290]. The flow domain consists of nine rectangular blocks (see Fig. 8.18). The geometry is defined by $L_1 = 50$, $L_2 = 10$, $H = 2$,

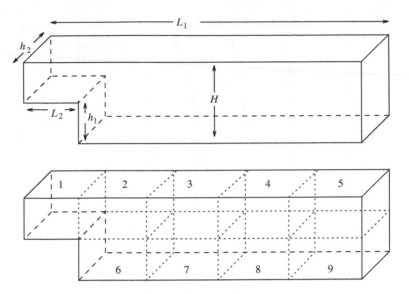

**Figure 8.18.** Domain for flow over a 3D backward-facing step and its division into nine blocks.

$h_1 = 1, h_2 = 1$. At the left boundary, a fully developed 2D velocity profile is prescribed. At the right boundary, Neumann boundary conditions are applied. We consider this problem for $Re = 200, \ 400, \ 600$ and $800$.

Here, we are interested in the length $(x_r)$ of the recirculation zone at the bottom of the channel. This length depends on the Reynolds number.

The shape of the recirculation length along the $y$-axis is shown in Fig. 8.19. In Fig. 8.20 two selected streamlines show the recirculation at the step for $Re = 400$. The recirculating region is clearly visible. The recirculation due to the step results *in a real 3D effect*; the flow direction moves towards the channel centerline. In 3D, recirculation zones are generally not determined by closed characteristics (closed streamlines), which is the case in 2D.

With respect to the multigrid convergence, this example is easy. The recirculation is a harmless one; it is not of convection dominating type. Correspondingly the convergence is fast and defect correction works well enough to obtain the second-order accuracy. The *defect correction convergence* shown in Fig. 8.21. is satisfactory for all Reynolds numbers considered.                                                                                      △

## 8.9 COMPRESSIBLE EULER EQUATIONS

In this section, we will discuss the multigrid solution of the compressible Euler equations. These equations are an important basis for many industrial applications. Physically, they model inviscid compressible flow.

Section 8.9.1 introduces the PDE system and gives a brief survey of some of its properties. In Section 8.9.2, we describe the idea of one particular finite volume discretization,

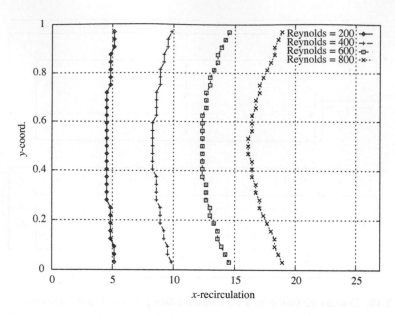

**Figure 8.19.** The shape of the recirculation length along the $y$-axis for different Reynolds numbers.

**Figure 8.20.** 3D flow over a backward-facing step (the different $16^3$ blocks are also visible) at $Re = 400$: two selected streamlines showing the recirculation.

which is based on the so-called Godunov upwind approach with Osher's flux difference splitting for the convective terms. This is one example of many different discretizations that have been used in multigrid algorithms for the compressible Euler equations. A general survey on discretization schemes for compressible flow equations is given in [196].

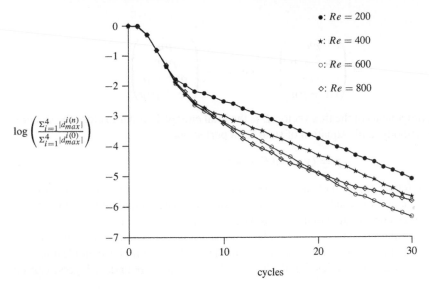

**Figure 8.21.** Nine-block defect correction convergence for 3D backward-facing step flow; grid $164 \times 32 \times 32$ cells.

In Section 8.9.2, we restrict ourselves to the *stationary* Euler equations to find steady-state solutions. The generalization to implicit discretizations of the time-dependent case is then straightforward and can be done as outlined in Section 2.8.2.

In Section 8.9.3, we give some examples for the multigrid treatment of the Euler equations, including an example with shocks.

In Section 8.9.4, we briefly discuss the application of the multistage Runge–Kutta approach, as introduced in Section 7.4.2, to the Euler equations.

We also refer to Appendix C, which gives general guidelines on how to apply multigrid efficiently to CFD problems including the compressible Euler equations.

### 8.9.1 Introduction

In 2D, the time-dependent compressible Euler equations can be written as

$$\frac{\partial \boldsymbol{u}}{\partial t} + N\boldsymbol{u} := \frac{\partial \boldsymbol{u}}{\partial t} + \frac{\partial \boldsymbol{f}}{\partial x} + \frac{\partial \boldsymbol{g}}{\partial y} = 0 \tag{8.9.1}$$

with

$$\boldsymbol{u} = \begin{pmatrix} \rho \\ \rho u \\ \rho v \\ E \end{pmatrix},$$

and where

$$f = \begin{pmatrix} \rho u \\ \rho u^2 + p \\ \rho uv \\ (E + p)u \end{pmatrix} \quad \text{and} \quad g = \begin{pmatrix} \rho v \\ \rho uv \\ \rho v^2 + p \\ (E + p)v \end{pmatrix}$$

are the two components of the flux vector. $E$ is the total energy. In order to close the system, we use the thermodynamic equation of state for a perfect gas

$$p = (\gamma - 1)(E - \tfrac{1}{2}\rho(u^2 + v^2)), \tag{8.9.2}$$

where $\gamma$ is the (constant) ratio of specific heats at constant pressure and constant volume.

The first equation represents the conservation of mass. The second and the third equations represent the conservation of momentum (neglecting viscosity). The vector $u$ is represented in the variables of the conservation laws: mass (or density), momentum and energy per unit volume. The equations are a reasonable model for flows at high Reynolds numbers away from solid wall boundaries, i.e. if viscous effects (such as boundary layers) can be neglected.

The Mach number $M_\infty$ is defined by

$$M_\infty = \frac{|u_\infty|}{c_\infty},$$

where $c_\infty$ is the speed of sound and $u_\infty$ is the velocity of the undisturbed flow ("at infinity"). In the literature, often also $(\rho, u, v, p)$ or $(u, v, c, z)$ are used as the unknown functions, where

$$c = \sqrt{\frac{\gamma p}{\rho}} \tag{8.9.3}$$

is the local speed of sound and

$$z = \ln \frac{p}{\rho^\gamma} \tag{8.9.4}$$

is a measure for the specific entropy.

Essential features of the Euler equations are the following: First, the Euler equations are a *first-order system* (only first derivatives are present). Although the physical assumptions for the compressible Euler equations are quite different compared to the incompressible Navier–Stokes equations (compressible versus incompressible flow, viscid versus inviscid flow), some formal relations between the first three of the compressible Euler equations and the incompressible Navier–Stokes equations are easily recognized (assuming constant density $\rho$).

The compressible Euler equations are an example of a *hyperbolic* system of PDEs. In *nonconservative* (differentiated) form, they read

$$\frac{\partial u}{\partial t} + \frac{\partial f}{\partial u} \frac{\partial u}{\partial x} + \frac{\partial g}{\partial u} \frac{\partial u}{\partial y} = 0. \tag{8.9.5}$$

The Jacobian matrix

$$n_1 A_1 + n_2 A_2 = n_1 \frac{\partial f}{\partial u} + n_2 \frac{\partial g}{\partial u}$$

has *real* eigenvalues for all directions $(n_1, n_2)$. The corresponding eigenvalues are $(n_1 u + n_2 v) + c$, $(n_1 u + n_2 v) - c$ and the double eigenvalue $n_1 u + n_2 v$. The sign of the eigenvalues determines the direction in which the information on the solution is propagated in time along the characteristics governed by $(n_1, n_2)$.

Because of the nonlinear terms, solutions of the Euler equations may develop *discontinuities* like shocks and contact discontinuities, even if the initial flow solution at $t = t_0$ is smooth. Formally, discontinuities are allowed if a weak formulation of the PDE system is used instead of (8.9.1). The weak formulation is known to give nonunique solutions. The physically relevant solution, which is the limit solution of the flow with disappearing viscosity, satisfies the so-called *entropy condition* [229].

Here, we depart from the integral form

$$\frac{\partial}{\partial t} \iint_\Omega u \, dx \, dy + \int_{\partial \Omega} (n_x f + n_y g) \, dS = 0, \qquad (8.9.6)$$

where $\partial \Omega$ is the boundary of $\Omega$ and $(n_x, n_y)$ is the outward normal unit vector at the boundary $\partial \Omega$.

### 8.9.2 Finite Volume Discretization and Appropriate Smoothers

Much progress has been made in the discretization and in the multigrid-based solution for complex compressible flow problems [196]. At first sight, due to the fact that we are dealing with a hyperbolic system, it may be surprising that multigrid can contribute to efficient solution methods. Heuristically, this can be understood since one basically deals with several "equations of convection type". We have seen for the example of the convection–diffusion equation, but also for the incompressible Navier–Stokes equations discretized by flux splitting discretizations, that some $h$-ellipticity is introduced by an appropriate discretization of convective terms.

For the finite volume discretization, the domain $\Omega$ is divided into quadrilaterals $\Omega_{i,j}$. For each quadrilateral, (8.9.1) holds in integral form

$$\oint_{\partial \Omega_{i,j}} (n_x f(u) + n_y g(u)) \, dS = 0 \qquad (8.9.7)$$

for each $(i, j)$, where $n_x$ and $n_y$ are the components of the unit outward normal vector on $\partial \Omega_{i,j}$ and where we assume steady-state flow.

The flux splitting concepts, that we have described in Section 8.8.4 were *originally developed for the Euler equations*. Proceeding as for the incompressible Navier–Stokes equations, the discretization results in (8.8.14), with $f_h$ and $g_h$ as defined in the Euler case. Again, $F_{i+i_0, j+j_0}$ is a suitable approximate Riemann solver, which approximates $(n_x f_h + n_y g_h)_k$ at the corresponding side $S_k$ of $\Omega_{i,j}$. The discretization requires a calculation of the convective flux at each control volume side $S_k$.

Flux difference splitting, which has been described in detail for the discretization of the inviscid terms of the incompressible Navier–Stokes equations, can also be applied here. Of course, the definition of the matrices $A_1$ and $A_2$ in (8.8.19) has to be adapted accordingly (see, for example [124]).

In the examples in this section, we use the Godunov upwind approach [158]. An approximate solution $F_{i+1/2,j}$ of the 1D Riemann problem is obtained by an approximate Riemann solver proposed in [288] in the so-called P-variant [188]:

$$F_{i+1/2,j} = \frac{1}{2}\left( F(u_{i,j}) + F(u_{i+1,j}) - \int_{u_{i,j}}^{u_{i+1,j}} |A(u)|\, du \right). \qquad (8.9.8)$$

Here, $u = (u, v, c, z)^T$ is the state vector (see (8.9.3) and (8.9.4)). $|A(u)|(= A^+(u) - A^-(u))$ is a splitting of the Jacobian matrix $A$ into matrices with positive and negative eigenvalues, $F = n_x f_h + n_y g_h$ is again the flux along the normal vector, and the integral corresponds to a special choice of the dissipative term $d$ in (8.8.16). This approximation is first-order accurate. Details on the discretization and on boundary conditions are given in [145, 188, 220, 372].

With respect to the multigrid solution, the FAS is commonly used for the discrete Euler equations. Since the unknowns occur in all equations, one can, in principle, use well-known collective relaxation methods. The equations are of "convection type", so, as in the flux splitting for the incompressible Navier–Stokes equations, Gauss–Seidel-type relaxations can very efficiently be used *for the first-order discretization*. Typical relaxation schemes for this nonlinear system for $u_h^{m+1}$ are the lexicographic collective Gauss–Seidel (CGS-LEX) point or line smoothers [188, 220, 274, 371]. For the (approximate) solution of the nonlinear system, one local Newton linearization per grid point or grid line is usually sufficient.

Symmetric CGS-LEX (forward followed by backward ordering) or even four-direction CGS-LEX and the corresponding alternating symmetric line smoothers are robust smoothing variants.

Second-order accurate discretizations can be obtained by the use of van Leer's $\kappa$-scheme [233]. The vectors $u_{i,j}$ and $u_{i+1,j}$ in (8.9.8) are replaced accordingly. TVD schemes (as described in Section 7.3.2) need to be employed if the solution has discontinuities.

As discussed for the convection–diffusion equation, the CGS-type smoothers are well-suited for the first-order discretization, but not for second-order schemes. For second-order discretizations, there are again three possibilities: first, we can combine a multigrid algorithm described above for the first-order scheme with defect correction (as described in Section 5.4.1), which has been done, for example, in [124, 188, 220, 371]. Another possibility is to use the KAPPA smoothers as introduced in Section 7.4.1, which have also been applied successfully to the Euler equations in [293]. The third possibility is to use multistage smoothers (see Section 8.9.4). This last option is actually most commonly used in practice when solving the Euler equations.

### 8.9.3 Some Examples

Here, we will present two examples of multigrid as described above for subsonic and transonic compressible flow.

**Example 8.9.1 (KAPPA smoothers for transonic flow)**   We consider a transonic Euler problem at Mach number $M_\infty = 0.85$ in a channel with a small circular bump; the height of the channel is 2.1, its length is 5 and the bump length is 1. For this transonic channel flow problem, we measured the multigrid convergence for the first- and the second-order discretization on three grids, a $24 \times 16$, a $48 \times 32$ and a $96 \times 64$ grid. The grids are moderately stretched (see Fig. 8.22). The pressure distribution is presented in Fig. 8.23. Obviously, the solution of this problem has a shock starting near the end of the bump. Correspondingly, we use van Leer's $\kappa$-scheme as described above together with the van Leer limiter (7.3.4) in order to avoid oscillations that may appear near shocks.

The V(2,1) cycle convergence for the first-order discretization with alternating line Gauss–Seidel smoothing is very fast: it is about 0.1 on all three grids.

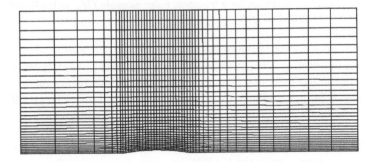

**Figure 8.22.**  The $48 \times 32$ grid in a channel with a bump.

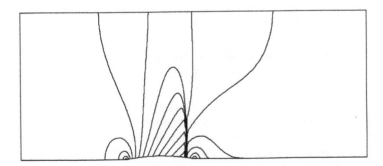

**Figure 8.23.**  The pressure distribution for the transonic test $M_\infty = 0.85$ on a $96 \times 64$-grid.

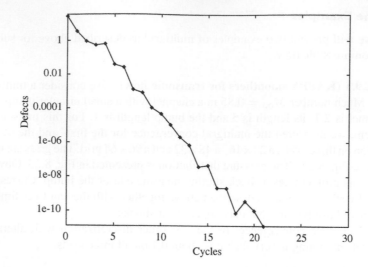

**Figure 8.24.** Multigrid convergence for a transonic Euler example ($M_\infty = 0.85$) with the KAPPA smoother on a $96 \times 64$ stretched grid.

The second-order discretization of the Euler equations is solved directly with V(2,1)-cycles using the (alternating symmetric line) KAPPA smoother ($\omega = 0.7$). With the KAPPA smoother, we find defect reduction factors of 0.3–0.4. The convergence on the finest grid is presented in Fig. 8.24. It is similar to the convergence obtained for scalar problems with the van Leer limiter (7.3.4). A single-grid version, however, does not lead to fast convergence in this case.

Although the multigrid convergence for the first-order discretization is excellent and the multigrid convergence with the KAPPA smoother is very satisfactory, the algebraic convergence of the defect correction iteration is slow.                                △

In the following example, we will see that for the defect correction approach, the convergence of relevant physical quantities such as the drag or the lift coefficient may be fast even if the algebraic convergence is slow. The "differential convergence" is faster than the algebraic convergence indicated by the defects.

**Example 8.9.2 (flow around an airfoil)**   A classical test case for 2D compressible Euler discretizations is the flow around a NACA0012 airfoil. Figure 8.25 shows a part of the $128 \times 24$ grid used in the computations. A well-known case is the Euler flow at $M_\infty = 0.63$ with a (flow) angle of attack $\alpha = 2°$. The corresponding flow is *subsonic*. Shocks are not present in the solution. The resulting pressure distribution (isobars) near the airfoil are presented in Fig. 8.26(a). The pressure at the airfoil is usually described by the pressure coefficient $c_P$. It is presented in Fig. 8.26(b). For a multigrid Euler solver, this is an easy test. The measured multigrid convergence with the KAPPA smoother is 0.28 for an F(1,1) cycle.

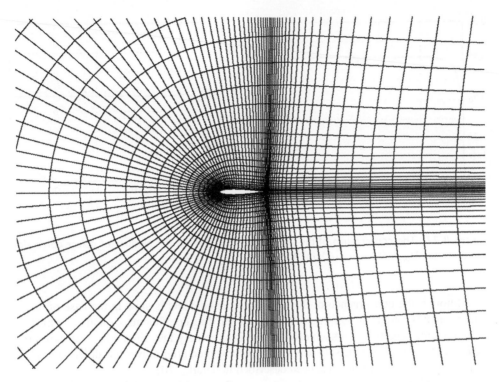

**Figure 8.25.** Part of the computational grid for the flow around an airfoil.

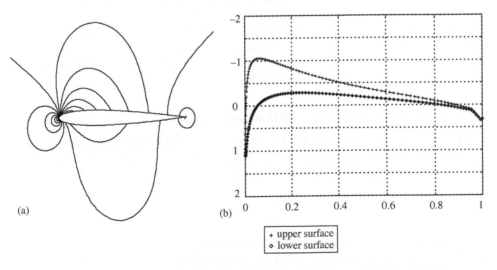

(a)

(b)

| + upper surface |
| ∘ lower surface |

**Figure 8.26.** Subsonic Euler flow around a NACA0012 airfoil, $M_\infty = 0.63$, $\beta = 2°$. (a) pressure distribution; (b) pressure coefficient, $c_P$.

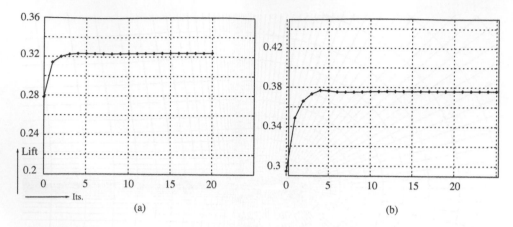

**Figure 8.27.** The convergence of the lift coefficient $c_L$ versus the number of defect correction iterations. (a) $M_\infty = 0.63$, $\alpha = 2°$; (b) $M_\infty = 0.85$, $\alpha = 1°$.

Although the convergence of the (second-order) defects is slow in the case of defect correction, it is found that convergence of the lift and drag coefficients, $c_L$ and $c_D$, which are the interesting quantities for such calculations is extremely fast. Figure 8.27(a) shows that the *lift* coefficient $c_L$ for the subsonic case has converged after only a few defect correction iterations. This phenomenon has been described in detail [220].

A second classical test case is the transonic flow at $M_\infty = 0.85$ with angle of attack $\alpha = 1°$. In this case two shocks appear in the flow solution; one at each side of the airfoil (see both parts of Fig. 8.28). The multigrid convergence with the van Leer limiter remains satisfactory. The F(1,1) cycle with the KAPPA smoother converges with an average reduction factor of 0.59. However, in this case, an underrelaxation parameter $\omega = 0.5$ needs to be used for convergence. The defect correction iteration also shows a good convergence, for example, of the lift coefficient (see Fig. 8.27(b)).                                    △

### 8.9.4 Multistage Smoothers in CFD Applications

In computational fluid dynamics (CFD), when complicated steady-state problems are to be solved, the following approach is often used. Instead of solving the discrete problem $N_h u_h = 0$ directly, a time variable is introduced and

$$\frac{\partial u_h(t)}{\partial t} = -N_h u_h \tag{8.9.9}$$

is considered, a system of ordinary differential equations (ODEs). An advantage of using this time-dependent formulation is that the corresponding initial value problem is "well posed", irrespective of the particular type of flow considered, sub- or supersonic, invicid or viscous. This initial value problem can be solved by a suitable ODE solver.

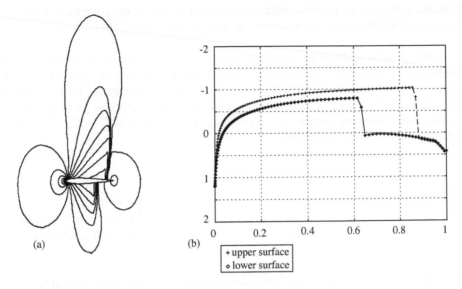

**Figure 8.28.** Transonic Euler flow around a NACA0012 airfoil, $M_\infty = 0.85$, $\beta = 1°$. (a) Pressure distribution; (b) pressure coefficient, $c_P$.

This approach is the starting point for the *multistage Runge–Kutta smoothers*. Certain time-integration schemes like the classical Runge–Kutta methods are slowly convergent (because of the *stiffness of the ODE system*) but have reasonable *smoothing properties* with respect to the original steady-state problem $N_h u_h = 0$. In fact, the multistage Runge–Kutta smoothers turn out to be equivalent to the multistage smoothers introduced in Section 7.4.2. (The respective parameters $\beta_j$ in (7.4.3) can consequently be interpreted as Runge–Kutta coefficients scaled by the time step $\tau$ and the mesh size $h$.)

This idea is the basis of the work and software development of Jameson et al. [202, 203]. Jameson successfully solved complicated 3D inviscid and viscous flow problems around aircraft [204] using four- and five-stage Runge–Kutta schemes for smoothing. This success for inviscid flow problems and the simplicity of the multistage approach has motivated many groups in industry to choose this approach for compressible flow problems.

Multistage smoothers can be applied directly to second-order upwind discretizations of the Euler equations. Their smoothing properties depend on the choice of the coefficients $\beta_j$ (see Section 7.4.2).

**Remark 8.9.1** Using LFA smoothing analysis, it is possible to find optimal smoothing parameter sets for different equations (replacing the classical Runge–Kutta parameters). Since these smoothers are often used for solving inviscid CFD equations, a simple reference equation, for which the optimal multistage parameters are calculated, is the limit case $\varepsilon \to 0$ of the convection–diffusion equation (see, for example, [96])

$$au_x + bu_y = 0 \quad (\Omega).$$

Often, the multistage parameters obtained are then also used for more complicated problems. For systems of PDEs, they are, however, not suitable in all cases [234].          ≫

A very popular multistage smoother for central discretization schemes with artificial viscosity (also called artificial dissipation) is the Jameson–Schmidt–Turkel scheme [202]. A variety of modifications in the form of *preconditioners* has been proposed for the multistage smoother from [202] in order to make it more efficient. An overview is given in [205].

The idea of these modifications is to introduce a preconditioner $C = C_h$ [234, 385] so that the system (8.9.9) is replaced by

$$\frac{\partial \tilde{u}_h(t)}{\partial t} + C_h N_h u_h = 0. \tag{8.9.10}$$

The preconditioning can, for instance, be based on collective Jacobi iteration [4, 126, 127, 318]. This leads to a clustering of eigenvalues, for which the optimal coefficients are not significantly different from those obtained for the convection equation [385] (which then can also be applied).

**Remark 8.9.2**   The multiple semicoarsening approach has been pioneered in [275, 276] for the Euler equations. The idea is to make the multigrid method with pointwise smoothers more robust, in particular in the case that the flow is aligned with the grid. The combination of multistage smoothing, preconditioning and semicoarsening has been shown to work well for the 2D Euler equations [118].          ≫

### 8.9.5 Towards Compressible Navier–Stokes Equations

In this section, we will survey multigrid for the compressible Navier–Stokes equations instead of giving a detailed discussion. These equations describe viscous and heat-conducting fluids. In contrast to the compressible Euler equations, they take, in particular, all the shear stresses into account.

Generally speaking, efficient multigrid based solution methods for the compressible Navier–Stokes equations remain a topic of current research. A survey of multigrid approaches for this PDE system can be found in [416]. In Section 10.5, we will outline the difficulties that arise when dealing with the compressible Navier–Stokes equations in an industrial aerodynamic design environment.

A typical boundary condition for the compressible Navier–Stokes equations at solid walls is the "no flow" boundary condition $u = v = 0$. As a result, *sharp boundary layers appear in the physical flow solutions which is an essential difference compared to the compressible Euler equations. In order to resolve such thin boundary layers, highly stretched cells (with aspect ratios of up to $10^4$) need to be employed.*

As usual, the anisotropies resulting from such grids can be dealt with by line relaxations and/or semicoarsening approaches. For compressible Navier–Stokes applications, semicoarsening techniques are starting to be accepted. A semicoarsening variant for the Navier–Stokes equations is presented in [307]. Smoothing results for semicoarsening are given in [4, 303, 305].

Due to these anisotropies, the "straightforward" application of multistage smoothers does not lead to efficient multigrid solution methods in this case, in contrast to the case of the Euler equations.

However, the multistage scheme can be made more suitable for compressible Navier–Stokes by using special preconditioners $C_h$ which take care of such anisotropies [304, 404]. For example, one may choose $C_h$ corresponding to collective line-Jacobi iteration, with lines chosen perpendicular to the boundary layer.

Significant improvements in the efficiency can be obtained by combining the multistage smoother with Jacobi preconditioning and semicoarsening.

Approaches, that also work on unstructured meshes, are decribed in [257–260, 271, 272]. The grid coarsening is done algebraically in that case (AMG-like, see Appendix A). A coarse grid operator for the compressible Navier–Stokes equations is then constructed by a Galerkin coarse grid approximation.

Turbulence modeling brings additional difficulties, that are typical for reactive flows as well. Some references for this topic are [15, 128, 156, 238, 245, 353].

Appendix C gives some general guidelines on how to apply multigrid to CFD problems including the compressible Navier–Stokes equations and the modeling of turbulence.

# 9

# ADAPTIVE MULTIGRID

Adaptivity is an important means to improve the efficiency of numerical methods for solving PDEs. In an adaptive method, the grid and/or the discretization (type, order) are *adapted* to the behavior of the solution in order to solve a given problem more efficiently and/or more accurately.

In this book, we restrict ourselves mainly to *adaptive grid refinement*. In principle, we distinguish two types of adaptive approaches:

- *predefined* (*also called static*) *refinement*: predefined refinement refers to grid structures where the refinement is determined in advance, i.e. before the solution process is started.
- *self-adaptive* (*also called dynamic*) *refinement*: in self-adaptive approaches, the grid refinements are carried out dynamically during the solution process, controlled by some appropriate adaptation criteria.

In practice, the approaches may, of course, be combined.

Most of our description of adaptive multigrid is independent of whether the grid is refined in a predefined way or self-adaptively and *applies to both static and dynamic* refinement.

The actual choice of the refinement criterion that has to be provided in the dynamic case depends on the application, more concretely, on the objective of the grid adaptation. One may be interested in balancing the global discretization error, in minimizing some error functionally or in improving the error locally. In our examples, we will mainly aim at balancing the global discretization error or at minimizing it in some norm (see [308] for a more general treatment of adaptivity objectives).

Self-adaptive refinements are particularly useful if the solution shows a certain singular behavior which is not (exactly) known in advance but detected during the solution process. The singularity may be a non-smooth behavior of the solution or of its derivatives, a rapid change like a boundary or interior layer, a shock, some turbulent behavior etc. In many cases, these singularities occur locally so that a *local* adaptation of the grid (and/or of the discretization) is appropriate. In addition to the local effects of singularities, typically a global impact on the discretization accuracy is also observed. This global effect, also called *pollution*, is an important phenomenon in the context of adaptivity.

There are different reasons for singular behavior of the solution. Peculiarities of the shape of the domain (like reentrant corners or other singularities), special features of the PDE data (discontinuous or nonsmooth coefficients, jumps or other singularities in the right-hand side or in the boundary conditions), but also more intrinsic features like changes of the PDE type (from elliptic to hyperbolic), hyperbolic nonlinearities leading to shocks etc.

To be more specific, we will give an example of adaptive refinements for singularities, which are caused by the shape of the domain, especially due to reentrant corners (see Section 9.1).

In this book, we are interested in the *combination of adaptivity and multigrid*. Actually, adaptive grid refinement turns out to be fully compatible with the multigrid structure and philosophy. Realizing this idea, Brandt [58, 59] has introduced the so-called multilevel adaptive technique (MLAT). The fast adaptive composite grid method (FAC) [252, 262], can be regarded as an alternative to the MLAT approach. In our view, MLAT and FAC are closely related and we will make use of both approaches.

Typically, adaptive grid refinement techniques start with a global (possibly coarse) grid covering the whole computational domain (or with a hierarchy of such global grids). Finer grids are then introduced locally only in parts of the domain in order to improve the accuracy of the solution. Starting with a representative example, we will present some notation for the global, the local and "composite" grids in Section 9.1. In Section 9.2, we will introduce the idea of adaptive multigrid and formulate the problem on the composite grid, i.e. the problem which is actually solved in the adaptive approach. In Section 9.3, we will discuss appropriate *conservative* discretizations at the boundaries of the local refinement areas.

There, we start with a consideration, showing how the conservative discretization at the interfaces can easily be dealt with in the adaptive multigrid algorithm. The main objective of this section is to describe an adaptive multigrid cycle in detail, assuming, for simplicity, a *static* (i.e. predefined) hierarchy of grids. Provided some automatic refinement criterion is given, this procedure immediately can be used in a dynamic (self-adaptive) way if combined with adaptive FMG.

Such a criterion has to answer the important question, where and how the refinement is to be constructed in the dynamic case. We will discuss refinement criteria in Section 9.4.

In Section 9.5, we will consider the parallelization of adaptive methods. Adaptivity and parallelism are numerically important principles, which, however, partly conflict with each other. Nevertheless, we will see in Section 9.6 that very satisfactory results are obtained (for systems of PDEs).

## 9.1 A SIMPLE EXAMPLE AND SOME NOTATION

### 9.1.1 A Simple Example

We will take a look again at one of the examples with a reentrant corner discussed in detail in Section 5.5. Here, we consider the classical example of Poisson's equation in an L-shaped domain $\Omega$:

$$
\begin{aligned}
-\Delta u &= f^{\Omega} \quad (\Omega = (-1, 1)^2 \backslash \{(0, 1) \times (-1, 0)\}) \\
u &= f^{\Gamma} \quad (\Gamma = \partial\Omega).
\end{aligned}
\tag{9.1.1}
$$

As we know from Section 5.5, the problem has, in general, a singularity at (0,0). The typical singular behavior is reflected by the function $u_s(r, \phi) = r^{2/3} \sin 2\phi/3$, where $(r, \phi)$ denote polar coordinates. We assume smooth data $f^{\Omega}$ and $f^{\Gamma}$ such that the solution $u$ of the problem (9.1.1) can be written as

$$u = \tilde{u} + \text{const } u_s,$$

where $\tilde{u}$ is a smooth function.

In order to obtain an impression of the potential of local refinement, we consider (9.1.1) with the specific solution $u_s$ and discretize this problem with the standard five-point stencil $\Delta_h$. Table 9.1 gives the maximum errors $\|u - u_h\|_{\infty}$ obtained on a uniform grid and on a locally refined grid, as illustrated in the left and the right pictures of Fig. 9.1, respectively. For $h = 1/128$, i.e. on the uniform grid with about 50 000 grid points, the maximum error $\|u-u_h\|_{\infty}$ is about $3.3 \times 10^{-3}$. On an appropriate locally refined grid, however, it is possible to obtain a similar accuracy with less than 700 grid points and thus with significantly less computational work.

**Table 9.1.** Comparison of the number of grid points and of the errors for a uniform and an adaptive grid for Problem 9.1.1.

| Grid | Uniform | Locally refined (adaptive) |
| --- | --- | --- |
| Grid points | 49665 | 657 |
| $\|u - u_h\|_{\infty}$ | 3.3(−3) | 3.8(−3) |

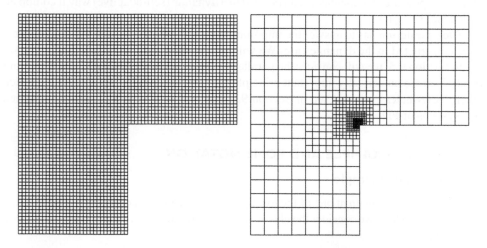

**Figure 9.1.** Uniform grid (left) and adaptive (composite) grid (right) for Problem (9.1.1).

In this simple example the full information about the behavior of $u$ close to the singularity is available (but was not used). *Nevertheless, the example shows the large potential of adaptive grids.* If the required accuracy is higher than in this example, the benefits of local refinements are even more spectacular. The same is true for 3D problems. More realistic examples will be treated in Section 9.6.

**Remark 9.1.1 (second-order accuracy)**   As we have discussed in Section 5.5, the accuracy in the above example is no longer $O(h^2)$ for the standard five-point discretization on the uniform grid. At a fixed distance from the singularity, the accuracy behaves like $O(h^{4/3})$. Near the singularity it is even worse, namely $O(h^{2/3})$. So, the singularity in the reentrant corner *leads to an increase of the discretization error in all of* $\Omega$. This is a typical example of *pollution*. (Theoretically, the pollution effect can be analyzed using Green's function of the underlying problem.)

It may be interesting to know whether the grid can be structured in such a way that an overall "second-order accuracy" is obtained by adaptation and how this can be achieved. In other words: how does the adaptive grid structure have to be arranged in order to improve the accuracy of the discrete solution by a factor of four if the mesh size of the global grid is divided by two?

This question has been studied systematically in [208], where a general formula giving "optimal" global and local refinement grid sizes for problems with reentrant corners was given. For the L-shaped domain, for instance, the result illustrated in Fig. 9.2 can be derived from that formula. Given a locally refined grid with global mesh size $h$ (the left picture in Fig. 9.2), we achieve a reduction of the overall error $u - u_h$ by a factor of four, under natural

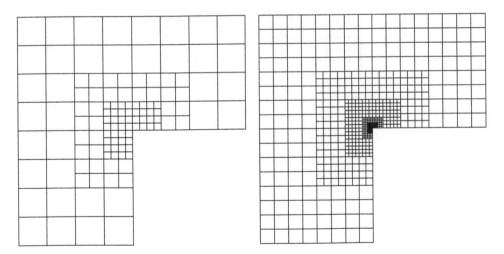

**Figure 9.2.** Adaptive refinement structures, giving an error reduction by a factor of four for Problem 9.1.1.

assumptions on $u$, by the following procedure. The mesh sizes of the entire locally refined grid are divided by 2 ($h \to h/2$ etc.), but, in addition *two* more refinement steps need to be added to the overall refinement structure (see right picture in Fig. 9.2).      ≫

### 9.1.2 Hierarchy of Grids

In nonadaptive multigrid (on uniform grids), we have used a hierarchy of global grids $\Omega_0, \Omega_1, \ldots, \Omega_\ell$. In the adaptive approach, we maintain the hierarchy of global grids, but we also introduce a hierarchy of increasingly finer grids restricted to smaller and smaller subdomains. *Adaptive* multigrid differs from standard multigrid only in the sense that we consider a hierarchy of grids

$$\Omega_0, \ldots, \Omega_\ell, \Omega_{\ell+1}, \ldots, \Omega_{\ell+\ell^*}, \tag{9.1.2}$$

where $\Omega_0, \ldots, \Omega_\ell$ are global, whereas $\Omega_{\ell+1}, \ldots, \Omega_{\ell+\ell^*}$ are local grids.

**Example 9.1.1**   A typical hierarchy of grids for the solution of the L-shaped problem is shown in Fig. 9.3. In this case, $\Omega_0$ and $\Omega_1$ are global grids ($\ell = 1$) and the refined grids $\Omega_{1+1}, \Omega_{1+2}$ ($\ell^* = 2$) cover increasingly smaller subdomains as indicated in Fig. 9.4.      △

We will call the boundaries of the locally refined domains in the interior of the domain $\Omega$ *interfaces* of the locally refined domains in the following and denote them by $\partial I_{\ell+1}, \ldots, \partial I_{\ell+\ell^*}$ (see Fig. 9.4).

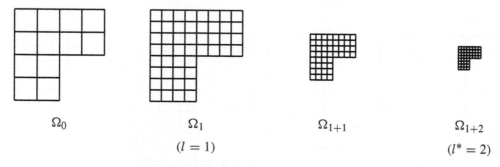

$\Omega_0$                $\Omega_1$                $\Omega_{1+1}$                $\Omega_{1+2}$

$(l = 1)$                            $(l^* = 2)$

**Figure 9.3.** Hierarchy of grids.

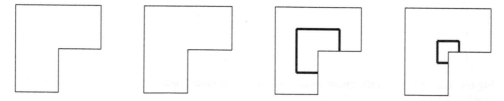

**Figure 9.4.** Subdomains and interfaces $\partial I_{\ell+k}$ corresponding to the hierarchy of grids in Fig. 9.3.

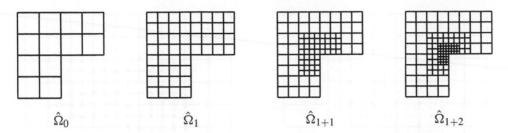

**Figure 9.5.** Composite grids corresponding to the hierarchy of grids in Fig. 9.3.

In addition to the global and the local grids, we consider their "composition". The corresponding sequence of *composite grids* (see Fig. 9.5) is defined by

$$\hat{\Omega}_k := \Omega_k \quad (k = 0, \ldots, \ell) \quad \text{and} \quad \hat{\Omega}_{\ell+k} := \Omega_\ell \cup \bigcup_{j=1}^{k} \Omega_{\ell+j} \quad (k = 1, \ldots, \ell^*).$$

$$(9.1.3)$$

## 9.2 THE IDEA OF ADAPTIVE MULTIGRID

Using the notation from the previous section, we are able to give a simple description of adaptive multigrid.

We will start with the two-grid case (for cycle and FMG) Section 9.2.1. In Section 9.2.2, we discuss the generalization of the two-grid cycle to multigrid. In these two sections, we assume (for simplicity and without loss of generality) that the locally refined grids are defined a priori.

Section 9.2.3 describes, how FMG can be applied in the context of *dynamic* local refinements.

### 9.2.1 The Two-grid Case

In this section, we start with the two-grid case of one global and one locally refined grid $\Omega_1$ and $\Omega_{1+1}$ as sketched in Fig. 9.3. If we know how to apply adaptive multigrid (FMG and the two-grid cycle) to this situation, the recursive generalization to more grid levels is straightforward. We will thus briefly describe an adaptive two-grid algorithm.

Assume that we have computed an approximation $u_1$ on the (global) coarse grid. As usual in FMG, we can interpolate this approximation to $\Omega_{1+1}$, for example by bicubic interpolation. The fact that $\Omega_{1+1}$ covers only a subgrid of $\Omega_1$ is not a problem at all and simply means that the interpolation is restricted to those parts of the computational domain where $\Omega_{1+1}$ is defined.

On $\Omega_{1+1}$, we can now perform a two-grid cycle in a straightforward way. First, standard smoothing schemes like GS-LEX, GS-RB or line smoothers can be applied on $\Omega_{1+1}$ without any trouble, keeping the approximations of $u_{1+1}$ at the interface points constant (i.e. these values are interpreted as "Dirichlet values").

For the coarse grid correction, we choose the FAS.

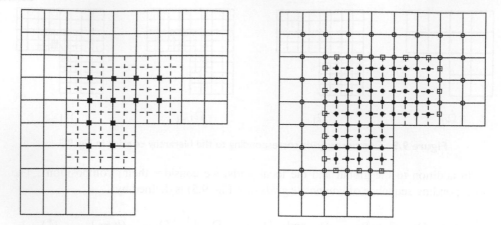

**Figure 9.6.** Two levels $\Omega_l$ and $\Omega_{l+1}$ of a locally refined grid. Left: coarse grid points inside the domain of local refinement; right: points on the composite grid $\hat{\Omega}_{l+1}$.

**Remark 9.2.1** It would not make much sense to use the multigrid *correction scheme* since in those parts of the domain, which are not covered by the finer grid $\Omega_{l+1}$, on $\Omega_l$, one has to work with the full solution and not just with a correction.                                           $\gg$

Of course, the FAS is applied only at those coarse grid points belonging to $\Omega_{l+1}$ which are not interface points. These coarse grid points are marked by ∎ in Fig. 9.6. The resulting problem on the coarse grid $\Omega_l$ is then defined by $L_l w_l = f_l$ with

$$
f_l := \begin{cases} I_{l+1}^l f_{l+1} + \tau_{l+1}^l & \text{at the points marked by ∎ in Fig. 9.6} \\ f^\Omega & \text{on the remaining part of } \Omega_l, \end{cases}
\tag{9.2.1}
$$

where $\tau_{l+1}^l$ is the $\tau$-correction as defined in Section 5.3.7 (applied to the current approximation).

When a solution $w_l$ (or a new approximation) of the coarse grid problem on $\Omega_l$ has been computed, the coarse grid corrections can be interpolated to $\Omega_{l+1}$ as usual, e.g. by bilinear interpolation, including the interface points (marked by □ and ∘). After the coarse grid correction, we can again smooth the current approximation on $\Omega_{l+1}$.

Of course, this two-grid cycle can be applied iteratively. What we have described so far, is essentially the two-grid case of the MLAT introduced in [58, 59].

**Remark 9.2.2 (the discrete problem solved by the two-grid algorithm)** We have described the above two-grid method as a generalization of the two-grid case on globally defined grids. We have, however, not defined the discrete problem which is to be solved if we iterate the two-grid cycle until convergence.

Obviously, we compute a solution on the composite grid $\hat{\Omega}_{1+1}$ though this composite grid is not addressed explicitly in the above description of the algorithm. This discrete solution is characterized

- by the discrete equations $L_{1+1}u_{1+1} = f_{1+1}$ at the points marked by • in Fig. 9.6 (i.e. on the locally refined grid $\Omega_{1+1}$ without interface points),
- by the discrete equations $L_1 u_1 = f_1$ at the points marked by ○ in Fig. 9.6 (i.e. at the interface points which also belong to the coarse grid and at those points of $\Omega_1$, which do not belong to $\Omega_{1+1}$) and
- by implicitly defined relations at the points marked by □ (i.e. at the interface points which do not belong to the coarse grid). These points are also called *hanging nodes*. In fact, the relations at the hanging nodes are not directly available but only defined by the above two-grid algorithm. In our description, a first approximation has been obtained by a cubic interpolation in FMG and, in each two-grid cycle, corrections from coarse grid points have been added using linear interpolation.

If we want to postulate that *cubic interpolation is to be fulfilled at the hanging nodes*, we have to modify the above two-grid algorithm. The easiest way is then to also use bicubic interpolation for the coarse grid correction at the hanging nodes.

In general, a "conservative interpolation" is recommended, which corresponds to a conservative discretization (see Section 9.3).                                      ≫

### 9.2.2 From Two Grids to Multigrid

The generalization from the two-grid cycle to multigrid by recursion is straightforward (see Fig. 9.7 for an illustration of adaptive multigrid V-cycles). MLAT works on the hierarchy of grids as presented in Fig. 9.3, *not* on the hierarchy of composite grids (Fig. 9.5).

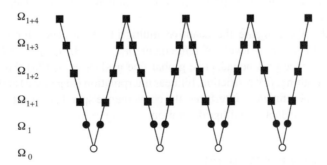

**Figure 9.7.** Cycles in the multilevel adaptive technique: ■, smoothing on locally refined grid levels; •, smoothing on global grid levels; ○, solution on coarsest grid.

Actually, MLAT can simply be regarded as an *adaptive generalization of FAS*, in the following sense.

---

- The sequence of the computation with respect to the grid levels is not affected, i.e. the grid levels are run through as in the nonrefined case.
- On level $\ell + k$, smoothing takes place in the *refined* parts of the domain $\Omega$, i.e. at all points of $\Omega_{\ell+k}$.
- In the interior parts of the locally refined grids, interpolation and restriction can be carried out as usual.
- At the interfaces of the locally refined regions suitable interpolations and restrictions of the solution have to be chosen. Whenever a coarse grid correction is interpolated or a restriction of the current approximation is performed in the cycle, then corresponding operations are carried out at the interfaces.

---

On each grid level $\Omega_k$ with $k = 0, \ldots, \ell + \ell^*$, we can apply any reasonable *smoothing procedure* to the discrete equations $L_k u_k = f_k$, where $f_k$ is defined by exactly the same formula as in (9.2.1), the index 1 being replaced by $k$ and the index $1 + 1$ being replaced by $k + 1$, i.e.

$$
f_k := \begin{cases} I_{k+1}^k f_{k+1} + \tau_{k+1}^k & \text{at those points where } \tau_{k+1}^k \text{ can be computed} \\ & \hspace{2em} \text{(marked by } \square \text{ in Fig. 9.6)} \\ f^\Omega & \text{on the remaining part of } \Omega_k. \end{cases} \tag{9.2.2}
$$

The unknowns at the interfaces $\partial I_k$ are *not* included in the smoothing iteration on the locally refined grid $\Omega_k$. On $\Omega_k$, they are treated as Dirichlet points. Their values are only updated by the coarse grid correction.

At the hanging nodes (marked by $\square$ in Fig. 9.6), we have the same options as in the two-grid case.

Summarizing, the idea of adaptive multigrid is a straightforward generalization of multigrid on uniform grids. The main difference is the treatment of the interface points.

**Example 9.2.1**  We compare the adaptive multigrid convergence for the example in Section 9.1.1 with $\ell^* = 4$ levels of refinement ($h_\ell = 1/8$, $h_{\ell+4} = 1/128$) with the multigrid convergence on a nonadaptive global fine grid ($h = 1/128$). For F(1,1) cycles with GS-RB smoothing, FW restriction, bilinear interpolation (in general) and bicubic interpolation at the local refinement interfaces, the convergence speed is almost identical for the adaptive and the nonadaptive multigrid iteration (see Fig. 9.8).                    $\triangle$

### 9.2.3 Self-adaptive Full Multigrid

So far, the discussion of adaptive multigrid has been essentially based on locally refined grids, which were defined a priori. In an advanced multigrid philosophy, however, the grid generation, the discretization and the solution process should not be seen as three separate

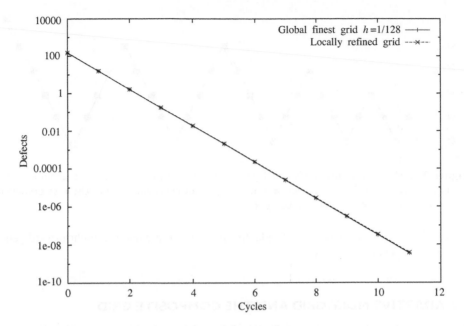

**Figure 9.8.** Multigrid convergence on a globally refined grid ($h = 1/128$) and on the locally refined grid, $h_0 = 1/8$ with 4 levels of refinement ($||\cdot||_\infty$).

processes but as closely connected. A close interaction between the grid, the discretization and the multigrid solution is realized in *self-adaptive FMG*.

Although the FMG approach can, of course, be applied to predefined locally refined grid structures, it is more naturally combined with the self-adaptive technique. FMG on adaptive grids is sketched in Fig. 9.9 and proceeds as follows:

(1) Perform FMG on a series of global grids until an approximation on the finest global level has been computed with satisfactory accuracy.
(2) Based on some appropriate criteria (see Section 9.4.1), determine whether the current grid should be locally refined. If yes, generate a locally refined grid in the subregions flagged for refinement. If not, stop.
(3) Interpolate the current approximation to the next finer grid level by some suitable FMG interpolation.
(4) Perform $r$ adaptive multigrid cycles on this level.
(5) Continue with Step (2) if the required accuracy is not yet achieved. Otherwise, the algorithm stops.

This procedure should terminate automatically if proper criteria for adaptive refinement are used. To be on the safe side, one may additionally prescribe a maximum number of grid levels.

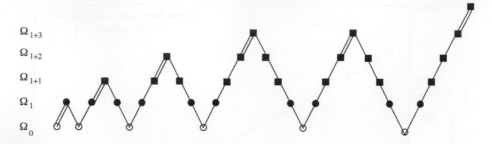

**Figure 9.9.** FMG on adaptive grids: starting on a hierarchy of global coarse grids, additional levels are introduced. //, FMG interpolation; ■, smoothing on locally refined grid levels; •, smoothing on global grid levels; ○, solution on coarsest grid.

Obviously, the basic idea of self-adaptive FMG is simple and a straightforward generalization of nonadaptive FMG.

## 9.3 ADAPTIVE MULTIGRID AND THE COMPOSITE GRID

Adaptive multigrid as sketched in Section 9.2 works on the hierarchy of locally refined grids as shown in Fig. 9.3. Of course, the problem which is finally solved, *depends on the discretization on the finest composite grid* (see Fig. 9.5) and may depend on algorithmic details of MLAT at the interface points.

McCormick *et al.* [252, 262] have proposed the so-called FAC method and outlined the importance of conservative discretization at the interfaces of local refinement areas. The main difference between MLAT and FAC is the composite grid orientation of FAC. Whereas the composite grid is only implicitly constructed in MLAT, FAC addresses the composite grid explicitly. Once the sequence of local refinement regions is defined and constructed, FAC can be regarded as a multigrid algorithm that works on the corresponding composite grid.

At first sight, FAC may look conceptually different from MLAT. In concrete MLAT and FAC algorithms, however, both approaches turn out to be closely related. Relations between MLAT and FAC have been studied for Poisson's equation in [235] with respect to parallel computing.

In this section, we follow the FAC approach in that we start with the formulation of the discrete problem on the composite grid. In Section 9.3.1, we describe a *conservative discretization* approach for points at the interfaces which can be considered as an example for an FAC discretization on the composite grid. In Section 9.3.2, we show that one way to obtain a conservative discretization at the interface points is to introduce ghost points near the interfaces and perform a so-called conservative interpolation at these points. In Section 9.3.3, we give a formal description of an adaptive multigrid cycle based on this technique. Here, we remain with the MLAT philosophy and *work on the hierarchy of locally refined grids*. In Section 9.3.4, we make some brief comments on some related approaches.

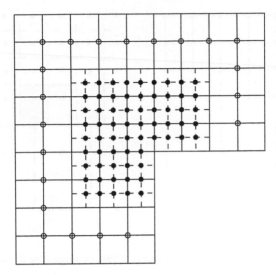

**Figure 9.10.** Regular points (marked by • and ○) on the composite grid.

In order to formulate the discrete problem on the composite grid, we assume operators $L_k$ ($k = \ell, \ldots, \ell + \ell^*$) and, for simplicity, a predefined local refinement. For convenience, we assume Dirichlet boundary conditions so that we do not have to bother about boundary points.

The discrete equations on a composite grid $\hat{\Omega}_{\ell+\ell^*}$ are now defined in a straightforward way at all *regular* points, i.e. at grid points which are not interface points. Here, we have the natural discrete equations

$$L_k u_k = f_k (:= f|_{\Omega_k}) \quad (k = \ell, \ell + 1, \ldots, \ell + \ell^*), \tag{9.3.1}$$

at those points of $\Omega_k$, which do not belong to a finer grid $\Omega_{k+1}$. The situation is illustrated by Fig. 9.10. At the points marked by •, a "regular" discretization with grid size $h_k$ is used. At the points marked by ○, we apply the "regular" discretization based on the mesh size $h_{k-1} = 2h_k$. For the interface points, more specific considerations are needed (see the next section).

### 9.3.1 Conservative Discretization at Interfaces of Refinement Areas

We have mentioned in Section 8.8.4 that conservative discretizations are important for many fluid dynamics problems. Conservation is then guaranteed independently of the grid (global or locally refined) under consideration. A general approach to define discrete equations is the finite volume discretization, resulting in conservative discretizations on uniform but also on locally refined grids. In particular, at the *interfaces* the finite volume discretization can be defined such that the "composite grid discretization" is conservative.

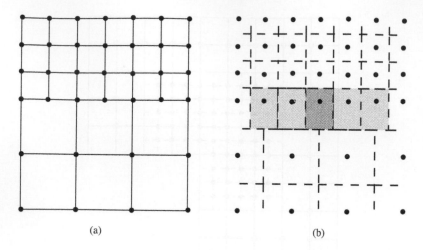

**Figure 9.11.** (a) Part of a composite grid; (b) corresponding finite volumes.

In this section, we present one possibility for a conservative finite volume discretization on a composite Cartesian grid. As a reference example, we use the 2D problem

$$-\nabla \cdot (a\nabla u) = f^{\Omega} \quad (\Omega)$$

with some boundary conditions on $\partial\Omega$ which we do not specify in detail here. Figure 9.11(a) shows a part of a composite grid on which the problem is discretized. Figure 9.11(b) shows the finite volumes around the grid points.

The domain is covered by *square* finite volumes, except for the grid points at the local refinement interface. Here, the volumes are enlarged, so that they are adjacent to the coarse volumes. For the square volumes, we apply the usual conservative finite volume discretization as described in Section 5.7. We consider the discretization for the enlarged nonsquare (shaded) volumes. The situation is shown in more detail in Fig. 9.12.

For the dark-shaded volume, *the fluxes* $F = a\nabla u$ are used at three faces, ($w$) west, ($e$) east and ($n$) north,

$$F_{h,w}\left(x - \frac{h}{2}, y\right) = a\left(x - \frac{h}{2}, y\right)[u_h(x, y) - u_h(x - h, y)]/h \tag{9.3.2}$$

$$F_{h,e}\left(x + \frac{h}{2}, y\right) = a\left(x + \frac{h}{2}, y\right)[u_h(x + h, y) - u_h(x, y)]/h \tag{9.3.3}$$

$$F_{h,n}\left(x, y + \frac{h}{2}\right) = a\left(x, y + \frac{h}{2}\right)[u_h(x, y + h) - u_h(x, y)]/h. \tag{9.3.4}$$

We assume that the values of the coefficients $a$ are known. In order to satisfy the conservation of flux at the local refinement boundary, the flux $F_{h,s}$ at the southern face is determined by linear interpolation

$$F_{h,s} = \tfrac{1}{2}(F_{H,l} + F_{H,r}) \tag{9.3.5}$$

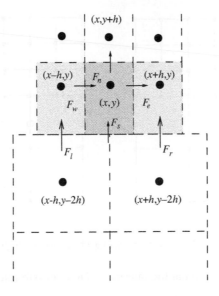

**Figure 9.12.** Part of the composite grid in detail, with the dark-shaded volume $\Omega_{xy}$ around $(x, y)$.

where $F_{H,l}$ and $F_{H,r}$ are the fluxes of the coarse (square) volumes (see Fig. 9.12). They are determined by

$$F_{H,l}(x - h, y - h) = \frac{a(x - h, y - h)(u_h(x - h, y) - u_h(x - h, y - 2h))}{2h} \quad (9.3.6)$$

$$F_{H,r}(x + h, y - h) = \frac{a(x + h, y - h)(u_h(x + h, y) - u_h(x + h, y - 2h))}{2h}. \quad (9.3.7)$$

The discrete equation at the point $(x, y)$ in the shaded volume $\Omega_{x,y}$ is given by

$$\frac{3h}{2}\left(F_{h,e}\left(x + \frac{h}{2}, y\right) - F_{h,w}\left(x - \frac{h}{2}, y\right)\right) + h\left(F_{h,n}\left(x, y + \frac{h}{2}\right) - F_{h,s}\left(x, y - \frac{h}{2}\right)\right)$$
$$= f(x, y)|\Omega_{x,y}|,$$

where the left-hand side of this flux balance equation is the *net flux* of the shaded volume $\Omega_{x,y}$. The resulting discretization can be generalized to 3D without any difficulties. This is a simple example of a conservative discretization on a composite grid since fluxes are preserved across the interface of a locally refined region.

### 9.3.2 Conservative Interpolation

We will discuss a simple procedure, by which a conservative discretization can easily be treated in the adaptive multigrid context. For this purpose, we introduce a so-called *conservative interpolation*, which can be interpreted as a connection between interpolation

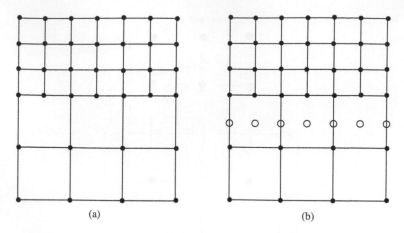

**Figure 9.13.** (a) Part of a composite grid; (b) grid with ghost points (○).

and conservative discretization at the interfaces. Other possibilities are outlined at the end of this section.

Since all the refinement levels are treated separately in adaptive multigrid, it is useful to rewrite the composite discretization in such a way that only grid points in the refined region are addressed when smoothing is carried out on the fine grid. For that purpose, we reformulate the composite conservative discretization in a setting with "Dirichlet-type local refinement ghost points" [412] (see Fig. 9.13).

With such ghost points, adaptive multigrid can proceed in a straightforward way. Standard smoothing techniques can be used at all grid points in the refined region including the interfaces. The ghost points are the points at which the values need to be determined by interpolation. They serve as Dirichlet boundary points of the locally refined grid. This situation is illustrated in Fig. 9.14. The equivalent conservative discretization based on the composite grid extended with these ghost points is obtained as follows. The net flux in the shaded square volume $\hat{\Omega}_{x,y}$ in Fig. 9.14(b) is determined by

$$\mathrm{Flux}(\hat{\Omega}_{x,y}) = h(F_{h,e} - F_{h,w}) + h(F_{h,n} - G_s)$$

with $G_s$ depicted in Fig. 9.14(b). The fluxes $F_{h,w}$, $F_{h,e}$ and $F_{h,n}$, are the same as in (9.3.2)–(9.3.4). $G_s$ is defined by

$$G_s = a(x, y - h/2)[u_h(x, y) - u_h(x, y - h)]/h \qquad (9.3.8)$$

where $u_h(x, y - h)$ is a Dirichlet value on the local refinement boundary. In order to regain the composite grid conservative discretization in the situation of Figs 9.14(a) and 9.14(b), the two net fluxes in $\Omega_{x,y}$ and $\hat{\Omega}_{x,y}$ need to be equal:

$$\frac{\mathrm{Flux}(\Omega_{x,y})}{3h^2/2} = \frac{\mathrm{Flux}(\hat{\Omega}_{x,y})}{h^2}. \qquad (9.3.9)$$

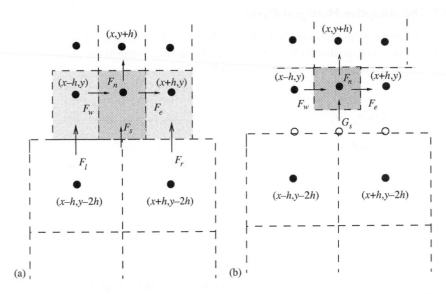

**Figure 9.14.** Part of the composite grid (a) composite grid finite volumes with $\Omega_{xy}$; (b) finite volumes with $\Omega_{xy}$ and $\circ$, ghost points.

From (9.3.9), we obtain an interpolation formula for the flux $G_s$:

$$G_s = \frac{F_{h,n} + 2F_{h,s}}{3}.$$

Now, it is possible to compute the boundary value $u_h$ at the ghost point $(x, y-h)$ from (9.3.8), namely

$$u_h(x, y - h) = u_h(x, y) - h\frac{G_s}{a(x, y - h/2)}. \qquad (9.3.10)$$

With constant coefficients $a$, the interpolation (9.3.10) coincides with the well-known quadratic interpolation. Hence, the order of the interpolation ($O(h^3)$) is lower than that of cubic interpolation ($O(h^4)$), but it is a conservative interpolation. Numerical experiments [412] confirm that this aspect is often more important than the order of the interpolation with respect to accuracy.

During a smoothing iteration on a locally refined grid, the unknowns from which the ghost points are interpolated are updated. In our multigrid algorithm after this update, before a restriction, the ghost points are updated again, i.e. an additional interpolation of the unknowns at the ghost points is carried out.

Other interpolations at interior locally refined boundary points can be found in [39, 390].

**Remark 9.3.1** Instead of conservative interpolation schemes one can also use proper $\tau_{k+1}^k$-*corrections* at those interface points which also belong to the coarse grid in order to enforce conservation ("conservative fine to coarse transfer" [14]). One has to analyze which corrections are missing, when cubic interpolation is used, in order to make the discretization conservative. These quantities are then added as a $\tau_{k+1}^k$-correction to the discrete equations.

### 9.3.3 The Adaptive Multigrid Cycle

Using the above definitions on all levels $k = 0, 1, \ldots, \ell, \ell + 1, \ldots, \ell + \ell^*$, an adaptive multigrid cycle as described in Section 9.3.2 (including an update at the ghost points within each smoothing step), is summarized in the following box.

---

**Adaptive cycle**   $u_k^{m+1} = \text{ADAPCYC}(k, \gamma, u_k^m, u_{k-1}^m, L_k, f_k, \nu_1, \nu_2)$

(1) Presmoothing
   – Compute $\bar{u}_k^m$ by applying $\nu_1 \, (\geq 0)$ smoothing steps to $u_k^m$ on $\Omega_k$.

(2) Coarse-grid correction
   – Compute the defect on $\Omega_k$        $\bar{d}_k^m = f_k - L_k \bar{u}_k^m$.
   – Restrict the defect from $\Omega_k$      $\bar{d}_{k-1}^m = I_k^{k-1} \bar{d}_k^m$.
   – Compute

$$\bar{u}_{k-1}^m = \begin{cases} \hat{I}_k^{k-1} \bar{u}_k^m & \text{on } \Omega_{k-1} \cap \Omega_k, \\ u_{k-1}^m & \text{on the remaining part of } \Omega_{k-1}. \end{cases}$$

   – Compute the right-hand side $f_{k-1}$ according to (9.2.2).
   – Compute an approximate solution $\hat{w}_{k-1}^m$ of the coarse grid equation on $\Omega_{k-1}$

$$L_{k-1} w_{k-1}^m = f_{k-1}. \tag{9.3.11}$$

   If $k = 1$ employ a direct or fast iterative solver for this purpose.
   If $k > 1$ solve (9.3.11) by performing $\gamma \, (\geq 1)$ adaptive multigrid $k$-grid cycles to (9.3.11) using $\bar{u}_{k-1}^m$ as initial approximation

$$\hat{w}_{k-1}^m = \text{ADAPCYC}^\gamma (k - 1, \gamma, \bar{u}_{k-1}^m, u_{k-2}^m, L_{k-1}, f_{k-1}, \nu_1, \nu_2).$$

   – Compute the correction at $\Omega_{k-1} \cap \Omega_k$      $\hat{v}_{k-1}^m = \hat{w}_{k-1}^m - \bar{u}_{k-1}^m$.
   – Set the solution at the other points of $\Omega_{k-1}$   $u_{k-1}^{m+1} = \hat{w}_{k-1}^m$.
   – Interpolate the correction to $\Omega_k$      $\hat{v}_k^m = I_{k-1}^k \hat{v}_{k-1}^m$.
   – Compute the corrected approximation on $\Omega_k$

$$u_k^{m, \text{ after CGC}} = \bar{u}_k^m + \hat{v}_k^m.$$

   – Carry out a conservative interpolation
     at the ghost points.

(3) Postsmoothing
   – Compute $u_k^{m+1}$ by applying $\nu_2 \, (\geq 0)$ smoothing steps to $u_k^{m, \text{ after CGC}}$ on $\Omega_k$.

---

Note that, in general, the current approximation on the composite grid $\hat{\Omega}_k$ is required for the computation of a new approximation $u_k^{m+1}$. However, for ease of presentation, we have included only the approximations on two levels ($u_k^m$ on $\Omega_k$ and $u_{k-1}^m$ on $\Omega_{k-1}$) in the argument list of ADAPCYC. (The adaptive cycle implicitly updates the approximation not only on the finest grid, but also on the entire composite grid.)

**Remark 9.3.2** For *time-dependent* applications, it is natural to use local refinements which follow certain time-dependent features in the solution, i.e. a subdomain which is refined in some time step may not be locally refined in a later one [38, 206, 392]. Furthermore, the extension of adaptive multigrid to discretizations on *staggered* grids is straightforward (see [390] for an example).                                                          ≫

### 9.3.4 Further Approaches

In addition to MLAT and FAC several related, but formally different, approaches for adaptivity have been proposed, in particular in the context of finite elements [18, 55, 425]. In the finite element context, local refinement techniques (which may be based on criteria similar to those discussed below) will typically lead to a discrete problem defined on the composite grid, for each level of refinement.

Finally, we would like to mention that additive versions of adaptive multigrid have also been proposed and are in use. A comparison of additive and multiplicative adaptive multigrid is given in [32] (in a finite element framework). One essential distinction in this context refers to the basic smoothing pattern. If the smoothing is applied only to unknowns corresponding to the "new" locally refined points ($\Omega_{k+1} \setminus \Omega_k$) in the refined regions, the respective method corresponds to Yserentant's hierarchical basis method (HB) [432, 434] in the additive case and to the so-called hierarchical basis multigrid method (HBMG) [18] in the multiplicative case. These approaches do *not* lead to level independent convergence, but to $O(N \log N)$ complexity instead. Only if the smoothing procedure is applied to all grid points in a refined region $\Omega_{k+1}$ as in MLAT, can level independent convergence be achieved. The corresponding additive version is the adaptive BPX method [55]. These methods are surveyed in [32].

**Remark 9.3.3 (AFAC)** The AFAC (asynchronous fast adaptive composite grid method) has been proposed [262] for parallel computing. It can be regarded as an additive version of FAC. Comparisons of AFAC with parallel versions of MLAT can be found in [235, 323].

## 9.4 REFINEMENT CRITERIA AND OPTIMAL GRIDS

In this section, we briefly discuss where and how locally refined grids are set up. Since we are interested in a fully self-adaptive technique, *refinement criteria* have to be provided, on the basis of which the algorithm can work and, in particular, also terminates *automatically*. A lot of research and experimental work deals with the definition of efficient and reliable refinement criteria. Although many interesting results are available [25, 92, 137, 308], the field of refinement strategies is far from being settled. Since different singular phenomena occur in different situations (geometric singularities in elliptic equations versus hyperbolic shocks, scalar equations versus systems, for instance), it is difficult to find refinement criteria which are reliable and efficient in all cases.

The general requirement is to use an indicator for the determination of refinement regions, which is based on calculated quantities. Usually, the indicator is oriented to an (a posteriori) estimate of the (local or global) discretization error.

With respect to adaptive multigrid, any refinement criterion can, in principle, be chosen. Most of our presentation is, however, based on the so-called $\tau$-criterion, which we regard as a *natural indicator in the context of multigrid*. This criterion is based on the $\tau_h^H$ quantity introduced in Section 5.3.7.

Other error estimators and error functionals are natural in the finite element context. A survey on such criteria is presented in [308]. These techniques can also be applied in the finite difference and finite volume context (see Remark 9.4.3 below).

A different class of criteria is based on the behavior of the solution itself, e.g. on physical features such as flow gradients in CFD applications. This type of criteria is somewhat problematic if the refinement strategy is solely based on it.

Before we return to the $\tau$-criterion and give examples, we would like to make the following trivial, but fundamental remark on local refinements.

**Remark 9.4.1** Consider an elliptic problem in $\Omega$ to which we want to apply a trivial local refinement technique around a point $P_0 \in \Omega$. If we repeat the local refinement process recursively, we will obtain a series of locally refined grids

$$\Omega_{\ell+k}; \quad k = 0, 1, 2, \ldots$$

covering smaller and smaller subdomains of $\Omega$ and correspondingly a local truncation error $\tau_k = L_k u - Lu$ at $P_0$ that tends to 0 for $k \to \infty$:

$$\tau_k(P_0) \to 0 \quad (k \to \infty).$$

This does, of course, *not* mean that the real error $u_k(P_0) - u(P_0)$ also tends to zero for $k \to \infty$. According to the elliptic nature of the problem, the error of $u_k(P_0)$, in general, depends on the discretization error at *all other* grid points of the composite grid. Therefore, even for an infinite number of purely local refinement steps, the accuracy that can be achieved is, in general, limited. $\gg$

## 9.4.1 Refinement Criteria

In this section, we discuss first the $\tau$-criterion (in the context of standard coarsening). As we know from Section 5.3.7, $\tau_h^{2h}$ is the quantity that is to be added to the right-hand side of the discrete problem on the coarse grid $\Omega_{2h}$ in order to obtain the accuracy of the fine grid (up to interpolation). Thus, $\tau_h^{2h}$ is a measure of the extent to which the local introduction of the grid $\Omega_h$ has influenced the global solution. Also, $\tau_h^{2h}$ is an approximation to $\tau_{2h} - \tau_h$

where $\tau_h$ and $\tau_{2h}$ are the respective truncation errors

$$\tau_h = L_h u - Lu, \qquad \tau_{2h} = L_{2h} u - Lu.$$

Heuristically, if $\tau_{2h} - \tau_h$ (and thus also $\tau_{2h}$) is large at some grid point, the discretizations on $\Omega_h$ and $\Omega_{2h}$ differ significantly, which indicates that an even finer grid should be used near this grid point. If $\tau_{2h}$ is small, the benefit of a finer grid will be much smaller.

*A useful local refinement criterion based on these quantities is to compare $h^d \tau_h^{2h}$, with a given tolerance $\varepsilon$, where $d$ denotes the dimension of the application.* Here, $\tau_h^{2h}$ is evaluated using the current approximation on the finest grid.

Note that any $h$-dependent scaling in the discrete equations has to be taken into account. In finite volume discretizations, for example, the discrete equations (and thus also $\tau_h^{2h}$) have been implicitly multiplied by $h^d$ already. In this case, the corresponding "$\tau_h^{2h}$" has then to be compared with the given tolerance $\varepsilon$.

This error indicator and variants are often used for scalar equations but also for systems of equations. (See, for example, [206] for an application in process simulation, [266] and many others for Euler applications or [270] for an application in semiconductor device simulation.)

In the simplest approach, the grid is refined locally if (and where) the estimator indicates values above $\varepsilon$. In practice, the local refinement areas then have to be determined (i.e., extended) in such a way that the refined regions are *consistent with the overall grid structure*.

For instance, if block-structured grids are used, this means that the refined grid should also have an appropriate block structure. Since this structural question is important for *parallel* adaptive methods, we will return to it in Section 9.5 and in the examples (see Section 9.6).

**Example 9.4.1** Figure 9.15 shows a refined grid for the convection–diffusion problem with a boundary layer (at the right and upper boundaries) as discussed in Example 7.1.2. The refinement regions have been found with the $\tau$-criterion. It can be seen that the boundary layers are well detected, and that the locally refined grid contains many fewer grid points than a corresponding globally refined grid. △

**Example 9.4.2 (nonlinear problem with a shock)** We consider the nonlinear convection-dominated conservation law (see also [359])

$$-\varepsilon \Delta u + \left(\frac{u^2}{2}\right)_x + u_y = 0 \qquad (9.4.1)$$

with $\varepsilon = 10^{-6}$ on $\Omega = \{(x, y); 0 \le x \le 3, 0 \le y \le 2\}$ and the boundary conditions

$$u = u_0 = \tfrac{1}{2}(\sin(\pi x) + 1) \qquad (9.4.2)$$

along the $x$-axis. For the other boundary conditions which are derived from the solution of the reduced equation $(u^2/2)_x + u_y = 0$, see [359]. The exact solution is constant along the characteristic lines $(u, 1)^T$. This example is related to Burger's equation.

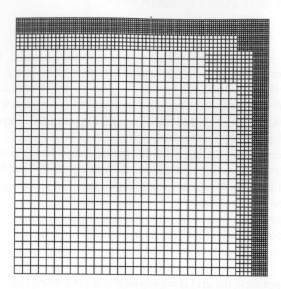

**Figure 9.15.** Locally refined grid near the boundary layer for a convection–diffusion problem.

The solution obtained by a conservative finite volume discretization is shown in Fig. 9.16. It contains a shock wave along the line $y = 2x - 2$. Limiters are necessary for an accurate solution of this problem. Here, we have applied the van Leer limiter (7.3.4). Figure 9.16 shows two levels of grid refinement near the shock detected by the $\tau$-criterion for adaptive refinement.                                                                △

So far, we have considered only grid refinement based on (2D) standard coarsening, i.e. grids which are refined by a factor of two in each direction. The two examples above represent cases, however, in which more efficient refinement structures are feasible. The following remark refers to more sophisticated refinements.

**Remark 9.4.2 (anisotropic refinements)**   In certain cases, in particular when boundary layers occur, a *1D local refinement* (corresponding to semicoarsening) may be sufficient. Compared to the grid shown in Fig. 9.15, such a 1D refinement (parallel to the upper and right boundaries) would give a similar accuracy with substantially fewer grid points.

1D refinements are also possible in more general situations. In Example 9.4.2, a "1D local refinement" relative to *local coordinates fitting to the shock position* would be sufficient to resolve the shock. Such possibilities have been considered in [66].          ≫

The following three remarks refer to other refinement criteria. In Remark 9.4.3, we describe a finite element-based criterion which we will use in one application at the end of this chapter. The other two remarks refer to the global discretization error and to the behavior of the solution, respectively.

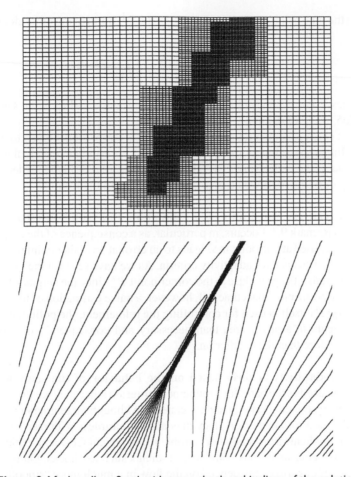

**Figure 9.16.** Locally refined grid near a shock and isolines of the solution.

**Remark 9.4.3 (FE residual)**  In the context of finite element discretizations, the finite element residual is a natural error indicator [308]. But for finite difference or finite volume discretizations, the finite element residual is also an option (proposed in [365] for the Euler equations). In that case, the discrete approximation $u_h$, which is defined only on the respective grid points of $\Omega_h$, is extended to the whole domain $\Omega$ by some suitable interpolation giving $\tilde{u}_h$. In the case of a PDE system $\boldsymbol{Lu} = \boldsymbol{f}$ (consisting of $q$ PDEs), the finite element residual

$$\tilde{\boldsymbol{r}}_h := \boldsymbol{L}\tilde{\boldsymbol{u}}_h - \boldsymbol{f}$$

can be calculated and evaluated.

In the examples below, we have used the finite element residual based on linear shape functions for the first-order Euler equations discretized by finite volumes in the following way: the rectangular control volume around a point $(x_i, y_j)$ is subdivided into two triangular elements $\Delta_1$ and $\Delta_2$. The values at the four corners of the two triangles can be interpolated linearly from the nine neighbor nodes. At $(x_i, y_j)$, the $L^1$-norm of the finite element residual is

$$\psi_{L^1}(x_i, y_j) = \sum_{\kappa=1,2} \sum_{k=1,...,q} \int_{\Delta_\kappa} |\tilde{r}_h^k| \, d\Omega, \qquad (9.4.3)$$

where the index $\kappa$ corresponds to the number of the triangle and $r_h^k$ is the $k$th component of $\boldsymbol{r}_h$ (see [365, 323] for details).

The grid at $(x_i, y_j)$ is then refined if $\psi_{L^1}(x_i, y_j)$ is larger than a given tolerance $\varepsilon$.

Note that linear shape functions are, in general, not appropriate for calculating the finite element residual if the underlying PDE (system) is of second or higher order. In that case, the principal part of the PDE would give no contribution to the residual.        $\gg$

**Remark 9.4.4 (criteria based on $u_h - u$ estimators and pollution effects)** Pollution as mentioned in Remark 9.1.1 occurs, in particular, as a consequence of the elliptic nature of a PDE. An error caused at one point in $\Omega$ leads to an error at any other point of $\Omega$. This effect has an obvious significance for adaptivity criteria. As we have seen for the example (9.1.1), it is sufficient to refine the grid close to the corner point $P^*$ since the singularity in $P^*$ is the reason for the discretization inaccuracy. If, however, a naive estimation of the discretization error $u(P) - u_h(P)$ itself is used for controlling the refinement, the pollution effect would lead to refinements in a larger area than necessary and consequently to a less efficient refinement procedure than the $\tau$-criterion.        $\gg$

**Remark 9.4.5 (adaptivity criteria based on the behavior of the solution)** Physical features are often used as the basis for a refinement criterion since they are directly available. However, the corresponding criteria are not general. Although large gradients of a solution may be an indication for large errors, they may also be harmless with respect to the errors. (Constant large gradients are usually not a problem.) In typical CFD applications, however, the largest errors occur in regions with large gradients such as shocks. On the other hand, the *location of a shock wave* may depend on the flow in a much more "smooth region" and this region should be detected by the refinement criterion.

Sometimes different sensors like shock, entropy and density sensors are applied simultaneously in order to identify crucial flow regions which require a finer resolution [193].
       $\gg$

In practice, one may well think of combining some of the criteria discussed above. For example, it may be useful to refine the grid in regions in which a particular nonsmooth behavior of the solution is detected and additionally in regions in which the $\tau$-tolerance is violated. Furthermore, if some a priori information is available about the solution (about the approximate position of shocks, singularities and the like), this information can be used in the construction of the global grid.

### 9.4.2 Optimal Grids and Computational Effort

The refinement criteria discussed above may, in principle, lead to an automatic refinement of the grids and to algorithms which no longer have "$O(N)$ complexity".

**Remark 9.4.6**  We illustrate the question of $O(N)$ complexity here for the example in Section 9.1. Assume that for this problem *on each refinement level* the size of the refinement region is approximately 1/4 of the size of the previously refined subregion (as illustrated in Fig. 9.3). In that case, the number of grid points of the refined regions is approximately the same *for all levels*. If, for $h \to 0$, this process of refinement is continued in the same way, we will have $\#\Omega_{\ell+k} \doteq \#\Omega_\ell$ ($k = 0, 1, 2, \ldots$), where $\doteq$ means equality up to lower order terms. For the number $N = N(k)$ of grid points of the corresponding composite grid $\hat{\Omega}_{\ell+k}$, we obtain $\#\hat{\Omega}_{\ell+k} \doteq (1 + 3k/4)\#\Omega_\ell = kO(1)$. Obviously, the above assumption that the refined region has 1/4 the size of the previously defined region is a limit case. If this ratio is larger than 1/4, $N$ will grow exponentially with $k$; if it is smaller than 1/4, $N$ will be bounded. In the first case, one may speak of a *generous* refinement, in the second case of a *contracting* refinement.                                                       ≫

If $N$ is the number of grid points of the composite grid, the adaptive FMG based on V-cycles leads to an $O(N)$ algorithm for generous refinements, whereas the complexity $O(N)$ is no longer maintained for contracting refinements. On the other hand, the algorithm with contracting refinement will be cheaper (less overall operations) than the "$O(N)$ optimal" algorithm with generous refinement. Trivially, the *nonadaptive* FMG algorithm is always $O(N)$ optimal (according to the considerations and under the assumptions in Section 3.2.2) whereas the adaptive FMG algorithm may not be (for contracting refinement).

Such a paradoxical phenomenon, that a "nonoptimal" algorithm may be much faster and more efficient than an "optimal" one, can be found in many other situations, in particular, if the term of optimality is oriented only to the algebraic convergence and if the *size/order* of the discretization error is not taken into account.

Looking at the *differential convergence* rather than looking at the algebraic convergence of adaptive multigrid, is the basis of the so-called λ-FMG strategy proposed in [66].

**Remark 9.4.7 (λ-FMG)**   The idea of the λ-FMG strategy is to define the grid refinement strategy within FMG in such a way that an optimal accuracy can be achieved *within a given amount of work*. Some numerical experiments based on the λ-FMG strategy have been made [320] for the Poisson equation in regions with reentrant corners. Here, the numerical efficiency was additionally improved *by the use of local relaxations* (see also Section 5.5). The construction of optimal grids is also an option in the context of the so-called "dual method" [308].                                                       ≫

## 9.5 PARALLEL ADAPTIVE MULTIGRID

### 9.5.1 Parallelization Aspects

If the adaptive multigrid techniques as described in Sections 9.2 and 9.3.3 are to be applied on a parallel computer with distributed memory, we face a specific difficulty. As we do not know in advance where the local refinements will occur, load balancing becomes crucial. Simplifying the situation to some extent, we have, in principle, two choices.

(1) We can use the grid partitioning approach as described in Section 6.3 in order to maintain the boundary-volume effect in communication. Clearly, we have to be aware

that some subdomains will be refined (by the refinement criteria used) whereas others will not. This results in a *load imbalance* which will, in general, be unacceptable.

(2) In order to avoid this load imbalance, one has to suitably redistribute the subgrids on each refined level. Of course, this means, that *volumes of data* must be redistributed on each level of local refinement leading to a possibly high amount of communication. *The boundary-volume effect will no longer be valid.* Instead we have to expect that the communication complexity is of the same order as the computing complexity.

This clearly shows that there is a *conflict between ideal parallelism and ideal adaptivity.*

> However, we will see below (see also [323]) that this conflict is not so crucial in practice, at least not in real-life problems. Whereas the first choice with its immanent load imbalance turns out to be unacceptable in many cases, the second choice is usually acceptable.

Since the different multigrid levels are treated sequentially in adaptive multigrid, a straightforward distribution strategy is to map the locally refined grid on each level uniformly to as many processors as possible. Again, the efficiency of this distribution strategy depends on the problem, the size of the refined grid regions and on the hardware properties. The *grid partitioning* employed here for locally refined grids is analogous to the strategy of parallelizing nonadaptive multigrid. The well-known deficiencies—the increase of the communication/computation ratio (for very small or rapidly contracting refined grids) and, eventually, fewer points than processors—not only apply to the global coarse levels but to the locally refined ones as well.

**Remark 9.5.1** One can also consider changing the adaptive features of the algorithm in order to establish a good compromise between the parallel efficiency requirements and the numerically ideal adaptivity. We are not going to discuss such modifications in detail. ≫

In the following, we consider *block-structured grids*. In block-structured grid applications, the number of blocks may be much lower than the number of available processors, $P$. Consequently, on a parallel machine, large blocks are further subdivided in order to provide the basis for a good *load balancing* which typically means that each "subblock" contains roughly the same number of grid points.

**Remark 9.5.2** The combination of adaptivity, multigrid and parallelism has been investigated for unstructured grids [31] including the additive version of adaptive multigrid. Depending on the application, additive multigrid may be of some interest in the case of locally refined grids, since smoothing at different levels can be performed simultaneously. Convincing examples of the benefits of additive methods over multiplicative methods for real-life applications are, however, not yet known. ≫

## 9.5.2 Distribution of Locally Refined Grids

During the refinement phase, the grid data of each new refinement level is initially available to only some of the processes. The parallel adaptive multigrid method cannot continue in a load-balanced way with the new level, until the new refined grid has been redistributed to all processors. (This distribution does not affect the distribution of previous refinement levels.) Generally, obtaining *optimal* load balancing at each stage of the adaptive multigrid process is too complicated and costly. An appropriate solution of this load-balancing problem is usually based on heuristic considerations. What is required is an algorithm that *rapidly* remaps locally refined block structures to *reasonably* load-balanced ones. One such strategy has been implemented, for example, in the adaptive routines of the CLIC library [321]. Other remapping strategies and corresponding systematic discussions can be found in [268, 446]. Here, we describe one practicable way of distributing (mapping) a locally refined block-structured grid to the processors of a parallel computer with distributed memory.

We assume that a refined grid $\Omega_{\ell+k}$ already exists and that the next refined grid, $\Omega_{\ell+(k+1)}$, has to be created and distributed.

(1) Based on a given refinement criterion, each process checks for refinement areas independently of the others. Since we deal with block-structured grids, each process has to embed its local refinement areas into structured (i.e. in 2D, logically *rectangular*) subgrids. If no process finds new refinement areas, the refinement procedure is finished.

(2) If refinement areas have been detected, communication is required to analyze the resulting block structure and to set up the corresponding data structure. At this point, local "process blocks" may be joined to larger "superblocks" whenever possible in order to obtain a final block structure with as few blocks as possible. The optimal number of grid points each processor can work on, $N^{(P)} = N_{\ell+(k+1)}/P$, is computed and broadcast.

(3) Based on a fast heuristic load balancing strategy, the blocks with the local refinement regions are redistributed among the processors. Very small blocks may share the same processor so that the total number of points treated by a processor is close to $N^{(P)}$. Blocks containing significantly more than $N^{(P)}$ points are subdivided for satisfactory load balancing.

The result of this procedure is a new block structure and a new (satisfactorily load-balanced) mapping of blocks to processes. Each of the essential mapping steps can be performed in parallel.

**Remark 9.5.3** Due to the required remapping of blocks, data has to be redistributed whenever locally refined levels are processed during an adaptive multigrid cycle. Let us assume that we have finished a relaxation sweep on grid $\Omega_{\ell+k}$ and that we need to transfer corrections from this level to the next finer one, $\Omega_{\ell+(k+1)}$. At this point, the relevant correction data is contained in the part of the coarse-level subgrid $\Omega_{\ell+k}$, which is being refined. Since this subgrid data is only available to some of the processors, one distributes the data *first* to the processes of grid $\Omega_{\ell+(k+1)}$ and *only then* performs the actual interpolation and correction. Similarly, during the fine-to-coarse transfer, all necessary computations (evaluation of defects, application of the FW operator etc.) are performed on the *fine* level; only the data which is really relevant for the coarser level are then redistributed. (For the fine-to-coarse transfer in the FAS, two types of grid functions have to be redistributed, defects and current approximations.) In addition to load-balanced computations, this way of redistributing data has another advantage: the amount of data to be redistributed on the coarse grid is smaller than the amount of data on the fine grid.                                  $\gg$

## 9.6 SOME PRACTICAL RESULTS

In this section, we give some results for two 2D problems and one 3D problem from computational fluid dynamics. The results have been obtained by using the remapping strategy as described in Section 9.5.2.

The two examples described in Sections 9.6.1 and 9.6.2 are the Euler equations for a flow around an airfoil and the incompressible Navier–Stokes equations for a double hole problem. Both examples and their treatment with parallel adaptive multigrid have been discussed in more detail in [323] and [424], respectively. Our intention here is only to demonstrate the improvement of computing times due to the adaptive multigrid technique and to show that adaptivity and parallelism can be combined. Although a full synergy of these somewhat conflicting principles cannot be expected, the gain in CPU time and storage achieved in practice is impressive.

In the first example, the refinement criterion based on the finite element residual (see Remark 9.4.3) has been used.

For nonadaptive multigrid applications corresponding to the examples discussed in Sections 9.6.1 and 9.6.2, the parallel efficiencies $E(P)$ on the IBM SP2-systems are higher than 90%. The utilization of parallel adaptive refinements at run-time requires additional communication and organization, and the communication among processes is no longer restricted to the exchange of boundary data but requires a redistribution of volumes of data. Nevertheless, the numerical results clearly show *considerable reductions of computing times and memory requirements* for the applications under consideration. These improvements by adaptive refinement are demonstrated by comparing the parallel computing times (wall-clock), the number of employed processors and the number of required grid points (the memory requirements being approximately proportional to the number of grid points).

### 9.6.1 2D Euler Flow Around an Airfoil

We consider a compressible Euler flow around the NACA0012 airfoil at Mach number $M_\infty = 0.85$ and an angle of attack of $1.0°$. The finite-volume discretization used is Osher's flux-difference splitting [220, 287] with a vertex-centered location of the variables.

Figure 9.17 shows the computational flow domain around the airfoil, which is partitioned into 16 blocks. Figure 8.25 presents a part of the computational grid before local refinements. Obviously, there is already a significant inherent refinement of this global grid towards the profile. Each block contains the same number of grid points and the blocks adjacent to the profile cover much smaller geometric domains than the others.

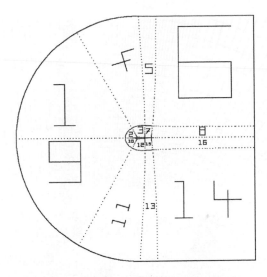

**Figure 9.17.** Computational domain around NACA0012 wing, 16 blocks.

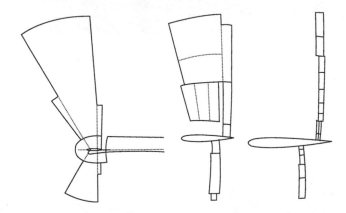

**Figure 9.18.** Geometric regions of adaptive refinement with block structures (16 blocks on each refinement level).

Figure 9.18 shows the geometric regions of refinement (with the corresponding block structures) of the three refinement levels. Note that the scale of each picture varies (as can be seen from the size of the profile).

A part of the adaptive composite grid, after the three steps of local refinement have been applied, is shown in Fig. 9.19. Obviously, the refinement regions are concentrated around the position of the shocks at the upper and the lower surface of the airfoil. Figure 9.20 shows that the shocks are determined much more sharply than on the global (coarse) grid without local refinement.

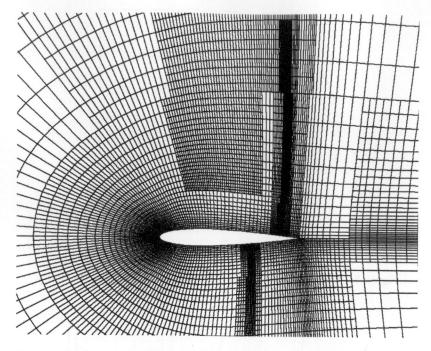

**Figure 9.19.** Part of adaptively refined (composite) grid around NACA0012 airfoil.

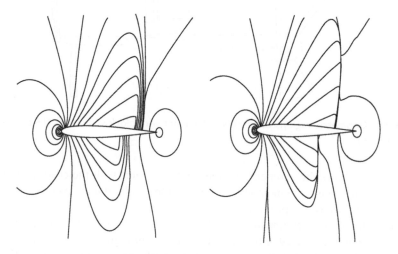

**Figure 9.20.** (Left) Pressure contours on global grid without refinement; (right) pressure contours on grid with adaptive refinement.

Table 9.2 shows that an improvement of more than a factor of ten is achieved with respect to computing time and memory for this example using the adaptive approach (three levels of refinement, approximately 14 000 grid points) as compared to a global fine grid (approximately 200 000 grid points). The accuracy achieved is approximately the same in both cases.

**Remark 9.6.1**  In [323] it is emphasized that a conservative discretization at the interface points is essential. The use of nonconservative formulas (e.g. cubic interpolation) may cause a deterioration in the accuracy and in the multigrid convergence (see also [100]).  $\gg$

### 9.6.2 2D and 3D Incompressible Navier–Stokes Equations

In the following example the incompressible Navier–Stokes equations have been solved based on a similar adaptive multigrid technique as used for the Euler equations [220]. Here, the finite volume discretization on the nonstaggered grid is based on Dick's flux difference splitting (see Section 8.8.4). Second-order accuracy is obtained using van Leer's $\kappa$-scheme in an outer defect correction iteration [220].

The example is shown in Fig. 9.21 (double-hole flow, in which the flow in a channel with height 1.0 is disturbed by two cavities; eleven blocks are used for the computation). The Reynolds number is $Re = 200$. Figure 9.22 shows clearly that the local refinement regions of the grid are located near the singularities (the upper corners of the cavities). Again, three refinement levels are used.

As can be seen from the results in Table 9.3, the global fine grid requires more than 23 times as many grid points as the adaptive grid in order to achieve comparable accuracy.

**Table 9.2.** Computing time, grid points and improvement factors for the flow around the NACA0012 profile.

|  | Global fine grid | Adaptive grid | Improvement factor |
|---|---|---|---|
| Processors | 16 | 16 |  |
| Grid points | 197632 | 13866 | 14 |
| Computing time (sec) | 6115 | 590 | 10 |

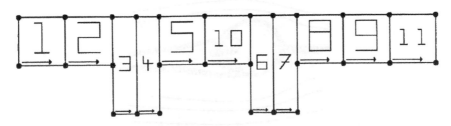

**Figure 9.21.** Block structure of double-hole flow problem.

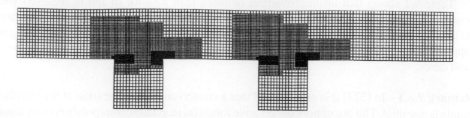

**Figure 9.22.** Refined grid of double-hole flow problem.

**Table 9.3.** Computing time, grid points and improvement factors for the double-hole flow problem.

|                       | Global fine grid | Adaptive grid | Improvement factor |
|-----------------------|:----------------:|:-------------:|:------------------:|
| Processors            | 11               | 11            |                    |
| Grid points           | 181761           | 7725          | 23                 |
| Computing time (sec)  | 656              | 49            | 13                 |

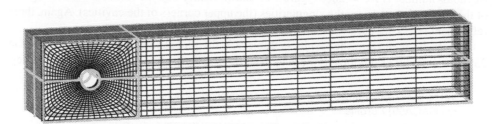

**Figure 9.23.** Domain and block structure for 3D flow around cylinder.

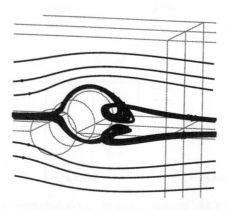

**Figure 9.24.** A streamline picture for steady flow around a cylinder at $Re = 20$.

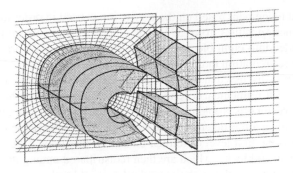

**Figure 9.25.** The first level of refinement around the cylinder after reblocking.

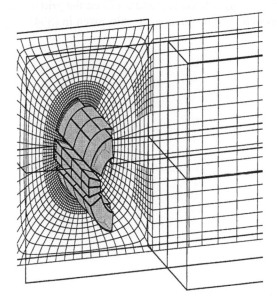

**Figure 9.26.** The second level of refinement around the cylinder after reblocking.

Using 11 processors of an IBM SP2 system for the computations, the computing time is reduced by a factor of 13 due to the adaptive refinement.

As a 3D example, we consider 3D steady incompressible Navier–Stokes flow around a circular cylinder. This is a reference problem, discussed in [394], where all the details of the domain and the boundary conditions are presented. The Reynolds number for the reference problem is $Re = 20$. One reference quantity to be determined is the pressure difference, $\delta p$ at the cylinder's centerline points in front of and behind the cylinder. We use a 3D finite volume discretization based on flux difference splitting for the convective terms [290]. Local refinement based on the $\tau$-criterion is used in order to obtain a solution close to the reference $\delta p$ values from [394]. The global 3D grid consisting of about 49 000 grid points is divided into eight blocks (see Fig. 9.23). Figure 9.24 presents a streamline picture with some recirculation in the flow.

Many of the topics discussed in this book, including parallelization, self-adaptive multigrid with an appropriate refinement criterion and the KAPPA smoother, have been employed in the multigrid solver for this problem. After FMG with $r = 3$ cycles on each global grid, the first local refinement region is determined self-adaptively and the refined grid is processed. Figure 9.25 presents the first level of refined blocks for this flow at $Re = 20$. The refined region is divided into 20 blocks, which are then redistributed among the eight processors. Figure 9.26 shows the 16 blocks on the second refinement level.

The corresponding multigrid convergence (with the KAPPA smoother) is excellent for this low Reynolds flow. The reference quantity $\delta p$ is found to be 0.178 on the global (coarse) grid, 0.174 on the first locally refined grid and 0.173 on the grid with two local refinement levels. This value fits well into the reference results given in [394].

# 10

# SOME MORE MULTIGRID APPLICATIONS

In this chapter, we summarize several multigrid applications. Some of the applications are difficult problems, some are less ambitious, but have particular features which were not covered in the earlier chapters.

Section 10.1 deals with appropriate multigrid techniques for the Poisson equation on the sphere. In Section 10.2, we discuss how multigrid and continuation techniques can be combined. As an example we use a nonlinear problem with a so-called turning point. In Section 10.3, we show how multigrid can be applied for the generation of curvilinear grids.

Section 10.4 describes the parallel multigrid software package $L_i SS$ [322], which is suited to a rather general class of PDEs on 2D block-structured grids. Many of the numerical results presented throughout this book, including those in Chapter 9, have been obtained with this package and its adaptive extension.

Section 10.5 is dedicated to the multigrid solution of compressible flow equations in industrial aerodynamic design. We describe some of the problems which occur for complicated flow problems in aircraft design, report some results and point out some topics of future research.

## 10.1 MULTIGRID FOR POISSON-TYPE EQUATIONS ON THE SURFACE OF THE SPHERE

The solution of elliptic equations in or on spherical geometries is of interest for several applications, for example, in astro- or geophysics and meteorology. We describe efficient multigrid solvers for 2D Poisson- and Helmholtz-type equations on the *surface* of the sphere.

In Section 10.1.1, we will briefly describe a discretization of the Laplacian on the sphere including the poles. This discretization is inherently anisotropic with varying directions of the anisotropy. Multigrid-related issues of the discrete problem will be discussed in Section 10.1.2.

Section 10.1.3 contains a discussion of a *special segment relaxation scheme*, proposed in [28], which takes into account the locality of the anisotropies of the discrete operator. The relaxation obtained is as cheap as usual line relaxation and has good smoothing properties.

### 10.1.1 Discretization

Here, we consider Poisson's equation in spherical coordinates:

$$x(r, \phi, \theta) = r \sin \phi \, \cos \theta$$
$$y(r, \phi, \theta) = r \sin \phi \, \sin \theta$$
$$z(r, \phi, \theta) = r \cos \phi$$

giving

$$\Delta = \frac{1}{r^2} \frac{\partial}{\partial r} \left( r^2 \frac{\partial}{\partial r} \right) + \frac{1}{r^2 \sin \phi} \frac{\partial}{\partial \phi} \left( \sin \phi \frac{\partial}{\partial \phi} \right) + \frac{1}{r^2 \sin^2 \phi} \frac{\partial^2}{\partial \theta^2}. \tag{10.1.1}$$

On the surface of a sphere, i.e. $r = 1$, $\Delta_{(\phi, \theta)}$ becomes

$$\Delta_{(\phi, \theta)} = \frac{1}{\sin \phi} \frac{\partial}{\partial \phi} \left( \sin \phi \frac{\partial}{\partial \phi} \right) + \frac{1}{\sin^2 \phi} \frac{\partial^2}{\partial \theta^2}, \quad \text{and}$$

$$-\Delta_{(\phi, \theta)} u = f \, (\Omega) \tag{10.1.2}$$

is the problem under consideration, where we denote the surface of the sphere in the coordinates $(\phi, \theta)$ by $\Omega$. On $\Omega$ we define the computational grid $\mathbf{G}_h$ by

$$\mathbf{G}_h = \{ (\phi_j, \theta_k) \colon j = 0, \ldots, n_\phi, k = 0, \ldots, n_\theta \}, \tag{10.1.3}$$

with $\phi_j = j h_\phi$, $\theta_k = k h_\theta$ and the mesh sizes $h_\phi = \pi / n_\phi$ and $h_\theta = 2\pi / n_\theta$. The quantities $n_\phi$ and $n_\theta$ are chosen so that standard coarsening strategies can be applied. This grid contains two singular lines which correspond to the north and the south pole (see Fig. 10.1).

A second-order accurate *finite volume discretization* [27] of (10.1.2) is

$$- \left[ \begin{array}{ccc} & h_\phi / (h_\theta \sin \phi_j) & \\ (h_\theta / h_\phi) \sin \phi_{j-1/2} & -\sum & (h_\theta / h_\phi) \sin \phi_{j+1/2} \\ & h_\phi / (h_\theta \sin \phi_j) & \end{array} \right] u_{j,k} \tag{10.1.4}$$
$$= h_\theta h_\phi \sin(\phi_j) f_{j,k}$$

$j = 1, \ldots, n_\phi - 1, k = 0, \ldots, n_\theta$, where $\sum$ denotes the sum of the four neighbor stencil elements. This discretization of the Poisson equation is equivalent to a finite difference form used in [380].

For the discretization at the poles, the finite volume integration of the equation is performed over all adjacent cells at once. At the north pole $P_N$ (where $\phi = 0$), we obtain [28]

$$\int_{\Omega_{P_N}} \Delta_{(\phi, \theta)} u \, d\Omega \approx \sum_{k=1}^{n_\theta} h_\theta \sin \phi_{1/2} \frac{u(\phi_1, \theta_k) - u(P_N)}{h_\phi}.$$

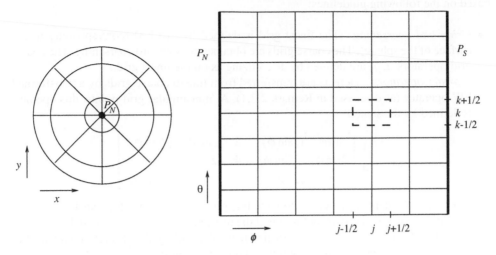

**Figure 10.1.** Spherical coordinates with poles $P_N$ and $P_S$ and a control volume at $(\phi_j, \theta_k)$.

By discretizing the right-hand side as

$$\int_{\Omega_{P_N}} f \, d\Omega \approx f(P_N)\pi \frac{h_\phi}{2} \sin \phi_{1/2},$$

we obtain

$$\sum_{k=1}^{n_\theta} \frac{h_\theta}{h_\phi} \sin \phi_{1/2}(u(\phi_1, \theta_k) - u(P_N)) = f(P_N)\pi \frac{h_\phi}{2} \sin \phi_{1/2}. \tag{10.1.5}$$

The south pole discretization is analogous. The resulting discretization is *conservative* and *symmetric*.

**Remark 10.1.1 (compatibility condition)**  The Poisson equation has a solution on the surface of the sphere only if the right-hand side $f$ satisfies the compatibility condition $\int_\Omega f \, d\Omega = 0$. The solution is then determined up to a constant. In general, the discrete compatibility condition is not fulfilled, even if the continuous condition is. The discrete right-hand side $f_h$ has then to be modified in order to ensure the existence of a solution to the discrete problem [28] (see also Section 5.6). $\qquad\qquad \gg$

### 10.1.2 Specific Multigrid Components on the Surface of a Sphere

Here, we discuss an efficient multigrid process for the Poisson equation on the surface of the sphere. We assume the discretization according to (10.1.4) and (10.1.5) and assume further that the discrete compatibility condition is fulfilled. The multigrid scheme is constructed based on the following guidelines:

- *Grids and coarsening*   The finest grid is defined by (10.1.3) (corresponding to the surface of the sphere). The coarse grids are obtained by standard coarsening. The coarse grid operators, $L_H$, are defined by discretizing as on the finest grid.
- *Transfer operators*   Defects are transferred from fine to coarse grids by a transformed FW operator (as discussed in Remark 2.3.1). At nonsingular grid points, this operator is defined by the stencil

$$\frac{1}{4} \begin{bmatrix} \sin\phi_{j-1}/\sin\phi_j & 2 & \sin\phi_{j+1}/\sin\phi_j \\ 2\sin\phi_{j-1}/\sin\phi_j & 4 & 2\sin\phi_{j+1}/\sin\phi_j \\ \sin\phi_{j-1}/\sin\phi_j & 2 & \sin\phi_{j+1}/\sin\phi_j \end{bmatrix} \qquad (10.1.6)$$

where $(\phi_j, \theta_k)$ denotes the corresponding fine grid point at which the transfer is carried out. At the poles $P$, a special definition of the restriction operator is needed. It is defined such that the transferred defect will automatically satisfy the discrete compatibility condition on the coarse grid if it is fulfilled on the fine grid:

$$d_{2h}(P) := 2\frac{\sin(h_\phi/2)}{\sin h_\phi}d_h(P) + \frac{4}{n_\theta}\sum_{k=1}^{n_\theta} d_h(\bar{\phi}, \theta_k), \qquad (10.1.7)$$

where $d_h$ denotes the defect on the fine grid, with $\bar{\phi} = h_\phi$ if $\phi = 0$ (north pole) or $\bar{\phi} = \pi - h_\phi$ if $\phi = \pi$ (south pole). $n_\theta$ is the number of grid points on the fine grid in the $\theta$-direction. The sum in (10.1.7) can also be interpreted as an average of all defects surrounding the pole.

Bilinear interpolation is employed to transfer the corrections from coarse to fine grids.

- *Relaxation*   Due to the strong anisotropy of the Laplacian in spherical coordinates, especially near the poles where its coefficients in the $\theta$-direction are orders of magnitude larger than in the $\phi$-direction, pointwise relaxation methods are not suitable as they have poor smoothing properties.

In order to design an appropriate smoothing procedure, we apply LFA to the problem obtained by freezing the (variable) coefficients of the discrete operator at each point. The LFA smoothing factors $\mu_{\text{loc}}$ depend on $\phi$, $\theta$ and on the quantity

$$q_l = h_\phi/h_\theta. \qquad (10.1.8)$$

As expected, GS-RB ($\omega = 1$) and the $\phi$-zebra relaxation give poor smoothing factors. Actually, they tend to one for points near the poles. On the other hand, it can be seen that $\theta$-zebra-line relaxation has good smoothing properties if $q_l \geq 1$: We find $\mu_{\text{loc}}$ ($\nu = 2$) $\leq 0.34$, so that we can expect convergence factors of about 0.1 for multigrid cycles employing two relaxation steps. For $q_l < 1$, the situation is more involved since in that case there is a change in the anisotropy direction inside the domain and none of the relaxations considered so far have good smoothing factors. As expected, *alternating zebra line relaxation* provides good

**Table 10.1.** V(1, 1) multigrid convergence factors using zebra line relaxation ($q_l = 2$).

| $n_\theta, n_\phi$ | Relaxation | |
|---|---|---|
| | $\theta$-line relax. | $\phi$-line relax. |
| 32, 8 | 0.020 | 0.76 |
| 64, 16 | 0.030 | 0.93 |
| 128, 32 | 0.040 | 0.97 |
| 256, 64 | 0.045 | 0.98 |

**Table 10.2.** V(1, 1) multigrid convergence factors using zebra line relaxation ($q_l = 0.25$).

| $n_\theta, n_\phi$ | Relaxation | | |
|---|---|---|---|
| | $\theta$-line relax. | $\phi$-line relax. | Altern. line |
| 8, 16 | 0.68 | 0.34 | 0.026 |
| 16, 32 | 0.72 | 0.53 | 0.031 |
| 32, 64 | 0.75 | 0.76 | 0.042 |
| 64, 128 | 0.77 | 0.93 | 0.051 |

smoothing in all cases. Tables 10.1 and 10.2 show measured V(1, 1)-cycle convergence factors using various relaxations. They are in agreement with the LFA results. Note that the simplest and most efficient choice is to apply $\omega$-GS-RB with an appropriate $\omega > 1$ for smoothing (see Remark 5.1.2) if the anisotropies are not too large. In the following section, we will discuss another efficient approach, which works well for any anisotropy.

### 10.1.3 A Segment Relaxation

As seen above, the line relaxation using lines in only one direction is not sufficient to achieve good multigrid convergence in cases with $q_l < 1$. There is actually no sudden deterioration of the convergence factors when crossing the value $q_l = 1$, but they will gradually worsen with decreasing values of $q_l$. In these cases, the $\theta$-line relaxation smoothes well near the poles but cannot provide good smoothing near the equator, where the anisotropy has a different direction.

Alternating line relaxation is a good smoother in any of these cases, but it requires twice the computational work of a single line relaxation. Half of the cost of alternating line relaxation can, however, be saved by a *segment* relaxation. The idea is to split the domain into some (nonoverlapping) regions, such that the anisotropy has a fixed direction inside them. In other words, we apply each type of line relaxation only in those regions in which its smoothing properties are good. As can be seen by LFA, the $\theta$-line relaxation does an excellent job in regions where $q_l \geq \sin \phi$, while for $q_l < \sin \phi$ the $\phi$-line relaxation is more effective. We then split the domain into regions according to whether or not $q_l \leq \sin \phi$. The pattern of the segment relaxation is illustrated in Fig. 10.2.

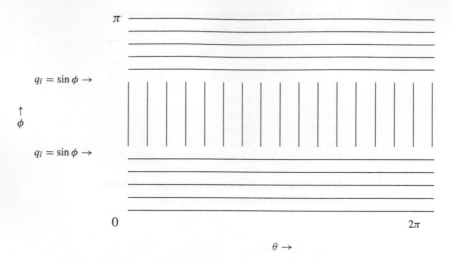

**Figure 10.2.** Pattern of the segment relaxation for cases with $q_l < 1$.

**Table 10.3.** V(1,1) multigrid convergence factors with the segment and with the alternating line relaxation.

| | $q_l = 0.5$ | | | $q_l = 0.25$ | |
|---|---|---|---|---|---|
| | Relaxation | | | Relaxation | |
| $n_\theta, n_\phi$ | Altern. | Segment | $n_\theta, n_\phi$ | Altern. | Segment |
| 16, 16 | 0.031 | 0.044 | 8, 16 | 0.026 | 0.026 |
| 32, 32 | 0.031 | 0.054 | 16, 32 | 0.031 | 0.059 |
| 64, 64 | 0.040 | 0.060 | 32, 64 | 0.042 | 0.073 |
| 128, 128 | 0.047 | 0.064 | 64, 128 | 0.051 | 0.082 |

A practical implementation of the segment *zebra* relaxation is done as follows: we perform $\theta$-line relaxation zebrawise (odd lines first, then even lines) on the regions around the poles (where $q_l < \sin \phi$) followed by $\phi$-zebra-line relaxation near the equator (where $q_l < \sin \phi$). Table 10.3 shows the convergence obtained with the segment relaxation.

**Remark 10.1.2 (3D generalization)**   The strong and varying anisotropy of the equations on the sphere causes difficulties for the realization of efficient 3D multigrid schemes. The use of segment based plane relaxation, implemented on the basis of 2D multigrid solvers, is one option for achieving an effective method. Of course, LFA can be used to provide guidelines on designing an efficient 3D relaxation. If the anisotropies are not too strong, suitable $\omega$-GS-RB relaxations can be applied efficiently.                                    $\gg$

**Remark 10.1.3 (icosahedral grids)**   The problems of the singularities caused by the poles and the anisotropies introduced by the grid are avoided if another type of grid is used. The

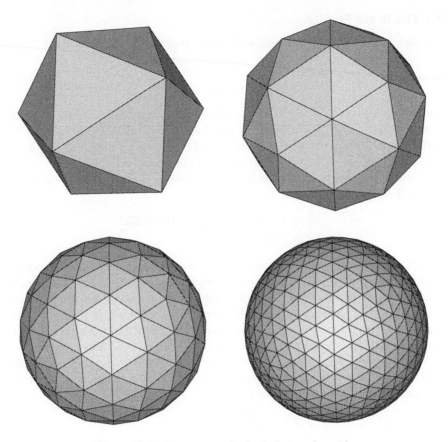

**Figure 10.3.** A sequence of spherical triangular grids.

so-called icosahedral grid, consisting of spherical triangular grid cells (see Fig. 10.3) [35] is of special interest for weather forecast models.

Depending on the place on the sphere, the grid points are surrounded by five or six triangular cells. A finite element or finite volume discretization can easily be applied on this icosahedral grid. Due to the regularity of cells and the hierarchical structure of the grids, shown in Fig. 10.3, multigrid can be applied in a natural way. Moreover, the need for line relaxation disappears and pointwise relaxation can be used [41]. Furthermore, the grid is well-suited for adaptive and parallel versions of multigrid. ≫

## 10.2 MULTIGRID AND CONTINUATION METHODS

In this section, we outline how multigrid methods can be used for the solution of nonlinear bifurcation problems of the form

$$N(u, \lambda) = 0. \tag{10.2.1}$$

Here $N$ is a nonlinear differential operator and $\lambda$ a parameter $\in \mathbb{R}$.

### 10.2.1 The Bratu Problem

As an example, we consider the classical 2D *Bratu* problem on the unit square

$$N(u, \lambda) = -\Delta u - \lambda e^u = 0 \quad (\Omega = (0, 1)^2)$$
$$u = 0 \quad (\partial \Omega) \tag{10.2.2}$$

and look for solutions $(u, \lambda)$, $\lambda \geq 0$. The structure of the solutions for the above problem is well known (see Fig. 10.4). Solutions exist only for certain values of $\lambda$, actually for $\lambda \leq \lambda^*$ ($\approx 6.8$). For $\lambda = 0$ only the trivial solution exists. For $0 < \lambda < \lambda^*$, there are two solutions, and for $\lambda = \lambda^*$ we have a "turning point" with only one solution.

We assume that the above problem is discretized using the standard second-order five-point stencil $\Delta_h$ on the square grid $\Omega_h$

$$N_h u_h = -\Delta_h u_h - \lambda e^{u_h} = 0 \quad (\Omega_h)$$
$$u_h = 0 \quad (\partial \Omega_h). \tag{10.2.3}$$

In principle, for fixed $\lambda$, we can either employ the nonlinear FAS multigrid approach or globally linearize the problem with an outer Newton-type iteration and use linear multigrid in each Newton step. Indeed, the straightforward FAS, for example, has been successfully used without any difficulties for finding the *lower* solution branch for $\lambda$ up to $\lambda \approx 6.5$. In this case, FAS is particularly efficient in combination with FMG, starting from a zero initial approximation on the coarsest grid, which may consist of only one interior point.

However, if $\lambda$ approaches the crucial value $\lambda^*$, standard multigrid methods fail to converge [350]. One reason is rather fundamental and will cause problems to any solver. At the turning point $\lambda^*$, the fundamental matrix

$$\frac{\partial N_h(u_h^*, \lambda_h^*)}{\partial u_h}$$

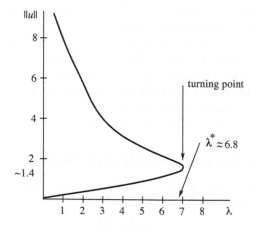

**Figure 10.4.** Bifurcation diagram for the Bratu problem, showing $||u||_\infty$ as a function of $\lambda$.

becomes singular. A second reason is specific for multigrid: if $\lambda_h$ is close to $\lambda^*$, certain *low* frequency error components are not reduced by coarse grid corrections.

This effect is also observed if the (linear) indefinite Helmholtz equation is solved by a multigrid method for a Helmholtz constant which is close to an eigenvalue of the discrete Laplacian (see also, for example, [84] and Appendix A, Section A.8.5). The fine grid problem is not approximated well on very coarse grids if standard multigrid is applied for these values of $\lambda$.

**Remark 10.2.1 (local Kaczmarz relaxation)** On very coarse grids, additional difficulties arise due to the fact that the diagonal elements of the corresponding coarse grid discrete operator may be close to 0 or may even change sign. When using one of the pointwise relaxation schemes, one has to divide by these very small coefficients. This difficulty can be overcome by a modification of the relaxation method used on these coarse grids. For example, a combination of the nonlinear Gauss–Seidel method with the Kaczmarz relaxation (see Section 4.7) method is suitable here [350]. $\gg$

## 10.2.2 Continuation Techniques

The question that now arises is how to find solutions on the upper branch in Fig. 10.4 or near the turning point.

A well-known process to deal with such problems is to apply a so-called *continuation technique*. This means that a sequence of auxiliary problems is constructed that converge towards the target problem. The auxiliary problems are constructed in such a way that a major difficulty of the original problem is overcome. This can be, for example, a small attraction basin for strongly nonlinear problems, for which a sequence of problems with a less strong nonlinear term can be used to provide a series of approximations for the actual problem to be solved.

In our example, we have a similar problem with multiple solutions (for one value of $\lambda$, two solutions may exist). One solution can usually be obtained easily, the other one can be obtained by a special continuation technique described in the following.

If we want to solve (10.2.3) on the upper solution branch or near $\lambda^*$ with a multigrid method, we have several choices and options. A first (naive) continuation idea is to apply multigrid for a fixed $\lambda$, for which multigrid converges well and to vary $\lambda$ systematically. We can increase $\lambda$ step-by-step and for example use the solution corresponding to the previous $\lambda$ as a first approximation for the next value of $\lambda$. This approach, however, has some deficiencies.

First, it is not known how to prescribe new values of $\lambda$ such that the problem still has a solution. ($\lambda^*$ is not known in more general situations.) Moreover, it is not clear how one can pass the turning point, i.e. switch the solution branch, with this approach. In addition, the multigrid convergence problems described in Section 10.2.1 will arise near $\lambda^*$ due to the singularity of the fundamental matrix.

A more sophisticated way to overcome these problems is to change the parametrization of the problem and of the continuation process. The problem parameter $\lambda$ is replaced by a new independent variable, $s$, in terms of which the solution is unique, $u$ and $\lambda$ are then regarded as unknowns. In other words, we form an augmented (extended) problem

$$N(u(s), \lambda(s)) = 0 \quad (\Omega)$$
$$u(s) = 0 \quad (\partial\Omega) \tag{10.2.4}$$
$$C(u(s), \lambda(s), s) = 0.$$

Here, the last equation gives an (implicit) definition of the relation between $u$, $s$ *and* $\lambda$. The procedure discussed above corresponds to the (inappropriate) choice $C(u, \lambda, s) = \lambda - s$. The operator $C(u, \lambda, s)$ can be interpreted as a *constraint* to the original system. A reasonable choice which works very well for many problems is given by the *arc-length continuation method* [101, 209] where the arc length is used for $s$, leading with the $\| \cdot \|_2$ norm to

$$C(u, \lambda, s) := ||du/ds||_2{}^2 + |d\lambda/ds|^2 - 1 = 0.$$

If we assume a solution $(u_0, \lambda_0) = (u(s_0), \lambda(s_0))$ and additionally $(\dot{u}(s_0), \dot{\lambda}(s_0)) = ((du/ds)(s_0), (d\lambda/ds)(s_0))$ to be known (exactly or approximately), we can use

$$C(u, \lambda, s) := \langle \dot{u}(s_0), u(s) - u(s_0)\rangle_2 + \dot{\lambda}(s_0)(\lambda(s) - \lambda(s_0)) - (s - s_0) = 0 \tag{10.2.5}$$

instead of the arc length itself (for sufficiently small $|s - s_0|$). Here $\langle \cdot, \cdot \rangle_2$ denotes the Euclidean inner product.

In [119] it is proved, under general conditions, that a unique solution $(u(s), \lambda(s), s)$ of (10.2.4) with (10.2.5) exists on a curve through $(u(s_0), \lambda(s_0))$ for all $|s - s_0| < \delta$ ($\delta$ sufficiently small) if $(u(s_0), \lambda(s_0))$ is a regular solution or a simple turning point of $N(u, \lambda)$. An important property of this continuation method is that along this curve, the derivative

$$\begin{pmatrix} \partial N(u(s), \lambda(s))/\partial u & \partial N(u(s), \lambda(s))/\partial \lambda \\ \partial C(u(s), \lambda(s), s)/\partial u & \partial C(u(s), \lambda(s), s)/\partial \lambda \end{pmatrix} \tag{10.2.6}$$

is not singular (not even at the turning point).

**Remark 10.2.2** In the case of the above Bratu problem, for which the numerical solution is known for all values of $\lambda$, simpler constraints are available. It is, for example, also possible to prescribe $||u_h||_2$ or $u_h(1/2, 1/2)$ as the additional constraint, i.e. one may set $C(u, \lambda, s) = ||u_h||_2 - s$ or $C(u, \lambda, s) = u_h(0.5, 0.5) - s$.                    $\gg$

## 10.2.3 Multigrid for Continuation Methods

In this section, we briefly describe one approach of how the FAS multigrid method (in particular the relaxation) can be modified in order to solve the augmented system (10.2.4) (for other approaches, see [19, 44, 269]).

Let us assume that a discretization of (10.2.4), (10.2.5) on $\Omega_\ell$ is given by

$$N_\ell(u_\ell, \lambda_\ell) = f_\ell \quad (\Omega_\ell)$$
$$u_\ell = 0 \quad (\partial\Omega_\ell) \tag{10.2.7}$$
$$C_\ell(u_\ell, \lambda_\ell, s) = \delta_\ell$$

with

$$C_\ell(u_\ell, \lambda_\ell, s) = \langle u_\ell(s_0) - u_\ell(s_{-1}), u_\ell(s) - u_\ell(s_0) \rangle_2$$
$$+ (\lambda_\ell(s) - \lambda_\ell(s_0))(\lambda_\ell(s_0) - \lambda_\ell(s_{-1}))$$

where $\langle \cdot, \cdot \rangle_2$ is the discrete Euclidean inner product and

$$\delta_\ell = (s - s_0)(s_0 - s_{-1}).$$

Here, we assume that $(u_k(s_0), \lambda_k(s_0))$ and $(u_k(s_{-1}), \lambda_k(s_{-1}))$ are obtained solutions.

For the continuation along the curve of the solution, we prescribe the distance (size $s - s_0$) on the secant with $|s - s_0|$ sufficiently small [350].

On the coarse grids $\Omega_k$, $k \neq \ell$, we have (10.2.7) with $\ell$ replaced by $k$. These coarse grid systems $(N_k, f_k, C_k, \delta_k)$ are defined as usual in the FAS process.

Note that $\delta_k$ is just a number, not a grid function and that a first approximation to the coarse grid $\lambda_k$-value can be obtained by straight injection.

One smoothing step on $\Omega_k$ for the augmented system consists of two parts, since both $\lambda$ and $u$ have to be updated.

(1) Apply one (nonlinear) "standard" relaxation step to the current approximation of $u_k(s)$ with a fixed $\lambda(s)$. This gives $\tilde{u}_k(s)$.
(2) Modify the relaxed approximation in such a way that the constraint is satisfied afterwards. This can be achieved by the corrections

$$\bar{u}_k(s) \leftarrow \tilde{u}_k(s) + \beta(\tilde{u}_k(s) - u_k(s_0))$$
$$\bar{\lambda}_k(s) \leftarrow \lambda_k(s) + \alpha(\lambda_k(s) - \lambda_k(s_0)), \tag{10.2.8}$$

where the parameters $\alpha$ and $\beta$ are obtained from the $2 \times 2$ system

$$C_k(\bar{u}_k(s), \bar{\lambda}_k(s)) = \delta_k$$
$$\widetilde{\sum}_{\Omega_k}(N_k(\bar{u}_k(s), \bar{\lambda}_k(s))) = \widetilde{\sum}_{\Omega_k} f_k.$$

Here, $\widetilde{\sum}$ characterizes an average over all equations. The $2 \times 2$ system can be approximately solved, for instance, by a few (even just one) steps of a Newton-type method.

Employing the FAS with such a relaxation procedure on all grids, one can march along the curve in Fig. 10.4 starting somewhere on the lower branch. Taking the previous solution (and the corresponding value of $\lambda$) as a first approximation and updating $\delta$, one can pass the turning point without any difficulty provided that local Kaczmarz relaxations are applied on coarse grids (as mentioned in Remark 10.2.1) if necessary.

**Remark 10.2.3** In general it is sufficient to update $\lambda$ (Step (2)) in the above relaxation *only on the coarsest grid* (or on a few very coarse grids). In this way, the cost of the relaxation can be reduced. $\gg$

**Remark 10.2.4 (embedding continuation into FMG)** An efficient variant of this multigrid solution method is obtained by incorporating the continuation process in a nested iteration process (in FMG): $s$ is increased already on the coarse grids in such a way that the target value of $s$ is already obtained when the finest grid is reached. $\gg$

### 10.2.4 The Indefinite Helmholtz Equation

If the Bratu problem is globally linarized, one obtains a Helmholtz-type equation of the form

$$-\Delta u + cu = f \quad (c = c(x, y) < 0).$$

Different from the harmless case $c(x, y) \geq 0$ (discussed in Section 2.8.1), the multigrid treatment of the case $c < 0$ can become rather involved and will be a real difficulty if $-c$ gets very large.

We will not describe these complications here. The situation and the typical phenomena are discussed in Appendix A (Section A.8.5). What is said there with respect to the AMG convergence of the (indefinite) Helmholtz equation is similarly valid for the standard multigrid treatment: Difficulties occur if $-c$ is close to one of the eigenvalues of the discrete Laplacian since the discrete equation becomes singular there (see [84, 386]).

If $-c$ is larger than the first eigenvalue, smoothers like Gauss–Seidel relaxation will no longer be convergent. But since their smoothing properties are still satisfactory, the multigrid convergence will deteriorate gradually for $-c$ increasing (see Figure A.8.22), provided that a direct solver is employed on the coarsest grid. However, if $-c$ gets larger and larger, the coarsest grid has to be chosen finer and finer (explained in A.8). Finally, if $-c \to \infty$ the $h$-ellipticity $E_h(L_h) \to 0$.

An efficient multigrid approach that is nevertheless feasible in this situation is the wave-ray multigrid [79]. Another multigrid treatment is proposed in [136].

## 10.3 GENERATION OF BOUNDARY FITTED GRIDS

A common way to generate boundary fitted grids is to solve quasilinear elliptic systems of PDEs, i.e. to solve transformed systems of Poisson or biharmonic equations [135, 354, 391]. Efficient multigrid solvers have been developed for such problems [244]. In the following, we will describe typical situations which may occur in grid generation and discuss the proper multigrid treatment. We start with a brief review of second- and fourth-order elliptic grid generation equations. Several aspects that have been discussed separately in previous sections in this book (like systems of equations, mixed derivatives, first-order derivatives etc.) come together here. For convenience, we restrict ourselves to 2D. The generalization to 3D is straightforward.

### 10.3.1 Grid Generation Based on Poisson's Equation

For a simply connected domain $\Omega$ in the $(x, y)$-plane with boundary $\partial\Omega$, boundary fitted coordinates $\xi = \xi(x, y)$ and $\eta = \eta(x, y)$ can be defined as the solution of the decoupled system of Poisson equations

$$\Delta\xi = s_1(x, y), \quad \Delta\eta = s_2(x, y) \qquad \text{on } \Omega$$
$$\xi = f(x, y), \quad \eta = g(x, y) \qquad \text{on } \partial\Omega. \qquad (10.3.1)$$

With points $A$, $B$, $C$ and $D$ on $\partial\Omega$ and constants $\xi_0 < \xi_1$ and $\eta_0 < \eta_1$, the boundary values are of the form (see Fig. 10.5)

$$f \equiv \xi_0 \ (\partial\Omega_{AD}), \quad f \equiv \xi_1 \ (\partial\Omega_{BC}), \quad g \equiv \eta_0 \ (\partial\Omega_{AB}) \quad \text{and} \quad g \equiv \eta_1 \ (\partial\Omega_{DC}).$$

On the remainder of the boundary, $f$ and $g$ are assumed to be monotonically increasing (surjective) functions. By choosing proper boundary values $f$ and $g$ and right-hand sides $s_1$ and $s_2$, one has a flexible control over the resulting $(\xi, \eta)$-coordinate system near the boundary and in the interior, respectively. (However, nonzero right-hand sides $s_1$ and $s_2$ have to be chosen with care in order to avoid overlapping coordinate lines.)

The required boundary fitted grid is finally defined by the discrete contour lines of $\xi$ and $\eta$. These are most conveniently obtained by considering the inverse transformation functions $x = x(\xi, \eta)$ and $y = y(\xi, \eta)$ which are the solutions of the following Dirichlet problem on the rectangle $\hat\Omega = [\xi_0, \xi_1] \times [\eta_0, \eta_1]$ in the $(\xi, \eta)$-plane:

$$\alpha x_{\xi\xi} - 2\beta x_{\xi\eta} + \gamma x_{\eta\eta} + J^2(s_1 x_\xi + s_2 x_\eta) = 0$$
$$\alpha y_{\xi\xi} - 2\beta y_{\xi\eta} + \gamma y_{\eta\eta} + J^2(s_1 y_\xi + s_2 y_\eta) = 0, \qquad (10.3.2)$$

where $x = \hat{f}(\xi, \eta)$, $y = \hat{g}(\xi, \eta)$ on $\partial\hat\Omega$ and

$$\alpha = x_\eta^2 + y_\eta^2, \qquad \beta = x_\xi y_\eta + y_\xi x_\eta, \qquad \gamma = x_\xi^2 + y_\xi^2$$
$$\text{and the Jacobian} \quad J = x_\xi y_\eta - x_\eta y_\xi.$$

The boundary values are such that, for example, $(\hat{f}(\xi, \eta_0), \hat{g}(\xi, \eta_0))$ $(\xi_0 \le \xi \le \xi_1)$ is a parametric representation of $\partial\Omega_{AB}$ (see Fig. 10.5). The system (10.3.2) is the system of interest here.

### 10.3.2 Multigrid Solution of Grid Generation Equations

An efficient multigrid treatment for (10.3.2) is guided by many of the considerations already touched on in this book, some of which we will list here. Two properties can immediately be seen. First, (10.3.2) is a *system* of differential equations and second, it is *nonlinear*. The nonlinearity does not lead to any essential problems if we apply the FAS (see Section 5.3) and use nonlinear variants of standard relaxation methods for smoothing. A straightforward generalization of Gauss–Seidel relaxation (pointwise, linewise, etc.) to systems of PDEs is the corresponding *collective Newton–Gauss–Seidel relaxation*.

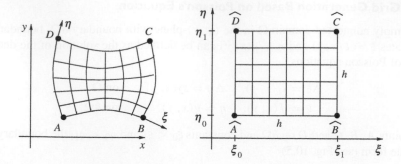

**Figure 10.5.** (Left) The physical domain; (right) the transformed rectangular computational domain.

**Remark 10.3.1** Note that, for (10.3.2) and collective point relaxation, both the Newton and Picard iteration (freezing of $\alpha, \beta, \gamma, J$) coincide if central differencing is used. At each point the corresponding systems of equations within the relaxation scheme are linear and the different unknowns are decoupled. The situation is different, however, if one-sided differencing is used for the first-order derivatives, and also in the case of collective line relaxation.                                                                                      $\gg$

Another issue is the occurrence of mixed derivatives in (10.3.2). In Section 7.6.1, we have seen that, if mixed derivatives become too large (i.e. if $\beta^2 - \alpha\gamma = -J^2 \approx 0$), the problem is close to being nonelliptic and special multigrid components (e.g. a specific smoother) need to be chosen for fast convergence.

The choice of a proper smoothing procedure depends essentially on the relative size of the (solution dependent) coefficients of the principle terms, i.e. on the ratio $\alpha/\gamma$. It is important to know how this ratio behaves asymptotically. As long as it is between 1/2 and 2, say, we may safely use (collective) point Gauss–Seidel relaxation for smoothing with standard grid coarsening. Geometrically, this is the case if $\xi$ and $\eta$ are, roughly, arc-length parameters of the corresponding $(x, y)$-curves (up to a scaling factor). Reasonable boundary fitted grids, however, should not be too far away from being orthogonal grids (at least not globally). Consequently, for relevant situations, the mixed derivative coefficient $\beta$ can be expected to be relatively small and should not have any negative influence on the multigrid convergence.

However, any kind of essential grid line concentration in the physical domain will cause anisotropies in the principal parts of the equations. For large anisotropies, line relaxation is required in combination with standard coarsening, or, alternatively, point relaxation with semicoarsening.

The first-order derivatives are usually discretized with central differences [391]. Even if this does not cause any difficulties on the finest grid, this discretization may still become unstable on some coarser grids and may even lead to divergence (see the discussion in Section 7.1.2). This is likely to happen if significant inhomogeneous source terms $s_1$ and $s_2$ are used. However, even for $s_1 \equiv s_2 \equiv 0$, the discretization of the first-order derivatives (in $\alpha, \beta, \gamma$) may become important. By linearizing (10.3.2) around a solution $(x, y)$, we see that a perturbation $\boldsymbol{\varepsilon} = (\varepsilon_1, \varepsilon_2)$ of the right-hand side causes a change $\boldsymbol{\delta} = (\delta_1, \delta_2)$ of $(x, y)$ which is approximately the solution of the linear differential system

$$\begin{pmatrix} L_1(x, y) & L_2(x, y) \\ L_2(y, x) & L_1(y, x) \end{pmatrix} \begin{pmatrix} \delta_1 \\ \delta_2 \end{pmatrix} = \begin{pmatrix} \varepsilon_1 \\ \varepsilon_2 \end{pmatrix} \tag{10.3.3}$$

where $L_1$ and $L_2$ are the differential operators

$$L_1(x, y) = \alpha \partial_{\xi\xi} - 2\beta \partial_{\xi\eta} + \gamma \partial_{\eta\eta} + 2(x_\eta x_{\eta\eta} - x_\eta x_{\xi\eta})\partial_\xi + 2(x_\eta x_{\xi\xi} - x_\xi x_{\xi\eta})\partial_\eta$$
$$L_2(x, y) = 2(y_\xi x_{\eta\eta} - y_\eta x_{\xi\eta})\partial_\xi + 2(y_\eta x_{\xi\xi} - y_\xi x_{\xi\eta})\partial_\eta.$$

The discretization of the first-order terms can be oriented to the stability properties of this linear system, i.e. the discretization of (10.3.2) has to be such (central, first-order upwind or higher order upwind, see Section 7.3.2) that the corresponding discretization of (10.3.3) is stable for all mesh sizes used in the corresponding multigrid process.

### 10.3.3 Grid Generation with the Biharmonic Equation

Often, in grid generation, control of the grid point distribution *and* of the angle of intersecting grid lines at the boundary $\partial\Omega$ is desirable, especially when separately generated grids are to be patched (in order to obtain smooth interfaces). Moreover, one would often like to generate grids that are orthogonal to the boundary. This, for example, simplifies the correct numerical treatment of boundary conditions. The approach (10.3.2) uses a system of second-order equations. In that case, it is impossible to control both the grid point distribution *and* the angle of intersecting grid lines at the same time because this would lead to an overdetermination: two boundary conditions per boundary for each grid function cannot be prescribed for a $2 \times 2$ system of Poisson-type equations.

One way to construct such grids is to use the *biharmonic equations*

$$\Delta\Delta\xi = s_1(x, y), \quad \Delta\Delta\eta = s_2(x, y) \qquad \text{on } \Omega \tag{10.3.4}$$

instead of Poisson equations [36, 349, 370]. As discussed in Section 8.4, it is convenient to write the above pair of biharmonic equations as two systems, each consisting of two Poisson equations:

$$\Delta\xi = p, \quad \Delta p = s_1 \quad \text{and} \quad \Delta\eta = q, \quad \Delta q = s_2. \tag{10.3.5}$$

By transforming the dependent to independent variables as in the previous section, the quasilinear system

$$\begin{aligned} \alpha x_{\xi\xi} - 2\beta x_{\xi\eta} + \gamma x_{\eta\eta} + J^2(px_\xi + qx_\eta) &= 0 \\ \alpha y_{\xi\xi} - 2\beta y_{\xi\eta} + \gamma y_{\eta\eta} + J^2(py_\xi + qy_\eta) &= 0 \\ \alpha p_{\xi\xi} - 2\beta p_{\xi\eta} + \gamma p_{\eta\eta} + J^2(pp_\xi + qp_\eta) - J^2 s_1 &= 0 \\ \alpha q_{\xi\xi} - 2\beta q_{\xi\eta} + \gamma q_{\eta\eta} + J^2(pq_\xi + qq_\eta) - J^2 s_2 &= 0 \end{aligned} \tag{10.3.6}$$

is derived from (10.3.5). Apart from specifying $x$ and $y$ at the boundary $\partial\hat{\Omega}$, two additional boundary conditions can now be imposed along $\partial\hat{\Omega}$. In terms of the physical domain, an

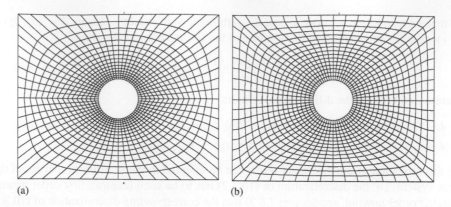

(a)                                                  (b)

**Figure 10.6.** Grids around a cylinder on a two-block domain, with (a) a second-order (10.3.2) and (b) a fourth-order (10.3.6) grid generation process.

obvious possibility is to prescribe the slope of the intersecting grid lines at the boundary and the mesh size orthogonal to the boundary.

Equation (10.3.6) can be discretized and solved by multigrid methods with a similar efficiency to (10.3.2) because all the equations have essentially the same structure as before. All remarks made in the previous section on the proper choice of the multigrid components carry over to this system. However, the peculiarity that is discussed in Section 8.4.2 for the biharmonic system has to be taken into account. We actually have two boundary conditions for $x$ and $y$, respectively, but no boundary condition for $p$ and $q$. These boundary conditions have to be handled with care especially when using V-cycles (see Section 8.4.2 and Remark 8.4.3).

### 10.3.4 Examples

We conclude this section with two examples of grid generation. Figure 10.6 presents grids generated in a two-block domain (with an upper and a lower block) around a cylinder. The left figure shows the grid generated with a straightforward application of the Poisson-type system (10.3.2), the grid in the right figure is generated with the biharmonic system (10.3.6). In this situation the biharmonic grid generator generates a smoother and more homogeneous grid. Near the physical and "interior" block boundaries the size and the shape of the grid cells are more suitable for numerical computations. Both grids are generated by solving discretizations of the respective equations (10.3.2) and (10.3.6) with standard multigrid solvers using alternating zebra line smoothers. A decoupled smoothing is sufficient here.

### 10.4 $L_i SS$: A GENERIC MULTIGRID SOFTWARE PACKAGE

In this and the following section, we will discuss two, in some sense typical, multigrid-based software packages. The first is the $L_i SS$ package, which has been designed to develop robust and efficient multigrid components for various applications. The second is the FLOWer code (developed by the German Aerospace Center DLR, Institute of Design Aerodynamics,

Braunschweig) which has been designed to solve complicated industrial aerodynamic flow problems and which also uses multigrid techniques.

Both codes are based on block-structured grids. The general procedure for solving a problem on block-structured grids can be summarized as follows: first, one starts with the definition of the problem geometry, i.e. one defines the computational domain. This domain is subdivided into subdomains (also called blocks), in each of which a logically rectangular boundary-fitted grid is generated (for example, by solving a system of transformed PDEs by a multigrid algorithm as described in Section 10.3).

The next step is the discretization of the PDE to be solved, and correspondingly the boundary conditions are discretized. After this, the problem description is finished and the solution process of the problem can start.

The PDE is then solved by a fast (iterative) solution method. Typically, the iterations are terminated if the defects are sufficiently small. In an advanced solution process, adaptivity is included so that an error indicator determines refinement regions and additional grid points are automatically added in these regions. After the solution process, the output is analyzed. In particular, the solution is visualized.

These steps of a solution process for PDEs are realized in the $L_i SS$ package [248, 322], which is a general tool for solving PDEs with multigrid. It combines the use of (optionally adaptive) multigrid with block-structured grids in a grid partitioning environment. Both sequential *and* parallel versions are included in a single code. The multigrid solver is portable among sequential and parallel architectures.

$L_i SS$ is designed for the solution of a system of linear or nonlinear PDEs

$$Nu = f$$

on a general bounded 2D domain $\Omega$ together with boundary conditions on $\partial\Omega$. It also allows to solve time-dependent problems

$$u_t = Nu - f$$

with initial and boundary conditions.

For the solution of a new system of PDEs, only a few interface subroutines of $L_i SS$ have to be provided by the user, for example those describing the discretization of a new application. Some example sets, containing all necessary interface routines are available in the package. These example sets currently include

- Poisson's equation,
- the incompressible Navier–Stokes equations using Dick's flux difference splitting for discretization [125] (see also Section 8.8.4),
- the Euler equations using Dick's or Osher's flux difference splitting for discretization [99, 124, 145].

### 10.4.1 Pre- and Postprocessing Components

#### Preprocessing

*Definition of the geometry*  The user has to specify the contour of the domain, its block structure and the boundary point distribution which is used by the grid generator to create the block-structured grid.

*Grid generation*    Grid generators are available which solve a system of transformed Poisson equations or, alternatively, of biharmonic equations with multigrid as described in Section 10.3.

*Subdivision of the block structure*    Often the user-defined block structure of an application consists of a possibly small number of blocks. The number of blocks required for a parallel computation is, however, typically larger, at least if many processors have to be employed. An interactive tool produces a further subdivision of the user-defined blocks in order to allow a full and load-balanced utilization of the available processors.

### Postprocessing

Postprocessing tools provide plots of the domain, the block structure and the corresponding grid and contour plots of all components of the discrete solution. For flow problems, the velocity of the flow and streamlines can be visualized and particle tracing is an option for time-dependent solutions.

### 10.4.2 The Multigrid Solver

The solver for a given PDE (system) is the kernel of the $L_i SS$ package. It is designed for *vertex-centered discretizations*.

Subroutines to be provided by the user define the discretization of the PDE and the boundary conditions.

The resulting (nonlinear) system of algebraic equations is solved by the FAS version of multigrid. The relaxation schemes implemented are GS-LEX, GS-RB and various line relaxations of Gauss–Seidel type. Nonlinear relaxation methods are also supported. Furthermore, the relaxation is performed collectively when dealing with systems of equations.

A defect correction iteration with a high order accurate discretization (as described in Section 5.4.1) is optional. High order discretization can also be dealt with directly using *smoothers* based on the defect correction idea (KAPPA smoothers, see Section 7.4.1).

Concerning the discretization in time, the user has the choice between the (implicit) backward Euler and the Crank–Nicolson scheme. The program supports a time marching method in which the time steps are determined by fixed time steps or by a time step size control.

Grid partitioning (see Section 6.3), where the subgrids (corresponding to the blocks) are mapped to different processes, is used as the parallelization strategy.

The complicated communication tasks in the parallel solution of grid-partitioned block-structured applications are independent of the PDE to be solved. In $L_i SS$, all the communication tasks are performed by routines of the GMD Communications Library [191] for block-structured grids. This library is based on the portable message-passing interface MPI [190]. Thus, it is portable among all parallel machines on which this interface is available.

For problems whose solution only requires a high resolution in parts of the domain, which may be solution-dependent and unknown in advance, fixed block-structured grids may still be too restrictive. Dynamic locally refined grids are adequate for such problems.

The adaptive multigrid approach (see Section 9.2) is used for these problems in $L_i SS$. Here, the locally refined grids are determined in such a way that they form a block structure themselves, so that the advantages of block-structured grids can be maintained on refinement levels. Since the different levels are treated sequentially, load balancing is an important issue at each level. In order to achieve a satisfactory load balancing, the blocks of a refined level are distributed to all processes (processors).

From the user's point of view, a locally refined level is created as follows. In each block, the user or the program determines point sets in which a locally refined grid should be introduced. This information is passed to the communications library. From then on, the library takes over: transparent for the user, it embeds the point sets in logically rectangular blocks, processes the local (blockwise) information, creates a new block structure and maps the blocks to the processors in a load-balanced way. The necessary data transfer is kept to a minimum. All necessary communication between different levels of the hierarchy is supported by the library. The use of the library for adaptive grids essentially reduces the programming effort to what one would have to do for a single block grid on a sequential machine.

**Remark 10.4.1 (generalization to 3D)**   The current features of $L_i SS$ have been developed for the solution of 2D systems of PDEs. One ongoing activity is to generalize $L_i SS$ to 3D block-structured grids (see Section 9.6.2 for an example).                                              ≫

## 10.5 MULTIGRID IN THE AERODYNAMIC INDUSTRY

In this section, we will give an example for multigrid in industrial practice. We would like to emphasize that a solver for industrial applications has to take into account many specific conditions. For example, in industry the requirements on robustness and flexibility of a solver with respect to grid structures, the number of equations in a PDE system, the complexity of the model etc. are often very important. Instead of an optimal method for one application, a satisfactory method for many applications is usually preferred. We will outline this for the particular case of the development of modern aircraft. Here, the numerical simulation is of high economic importance. Objectives in the design of an aircraft are, for example: *drag minimization, minimization of energy consumption* and *noise reduction*. Many aerodynamic phenomena are too expensive and too time consuming to be dealt with experimentally, if they can be treated at all. The numerical simulation of the full flow around an aircraft is a way out. This was recognized many years ago; it is thus not surprising that (theoretical and) computational fluid dynamics has a long history.

Corresponding to the development of modern computer hardware and fast numerical algorithms during recent decades, increasingly complicated flow problems have been solved numerically. Having started in the 1950s with simple flow models around circles and 2D airfoils and having continued with the Euler equations, it is now possible to solve the compressible Reynolds-averaged Navier–Stokes equations (RANS) for the 3D flow around a full aircraft with complicated geometry. *This has become feasible because of the development of appropriate parallel multigrid methods and the use of high performance parallel computers.* Typically, the grids required for such numerical computations consist of many millions of grid points.

For such computations, the German aircraft industry and others use the FLOWer code, a parallel 3D multigrid program [224].

### 10.5.1 Numerical Challenges for the 3D Compressible Navier–Stokes Equations

Before we report on some results obtained with the FLOWer code, we will list some problems, which have to be tackled when solving the compressible Navier–Stokes equations for the flow around an aircraft.

- The PDE systems are *nonlinear*.
- The derivation of an appropriate set of *boundary conditions*, their discretization and multigrid treatment requires a careful study.
- The operator determinant of the PDE system is, in general, a product of several scalar operators, each with different properties. For a fully efficient multigrid solver, each of these properties should, in principle, be taken into account (see also Appendix C).
- Realistic geometries are very *complicated*. The generation of suitable *grids* for such geometries becomes crucial. One possibility is to use *block-structured grids* (see Fig. 10.7 for a wing–body–engine aircraft configuration). In practical aircraft

**Figure 10.7.** Block-structured boundary fitted grid around a DLR-F6 wing–body–engine configuration (courtesy of DLR).

simulations, these grids often have singularities: whole edges (or even faces) of the blocks may collapse into one single geometrical point.

- *Boundary layers* may require the use of extremely fine and anisotropic grids near certain boundaries of the computational domain. As a consequence, the discrete systems may have *large anisotropies* (see Sections 5.1 and 5.2).
- The number of grid points required is so large that only with highly *parallel computers* does one have a chance to solve the problem. Therefore, parallel multigrid is needed (see Chapter 6), preferably combined with a robust adaptive grid approach.
- *Conservative discretization* is needed. Appropriate discretizations can be achieved with finite volumes (similar to Sections 8.8.4 and 8.9).
- *Dominance of convection* and corresponding singular perturbations are common features of many flow problems. They require special treatment (see Sections 7.1–7.4). Sonic lines and recirculation phenomena should also be taken into account (see Appendix C).
- The properties of the PDE systems and the appropriate multigrid treatment depend strongly on the amount of *viscosity* (added artificially).
- Typically, *second-order accuracy* is necessary to obtain sufficiently accurate approximations. If first-order discretizations were used, the grids would have to be so fine that the computing times would become unacceptable.

    Second-order discretizations for systems like the Euler or the Navier–Stokes equations require the development of advanced multigrid components. *Reliable defect correction techniques or the development of special smoothers* are necessary (see Sections 5.4.1 and 7.4).
- The development of optimal multigrid algorithms for turbulent flow, in particular for advanced turbulence models such as the so-called two-equation models (e.g. $k - \varepsilon$ or $k - \omega$) or even large eddy simulation (LES), is a topic of current research.
- A further area which requires special numerical treatment is that of chemically reacting flows (e.g. in hypersonic cases).
- One finally wants to *design* the aircraft and not just calculate the flow around it. Therefore, one is interested in *optimizing* certain parameters (design variables), which leads to (possibly ill-posed) problems of differential optimization [382–384].

### 10.5.2 FLOWer

The combination of the numerical problems listed above makes it very difficult to develop an optimized multigrid solver. Correspondingly, various of the above problems and their optimal multigrid treatment are still topics of current research (see Appendix C). However, quite robust multigrid techniques have been developed and are in practical use. The FLOWer code [224], which is used to compute the *compressible viscous flow around complex 3D aircraft configurations* based on RANS, is a parallel multigrid code for 3D block-structured grids and finite volume discretization using standard coarsening. Jameson's multistage Runge–Kutta schemes are employed as smoothers [202–205] (see also Section 8.9.4), and are improved by various techniques (for details, see [224, 225]). We refer to [348], for a detailed description of FLOWer and its performance.

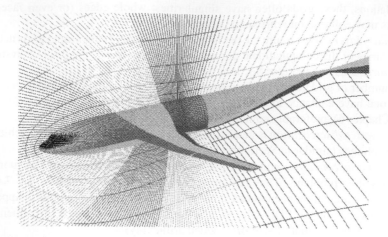

**Figure 10.8.** Block-structured boundary fitted grid around a generic DLR-F4 wing–body configuration (courtesy of DLR).

**Example 10.5.1**   As an example, we want to cite a result [42] for the viscous flow around a generic DLR-F4 wing–body configuration (see Fig. 10.8).

A grid consisting of 6.64 million grid points has been used for the computation of the flow. The memory required for this calculation is over 5 Gbytes. Using four levels, 800 multigrid cycles have been carried out (see the left picture in Fig. 10.9 for the convergence history). For the parallel solution, the whole grid has been partitioned into 128 equally sized blocks. This calculation takes about 3 hours on an IBM-SP2 with 128 Thin2-nodes (or 1.25 hours on 16 processors of a NEC-SX-4 with the grid partitioned into 16 equally sized blocks).

The relatively large number of 800 multigrid cycles indicates that an "optimal" multigrid treatment of all the difficulties described in Section 10.5.1 is not yet available. Nevertheless, FLOWer is *by far* more efficient than any known single grid solver for such applications and has thus become an important design tool in industrial aerodynamics.

The right picture in Figure 10.9 shows how the computed lift coefficient $C_L$ depends on the grid size. Grid 4 denotes the finest grid. The coarser grids (Grids 3, 2, 1) are obtained by standard coarsening. The necessity for fine grid resolution is obvious if the lift coefficient has to be predicted to within 2% accuracy. Here, the scale $1/N^{2/3}$ has been chosen since on this scale a linear behavior of $c_L$ is expected for $N \to \infty$ (i.e. $1/N^{2/3} \to 0$). Note that $h = 1/N^{1/3}$ and we have $O(h^2)$ accuracy.                                           △

### 10.5.3 Fluid–Structure Coupling

Above, we have reported that the computation of the compressible flow around a full aircraft is possible within a few hours. Implicitly, we have assumed a steady flow and a fixed, i.e. a nonflexible geometry of the aircraft.

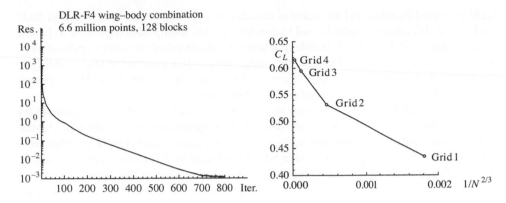

**Figure 10.9.** Multigrid convergence history and "grid convergence".

However, such a steady-state flow around a rigid aircraft configuration describes the real flight behavior of the aircraft only to some degree. This assumption neglects, in particular, the interaction of air flow and structure.

In reality, wings are not rigid but flexible to some extent. This can easily be observed. While the wings hang down before the start of an aircraft due to their own weight, they bend themselves upward during the flight because of the lift. Moreover, the air flow produces forces which influence the position of the wings and, in particular, the effective angle of attack of the flow. A modification of the angle of attack, however, causes a change in the flow, a change of forces acting on the wing and, thus, a further change of the wing position and of the angle of attack, and so on.

The knowledge of the precise situation during the flight is important for aircraft design. Even relatively small changes in the angle of attack can lead to reduced flight performance with larger drag and larger fuel consumption.

In order to solve the complete problem, the dynamic interaction of the wing structure and the aerodynamics has to be taken into account. This is a typical, but still simple, example of a *coupled or multidisciplinary application*. (For example, the propulsion and corresponding interactions may also have to be considered.) Similar multidisciplinary problems have to be solved in many other fields, e.g. in the design of artificial heart valves, which also has to take a dynamic interaction of structural mechanics and fluid dynamics into account.

Multidisciplinary applications are, of course, much more involved and time-consuming than the underlying monodisciplinary subproblems. The development of optimal multigrid solvers for multidisciplinary applications is a new challenge for scientific computing. Some software tools for this purpose are already available (e.g. the MpCCI coupling interface, see http://www.mpcci.org).

## 10.6 HOW TO CONTINUE WITH MULTIGRID

As seen in the above example, the typical multigrid efficiency has not yet been reached for all of the difficulties listed in Section 10.5.1. In Appendix C, A. Brandt sketches some ideas for tackling such difficulties and the status of current research. The general strategy is to study each of the difficulties separately and to develop specific multigrid components to handle each complication. For real-life problems where several complication occur simultaneously, an appropriate combination of the specific components is then expected to lead to multigrid algorithms with the typical multigrid efficiency for a large class of problems. This approach is in accordance with the general philosophy of this book.

In typical real-life applications, one may not be free to choose optimal components for multigrid since, for example, grids and discretization approaches are typically predefined (like in most numerical simulation codes). For such reasons, practitioners are often very interested in robust multigrid approaches which work well for large classes of applications. The AMG approach is systematically introduced in Appendix A of this book. The aim for robustness is fundamental to AMG. In many applications, an acceleration of existing programs can often be achieved by using AMG as a black-box solver, replacing a conventional solver.

In our view, multigrid will remain an interesting topic for future research. The development of robust and efficient multigrid, multilevel and multiscale solvers is still a challenge in some of the traditional application fields, but even more in new areas. The combination of multigrid with adaptivity and parallelism still has a large potential for breakthroughs in numerical simulation.

# Appendix A

# AN INTRODUCTION TO ALGEBRAIC MULTIGRID

### K. Stüben

German National Research Centre for Information Technology (GMD)
Institute for Algorithms and Scientific Computing (SCAI)
Schloss Birlinghoven, D-53757 St. Augustin, Germany

## A.1 INTRODUCTION

In contrast to geometrically based multigrid, algebraic multigrid (AMG) does not require a given problem to be defined on a grid but rather operates directly on (linear sparse) algebraic equations

$$Au = f \quad \text{or} \quad \sum_{j=1}^{n} a_{ij}u_j = f_i \quad (i = 1, 2, \ldots, n). \tag{A.1.1}$$

If one replaces the terms *grids, subgrids* and *grid points* by *sets of variables, subsets of variables* and *single variables*, respectively, one can describe AMG in formally the same way as a geometric multigrid method. In particular, coarse-grid discretizations used in geometric multigrid to reduce low-frequency error components now correspond to certain matrix equations of reduced dimension. However, no multigrid[1] hierarchy needs to be known a priori. In fact, the construction of a (problem-dependent) hierarchy — including the coarsening process itself, the transfer operators as well as the coarse-grid operators — is part of the AMG algorithm, based solely on algebraic information contained in the given system of equations.

Although the central ideas behind AMG and its range of applicability are more general, in this introduction, the focus is on the solution of *scalar* elliptic partial differential equations of

---

[1] We should actually use the term multi*level* rather than multi*grid*. For historical reasons we use the term multi*grid*.

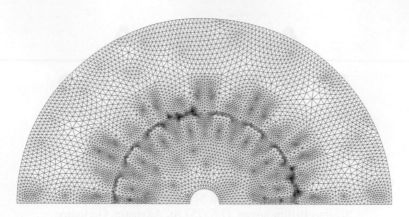

**Figure A.1.** Unstructured finite element mesh (see also Fig. A.31).

second-order. Moreover, we mostly consider *symmetric, positive (semi-) definite* problems. This is because AMG is best developed for such problems. Various recent research activities aim to apply AMG to *systems* of partial differential equations (such as Navier–Stokes equations or structural mechanics problems). However, although important progress has been achieved for different types of systems, major research is still ongoing and there is no well-settled approach.

We will see that AMG provides very robust solution methods. However, the real practical advantage of AMG is that it can be applied directly to structured as well as unstructured grids (see Fig. A.1), in 2D as well as in 3D. In order to point out the similarities and differences of geometric and algebraic multigrid, we will first give a brief review of some major steps in the development of robust geometric approaches.

### A.1.1 Geometric Multigrid

In the early days of multigrid, coarse-grid correction approaches were based on simple coarsening strategies (typically by doubling the mesh size in each spatial direction, that is, by $h \rightarrow 2h$ coarsening), straightforward geometric grid transfer operators (standard interpolation and restriction) and coarse-grid operators being natural analogs of the one given on the finest grid. Later, it was realized that such simple "coarse-grid components" were not appropriate for various types of more complex problems such as diffusion equations with strongly varying or even discontinuous coefficients. The so-called *Galerkin operator* [184] was introduced as an alternative to the "natural" selection of the coarse-grid operators mentioned before. From a practical point of view, it is advantageous that this operator can be constructed purely algebraically. This makes it very convenient for the treatment of, for instance, differential operators with strongly varying coefficients. From a theoretical point of view, the major advantage of Galerkin-based coarse-grid correction processes is that they satisfy a variational principle (for symmetric and positive definite problems). This opened new perspectives for theoretical convergence investigations.

The introduction of *operator-dependent interpolation* [3, 439], interpolation which directly relies on the discretization stencils, was equally important. Together with the Galerkin operator, this kind of interpolation allowed the treatment of larger classes of problems including problems with strongly discontinuous coefficients. The main trouble with such problems is that, after applying a typical smoothing process (relaxation), the error is *no* longer geometrically smooth. The smoothed error exhibits the same discontinuous behavior across discontinuities as the solution itself. Galerkin-based coarsening, however, requires interpolation which correctly operates on such error. While geometric interpolation (which can be accurately applied only to corrections with continuous first derivatives) does not correctly transfer such corrections to finer levels, the discretization stencils themselves *do* reflect the discontinuities and, if used for interpolation, also correctly transfer the discontinuities. These Galerkin-based coarse-grid correction processes with operator-dependent interpolation have become increasingly popular.

All geometric multigrid approaches operate on predefined grid hierarchies. That is, the *coarsening process* itself is fixed and kept as simple as possible. Fixing the hierarchy, however, puts particular requirements on the smoothing properties of the smoother used in order to ensure an efficient interplay between smoothing and coarse-grid correction. Generally speaking, error components which cannot be corrected by appealing to a coarser-grid problem, must be effectively reduced by smoothing (and vice versa). For instance, assuming the coarser levels to be obtained by $h \rightarrow 2h$ coarsening, pointwise relaxation is very efficient for essentially isotropic problems. For anisotropic problems, however, pointwise relaxation exhibits good smoothing properties only "in the direction of strong couplings" (cf. Section A.1.3). Consequently, more complex smoothers, such as alternating line-relaxation or ILU-type smoothers, are required in order to maintain fast multigrid convergence. Multigrid approaches for which the interplay between smoothing and coarse-grid correction works efficiently for large classes of problems are often called "robust".

While the implementation of efficient and robust smoothers was not difficult in 2D model situations, for 3D applications on complex meshes their realization tended to become rather cumbersome. For instance, the robust 3D analog of alternating line relaxation is alternating *plane* relaxation (realized by 2D multigrid within each plane) which, in complex geometric situations, becomes very complicated, if possible at all. ILU smoothers, on the other hand, lose much of their smoothing property in general 3D situations.

It is therefore not surprising that a new trend arose which aimed to simplify the smoother without sacrificing convergence. However, in order to maintain an efficient interplay between smoothing and coarse-grid correction, this required putting more effort into the coarse-grid correction process. More sophisticated coarsening techniques were developed, for example, employing more than one coarser grid on each level of the multigrid hierarchy such as the *multiple semicoarsening* technique (semicoarsening in multiple directions) [122, 275, 277, 411].

### A.1.2 Algebraic Multigrid

Regarding the interplay between smoothing and coarse-grid correction, AMG can be regarded as the most radical attempt to maintain simple smoothers but still achieve robust convergence. Its development started in the early 1980s [67, 80, 81] when Galerkin-based coarse-grid correction processes and, in particular, operator-dependent interpolation were introduced into geometric multigrid (see previous section). One of the motivations for AMG was the observation that reasonable operator-dependent interpolation and the Galerkin operator can be derived directly from the underlying matrices, without any reference to the grids. To some extent, this fact had already been exploited in the first "black-box" multigrid code [120]. However, regarding the selection of coarser levels, this code was still geometrically based. In a purely algebraic setting, the coarsening process itself also needs to be defined solely on the basis of information contained in the given matrix.

This leads to the most important conceptual difference between geometric and algebraic multigrid (cf. Fig. A.2). Geometric approaches employ fixed grid hierarchies and, therefore, an efficient interplay between smoothing and coarse-grid correction has to be ensured by selecting appropriate smoothing processes. In contrast to this, AMG fixes the smoother to some simple relaxation scheme such as plain point Gauss–Seidel relaxation, and enforces an efficient interplay with the coarse-grid correction by choosing the coarser levels and interpolation appropriately. Geometrically speaking, AMG attempts to coarsen only in directions in which relaxation really smoothes the error for the problem at hand. However, since the relevant information is contained in the matrix itself (in terms of size and sign of coefficients), this process can be performed based only on matrix information, producing coarser levels which are *locally* adapted to the smoothing properties of the given smoother. The guiding principle in constructing the operator-dependent interpolation is to *force* its range to approximately contain those "functions" which are unaffected by relaxation. It will turn out that this is the crucial condition for obtaining efficient coarse-grid correction processes.

The coarsening process is fully automatic. This automation is the major reason for AMG's flexibility in adapting itself to specific requirements of the problem to be solved and is the main reason for its robustness in solving large classes of problems *despite using very*

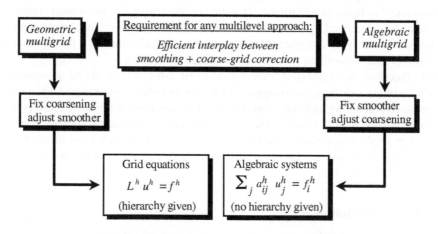

**Figure A.2.** Geometric versus algebraic multigrid.

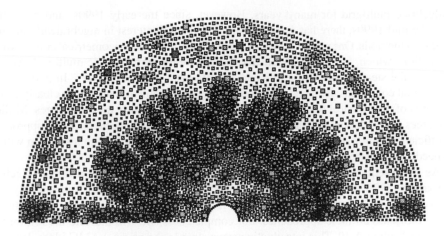

**Figure A.3.** Standard AMG coarsening.

*simple pointwise smoothers*. There is no need for something like multiple semicoarsened grids. Figure A.3 visualizes the hierarchy of grids created by AMG if applied to a diffusion equation discretized on the grid depicted in Fig. A.1. See Section A.1.3 for an explanation of this type of picture and a more detailed example of AMG's coarsening strategy.

The flexibility of AMG and its simplicity of use, of course, have a price: A *setup phase*, in which the given problem (A.1.1) is analyzed, the coarse levels are constructed and all operators are assembled, has to be concluded before the actual *solution phase* can start. This extra overhead is one reason why AMG is usually less efficient than geometric multigrid approaches (if applied to problems for which geometric multigrid *can* be applied efficiently). Another reason is that AMG's components can, generally, not be expected to be "optimal", they will always be constructed on the basis of compromises between numerical work and overall efficiency. Nevertheless, if applied to standard elliptic test problems, the computational cost of AMG's solution phase (ignoring the setup cost) is typically comparable to the solution cost of a *robust* geometric multigrid solver. However, AMG should not be regarded as a competitor to geometric multigrid. The strengths of AMG are its robustness, its applicability in complex geometric situations, and its applicability to solving certain problems which are out of the reach of geometric multigrid, in particular, problems with no geometric or continuous background at all (as long as the given matrix satisfies certain conditions). That is, AMG provides an attractive multilevel variant whenever geometric multigrid is either too difficult to apply or cannot be used at all. In such cases, AMG should be regarded as an efficient alternative to standard numerical methods such as conjugate gradient accelerated by typical (one-level) preconditioners. We will see that AMG itself also provides a very efficient preconditioner. In fact, we will see that simplified AMG variants, used as preconditioners, are often better than more complex ones applied as stand-alone solvers.

The first fairly general algebraic multigrid program was described and investigated in [333, 334, 376], see also [109]. Since the resulting code, AMG1R5, was made publically available in the mid 1980s, there had been no substantial further research and development in algebraic multigrid for many years. However, since the early 1990s, and even more since the mid 1990s, there has been a strong increase of interest in algebraically oriented multilevel methods. One reason for this is certainly the increasing geometrical complexity of applications which, technically, limits the immediate use of geometric multigrid. Another reason is the steadily increasing demand for efficient "plug-in" solvers. In particular, in commercial codes, this demand is driven by increasing problem sizes which clearly exhibit the limits of the classical one-level solvers which are still used in most packages. Millions of degrees of freedom in the underlying numerical models require hierarchical approaches for efficient solution and AMG provides a possibility of obtaining such a solution without the need to completely restructure existing software packages.

As a consequence of this development, there now exist various different algebraic approaches [377], all of which are hierarchical but some of which differ substantially from the original AMG ideas as outlined above. It is beyond the scope of this introduction to AMG to discuss all these approaches. For completeness, we will indicate the relevant literature in Section A.10. This introduction stays close to the original AMG ideas described in [334]. In particular, AMG as we understand it, is structurally completely analogous to standard multigrid methods in the sense that algorithmic components such as smoothing and coarse-grid correction play a role in AMG similar to the one they play in standard multigrid. Nevertheless, there is no unique AMG algorithm and one may think of various modifications and improvements in the concrete realization of AMG's coarsening strategy. We here refer to an approach which, in our experience, has turned out to be very flexible, robust and efficient in practice. It has been implemented in the code RAMG05[2], which is a successor of the original code AMG1R5. However, RAMG05 is completely new and, in particular, incorporates more efficient and more flexible interpolation and coarsening strategies.

### A.1.3 An Example

The flexibility of AMG in adjusting its coarsening process locally to the requirements of a given problem is demonstrated in Fig. A.4. The underlying problem is the differential equation

$$-(au_x)_x - (bu_y)_y + cu_{xy} = f(x, y) \tag{A.1.2}$$

defined on the unit square (with Dirichlet boundary conditions). We set $a = b = 1$ everywhere except in the upper left quarter of the unit square (where $b = 10^3$) and in the lower right quarter (where $a = 10^3$). The coefficient $c$ is zero except for the upper right quarter where we set $c = 2$.

The diffusion part is discretized by the standard fine-point stencil and the mixed derivative by the (left-oriented) seven-point stencil (A.8.23). The resulting discrete system is isotropic in the lower left quarter of the unit square but strongly anisotropic in the remaining quarters. In the upper left and lower right quarters we have strong connections in the $y$- and $x$-directions, respectively. In the upper right quarter strong connectivity is in the diagonal direction. Figure A.4(a) shows what a "smooth" error looks like on the finest level

---

[2]The development of RAMG05 has partly been funded by Computational Dynamics Ltd., London.

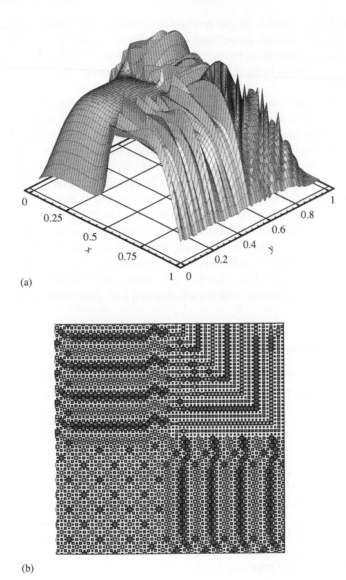

(a)

(b)

**Figure A.4.** (a) "Smooth" error in case of problem (A.1.2); (b) finest and three consecutive levels created by standard AMG coarsening algorithm.

after applying a few point relaxation steps to the homogeneous problem, starting with a random function. The different anisotropies as well as the discontinuities across the interface lines are clearly reflected in the picture.

It is heuristically clear that such error can only be effectively reduced by means of a coarser grid if that grid is obtained by essentially coarsening in directions in which the error really changes smoothly in the geometric sense and if interpolation treats the discontinuities correctly. Indeed, as outlined before, this is exactly what AMG does. First, the operator-based interpolation ensures the correct treatment of the discontinuities. Secondly, AMG coarsening is in the direction of strong connectivity, that is, in the direction of smoothness.

Figure A.4(b) depicts the finest and three consecutive grids created by using standard AMG coarsening and interpolation (cf. Section A.7). The smallest dots mark grid points which are contained *only* on the finest grid, the squares mark those points which are also contained on the coarser levels (the bigger the square, the longer the corresponding grid point stays in the coarsening process). The picture shows that coarsening is uniform in the lower left quarter where the problem is isotropic. In the other quarters, AMG adjusts itself to the different anisotropies by locally coarsening in the proper direction. For instance, in the lower right quarter, coarsening is in the $x$-direction only. Since AMG takes only *strong* connections in coarsening into account and since all connections in the $y$-direction are *weak*, the individual lines are coarsened *independently of each other*. Consequently, the coarsening of neighboring $x$-lines is not "synchronized"; it is actually a matter of "coincidence" where coarsening starts within each line. This has to be observed in interpreting the coarsening pattern in the upper right quarter: within each diagonal line, coarsening is essentially in the direction of this line.

### A.1.4 Overview of the Appendix

The intention of this appendix is to give an elementary, self-contained introduction to an algebraic multigrid approach which is suited, in particular, for the treatment of large classes of scalar elliptic differential equations and problems whose matrices have a similar structure. Although the theoretical considerations are in the framework of positive definite problems, the algorithm presented does not exploit symmetry and can, to some extent, also be applied to certain nonsymmetric and indefinite problems.

We assume that readers will have some basic knowledge of standard (geometric) multigrid. In particular, they should be familiar with the basic principles (smoothing and coarse-grid correction) and with the recursive definition of multigrid cycles (such as V- or F-cycles). This is because, for simplicity, we limit all our descriptions to just two levels. Accordingly, whenever we talk about the efficiency of a particular approach, we always implicitly assume the underlying two-level approach to be recursively extended to full cycles. (Clearly, a mere two-level method is hardly ever practical.)

Section A.2 describes our notation and contains basic theoretical aspects. In particular, it summarizes well-known properties of Galerkin-based coarse-grid correction approaches and shows that AMG, in certain limit cases, degenerates to direct solvers (Section A.2.3). Although, generally, these direct solvers are extremely inefficient in terms of computational work and memory requirement, they can be approximated by more realistic (iterative) approaches in various ways, indicating the formal generality of the approach.

The AMG method as efficiently used in practice, is largely heuristically motivated. However, under certain assumptions, in particular symmetry and positive definiteness, a two-level theory is available showing that convergence can be expected to be independent of the size of the problem and as fast (and expensive) as we wish. Actually, the *convergence* of AMG is not generally the problem (in fact, AMG can always be *forced* to converge rapidly) but rather the trade-off between convergence and numerical work which is also directly related to the memory requirements. Note that this is, in a sense, the opposite of standard multigrid approaches where the numerical work per cycle is known and controllable but the convergence may not be satisfactory.

This appendix covers both theoretical and practical aspects. While the theoretical investigations are contained in Sections A.3–A.6, practical aspects (a concrete algorithm and a discussion of its performance for various types of problems) are presented in Sections A.7–A.8. Although the theoretical results form the basis for the details of the algorithm presented, we have tried to keep the practically oriented sections as independent from the theoretical sections as possible. Readers not interested in the theory may thus decide to skip the corresponding sections. Whenever necessary, we will make reference to relevant theoretical aspects.

In Section A.3, we first introduce the basic concept of algebraic smoothness [67]. This will be used in Section A.4 to prove the convergence of two-level methods using *postsmoothing*. While the main approach is the same as in [67, 334], the definition of interpolation has been modified and extended. The case of *presmoothing* is considered in Section A.5. In both cases, it turns out that it is crucial to define coarsening and interpolation so that the "interpolation error", in some algebraic sense, is uniformly bounded. A realistic extension of the two-level theory to complete V-cycles is not yet available (cf. Section A.6). Moreover, while the underlying AMG code has been successfully applied to various nonsymmetric problems, there is no comparable theory to date for the nonsymmetric case.

The algorithm used in the code RAMG05 mentioned above is described in some detail in Section A.7. Although one can imagine several modifications and improvements, the approach presented has turned out to be very flexible, robust and efficient in practice. Various applications and a discussion of RAMG05's performance are presented in Section A.8. We investigate standard model cases and some industrially relevant cases, for instance, from computational fluid dynamics.

Section A.9 outlines so-called "aggregation-based" AMG variants and points out their relation to the "standard" approach considered in the other parts of this appendix. Finally, in Section A.10, we summarize important further developments and draw some conclusions. Although we try to cover the most important references, the list is certainly not complete in this rapidly developing field of research.

**Remark A.1.1**   Many considerations in the theoretical parts of this appendix refer to a given matrix $A$. However, it should be clear that we are not really interested in, for instance, convergence estimates for one particular $A$ only but rather in having *uniform* convergence if $A$ ranges over some reasonable *class* of matrices, $\mathcal{A}$. A typical class is the class consisting of all M-matrices. However, a reasonable class may also consist of the discretization matrices of a particular elliptic differential equation discretized on a series of grids with mesh size $h \rightarrow 0$. Uniform convergence for $A \in \mathcal{A}$ then means that AMG convergence does not depend on the mesh size (a typical property of geometric multigrid methods). In this sense, we sometimes say that convergence does not depend on the *size of the matrix*.     $\gg$

**Remark A.1.2** All results which refer to positive definite matrices carry over also to the semidefinite zero row-sum case. One just has to exclude the constant vectors from all considerations. Of course, besides treating the coarsest level properly, it is crucial to transfer constants exactly between levels. Since this will be ensured by all the interpolation processes discussed, we will not discuss the semidefinite case explicitly any further.                    ≫

**Acknowledgments** I would like to thank A. Brandt, K. Witsch, R. Lorentz and A. Krechel for various discussions concerning specific aspects of this appendix. I also would like to point out that significant parts of this appendix rely on the early work [334] which was the result of a close cooperation with A. Brandt, S. McCormick, and J. Ruge. Finally, I would like to thank Computational Dynamics Ltd., London, for funding substantial parts of the work on AMG.

## A.2 THEORETICAL BASIS AND NOTATION

As mentioned in the introduction, AMG is based on the Galerkin approach. The formal structure of AMG, with all its components, is described in Section A.2.1, some additional notation is contained in Section A.2.2. The remainder of the section summarizes fundamental aspects related to Galerkin-based coarse-grid correction processes. This includes the discussion of certain limit cases in Section A.2.3 for which AMG degenerates to a direct solver. Since these (unrealistic) limit cases are presented mainly for reasons of motivation—in particular, to indicate the formal generality of the overall approach—this section may well be skipped in first reading. Section A.2.4, which recalls the variational principle of Galerkin-based coarse-grid correction processes for symmetric and positive definite matrices $A$, is more important for the concrete approaches investigated in this appendix.

### A.2.1 Formal Algebraic Multigrid Components

Since the recursive extension of any two-level process to a real multilevel process is formally straightforward, we describe the components of AMG only on the basis of two-level methods with indices $h$ and $H$ distinguishing the fine and coarse level, respectively. In particular, we rewrite (A.1.1) as

$$A_h u^h = f^h \quad \text{or} \quad \sum_{j \in \Omega^h} a_{ij}^h u_j^h = f_i^h \quad (i \in \Omega^h) \tag{A.2.1}$$

with $\Omega^h$ denoting the index set $\{1, 2, \ldots, n\}$. We implicitly assume always that $A_h$ corresponds to a *sparse* matrix. The particular indices, $h$ and $H$, have been chosen to have a formal similarity to geometric two-grid descriptions. In general, they are not related to a discretization parameter.

In order to derive a coarse-level system from (A.2.1), we first need a splitting of $\Omega^h$ into two disjoint subsets $\Omega^h = C^h \cup F^h$ with $C^h$ representing those variables which are to be contained in the coarse level (*C-variables*) and $F^h$ being the complementary set

(*F-variables*). Assuming such a splitting to be given and defining $\Omega^H = C^h$, coarse-level AMG systems,

$$A_H u^H = f^H \quad \text{or} \quad \sum_{l \in \Omega^H} a_{kl}^H u_l^H = f_k^H \quad (k \in \Omega^H), \tag{A.2.2}$$

will be constructed based on the *Galerkin principle*, i.e. the matrix $A_H$ is defined as the *Galerkin operator*

$$A_H := I_h^H A_h I_H^h \tag{A.2.3}$$

where $I_H^h$ and $I_h^H$ denote *interpolation* (or *prolongation*) and *restriction operators*, respectively, mapping coarse-level vectors into fine-level ones and vice versa. We always assume that both operators have *full rank*.

Finally, as with any multigrid method, we need a *smoothing process* with a corresponding linear *smoothing operator* $S_h$. That is, one smoothing step is of the form

$$u^h \longrightarrow \bar{u}^h \quad \text{where } \bar{u}^h = S_h u^h + (I_h - S_h) A_h^{-1} f^h \tag{A.2.4}$$

($I_h$ denotes the identity operator). Consequently, the error $e^h = u_\star^h - u^h$ ($u_\star^h$ denotes the exact solution of (A.2.1)) is transformed according to

$$e^h \longrightarrow \bar{e}^h \quad \text{where } \bar{e}^h = S_h e^h. \tag{A.2.5}$$

Note that we normally use the letter $u$ for *solution* quantities and the letter $e$ for *correction* or *error* quantities.

As mentioned before, AMG employs simple smoothing processes. In this introduction to AMG, we consider only plain *Gauss–Seidel relaxation* (i.e. $S_h = (I_h - Q_h^{-1} A_h)$ with $Q_h$ being the lower triangular part of $A_h$, including the diagonal) or *$\omega$-Jacobi relaxation* (i.e. $S_h = I_h - \omega D_h^{-1} A_h$ with $D_h = \text{diag}(A_h)$). Clearly, unless $A_h$ is positive definite (which we assume most of the time), the use of such variablewise relaxation methods implicitly requires additional assumptions on $A_h$, in particular, its diagonal elements should be sufficiently large compared to the off-diagonal elements.

For completeness, we want to mention that, particularly in connection with theoretical investigations, we also consider *partial* relaxation steps, namely, Gauss–Seidel and Jacobi relaxation applied only to F-variables (with frozen values for the C-variables). Since such partial relaxation will then formally play the role of smoothing, we will refer to it as *F-smoothing*. Note, however, that F-smoothing by itself has no real smoothing properties in the usual sense.

---

**Remark A.2.1** The coarse-level system (A.2.2) formally plays the same role as coarse-grid correction equations in geometric multigrid. In particular, $f^H$ and $u^H$ actually correspond to *residuals* (i.e. defects) and *corrections*, respectively. More precisely, one two-level correction step is defined as

$$u_{\text{new}}^h = u_{\text{old}}^h + I_H^h e^H \quad \text{where } A_H e^H = I_h^H (r_{\text{old}}^h) = I_h^H (f^h - A_h u_{\text{old}}^h). \quad \text{(A.2.6)}$$

For the corresponding errors, this means

$$e_{\text{new}}^h = K_{h,H} e_{\text{old}}^h \quad \text{with } K_{h,H} := I_h - I_H^h A_H^{-1} I_h^H A_h, \quad \text{(A.2.7)}$$

$K_{h,H}$ being the so-called *coarse-grid correction operator*. Consequently, error reduction by one complete two-grid iteration step, including $\nu_1$ and $\nu_2$ pre- and postsmoothing steps, respectively, is described by the *two-grid iteration operator* (cf. (A.2.5)):

$$e_{\text{new}}^h = M_{h,H} e_{\text{old}}^h \quad \text{with } M_{h,H}(\nu_1, \nu_2) = S_h^{\nu_2} K_{h,H} S_h^{\nu_1}. \qquad \gg$$

---

Summarizing, the C/F-splitting and the transfer operators $I_H^h$ and $I_h^H$ need to be *explicitly constructed* in order to formally set up a two-level (and by recursive application a multilevel) process. The construction of these components, which forms the major task of AMG's setup phase, involves closely related processes and, whenever we talk about transfer operators, we always implicitly assume a suitable C/F-splitting to be given. These components need to be selected so that an efficient interplay between smoothing and coarse-grid correction, and consequently good convergence, is achieved. It is equally important that the splitting and the transfer operators are such that $A_H$ is still reasonably sparse and much smaller than $A_h$. In Section A.7, we will describe a practical algorithm. From a more theoretical point of view, we consider AMG components in Sections A.3–A.5.

Except for Section A.2.3, the theoretical parts of this appendix refer to *symmetric* and *positive definite* matrices $A_h$ for which we always define the restriction as the transpose of interpolation,

$$I_h^H = (I_H^h)^T. \qquad \text{(A.2.8)}$$

It then immediately follows that $A_H$ is also symmetric and positive definite, independent of the concrete choice of $I_H^h$ (as long as it has full rank):

$$(A_H u^H, u^H)_E = (I_h^H A_h I_H^h u^H, u^H)_E = (A_h I_H^h u^H, I_H^h u^H)_E = (u^H, A_H u^H)_E$$

where $(.,.)_E$ denotes the *Euclidean inner product*. (Unless explicitly stated otherwise, the terms symmetric and positive definite as well as the transpose of a matrix always refer to the Euclidean inner product.) Moreover, the coarse-grid correction operator $K_{h,H}$ turns out to satisfy a *variational principle* (cf. Section A.2.4).

Finally, all interpolations $e^h = I_H^h e^H$ considered in this appendix are of the form

$$
e_i^h = (I_H^h e^H)_i = \begin{cases} e_i^H & \text{if } i \in C^h \\ \sum_{k \in P_i^h} w_{ik}^h e_k^H & \text{if } i \in F^h \end{cases} \tag{A.2.9}
$$

where $P_i^h \subset C^h$ is called the set of *interpolatory variables*. Clearly, for reasons of efficiency, $P_i$ should be a reasonably small subset of C-variables "near" $i$. Note that any such interpolation has *full rank*.

**Remark A.2.2** According to the above description, we regard the set of coarse-level variables as a subset of the fine-level ones. In particular, (A.2.9) expresses the fact that the coarse-level correction $e_k^H$ is used to directly correct the corresponding fine-level variable, $u_k^h$. Note that this is formally different from algebraic multigrid approaches based on "aggregation" [51, 104, 250, 398]. However, we will see in Section A.9 that aggregation-type approaches can be regarded as a special case of the approach considered here.    $\gg$

### A.2.2 Further Notation

AMG is set up in an algebraic environment. However, rather than using vector–matrix terminology, it is often convenient to formally stick to the grid terminology by introducing *fictitious grids* with grid points being simply the nodes of the directed graph which can be associated with the given matrix. In this sense, we identify each $i \in \Omega^h$ with a *point* and define connections between points in the sense of the associated graph. That is, point $i \in \Omega^h$ is defined to be (directly) *coupled* (or *connected*) to point $j \in \Omega^h$ if $a_{ij}^h \neq 0$. Correspondingly, we define the (direct) *neighborhood* of a point $i$ by

$$
N_i^h = \{j \in \Omega^h : j \neq i, \ a_{ij}^h \neq 0\} \quad (i \in \Omega^h). \tag{A.2.10}
$$

Referring to a point $i \in \Omega^h$ means nothing other than referring to the variable $u_i^h$. Using grid terminology, we can formally interpret the equations $A_h u^h = f^h$ as *grid equations* on the fictitious *grid* $\Omega^h$. Analogously, coarser level equations $A_H u^H = f^H$ can be interpreted as grid equations on *subgrids* $\Omega^H \subset \Omega^h$.

In the course of this appendix, we will use both grid and vector–matrix terminology whichever is more convenient for the given purpose. Moreover, we will usually omit the indices $h$ and $H$, writing for instance, $A$, $e$, $C$ and $K$ instead of $A_h$, $e^h$, $C^h$ and $K_{h,H}$, respectively. We only use these indices if we explicitly need to distinguish between two consecutive levels.

For theoretical investigations, it is often convenient to assume vectors and matrices to be reordered so that, w.r.t. a given C/F-splitting, the set of equations (A.2.1) can be written in block form,

$$
A_h u = \begin{pmatrix} A_{FF} & A_{FC} \\ A_{CF} & A_{CC} \end{pmatrix} \begin{pmatrix} u_F \\ u_C \end{pmatrix} = \begin{pmatrix} f_F \\ f_C \end{pmatrix} = f. \tag{A.2.11}
$$

Correspondingly, the intergrid transfer operators are then written as

$$I_H^h = \begin{pmatrix} I_{FC} \\ I_{CC} \end{pmatrix}, \qquad I_h^H = (I_{CF}, I_{CC}) \qquad (A.2.12)$$

with $I_{CC}$ being the identity operator. Instead of $e^h = I_H^h e^H$ and (A.2.10) we simply write $e_F = I_{FC} e_C$ and

$$e_i = \sum_{k \in P_i} w_{ik} e_k \quad (i \in F), \qquad (A.2.13)$$

respectively. This is for simplicity and should not lead to any confusion.

We finally list some more specific notation. The *range* and *null space* of any operator $Q$ are denoted by $\mathcal{R}(Q)$ and $\mathcal{N}(Q)$, respectively. For any square matrix $Q$, its *spectral radius* is denoted by $\rho(Q)$. At several places in this appendix, we make use of the fact that, for any two matrices, $Q_1$ and $Q_2$, we have

$$\rho(Q_1 Q_2) = \rho(Q_2 Q_1). \qquad (A.2.14)$$

We write $A > 0$ if $A$ is symmetric and positive definite. Correspondingly, $A > B$ stands for $A - B > 0$. For vectors, $u > 0$ and $u \geq 0$ mean that the corresponding inequalities hold componentwise.

If $A_h > 0$, we use the following three inner products in addition to the Euclidean one:

$$(u, v)_0 = (D_h u, v)_E, \quad (u, v)_1 = (A_h u, v)_E \quad \text{and} \quad (u, v)_2 = (D_h^{-1} A_h u, A_h v)_E, \qquad (A.2.15)$$

along with their associated norms $\|.\|_i$ $(i = 0, 1, 2)$. Here, $D_h = \text{diag}(A_h)$. $(., .)_1$ is the so-called *energy inner product* and $\|.\|_1$ the *energy norm*. Moreover, given any C/F-splitting, we will use the analogs of the first two inner products applied to $A_{FF}$ (A.2.11) instead of $A_h$,

$$(u_F, v_F)_{0,F} = (D_{FF} u_F, v_F)_E \quad \text{and} \quad (u_F, v_F)_{1,F} = (A_{FF} u_F, v_F)_E, \qquad (A.2.16)$$

and the associated norms $\|.\|_{i,F}$ $(i = 0, 1)$ where $D_{FF} = \text{diag}(A_{FF})$. (Note that $A_{FF}$ is positive definite.)

Important parts of our theoretical discussion refer to the model class of *symmetric M-matrices*, where a symmetric matrix is defined to be an M-matrix if it is *positive definite* and *off-diagonally nonpositive*. Such matrices often arise from second-order discretizations of scalar elliptic differential equations. If a matrix $A$ contains both negative and positive off-diagonal entries, we use the notation

$$a_{ij}^- = \begin{cases} a_{ij} & (\text{if } a_{ij} < 0) \\ 0 & (\text{if } a_{ij} \geq 0) \end{cases} \quad \text{and} \quad a_{ij}^+ = \begin{cases} 0 & (\text{if } a_{ij} \leq 0) \\ a_{ij} & (\text{if } a_{ij} > 0). \end{cases} \qquad (A.2.17)$$

Correspondingly, we write

$$N_i^- = \{j \in N_i : a_{ij}^h < 0\} \quad \text{and} \quad N_i^+ = \{j \in N_i : a_{ij}^h > 0\}. \qquad (A.2.18)$$

### A.2.3 Limit Case of Direct Solvers

In this section, we will see that, for very specific (impractical) definitions of the smoothing and transfer operators, the two-level methods corresponding to pre- and postsmoothing degenerate to direct solvers, that is, we have either $K_{h,H}S_h = 0$ or $S_h K_{h,H} = 0$. This is true under the mere assumption that $A_h$ is *nonsingular*. In order to show this, let us first state some basic properties of the coarse-grid correction operator, $K_{h,H}$. The transfer operators $I_H^h$ and $I_h^H$ are required to have full rank but are not required to be the transpose of each other.

**Lemma A.2.1** *Let $A_h$ be any nonsingular matrix and assume the C/F-splitting and the transfer operators to be given such that $A_H^{-1}$ exists. We then have*

$$K_{h,H} I_H^h e^H \equiv 0, \quad K_{h,H}^2 = K_{h,H} \quad \text{and} \quad I_h^H A_h K_{h,H} e^h \equiv 0$$

*which implies $\mathcal{N}(K_{h,H}) = \mathcal{R}(I_H^h)$ and $\mathcal{R}(K_{h,H}) = \mathcal{N}(I_h^H A_h)$. Consequently, given any smoothing operator $S_h$, the following holds:*

$$K_{h,H}S_h = 0 \iff \mathcal{R}(S_h) \subseteq \mathcal{R}(I_H^h) \quad \text{and} \quad S_h K_{h,H} = 0 \iff \mathcal{N}(I_h^H A_h) \subseteq \mathcal{N}(S_h).$$

*Proof.* All statements are immediate consequences of the fact that $A_H$ is the Galerkin operator (A.2.3). For instance, the first identity holds because

$$K_{h,H} I_H^h = I_H^h - I_H^h A_H^{-1} I_h^H A_h I_H^h = I_H^h - I_H^h = 0$$

which, in turn, implies $\mathcal{N}(K_{h,H}) \supseteq \mathcal{R}(I_H^h)$. The reverse relation, $\mathcal{N}(K_{h,H}) \subseteq \mathcal{R}(I_H^h)$, follows directly from the definition (A.2.7) of $K_{h,H}$. The proof of the remaining statements is similarly straightforward. $\square$

In the following, we use the notation (A.2.11) and (A.2.12) and, for the purpose of this section, define a very specific "smoothing process" as follows:

$$u \longrightarrow \bar{u} \quad \text{where} \quad A_{FF}\bar{u}_F + A_{FC}u_C = f_F, \quad \bar{u}_C = u_C. \quad \text{(A.2.19)}$$

(Although this is not a practical smoothing process, we formally stick to the standard multigrid terminology.) In terms of the error, $e = u^\star - u$, this means

$$e \longrightarrow \bar{e} \quad \text{where} \quad A_{FF}\bar{e}_F + A_{FC}e_C = 0, \quad \bar{e}_C = e_C \quad \text{(A.2.20)}$$

and the "smoothing operator" is seen to be

$$\hat{S}_h = \begin{pmatrix} 0 & -A_{FF}^{-1}A_{FC} \\ 0 & I_{CC} \end{pmatrix} \quad \text{(A.2.21)}$$

which has the properties

$$\mathcal{R}(\hat{S}_h) = \{e: e_F = -A_{FF}^{-1}A_{FC}\,e_C\} \quad \text{and} \quad \mathcal{N}(\hat{S}_h) = \{e: e_C = 0\}. \quad \text{(A.2.22)}$$

In addition, we define very specific transfer operators by

$$\hat{I}_{FC} = -A_{FF}^{-1}A_{FC} \quad \text{and} \quad \hat{I}_{CF} = -A_{CF}A_{FF}^{-1}. \quad \text{(A.2.23)}$$

We then obtain the following theorem:

**Theorem A.2.1** [222]    *Let $A_h$ be nonsingular and let a C/F-splitting be given such that $A_{FF}^{-1}$ exists. Furthermore, use (A.2.19) as "smoothing process". Then the following statements hold:*

*(1) For $I_{FC} = \hat{I}_{FC}$ and arbitrary $I_{CF}$, $A_H^{-1}$ exists and $K_{h,H} \hat{S}_h = 0$.*
*(2) For $I_{CF} = \hat{I}_{CF}$ and arbitrary $I_{FC}$, $A_H^{-1}$ exists and $\hat{S}_h K_{h,H} = 0$.*
*(3) In either of the two cases, the Galerkin operator (A.2.3) is just the Schur complement corresponding to (A.2.11), that is, $A_H = C_H$ where*

$$C_H := A_{CC} - A_{CF} A_{FF}^{-1} A_{FC}. \tag{A.2.24}$$

*Proof.* If $I_{FC} = \hat{I}_{FC}$, a straightforward computation shows that, independent of $I_{CF}$, the Galerkin operator equals the Schur complement:

$$A_H = I_h^H A_h I_H^h = \begin{pmatrix} I_{CF}, & I_{CC} \end{pmatrix} \begin{pmatrix} A_{FF} & A_{FC} \\ A_{CF} & A_{CC} \end{pmatrix} \begin{pmatrix} -A_{FF}^{-1} A_{FC} \\ I_{CC} \end{pmatrix}$$

$$= \begin{pmatrix} I_{CF}, & I_{CC} \end{pmatrix} \begin{pmatrix} 0 \\ A_{CC} - A_{CF} A_{FF}^{-1} A_{FC} \end{pmatrix} = C_H.$$

Since both $A_h$ and $A_{FF}$ are assumed to be nonsingular, $C_H$ is also nonsingular. Hence, $A_H^{-1}$ exists. By definition, we have $\mathcal{R}(I_H^h) = \mathcal{R}(\hat{S}_h)$ which, according to Lemma A.2.1, implies $K_{h,H} \hat{S}_h = 0$. Regarding the second statement, one can see by analogous arguments as above that $A_H$ equals the Schur complement also in this case. Because of

$$I_h^H A_h = \begin{pmatrix} -A_{CF} A_{FF}^{-1}, & I_{CC} \end{pmatrix} \begin{pmatrix} A_{FF} & A_{FC} \\ A_{CF} & A_{CC} \end{pmatrix} = \begin{pmatrix} 0, & A_H \end{pmatrix}$$

we have $I_h^H A_h e = A_H e_C$ for all $e$. Hence, $\mathcal{N}(I_h^H A_h) = \{e: e_C = 0\} = \mathcal{N}(\hat{S}_h)$ which, according to Lemma A.2.1, implies $\hat{S}_h K_{h,H} = 0$.    □

According to the theorem, only one of the transfer operators has to be explicitly defined in order to obtain a direct method. For the two-level method to be a direct solver independent of whether pre- or postsmoothing is used, both operators have to be specified accordingly.

**Remark A.2.3**   Note that interpolation $e_F = \hat{I}_{FC} e_C$ is defined by exactly solving the homogeneous F-equations

$$A_{FF} e_F + A_{FC} e_C = 0. \tag{A.2.25}$$

That is, interpolation and "smoothing" (A.2.20) are based on the same set of equations. Note furthermore that, for *symmetric* matrices $A_h$, we have $\hat{I}_{CF} = \hat{I}_{FC}^T$, that is, the restriction is just the transpose of interpolation. In contrast to this, for *nonsymmetric* matrices, $\hat{I}_{CF} = \tilde{I}_{FC}^T$ where $\tilde{I}_{FC} = -(A_{FF}^{-1})^T A_{CF}^T$ which is related to solving

$$A_{FF}^T e_F + A_{CF}^T e_C = 0 \qquad\qquad (A.2.26)$$

instead of (A.2.25). Thus, $\hat{I}_{CF}$ is just the transpose of another interpolation, namely, the one corresponding to $A_h^T$.                                                                              $\gg$

---

> The specific two-level approaches defined above can be extended to full V-cycles in a straightforward way by recursively applying the same strategy to the coarse-level Galerkin problems (A.2.2). Assuming the coarsest level equations to be solved exactly, the resulting V-cycles then also converge in just one iteration step. However, such cycles are far from being practical, the obvious reason being that $A_{FF}^{-1}$ is involved in computing both the smoothing and the transfer operators. Generally, the explicit computation of $A_{FF}^{-1}$ is much too expensive and, moreover, a recursive application in a multilevel context would be prohibitive due to fill-in on coarser levels.

---

Of course, the complexity of the matrix $A_{FF}$ strongly depends on the selected C/F-splitting, and by just choosing the splitting appropriately, one may *force* $A_{FF}$ to become simple and easy to invert. For instance, on each level, the splitting can be selected so that all corresponding matrices $A_{FF}$ simply become *diagonal* (assuming nonzero diagonal entries). In some exceptional situations, this indeed leads to an efficient method. For instance, if $A_h$ corresponds to a tridiagonal matrix, the resulting V-cycle can easily be seen to coincide with the well-known method of *total reduction* [342]. Nevertheless, in general, the resulting method will still become extremely inefficient: although the selection of such special C/F-splittings often makes sense in constructing the *second* level, further coarsening rapidly becomes extremely slow causing the corresponding Galerkin matrices to become dense very quickly. This is illustrated in the following example.

**Example A.2.1**   Consider any standard five-point discretization on a rectangular mesh, for instance, the five-point discretization of the 2D Poisson equation. Then, obviously, $A_{FF}$ becomes diagonal if we select the C/F-splitting so that, for each $i \in F$, *all* of its neighbors are in C, that is, if we select *red–black coarsening*, yielding a grid coarsening ratio of 0.5. The coarse-grid operator on the second level (consisting of the black points, say) can be seen to correspond to nine-point stencils. That is, although the reduction of points is substantial, the overall size of the second-level matrix is still close to the finest-level one. Proceeding analogously in creating the third level, will now reduce the grid size only by a factor of 3/4. At the same time, the Galerkin operator grows further: the largest matrix rows on level 3 correspond to 21-point stencils. Clearly, continuing this process will lead to a completely impractical coarsening. For corresponding 3D problems, the situation is even more dramatic.                                                                              $\triangle$

The above V-cycles actually correspond to specific variants of Gauss elimination rather than real multigrid processes. Clearly, in order to obtain more practical *iterative* approaches, the explicit inversion of $A_{FF}$ has to be avoided. From the multigrid point of view, it is most natural to approximate the operators $\hat{I}_{FC}$ and $\hat{S}_h$ by more realistic interpolation and smoothing operators, $I_{FC}$ and $S_h$, respectively (and similarly $\hat{I}_{CF}$ by some $I_{CF}$ if $A_h$ is nonsymmetric). According to Remark A.2.3, all $e \in \mathcal{R}(I_H^h)$ and all $e \in \mathcal{R}(S_h)$ should approximately satisfy equation (A.2.25). We do not want to quantify this here any further but rather refer to Sections A.4 and A.5.

### A.2.4 The Variational Principle for Positive Definite Problems

In the following, we will summarize the basic properties of Galerkin-based coarse-grid correction processes for symmetric, positive definite matrices $A_h$. For symmetric matrices, we always assume (A.2.8). We have already mentioned that the Galerkin operator $A_H$ (A.2.3) is then also symmetric and positive definite. Moreover, a variational principle for the coarse-grid correction operator $K_{h,H}$ (A.2.7) is implied (see the last statement of Corollary A.2.1 below) which simplifies theoretical investigations substantially. This principle follows from well-known facts about orthogonal projectors which, for completeness, are summarized in the following theorem.

**Theorem A.2.2** *Let $(.,.)$ be any inner product with corresponding norm $\|.\|$ and let the matrix $Q$ be symmetric w.r.t. $(.,.)$. Furthermore, let $Q^2 = Q$. Then $Q$ is an orthogonal projector. That is, we have*

*(1) $\mathcal{R}(Q) \perp \mathcal{R}(I - Q)$.*
*(2) For $u \in \mathcal{R}(Q)$ and $v \in \mathcal{R}(I - Q)$ we have $\|u + v\|^2 = \|u\|^2 + \|v\|^2$.*
*(3) $\|Q\| = 1$.*
*(4) For all $u$: $\|Qu\| = \min_{v \in \mathcal{R}(I-Q)} \|u - v\|$.*

*Proof.* The first statement follows immediately since $Q$ is symmetric and $Q^2 = Q$:

$$(Qu, (I - Q)v) = (u, Q(I - Q)v) = (u, 0) = 0.$$

This, in turn, implies the second statement. Regarding the third statement, we obtain by decomposing $u = Qu + (I - Q)u$,

$$\|Q\|^2 = \sup_{u \neq 0} \frac{\|Qu\|^2}{\|u\|^2} = \sup_{u \neq 0} \frac{\|Qu\|^2}{\|Qu\|^2 + \|(I - Q)u\|^2} \leq 1$$

which shows $\|Q\| \leq 1$. Selecting any $u \in \mathcal{R}(Q)$, proves that $\|Q\| = 1$. Regarding the last statement, again by decomposing $u$ as before, we obtain

$$\min_{v \in \mathcal{R}(I-Q)} \|u - v\|^2 = \min \|Qu + (I - Q)u - v\|^2 = \min \|Qu - v\|^2$$

$$= \min(\|Qu\|^2 + \|v\|^2) = \|Qu\|^2. \qquad \square$$

To apply this theorem to $K_{h,H}$, we observe that $A_h K_{h,H}$ corresponds to a symmetric matrix, that is, $K_{h,H}$ itself is symmetric w.r.t. the *energy inner product* (A.2.15):

$$(K_{h,H} u^h, v^h)_1 = (A_h K_{h,H} u^h, v^h)_E = (u^h, A_h K_{h,H} v^h)_E = (u^h, K_{h,H} v^h)_1 .$$

Since we also have $K^2_{h,H} = K_{h,H}$ (see Lemma A.2.1), $K_{h,H}$ is an orthogonal projector. By finally observing that

$$\mathcal{R}(I_h - K_{h,H}) = \mathcal{R}(I^h_H), \tag{A.2.27}$$

we obtain the following corollary.

**Corollary A.2.1** *Let $A_h > 0$ and let any C/F-splitting and any full rank interpolation $I^h_H$ be given. Then the coarse-level correction operator $K_{h,H}$ is an orthogonal projector w.r.t. the energy inner product $(.,.)_1$. In particular, we have:*

*(1) $\mathcal{R}(K_{h,H}) \perp_1 \mathcal{R}(I^h_H)$, i.e. $(A_h K_{h,H} u^h, I^h_H v^H)_E = 0$ for all $u^h, v^H$.*
*(2) For $u^h \in \mathcal{R}(K_{h,H})$ and $v^h \in \mathcal{R}(I^h_H)$ we have $\|u^h + v^h\|^2_1 = \|u^h\|^2_1 + \|v^h\|^2_1$.*
*(3) $\|K_{h,H}\|_1 = 1$.*
*(4) For all $e^h$: $\|K_{h,H} e^h\|_1 = \min_{e^H} \|e^h - I^h_H e^H\|_1$.*

The last statement of the corollary expresses the variational principle mentioned above: Galerkin-based coarse-grid corrections minimize the *energy norm* of the error w.r.t. all variations in $\mathcal{R}(I^h_H)$. As a trivial consequence, a two-level method can never diverge if the smoother satisfies $\|S_h\|_1 \leq 1$ (e.g. Gauss–Seidel relaxation or $\omega$-Jacobi relaxation with a suitably selected underrelaxation parameter $\omega$). That this also holds for complete V-cycles, assuming *any* hierarchy of C/F-splittings and (full rank) interpolation operators to be given, follows immediately by a recursive application (replacing exact coarse-grid corrections by V-cycle approximations with zero initial guess) of the following lemma.

**Lemma A.2.2** *Let the exact coarse-level correction $e^H$ in (A.2.6) be replaced by any approximation $\tilde{e}^H$ satisfying $\|e^H - \tilde{e}^H\|_1 \leq \|e^H\|_1$ (where $\|.\|_1$ is taken w.r.t. $A_H$). Then the approximate two-level correction operator still satisfies $\|\tilde{K}_{h,H}\|_1 \leq 1$.*

*Proof.* For the approximate two-level correction operator

$$\tilde{K}_{h,H} e^h = e^h - I^h_H \tilde{e}^H = K_{h,H} e^h + I^h_H (e^H - \tilde{e}^H)$$

we obtain

$$\|\tilde{K}_{h,H} e^h\|^2_1 = \|K_{h,H} e^h\|^2_1 + \|I^h_H (e^H - \tilde{e}^H)\|^2_1 .$$

Since $\|I^h_H v^H\|_1 = \|v^H\|_1$ holds for all $v^H$, we have

$$\|I^h_H (e^H - \tilde{e}^H)\|^2_1 = \|e^H - \tilde{e}^H\|^2_1 \leq \|e^H\|^2_1 = \|I^h_H e^H\|^2_1 .$$

Hence,

$$\|\tilde{K}_{h,H} e^h\|^2_1 \leq \|K_{h,H} e^h\|^2_1 + \|I^h_H e^H\|^2_1 = \|e^h\|^2_1$$

and, therefore, $\|\tilde{K}_{h,H}\|_1 \leq 1$. $\qquad\square$

Although this does not say anything about the efficiency of a V-cycle, from a practical point of view, it ensures at least some kind of minimum robustness. Based on the properties in Corollary A.2.1, one can easily formulate concrete conditions which imply V-cycle convergence at a rate which is independent of the size of $A_h$ (see, for example, Theorem 3.1 in [334]). Since, unfortunately, these conditions are not suited for the explicit construction of realistic AMG processes, we here just refer to related discussions in [334].

---

**Remark A.2.4** In our final algorithm (see Section A.7), we will employ certain truncation mechanisms in order to limit the growth of the Galerkin operators towards increasingly coarser levels. According to the variational principle and the above remarks, the truncation of *interpolation* (before computing the corresponding Galerkin operator) is a "safe process": in the worst case, overall convergence may slow down, but no divergence can occur. On the other hand, a truncation of the Galerkin operators themselves may be dangerous since this violates the validity of the variational principle and, if not applied with great care, may cause strong divergence in practice.                                                                      $\gg$

---

## A.3 ALGEBRAIC SMOOTHING

In algebraic multigrid, smoothing and coarse-grid correction play formally the same role as in geometric multigrid. However, the meaning of the term "smooth" is different.

- In a geometric environment, the term "smooth" is normally used in a restrictive (viz. the "natural") way. Moreover, in the context of multigrid, an error is regarded as smooth only if it *can be* approximated on some predefined coarser level. That is, smoothness in geometric multigrid has always to be seen relative to a coarser level. For example, an error may be smooth with respect to a semicoarsened grid but not with respect to a standard $h \rightarrow 2h$ coarsened grid. Correspondingly, the "smoothing property" of a given smoother always involves two consecutive levels.

- In contrast to this, in algebraic multigrid, there are no predefined grids and a smoothing property in the geometric sense becomes meaningless. Instead, we *define* an error $e$ to be *algebraically smooth* if it is slow to converge with respect to $S_h$, that is, if $S_h e \approx e$. In other words, we call an error "smooth" if it *has to be* approximated by means of a coarser level (which then needs to be properly constructed) in order to speed up convergence. From an algebraic point of view, this is the important point in distinguishing smooth and nonsmooth errors.

In this section, assuming $A_h$ to be symmetric and positive definite ($A_h > 0$), we will consider algebraic smoothing by relaxation and introduce a concept [67] of how to characterize it. For typical types of matrices, we give some heuristic interpretation of algebraic smoothness which is helpful in finally constructing the coarsening and interpolation.

### A.3.1 Basic Norms and Smooth Eigenvectors

In investigating the smoothing properties of relaxation, we use the inner products and norms defined in (A.2.15). The following lemma (omitting the index $h$) summarizes some basic relations which will be needed later. Note that we can assume $\rho(D^{-1}A)$ to be uniformly bounded for all important classes $\mathcal{A}$ of matrices under consideration.

**Lemma A.3.1**  *Let $A > 0$. Then the following inequalities hold for all $e$:*

$$\|e\|_1^2 \le \|e\|_0 \|e\|_2, \qquad \|e\|_2^2 \le \rho(D^{-1}A) \|e\|_1^2, \qquad \|e\|_1^2 \le \rho(D^{-1}A) \|e\|_0^2.$$
(A.3.1)

*Applying these norms to the eigenvectors of $D^{-1}A$, we have*

$$D^{-1}A \phi = \lambda \phi \implies \|\phi\|_2^2 = \lambda \|\phi\|_1^2 \quad and \quad \|\phi\|_1^2 = \lambda \|\phi\|_0^2.$$
(A.3.2)

*Proof.* The first inequality in (A.3.1) follows from Schwarz' inequality:

$$\|e\|_1^2 = (Ae, e)_E = (D^{-1/2}Ae, D^{1/2}e)_E \le \|e\|_2 \|e\|_0.$$

The other inequalities follow from the equivalence

$$(B_1 e, e)_E \le c (B_2 e, e)_E \iff \rho(B_2^{-1} B_1) \le c$$
(A.3.3)

which holds for all $B_1 > 0$ and $B_2 > 0$. The verification of (A.3.2) is straightforward. $\square$

---

**Remark A.3.1**  The eigenvectors $\phi$ of $D^{-1}A$ play a special role. In particular, eigenvectors corresponding to the *smallest* eigenvalues $\lambda$ are those which typically cause slowest convergence of relaxation and, therefore, correspond to what we defined as an algebraically smooth error. This can most easily be verified for $\omega$-Jacobi relaxation (using proper underrelaxation) by observing that small $\lambda$s correspond to eigenvalues of the $\omega$-Jacobi iteration operator $S = (I - \omega D^{-1}A)$ close to one. This is also true for related schemes such as Gauss–Seidel relaxation, but is not so easy to see.

Clearly, for all relevant applications, we can assume the smallest eigenvalues $\lambda$ to approach zero (otherwise, standard relaxation methods converge rapidly on their own and no multilevel improvement is required). For instance, for standard elliptic problems of second-order, discretized on a square grid with mesh size $h$, the smallest eigenvalues satisfy $\lambda = O(h^2)$ and, for isotropic (e.g. Poisson-like) problems, correspond to just those eigenfunctions $\phi$ which are very smooth geometrically in all spatial directions. On the other hand, the largest eigenvalues satisfy $\lambda = O(1)$ and correspond to geometrically nonsmooth eigenvectors. For an illustration, see Example A.3.1 below.

Generally, however, whether or not slow-to-converge error really corresponds to geometrically smooth error (assuming a geometric background to exist), depends on $A$ (see Example A.3.2 in the next section).                                              $\gg$

**Example A.3.1**   To illustrate the previous remark, consider the matrix $A$ which corresponds to the Laplace operator, discretized on the unit square with mesh size $h = 1/N$,

$$\frac{1}{h^2} \begin{bmatrix} & -1 & \\ -1 & 4 & -1 \\ & -1 & \end{bmatrix}_h .$$    (A.3.4)

Assuming Dirichlet boundary conditions, the eigenvalues and eigenfunctions of $D^{-1}A = h^2 A/4$ are known to be

$$\lambda_{n,m} = (2 - \cos n\pi h - \cos m\pi h)/2 \quad \text{and} \quad \phi_{n,m} = \sin(n\pi x)\sin(m\pi y)$$    (A.3.5)

where $n, m = 1, 2, \ldots, N - 1$. Obviously, we have

$$\lambda_{\min} = \lambda_{1,1} = 1 - \cos \pi h = O(h^2)$$

and

$$\lambda_{\max} = \lambda_{N-1,N-1} = 1 + \cos \pi h = O(1),$$

corresponding to the lowest and highest frequency eigenfunctions, respectively,

$$\phi_{\min} = \sin(\pi x)\sin(\pi y) \quad \text{and} \quad \phi_{\max} = \sin((N-1)\pi x)\sin((N-1)\pi y). \qquad \triangle$$

The previous discussion on $\lambda$ and the corresponding eigenvectors $\phi$ of $D^{-1}A$, together with the relations (A.3.2), motivate the significance of the above norms, in particular, in the context of algebraic smoothing: if applied to a slow-to-converge error $e = \phi$ ($\lambda$ close to zero), all three norms are largely different in size,

$$\|\phi\|_2 \ll \|\phi\|_1 \quad \text{and} \quad \|\phi\|_1 \ll \|\phi\|_0.$$    (A.3.6)

On the other hand, if applied to algebraically nonsmooth error, all three norms are comparable in size. This different behavior makes it possible to identify slow-to-converge error by simply comparing different norms and gives rise to the characterization of algebraic smoothness in the next section.

## A.3.2 Smoothing Property of Relaxation

We say that a relaxation operator $S$ satisfies the *smoothing property* w.r.t. a matrix $A > 0$ if

$$\|Se\|_1^2 \leq \|e\|_1^2 - \sigma \|e\|_2^2 \quad (\sigma > 0)$$    (A.3.7)

holds with $\sigma$ being independent of $e$. This implies that $S$ is efficient in reducing the error $e$ as long as $\|e\|_2$ is relatively large compared to $\|e\|_1$. However, it will generally become very inefficient if $\|e\|_2 \ll \|e\|_1$. In accordance with the motivations given before, such error is called *algebraically smooth*. We say that $S$ satisfies the smoothing property w.r.t. a class $\mathcal{A}$ of matrices if (A.3.7) holds *uniformly* for all $A \in \mathcal{A}$, that is, with the same $\sigma$.

Below, we will show that Gauss–Seidel and $\omega$-Jacobi relaxation satisfy (A.3.7) uniformly for all matrices which are of interest here. First, however, we want to make some further remarks on algebraic smoothness.

**Remark A.3.2** Note that $\sigma \|e\|_2^2 \leq \|e\|_1^2$ is necessary for (A.3.7) to hold which, because of (A.3.3), is equivalent to $\rho(D^{-1}A) \leq 1/\sigma$. Consequently, a necessary condition for (A.3.7) to hold uniformly for all $A \in \mathcal{A}$ is the uniform boundedness of $\rho(D^{-1}A)$ in $\mathcal{A}$ which, as mentioned before, is satisfied for all important classes $\mathcal{A}$ under consideration. $\gg$

We have already indicated that the term "algebraically smooth" in the above sense is not necessarily related to what is called smooth in a geometric environment. In order to illustrate this, we give two examples.

**Example A.3.2** As an extreme case, consider the matrix $\hat{A} > 0$ which corresponds to the (somewhat artificial) stencil

$$\frac{1}{h^2} \begin{bmatrix} & 1 & \\ 1 & 4 & 1 \\ & 1 & \end{bmatrix}_h \tag{A.3.8}$$

with Dirichlet boundary conditions and $h = 1/N$. That is, $\hat{A}$ is similar to $A$ in Example A.3.1 (Poisson equation) except that the sign of all off-diagonal entries has changed from negative to positive. As a consequence of this, compared to the Poisson case, the role of geometrically smooth and nonsmooth error is completely interchanged: algebraically smooth error is actually highly oscillatory geometrically and algebraically nonsmooth error is very smooth geometrically. In order to see this more clearly, observe that

$$\hat{A} = -(A - cI) \quad \text{with } c = 8/h^2.$$

That is, the eigenvalues and eigenfunctions of $\hat{D}^{-1}\hat{A} = (h^2/4)\hat{A}$ are directly related to those of $D^{-1}A = (h^2/4)A$ (cf. Example A.3.1), namely,

$$\hat{\lambda}_{n,m} = -\lambda_{n,m} + 2 \quad \text{and} \quad \hat{\phi}_{n,m} = \phi_{n,m}$$

where $n, m = 1, 2, \ldots, N - 1$. A straightforward computation shows that the smallest and largest eigenvalues of $\hat{D}^{-1}\hat{A}$ and $D^{-1}A$ are the same,

$$\hat{\lambda}_{\min} = \hat{\lambda}_{N-1,N-1} = -\lambda_{\max} + 2 = \lambda_{\min}$$

and

$$\hat{\lambda}_{\max} = \hat{\lambda}_{1,1} = -\lambda_{\min} + 2 = \lambda_{\max},$$

but the corresponding eigenfunctions are interchanged. $\triangle$

**Example A.3.3** For certain matrices, there is no algebraically smooth error at all. For instance, assume $A$ to be strongly diagonally dominant, that is, $a_{ii} - \sum_{j\neq i} |a_{ij}| \geq \delta a_{ii}$ with $\delta > 0$. The latter immediately implies $\rho(A^{-1}D) \leq 1/\delta$ which, because of (A.3.3), is equivalent to $\|e\|_2^2 \geq \delta \|e\|_1^2$ for all $e$. That is, if $\delta$ is of significant size, there is no algebraically smooth error. Clearly, such cases are not really interesting here since they do not require any multilevel improvement. In fact, (A.3.7) implies rapid convergence for all $e$. In the following, we will tacitly exclude such cases. $\triangle$

> As seen from the above considerations, the term "smooth" is sometimes misleading and should better be replaced by, for instance, "slow-to-converge". However, for historical reasons, we stick to the term "smooth".

The following lemma is used for proving the subsequent theorems which refer to the smoothing properties of Gauss–Seidel and Jacobi relaxation, respectively.

**Lemma A.3.2** [334]   *Let $A > 0$ and let the smoothing operator be of the form $S = I - Q^{-1}A$ with some nonsingular matrix $Q$. Then the smoothing property (A.3.7) is equivalent to*

$$\sigma \, Q^T D^{-1} Q \le Q + Q^T - A.$$

*Proof.* Using the particular form of $S$, a straightforward calculation shows

$$\|Se\|_1^2 = \|e\|_1^2 - ((Q + Q^T - A)Q^{-1}Ae, \, Q^{-1}Ae)_E.$$

Hence, (A.3.7) is equivalent to

$$\sigma \, \|e\|_2^2 \le ((Q + Q^T - A)Q^{-1}Ae, \, Q^{-1}Ae)_E$$

which, in turn, is equivalent to

$$\sigma \, (D^{-1}Qe, \, Qe)_E \le ((Q + Q^T - A)e, \, e)_E. \qquad \square$$

**Theorem A.3.1** [67, 334]   *Let $A > 0$ and define, with any vector $w = (w_i) > 0$,*

$$\gamma_- = \max_i \left\{ \frac{1}{w_i a_{ii}} \sum_{j<i} w_j |a_{ij}| \right\}, \qquad \gamma_+ = \max_i \left\{ \frac{1}{w_i a_{ii}} \sum_{j>i} w_j |a_{ij}| \right\}.$$

*Then Gauss–Seidel relaxation satisfies (A.3.7) with $\sigma = 1/(1 + \gamma_-)(1 + \gamma_+)$.*

*Proof.* Gauss–Seidel relaxation satisfies the assumptions of Lemma A.3.2 with $Q$ being the lower triangular part of $A$ (including the diagonal) and we have $Q + Q^T - A = D$. Thus, (A.3.7) is equivalent to $\sigma \, (Q^T D^{-1} Qe, e)_E \le (De, e)_E$ which, because of (A.3.3), is equivalent to $\sigma \le 1/\rho(D^{-1} Q^T D^{-1} Q)$. A sufficient condition for the latter inequality is given by

$$\sigma \le 1/|D^{-1}Q^T| \, |D^{-1}Q|$$

where $|.|$ stands for an arbitrary matrix norm which is induced by a vector norm (i.e., which is the corresponding operator norm). For the special choice

$$|L| = |L|_w = \max_i \left\{ \frac{1}{w_i} \sum_j w_j |l_{ij}| \right\} \tag{A.3.9}$$

we have $|D^{-1}Q| = 1 + \gamma_-$ and $|D^{-1}Q^T| = 1 + \gamma_+$ which proves the theorem. $\qquad \square$

From this theorem we conclude that Gauss–Seidel relaxation satisfies the smoothing property uniformly for all important classes $\mathcal{A}$ of matrices under consideration.

- For all *symmetric M-matrices*, the smoothing property is satisfied with $\sigma = 1/4$. This can be seen by observing that, for any such matrix, there exists a vector $z > 0$ with $Az > 0$ [341]. By choosing $w = z$ in Theorem A.3.1, we obtain

$$\gamma_- = \max_i \left\{ \frac{1}{z_i a_{ii}} \sum_{j<i} z_j |a_{ij}| \right\} = \max_i \left\{ 1 - \frac{1}{z_i a_{ii}} \sum_{j\leq i} z_j a_{ij} \right\} < 1.$$

Similarly, we obtain $\gamma_+ < 1$.
- The previous result, trivially, carries over to all $A > 0$ which are obtained from a symmetric M-matrix by symmetrically flipping some or all off-diagonal signs.
- For any $A > 0$ with $\leq \ell$ nonvanishing entries per row, the smoothing property is satisfied with $\sigma = 1/\ell^2$. This can be seen by selecting $w_i = 1/\sqrt{a_{ii}}$. Because of $a_{ij}^2 < a_{ii} a_{jj}$ ($j \neq i$), it follows that $\gamma_-, \gamma_+ < \ell - 1$.
- From a practical point of view, the previous result is far too pessimistic. We typically have $\sum_{j \neq i} |a_{ij}| \approx a_{ii}$, which means that, by selecting $w_i \equiv 1$, we can expect $\gamma_-$ and $\gamma_+$ to be close to or even less than 1. That is, $\sigma \approx 1/4$ is typical for most applications we have in mind here.

**Theorem A.3.2** [67, 334] *Let $A > 0$ and $\eta \geq \rho(D^{-1}A)$. Then Jacobi relaxation with relaxation parameter $0 < \omega < 2/\eta$ satisfies (A.3.7) with $\sigma = \omega(2 - \omega\eta)$. In terms of $\eta$, the optimal parameter (which gives the largest value of $\sigma$) is $\omega^\star = 1/\eta$. For this optimal parameter, the smoothing property is satisfied with $\sigma = 1/\eta$.*

*Proof.* Jacobi relaxation satisfies the assumptions of Lemma A.3.2 with $Q = (1/\omega)D$. Hence, (A.3.7) is equivalent to $(Ae, e)_E \leq (2/\omega - \sigma/\omega^2)(De, e)_E$ which, because of (A.3.3), is equivalent to $\rho(D^{-1}A) \leq 2/\omega - \sigma/\omega^2$. Replacing $\rho(D^{-1}A)$ by the upper bound $\eta$, leads to the sufficient condition $\eta \leq 2/\omega - \sigma/\omega^2$, or, in terms of $\sigma$, $\sigma \leq \omega(2 - \omega\eta)$. Obviously, $\sigma$ is positive if $0 < \omega < 2/\eta$. This proves the theorem. $\qquad\square$

This theorem shows that Jacobi relaxation has smoothing properties similar to Gauss–Seidel relaxation. However, as in geometric multigrid, some relaxation parameter, $\omega$, is required. Using $\eta = |D^{-1}A|_w$ as an upper bound for $\rho(D^{-1}A)$ (see (A.3.9)), one obtains, for instance, $\eta = 2$ for all symmetric M-matrices. More generally, for all typical scalar PDE applications satisfying $\sum_{j \neq i} |a_{ij}| \approx a_{ii}$ we have $\eta \approx 2$. That is, using the relaxation parameter $\omega = 1/2$, we have $\sigma \approx 1/2$.

Finally we note that Gauss–Seidel and $\omega$-Jacobi relaxation also satisfy the following variant of the smoothing property (A.3.7),

$$\|Se\|_1^2 \leq \|e\|_1^2 - \tilde{\sigma} \|Se\|_2^2 \quad (\tilde{\sigma} > 0). \tag{A.3.10}$$

Regarding the proof, we refer to [334]. Further discussions on smoothing properties of different relaxation schemes can be found in [67].

### A.3.3 Interpretation of Algebraically Smooth Error

We have seen in the previous section that, in the sense of (A.3.7), Gauss–Seidel and $\omega$-Jacobi relaxation have smoothing properties for all matrices $A > 0$ under consideration. This smoothness needs to be exploited in order to finally construct reasonable C/F-splittings and interpolation (see Section A.4.2). Therefore, in this section, we (heuristically) interpret algebraic smoothness for some typical cases.

Algebraically smooth error is characterized by $Se \approx e$ which, according to (A.3.7), implies $\|e\|_2 \ll \|e\|_1$ (see also (A.3.6)). In terms of the residual, $r = Ae$, this means

$$(D^{-1}r, r)_E \ll (e, r)_E$$

which indicates that, on the average, algebraically smooth error is characterized by (scaled) residuals which are much smaller than the error itself. This can also be seen directly. For instance, Gauss–Seidel relaxation, performed at point $i$, corresponds to replacing $u_i$ by $\bar{u}_i$ where

$$\bar{u}_i = \frac{1}{a_{ii}} \left( f_i - \sum_{j \neq i} a_{ij} u_j \right) = \frac{1}{a_{ii}} \left( a_{ii} u_i + f_i - \sum_j a_{ij} u_j \right) = u_i + \frac{r_i}{a_{ii}}$$

or, in terms of the corresponding error,

$$\bar{e}_i = e_i - \frac{r_i}{a_{ii}}.$$

Here, $r_i$ denotes the residual *before* relaxation at point $i$. From this we can heuristically conclude that, for algebraically smooth error (i.e. $\bar{e}_i \approx e_i$),

$$|r_i| \ll a_{ii} |e_i|.$$

That is, although the error may still be quite large globally, locally we can approximate $e_i$ as a function of its neighboring error values $e_j$ by evaluating

$$(r_i =) \quad a_{ii} e_i + \sum_{j \in N_i} a_{ij} e_j = 0. \tag{A.3.11}$$

In this sense, algebraically smooth error provides some rough approximation to the solution of the basic equations (A.2.25).

The fact that (scaled) residuals are much smaller than the errors themselves, is, algebraically, the most important characteristic of smooth error. However, for some specific classes of matrices, we can give algebraic smoothness a more intuitive interpretation.

### A.3.3.1 M-matrices

An algebraically smooth error $e$ satisfies $\|e\|_2 \ll \|e\|_1$ which, because of the first inequality in (A.3.1), implies $\|e\|_1 \ll \|e\|_0$ (see also (A.3.6)) or, equivalently,

$$\frac{1}{2} \sum_{i,j} (-a_{ij})(e_i - e_j)^2 + \sum_i s_i e_i^2 \ll \sum_i a_{ii} e_i^2. \tag{A.3.12}$$

This follows immediately from the equality

$$\|e\|_1^2 = (Ae, e)_E = \sum_{i,j} a_{ij}\, e_i\, e_j = \frac{1}{2} \sum_{i,j} (-a_{ij})(e_i - e_j)^2 + \sum_i s_i e_i^2 \qquad (A.3.13)$$

which can easily be seen to hold for all symmetric matrices $A$. Here and in the sequel, $s_i = \sum_j a_{ij}$ denotes the $i$-th row sum of $A$.

For symmetric M-matrices (see Section A.2.2), we have $a_{ij} \leq 0$ ($j \neq i$) and, in the most important case of $s_i \approx 0$, (A.3.12) means that, on the average for each $i$,

$$\sum_{j \neq i} \frac{|a_{ij}|}{a_{ii}} \frac{(e_i - e_j)^2}{e_i^2} \ll 1. \qquad (A.3.14)$$

> That is, *algebraically smooth error varies slowly in the direction of large (negative) connections*, i.e., from $e_i$ to $e_j$ if $|a_{ij}|/a_{ii}$ is relatively large. In other words, relaxation schemes which satisfy the smoothing property (A.3.7), smooth the error along *strong (negative) connections*.

**Example A.3.4** The most typical example which illustrates the previous statement is given by matrices derived from the model operator $-\varepsilon u_{xx} - u_{yy}$, discretized on a uniform mesh. While, for $\varepsilon \approx 1$, algebraically smooth error changes slowly in both spatial directions, for $\varepsilon \ll 1$ (the anisotropic case) this is true only for the $y$-direction (cf. Fig. A.4 in Section A.1.3 for an example with varying directions of anisotropies).

Another example is illustrated in Fig. A.5. Here $A$ is derived by discretizing

$$-(\varepsilon u_x)_x - (\varepsilon u_y)_y = f(x, y) \qquad (A.3.15)$$

on the unit square with mesh size $h$ and using Dirichlet boundary conditions. The coefficient function $\varepsilon$ is piecewise constant and defined as indicated in Fig. A.5(a). Using standard five-point differencing (for the definition, see Section A.8.4), we obtain uniform stencils away from the interface of discontinuity and, consequently, in this area, algebraically smooth error changes smoothly in both coordinate directions. At the interface itself, however, the discretization stencil (depicted in Fig. A.5(a)) clearly shows that the inner subsquare is virtually decoupled from the rest of the domain: $\varepsilon_{\text{out}}$ is negligible compared to $\varepsilon_{\text{in}}$. Consequently, the error inside the subsquare is unaffected by the error in the rest of the domain and we cannot expect an algebraically smooth error to change smoothly across the interface. In fact, it generally exhibits a sharp discontinuity. This is depicted in Fig. A.5(b) which shows a typical algebraically smooth error obtained after the application of a few Gauss–Seidel relaxation steps to the homogeneous equations (A.3.15). △

### A.3.3.2 Essentially positive-type matrices

A positive definite matrix is of *essentially positive type* [67] if there exists a constant $c > 0$ such that, for all $e$,

$$\sum_{i,j} (-a_{ij})(e_i - e_j)^2 \geq c \sum_{i,j} (-a_{ij}^-)(e_i - e_j)^2. \qquad (A.3.16)$$

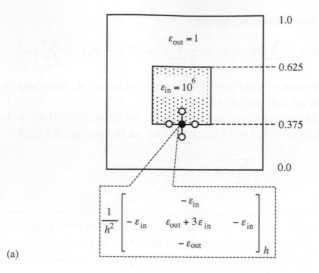

(a)

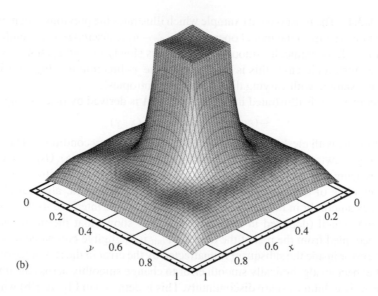

(b)

**Figure A.5.** (a) Coefficient $\varepsilon$ for problem (A.3.15) and discretization stencil at the inner interface; (b) algebraically smooth error obtained after a few Gauss–Seidel relaxation steps.

(Here and further below, we make use of the notation $a_{ij}^-$ and $a_{ij}^+$ as defined in (A.2.17).) The main conclusion for M-matrices carries over to essentially positive type matrices. In particular, instead of (A.3.12), algebraically smooth error satisfies

$$\frac{c}{2} \sum_{i,j} (-a_{ij}^-)(e_i - e_j)^2 + \sum_i s_i e_i^2 \ll \sum_i a_{ii}\, e_i^2 \qquad (A.3.17)$$

which still leads to the conclusion that an algebraically smooth error varies slowly in the direction of large (negative) connections.

Higher order difference approximations to second-order elliptic problems or problems involving mixed derivatives often lead to essentially positive-type matrices. Such and similar matrices have the property that, for each $a_{ij} > 0$, there exist paths of length two (or more) from $i$ to $j$ corresponding to relatively large *negative* connections (we call such paths "strong negative paths"). For instance, we may have $a_{ik} < 0$ and $a_{kj} < 0$ with $|a_{ik}|$, $|a_{kj}|$ being sufficiently large compared to $a_{ij}$. In such cases, (A.3.16) can explicitly be verified by using simple estimates like

$$\frac{\alpha\beta}{\alpha + \beta}(a + b)^2 \le \alpha a^2 + \beta b^2 \quad (\alpha, \beta > 0).\tag{A.3.18}$$

**Example A.3.5** Ignoring boundary conditions, the fourth order discretization of $-\Delta u$ leads to the stencil

$$\frac{1}{12h^2}\begin{pmatrix} & & 1 & & \\ & & -16 & & \\ 1 & -16 & 60 & -16 & 1 \\ & & -16 & & \\ & & 1 & & \end{pmatrix}.$$

Using (A.3.18) with $\alpha = \beta = 1$, one can easily verify (A.3.16) with $c = 3/4$. Similarly, the nine-point discretization of $-\Delta u + u_{xy}$,

$$\frac{1}{h^2}\begin{pmatrix} -1/4 & -1 & +1/4 \\ -1 & 4 & -1 \\ +1/4 & -1 & -1/4 \end{pmatrix},\tag{A.3.19}$$

satisfies (A.3.16) with $c = 1/2$. $\triangle$

Figure A.6 compares algebraically smooth error corresponding to (A.3.19) with that corresponding to the standard five-point Poisson stencil (A.3.4), obtained after the same number of Gauss–Seidel relaxation steps and starting with the same random function. The result is virtually the same, indicating that the positive matrix entries do not significantly influence the smoothing behavior of Gauss–Seidel.

**Remark A.3.3** Note that, for an essentially positive-type matrix, each row containing off-diagonal elements has at least one negative off-diagonal entry. For the $k$-th row, this follows immediately by applying (A.3.16) to the special vector $e = (e_i)$ with $e_i = \delta_{ik}$ (Kronecker symbol). $\gg$

### A.3.3.3 Large positive connections

For essentially positive-type matrices, positive off-diagonal entries are relatively small. If there exist strong negative paths as in the previous examples, algebraically smooth error

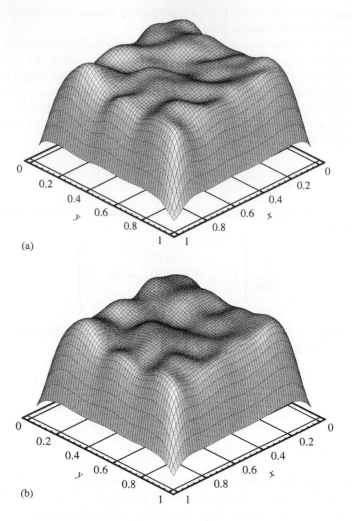

**Figure A.6.** (a) Algebraically smooth error for the stencil (A.3.19) and (b) standard five-point Poisson stencil (A.3.4), respectively.

still varies slowly *even in the direction of positive connections*. However, this cannot be expected to be true any more if positive connections exceed a certain size, in particular not, if $a_{ij} > 0$ and there exist *no* strong negative paths from $i$ to $j$. We demonstrate this for matrices $A > 0$ which are close to being weakly diagonally dominant [200].

To characterize algebraically smooth error analogously as before, observe first that $s_i = t_i + 2 \sum_{j \neq i} a_{ij}^+$ where $t_i := a_{ii} - \sum_{j \neq i} |a_{ij}|$. Using this, one can evaluate (A.3.13) further:

$$
\begin{aligned}
(Ae, e)_E &= \frac{1}{2} \sum_{i,j} |a_{ij}^-|(e_i - e_j)^2 - \frac{1}{2} \sum_{i,j} a_{ij}^+(e_i - e_j)^2 + \sum_i s_i e_i^2 \\
&= \frac{1}{2} \sum_{i,j} |a_{ij}^-|(e_i - e_j)^2 + \sum_i \sum_{j \neq i} a_{ij}^+(2e_i^2 - (e_i - e_j)^2/2) + \sum_i t_i e_i^2 \\
&= \frac{1}{2} \sum_{i,j} |a_{ij}^-|(e_i - e_j)^2 + \frac{1}{2} \sum_i \sum_{j \neq i} a_{ij}^+(2e_i^2 + 2e_j^2 - (e_i - e_j)^2) + \sum_i t_i e_i^2 \\
&= \frac{1}{2} \sum_i \left( \sum_{j \neq i} |a_{ij}^-|(e_i - e_j)^2 + \sum_{j \neq i} a_{ij}^+(e_i + e_j)^2 \right) + \sum_i t_i e_i^2.
\end{aligned}
\tag{A.3.20}
$$

Assuming $t_i \approx 0$ (approximate weak diagonal dominance), $\|e\|_1 \ll \|e\|_0$ now leads to the conclusion that, on the average for each $i$, algebraically smooth error satisfies

$$
\sum_{j \neq i} \frac{|a_{ij}^-|}{a_{ii}} \frac{(e_i - e_j)^2}{e_i^2} + \sum_{j \neq i} \frac{a_{ij}^+}{a_{ii}} \frac{(e_i + e_j)^2}{e_i^2} \ll 1
\tag{A.3.21}
$$

instead of (A.3.14).

---

Consequently, as before, algebraically smooth error can be expected to change slowly in the direction of strong *negative* directions. However, $e_j$ tends to approximate $-e_i$ (relative to the size of $e_i$) if $a_{ij}$ is *positive* and $a_{ij}/a_{ii}$ is relatively large. In other words, algebraically smooth error tends to oscillate along strong positive connections.

---

**Example A.3.6** If $A$ corresponds to the following stencil

$$
\begin{bmatrix} & +1 & \\ -1 & 4 & -1 \\ & +1 & \end{bmatrix},
\tag{A.3.22}
$$

the algebraically smooth error is geometrically smooth only in the $x$-direction but strongly oscillatory in the $y$-direction. This is depicted in Fig. A.7(a). Note that the situation here is completely different from the anisotropic case $-u_{xx} - \varepsilon u_{yy}$ with $\varepsilon \ll 1$. While, in the latter case, the error between any two horizontal gridlines ($y \equiv$ const) is virtually unrelated, it is strongly related in case of (A.3.22). According to the oscillatory behavior in the $y$-direction, the error is actually rather smooth globally, if one considers only every other horizontal gridline.

As an example which is more typical for differential problems, consider the standard discretization of Poisson equation with *antiperiodic* boundary conditions. Compared to the corresponding *periodic* case (only negative off-diagonal entries), here certain off-diagonal entries have changed sign near the boundary. As a consequence, algebraically smooth error will generally exhibit a jump across the boundary. This is illustrated in Fig. A.7(b). Note

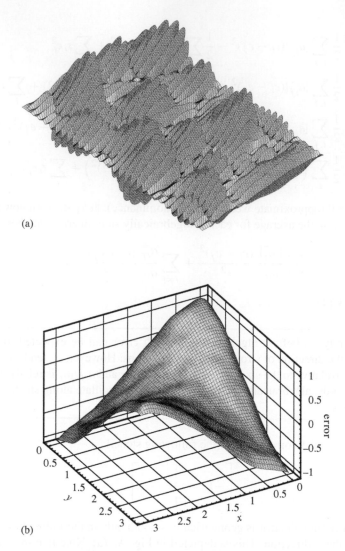

(a)

(b)

**Figure A.7.** Algebraically smooth error in case of problem (A.3.22) and the five-point Laplace operator with anti-periodic boundary conditions.

that AMG will not be able to detect the boundary: all equations with positive off-diagonals look like interior equations to AMG.                                                      △

## A.4 POSTSMOOTHING AND TWO-LEVEL CONVERGENCE

In this section, we investigate the two-level convergence for symmetric and positive definite problems. As mentioned before, using transfer operators satisfying (A.2.8), the corresponding Galerkin operators (A.2.3) are also symmetric and positive definite and the coarse-grid correction operators, $K = K_{h,H}$ (A.2.7), satisfy the variational principle described in Section A.2.4.

We here consider the case of *postsmoothing*, adopting the theoretical approach introduced in [67], also see [334]. For simplicity, we assume that only one smoothing step is performed per cycle, that is, the two-grid operator to be considered is $SK$. In Section A.4.1, we will derive a simple algebraic requirement on interpolation (which implicitly includes also a requirement on the C/F-splitting), in terms of a bound for its "error", which implies uniform two-level convergence w.r.t. the energy norm. In Sections A.4.2 and A.4.3, we will discuss concrete interpolation approaches satisfying this requirement for relevant classes of matrices. Compared to [334], interpolation has been modified and generalized.

Throughout this section and Section A.5 (which covers the case of presmoothing), we employ the inner products and norms defined in (A.2.15) and (A.2.16). Indices $h$ and $H$ will be used only if absolutely necessary.

### A.4.1 Convergence Estimate

For $SKe$ to become small, it is important that the smoothing operator $S$ efficiently reduces all vectors contained in $\mathcal{R}(K)$. Loosely speaking, the error after a coarse-grid correction step has to be "relaxable". Since, assuming property (A.3.7) to be satisfied, error reduction by smoothing becomes the less efficient the smaller $\|e\|_2$ is relative to $\|e\|_1$, the least we have to require is that, for all $e \in \mathcal{R}(K)$, $\|e\|_2$ is bounded *from below* by $\|e\|_1$. This leads to the following theorem.

**Theorem A.4.1** [334]   *Let $A > 0$ and let $S$ satisfy the smoothing property (A.3.7). Furthermore, assume the C/F-splitting and interpolation to be such that*

$$\|Ke\|_1^2 \leq \tau \, \|Ke\|_2^2 \tag{A.4.1}$$

*with some $\tau > 0$ being independent of $e$. Then $\tau \geq \sigma$ and $\|SK\|_1 \leq \sqrt{1 - \sigma/\tau}$.*

*Proof.* By combining (A.3.7) and (A.4.1) one immediately obtains

$$\|SKe\|_1^2 \leq \|Ke\|_1^2 - \sigma \, \|Ke\|_2^2 \leq (1 - \sigma/\tau) \, \|Ke\|_1^2 \leq (1 - \sigma/\tau) \, \|e\|_1^2$$

which proves the theorem.   $\square$

Condition (A.4.1) is not very practical. The following theorem gives a sufficient condition directly in terms of "accuracy" of interpolation.

**Theorem A.4.2** [334]   *If the C/F-splitting and interpolation $I_{FC}$ are such that, for all $e$,*

$$\|e_F - I_{FC}e_C\|_{0,F}^2 \leq \tau \, \|e\|_1^2 \tag{A.4.2}$$

*with $\tau$ being independent of $e$, then (A.4.1) is satisfied.*

*Proof.* Let any $e \in \mathcal{R}(K)$ be given. Then, for arbitrary $e^H$, the orthogonality properties of $K$ (Corollary A.2.1) imply

$$\|e\|_1^2 = (Ae, e - I_H^h e^H)_E$$

and a straightforward application of Schwarz' inequality yields

$$\|e\|_1^2 = (D^{-1/2}Ae, D^{1/2}(e - I_H^h e^H))_E \le \|e\|_2 \|e - I_H^h e^H\|_0 . \tag{A.4.3}$$

By selecting $e^H$ to be just the straight projection of $e$ to the coarse level, we obtain

$$\|e - I_H^h e^H\|_0 = \|e_F - I_{FC}e_C\|_{0,F}.$$

Hence, assumption (A.4.2) implies $\|e\|_1^2 \le \tau \|e\|_2^2$ for all $e \in \mathcal{R}(K)$ which proves (A.4.1). $\qquad\square$

We have already noted (Remark A.1.1) that we are not interested in convergence for one particular $A$ only but rather in having *uniform* convergence if $A$ ranges over some reasonable *class* of matrices, $\mathcal{A}$. Since, according to Section A.3.2, standard relaxation schemes satisfy the smoothing property (A.3.7) uniformly for all problems under consideration, Theorem A.4.1 actually implies uniform two-level convergence for $A \in \mathcal{A}$ if we can show that an (operator-dependent) interpolation can be constructed so that (A.4.2) holds uniformly for all such $A$ (i.e. with the same $\tau$). That this is possible for relevant classes $\mathcal{A}$, will be shown in the following section. First, however, we want to make some general remarks on the requirement (A.4.2).

---

**Remark A.4.1** In the limit case of zero row sum matrices, the right-hand side of (A.4.2) is zero for all constant vectors $e$. Hence, constants necessarily have to be interpolated exactly (cf. Remark A.1.2). Note that this is not necessary for nonsingular matrices. In fact, one may be able to satisfy (A.4.2) with smaller $\tau$-values if one does *not* force constants to be interpolated exactly (see the related discussion in Section A.4.2). $\qquad\gg$

---

**Remark A.4.2** Although (A.4.2) has to hold for all $e$, it implies a nontrivial condition only if $e$ is algebraically smooth. This is most easily seen by writing (A.4.2) in terms of the eigenvectors $\phi$ of $D^{-1}A$ and using (A.3.2),

$$\|\phi_F - I_{FC}\phi_C\|_{0,F}^2 \le \lambda \tau \|\phi\|_0^2. \tag{A.4.4}$$

Requiring this to be uniformly satisfied within a relevant class $\mathcal{A}$ of matrices implies a nontrivial condition only for those $\phi = \phi_A$ which correspond to the small eigenvalues $\lambda = \lambda_A$, in particular, those which approach zero if $A$ varies in $\mathcal{A}$. According to Remark A.3.1, these are just the algebraically smooth eigenvectors. If $\mathcal{A}$ consists of the $h$-discretization matrices corresponding to a standard and isotropic elliptic problem, we know that algebraically smooth eigenvectors are also geometrically smooth and their eigenvalues satisfy $\lambda = O(h^2)$. In such cases, obviously, (A.4.2) is closely related to the requirement of *first-order* interpolation. $\qquad\gg$

We finally note that the two-level approach considered here can be regarded as an approximation to the direct solver (postsmoothing variant) described in Section A.2.3. In particular, the requirement that the error after a coarse-grid correction step has to be "relaxable" (see beginning of this section) corresponds to the property of the direct solver that all vectors in $\mathcal{R}(K)$ are annihilated by smoothing (cf. Lemma A.2.1).

## A.4.2 Direct Interpolation

In the following, we consider basic approaches to automatically construct (operator-dependent) interpolation which, for relevant classes $\mathcal{A}$, can be shown to uniformly satisfy (A.4.2). For ease of motivation, we start with the class of M-matrices in Section A.4.2.1. Generalizations are given in Sections A.4.2.2 and A.4.2.3. A realization in practice, is given in Section A.7.

In order to motivate the general approach in constructing interpolation, we recall that (A.4.2) is a nontrivial condition only for algebraically smooth error (see Remark A.4.2). For such error, however, we have heuristically seen in Section A.3.3 that the basic equations (A.2.25) are approximately satisfied (cf. (A.3.11)). Consequently, the definition of interpolation will be based on the same equations.

That is, given a C/F-splitting and sets $P_i \subseteq C$ ($i \in F$) of interpolatory points, the goal is to define the interpolation weights $w_{ik}$ in

$$e_i = \sum_{k \in P_i} w_{ik} e_k \quad (i \in F) \tag{A.4.5}$$

so that (A.4.5) yields a reasonable approximation for any algebraically smooth $e$ which approximately satisfies

$$a_{ii} e_i + \sum_{j \in N_i} a_{ij} e_j = 0 \quad (i \in F). \tag{A.4.6}$$

Of course, the actual construction of the C/F-splitting and the interpolation itself are closely related processes. Generally, the splitting has to be such that each F-point has a "sufficiently strong connection" to the set of C-points. Although this connectivity does not necessarily have to be via *direct* couplings, in the following sections we only consider "direct" interpolation, that is, we assume the sets of interpolatory points to satisfy $P_i \subseteq C \cap N_i$ where, as before, $N_i$ denotes the direct neighborhood (A.2.10) of point $i$. This is for simplicity; some remarks on more general "indirect" interpolations are contained in Section A.4.3.

**Remark A.4.3** Variables which are not coupled to any other variable (corresponding to matrix rows with all off-diagonal entries being zero) will always become F-variables which, however, do not require any interpolation. For simplicity, we exclude such trivial cases in the following. $\gg$

### A.4.2.1 M-matrices

We have seen in Section A.3.3.1 that, for symmetric M-matrices, algebraically smooth error varies slowly in the direction of *strong* couplings. That is, the error at a point $i$ is essentially determined by a weighted average of the error at its strong neighbors. Consequently, assuming $\emptyset \neq P_i \subseteq C \cap N_i$, the more strong connections of any F-variable $i$ are contained in $P_i$, the better will

$$\frac{1}{\sum_{k \in P_i} a_{ik}} \sum_{k \in P_i} a_{ik} e_k \approx \frac{1}{\sum_{j \in N_i} a_{ij}} \sum_{j \in N_i} a_{ij} e_j \tag{A.4.7}$$

be satisfied for smooth error. This suggests approximating (A.4.6) by

$$a_{ii} e_i + \alpha_i \sum_{k \in P_i} a_{ik} e_k = 0 \quad \text{with } \alpha_i = \frac{\sum_{j \in N_i} a_{ij}}{\sum_{k \in P_i} a_{ik}} \tag{A.4.8}$$

which leads to an interpolation formula (A.4.5) with matrix-dependent, positive weights:

$$w_{ik} = -\alpha_i a_{ik}/a_{ii} \quad (i \in F, \ k \in P_i). \tag{A.4.9}$$

Note that the row sums of (A.4.6) and (A.4.8) are equal and we have

$$a_{ii} \left( 1 - \sum_{k \in P_i} w_{ik} \right) = s_i := \sum_j a_{ij} \tag{A.4.10}$$

showing that $\sum_{k \in P_i} w_{ik} = 1$ if $s_i = 0$. Consequently, in the limit case of zero row sum matrices, constants are interpolated exactly (cf. Remark A.4.1). For regular matrices, however, this is not the case. Instead, the weights are chosen so that $I_{FC} 1_C$ is an approximation to $\hat{I}_{FC} 1_C$ (see Section A.2.3). More precisely, $I_{FC} 1_C$ equals the result of one Jacobi step applied to the equations (A.4.6) with the vector $e = 1$ as the starting vector. (Here $1$ denotes the vector with all components being ones.)

The above interpolation approach can formally be applied to any M-matrix and any C/F-splitting provided that $C \cap N_i \neq \emptyset$ for each $i \in F$. The following theorem shows that (A.4.2) can be satisfied uniformly within the class of weakly diagonally dominant M-matrices whenever the sets $C \cap N_i$ are reasonably large.

**Theorem A.4.3** *Let $A$ be a symmetric M-matrix with $s_i = \sum_j a_{ij} \geq 0$. With fixed $\tau \geq 1$ select a C/F-splitting so that, for each $i \in F$, there is a set $P_i \subseteq C \cap N_i$ satisfying*

$$\sum_{k \in P_i} |a_{ik}| \geq \frac{1}{\tau} \sum_{j \in N_i} |a_{ij}|. \tag{A.4.11}$$

*Then the interpolation (A.4.5) with weights (A.4.9) satisfies (A.4.2).*

*Proof.* We first note that, according to Remark A.4.3, $P_i \neq \emptyset$ for all $i \in F$. Because of (A.3.13), we can estimate for all $e$

$$\|e\|_1^2 = (Ae, e)_E \geq \sum_{i \in F} \left( \sum_{k \in P_i} (-a_{ik})(e_i - e_k)^2 + s_i e_i^2 \right). \tag{A.4.12}$$

On the other hand, employing Schwarz' inequality, we can estimate

$$\|e_F - I_{FC}e_C\|_{0,F}^2 = \sum_{i \in F} a_{ii} \left( e_i - \sum_{k \in P_i} w_{ik}e_k \right)^2$$

$$= \sum_{i \in F} a_{ii} \left( \sum_{k \in P_i} w_{ik}(e_i - e_k) + \left( 1 - \sum_{k \in P_i} w_{ik} \right)e_i \right)^2 \qquad (A.4.13)$$

$$\leq \sum_{i \in F} a_{ii} \left( \sum_{k \in P_i} w_{ik}(e_i - e_k)^2 + \left( 1 - \sum_{k \in P_i} w_{ik} \right)e_i^2 \right). \qquad (A.4.14)$$

Observing (A.4.10), the previous two estimates imply (A.4.2) if $a_{ii}w_{ik} \leq \tau |a_{ik}|$ holds for $i \in F$, $k \in P_i$. According to the definition of the interpolation weights (A.4.9), this is equivalent to $\alpha_i \leq \tau$ $(i \in F)$ which, in turn, is equivalent to assumption (A.4.11). $\square$

The requirement of weak diagonal dominance in the previous theorem is sufficient but not necessary. The following generalization applies to the class of M-matrices whose row sums are uniformly bounded from below and whose eigenvalues are uniformly bounded away from zero.

**Theorem A.4.4** *Let the symmetric M-matrix A satisfy $s_i = \sum_j a_{ij} \geq -c$ with some $c \geq 0$ and assume $(Ae, e)_E \geq \varepsilon(e, e)_E$ for all e with some $\varepsilon > 0$. With fixed $\tau \geq 1$, select a C/F-splitting as in Theorem A.4.3. Then the interpolation (A.4.5) with weights (A.4.9) satisfies (A.4.2) with $\tau$ replaced by some $\tilde{\tau} = \tilde{\tau}(\varepsilon, c, \tau)$. As a function of $\varepsilon$ and $c$, we have $\tilde{\tau} \to \infty$ if either $c \to \infty$ or $\varepsilon \to 0$.*

*Proof.* Let us assume that $s_i < 0$ for (at least) one $i$. Instead of (A.4.12) we employ the following estimate with $\tilde{A} = A + cI$:

$$(\tilde{A}e, e)_E \geq \sum_{i \in F} \left( \sum_{k \in P_i} (-a_{ik})(e_i - e_k)^2 + (c + s_i)e_i^2 \right).$$

In order to estimate the interpolation error, we proceed as in the proof of the previous theorem. However, we need to modify the estimation of (A.4.13) for those $i \in F$ for which $s_i < 0$ (because $1 - \sum_{k \in P_i} w_{ik} < 0$, see (A.4.10)) by, for instance, inserting an additional estimate of the form $(a + b)^2 \leq 2(a^2 + b^2)$ for each such $i$. A straightforward computation, exploiting that $|s_i| \leq c$ and $a_{ii} \geq \varepsilon$, then yields an estimate of the form

$$\|e_F - I_{FC}e_C\|_{0,F}^2 \leq \tau_1(\tilde{A}e, e)_E + \tau_2(e, e)_E \leq \tilde{\tau} \|e\|_1^2$$

with $\tilde{\tau} = \tilde{\tau}(\varepsilon, c, \tau)$. The rest of the theorem follows from the explicit form of $\tilde{\tau}$. $\square$

While the smoothing property (A.3.7) is uniformly satisfied in the class of *all* symmetric M-matrices, the previous theorem indicates that (A.4.2) cannot be expected to uniformly hold in this class. In particular, the smaller the first eigenvalue of $A$ (i.e. the smaller $\varepsilon$) is, the higher is the required accuracy in interpolating the corresponding eigenvector. However, unless this eigenvector is constant (cf. Remark A.1.2), this cannot be achieved by the above interpolation. The following example illustrates this situation.

**Example A.4.1**  Consider the class of matrices $A_c$ ($0 \leq c < \lambda_0$) defined by discretizing the Helmholtz operator $-\Delta u - cu$ on the unit square with *fixed* mesh size $h$ and Dirichlet boundary conditions. Here $\lambda_0 > 0$ denotes the smallest eigenvalue of the corresponding discrete Laplace operator, $A_0$. If we select $e = \phi_0$ as the corresponding eigenfunction (normalized so that $\|e\|_E = 1$), we have for each $A = A_c$ that $\|e\|_1^2 = \lambda_0 - c$. Thus, for (A.4.2) to hold independently of $c$,

$$\|e_F - I_{FC} e_C\|_{0,F}^2 \leq \tau (\lambda_0 - c)$$

is required which means that the first eigenfunction of the Laplace operator has to be approximated with increasing accuracy if $c \rightarrow \lambda_0$. However, this is generally not true unless the interpolation formula is improved by special techniques (e.g. based on an approximate knowledge of $\phi_0$ [334]). Numerical results for this case, are given in Section A.8.5.3.   $\triangle$

According to the above theorems, the C/F-splitting should be selected so that, for each $i \in F$, a *fixed fraction* of its total strength of connection is represented in C (and used for interpolation). However, given any $\tau$, this leaves quite some freedom in realizing a concrete splitting algorithm. Although specific realizations cannot be distinguished within the framework of this algebraic two-level theory, the convergence that finally results may substantially depend on the details of this realization. We make some basic remarks.

**Remark A.4.4**  The concrete choice of $\tau$ is crucial. Clearly, the larger $\tau$ is, the weaker is the assumption (A.4.11). In particular, for large $\tau$, (A.4.11) allows for rapid coarsening, but the two-level convergence will be very slow (see Section A.4.1). On the other hand, the choice $\tau = 1$ gives best convergence, but will force *all* neighbors of $i \in F$ to be in C. In fact, since the latter means that the submatrix $A_{FF}$ becomes diagonal, this results in a direct solver as described in Section A.2.3 (if combined with F-relaxation for smoothing) and we have already seen that this approach, if applied recursively, will be extremely inefficient (cf. Example A.2.1). A reasonable compromise is $\tau = 2$ which means that about 50% of the total strength of connections of every F-point has to be represented on the next coarser level. However, from a practical point of view, coarsening may still be too slow, in particular, for matrices which have many row entries of similar size. We will return to a practical algorithm in Section A.7.                              $\gg$

**Remark A.4.5**  To satisfy (A.4.11) with as few C-points as possible, one should arrange the splitting so that C-points are only chosen from the *strongest* connections of every F-point. This just means coarsening "in the direction of smoothness".  ≫

**Remark A.4.6**  Given an application with geometrical background, the algebraic condition (A.4.11) does not take the geometric locations of C-points relative to the F-points into account. In fact, this is the main reason for the limited accuracy of purely algebraically defined interpolation as mentioned in Remark A.4.2. In practice, however, the accuracy of interpolation, and through this the resulting convergence, can often be substantially improved by just arranging the C/F-distribution carefully. As a rule, it has been beneficial to arrange the C/F-splitting so that the set of C-points builds (approximately) a maximal set with the property that the C-points are not strongly coupled among each other ("maximally independent set") and that the F-points are "surrounded" by their interpolatory C-points. This can be ensured to a large extent by merely exploiting the connectivity information contained in the matrix. For an illustration, see the following example.  ≫

**Example A.4.2**  A careful arrangement of the C/F-splitting in the sense of the previous remark is particularly beneficial if the matrix $A$ corresponds to a geometrically posed problem with many off-diagonal entries being *of essentially the same size*. Consider, for instance, the nine-point discretization of the Poisson operator

$$\frac{1}{3h^2} \begin{bmatrix} -1 & -1 & -1 \\ -1 & 8 & -1 \\ -1 & -1 & -1 \end{bmatrix}_h . \tag{A.4.15}$$

Figure A.8 illustrates some local C/F-arrangements all of which are admissible (in the sense of (A.4.11)) if we set $\tau = 4$. Clearly, interpolation given by the left-most arrangement is worst: it just gives first-order accuracy, the best we can really *ensure* by purely algebraic means (cf. Remark A.4.2). Following the rule mentioned in Remark A.4.6, we would obtain a C/F-splitting for which the two right-most arrangements are characteristic. Both of them correspond to second-order interpolation which, as we know from geometric multigrid,

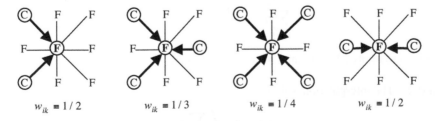

**Figure A.8.**  Different C/F-arrangements and corresponding interpolation formulas.

gives a much better performance for such balanced stencils as considered in this example. The second arrangement does not give second-order, but is still better than the first one.

This illustrates that a proper arrangement of the splitting may substantially enhance convergence. Ignoring this, may, in the worst case, not only cause relatively slow convergence of the two-level method, but also an $h$-dependent convergence behavior of full V-cycles. For an extreme example of such a situation, see Example A.6.1. Clearly, in general, there is no way to strictly ensure optimal interpolation and convergence by exploiting only algebraic information contained in the matrix. In practice, however, it is usually sufficient to avoid extreme one-sided interpolation (see the related discussion in Sections A.6 and A.9). $\triangle$

For completeness, we want to briefly mention a few other typical approaches to define interpolation which are, however, not in the focus of this appendix and may well be skipped in reading. In each case, we briefly discuss requirements which, instead of (A.4.11), ensure (A.4.2) for weakly diagonally dominant M-matrices. Note that any interpolation (A.4.5) with $P_i \subseteq C \cap N_i$ satisfies (A.4.2) if the following two inequalities hold:

$$0 \le a_{ii} w_{ik} \le \tau |a_{ik}|, \qquad 0 \le a_{ii}\left(1 - \sum_{k \in P_i} w_{ik}\right) \le \tau s_i. \tag{A.4.16}$$

This follows immediately from the proof of Theorem A.4.3 and will be used below.

**Variant 1** One variant, considered in [334], starts from the assumption

$$\sum_{j \notin P_i} a_{ij} e_j \approx \left(\sum_{j \notin P_i} a_{ij}\right) e_i$$

instead of (A.4.7). This corresponds to adding all noninterpolatory connections to the diagonal, leading to the weights

$$w_{ik} = -a_{ik} \bigg/ \sum_{j \notin P_i} a_{ij}. \tag{A.4.17}$$

A certain drawback of this approach is that the denominator in (A.4.17) may, in principle, become zero or even negative for matrices which contain rows with $s_i < 0$. Apart from this, however, this variant performs very comparably to the interpolation (A.4.9) (in fact, in the limit case of zero row sums, $s_i \equiv 0$, both interpolations are identical). For weakly diagonally dominant M-matrices, the requirement

$$\sum_{j \notin P_i} a_{ij} \ge \frac{1}{\tau} a_{ii} \quad (i \in F). \tag{A.4.18}$$

can be seen to imply (A.4.16) and, hence, also (A.4.2).

**Variant 2** The interpolation weights

$$w_{ik} = a_{ik} \bigg/ \sum_{j \in P_i} a_{ij}$$

are constructed so that $\sum_{k \in P_i} w_{ik} \equiv 1$ which forces constants to be *always* interpolated exactly. Compared to (A.4.9), no essential difference in behavior is expected as long as $s_i \approx 0$. However, if this is strongly violated for some $i$ (e.g. near boundaries), the approximation of algebraically smooth error (which is *not* constant) becomes less accurate which, in turn, may cause a certain reduction in convergence speed. According to (A.4.16), (A.4.2) is satisfied if

$$\sum_{k \in P_i} |a_{ik}| \geq \frac{1}{\tau} a_{ii} \quad (i \in F). \tag{A.4.19}$$

This also indicates that, in cases of strongly diagonally dominant rows, (A.4.19) may unnecessarily slow down the coarsening process compared to (A.4.11), or, alternatively, lead to worse bounds for the interpolation error.

**Variant 3**   As an alternative to the previous interpolation, it is sometimes proposed to simply use equal weights $w_{ik} = 1/n_i$ where $n_i = |P_i|$ just denotes the number of neighbors used for interpolation. Then (A.4.2) is satisfied if

$$n_i |a_{ik}| \geq \frac{1}{\tau} a_{ii} \quad (i \in F, \, k \in P_i). \tag{A.4.20}$$

Obviously, for this interpolation to be reasonable it is particularly crucial to interpolate *only* from strong connections (otherwise, $n_i$ has to be too large). Unless all interpolatory connections are of approximately the same size, this interpolation will be substantially worse than the previous ones.

**Variant 4**   Aggregation-based AMG approaches (see Section A.9) can be interpreted to use interpolation to any F-point from *exactly one* C-point only (with weight 1), that is, $|P_i| = 1$. This interpolation allows for a rapid coarsening and, moreover, the computation of the coarse-level Galerkin operators becomes extremely simple. On the other hand, this interpolation is the crudest possible by definition, leading to rather bad bounds for the interpolation error. In fact, (A.4.2) is satisfied only if

$$|a_{ik}| \geq \frac{1}{\tau} a_{ii} \quad (i \in F, \, k \in P_i). \tag{A.4.21}$$

Consequently, the smaller the (significant) off-diagonal entries are compared to the diagonal, the larger $\tau$ will be. In particular, (A.4.2) cannot be satisfied uniformly within the class of weakly diagonally dominant M-matrices any more. Generally, using this most simple interpolation, convergence will be rather poor (and $h$-dependent for complete V-cycles; see the related discussion in Section A.6).

In spite of this, aggregation-based AMG approaches have become quite popular, one reason being their simplicity and ease in programming. However, to become practical, they require additional acceleration by means of various techniques, for instance, by smoothing of interpolation ("smoothed aggregation"). In addition, aggregation-based AMG is typically used as a preconditioner rather than a stand-alone. We will return to such approaches in Section A.9.

### A.4.2.2 Essentially positive-type matrices

In the previous section we considered off-diagonally nonpositive matrices. However, the essential estimates carry over to certain matrices with relatively small positive off-diagonal entries such as the *essentially positive-type matrices* considered in Section A.3.3.2. The theorem below shows that, for such matrices, it is sufficient to base the splitting on the connectivity structure induced by the negative off-diagonal entries only. Accordingly, interpolation needs to be done only from neighbors with negative coefficients. In the following, we make use of the notation $a_{ij}^+$, $a_{ij}^-$ and $N_i^+$, $N_i^-$ as defined in Section A.2.2.

Using interpolatory points $\emptyset \neq P_i \subseteq C \cap N_i^-$ and recalling the heuristic considerations on algebraically smooth error in Section A.3.3.2, we might define interpolation exactly as before (A.4.8). However, for reasons explained in Remark A.4.8 below, we prefer to extend the interpolation to the case of positive entries by adding all such entries to the diagonal (to preserve row sums). That is, we use

$$\tilde{a}_{ii}e_i + \alpha_i \sum_{k \in P_i} a_{ik}^- e_k = 0 \quad \text{with } \tilde{a}_{ii} = a_{ii} + \sum_{j \in N_i} a_{ij}^+, \quad \alpha_i = \frac{\sum_{j \in N_i} a_{ij}^-}{\sum_{k \in P_i} a_{ik}^-} \quad \text{(A.4.22)}$$

instead of (A.4.8), yielding positive interpolation weights

$$w_{ik} = -\alpha_i a_{ik}^- / \tilde{a}_{ii} \quad (i \in F, \ k \in P_i). \quad \text{(A.4.23)}$$

Note that the row sums of (A.4.6) and (A.4.22) are equal and we have

$$\tilde{a}_{ii}\left(1 - \sum_{k \in P_i} w_{ik}\right) = s_i. \quad \text{(A.4.24)}$$

**Theorem A.4.5** *Let $A > 0$ be essentially positive-type (A.3.16) with $s_i = \sum_j a_{ij} \geq 0$. With fixed $\tau \geq 1$, select a C/F-splitting such that, for each $i \in F$, there is a set $P_i \subseteq C \cap N_i^-$ satisfying*

$$\sum_{k \in P_i} |a_{ik}^-| \geq \frac{1}{\tau} \sum_{j \in N_i} |a_{ij}^-|. \quad \text{(A.4.25)}$$

*Then the interpolation (A.4.5) with weights (A.4.23) satisfies (A.4.2) with $\tau/c$ rather than $\tau$.*

*Proof.* The proof runs analogously to that of Theorem A.4.3. We first note again that $P_i \neq \emptyset$ for all $i \in F$ (cf. Remark A.3.3). Since $c \leq 1$ and $s_i \geq 0$, (A.3.13) and (A.3.16) imply

$$(Ae, e)_E \geq \frac{c}{2} \sum_{i,j} (-a_{ij}^-)(e_i - e_j)^2 + \sum_i s_i e_i^2$$

$$\geq c \sum_{i \in F} \left( \sum_{k \in P_i} (-a_{ik})(e_i - e_k)^2 + s_i e_i^2 \right).$$

Using (A.4.14) and observing that $\tilde{a}_{ii} \geq a_{ii}$, we see that (A.4.2) is satisfied with $\tau/c$ rather than $\tau$ if $\alpha_i \leq \tau$ $(i \in F)$ which is equivalent to (A.4.25). $\qquad\square$

**Remark A.4.7** There is a straightforward generalization of this theorem to the case of negative row sums which is analogous to Theorem A.4.4. $\gg$

**Remark A.4.8** The reason for adding positive entries to the diagonal (rather than using the same formula as in the previous section) is mainly practical. In practice, we want to implement an interpolation which (at least formally) can also be employed to matrices which are not of strictly essentially positive type, for instance, which contain some particularly large positive entries. However, in such cases, $\sum_{j \in N_i} a_{ij}$ might become zero or even positive for certain $i \in F$ and, using (A.4.8), we would obtain zero or even negative interpolation weights. This is avoided by adding positive entries to the diagonal. Nevertheless, the approximation (A.4.22) is, obviously, only reasonable if we can assume that, for each $i \in F$, an algebraically smooth error satisfies

$$\sum_j a_{ij}^+ e_j \approx \sum_j a_{ij}^+ e_i \qquad (A.4.26)$$

which, for $j \neq i$, either requires $e_j \approx e_i$ or $a_{ij}^+$ to be small (relative to $a_{ii}$). According to the heuristic considerations in Section A.3.3, this can be assumed for essentially positive-type matrices as considered here. However, we will see in the next section that (A.4.22) becomes less accurate (in the sense of (A.4.2)) if (A.4.26) is strongly violated. $\gg$

### A.4.2.3 General case

Although the previous interpolation can, formally, always be used (provided each $i$ has, at least, one negative connection), it is well suited only if (A.4.26) can be assumed to hold for an algebraically smooth error. To demonstrate this, let us consider a problem already mentioned in Section A.3.3.3 (Example A.3.6).

**Example A.4.3** Consider the matrix $A$ which corresponds to the stencil

$$\begin{bmatrix} & +1 & \\ -1 & 4 & -1 \\ & +1 & \end{bmatrix}, \qquad (A.4.27)$$

applied on an $N \times N$ grid with mesh size $h = 1/N$ and Dirichlet boundary conditions.

We select the particular error vector $e$ with components defined by $+1$ $(-1)$ along even (odd) horizontal grid lines. For this vector, we have $e_i^2 = 1$ for all $i$, $(e_i - e_j)^2 = 0$ if $a_{ij} < 0$ and $(e_i - e_j)^2 = 4$ if $a_{ij} > 0$. Using (A.3.13), we therefore see that

$$(Ae, e)_E = -2 \sum_{i,j} a_{ij}^+ + \sum_i s_i = \sum_i \left( s_i - 2 \sum_j a_{ij}^+ \right) = \sum_i t_i$$

with $t_i := a_{ii} - \sum_{j \neq i} |a_{ij}|$. Since $t_i = 0$ in the interior and $t_i = 1$ along the boundary, we obtain

$$(Ae, e)_E = O(N). \qquad (A.4.28)$$

Assume now any C/F-splitting and any interpolation $I_{FC}$ with $P_i \subseteq C \cap N_i^-$ to be given. Using (A.4.13), we then see that

$$\|e_F - I_{FC}e_C\|_{0,F}^2 = \sum_{i \in F} a_{ii} \left(1 - \sum_{k \in P_i} w_{ik}\right)^2. \tag{A.4.29}$$

If $I_{FC}$ is the particular interpolation from the previous section, (A.4.24) implies that each term of the sum in (A.4.29) is nonzero and we obtain

$$\|e_F - I_{FC}e_C\|_{0,F}^2 = O(N^2). \tag{A.4.30}$$

Because of this and (A.4.28), inequality (A.4.2) cannot hold independently of $N$.          $\triangle$

The problem seen in this example is that an algebraically smooth error is geometrically smooth only in the $x$-direction but highly oscillatory in the $y$-direction (cf. Section A.3.3.3). (The particular error vector considered in the example is actually algebraically smooth.) That is, we have $e_j \approx -e_i$ if $a_{ij} > 0$ and $j \neq i$. Consequently, (A.4.26) is strongly violated which explains the high interpolation error observed above. A redefinition of $\tilde{a}_{ii}$ in (A.4.22) by *subtracting* all positive connections from the diagonal (rather than *adding* them), is a way out for situations such as the one considered here (in fact, this would give $O(N)$ in (A.4.30) rather than $O(N^2)$) but it is the wrong thing to do in other cases.

This indicates that the correct treatment of positive connections in interpolation is, in general, more critical than that of negative connections. We have seen in Section A.3.3 that, assuming $a_{ij} > 0$, algebraically smooth error $e$ can still be expected to change slowly from $i$ to $j$ for essentially positive-type matrices, in particular, if there exist strong negative paths from $i$ to $j$. On the other hand, for matrices which are approximately weakly diagonally dominant, we have to expect an oscillatory behavior (which is the stronger the larger $a_{ij}$ is relative to $a_{ii}$, see (A.3.21)). In *both* cases, however, we may expect that those $e_k$ which correspond to positive connections $a_{ik} > 0$, change slowly *among each other* (unless $a_{ik}$ is very small in which case we can ignore its influence). This gives rise to the following generalization of the interpolation approach (A.4.8) (M-matrix case) which is completely symmetric in the treatment of negative and positive connections.

Let us assume that some $i \in F$ has both negative and positive connections, that is, $N_i^- \neq \emptyset$ and $N_i^+ \neq \emptyset$. Then, assuming the C/F-splitting to be such that at least one connection of either sign is contained in C, we can select two sets of interpolation points, $\emptyset \neq P_i^- \subseteq C \cap N_i^-$ and $\emptyset \neq P_i^+ \subseteq C \cap N_i^+$. Setting $P_i = P_i^- \cup P_i^+$, we then use

$$a_{ii}e_i + \alpha_i \sum_{k \in P_i} a_{ik}^- e_k + \beta_i \sum_{k \in P_i} a_{ik}^+ e_k = 0 \tag{A.4.31}$$

instead of (A.4.8) with

$$\alpha_i = \frac{\sum_{j \in N_i} a_{ij}^-}{\sum_{k \in P_i} a_{ik}^-} \quad \text{and} \quad \beta_i = \frac{\sum_{j \in N_i} a_{ij}^+}{\sum_{k \in P_i} a_{ik}^+}.$$

At this point, we have assumed that, for an algebraically smooth error, approximations analogous to (A.4.7) hold *separately* for the negative and the positive connections. This leads to the following interpolation weights:

$$w_{ik} = \begin{cases} -\alpha_i \, a_{ik}/a_{ii} & (k \in P_i^-) \\ -\beta_i \, a_{ik}/a_{ii} & (k \in P_i^+). \end{cases} \tag{A.4.32}$$

Note that $w_{ik} > 0$ $(k \in P_i^-)$ and $w_{ik} < 0$ $(k \in P_i^+)$. If either $N_i^+ = \emptyset$ or $N_i^- = \emptyset$, these definitions are to be modified in a straightforward way by setting $P_i^+ = \emptyset$, $\beta_i = 0$ and $P_i^- = \emptyset$, $\alpha_i = 0$, respectively. In particular, for M-matrices, the above interpolation is identical to the one in (A.4.8). In any case, the row sums of (A.4.6) and (A.4.31) are equal and we have

$$a_{ii} \left( 1 - \sum_{k \in P_i} w_{ik} \right) = s_i. \tag{A.4.33}$$

Formally, this approach can always be applied even in cases where most, or even all, entries are positive. The following theorem, a direct generalization of Theorem A.4.3, shows that (A.4.2) can be uniformly satisfied in the class of weakly diagonally dominant matrices $A$ under completely symmetric conditions for positive and negative entries. Note that the above example belongs to this class. A more realistic application arises, for instance, in connection with *antiperiodic* boundary conditions (see also Section A.3.3.3).

**Theorem A.4.6** *Let $A > 0$ and $t_i = a_{ii} - \sum_{j \in N_i} |a_{ij}| \geq 0$. With fixed $\tau \geq 1$, select a C/F-splitting such that the following holds for each $i \in F$. If $N_i^- \neq \emptyset$, there is a set $P_i^- \subseteq C \cap N_i^-$ satisfying*

$$\sum_{k \in P_i^-} |a_{ik}| \geq \frac{1}{\tau} \sum_{j \in N_i^-} |a_{ij}| \tag{A.4.34}$$

*and, if $N_i^+ \neq \emptyset$, there is a set $P_i^+ \subseteq C \cap N_i^+$ satisfying*

$$\sum_{k \in P_i^+} a_{ik} \geq \frac{1}{\tau} \sum_{j \in N_i^+} a_{ij}. \tag{A.4.35}$$

*Then the interpolation (A.4.5) with weights (A.4.32) satisfies (A.4.2).*

*Proof.* Using (A.3.20), we can estimate

$$(Ae, e)_E \geq \sum_{i \in F} \left( \sum_{k \in P_i^-} |a_{ik}|(e_i - e_k)^2 + \sum_{k \in P_i^+} a_{ik}(e_i + e_k)^2 + t_i e_i^2 \right). \tag{A.4.36}$$

To obtain an estimate for $\| e_F - I_{FC} e_C \|_{0,F}^2$, we start from (A.4.13). We first note that

$$a_{ii} \left( 1 - \sum_{k \in P_i} w_{ik} \right) = s_i = 2 \sum_{j \in N_i^+} a_{ij} + t_i = 2\beta_i \sum_{k \in P_i^+} a_{ik} + t_i$$

$$= 2a_{ii} \sum_{k \in P_i^+} |w_{ik}| + t_i$$

which implies

$$\sum_{k \in P_i^+} w_{ik}(e_i - e_k) + \left(1 - \sum_{k \in P_i} w_{ik}\right) e_i = \sum_{k \in P_i^+} |w_{ik}|(e_i + e_k) + t_i/a_{ii} \, e_i \,.$$

The latter, inserted into (A.4.13), gives the following estimate by applying Schwarz's inequality

$$\|e_F - I_{FC} e_C\|_{0,F}^2 = \sum_{i \in F} a_{ii} \left( \sum_{k \in P_i^-} w_{ik}(e_i - e_k) + \sum_{k \in P_i^+} |w_{ik}|(e_i + e_k) + t_i/a_{ii} \, e_i \right)^2$$

$$\leq \sum_{i \in F} a_{ii} \left( \sum_{k \in P_i^-} w_{ik}(e_i - e_k)^2 + \sum_{k \in P_i^+} |w_{ik}|(e_i + e_k)^2 + t_i/a_{ii} \, e_i^2 \right).$$

$$(A.4.37)$$

Regarding this estimate, note that $\sum_{k \in P_i} |w_{ik}| + t_i/a_{ii} = 1$. The estimates (A.4.36) and (A.4.37) imply (A.4.2) if $a_{ii}|w_{ik}| \leq \tau |a_{ik}|$ $(i \in F, k \in P_i)$ which is equivalent to the assumptions of the theorem.                                                                                   □

**Remark A.4.9**  We note that there is a straightforward extension of this theorem to the case $t_i \geq -c$ with some $c \geq 0$ which is analogous to Theorem A.4.4.                    ≫

Although the above theorem applies only to weakly diagonally dominant matrices, it is heuristically clear that the approach (A.4.31) is also reasonable in other cases. In particular, if we can assume algebraically smooth error to vary slowly even along positive connections (as in the essentially positive-type case), (A.4.31) is as good an approximation as (A.4.22). In fact, replacing $e_k$ $(k \in P_i^+)$ in (A.4.31) by $e_i$ gives exactly (A.4.22).

The latter indicates that, in practice, there is no need to use the approach (A.4.31) for all $i \in F$ but rather only for those for which oscillatory behavior has to be expected. This simplifies the coarsening algorithm substantially in case of elliptic differential problems, where the largest couplings will usually be negative. In such cases, we will use the full interpolation as described above only when there really exist large positive couplings (comparable in size to the largest negative connection, say). Otherwise, we proceed as described in the previous section, that is, we do not interpolate from positive connections but rather add them to the diagonal (for more details, see Section A.7).

**Remark A.4.10**  A different interpolation has been considered in [200] (which is actually a direct generalization of the one used in [334]). Compared to (A.4.31), $\alpha_i = \beta_i = 1$ has been selected and all noninterpolatory connections are used to modify the diagonal element:

$$a_{ii} \longrightarrow \tilde{a}_{ii} = a_{ii} - \sum_{j \notin P_i, \, j \neq i} |a_{ij}|. \qquad (A.4.38)$$

The resulting interpolation is limited to (and actually has been developed for) weakly diagonally dominant matrices where algebraically smooth error really oscillates along positive connections (note that positive matrix entries are *subtracted* from the diagonal element rather than *added*; see the related comment further above). It is not suited for other applications. In particular, (A.4.38) does not preserve row sums and, consequently, constants are not interpolated exactly if $s_i \equiv 0$. Also, the denominator in (A.4.38) may become zero or negative. $\gg$

Finally we note that interpolation from positive connections is not always necessary even if an algebraically smooth error tends to oscillate in certain directions. We just have to ensure that the C/F-splitting is such that the C-variables can represent the oscillations sufficiently well and that interpolation along negative couplings is "accurate enough". We do not want to discuss this aspect any further but rather refer to Example A.4.3 where the situation is particularly simple. If we do not perform coarsening in the $y$-direction (i.e. in the direction of strong positive connections), we may, for instance, use

$$w_{ik} = a_{ik} / \sum_{j \in P_i} a_{ij} \quad \text{with } k \in P_i \subseteq C \cap N_i^-$$

for interpolation (cf. Variant 2 in Section A.4.2.1).

### A.4.3 Indirect Interpolation

In the previous sections, we considered approaches to construct interpolation based on *direct* connections, that is, interpolation to an F-point $i$ only from its direct neighbors. Correspondingly, the C/F-splittings had to be such that each $i \in F$ is sufficiently strongly connected to the set of C-points via direct connections.

Although a strong F-to-C connectivity is indeed crucial, it does not necessarily have to be via direct connections. In practice, this may limit the speed of coarsening. Too slow coarsening, however, may cause high memory requirements, which are often unacceptably high. Clearly, a faster coarsening will typically imply a slower convergence. However, the advantages in terms of less memory requirement and lower computational cost per cycle and for the setup, in many cases outweigh the disadvantage of slower convergence. Moreover, the use of AMG as a preconditioner (rather than stand-alone) is usually a very efficient means to bring the speed of convergence back up again. This will be seen in Section A.8.

In order to permit more rapid coarsening, one has to allow that F-points may be interpolated via sufficiently many of their strong *F-neighbors*. For an illustration, consider the typical geometric scenario of (isotropic) five-point discretizations on regular meshes. Interpolation based only on direct connections would not allow for the $h \to 2h$ coarsening which is typically used in geometric multigrid methods, the reason being that those F-points $i$ located in the center of a coarse-grid cell have no direct connection to the C-points (see Fig. A.9, left picture). However, all their direct F-neighbors, $j$, *do have* strong connections to the C-points and, thus, can interpolate directly.

This simple scenario gives rise to a straightforward generalization of interpolation. First interpolate the $j$-variables and then, via the resulting interpolation formulas, the $i$-variable

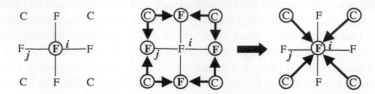

**Figure A.9.** Illustration of indirect interpolation in case of five-point stencils.

(see Fig. A.9, right picture). Since the details of a corresponding generalization of the theorems in the previous sections are elementary but rather involved, we here just outline the main additional step required for the generalization of Theorem A.4.3 (M-matrix case). We have a scenario in mind which is analogous to the one described above.

Assuming a given C/F-splitting to be such that there is some $i \in F$ which has no (strong) connection to any C-point, we select a set of points $P_i^F \subseteq F \cap N_i$ satisfying (A.4.11) with $P_i$ replaced by $P_i^F$. With the same arguments as at the beginning of Section A.4.2.1 we approximate in an intermediate step

$$e_i = \sum_{j \in P_i^F} w_{ij}^F e_j \quad \text{with } w_{ij}^F = -\alpha_i a_{ij}/a_{ii}, \ \alpha_i \leq \tau. \quad (A.4.39)$$

Assume now that each of the neighboring points $j \in P_i^F$ *can be* interpolated, that is, we have

$$e_j = \sum_{k \in P_j} w_{jk} e_k \quad \text{with } w_{jk} = -\alpha_j a_{jk}/a_{jj}, \ \alpha_j \leq \tau. \quad (A.4.40)$$

Inserting (A.4.40) into (A.4.39), yields

$$e_i = \sum_{j \in P_i^F} \sum_{k \in P_j} w_{ij}^F w_{jk} e_k =: \sum_{k \in P_i} w_{ik} e_k \quad \text{with } P_i = \bigcup P_j. \quad (A.4.41)$$

If, for simplicity, we finally assume that the pairwise intersection of all $P_j$s is empty (otherwise there will be several terms of the following form), we can estimate for $k \in P_i$

$$\begin{aligned}
a_{ii} w_{ik} (e_i - e_k)^2 &= a_{ii} w_{ij}^F w_{jk} (e_i - e_k)^2 \\
&\leq 2a_{ii} w_{ij}^F w_{jk} ((e_i - e_j)^2 + (e_j - e_k)^2) \\
&= 2\alpha_i \alpha_j \frac{a_{ij} a_{jk}}{a_{jj}} ((e_i - e_j)^2 + (e_j - e_k)^2) \\
&\leq c_1 |a_{ij}| (e_i - e_j)^2 + c_2 |a_{jk}| (e_j - e_k)^2,
\end{aligned}$$

where $c_1$ and $c_2$ depend only on $\tau$ and not on the given matrix.

By means of such simple estimates and by using the full representation (A.3.13) of $\|e\|_1^2$ in (A.4.12), one can immediately extend the proof of Theorem A.4.3 showing that (A.4.2) holds with some $\tilde{\tau}$ which is larger than $\tau$ but does not depend on $A$. Practical aspects will be considered in Section A.7.2.2 ("multipass interpolation").

## A.5 PRESMOOTHING AND TWO-LEVEL CONVERGENCE

In this section, we investigate two-level convergence if *presmoothing* is used rather than postsmoothing, that is, we consider the (energy) norm of the two-level operator $K S^\nu$ with some $\nu \geq 1$. Post- and presmoothing are treated separately because the *immediate* interaction between coarse-grid correction and smoothing is different for these cases. Clearly, regarding the *asymptotic* two-level convergence, it does not matter whether pre- or postsmoothing is used. In this sense, the conclusions for either of the two cases carry over to the other one. (In practice, we usually employ post- *and* presmoothing anyway.)

The theoretical approach used in this section is quite different from the one used in the previous section. In particular, it does not use the smoothing property (A.3.7) but is directly based on the following equivalence which is an immediate consequence of the variational principle (last statement in Corollary A.2.1):

**Corollary A.5.1** *The two-level estimate $\|K S^\nu\|_1 \leq \eta$ holds if and only if, for all $e$, there exists an $e^H$ such that*

$$\|S^\nu e - I_H^h e^H\|_1 \leq \eta \|e\|_1. \tag{A.5.1}$$

The interpretation of (A.5.1) is that the speed of convergence depends solely on the efficient interplay between *smoothing* and *interpolation*. To this end, $S$ does *not* necessarily have to satisfy the smoothing property as long as $S^\nu e$ is represented in $\mathcal{R}(I_H^h)$ sufficiently well. The better this is satisfied, the faster the convergence will be.

In fact, we will see in Section A.5.1 that we can force $\eta$ to be small (independently of $A \in \mathcal{A}$ for relevant classes $\mathcal{A}$) *by smoothing only at F-points*. Interpolation can be constructed as described in Section A.4.2. Additional Jacobi relaxation steps, applied to the interpolation operator, can be used to speed up convergence further. In Section A.5.2 we make some remarks on the use of complete (rather than just F-) relaxation steps for smoothing.

### A.5.1 Convergence using Mere F-Smoothing

In order to motivate our theoretical approach, let us briefly recall the results on direct solvers (presmoothing variant) described in Section A.2.3. The very specific smoothing and interpolation operators introduced there, $\hat{S}$ and $\hat{I}_H^h$ (see (A.2.21) and (A.2.23), respectively), have the property

$$\mathcal{R}(\hat{S}) = \mathcal{R}(\hat{I}_H^h) = \mathcal{E} := \{e: e_F = -A_{FF}^{-1} A_{FC} e_C\} \tag{A.5.2}$$

which, obviously, implies that the left-hand side of (A.5.1) is identically zero (for $\nu = 1$). This suggests trying to approximate $\hat{S}$ and $\hat{I}_H^h$ by some more realistic operators, $S$ and $I_H^h$.

One can imagine various ways to do this. Recalling that $\hat{S}$ and $\hat{I}_H^h$ have been defined by exactly solving the F-equations (A.2.19) and (A.2.25), respectively, we will here define $S$ and $I_H^h$ by solving these equations only approximately. This becomes particularly easy, if one assumes the submatrix $A_{FF}$ to be *strongly diagonally dominant*,

$$a_{ii} - \sum_{j \in F, \, j \neq i} |a_{ij}| \geq \delta a_{ii} \quad (i \in F) \tag{A.5.3}$$

with some fixed, predefined $\delta > 0$. This assumption is very natural for large classes of problems, in particular those considered in this appendix. It essentially means that a strong F-to-C connectivity is required which, however, has to be ensured by a reasonable C/F-splitting anyway (cf. Section A.4.2 and Remark A.5.3).

> All of the following is based on the assumption of strong diagonal dominance (A.5.3) and we consider the simplest means to approximate the solutions of the F-equations (A.2.19) and (A.2.25), namely, by straightforward relaxation involving only F-variables ("F-relaxation", either Gauss–Seidel or Jacobi).

**Remark A.5.1** Note that strong diagonal dominance (A.5.3) is assumed for simplicity. What we really require is that F-relaxation converges at a rate which is independent of the original matrix $A$. Strong diagonal dominance of $A_{FF}$, with predefined $\delta$, is sufficient to ensure this, but it is not necessary.                                                    ≫

Before we derive the main convergence estimate in Section A.5.1.4, we formally define the smoothing and interpolation operators in Sections A.5.1.2 and A.5.1.3. The following section contains an auxiliary result.

### A.5.1.1 An auxiliary result

In all of the following, the subspace $\mathcal{E}$, defined in (A.5.2), will play an essential role. Given any $e = (e_F, e_C)^T$, we denote its projection onto $\mathcal{E}$ by $\hat{e}$,

$$\hat{e} := (\hat{e}_F, e_C)^T \quad \text{where} \quad \hat{e}_F := \hat{I}_{FC} e_C = -A_{FF}^{-1} A_{FC} e_C. \tag{A.5.4}$$

The following lemma states some basic properties which will be needed below. In particular, it relates the energy norm of $e$ to that of $\hat{e}$.

**Lemma A.5.1** *Let $A > 0$ and any C/F-splitting be given. Then the Schur complement $C_H$ (A.2.24) is also positive definite and satisfies $\rho(C_H^{-1}) \leq \rho(A^{-1})$. For all $e$, we have*

$$(Ae, e)_E = (A_{FF}(e_F - \hat{e}_F), e_F - \hat{e}_F)_E + (C_H e_C, e_C)_E = \|e - \hat{e}\|_1^2 + \|\hat{e}\|_1^2 \tag{A.5.5}$$

*which immediately implies*

$$\|\hat{e}\|_1 \leq \|e\|_1 \quad \text{and} \quad \|e_F - \hat{e}_F\|_{1,F} = \|e - \hat{e}\|_1 \leq \|e\|_1. \tag{A.5.6}$$

*Proof.* The left equality in (A.5.5) follows by a straightforward computation. The positive definiteness of $C_H$ then follows from $(C_H e_C, e_C)_E = (A\hat{e}, \hat{e})_E$. Denoting the smallest eigenvalue of $A$ by $\varepsilon > 0$, we have $(Ae, e)_E \geq \varepsilon(e, e)_E$ for all $e$. We can then estimate for all $e_C$

$$(C_H e_C, e_C)_E = (A\hat{e}, \hat{e})_E \geq \varepsilon(\hat{e}, \hat{e})_E \geq \varepsilon(e_C, e_C)_E.$$

Hence, $\rho(C_H^{-1}) \leq 1/\varepsilon = \rho(A^{-1})$.                                              □

### A.5.1.2 F-smoothing

As motivated above, we define $\nu$ smoothing steps by applying $\nu$ F-relaxation steps to approximately solve (A.2.19), starting with the most recent approximation $u = (u_F, u_C)^T$ (keeping $u_C$ fixed). We will refer to this process as *F-smoothing*.

> We here use the term "smoothing" although mere F-relaxation (like any other partial relaxation) has no real smoothing properties in the usual sense. In particular, the considerations in Section A.3 on algebraically smooth error do not apply here. In fact, F-relaxation aims at approximately *solving* the F-equations (A.2.19) (with fixed $u_C$) rather than smoothing the error of the full system of equations.

Consequently, a relaxation parameter, $\omega$, may be used to speed up the convergence. However, since we require $A_{FF}$ to be strongly diagonally dominant, this is not really necessary and, for simplicity, we only consider the case $\omega = 1$. Having this in mind, one F-smoothing step consists of

$$u \longrightarrow \bar{u} \quad \text{where } Q_{FF}\bar{u}_F + (A_{FF} - Q_{FF})u_F + A_{FC}u_C = f_F, \quad \bar{u}_C = u_C. \quad \text{(A.5.7)}$$

$Q_{FF}$ is the lower triangular part of $A_{FF}$ (including the diagonal) in case of *Gauss–Seidel relaxation* and $Q_{FF} = D_{FF}$ in case of *Jacobi relaxation*. (In practice, we only use Gauss–Seidel relaxation.) To explicitly compute the corresponding smoothing operator, we rewrite this as

$$\bar{u}_F = S_{FF}u_F + (I_{FF} - S_{FF})A_{FF}^{-1}(f_F - A_{FC}u_C), \qquad \bar{u}_C = u_C \quad \text{(A.5.8)}$$

or, in terms of the error $e = u^\star - u$ (with $u^\star$ denoting the exact solution of $Au = f$),

$$\bar{e}_F = S_{FF}e_F - (I_{FF} - S_{FF})A_{FF}^{-1}A_{FC}e_C, \qquad \bar{e}_C = e_C.$$

Here, $S_{FF} = I_{FF} - Q_{FF}^{-1}A_{FF}$ denotes the iteration matrix corresponding to the relaxation of the F-variables. Consequently, the relaxation operator reads

$$S_h = \begin{pmatrix} S_{FF} & (S_{FF} - I_{FF})A_{FF}^{-1}A_{FC} \\ 0 & I_{CC} \end{pmatrix} \quad \text{(A.5.9)}$$

and a simple calculation shows that

$$S_h^\nu e = \begin{pmatrix} S_{FF}^\nu(e_F - \hat{e}_F) + \hat{e}_F \\ e_C \end{pmatrix}. \quad \text{(A.5.10)}$$

Obviously, we have rapid convergence $S_h^\nu e \to \hat{e}$ $(\nu \to \infty)$ for each $e$.

> **Remark A.5.2** Note that F-smoothing does *not* satisfy the smoothing property (A.3.7). In fact, (A.5.10) shows that $Se = e$ for all $e \in \mathcal{E}$. Hence, (A.3.7) cannot hold. $\gg$

### A.5.1.3 Jacobi interpolation

Given any $e_C$, we define interpolation by applying $\mu$ F-relaxation steps to approximately solve the F-equations (A.2.25). However, in order to keep the resulting operator as "local" as possible, we only consider Jacobi relaxation. That is, we iteratively define

$$e_F^{(\mu)} = P_{FF} e_F^{(\mu-1)} - D_{FF}^{-1} A_{FC} e_C \tag{A.5.11}$$

and set $I_{FC} e_C = I_{FC}^{(\mu)} e_C := e_F^{(\mu)}$. Here $P_{FF}$ denotes the Jacobi iteration operator, $P_{FF} = I_{FF} - D_{FF}^{-1} A_{FF}$. In contrast to the F-smoothing process described in the previous section, however, here no "natural" first approximation, $e_F^{(0)}$, is available to start the relaxation process with. Using zero as the first guess is not sufficient as will be seen in Remark A.5.6. For now, we assume any "first guess interpolation", $I_{FC}^{(0)}$, to be given and use $e_F^{(0)} = I_{FC}^{(0)} e_C$ as the first guess. (In the following section, we will derive a requirement on $I_{FC}^{(0)}$.)

Since the interpolation operator needs to be known *explicitly* in order to compute the Galerkin operator, we rewrite (A.5.11) in operator form,

$$I_{FC}^{(\mu)} = P_{FF} I_{FC}^{(\mu-1)} - D_{FF}^{-1} A_{FC} \tag{A.5.12}$$

starting with the first-guess interpolation operator, $I_{FC}^{(0)}$. (Regarding the practical computation of the interpolation, see Section A.7.2.3.) By subtracting these equations from the equality $\hat{I}_{FC} = P_{FF} \hat{I}_{FC} - D_{FF}^{-1} A_{FC}$, we obtain

$$J_{FC}^{(\mu)} = I_{FC}^{(\mu)} - \hat{I}_{FC} = P_{FF}^{\mu} (I_{FC}^{(0)} - \hat{I}_{FC}) = P_{FF}^{\mu} J_{FC}^{(0)} \tag{A.5.13}$$

where $J_{FC} = I_{FC} - \hat{I}_{FC}$ denotes the "interpolation error" of $I_{FC}$ relative to $\hat{I}_{FC}$.

We will later refer to this relaxation of interpolation as *Jacobi interpolation*. Clearly, for any given $e = (e_F, e_C)^T$, we have rapid convergence $(I_H^h)^{(\mu)} e_C \to \hat{e}$ ($\mu \to \infty$).

### A.5.1.4 Convergence estimate

The following theorem yields a requirement on the first-guess interpolation $I_{FC}^{(0)}$ which is sufficient to imply uniform two-level convergence. Further below, in Lemma A.5.2, we will see that this requirement is very much related to the corresponding one (A.4.2) in Section A.4.1.

**Theorem A.5.1** *Let $A > 0$ and assume the C/F-splitting to be such that $A_{FF}$ is strongly diagonally dominant (A.5.3) with fixed $\delta > 0$. Let smoothing be performed by $\nu \geq 1$ F-relaxation steps (A.5.7). Finally, let the interpolation be defined by $I_{FC} = I_{FC}^{(\mu)}$ with some $\mu \geq 0$ (see (A.5.12)) and assume that the first-guess interpolation, $I_{FC}^{(0)}$, satisfies*

$$\|(\hat{I}_{FC} - I_{FC}^{(0)}) e_C\|_{1,F} \leq \tau \|e\|_1 \tag{A.5.14}$$

*for all $e$ with some $\tau \geq 0$ being independent of $e$. Then the following estimate holds:*

$$\|K S^\nu e\|_1 \leq (\|S_{FF}\|_{1,F}^\nu + \tau \|P_{FF}\|_{1,F}^\mu) \|e\|_1. \tag{A.5.15}$$

*Proof.* Because of the variational principle (Corollary A.2.1) and exploiting the representation (A.5.10) of $S^\nu$, we can estimate for all $e$

$$\|KS^\nu e\|_1 = \min_{e^H} \|S^\nu e - I_H^h e^H\|_1 \leq \|S^\nu e - I_H^h e_C\|_1$$
$$= \|S_{FF}^\nu (e_F - \hat{e}_F) + \hat{e}_F - I_{FC} e_C\|_{1,F}.$$

Recalling that $\hat{e}_F = \hat{I}_{FC} e_C$, the application of the triangular inequality gives

$$\|KS^\nu e\|_1 \leq \|S_{FF}^\nu (e_F - \hat{e}_F)\|_{1,F} + \|(I_{FC} - \hat{I}_{FC}) e_C\|_{1,F}.$$

Finally, because of (A.5.13) and (A.5.6), assumption (A.5.14) implies

$$\|KS^\nu e\|_1 \leq \|S_{FF}\|_{1,F}^\nu \|e_F - \hat{e}_F\|_{1,F} + \|P_{FF}\|_{1,F}^\mu \|(I_{FC}^{(0)} - \hat{I}_{FC}) e_C\|_{1,F}$$
$$\leq (\|S_{FF}\|_{1,F}^\nu + \tau \|P_{FF}\|_{1,F}^\mu) \|e\|_1,$$

which proves the theorem. □

> Clearly, the norms of $S_{FF}$ and $P_{FF}$ in (A.5.15) are less than one and depend only on $\delta$, not on $A$. In particular, the larger $\delta$ is, the smaller these norms are. Consequently, the previous theorem shows that, in principle, we can enforce arbitrarily fast two-level convergence by selecting $\nu$ and $\mu$ accordingly. Moreover, the convergence is *uniform* for $A \in \mathcal{A}$ if we can construct the first-guess interpolation, $I_{FC}^{(0)}$, so that (A.5.14) is uniformly satisfied for all such $A$. We will see that this can be achieved for the same classes of matrices for which the related condition (A.4.2) can be uniformly satisfied (which has been discussed in detail in Section A.4.2). Lemma A.5.2 shows that (A.5.14) and (A.4.2) are essentially equivalent.

**Lemma A.5.2** *Consider the two estimates*

> (a) $\|e_F - I_{FC} e_C\|_{0,F}^2 \leq \tau_1 \|e\|_1^2$, (b) $\|(\hat{I}_{FC} - I_{FC}) e_C\|_{1,F}^2 \leq \tau_2 \|e\|_1^2$. (A.5.16)

*If (a) holds for all $e$ and if $\eta \geq \rho(D^{-1}A)$, then (b) holds for all $e$ with $\tau_2 = \eta\tau_1$. If (b) holds for all $e$ and if $A_{FF}$ is strongly diagonally dominant (A.5.3), then (a) holds for all $e$ with $\tau_1 = (1 + \sqrt{\tau_2})^2/\delta$.*

*Proof.* We first note that $\rho(D_{FF}^{-1}A_{FF}) \leq \rho(D^{-1}A)$:

$$\rho(D_{FF}^{-1}A_{FF}) = \sup_{e_F} \frac{(A_{FF}e_F, e_F)_E}{(D_{FF}e_F, e_F)_E} = \sup_{(e_F,0)} \frac{(Ae, e)_E}{(De, e)_E}$$
$$\leq \sup_{e} \frac{(Ae, e)_E}{(De, e)_E} = \rho(D^{-1}A).$$

Using this and assuming (a) to hold for all $e$, we obtain because of (A.3.3)

$$\|e_F - I_{FC} e_C\|_{1,F}^2 \leq \eta \|e_F - I_{FC} e_C\|_{0,F}^2 \leq \eta\tau_1 \|e\|_1^2.$$

Applying this to $\hat{e}$ rather than $e$ and using (A.5.6), gives

$$\|\hat{e}_F - I_{FC}e_C\|_{1,F}^2 = \|(\hat{I}_{FC} - I_{FC})e_C\|_{1,F}^2 \leq \eta\tau_1 \|\hat{e}\|_1^2 \leq \eta\tau_1 \|e\|_1^2$$

which proves the first statement. Regarding the second one, we first estimate for any $e$

$$\|e_F - I_{FC}\,e_C\|_{1,F} \leq \|e_F - \hat{e}_F\|_{1,F} + \|(\hat{I}_{FC} - I_{FC})\,e_C\|_{1,F} \leq (1 + \sqrt{\tau_2})\|e\|_1.$$

By observing that

$$\rho(A_{FF}^{-1}D_{FF}) = 1 / \min\{\lambda : \lambda \text{ eigenvalue of } D_{FF}^{-1}A_{FF}\} \leq 1/\delta,$$

we conclude that

$$\|e_F - I_{FC}e_C\|_{0,F}^2 \leq \rho(A_{FF}^{-1}D_{FF}) \|e_F - I_{FC}e_C\|_{1,F}^2 \leq \frac{1}{\delta}(1 + \sqrt{\tau_2})^2 \|e\|_1^2.$$

This proves the lemma. $\qquad\square$

According to this lemma, we can use the same interpolation approaches as described in Section A.4.2 to define $I_{FC}^{(0)}$. Regarding the practical realization, we want to make the following remark:

---

**Remark A.5.3** The requirement of strong diagonal dominance (A.5.3) can easily be satisfied. If the C/F-splitting and interpolation are constructed according to Theorems A.4.3 and A.4.6, strong diagonal dominance is automatically satisfied, namely, with $\delta = 1/\tau$. For instance, the assumptions of Theorem A.4.3 imply for all $i \in F$:

$$a_{ii} - \sum_{j \in F, \, j \neq i} |a_{ij}| = s_i + \sum_{j \in P_i} |a_{ij}| + \sum_{j \in C \backslash P_i} |a_{ij}|$$

$$\geq s_i + \frac{1}{\tau}\sum_{j \in N_i} |a_{ij}| = \frac{1}{\tau}a_{ii} + \left(1 - \frac{1}{\tau}\right)s_i \geq \frac{1}{\tau}a_{ii}.$$

A second remark refers to the parameter $\mu$. Although the two-level method converges for all $\mu \geq 0$, Theorem A.5.1 states fast convergence only if $\mu$ is sufficiently large. In practice, however, $\mu > 2$ is hardly ever required (at least if $\delta$ is not too small). Nevertheless, each additional relaxation step increases the "radius" of interpolation (causing additional fill-in in the resulting Galerkin operator). Most of the new entries, however, will be relatively small and can be ignored without seriously sacrificing convergence. Consequently, in order to keep the resulting Galerkin operator as sparse as possible, relaxation of interpolation should *always* be combined with a reasonable truncation, performed before the Galerkin operator is computed (cf. Remark A.2.4; see also Section A.7.2.4). We also note that, in practice, it is usually not necessary to perform F-relaxation with the complete matrix $A_{FF}$. Instead, one may well ignore all those entries of $A_{FF}$ which are relatively small (and add them to the diagonal, say, in order to preserve the row sums of interpolation). $\gg$

For completeness, the following remarks summarize some algebraic conditions which are equivalent to (A.5.14). However, they are not important for the remainder of the appendix.

**Remark A.5.4** Requirement (A.5.14) is equivalent to

$$\|(\hat{I}_{FC} - I_{FC}^{(0)}) e_C\|_{1,F}^2 \le \tau^2 (C_H e_C, e_C)_E \tag{A.5.17}$$

for all $e_C$. This follows immediately by applying (A.5.14) to $\hat{e}$ rather than $e$ and using Lemma A.5.1. (A.5.17), in turn, is equivalent to

$$\rho(C_H^{-1} J_{CF}^{(0)} A_{FF} J_{FC}^{(0)}) = \|A_{FF}^{1/2} J_{FC}^{(0)} C_H^{-1/2}\|_E^2 \le \tau^2 \tag{A.5.18}$$

where, as above, $J_{FC}^{(0)} = I_{FC}^{(0)} - \hat{I}_{FC}$ and $J_{CF}^{(0)} = (J_{FC}^{(0)})^T$ denote the "errors" of the first-guess interpolation and restriction, respectively. $\gg$

**Remark A.5.5** Denoting the Galerkin operator corresponding to the first-guess interpolation by $A_H^{(0)}$, the requirement that (A.5.14) holds uniformly for $A \in \mathcal{A}$ is equivalent to the *spectral equivalence* of $C_H$ and $A_H^{(0)}$ for $A \in \mathcal{A}$,

$$(C_H e_C, e_C)_E \le (A_H^{(0)} e_C, e_C)_E \le (1 + \tau^2) (C_H e_C, e_C)_E. \tag{A.5.19}$$

In order to see this, one first verifies by a straightforward computation that

$$A_H^{(0)} = C_H + J_{CF}^{(0)} A_{FF} J_{FC}^{(0)}.$$

This equality, together with (A.5.17), proves the statement. Note that, because of (A.3.3), (A.5.19) implies that the related (spectral) condition number is uniformly bounded,

$$\kappa((A_H^{(0)})^{-1} C_H) := \frac{\lambda_{\max}((A_H^{(0)})^{-1} C_H)}{\lambda_{\min}((A_H^{(0)})^{-1} C_H)} \le 1 + \tau^2. \qquad \gg$$

We conclude with a remark which stresses the importance of constructing the first-guess interpolation reasonably.

---

**Remark A.5.6** If we select the first-guess interpolation too simply, in general, we cannot expect uniform two-level convergence. For instance, if we just select $I_{FC}^{(0)} = 0$, (A.5.18) is equivalent to

$$\rho(C_H^{-1} A_{CF} A_{FF}^{-1} A_{FC}) = \|A_{FF}^{-1/2} A_{FC} C_H^{-1/2}\|_E^2 \le \tau^2. \tag{A.5.20}$$

For $h$-discretized elliptic problems, we typically have $\|A_{FF}^{-1}\|_E = O(h^2)$ (provided $A_{FF}$ is strongly diagonally dominant) and $\|A_{FC}\|_E = O(h^{-2})$. Hence, (A.5.14) cannot be expected to hold with $\tau$ being independent of $h \to 0$; we actually have $\tau = O(h^{-1})$. In order to still obtain uniform convergence, we would need to select $\mu = O(\log(h^{-1}))$. $\gg$

### A.5.2 Convergence using Full Smoothing

The approach discussed in the previous section is not really in the spirit of multigrid since smoothing in the usual sense is not exploited. Two-level convergence is actually obtained by purely algebraic means. In fact, as already pointed out before, the role of F-smoothing is merely to *force* $S^\nu e \approx \hat{e}$ rather than to smooth the error of the full system. This, together with Jacobi interpolation, is a "brute force" approach to make $\| S^\nu e - I_H^h e_C \|_1$ small for all $e$.

Although this brute force approach helps convergence, particularly for "tough problems", we will see in Section A.8 that the use of *full* relaxation steps for smoothing usually leads to cycles which are considerably more efficient if computational work is taken into account. The heuristic reason is that, assuming $S$ to satisfy the smoothing property (A.3.7), relatively simple interpolation of the type derived in Section A.4.2 is usually sufficient to approximate algebraically smooth error. However, if mere F-smoothing is employed, approximations of the type (A.4.7), as used in Section A.4.2, are too crude and, generally, additional effort needs to be invested to "improve" interpolation by Jacobi relaxation in order to cope with all those error components which are not affected by mere F-smoothing. In particular, recall that an error $e \in \mathcal{E}$ is not reduced at all by F-smoothing (cf. Remark A.5.2).

Unfortunately, Theorem A.5.1 does not carry over to the use of general smoothing processes based on full relaxation steps. This is because the proof is based on an estimate of $\| S^\nu e - \hat{e} \|_1$. The latter, however, can most easily be obtained for F-smoothing but not for smoothing by full relaxation steps, except if we perform relaxation in *CF-ordering*, that is, if we first relax all C-variables and afterwards all F-variables. In this case, Theorem A.5.1 trivially carries over, at least for $\nu = 1$, by simply ignoring the C-part of the relaxation. For completeness, we give this result here although it is unrealistic in the sense that it cannot explain the better performance mentioned above.

**Corollary A.5.2** *Under the same assumptions as in Theorem A.5.1, except that smoothing is replaced by one step of Gauss–Seidel relaxation in CF-order, we obtain the following estimate*

$$\| KSe \|_1 \leq (\| S_{FF} \|_{1,F} + \tau \| P_{FF} \|_{1,F}^\mu) \| e \|_1. \tag{A.5.21}$$

> Gauss–Seidel CF-relaxation has turned out to be a very efficient smoother in practice. In particular, for positive definite problems, it is usually more efficient than Gauss–Seidel relaxation without any specific order of variables. (Note that CF-relaxation is related to red–black relaxation in geometric multigrid.)

From a practical point of view, (A.5.21) is too pessimistic since it implies convergence only if $\mu$ is sufficiently large. However, since the *asymptotic* two-level convergence does not depend on whether pre- or postsmoothing is performed, we can conclude from the results in Section A.4.1 that the above two-level cycle asymptotically converges *even for $\mu = 0$* and convergence is uniform for $A \in \mathcal{A}$ provided (A.5.14) holds uniformly for all such $A$.

Compared to the results in Section A.4.1, the relevance of Theorem A.5.1 and Corollary A.5.2 is due to the complementary information. In particular, the fact that Jacobi interpolation with $\mu > 0$ provides a (purely algebraic) means to improve convergence and that (additional) F-smoothing steps can be used in case of "tough" problems. We will present examples in Section A.8 demonstrating the effect of relaxation of interpolation. Various numerical experiments employing F-smoothing and Jacobi interpolation can also be found in [222].

## A.6 LIMITS OF THE THEORY

The two-level investigations of the previous sections are the basis for the definition of our final *multilevel* method in Section A.7. Various results will be presented in Section A.8, showing that AMG's V-cycle convergence is, to a large extent, independent of the size of the problem, at least for the geometrically posed problems as considered here.

Unfortunately, the latter cannot be proven in a purely algebraic setting. Since the main reason for this implies some important additional objective which should be taken into regard in AMG's coarsening algorithm, already indicated in Remark A.4.6, we want to briefly discuss the limitations of the theoretical approach.

First, uniform two-level convergence has strictly been proven only for certain "ideal" classes of positive definite matrices such as M-matrices, essentially positive-type matrices, weakly diagonally dominant matrices and some perturbations thereof. Although it is plausible that we can still expect uniform convergence w.r.t. certain larger classes, the uniformity may get lost if the matrices considered are too far off. A special limit case has already been discussed in Example A.4.1. It is also clear that we have to expect a degradation if the given problem is still symmetric and positive definite but corresponds to a *system* of PDEs (e.g. from structural mechanics) rather than a scalar PDE. In such cases, generally, the overall approach requires modification: the specific connectivity between different physical quantities needs to be taken into account (at least, all these quantities should be kept separate) in order to still obtain an efficient solution method. Since such problems are not within the focus of this appendix, we will not discuss such modifications here but rather refer to the preliminary discussion in [334].

However, apart from such limit cases, AMG's performance in practice turns out to be fairly insensitive to deviations of the underlying matrices from the ideal types. This is important to know, in particular, in recursively extending two-level to real multilevel cycles: even if the given matrix $A$ belongs to one of the ideal classes mentioned above, the recursively defined coarse-level Galerkin operators will generally not. It is often possible to avoid this by particular coarsening strategies. For instance, if $A$ is a weakly diagonally dominant M-matrix, the corresponding Galerkin operator $A_H$ will also be a weakly diagonally dominant M-matrix if coarsening is performed according to Theorem A.4.3 with $\tau \leq 2$ (see [334] for a proof). A similar result can be shown for weakly diagonally dominant matrices $A > 0$ (cf. [200]). However, these results turned out to be not really relevant in practice since they put unnecessary restrictions on the coarsening strategy.

> From experience, allowing faster coarsening and accepting that the coarse-level matrices do not exactly remain in the respective classes, typically leads to much more efficient solution processes.

However, even if all coarse-level matrices belong to one of the ideal classes (for instance, if they are all weakly diagonally dominant M-matrices), the two-level theory does *not* carry over to a multilevel V-cycle theory. To demonstrate this, we consider the following very simple but characteristic counter-example.

**Example A.6.1** [67, 334]   Let $A_h$ be derived from discretizing $-u''$ on the unit interval with mesh size $h$, i.e., the rows of $A_h$ correspond to the difference stencil

$$\frac{1}{h^2}[-1 \quad 2 \quad -1]_h,$$

with Dirichlet boundary conditions. (However, since the concrete boundary conditions are irrelevant for this example, we may ignore the boundary in the following.) One possibility of satisfying the assumptions of Theorem A.4.3 with $\tau = 2$ is to assume $h \to 2h$ coarsening and define interpolation to each F-point *strictly one-sided* (with the interpolation weight being 1), see Fig. A.10.

The corresponding coarse-grid operator, $A_H$, is easily computed to correspond to the difference stencil

$$\frac{1}{(2h)^2}[-4 \quad 8 \quad -4]_{2h}$$

which, after proper scaling of the restriction operator by $1/2$, is seen to be off by a factor of 2 compared to the "natural" $2h$-discretization of $-u''$. Due to this, for a very smooth error frequency, $\sin(\pi x)$, say, we obtain $Ke \approx (1/2)e$. Consequently, as smoothing hardly affects this frequency (if $h$ is very small), *we cannot expect the asymptotic two-level convergence factor to be better than* $1/2$.

If the same strategy is now used recursively to introduce coarser and coarser levels, the above arguments carry over to each of the intermediate levels and, in particular, each coarser-grid operator is off by a factor of 2 compared to the previous one. A simple recursive argument, applied to the same error frequency as above, shows that errors are accumulated from grid to grid and the asymptotic V-cycle convergence factor cannot be expected to be better than $1 - 2^{-m}$ where $m$ denotes the number of coarser levels. *That is, the V-cycle convergence is h-dependent.*                                    ≫

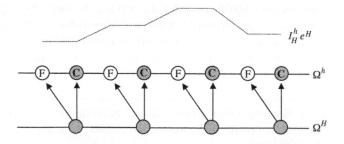

**Figure A.10.** Strictly one-sided interpolation (piecewise constant).

The major cause of the problem seen here is that the interpolation is only *piecewise constant* (first order) which, obviously, is insufficient to ensure $h$-independent V-cycle convergence. (We will consider piecewise constant interpolation again in the context of aggregation-based AMG in Section A.9, where the basic problem with piecewise constant interpolation will become clear; see Section A.9.1.) Note that one-sided interpolation has been artificially introduced in this particular example to demonstrate the consequences. In practice, each F-point should, of course, be interpolated from *both* its C-neighbors, leading to linear interpolation (which, in this 1D case, even gives a direct solver). In any case, Theorem A.4.3 formally allows such interpolations, and $h$-dependent V-cycle convergence always has to be expected whenever essentially one-sided interpolation is employed to solve analogous problems, in any dimension.

The main hurdle in extending the two-level to a V-cycle theory is due to the fact that the basic algebraic condition for interpolation, (A.4.2) (and hence also (A.5.14)), is too weak to imply uniform V-cycle convergence (cf. Remark A.4.2). In [334], the stronger requirement

$$\|e_F - I_{FC}e_C\|_{0,F}^2 \le \tau \, \|e\|_2^2 \tag{A.6.1}$$

has been discussed which is more suited to an algebraic V-cycle convergence theory. Indeed, following the same arguments as in Remark A.4.2, one sees that interpolation based on (A.6.1) is related to *second-order* interpolation (assuming an adequate geometric problem). Unfortunately, it hardly seems possible to construct interpolation so that (A.6.1) is satisfied exactly by using only algebraic information such as the matrix entries.

> In practice, however, it turned out that potential problems with interpolation can, to a large extent, easily be avoided, at least for all applications considered here. In fact, for geometrically posed applications, a sufficient improvement of the accuracy of interpolation (compared to the worst case as considered above) is generally obtained, by just arranging the C/F-splittings as uniformly as possible (based on the connectivity information contained in the matrices) so that each F-point is reasonably surrounded by its interpolatory neighbors. The importance of this has already been stressed in Remark A.4.6. Of course, if there is no geometrical background for a given problem or if the underlying connectivity structure is far from being local, there is no a priori guarantee of highest efficiency.

The application of one Jacobi relaxation step to a given interpolation is another simple (but generally more costly) way of improvement. In the above example, for instance, this would immediately "overwrite" the given piecewise constant interpolation by linear interpolation. Although, in general, relaxation of interpolation will not be able to increase the *order* of interpolation, it tends to substantially improve it.

The most popular way to overcome $h$-dependent V-cycle convergence is to use "better" cycles like F- or W-cycles. However, apart from the fact that such cycles are more expensive (which may be considerable in AMG, depending on the actual coarsening), they will, at best, have the same convergence factor as the corresponding two-level method which, in turn and for the same reasons which may lead to $h$-dependent V-cycle convergence, might not be fully satisfactory. That is, although better cycles may well be a pragmatic way to overcome a convergence problem, they tend to hide the true reasons for such problems and should only be considered as a second choice.

Although not needed for the type of applications considered here, we finally want to mention that one can imagine other strategies to improve interpolation. For instance, one could exploit a minimum amount of geometric information (e.g. point locations). More algebraically, local fitting of interpolation weights provides various possibilities to better approximate smooth error (cf. [334]), for instance, fitting based on some "test vector(s)" provided by the user upon calling AMG. In general, however, such "sophisticated" techniques are rather complicated and tend to be computationally expensive.

## A.7 THE ALGEBRAIC MULTIGRID ALGORITHM

The application of AMG to a given problem is a two-part process. The first part, a fully automatic *setup phase*, consists of recursively choosing the coarser levels and defining the transfer and coarse-grid operators. The second part, the *solution phase*, just uses the resulting components in order to perform normal multigrid cycling until a desired level of tolerance is reached (usually involving Gauss–Seidel relaxation for smoothing). The solution phase is straightforward and requires no explicit description.

This section describes algorithmic components used in the setup phase of the RAMG05 code mentioned in Section A.1.2. According to Section A.2.1, only the C/F-splitting and the interpolation, $I_{FC}$, need to be explicitly defined. These definitions closely follow the approaches suggested by the analysis contained in Sections A.4.2 and A.4.3. Restriction is taken as the transpose of interpolation (A.2.8) and the computation of the coarse-level Galerkin operators (A.2.3) is straightforward. There is still much scope for modifications and further enhancements, but the algorithmical components proposed here have been tested for a wide variety of problems and have been found to lead to robust and efficient solution processes. Typical results will be presented in Section A.8.

The algorithm described below does not exploit symmetry, except that restriction is *always* taken as the transpose of interpolation (which is not necessarily the best for non-symmetric problems, see Section A.2.3). Without any modification, the algorithm has been applied to various nonsymmetric problems. Practical experience indicates that, generally, the nonsymmetry by itself does not necessarily cause particular problems for AMG. Other properties of the underlying matrices, such as a strong violation of weak diagonal dominance, seem to influence the performance of AMG (as it stands) to a much larger extent.

In the following, we will first describe the splitting process (Section A.7.1) and afterwards the interpolation (Section A.7.2). The approach for constructing the splitting and the interpolation is the same for all levels of the AMG hierarchy. Therefore, the following description will be for any fixed level. Clearly, all of the following has to be repeated recursively for each level until the level reached contains sufficiently few variables to permit a direct solution. All quantities occurring will actually depend on the level, but the index will be omitted for convenience.

To make the current section easier to read and more self-contained, we keep references to previous sections to a minimum. Instead, we repeat the most relevant aspects.

### A.7.1 Coarsening

In order to achieve fast convergence, algebraically smooth error needs to be approximated well by the interpolation. As a rule, the stronger the F-to-C connectivity is, the better this can be achieved. On the other hand, the size of the coarse-level operator (and the time to compute it) strongly depends on the number of C-variables. Since the overall efficiency is determined by both the speed of convergence and the amount of work needed per cycle (which is also directly related to the total memory requirement), it is absolutely imperative to limit the number of C-variables while still guaranteeing that all F-variables are sufficiently strongly connected to the C-variables.

However, the goal should not just be to minimize the total number of C-points. According to Remark A.4.6 and the related discussion in Section A.6, an important objective is to create C/F-splittings which are as uniform as possible with F-variables being "surrounded" by C-variables to interpolate from. Although there is no algebraic proof, interpolation tends to be considerably better, resulting in much faster convergence, if this objective is considered. A simple algorithm is described in Section A.7.1.1.

Requiring strong F-to-C connectivity does not necessarily mean that all F-variables need to have strong *direct* connections to C-variables. In general, strong connectivity may be via strongly connected neighboring F-variables (cf. Section A.4.3). This leads to "aggressive" coarsening strategies as described in Section A.7.1.2. Such strategies allow for a drastic reduction of the setup and cycle cost, the complexity of the coarse-level operators as well as the memory requirement. Clearly, these benefits will be at the expense of a reduced convergence speed since smoothing becomes less efficient and since it becomes more difficult to "match the ranges" of the smoothing and the interpolation operators. In practice, however, this disadvantage is usually more than compensated for by the benefits, in particular, if AMG is used as a preconditioner rather than stand-alone (cf. Section A.7.3). We will present examples in Section A.8.

#### A.7.1.1 Standard coarsening

In this section, we consider C/F-splittings based on *direct* couplings: each F-variable $i$ is required to have a minimum number of those of its couplings $j \in N_i$ be represented in C which affect the error at $i$ most, that is, for which $|a_{ij}|$ is largest in some sense ("strong connections").

For all the applications we have in mind here, most of the strong couplings are *negative* and we first describe a fast procedure which generates a C/F-splitting taking only negative couplings into account (for positive couplings, see Section A.7.1). That is, the resulting C/F-splitting will be such that all F-variables have a substantial (direct) *negative* connectivity to neighboring C-variables. In other words, we essentially coarsen in directions *in which algebraically smooth error changes slowly* (cf. Section A.3.3).

To be more specific, let us define a variable $i$ to be *strongly negatively coupled* (or *strongly n-coupled*) to another variable, $j$, if

$$-a_{ij} \geq \varepsilon_{\text{str}} \max_{a_{ik} < 0} |a_{ik}| \quad \text{with fixed } 0 < \varepsilon_{\text{str}} < 1 \tag{A.7.1}$$

and let us denote the set of all strong n-couplings of variable $i$ by $S_i$,

$$S_i = \{j \in N_i : i \text{ strongly n-coupled to } j\}. \tag{A.7.2}$$

(Note that *all* positive connections are regarded as *weak* at this point.) According to practical experience, the concrete value of $\varepsilon_{str}$ is not critical, $\varepsilon_{str} = 0.25$ being a reasonable default value. Since the relation of variables being strongly n-coupled is generally nonsymmetric (even if $A$ is symmetric), we introduce the set $S_i^T$ of strong *transpose* n-couplings of $i$ consisting of all variables $j$ which are strongly n-coupled to $i$:

$$S_i^T = \{j \in \Omega : i \in S_j\}.$$

The proposed simple splitting algorithm corresponds to the "preliminary C-point choice" [334]. Essentially, one starts with defining some first variable, $i$, to become a C-variable. Then all variables, $j$, which are strongly n-coupled to $i$ (i.e. all $j \in S_i^T$) become F-variables. Next, from the remaining undecided variables, another one is defined to become a C-variable and all variables which are strongly n-coupled *to it* (and which have not yet been decided upon) become F-variables. This process is repeated until all variables have been taken care of.

The only problem is that, in order to avoid randomly distributed C/F-patches and instead obtain reasonably uniform distributions of C- and F-variables, we need to perform this process in a certain order. In order to ensure that there is a tendency to build the splitting starting from one variable and continuing "outwards" until all variables are covered, we introduce a "measure of importance", $\lambda_i$, of any undecided variable $i$ to become the next C-variable. We define

$$\lambda_i = |S_i^T \cap U| + 2|S_i^T \cap F| \quad (i \in U)$$

where $U$, at any stage of the algorithm, denotes the current set of undecided variables. (For any set $P$, $|P|$ denotes the number of elements it contains.) $\lambda_i$ acts as a measure of how valuable a variable $i \in U$ is as a C-variable, given the current status of C and F. Initially, variables with many others strongly n-coupled to them become C-variables, while later the tendency is to pick as C-variables those on which many *F-variables* strongly depend.

The complete algorithm is outlined in Fig. A.11. We point out that the measure $\lambda_i$ has to be computed globally only once, at the beginning of the algorithm. At later stages, it just needs to be updated locally. For isotropic five-point and nine-point stencils, the first coarsening steps are illustrated in Fig. A.12.

**Remark A.7.1**    Before the above algorithm starts, variables which have no connection at all (e.g. resulting from Dirichlet boundary points which have not been eliminated from the system) are filtered out and become F-variables. Trivially, such variables do not require interpolation. Similarly, variables which correspond to (very) strongly diagonally dominant rows of the matrix might be filtered out at this point.                                    ≫

**Remark A.7.2**    After termination of the above algorithm, all F-variables have (at least) one strong n-coupling to a C-variable (except for the "trivial" ones taken out at the very beginning, see Remark A.7.1). However, there may be a few U-variables left, in particular, in non-symmetric problems. Such particular variables have the property that they are not strongly n-coupled to any of the C-variables (otherwise they would have become F-variables earlier in the process). Moreover, such variables have no strong n-connection among themselves

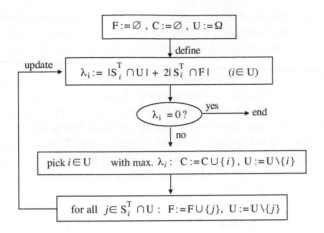

**Figure A.11.** Standard coarsening algorithm [334].

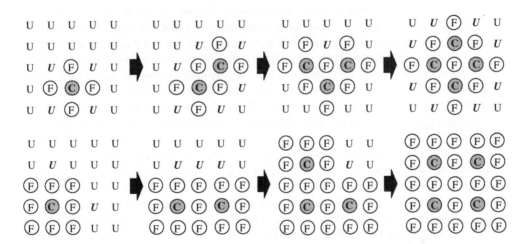

**Figure A.12.** First steps of the standard coarsening process in case of isotropic five-point (top) and nine-point stencils (bottom). At each stage, those undecided points with highest $\lambda$-value are shown in bold-italics.

nor is any F-variable strongly n-coupled to any of them (otherwise their measure $\lambda_i$ would be nonzero). However, each of these U-variables is strongly n-coupled to (at least) one of the F-variables. We therefore redefine all potentially remaining U-variables to become F-variables. In interpolation, they will be interpolated *indirectly* via their strong F-couplings (see Section A.7.2.1). $\gg$

**Remark A.7.3** None of the C-variables is strongly n-coupled to any of those C-variables created prior to itself in the coarsening process described above. However, since the relation of being strongly n-coupled is not necessarily symmetric, this may not fully be true the other way around. In any case, however, the resulting set of C-variables is close to a *maximal* set of variables which are not strongly n-connected to each other (see Remark A.4.6).     ≫

**Remark A.7.4**  The theoretical investigation of special processes such as F-smoothing (Section A.5.1.2) and relaxation of interpolation (Jacobi interpolation, Section A.5.1.3) was based on the assumption that the submatrices $A_{FF}$ are strongly diagonally dominant. Clearly, if required, this condition can be exactly satisfied during the coarsening step or, very easily, by adding a few C-points afterwards (if necessary). However, for those applications considered in this appendix, the above coarsening algorithm tends to ensure sufficient diagonal dominance without any modification. Therefore, by default, we do not explicitly check for strong diagonal dominance.     ≫

### A.7.1.2 Aggressive coarsening

In many PDE applications, we have to deal with small stencils. In such cases, the previous splitting algorithm, because it is based on direct connections, may cause a relatively high complexity (memory requirement due to the coarse-level Galerkin operators). For instance, isotropic seven-point stencils on regular 3D meshes, will cause the first coarser level to correspond to the black points of a red–black coarsened grid (as in the 2D case depicted in Fig. A.12, upper picture). One easily sees that the Galerkin operator on this level corresponds to a 19-point stencil. That is, the Galerkin *matrix* is larger than the original matrix by a factor of 1.36. Although subsequent coarsening will typically become faster (simply because the corresponding stencils are larger), the first coarsening step significantly contributes to the final complexity. Complexity can substantially be reduced by employing "aggressive coarsening".

In order to allow aggressive coarsening, we extend the definition of strong connectivity to also include variables which are not directly coupled. Following [334], we introduce the concept of *long-range strong n-connections*. A variable $i$ is said to be strongly n-connected to a variable $j$ *along a path of length* $\ell$ if there exists a sequence of variables $i_0, i_1, \ldots, i_\ell$ with $i = i_0$ and $j = i_\ell$ such that $i_{k+1} \in S_{i_k}$ for $k = 0, 1, 2, \ldots, \ell - 1$. With given values $p \geq 1$ and $\ell \geq 1$, we then define a variable $i$ to be *strongly n-connected to a variable* $j$ *w.r.t.* $(p, \ell)$ if at least $p$ paths of length $\leq \ell$ exist such that $i$ is strongly n-connected to $j$ along each of these paths (in the above sense).

In principle, for any given $p$ and $\ell$, the splitting algorithm described in the previous section immediately carries over if we apply it to the set

$$S_i^{p,\ell} = \{j \in \Omega : i \text{ strongly n-connected to } j \text{ w.r.t. } (p, \ell)\} \tag{A.7.3}$$

rather than $S_i$ (A.7.2). From a practical point of view, however, it does not pay to exploit strong n-connectivity in this generality. In fact, the cases $p = 2, \ell = 2$ and $p = 1, \ell = 2$ turn out to be the most useful. Moreover, it hardly ever pays to use aggressive coarsening on more than one level. (On all but the first level, standard coarsening is usually efficient enough.) We will refer to the coarse strategies corresponding to $S_i^{2,2}$ and $S_i^{1,2}$ as *A2*- and *A1-coarsening*, respectively.

**Remark A.7.5** Applying the splitting algorithm directly to $S_i^{p,\ell}$ instead of $S_i$, requires the storage of the complete connectivity information (and the corresponding transpose information contained in $(S_i^{p,\ell})^T$) for each variable $i$. Even for the cases $S_i^{2,2}$ and $S_i^{1,2}$ considered here, this may be quite substantial but can be avoided to a large extent by *applying the standard coarsening algorithm twice*. In the first step, it is applied exactly as described in the previous section. Then, instead of (A.7.3), we define strong n-connectivity *only between the resulting C-variables* (via neighboring F-variables). That is, for each variable $i \in C$, we define

$$\hat{S}_i^{p,\ell} = \{j \in C: i \text{ strongly n-connected to } j \text{ w.r.t. } (p, \ell)\}. \tag{A.7.4}$$

Using this definition, the standard coarsening algorithm is now applied a second time but restricted to the set of C-variables. The subset of "new" C-variables resulting from this second step will then be used as the next coarser level.          $\gg$

Clearly, A1- is faster than A2-coarsening. As an example, Fig. A.13 illustrates the second step of the two-step process described in the previous remark for isotropic five-point stencils. Generally, while A2-coarsening is effective only in (at least) "planewise" isotropic problems, A1 is also effective in strongly anisotropic cases. Figure A.14 shows the result of using A2 and A1 coarsening for the example discussed in Section A.1.3. In case of strategy A2 (left picture), the coarsening pattern is essentially the same as for standard coarsening (sec Fig. A.4) except in the lower left quarter where the problem is isotropic and coarsening is faster. Strategy A1, on the other hand, also speeds coarsening up in the anisotropic areas. Indeed, the right picture in the figure shows that coarsening is now substantially faster everywhere. It is still essentially in the direction of strong couplings. However, the coarse-level points are further apart than before.

**Remark A.7.6** If aggressive coarsening is used, strong diagonal dominance of the corresponding submatrices $A_{FF}$ (theoretically required if F-smoothing or Jacobi interpolation is

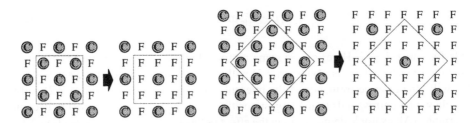

**Figure A.13.** Results of aggressive A2 (left) and A1 coarsening (right) in case of isotropic five-point stencils. The dashed boxes depict the range of strong connectivity in the sense of $\hat{S}_i^{2,2}$ and $\hat{S}_i^{1,2}$, respectively.

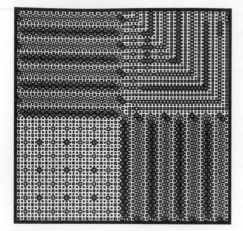 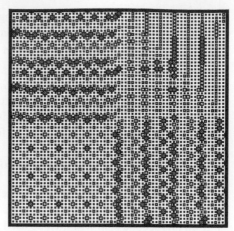

**Figure A.14.** The finest and three consecutive AMG levels if aggressive A2 (left) and A1 coarsening (right) is applied (only) on the first level.

to be employed in the sense of Section A.5.1), can no longer be assumed in the strict sense, at least not for each row.                                                                 ≫

### A.7.1.3 Strong positive connections

The previous approaches to constructing a C/F-splitting were based on negative couplings only. Provided that potentially existing *positive* couplings are relatively small, they can, indeed, be ignored in coarsening and interpolation (cf. the related discussion in Section A.4.2). However, this cannot always be assumed and we have to allow for matrices which also contain some strong *positive* entries.

According to Theorem A.4.6, a more general splitting process should ensure that, for all F-variables which have strong negative *and* positive couplings, a minimum number of *both* types of couplings is represented in C. However, to construct such a splitting within one step, turns out to be relatively complicated. Since, for all problems we have in mind here, most strong connections are negative, we propose a very simple alternative. After one of the coarsening processes described before has been applied, we test for all F-variables $i$ whether or not there are strong positive F–F couplings. For instance, we simply check whether

$$a_{ij} \geq \varepsilon_{str}^{+} \max_{k \neq i} |a_{ik}| \tag{A.7.5}$$

holds for some $j \neq i$. Here, $\varepsilon_{str}^{+}$ is some reasonable tolerance, for instance, $\varepsilon_{str}^{+} = 0.5$. If such a $j$ exists, *all* $j$s satisfying (A.7.5) will a posteriori be added to the set $S_i$ (A.7.2) (this will affect the performance of the interpolation routines, see Section A.7.2) and the variable which corresponds to the largest positive coupling, will be redefined to become a C-variable. (An example demonstrating the effect of this process, is given in Section A.8.4.)

Clearly, this a posteriori update of the C/F-splitting is suitable only if there are not too many strong positive connections. Otherwise, one has to change the original splitting algorithm as mentioned above.

### A.7.2 Interpolation

In defining interpolation to the currently finest level, we assume that a C/F-splitting has been constructed either by means of standard or aggressive coarsening. In the first case, interpolation is used as described in Section A.7.2.1 (*direct* or *standard interpolation*). In the second case, interpolation is used as described in Section A.7.2.2 (*multipass interpolation*). In both cases, interpolation can optionally be improved further by means of additional relaxation steps (Section A.7.2.3, *Jacobi interpolation*).

> Some of the interpolation variants described in the following exploit strong *indirect* C-couplings which will increase the "radius" of interpolation. In all such cases, it is important to reasonably truncate the resulting interpolation operator before computing the Galerkin operator (see Section A.7.2.4).

The following abbreviations will be needed below. Here, $S_i$ is as defined in (A.7.2), possibly modified a posteriori as described in Section A.7.1.3:

$$C_i = C \cap N_i, \qquad C_i^s = C \cap S_i,$$
$$F_i = F \cap N_i, \qquad F_i^s = F \cap S_i.$$

#### A.7.2.1 Direct and standard interpolation
The following procedures apply in case the C/F-splitting has been constructed by means of standard coarsening.

#### Direct interpolation
In the simplest case, the definition of interpolation, as described in Section A.4.2.3, is applied immediately. More precisely, for each $i \in F$, we define the set of interpolatory variables by $P_i = C_i^s$ and approximate

$$a_{ii}e_i + \sum_{j \in N_i} a_{ij}e_j = 0 \quad \Longrightarrow \quad a_{ii}e_i + \alpha_i \sum_{k \in P_i} a_{ik}^- e_k + \beta_i \sum_{k \in P_i} a_{ik}^+ e_k = 0 \qquad \text{(A.7.6)}$$

with

$$\alpha_i = \frac{\sum_{j \in N_i} a_{ij}^-}{\sum_{k \in P_i} a_{ik}^-} \quad \text{and} \quad \beta_i = \frac{\sum_{j \in N_i} a_{ij}^+}{\sum_{k \in P_i} a_{ik}^+}.$$

This immediately leads to the interpolation formula

$$e_i = \sum_{k \in P_i} w_{ik} e_k \quad \text{with } w_{ik} = \begin{cases} -\alpha_i\, a_{ik}/a_{ii} & (k \in P_i^-) \\ -\beta_i\, a_{ik}/a_{ii} & (k \in P_i^+). \end{cases} \qquad \text{(A.7.7)}$$

If $P_i^+ = \emptyset$, this formula is modified according to Section A.4.2.2, that is, we set $\beta_i = 0$ and add all positive entries, if there are any, to the diagonal. Since this interpolation involves only direct connections of variable $i$, we will refer to it as *direct interpolation*.

**Remark A.7.7** The above procedure can be applied as long as $C_i^s \neq \emptyset$. However, this is ensured by the standard coarsening algorithm for all F-points $i$ with the potential exception of just a few of them (see Remark A.7.2). Such "exceptional" F-points, however, necessarily have at least one strong connection to a "regular" F-point, and will be interpolated indirectly as described next. $\gg$

### Standard interpolation

The standard coarsening strategy ensures that there is a strong F–C connectivity. However, it does not strictly enforce what actually is required by the two-level theory, namely, that each F-variable should have a *fixed percentage* of its total connectivity reflected in C (defined by $\tau$, see Section A.4.2). Although this is usually not a problem in practice (since the coarsening algorithm by itself usually ensures sufficient F–C connectivity), we can easily make up for this. We modify the previous direct interpolation so that, for each $i \in F$, its strong *F-connections* are also (indirectly) included in interpolation.

That is, instead of immediately approximating the $i$th equation (left equation in (A.7.6)), we first (approximately) eliminate all $e_j$ ($j \in F_i^s$) by means of the corresponding $j$th equations. More specifically, for each $j \in F_i^s$, we replace

$$e_j \longrightarrow -\sum_{k \in N_j} a_{jk} e_k / a_{jj} \tag{A.7.8}$$

resulting in a new equation for $e_i$,

$$\hat{a}_{ii} e_i + \sum_{j \in \hat{N}_i} \hat{a}_{ij} e_j = 0 \quad \text{with } \hat{N}_i = \{j \neq i \colon \hat{a}_{ij} \neq 0\}. \tag{A.7.9}$$

By defining $P_i$ as the union of $C_i^s$ and all $C_j^s$ ($j \in F_i^s$), we now define interpolation exactly as in (A.7.6) and (A.7.7) with all $a$s replaced by $\hat{a}$s and $N_i$ replaced by $\hat{N}_i$.

This modification usually enhances the quality of interpolation substantially (see Example A.7.1 below), the main reason being that the type of approximation (A.7.6), if applied to the "extended" equation (A.7.9), introduces less error. Moreover, it further contributes to the objective of having F-variables largely "surrounded" by interpolatory variables. This modified interpolation will be referred to as *standard interpolation* below.

**Example A.7.1** Consider the same case as in Example A.6.1 except that the coarse grid is assumed to be created by $h \to 3h$ coarsening, see Fig. A.15. (This is just for ease of demonstration; the standard coarsening process described in Section A.7.1 would not really create this coarsening.)

Direct interpolation in this situation would obviously give piecewise constant interpolation (dashed line in the figure) which, as we know from Example A.6.1, is not quite satisfactory. In fact, the resulting Galerkin operator is off by a factor of 3 compared to the natural one. (Compare Example A.6.1 where the corresponding factor was 2.) Standard interpolation, on the other hand, can easily be seen to correspond to linear interpolation. For instance, the interpolated value for $e_0$ is computed from the equation $-e_{-1} + 2e_0 - e_1 = 0$

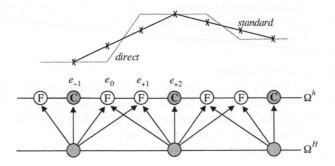

**Figure A.15.** Direct versus standard interpolation.

by substituting

$$e_1 \longrightarrow (e_0 + e_2)/2,$$

giving $e_0 = \frac{2}{3}e_{-1} + \frac{1}{3}e_2$. $\triangle$

Of course, direct and standard interpolation processes may also be mixed in a straightforward way. That is, standard interpolation is used only for variables $i$ for which, based on some reasonable criterion, the (direct) F–C connectivity appears to be too low. However, for simplicity, such mixed interpolation will not be considered here. Moreover, for critical F-variables $i$, one might be tempted to eliminate *all* $e_j$ ($j \in F_i$) (rather than just the *strong* F-neighbors) and to use the union of $C_i$ and all $C_j$ ($j \in F_i$) as $P_i$. However, taking computational work into account, this *extended interpolation* is rarely ever advantageous and will not be discussed further.

**Remark A.7.8** Apart from other minor differences, interpolation in AMG1R5 was a compromise between the direct interpolation and the standard interpolation described above. There, an attempt was made to replace $e_j$ ($j \in F_i^s$) by averages *involving only variables in $C_i^s$*. That is, the goal was to improve the direct interpolation *without* increasing the set of interpolatory variables $P_i$. If it turned out that this was not possible, based on certain criteria, new C-variables were added to the splitting, *increasing $C_i^s$ a posteriori*. Although this approach works quite well in many situations, it has two drawbacks.

First, an a posteriori introduction of additional C-variables may turn a fairly regular C/F-splitting (produced by the coarsening algorithm) into a quite disturbed one which, during subsequent coarsening steps, may lead to more irregular and more complex Galerkin operators. In fact, in complex 3D situations, many additional C-variables are typically introduced a posteriori often causing unacceptably high complexities (see Section A.8 for examples). Second, the above-mentioned replacement of the $e_j$s by averaged values was motivated by geometric arguments. It works very well in case of matrices which are close to M-matrices and are related to regular geometric situations. However, it may substantially deteriorate in other cases.

In practice, the standard interpolation as described above—extending the interpolation pattern on a *fixed* set C followed by a truncation (see Section A.7.2.4)—has turned out to be more robust and often considerably more efficient. $\gg$

### A.7.2.2 Multipass interpolation

The following interpolation procedure applies when the C/F-splitting has been constructed by means of aggressive coarsening. It proceeds in several passes, using direct interpolation whenever possible and, for the remaining variables, exploiting interpolation formulas at neighboring F-variables. (This corresponds to the approach described in Section A.4.3.) The individual passes are as follows:

(1) Use direct interpolation (Section A.7.2.1) to derive formulas for all $i \in F$ for which $C_i^s \neq \emptyset$ and define the set $F^\star$ to contain all these variables. If $F^\star = F$ stop, otherwise proceed.

(2) For all $i \in F \setminus F^\star$ for which $S_i \cap F^\star \neq \emptyset$ do the following: take the $i$th equation (left equation in (A.7.6)) and, for all $j \in S_i \cap F^\star$, replace

$$e_j \longrightarrow \sum_{k \in P_j} w_{jk} e_k$$

leading to a new equation (A.7.9) for $e_i$. Defining the set of interpolatory variables $P_i$ as the union of all $P_j$ for $j \in S_i \cap F^\star$, an interpolation formula is then computed exactly as in the case of standard interpolation. If all such variables $i$ have been processed, update $F^\star$ to also include all variables which have obtained an interpolation formula during this pass.

(3) If $F^\star = F$ stop. Otherwise go back to Step 2.

Using the aggressive (A1 or A2) coarsening strategy described in Remark A.7.5, this process can be shown to terminate after at most four passes. Note that the update of $F^\star$ is done in a Jacobi (not Gauss–Seidel) fashion. This is done to preserve the locality of interpolation. We will refer to this interpolation as *multipass interpolation*.

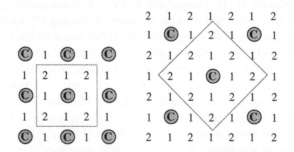

**Figure A.16.** Multipass interpolation for isotropic five-point problems. Left, A2-coarsening; right, A1-coarsening.

**Example A.7.2** Figure A.16 illustrates multipass interpolation in case of A2- and A1-coarsening, applied to the five-point Poisson stencil (cf. Fig. A.13). F-points marked by "1" and "2" are those which are interpolated in the first and second pass, respectively. In case of A2-coarsening (left), the resulting interpolation can easily be seen to be linear. For A1-coarsening (right), we obtain constant interpolation at all points marked by "1", and linear interpolation at the remaining points (see also Example A.7.3 below). $\triangle$

### A.7.2.3 Jacobi interpolation

Given any of the previous interpolation formulas, we can optionally improve it by a posteriori applying Jacobi relaxation as formally described in Section A.5.1.3. More explicitly, one step of this Jacobi relaxation, proceeding from iteration $\mu - 1$ to iteration $\mu$ (where $\mu = 0$ corresponds to the given interpolation), proceeds as follows.

For all variables $i \in F$ in turn, take the $i$th equation (left equation in (A.7.6)) and, for *all* $j \in F_i$, replace

$$e_j \longrightarrow \sum_{k \in P_j} w_{jk}^{(\mu-1)} e_k \tag{A.7.10}$$

which leads to a new equation (A.7.9) for $e_i$. Defining the set of interpolatory variables $P_i$ as the union of $C_i$ and all $P_j$ for $j \in F_i$, the $\mu$th interpolation formula is then obtained exactly as in the case of standard interpolation.

Clearly, only one or (at most) two steps are practical. Depending on the situation, relaxation of interpolation may considerably enhance convergence. Often, however, the additional effort for computing this interpolation does not pay if total computational cost is taken into account. Clearly, one may save a substantial amount of work by only applying relaxation of interpolation locally wherever it appears to be reasonable. This is not done in the following.

The interpolation as described above will be referred to as *fully relaxed Jacobi interpolation*. In most cases, it will be sufficient to use the replacement (A.7.10) only for $j \in F_i^s$, that is, only for the *strongly* connected F-variables. Accordingly, the set $P_i$ is then selected as the union of $C_i^s$ and all $P_j$ for $j \in F_i^s$. This will be referred to as *partially relaxed Jacobi interpolation*.

**Example A.7.3** Consider the same case as in Example A.7.2. We saw there that multipass interpolation, applied to the A1-coarsened grid (right-hand side of Fig. A.16), gives constant interpolation at those points marked by "1", and linear interpolation at the remaining points. One easily sees that, applying Jacobi relaxation just to the points marked by "1" yields linear interpolation everywhere. $\triangle$

### A.7.2.4 Truncation of interpolation

For both standard as well as Jacobi interpolation, the sets $P_i$ of interpolatory variables may become quite large. This is, in particular, true for the Jacobi interpolation since each relaxation step introduces, roughly, a "new layer" of additional C-variables to be used for interpolation. Consequently, even if only one Jacobi step is applied at each AMG level, the resulting Galerkin operators will substantially increase towards coarser levels. This process, without reasonable truncation, will generally be much too costly.

However, interpolation weights corresponding to variables "far away" from variable $i$ will usually be much smaller than the largest ones. Before computing the coarser-level Galerkin operator, we therefore always truncate the full interpolation operator by ignoring all interpolatory connections which are smaller (in absolute value) than the largest one by a factor of $\varepsilon_{tr}$ and rescale the remaining weights so that the total sum remains unchanged. In practice, a value of $\varepsilon_{tr} = 0.2$ is usually taken.

**Remark A.7.9** If interpolation contains substantial positive *and* negative weights, one should truncate and rescale positive and negative weights *separately* (analogously to the definition of interpolation (A.7.6)). Otherwise, convergence may substantially degrade. ≫

**Remark A.7.10** One might be tempted to truncate the *Galerkin operator* rather than the interpolation operator. This would formally give more control on the growth of the coarse-level operators. However, we have already pointed out in Remark A.2.4, that this may cause serious convergence problems if not applied with great care.                    ≫

### A.7.3 Algebraic Multigrid as Preconditioner

In order to increase the robustness of standard multigrid approaches, it has become very popular in recent years, to use multigrid not as a stand-alone solver but rather to combine it with acceleration methods such as conjugate gradient, BI-CGSTAB [397] or GMRES [335, 337]. In the simplest case, complete multigrid cycles are merely used as preconditioners [211, 292]; in more sophisticated approaches, acceleration is even used on the individual grids of the hierarchy [82, 294]. This development was driven by the observation that, it is often not only simpler but also more efficient to use accelerated multigrid approaches rather than to try to optimise the interplay between the various multigrid components in order to improve the convergence of stand-alone multigrid cycles.

This has turned out to be similar for AMG which was originally designed to be used stand-alone. Practical experience has clearly shown that AMG is also a very good pre-conditioner, much better than standard (one-level) ILU-type preconditioners, for example. Heuristically, the major reason is due to the fact that AMG, in contrast to any one-level preconditioner, operates efficiently on *all* error components, short-range as well as long-range. This has the implication that, instead of using AMG stand-alone, it is generally more efficient to put less effort into the (expensive) setup phase and use AMG as preconditioner, for example, by using aggressive coarsening strategies (cf. the applications in Section A.8).

In this context, we also point out that, although AMG tries to capture all relevant influences by proper coarsening and interpolation, its interpolation will hardly ever be optimal. It may well happen that error reduction is significantly less efficient for some very specific error components. This may cause a few eigenvalues of the AMG iteration matrix to be considerably closer to one than all the rest. If this happens, AMG's convergence factor is limited by the slow convergence of just a few exceptional error components while the majority of the error components are reduced very quickly. Acceleration by, for instance, conjugate gradient typically eliminates these particular frequencies very efficiently. The alternative, namely, to try to prevent such situations by putting more effort into the construction of interpolation, will generally be much more expensive. And even then, there is no final guarantee that such situations can be avoided. (We note that this even happens with "robust" geometric multigrid methods, see, for instance, Remark A.8.10.)

## A.8 APPLICATIONS

In this section we will demonstrate the efficiency and robustness of AMG in solving second-order elliptic differential equations. All results presented have been obtained by the code RAMG05 described in the previous section. Although the strength of RAMG05 is its direct applicability to geometrically complex problems, we will often consider selected model problems on simple geometries in quite some detail. Such model problems do not give the full picture, but they permit easy investigation of AMG's asymptotic behavior as well as its dependence on various specific aspects such as anisotropies, discontinuities, singular perturbations and the like. Practical experience has shown that AMG's performance in geometrically complex situations, in 2D as well as 3D, is very comparable to that in related model situations. We will present some typical examples to demonstrate this.

We have already pointed out that it is not sufficient to merely look at AMG's convergence behavior in order to judge its performance. Useful comparisons have to take both *computing times* and *memory requirements* into account. Having this in mind, we will compare the influence of different algorithmical components (such as type of interpolation and speed of coarsening) and solution approaches (stand-alone versus accelerated cycles) on the performance. Moreover, in order to quantify the overall efficiency, we will make some comparisons with well-known standard (one-level) solution methods such as ILU preconditioned conjugate gradient ("CG"). However, we want to point out that the purpose of these comparisons is merely to give a first indication, they do not give a final picture.

First, many variants and improvements of ILU preconditioned CG methods are available, here we focus on simple, well-known strategies. In particular, our comparisons are not meant to judge the performance of such classical methods in general. Secondly, RAMG05 is still under development and is continuously being enhanced and generalized. In particular, RAMG05 is far from being optimized. In fact, this code has not been designed for optimum efficiency but rather for flexibility in testing and extending the method. In particular, the setup cost may be substantially reduced. Depending on the concrete approach, 50% savings or even more seem realistic. Moreover, our main interest here is to demonstrate typical trends in the influence of different algorithmical components. For simplicity, these components are implemented as "fixed" strategies, that is, they are not locally adjusted to particular requirements of the given problem. For example, if Jacobi relaxation of interpolation is performed, it is always applied "globally". In many situations, however, a local application, controlled by some reasonable measure (e.g. based on the total strength of C-connectivity found at an F-point), may give similar convergence improvements at much lower cost and memory. Similarly, if aggressive coarsening is performed, it is done everywhere. Thus, there is much room for quite substantial optimizations. Nevertheless, the results indicate that the code is very efficient even as it stands.

In the sequel, for brevity, we will refer to "AMG" rather than to "RAMG05". However, one should keep in mind that there is rapid, ongoing development of new AMG approaches and variants, and that there is no unique and best AMG approach yet.

### A.8.1 Default Settings and Notation

Various parameters have to be set to define AMG's setup and cycle (see Section A.7). Unless explicitly stated otherwise, we use the following default settings and procedures.

- $\varepsilon_{str} = 0.25$ to define *strong connectivity* (Section A.7.1.1).
- $\varepsilon_{tr} = 0.2$ to define *truncation of interpolation* (Section A.7.2.4).
- Coarsening is terminated if the number of variables on the coarsest level drops below 40. The coarsest-level equations are solved by direct Gauss elimination.
- Smoothing is done by Gauss–Seidel relaxation, one pre- and one postsmoothing step being the default. Unless explicitly stated otherwise, the order of relaxation is "CF", that is, first all C-variables are relaxed and then all F-variables. (This corresponds to red–black relaxation in geometric multigrid.)

Other degrees of freedom in defining the concrete strategy will be varied in the experiments below and some notation is required to distinguish these cases.

- *Type of cycle and coarsening* The abbreviations VS and VA are used to distinguish V-cycles based on *standard* and *aggressive coarsening*, respectively (Section A.7.1). Aggressive coarsening is performed only in creating the second AMG level, and only the types A1 and A2 are used (see Section A.7.1.2). Correspondingly, we distinguish VA1 and VA2 cycles. For F-cycles, the "V" is replaced by "F".
- *Type of smoothing* As mentioned above, by default we use one Gauss–Seidel CF-relaxation step for pre- and postsmoothing. If this is not the case, we append the type of smoothing to the abbreviation of the cycle. For instance, VS-FF stands for a V-cycle using standard coarsening but employing two F-smoothing steps rather than one CF-step for pre- and postsmoothing (cf. Section A.5.1.2). SGS stands for symmetric Gauss–Seidel relaxation.
- *Type of interpolation* The type of interpolation used is appended in parentheses: the letters "D" and "S" stand for *direct* and *standard interpolation*, respectively (see Section A.7.2.1). Our standard AMG cycle is VS(S). Note that, if aggressive coarsening is employed, interpolation to the finest level is *always multipass interpolation* (Section A.7.2.2). For example, VA2(S) means that standard interpolation is performed on all *but the finest level*.

  If, in addition, Jacobi relaxation is applied to improve interpolation, the letters "F" and "P" refer to *fully* and *partially* relaxed interpolation, respectively (Section A.7.2.3). For instance, VS(S-2F) stands for a V-cycle using standard coarsening and standard interpolation improved by two full Jacobi relaxations. As mentioned above, truncation with $\varepsilon_{\mathrm{tr}} = 0.2$ is the default. Otherwise, the truncation parameter is also contained within the parentheses, for example, VS(S-1F,0.02).

- *Acceleration*  If a cycle is used as *preconditioner* rather than stand-alone, the type of accelerator is appended to the corresponding cycle abbreviation. For instance, VA1(D)/CG means that the VA1(D) cycle is used as preconditioner for CG. Note that, if a cycle is used as preconditioner for CG, pre- and postsmoothing will always be done in a symmetric way. For instance, if presmoothing is done by CF-relaxation, postsmoothing will be CF-relaxation *with the order of variables reversed*. For nonsymmetric problems, we will usually use AMG as preconditioner for BI-CGSTAB.

The following sections contain results on *asymptotic convergence, memory requirement* as well as *computational work*. The *asymptotic convergence factor*, $\rho$, is always computed numerically by applying a von Mises vector iteration to the homogeneous problem, usually starting with a random first approximation. Results on memory requirement will be given in terms of the *grid* and *operator complexity*, $c_G$ and $c_A$,

$$c_G = \sum_\ell n_\ell/n_1 \quad \text{and} \quad c_A = \sum_\ell m_\ell/m_1, \tag{A.8.1}$$

where $n_\ell$ and $m_\ell$ denote the number of variables and nonzero matrix entries, respectively, on level $\ell$. Note that $\ell = 1$ corresponds to the finest level. Although the true memory requirement of AMG is not fully reflected by these quantities (some extra work space still needs to be allocated), they are closely related.

Unless explicitly stated otherwise, all timings given have been obtained on a Pentium II/300 PC using the Lahey F90 Compiler (version 4.0). We point out that timings for a particular machine always have to be judged with care. Comparisons typically change from machine to machine and even from compiler to compiler. For instance, the Pentium II is relatively fast in integer (compared to floating point) computations. This is advantageous for substantial parts of the AMG algorithm which essentially require integer computations (in particular, during the setup phase). Consequently, comparisons of the setup and solution costs may give a different picture on machines for which floating point computations are more efficient than integer computations (such as on IBM RS6000 workstations).

### A.8.2 Poisson-like Problems

In the following, we investigate the performance of AMG in some detail if applied to the diffusion equation

$$-((1 + \sin(x + y))u_x)_x - (e^{x+y}u_y)_y = f(x, y), \tag{A.8.2}$$

defined on the unit square with $f(x, y) \equiv 1$ and homogeneous Dirichlet boundary conditions. Discretization is on a uniform grid of mesh size $h = 1/N$ using standard five-point stencils. Although this example is very simple, the resulting AMG behavior is typical for general "Poisson-like" problems and the relevant conclusions qualitatively carry over also to unstructured meshes (cf. the examples in Section A.8.3). We summarize our practical experience with AMG in the following remark.

**Remark A.8.1** Compared to problems on very regular meshes, a certain decrease of AMG convergence has to be expected in case of irregular meshes. This is essentially due to the fact that, on regular meshes, standard AMG interpolation tends to be close to geometrical interpolation (which is very good for Poisson-like problems as considered here). This cannot be expected to be satisfied to the same extent on irregular meshes. Similarly, convergence has to be expected to be somewhat slower in 3D than in 2D situations. In 3D, we have the additional effect that smoothing by Gauss–Seidel relaxation is (slightly) less efficient than in 2D problems (just like in geometric multigrid). By how much convergence will finally be influenced by the irregularity of the grid and its dimension, depends somewhat on the concrete problem. By experience, however, the effects mentioned are very limited and the results presented in the following exhibit the typical AMG behavior observed in many other cases. $\gg$

### A.8.2.1 Coarsening and complexity

Problem (A.8.2) has a slight anisotropy towards the upper right corner. Due to the setting $\varepsilon_{str} = 0.25$, however, AMG still treats all connections contained in the corresponding matrix as strong, at least on the finest level. Consequently, the first standard coarsening step of AMG corresponds to geometrical red–black coarsening. This is shown in Fig. A.17(a). Subsequent coarsening then becomes faster (here, grid size ratio $1:4$) simply because the Galerkin stencils become larger on coarser levels. For example, the Galerkin operator on level 2 corresponds to a nine-point stencil. There is a slight disturbance of the regular coarsening in the upper right corner where the anisotropy of the problem is largest. However, the coarsening pattern is still essentially the same.

This type of coarsening is typical for five-point stencils with all connections being strong, yielding grid complexities of $c_G \approx 1.7$. If interpolation were to be defined geometrically (i.e. linear interpolation), the Galerkin operators on all coarser levels would correspond

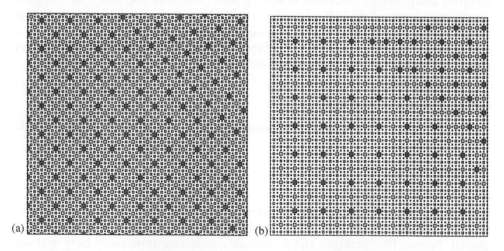

**Figure A.17.** The finest and three consecutive AMG levels created by: (a) standard coarsening; (b) aggressive A2-coarsening (applied only in the first coarsening step).

to nine-point stencils and the "ideal" operator complexity would be $c_A \approx 2.2$. For AMG, however, the situation is more involved, since the final values of $c_G$ and $c_A$ are influenced by the AMG interpolation operator which tends to cover more points than geometric interpolation, especially towards coarser levels. As a consequence, the AMG Galerkin operators will tend to become somewhat larger towards coarser levels. This effect, however, is normally limited and is more than compensated for by the decrease of grid points. In any case, one has to expect the final operator complexity of AMG to be larger than the ideal complexity by a certain factor. In the above example, we obtain $c_A \approx 2.38$ if standard coarsening and interpolation are used (see Table A.1).

**Remark A.8.2** Memory requirement is somewhat higher in corresponding 3D situations. For instance, if applied to seven-point stencils with all connections being strong, standard AMG coarsening again yields geometrical (3D) red–black coarsening in creating the first coarser level, with the Galerkin operator corresponding to a nineteen-point stencil. Subsequent coarsening, as before, will become faster. However, the "ideal" operator complexity (obtained if geometrical interpolation was used) is now $c_A \approx 2.8$. Consequently, the true AMG complexity, generally, has to be expected to be larger than 3.0. $\gg$

If memory requirement is an issue, aggressive coarsening may be used instead of standard coarsening. As mentioned earlier, it is usually sufficient to apply this type of coarsening in the first coarsening step and maintain standard coarsening for all subsequent levels. Figure A.17(b) shows the resulting first four levels if aggressive A2-coarsening is performed to create the first coarser level. The first AMG coarsening step now corresponds to geometrical $h \to 2h$ coarsening rather than red–black coarsening. Except for the upper right area of the domain, this also holds for the subsequent (standard) coarsening steps. Near the upper right corner, the situation is slightly different. Obviously, the Galerkin operator on level 2 is more anisotropic than it was on the finest level. AMG detects this and creates the third level by linewise coarsening in the $y$-direction. Since linewise coarsening essentially removes the anisotropy, the next coarsening step is again in both directions.

Ignoring the special coarsening near the upper right corner, the grid complexity now is only $c_G \approx 1.33$ and the ideal operator complexity becomes $c_A \approx 1.6$ (assuming linear interpolation, all Galerkin operators correspond to nine-point stencils). As before, the true AMG operator complexity will be somewhat larger; in the above example we obtain $c_A \approx 1.77$ (see Table A.1) which is very reasonable. In corresponding 3D situations, the gain in terms of memory reduction by means of aggressive coarsening is even higher. Memory usage can further be reduced either by also using A2-coarsening on the coarser levels (which usually does not pay, see above) or by using A1-coarsening to create the first coarser level (cf. the examples in Section A.8.3).

### A.8.2.2 Performance and comparisons

Figure A.18(a) shows asymptotic convergence factors, $\rho$, for several AMG strategies and increasing $N$. We first observe that our standard cycle, the VS(S)-cycle, exhibits very stable convergence behavior with $\rho < 0.15$ for increasing $N$. Investing more effort into the interpolation, by applying one Jacobi F-relaxation, improves convergence only marginally,

**Table A.1.** Complexities and computing times ($N = 512$).

| | Complexities | | Setup time | Stand-alone | | Conjugate gradient | |
|---|---|---|---|---|---|---|---|
| Method | $c_A$ | $c_G$ | | Cycle | $\varepsilon_0 = 10^{-10}$ | Cycle | $\varepsilon_0 = 10^{-10}$ |
| | | | | | Times (sec)/Pentium II, 300 MHz | | |
| ILU(0) | | | 0.87 | | | 1.07 | 628.6 (587) |
| AMG1R5 | 2.42 | 1.71 | 6.97 | 2.10 | 25.9 (9) | | |
| VS(S) | 2.38 | 1.67 | 11.8 | 2.32 | 37.4 (11) | 2.93 | 32.6 (7) |
| FS(S) | " | " | " | 3.87 | 31.2 (5) | 4.46 | 29.7 (4) |
| VS(D) | 2.20 | 1.67 | 8.51 | 2.22 | 48.5 (18) | 2.83 | 39.7 (11) |
| VA2(S) | 1.77 | 1.35 | 10.3 | 1.78 | 58.2 (27) | 2.38 | 41.3 (13) |
| VA1(S) | 1.50 | 1.19 | 8.07 | 1.47 | 65.4 (39) | 2.07 | 45.3 (18) |

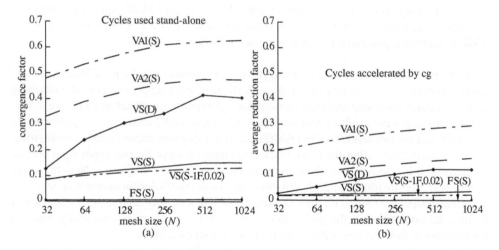

**Figure A.18.** (a) Convergence factors for cycles used stand-alone; (b) average reduction factors for accelerated cycles.

indicating that the standard interpolation is fairly good in this case (cf. VS(S-1F,0.02)-cycle). In contrast to this, investing more work into the cycle itself, by applying an F- instead of a V-cycle, causes extremely fast convergence (cf. Remark A.8.4).

The influence of aggressive coarsening is demonstrated by the low-memory VA2(S)- and VA1(S)-cycles. As expected, convergence becomes considerably slower but still approaches an upper limit for large $N$. As mentioned earlier, aggressive coarsening not only causes the smoothing to be less effective (we still use only one CF-relaxation step for pre- and postsmoothing) but also interpolation to be less accurate (multipass interpolation from level 2 to level 1). Clearly, convergence can be improved, for instance, by using twice as many smoothing steps and using Jacobi relaxation to improve interpolation to the finest level. However, this would substantially increase the cost of the cycle and, more importantly, tend to reduce the advantage of low operator complexity which was the major purpose for using aggressive coarsening to begin with.

A simpler and very effective way of improving the convergence of any cycle, in particular those using aggressive coarsening, is to use them as preconditioners rather than stand-alone. This is demonstrated in Fig. A.18(b) which shows the AMG convergence if used as preconditioner for CG. For the same cycles as before, the average residual reduction factors are shown, obtained over (at most) 50 iterations in solving the homogeneous system starting with a random first approximation. At relatively little extra cost, convergence is substantially enhanced.

**Remark A.8.3**    Figure A.18 also depicts the convergence of the VS(D)-cycle (where standard interpolation is replaced by the simpler direct interpolation). Although this cycle also gives good convergence in most cases, its convergence behavior is often not as stable as that of the VS(S)-cycle. This is why the VS(S)-cycle is our standard cycle.    ≫

**Remark A.8.4**    The reason for using F-cycles rather than V-cycles is to solve the coarse-level correction equations more accurately and to reduce the accumulation of errors from the individual levels of the AMG hierarchy. Clearly, in general, F-cycle convergence cannot be better than the corresponding two-level convergence. The drastic improvement in convergence shown in Fig. A.18 (cf. VS(S)- and FS(S)-cycles) is due to a particular situation which cannot be expected in general. Since standard coarsening, applied to a five-point stencil with all connections being strong, corresponds to red–black coarsening, there are no F–F connections on the finest level (i.e., $A_{FF}$ is a diagonal matrix). Therefore, according to Section A.2.3, the two-level method involving the first two levels corresponds to a direct solver. Consequently, any increase of accuracy in solving the correction equations on level 2, directly improves the AMG convergence by a corresponding amount. Since, compared to the V-cycle, the F-cycle solves the second-level correction equation approximately twice as accurately, this immediately causes the overall convergence to be twice as fast. (Note that we have a similar effect for isotropic seven-point stencils in 3D.)    ≫

Figure A.19(a) shows the convergence histories of the VS-, VA2- and VA1-cycles, with and without acceleration by CG, for solving the given problem ($N = 512$ and using $u \equiv 1$ as first approximation)[3]. The figure also shows the convergence history of ILU(0) preconditioned CG for the first 30 iterations.

However, the computational time, and how it increases with increasing $N$, is more important than convergence histories. Figure A.19(b) shows computational times to solve (A.8.2) for varying mesh sizes up to a residual reduction by ten orders of magnitude. Times are given in milliseconds *per finest grid point* and include the setup cost. For all AMG variants discussed here, the total cost approaches an upper limit for increasing mesh sizes which demonstrates their computational optimality for solving problems of the kind at hand. The figure also depicts the corresponding increase in cost for ILU(0) preconditioned CG.

---

[3]For ease of reading, we always use thin lines for stand-alone AMG cycles and corresponding bold lines for their accelerated analogs.

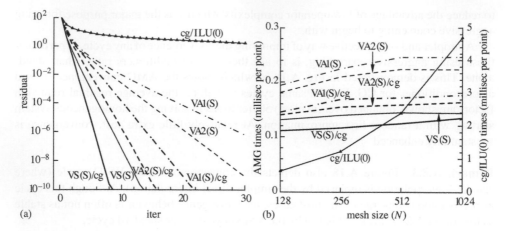

**Figure A.19.** (a) Convergence histories for $N = 512$; (b) total time in millisec per finest grid point to reduce the residual by ten orders of magnitude.

Since the convergence speed of CG/ILU(0) depends on $N$, the advantage of AMG over CG/ILU(0) substantially increases with increasing problem size. For $N = 1024$, it can be seen that the accelerated standard cycle, VS(S)/CG, is about 37 times faster than CG/ILU(0).

**Remark A.8.5** The computational work of AMG is essentially determined by the operator complexity $c_A$ and the convergence factor $\rho$. Only if both quantities are bounded as a function of $h$, do we have an asymptotically optimal performance. We note that $c_A$ is indeed virtually independent of $N$ for all AMG variants shown. The slight increase of the total cost for medium sized meshes, seen in Fig. A.19(b), is caused by the small increase of the convergence factors in this area (cf. Fig. A.18).                                    $\gg$

Detailed measurements are given in Table A.1 for $N = 512$ including the complexity values, $c_G$ and $c_A$. In addition to the computational times for the setup phase and single cycles, total execution times (including setup) are given for the reduction of the residual by a factor of $\varepsilon_0 = 10^{-10}$ (the values in parentheses indicate the number of iterations required). The table shows that the standard cycle, VS(S)/CG, is nearly 20 times faster than standard ILU(0) preconditioned CG. The lowest memory cycle, VA1(S), reduces the memory overhead for storing the coarse-level matrices by approximately 64%. If used as a preconditioner, it is still approximately 14 times faster than CG/ILU(0).

Table A.1 also shows complexity values and timings for the original AMG solver, AMG1R5, which should be compared with the VS(S)-cycle. Obviously, for the current problem, AMG1R5 converges somewhat faster (nine iterations instead of 11) and the total execution time is lower[4]. The faster convergence is due to the particularly simple geometrical

---

[4]This lower computational time is, to some extent, due to the fact that AMG1R5 is a FORTRAN77 code using only static arrays while RAMG05 uses dynamic FORTRAN90 arrays (which decreases the performance in case of the Lahey compiler used here).

situation which benefits the interpolation used in AMG1R5 (cf. Remark A.7.8). We will see later that this advantage gets lost in more complex geometric situations or for more complicated problems.

**Remark A.8.6**  The cost of "better" cycles such as F-cycles (and even more for W-cycles) is usually substantially higher than that of the simpler V-cycles, at least in connection with standard coarsening. While, in V-cycles, each level is visited just once, in F-cycles, level $n$ is visited $n$ times. This increase in cost is, generally, more critical in AMG than it is for comparable geometric multigrid cycles. This is mainly due to the more complex AMG coarse-level operators. Thus, although the convergence of F-cycles may be faster than that of their V-cycle analog, the total computational time is usually higher. That, in Table A.1, the total times for the FS(S)-cycle are (slightly) *lower* than those for the VS(S)-cycle is a consequence of the extremely fast F-cycle convergence which, in turn, is a consequence of the particular situation mentioned in Remark A.8.4.                                          $\gg$

### A.8.2.3 F-smoothing and Jacobi interpolation
According to the theoretical results of Section A.5.1.4, it is possible to employ mere F-smoothing instead of full smoothing except that standard interpolation might then not be sufficient any more. Indeed, in general, additional work needs to be invested in improving interpolation by Jacobi F-relaxation in order to cope with all those error components which cannot efficiently be reduced by mere F-smoothing. That just one Jacobi step is enough is demonstrated in Fig. A.20 which compares the convergence factors of the standard cycle, VS(S), with that of various cycles using mere F-smoothing[5].

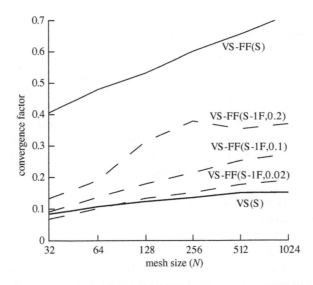

**Figure A.20.**  Convergence factors of cycles using F-smoothing.

[5]We have forced strong diagonal dominance (A.5.3) here with $\delta = 0.75$.

**Table A.2.** Complexities and computing times for cycles using F-smoothing ($N = 512$).

| Method | Complexities | | Times (sec)/Pentium II, 300 MHz | | | |
| --- | --- | --- | --- | --- | --- | --- |
| | $c_A$ | $c_G$ | Setup | Cycle | $\varepsilon_0 = 10^{-10}$ | $\rho$ |
| VS(S) | 2.38 | 1.67 | 11.8 | 2.32 | 37.4 (11) | 0.151 |
| VS-FF(S) | 2.45 | 1.68 | 13.1 | 2.65 | 113.6 (38) | 0.656 |
| VS-FF(S-1F,0.2) | 2.82 | 1.67 | 22.0 | 2.91 | 65.6 (15) | 0.354 |
| VS-FF(S-1F,0.1) | 2.95 | 1.67 | 24.7 | 3.02 | 60.9 (12) | 0.253 |
| VS-FF(S-1F,0.02) | 3.24 | 1.67 | 39.6 | 3.20 | 71.9 (10) | 0.177 |

Obviously, the VS-FF(S) cycle (i.e., the VS(S)-cycle with each CF-relaxation step replaced by two F-relaxation steps), is substantially inferior to the VS(S)-cycle and $h$-dependent. However, one Jacobi F-relaxation step, applied to the interpolation, enhances convergence substantially. If truncation with $\varepsilon_{tr} \leq 0.1$ is used, convergence approaches that of the VS(S)-cycle. (Truncation with the default value, $\varepsilon_{tr} = 0.2$, is not quite sufficient.) Unfortunately, this is at the expense of an increase of the total solution cost (mainly because of a strong increase of the setup cost). In addition, operator complexities substantially increase. Detailed results are shown in Table A.2 for the case $N = 512$. Although we have already mentioned ways of improvement, the trend indicated is typical: skipping relaxation of the C-equations may unnecessarily increase cost and memory requirement (cf. Section A.5.1.1).

### A.8.3 Computational Fluid Dynamics

Industrial CFD applications involve very complicated flow problems. For instance, in the car industry, flows through heating and cooling systems, complete vehicle underhood flows or flows within passenger compartments are computed on a regular basis. Large complex meshes, which are normally unstructured, are used to model such situations. Requirements on the achievable accuracy are ever increasing, leading to finer and finer meshes. Locally refined grid patches are introduced to increase the accuracy with as few additional mesh points as possible. Figures A.21 and A.22 show two exemplary meshes used to model the flow through a down-shot coal furnace and the cooling jacket of a four-cylinder engine, respectively[6].

The software industry is continuously improving the generality and efficiency of its codes. The incorporation of multigrid methods would be one way to improve performance. Geometrically oriented approaches, however, can hardly cope with the complex geometries under consideration. Generally, there is no natural grid hierarchy which could easily be exploited. But even if there was such a hierarchy, the coarsest level would still be required to be fine enough to resolve the geometry to some extent. For industrially relevant configurations, such coarsest grids would still be much too fine for an efficient multilevel solution.

---

[6]All examples have been provided by Computational Dynamics Ltd., London, UK.

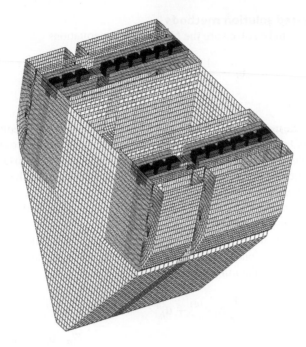

**Figure A.21.** View into the interior of the bottom part of a coal furnace model (325 000 mesh cells; for simplicity, only the mesh surface is visualized).

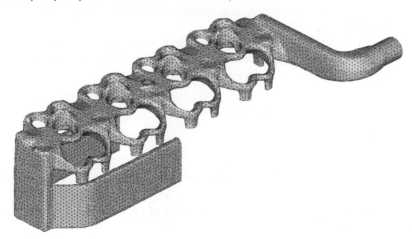

**Figure A.22.** Cooling jacket of a four-cylinder engine (100 000 cells).

AMG is of particular interest here since it can be used as a "plug-in solver" for existing codes.

In this section, we present some examples of AMGs performance if applied to industrial CFD applications based on segregated solution methods.

### A.8.3.1 Segregated solution methods

The basic equations to be solved are the Navier–Stokes equations

$$\mathbf{u}_t - \frac{1}{Re}\Delta\mathbf{u} + \mathbf{u}\cdot\nabla\mathbf{u} + \nabla p = f \tag{A.8.3}$$

$$\nabla\cdot\mathbf{u} = 0 \tag{A.8.4}$$

where (A.8.3) are the momentum equations and (A.8.4) is the continuity equation. $\mathbf{u}$ denotes the velocity vector, $p$ the pressure and $f$ the body force. $Re$ is the Reynolds number.

*Segregated solution methods* (or *pressure-correction type* methods) to tackle the solution of the Navier–Stokes equations belong to the most established approaches used in general-purpose commercial CFD codes. Their major advantage is that, at each time step, the approximate solution of just a series of scalar equations is required rather than the solution of the coupled Navier–Stokes system.

Assuming an implicit time-stepping scheme (here backward Euler for simplicity) and a stable discretization in space (indicated by the subscript $h$), equations of the following form have to be solved at the $n$th time step:

$$\frac{1}{\delta t}(\mathbf{u}_h^{(n)} - \mathbf{u}_h^{(n-1)}) - \frac{1}{Re}\Delta_h\mathbf{u}_h^{(n)} + \mathbf{u}_h^{(n-1)}\cdot\nabla_h\mathbf{u}_h^{(n)} + \nabla_h p_h^{(n)} = f_h^{(n)} \tag{A.8.5}$$

$$\nabla_h\cdot\mathbf{u}_h^{(n)} = 0. \tag{A.8.6}$$

Here, the convective part has been linearized. (If required, the solution of the nonlinear equations can be computed iteratively in a straightforward way.) For ease of reading, we omit the subscript $h$ in the following.

There are several variants of pressure-correction type approaches all of which proceed in two steps: first, an intermediate velocity approximation, $\mathbf{u}^\star$, is computed by replacing $p^{(n)}$ in (A.8.5) by values from the previous time step:

$$\frac{1}{\delta t}(\mathbf{u}^\star - \mathbf{u}^{(n-1)}) - \frac{1}{Re}\Delta\mathbf{u}^\star + \mathbf{u}^{(n-1)}\cdot\nabla\mathbf{u}^\star + \nabla p^{(n-1)} = f^{(n)}. \tag{A.8.7}$$

Second, corrections

$$\mathbf{u}^{(n)} = \mathbf{u}^\star + \mathbf{u}' \quad \text{and} \quad p^{(n)} = p^{(n-1)} + p' \tag{A.8.8}$$

are computed such that $\mathbf{u}^{(n)}$ is an improved solution of (A.8.5) satisfying the continuity equation (A.8.6). To strictly satisfy (A.8.5), one would have to solve

$$\frac{1}{\delta t}\mathbf{u}' - \frac{1}{Re}\Delta\mathbf{u}' + \mathbf{u}^{(n-1)}\cdot\nabla\mathbf{u}' + \nabla p' = 0. \tag{A.8.9}$$

However, since the correction $\mathbf{u}'$ is assumed to be relatively small and change little in space, the $\mathbf{u}'$-dependent part in (A.8.9) is approximated by $A(\mathbf{u}^{(n-1)})\mathbf{u}'$ with some simple (invertible) matrix $A$, depending on old velocity values. Usually $A$ is assumed to be diagonal, in the simplest case just $A = (1/\delta t)I$. Well-known methods such as SIMPLE and SIMPLEC are based on such approximations. Consequently, the velocity correction $\mathbf{u}'$ is computed via

$$\mathbf{u}' = -[A(\mathbf{u}^{(n-1)})]^{-1}\nabla p' \qquad (A.8.10)$$

where $p'$ is the solution of the so-called *pressure-correction equation*

$$\nabla \cdot [A(\mathbf{u}^{(n-1)})]^{-1}\nabla p' = \nabla \cdot \mathbf{u}^{\star}. \qquad (A.8.11)$$

The latter follows immediately from (A.8.10) because of the requirement that the velocity, after its correction (A.8.8), has to satisfy the continuity equation (A.8.6). Typically, the pressure-correction equation has to be solved several times at each time step.

In practice, segregated solution methods are used to solve time-dependent as well as steady-state problems with well-known approaches being PISO and SIMPLER, respectively (for more information on segregated solution methods, see [107, 216, 281, 299–301]).

Summarizing, two different types of scalar equations have to be solved within each step: a set of (decoupled) convection–diffusion equations (A.8.7) and the pressure-correction equation (A.8.11), a Poisson-like equation with coefficients which, in general, depend on known velocity values. AMG can be used efficiently to solve both types of scalar equation. Regarding Poisson-like equations, we have demonstrated this in Section A.8.2 by means of a simple model problem. Convection–diffusion problems will be considered in Section A.8.5.2. However, the pressure-correction equation is generally by far the most expensive to solve. We therefore focus on this equation. For a further discussion of segregated solution methods, in particular in the context of AMG, see also [166].

**Remark A.8.7** The fact that the pressure-correction equation is just one component within an outer iteration has two implications. First, there is no need to solve it too accurately, in particular not in steady-state computations. We will therefore consider the efficiency of AMG not only for obtaining high- but also low-accuracy approximations. Secondly, potential performance gains through the use of an efficient solver solely for solving the pressure-correction equations is limited by the cost of the remaining components. But, since the solution of the pressure-correction equations typically makes up the largest part of the overall computation, potential benefits may be substantial. $\gg$

**Remark A.8.8** Currently, there is a trend in commercial code development towards solving the Navier–Stokes equations directly as a fully coupled system. However, this has several drawbacks, for instance, regarding overall memory requirements (which is a major concern for all commercial software providers). On the other hand, having efficient solvers available for directly solving coupled systems on complex geometries, would increase the overall performance substantially. AMG solvers can play a very important role here. Extensions of the AMG approach which can handle such coupled systems are under development. $\gg$

### A.8.3.2 Industrial test cases

In order for any new solver to be suitable for industrial use, it has to be *fast*, *robust* and require only a *low amount of memory*. Whether or not this is satisfied, has to be judged by comparison with typical industrial solvers such as the ILU preconditioned conjugate gradient.

In terms of performance, AMG is generally much more efficient than any one-level method, in particular, if the underlying meshes are large and complex, if there are thin substructures in different directions, or if coefficients are not smoothly varying. The robustness has turned out to be extraordinarily high for all industrial problems solved so far. Regarding memory requirement, AMG can not compete with a simple one-level method. Any hierarchical solver, and hierarchical approaches are *necessary* to obtain fast solution, requires additional memory. Memory requirement, however, is a major concern for any commercial software provider. Industrial users of commercial codes always drive their simulations to the limits of their computers, shortage of memory being a serious one. In fact, most industrial users would prefer to wait longer for the results if they would otherwise not be able to solve what they really want to solve.

For these reasons, low-memory AMG approaches are of particular interest, even if the reduced memory requirement causes an increase in the total computational time. A memory overhead of some tens of percents is certainly acceptable. In any case, however, the operator complexity $c_A$ must not be significantly larger than say, 2.0. We will see that the low-memory cycles, VA1(S) and VA2(S), will satisfy the industrial requirements in all cases considered.

In this section, our focus will be on solving pressure-correction equations to a *high accuracy*, namely, by reducing the residual by ten orders of magnitude. (See Section A.8.3.3 for *low-accuracy* approximations.) We consider problems with different types of meshes. Discretization is based on a standard finite-volume approach. In all cases, the concrete data used correspond to one particular time step taken from a normal production run.

The first problem corresponds to the 2D simulation of the flow through a fan model. The core part of the corresponding mesh is outlined in Fig. A.23(a). The mesh consists mostly of quadrilaterals and some triangles.

Figure A.23(b) shows the convergence histories for the standard VS(S) cycle as well as for the low-memory variants VA1(S) and VA2(S), used with and without acceleration by conjugate gradient. We observe that the convergence behavior is comparable to that of the simple model equation (A.8.2) except that convergence is slightly slower here (cf. Fig. A.19(a)). As before, the VA-cycles are not supposed to be used stand-alone but rather as preconditioners.

The fastest cycle, VS(S)/CG, needs 11 steps to reduce the residual by ten orders of magnitude. It also provides the best method in terms of computational time as can be seen from Table A.3. It requires a total time of 3.68 sec which is about 8.5 times less than the time required by ILU(0)/CG. Memory overhead is reasonable for this 2D problem, however, relative to our above requirements, it remains too high. The VA2- and VA1-cycles are still much faster than ILU(0)/CG but require substantially less memory. In particular, the operator complexity of the VA1/CG-cycle is only $c_A = 1.39$. That is, its memory overhead is smaller than that of the standard cycle by 72% at the expense of some 25% increase in total execution time.

The following two examples correspond to 3D flow computations with largely different unstructured meshes, namely, the flows through the cooling jacket of a four-cylinder engine (Fig. A.22) and through a coal furnace (Fig. A.21), respectively. While the first mesh is a fairly uniform tetrahedral mesh, the second one consists mainly of hexahedra and a few thousand pentahedra, including many locally refined grid patches. According to this, the discretized problems employ mostly five-point stencils in the first case and varying stencil sizes in the second case (ranging from four- to 11-point stencils).

The convergence histories for both problems are depicted in Fig. A.24. The difference in size and structure of the above meshes hardly influences the convergence of AMG. Compared to the 2D problem (Fig. A.23(b)), however, convergence is somewhat slower here, which is typical for 3D applications (cf. Remark A.8.1). The only major difference between

(a)

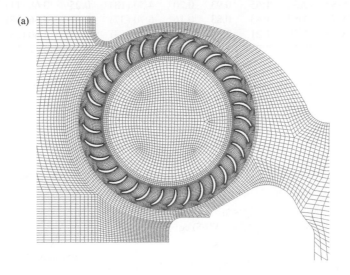

(b)

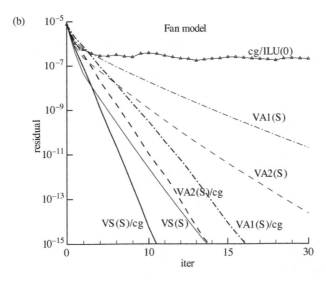

**Figure A.23.** (a) Core part of a fan model; (b) convergence histories.

**Table A.3.** Complexities and computing times (fan model).

| | | | | Times (sec)/Pentium II, 300 MHz | | | |
| Method | Complexities | | Setup time | Stand-alone | | Conjugate gradient | |
| | $c_A$ | $c_G$ | | Cycle | $\varepsilon_0 = 10^{-10}$ | Cycle | $\varepsilon_0 = 10^{-10}$ |
|---|---|---|---|---|---|---|---|
| ILU(0) | | | 0.11 | | | 0.09 | 32.0 (354) |
| VS(S) | 2.38 | 1.65 | 0.93 | 0.20 | 4.50 (18) | 0.25 | 3.68 (11) |
| VA2(S) | 1.87 | 1.43 | 0.83 | 0.16 | 6.70 (37) | 0.21 | 4.56 (18) |
| VA1(S) | 1.39 | 1.21 | 0.60 | 0.13 | 8.44 (63) | 0.17 | 4.60 (23) |

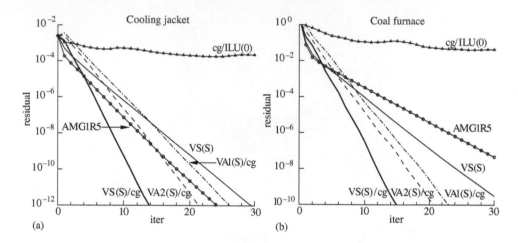

**Figure A.24.** Convergence histories: (a) cooling jacket; (b) coal furnace.

the 2D and 3D problems is that the standard VS(S)-cycle, used stand-alone, converges significantly slower in the 3D cases. As a consequence, the accelerated VA-cycles converge faster than the standard cycle. For both meshes, the accelerated standard cycle exhibits fastest convergence and requires 15 iterations to reduce the residual by ten orders of magnitude.

Table A.4 shows that, in terms of total execution time, the accelerated standard cycle is nearly 20 times faster than ILU(0)/CG for the cooling jacket, and around 6.5 times for the coal furnace. The lowest-memory cycle, VA1(S)/CG, is still over 17 times faster than CG/ILU(0) for the first case. For the second case, it is even cheaper than the accelerated standard cycle although it requires eight additional iterations.

Generally, standard coarsening requires significantly more memory in 3D than in 2D. According to the table, the operator complexity of the VS(S)-cycle is given by $c_A = 2.77$ and $c_A = 3.39$ for the two problems (which is still practical but too high w.r.t. our requirements). The reasons for this increase in complexity are similar to those stated in Remark A.8.2 for the model problem. The first standard coarsening step tends to be relatively slow (in terms

**Table A.4.** Complexities and computing times.

| Method | Complexities | | Times (sec)/Pentium II, 300 MHz | | | | |
| | | | | Stand-alone | | Conjugate gradient | |
| | $c_A$ | $c_G$ | Setup time | Cycle | $\varepsilon_0 = 10^{-10}$ | Cycle | $\varepsilon_0 = 10^{-10}$ |
|---|---|---|---|---|---|---|---|
| | | | | Cooling jacket | | | |
| ILU(0) | | | 0.39 | | | 0.40 | 434.1 (1084) |
| AMG1R5 | 5.35 | 1.98 | 7.25 | 1.29 | 42.1 (27) | | |
| VS(S) | 2.77 | 1.55 | 5.77 | 0.90 | 34.4 (32) | 1.11 | 22.6 (15) |
| VA2(S) | 2.25 | 1.30 | 4.94 | 0.73 | 55.6 (69) | 0.96 | 27.2 (23) |
| VA1(S) | 1.44 | 1.14 | 3.18 | 0.56 | 59.2 (100) | 0.76 | 24.5 (28) |
| | | | | Coal furnace | | | |
| ILU(0) | | | 1.86 | | | 1.70 | 743.8 (436) |
| AMG1R5 | 7.06 | 1.90 | 72.6 | 6.76 | 370.0 (44) | | |
| VS(S) | 3.39 | 1.59 | 33.9 | 4.36 | 174.2 (32) | 5.27 | 113.5 (15) |
| VA2(S) | 2.12 | 1.27 | 21.8 | 3.00 | 173.5 (51) | 3.93 | 104.6 (21) |
| VA1(S) | 1.47 | 1.14 | 14.4 | 2.33 | 170.2 (67) | 3.27 | 89.5 (23) |

of a reduction of grid points) while, at the same time, the size of the Galerkin stencils on the second level becomes substantially larger than on the finest one. As can be seen from Table A.4, aggressive coarsening avoids this problem very efficiently in both cases. In particular, A1-coarsening reduces the memory overhead by around 80% in both test cases. Although this significantly increases the number of iterations required, the resulting method is still very efficient (even the most efficient in some cases).

**Remark A.8.9** For comparison, Fig. A.24 and Table A.4 also show the performance of the original code AMG1R5. Note first that the convergence of AMG1R5 is comparable to that of RAMG05 for the cooling jacket but significantly slower for the coal furnace. More importantly, however, the complexity values of AMG1R5, $c_A = 5.35$ and $c_A = 7.06$, respectively, indicate an unacceptably high memory requirement of AMG1R5 in both cases. This confirms what has already been mentioned in Remark A.7.8: the coarsening strategy used in AMG1R5 may become very inefficient for nonregular 3D meshes such as those considered here. Since $c_A$ is directly related to the computational time, AMG1R5 is more expensive, particularly for the coal furnace case. ≫

### A.8.3.3 Low-accuracy approximations
Particularly in steady-state computations, the pressure-correction equation usually needs to be solved only to a low accuracy of one or two digits. One might expect that the use of AMG solvers is an "overkill" (because of the high setup cost involved) and simple one-level methods would be more efficient.

**Table A.5.** Total computing times to reach a low *residual* and *error* reduction, respectively.

| Method | Fan model | | | | | Cooling jacket | | | | |
| | Setup cost | $\varepsilon_0 = 10^{-1}$ | | $\varepsilon_0 = 10^{-2}$ | | Setup cost | $\varepsilon_0 = 10^{-1}$ | | $\varepsilon_0 = 10^{-2}$ | |
| | | Resid. | Error | Resid. | Error | | Resid. | Error | Resid. | Error |
|---|---|---|---|---|---|---|---|---|---|---|
| CG/ILU(0) | 0.11 | 0.28 | 9.98 | 11.1 | 12.2 | 0.39 | 7.64 | 114.8 | 202.8 | 177.4 |
| VS(S) | 0.93 | 1.15 | 1.32 | 1.32 | 1.69 | 5.77 | 7.62 | 8.50 | 10.2 | 11.2 |
| VS(S)/CG | " | 1.20 | 1.45 | 1.70 | 1.45 | " | 8.99 | 8.99 | 10.3 | 8.99 |
| VA2(S) | 0.83 | 1.10 | 1.23 | 1.41 | 1.54 | 4.94 | 8.84 | 9.50 | 13.2 | 13.9 |
| VA2(S)/CG | " | 1.20 | 1.20 | 1.65 | 1.20 | " | 8.95 | 7.00 | 10.8 | 8.95 |
| VA1(S) | 0.60 | 0.86 | 1.22 | 1.47 | 1.83 | 3.18 | 7.53 | 8.62 | 12.9 | 14.1 |
| VA1(S)/CG | " | 0.94 | 0.94 | 1.46 | 1.46 | " | 4.70 | 4.70 | 8.49 | 6.22 |

Indeed, this seems to be true if one compares the total computing times needed by AMG and CG/ILU(0) to reduce the residual by only *one order of magnitude*. Table A.5 shows corresponding timings (in the columns labeled "Resid.") for both the fan model and the cooling jacket. The results show that the AMG execution time is still comparable to that of CG/ILU(0) for the cooling jacket but is considerably higher for the fan model, where CG/ILU(0) appears to be up to four times faster. Note that, in this case, AMG's setup time alone is already two to three times higher than the total execution time of CG/ILU(0)! If, however, the residual is required to be reduced by *two orders of magnitude* instead, the execution time of CG/ILU(0) again becomes higher than that of AMG (accelerated VA1(S)-cycle) by factors of approximately 8 and 23 for the two problems.

Generally, however, one has to be very careful in drawing conclusions from small residual reductions to corresponding reductions in the error. Indeed, if one compares AMG with CG/ILU(0) on the basis of *error* rather than *residual* reductions, the picture looks completely different. To demonstrate this, Table A.5 contains total computing times on the basis of the true error reduction (in the columns labeled "Error"). According to these results, AMG is *always* faster than CG/ILU(0). For instance, even if the requirement on the error reduction is merely one order of magnitude, the execution time of VA1(S)/CG is lower than that of CG/ILU(0) by factors of approximately 10 and 25 for the first and second test case, respectively. Note that these results depend, to some extent, on the used norm (here, we used the Euclidean norm). The tendency, however, will be similar for other norms.

This advantageous behavior of AMG in terms of error reduction is related to its property to *globally* reduce errors much more effectively than a one-level method (such as ILU(0)). To illustrate this further, Fig. A.25 shows separate convergence histories for residuals and errors for both CG/ILU(0) and VA2(S)/CG. Obviously, during the first iterations, AMG reduces errors much more effectively than CG/ILU(0). This unsatisfactory behavior of CG/ILU(0) makes the use of termination criteria, based merely on the residual reduction, very unpractical. If the given tolerance is too large, CG/ILU(0) may stop after only a few iterations although the error may still be far too large. On the other hand, selecting a (slightly) smaller tolerance, may drastically increase computing cost.

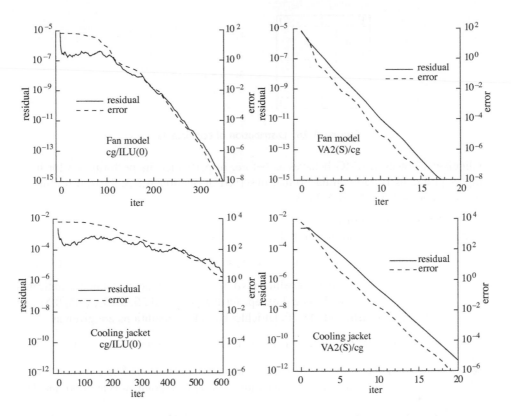

**Figure A.25.** Convergence histories: residual versus error for fan model and cooling jacket.

Nevertheless, in computing low-accuracy approximations, AMG's setup cost becomes quite substantial. In fact, most of the total computing time may be spent in the setup routines. However, in the situations as described here, typically chains of (often many hundreds or thousands of) problems have to be solved for which the underlying matrices usually change only slowly from one step to the next. Consequently, AMG's setup phase only needs to be performed once in a while. For most of the problems, the complete setup can be "frozen" (or "updated" by fixing the interpolation and just recomputing the Galerkin operators). To control such an optimized use of AMG by an efficient and automatic strategy is straightforward. In this way, AMG's total setup overhead can be drastically reduced. For those examples considered here, new setups are, on average, needed only after every fifth time step. Clearly, this substantially enhances the efficiency of AMG.

### A.8.4 Problems with Discontinuous Coefficients

In this section, we consider problems with strongly discontinuous coefficients, again beginning with the investigation of a typical model problem. We will see that, compared to Poisson-like problems, the overall performance of AMG decreases to some extent.

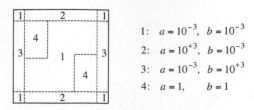

**Figure A.26.** Distribution of coefficients.

Qualitatively, however, AMG behaves as before. In particular, its performance for more complex problems is very similar to that observed in Section A.8.3. Typical results are given in Sections A.8.4.2 and A.8.4.3.

### A.8.4.1 A model problem

We consider the diffusion problem [292]

$$-(au_x)_x - (bu_y)_y = f(x, y) \tag{A.8.12}$$

on the unit square with discontinuous coefficients $a > 0$ and $b > 0$ being defined as indicated in Fig. A.26. $f(x, y)$ is defined to be 0 except for the points $(0.25, 0.25)$, $(0.5, 0.5)$ and $(0.75, 0.75)$ where it is defined to be 10. Dirichlet boundary conditions are given as

$$u = 1 \quad \text{for } x \le 0.5, \, y = 0 \text{ and } x = 0, \, y \le 0.5; \quad \text{otherwise: } u = 0.$$

Discretization is assumed to be done by the standard five-point stencil on a regular grid of mesh size $h = 1/N$. For instance, the $x$-derivative $-(au_x)_x$ at point $x_0$ is approximated by

$$\frac{1}{h^2}\big(-a(x_0 - h/2)u(x_0 - h) + cu(x_0) - a(x_0 + h/2)u(x_0 + h)\big) \tag{A.8.13}$$

with $c = a(x_0 - h/2) + a(x_0 + h/2)$.

Besides discontinuous changes in the size of the coefficients by orders of magnitude, the resulting equations are strongly anisotropic near the boundary, with the strong connectivity being in the direction of the boundary. Regarding the treatment of such problems by geometric multigrid methods, both properties require special attention. In order to see the similarities between geometric and algebraic multigrid if applied to such problems, we briefly recall a typical geometric approach.

First, assuming usual $h \rightarrow 2h$ coarsening, smoothing needs to be done by "robust" smoothers such as alternating line relaxation. Secondly, geometric (linear) interpolation of corrections is no longer appropriate. This is because linear interpolation for $u$ at a grid point $x_0$ requires the continuity of its first derivatives. However, in our example, $u_x$ and $u_y$ are not continuous but rather $au_x$ and $bu_y$. Since corresponding corrections exhibit the same discontinuous behavior, proper interpolation has to approximate the continuity of $au_x$ and $bu_y$ rather than that of $u_x$ and $u_y$. Consequently, we obtain a better interpolation, for instance in the $x$-direction, if we start from the equation

$$(au_x)(x_0 - h/2) = (au_x)(x_0 + h/2)$$

and approximate this by

$$a(x_0 - h/2)(u(x_0) - u(x_0 - h)) = a(x_0 + h/2)(u(x_0 + h) - u(x_0)).$$

This yields the interpolation formula

$$c\, u(x_0) = a(x_0 - h/2)u(x_0 - h) + a(x_0 + h/2)u(x_0 + h) \qquad (A.8.14)$$

for computing $u(x_0)$ from its neighbors.

The extension of such relations to both space dimensions forms the basis for the definition of "operator-dependent" interpolation in geometric multigrid. Clearly, by means of some additional approximations, the "interpolation pattern" has to be modified in order to match the coarse-grid points that are available. Such an interpolation was first investigated in [3]. In that paper, it was also shown that the use of operator-dependent interpolation gives the most robust multigrid convergence if, in addition, Galerkin operators are used on coarser levels (rather than the coarse-level stencils corresponding to (A.8.12)).

The development of AMG can be regarded as the attempt to generalize the ideas contained in [3]. In fact, geometrically motivated operator-dependent interpolation (A.8.14) is simply an approximation to the (homogeneous) difference equations (A.8.13). This is exactly the way AMG attempts to define interpolation. Clearly, in contrast to the geometric approach, line relaxations are not required in AMG since anisotropies are "resolved" by coarsening essentially in the directions of strong connectivity (see Fig. A.28(a)).

Figure A.27(a) shows the convergence factors of the VS- and VA-cycles if applied to the above problem. The standard VS(S)-cycle converges somewhat slower than for the Poisson-like model problem and its rate exhibits a (slight) $h$-dependence. But even for the finest grid, it still converges at a rate of 0.38 per cycle. In contrast to the Poisson-like example, improving interpolation by one additional Jacobi relaxation step (VS(S-1F)-cycle) substantially speeds

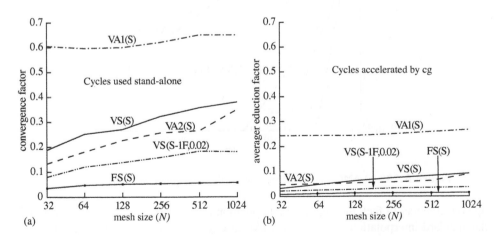

**Figure A.27.** (a) Convergence factors of cycles used stand-alone; (b) average reduction factors of accelerated cycles.

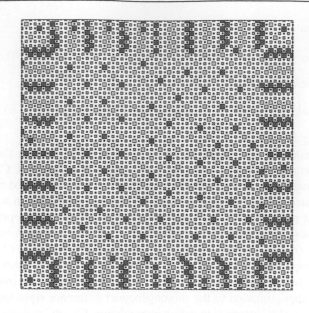

(a)

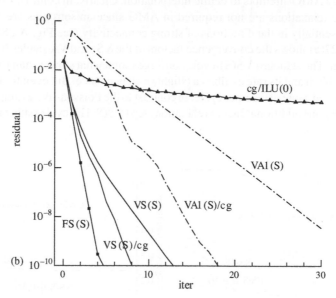

(b)

**Figure A.28.** (a) AMG standard coarsening; (b) convergence histories ($N = 512$).

up convergence. In fact, the improved interpolation restores the convergence observed for the standard interpolation in the Poisson-like case (cf. VS(S)-cycle in Fig. A.18(a)). The F-cycle convergence factor is around 0.05 and virtually constant for all grids considered. The VA1(S)-cycle converges asymptotically at about the same rate as in the Poisson case.

**Table A.6.** Complexities and computing times ($N = 512$).

| Method | Complexities | | Setup time | Times (sec)/Pentium II, 300 MHz | | | |
|---|---|---|---|---|---|---|---|
| | | | | Stand-alone | | Conjugate gradient | |
| | $c_A$ | $c_G$ | | Cycle | $\varepsilon_0 = 10^{-8}$ | Cycle | $\varepsilon_0 = 10^{-8}$ |
| ILU(0) | | | 0.93 | | | 1.07 | 693.4 (647) |
| AMG1R5 | 2.58 | 1.81 | 6.87 | 2.19 | 41.9 (16) | | |
| VS(S) | 2.52 | 1.79 | 9.43 | 2.44 | 41.1 (13) | 3.03 | 33.8 (8) |
| FS(S) | " | " | " | 4.67 | 32.8 (5) | 5.20 | 30.2 (4) |
| VA2(S) | 2.14 | 1.58 | 8.78 | 2.08 | 29.6 (10) | 2.68 | 30.3 (8) |
| VA1(S) | 1.78 | 1.32 | 7.52 | 1.67 | 65.8 (35) | 2.29 | 48.6 (18) |
| VS(S-1F,0.02) | 3.20 | 1.79 | 25.9 | 2.79 | 48.2 (8) | 3.40 | 46.3 (6) |
| VS(S-1P,0.02) | 3.05 | 1.79 | 19.0 | 2.82 | 47.2 (10) | 3.30 | 42.1 (7) |

Finally, the VA2-cycle converges even faster than the VS-cycle. However, this is unusual and can not be expected in most cases.

As before, acceleration by conjugate gradient improves convergence substantially (see Fig. A.27(b)). Moreover, convergence speed becomes virtually independent of the mesh size. How this improvement translates into number of iteration steps and computational time needed to solve (A.8.12) by eight orders of magnitude (for $N = 512$ and starting with $u \equiv 0$ as first approximation), is shown in Fig. A.28(b) and Table A.6. The accelerated VS-cycle requires just eight iterations and is over 20 times faster than CG/ILU(0). The corresponding F-cycle even solves this problem in only four cycles and is approximately 23 times faster than CG/ILU(0). However, this extraordinarily rapid convergence occurs for reasons similar to those mentioned in Remark A.8.4. Although the two-level method involving the first two levels no longer strictly corresponds to a direct solver (due to the one-dimensional coarsening near the boundary, see Fig. A.28(a)), it is still very close since the F-points of the finest level are only (very) weakly connected.

We note that the memory requirement for strongly anisotropic problems is typically higher than that for isotropic problems, the reason being that AMG will essentially perform one-dimensional coarsening (in the direction of strong connectivity). To a limited extent, this is also observed in Table A.6. To reduce the memory requirement in anisotropic areas, A1-coarsening needs to be employed and is quite effective. From the table we see that, compared to the VS-cycle, the VA1-cycle requires 50% less overhead memory at the expense of a 50% increase in execution time.

We have seen above that, relaxation of interpolation substantially improves convergence. Although the mere solution time of VS(S-1F,0.02) (*not* counting the setup) is significantly lower than that of VS(S), the table shows that this advantage is eaten up by a much higher setup cost. Unfortunately, this is also typical for more general situations. Moreover, relaxation of interpolation naturally increases the memory requirement. As can be seen in Table A.6, the VA1-cycle requires only one-third of the memory overhead of the VS(S-1F,0.02)-cycle. As already mentioned at the beginning of Section A.8, there is much room for optimizing the application of relaxation of interpolation, for instance, by only applying it locally where really needed and by locally optimizing truncation of interpolation.

**Remark A.8.10**   The above test case was constructed as a worst case example for the geometric multigrid code MG2 [292]. Although MG2 is normally very efficient in solving problems with discontinuous coefficients, for this particular problem there exist eigenvalues of the MG2 iteration matrix which are very close to one. As a consequence, MG2 V-cycle convergence becomes extremely slow. Using MG2 as preconditioner substantially improves convergence. But about 60 MG2 V-cycles have to be performed to reduce the residual by eight orders of magnitude. Even the F-cycle still requires about 40 cycles [292]. Compared to this, AMG does not show any particular problems for this test case and converges much faster.                                                                                                 ≫

### A.8.4.2 Oil reservoir simulation

In oil reservoir simulation, the basic task is to solve complex multiphase flows in porous media. For each phase, $\ell$, of a multiphase problem, the governing equations are the *continuity equation*

$$-\nabla \cdot (\rho_\ell \mathbf{u}_\ell) = \frac{\partial}{\partial t}(\rho_\ell \phi S_\ell) + q_\ell \qquad (A.8.15)$$

and *Darcy's law*

$$\mathbf{u}_\ell = -K\frac{k_{r\ell}}{\mu_\ell}(\nabla p_\ell - \rho_\ell g \nabla z). \qquad (A.8.16)$$

The continuity equation describes the mass conservation. For each phase, $\mathbf{u}_\ell$ denotes the velocity vector, $S_\ell$ the saturation, $\rho_\ell$ the density distribution (depending on $p_\ell$) and $q_\ell$ represents injection or production wells. $\phi$ denotes the porosity of the medium. Darcy's law essentially describes the velocity–pressure dependence. Here, $p_\ell$ denotes the pressure, $\mu_\ell$ the viscosity and $k_{r\ell}$ the relative permeability (depending on $S_\ell$). $g$ is the gravity acceleration constant (we here assume that gravity acts in the direction of the $z$-axis). Finally, $K$ is a tensor (absolute permeability). The absolute permeability varies in space by, typically, several orders of magnitude in a strongly discontinuous manner. The gray scale in Fig. A.29 indicates the variation of the permeability as a function of space for a typical case.

The phase velocities can be eliminated by inserting (A.8.16) into (A.8.15). For *incompressible flows* we can assume $\partial \phi/\partial t = 0$ and $\partial \rho_\ell/\partial t = 0$, and one obtains the following equations involving pressures and saturations:

$$\nabla \cdot \left( K\frac{k_{r\ell}}{\mu_\ell}(\nabla p_\ell - \rho_\ell g \nabla z) \right) = \phi \frac{\partial}{\partial t} S_\ell + q_\ell/\rho_\ell. \qquad (A.8.17)$$

From this set of equations, pressures and saturations can be computed if one takes into account that $\sum_\ell S_\ell \equiv 1$ and that the individual phase pressures are directly interrelated by means of simple non-PDE relations involving known (but saturation-dependent) capillary pressures.

It is rather expensive to solve the resulting nonlinear system in a *fully implicit* manner. This limits the size of the problems that can be handled. The more classical IMPES approach (implicit in pressure, explicit in saturation) treats (A.8.17) by an *explicit* time-stepping. Consequently, in each time step, the pressures need to be computed with all saturations being known from the previous time step. Exploiting the interrelation of the individual phase pressures mentioned above, only *one* pressure (for instance, the oil pressure) requires the solution of a partial differential equation of the form

$$-\nabla \cdot (T \nabla p) = Q \qquad (A.8.18)$$

which is obtained by adding the individual equations (A.8.17) (and using $\sum_\ell S_\ell \equiv 1$). The tensor $T$ is directly related to $K$. According to the assumption of incompressibility, both $T$ and $Q$ depend only on the saturations and given quantities.

Clearly, as with any explicit time-stepping method, the major drawback of the IMPES approach is the serious restriction in the maximally permitted time step size (CFL (Courant–Friedrichs–Lewy) condition). Since this restriction becomes stronger with decreasing spatial mesh size or increasing variation in the magnitude of $T$, in practice, the classical IMPES method also strongly limits the treatment of large problems.

Recently, however, a new IMPES-type approach has become quite popular which eliminates the time-step restriction due to the CFL condition. Rather than updating the saturations directly on the grid based on (A.8.17), a *streamline method* is used instead [34, 387]. By transporting fluids along periodically changing streamlines, the streamline approach is actually equivalent to a dynamically adapting grid that is decoupled from the underlying, static, grid used to describe the reservoir geology (and to compute the pressure). The 1D nature of a streamline allows the 3D problem to be decoupled into multiple 1D problems. The main advantage of this approach is that the CFL conditions are eliminated from the fluid transport, allowing global time-step sizes that are independent of the underlying grid constraints.

Although this approach cannot (yet) be applied to all relevant situations occurring in oil-reservoir simulation, it is well suited for large heterogeneous multiwell problems that are convectively dominated. This has been demonstrated [34] for a problem consisting of one million mesh cells. Cases of this size could not be solved as easily and quickly by standard implicit methods. Clearly, an efficient solver for the pressure equation (A.8.18) then becomes highly important.

The one million cell case previously mentioned, provided by StreamSim Technologies (CA, USA), has been used as a test case for AMG. The variation of the absolute permeability, which directly corresponds to a discontinuous variation of the coefficients in the resulting matrix by four orders of magnitude, is shown in Fig. A.29. Figure A.30 shows the convergence histories of the typical AMG cycles for this case (starting with the zero first approximation). We see that all cycles presented show essentially the same convergence behavior as for the Poisson-like problems considered in Section A.8.3.2 (cf. Fig. A.24). This demonstrates the robustness of AMG with respect to strong, discontinuous variations in the matrix coefficients.

Table A.7 presents some detailed measurements. For all cycles considered, we see a substantial benefit by using them as preconditioners rather than stand-alone. The accelerated standard VS(S)-cycle takes 16 iterations to reduce the residual by ten orders of magnitude.

**Figure A.29.** Distribution of permeability as a function of space (logarithmic gray scale).

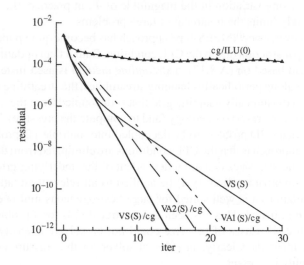

**Figure A.30.** Convergence histories (one million cell case).

Although the lowest-memory cycle, VA1(S)/CG, converges more slowly (24 iterations), in terms of total computation time it is fastest and about 16 times faster than CG/ILU(0). Its complexity value $c_A = 1.41$ is very reasonable. Note, however, that the memory reduction by aggressive A2-coarsening is not very effective, the reason being the strong anisotropies in the problem. On the whole, the AMG performance is strongly comparable to that shown in Table A.4 for Poisson-like cases.

The performance of AMG1R5 shows that the original interpolation (cf. Remark A.7.8) leads to unacceptably high memory requirements for this example: $c_A = 7.66$ as compared to $c_A = 2.85$ for VS(S). As a consequence, although AMG1R5 converges faster than VS(S), its efficiency is substantially lower.

**Table A.7.** Complexities and computing times (one million cell case).

| | | | | Times (sec)/IBM Power PC, 333 MHz | | | |
| | Complexities | | Setup | Stand-alone | | Conjugate gradient | |
| Method | $c_A$ | $c_G$ | time | Cycle | $\varepsilon_0 = 10^{-10}$ | Cycle | $\varepsilon_0 = 10^{-10}$ |
|---|---|---|---|---|---|---|---|
| ILU(0) | | | 3.89 | | | 3.74 | 3376. (902) |
| AMG1R5 | 7.66 | 2.21 | 167. | 22.7 | 758.6 (26) | | |
| VS(S) | 2.85 | 1.56 | 43.8 | 10.5 | 401.2 (34) | 12.6 | 245.1 (16) |
| VA2(S) | 2.56 | 1.39 | 41.6 | 9.20 | 492.2 (49) | 11.3 | 267.3 (20) |
| VA1(S) | 1.41 | 1.13 | 26.7 | 5.56 | 476.4 (81) | 7.67 | 210.7 (24) |

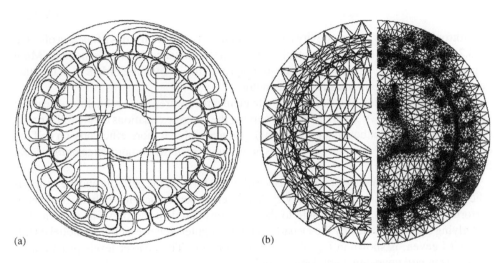

(a)　　　　　　　　　　　　　(b)

**Figure A.31.** Synchronous line-start motor: (a) magnetic field plot; (b) initial and locally refined mesh [265].

### A.8.4.3 Electromagnetic systems

In this section we consider a synchronous line-start motor excited with permanent magnets. The knowledge of the magnetic field inside such a motor, induced by the currents in the stator and the magnets in the rotor, allows its optimization w.r.t. functionality and efficiency. For an example, see Fig. A.31(a).

The governing equation is the *magnetostatics Maxwell equation* (also known as *Ampere's law*)

$$\nabla \times H = J, \tag{A.8.19}$$

where $H$ denotes the *magnetic field intensity* and $J$ the *electric current density*. According to Maxwell's equation for the *magnetic flux density*,

$$\nabla \cdot B = 0,$$

we know that there is a *magnetic vector potential*, $A$, such that $B = \nabla \times A$. Finally, we observe that the constitutive relation $B = \mu H$, with $\mu$ being the *permeability*, (A.8.19) can be rewritten as

$$\nabla \times (\nu \nabla \times A) = J \tag{A.8.20}$$

where $\nu = 1/\mu$ is the *reluctivity*.

We here consider only 2D intersections (Cartesian coordinates) and, for reasons of symmetry, we can assume $A$ and $J$ to be of the special form

$$A = (0, 0, u(x, y)), \quad J = (0, 0, f(x, y)).$$

Hence, (A.8.20) can be seen to correspond to a scalar diffusion equation for the ($z$-component of the) magnetic potential, namely,

$$-\nabla \cdot (\nu \nabla u) = f. \tag{A.8.21}$$

For isotropic materials, the reluctivity $\nu$ is a scalar quantity. Normally, it is a function of $u$ and (A.8.21) has to be solved by some outer linearization (for instance, Newton's method). More importantly, however, $\nu$ is strongly discontinuous and differs by three orders of magnitude between the steel and air areas inside the motor.

An accurate solution of (A.8.21) by finite elements requires local refinements near the critical areas (see Fig. A.31(b)). Instead of solving the discrete equations on the full circular domain (with Dirichlet boundary conditions), one may solve it more efficiently on half the domain using *periodic* boundary conditions or on a quarter of the domain using *antiperiodic* boundary conditions.

Figure A.32(a), (b) show the performance of AMG for both the periodic and the antiperiodic case. Except that the low-memory cycle converges somewhat slower here, the overall performance is strongly comparable to the case considered in the previous section. The underlying mesh, provided by the Department of Computer Science of the Katholieke Universiteit Leuven, is shown in Fig. A.1 (half the domain). The coarser levels produced by AMG's standard coarsening are depicted in Fig. A.3.

It should be noted that *antiperiodic* boundary conditions cause strong *positive* connections to occur in the underlying matrix (in all equations which correspond to points near the antiperiodic boundary). As discussed theoretically in Section A.4.2.3, interpolation should take such connections into account, for instance, in the way as described in Section A.7.1.3. If this is ignored, that is, if interpolation is done only via strong *negative* couplings, the discontinuous behavior of corrections across the antiperiodic boundary is not properly reflected by the interpolation, leading to a substantial degradation of the method.

This is demonstrated in Fig. A.32(c). We observe that the standard VS(S)-cycle, without acceleration, hardly converges. In fact, the convergence speed is limited by the convergence of the smoother on the finest level. It is heuristically clear that this slow convergence is only caused by particular error components, namely, those which really exhibit the discontinuity. All other components are very effectively reduced. Consequently, the use of accelerated cycles "cures" this problem to some extent as can be seen from the figure. However, it takes ten "wasted" cycles before the critical error components are sufficiently reduced by conjugate gradient and the AMG performance becomes visible again. Although this demonstrates

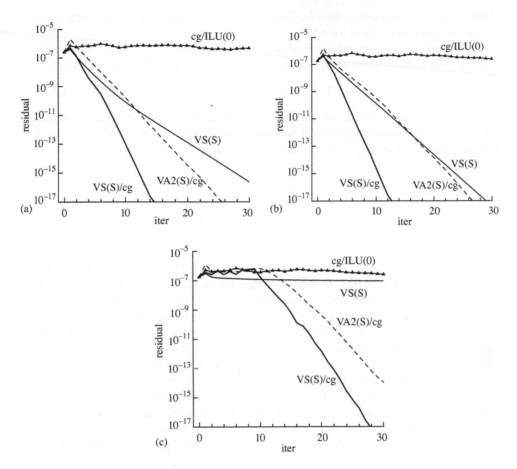

**Figure A.32.** Convergence histories: (a) periodic case; (b) antiperiodic case; (c) antiperiodic case (positive connections ignored).

that the use of accelerators such as conjugate gradient helps to stabilize convergence, a situation like that shown here should clearly be avoided since it demonstrates that something is wrong conceptually. Moreover, the area of "stalling" convergence (here just the first ten iterations) strongly depends on the problem and the size of the grid (more precisely, on the distribution of those eigenvalues of the AMG iteration matrix which are close to one).

## A.8.5 Further Model Problems

### A.8.5.1 Special anisotropic problems

We have seen before that AMG treats anisotropies by coarsening in the proper direction. This works perfectly if the anisotropies are essentially aligned with the grid. Moreover, since

AMG adjusts its coarsening *locally*, diffusion equations (A.8.12) with strongly varying anisotropies are no problem for AMG.

However, one has to expect certain difficulties if strong anisotropies are *not* aligned with the grid. The following model problem is a well-known test case for such a situation:

$$-(c^2 + \varepsilon s^2)u_{xx} + 2(1 - \varepsilon)scu_{xy} - (s^2 + \varepsilon c^2)u_{yy} = f(x, y) \qquad (A.8.22)$$

with $s = \sin \alpha$ and $c = \cos \alpha$. We consider this differential operator on the unit square with $f(x, y) \equiv 1$, homogeneous Dirichlet boundary conditions, $\varepsilon = 10^{-3}$ and $0° \leq \alpha \leq 90°$. For such values of $\alpha$, $u_{xy}$ is most naturally discretized by the *left-oriented* seven-point stencil

$$\frac{1}{2h^2} \begin{bmatrix} -1 & 1 & \\ 1 & -2 & 1 \\ & 1 & -1 \end{bmatrix} \qquad (A.8.23)$$

where $h = 1/N$.

The main difficulty with the differential operator (A.8.22) is that it corresponds to the operator $-u_{ss} - \varepsilon u_{tt}$ in an $(s, t)$-coordinate system obtained by rotating the $(x, y)$-system by an angle of $\alpha$ ("rotated anisotropic diffusion equation" [415]). That is, (A.8.22) is strongly anisotropic with the direction of strong connectivity given by the angle $\alpha$ (see Fig. A.33).

In particular, for $\alpha = 0°$ and $\alpha = 90°$, (A.8.22) becomes $-u_{xx} - \varepsilon u_{yy} = f$ and $-\varepsilon u_{xx} - u_{yy} = f$, respectively, and the anisotropies are just aligned with the axes. Geometric multigrid methods solve these equations very efficiently by employing $h \to 2h$ coarsening and using line relaxations (in the direction of strong connectivity) for smoothing. In contrast to this, AMG uses point relaxation for smoothing but coarsens in the direction of strong connectivity. The standard VS(S)-cycle, for instance, converges at a rate of 0.1 per cycle, independent of the grid size. This can be seen from Fig. A.34(a). Using acceleration by conjugate gradient gives a convergence factor better than 0.01 per cycle (see Fig. A.34(b)).

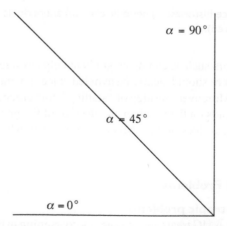

**Figure A.33.** Direction of strong connectivity ($\varepsilon \ll 1$).

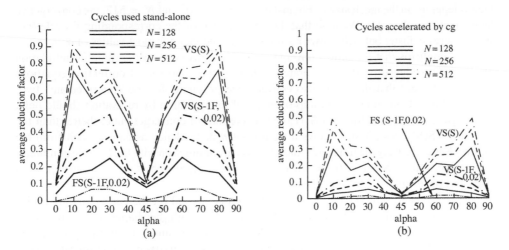

**Figure A.34.** (a) Convergence factors of cycles used stand-alone; (b) average reduction factors of accelerated cycles.

The figure shows that AMG also performs well for $\alpha = 45°$. In this case, the discretization of (A.8.22) corresponds to the stencil

$$\frac{1}{h^2} \begin{bmatrix} -(1-\varepsilon)/2 & -\varepsilon & \\ -\varepsilon & 1+3\varepsilon & -\varepsilon \\ & -\varepsilon & -(1-\varepsilon)/2 \end{bmatrix}, \tag{A.8.24}$$

which exhibits a strong connectivity in the diagonal direction. It has only nonpositive off-diagonal entries and essentially degenerates to a three-point stencil for small $\varepsilon$. Since the anisotropy is still aligned with the grid, AMG can cope with this anisotropy as efficiently as in the previous cases by coarsening in the diagonal direction. Nevertheless, solving this case with geometric multigrid (using $h \to 2h$ coarsening), brings a problem. Even alternating line relaxation no longer has good smoothing properties. In fact, it is no better than point relaxation (since the connections in both coordinate directions are very weak).

For other values of $\alpha$, the strong anisotropies are no longer aligned with the grid. (Note also that, generally, the resulting discretization matrices are not M-matrices.) This causes particular difficulties for *any* multigrid method. In geometric multigrid, as above, neither point- nor line-relaxation schemes have good smoothing properties with respect to $h \to 2h$ grid coarsening. More importantly, however, the extent to which the anisotropy is captured by grid points, strongly depends on $\alpha$ and is different on different grid levels. This substantially reduces the effectiveness of coarse-grid correction processes and, through this, the overall cycle convergence.

Since AMG cycles also obtain their correction quantities from points which form sub-grids of the given grid, the nonalignment also influences the AMG performance. This is demonstrated in Fig. A.34(a) which, for $N = 128, 256, 512$ and different cycles, shows convergence factors as a function of $\alpha$. One sees that, for certain values of $\alpha$, the standard VS(S)-cycle (upper three curves in the figure) converges very slowly and the convergence factor depends on the mesh size $h$. For instance, for $\alpha = 10°$ and $N = 512$, the convergence factor is worse than 0.9. We note that the convergence of the corresponding F-cycle (not shown here) is faster but still shows a similar $h$-dependency for most values of $\alpha$.

This confirms that the slow convergence is not (only) due to the mere accumulation of errors (introduced by the inaccurate solution of the coarse-level correction equations in a V-cycle) but that there are also convergence problems for the "intermediate" two-level methods. Consequently, an improvement of interpolation by relaxation should help. Although the corresponding VS(S-1F,0.02)-cycle indeed converges much better than the standard VS(S)-cycle, it still shows a significant $h$-dependency as can be seen from the figure. However, using the improved interpolation in conjunction with the F-cycle eliminates this problem: the FS(S-1F,0.02)-cycle converges at a rate which is the same for all mesh sizes and better than 0.1 for all $\alpha$ considered.

Figure A.34(b) shows average reduction factors obtained for the corresponding cycles if used as a preconditioner for conjugate gradient. Although the shapes of the curves are similar to the previous ones, the accelerated cycles converge much faster. In particular, the accelerated F-cycle with improved interpolation converges at a rate $\leq 0.02$ for all $\alpha$ and $h$ considered.

Finally, Fig. A.35 shows convergence histories for $N = 512$ and $\alpha = 20°$, starting with $u = 0$ as first approximation. One observes that the VS(S)-cycle converges rapidly for the first few cycles before convergence levels off and reaches its slow asymptotic value. This effect is virtually eliminated for the corresponding accelerated cycle. To a lesser extent, a similar effect occurs also for the FS(S)-cycle. The accelerated F-cycle with improved interpolation reduces the residual by ten orders of magnitude in only six iteration steps.

We see that it is easy to obtain fast convergence even in this example which can be regarded as very difficult for geometric multigrid methods. However, as can be seen from Table A.8, cost and memory requirement become rather high if relaxation of interpolation is employed. An operator complexity $c_A$ of over 6 is unacceptably high for practical applications. The setup cost is also much higher than for the other cycles. The table shows that,

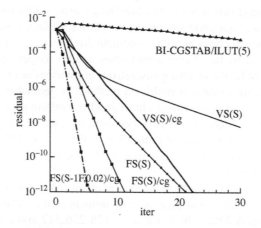

**Figure A.35.** Convergence histories ($N = 512$, $\alpha = 20°$).

**Table A.8.** Complexities and computing times ($N = 512$, $\alpha = 20°$).

| Method | Complexities | | Setup time | Times (sec)/Pentium II, 300 MHz | | | |
| | $c_A$ | $c_G$ | | Stand-alone | | Conjugate gradient | |
| | | | | Cycle | $\varepsilon_0 = 10^{-10}$ | Cycle | $\varepsilon_0 = 10^{-10}$ |
|---|---|---|---|---|---|---|---|
| AMG1R5 | 3.02 | 1.84 | 12.1 | 3.07 | 414.0 (131) | | |
| VS(S) | 3.24 | 1.84 | 18.6 | 3.55 | 263.6 (69) | 4.22 | 119.7 (24) |
| FS(S) | " | " | " | 6.87 | 183.4 (24) | 7.49 | 108.4 (12) |
| VA2(S) | 1.72 | 1.41 | 11.6 | 2.25 | 153.1 (63) | 2.92 | 75.8 (22) |
| VS(S-1F,0.02) | 6.05 | 1.84 | 91.9 | 5.65 | 199.2 (19) | 6.45 | 162.9 (11) |
| FS(S-1F,0.02) | " | " | " | 13.1 | 210.0 (9) | 13.8 | 174.4 (6) |
| VS(D-1F,0.5) | 2.54 | 1.88 | 15.4 | 3.13 | 125.1 (35) | 3.82 | 76.5 (16) |
| FS(D-1F,0.5) | " | " | " | 5.91 | 109.9 (16) | 6.53 | 74.2 (9) |

although the cycles with standard interpolation converge much slower, in terms of total computational work they are still more efficient, the best one being the accelerated VA2-cycle. Remark A.8.11 outlines the main reason for the particularly high memory requirement observed in this example.

**Remark A.8.11**  We have already mentioned that, compared to isotropic problems, the memory requirement is generally higher for anisotropic problems. The memory requirement increases further for problems where the anisotropies are not aligned with the grid. Table A.8 shows that, using standard coarsening and standard interpolation, the operator complexity is $c_A = 3.24$ while for the Poisson-like problem discussed earlier, it was only $c_A = 2.38$ (see Table A.1). The major reason is that the nonalignment causes the strong connections to "fan out" so that each point is strongly connected to an increasing number of points on both sides of the "true" anisotropic line. Thus interpolation stencils become larger on coarser levels and, as an immediate consequence, so do the Galerkin operators. Clearly, this effect is strongly amplified by relaxation of interpolation.                    $\gg$

The fan-out effect mentioned in this remark may be reduced by choosing a larger truncation value for interpolation, $\varepsilon_{tr}$. Indeed, this dramatically improves efficiency as seen from the last two rows in Table A.8 where we have chosen $\varepsilon_{tr} = 0.5$. Compared to the case $\varepsilon_{tr} = 0.02$, the operator complexity is reduced from 6.05 to 2.54 and the total execution time is reduced by more than a factor of two! (Note that we have also used the simpler direct interpolation instead of the standard one.)

> In this context we want to recall that our main goal in this appendix on applications is to demonstrate how different AMG components may influence the overall performance. We have not tried to find optimal components or parameters but rather confined ourselves to a few typical ones. The above results clearly show that optimized parameter settings may, depending on the application, substantially improve the performance.

**Remark A.8.12** If, instead of the seven-point discretization (A.8.23), we use the standard nine-point discretization, the AMG convergence behavior is qualitatively the same (except that $\alpha = 45°$ does not play such a particular role any more). Generally, the performance of AMG suffers from the nonalignment of the anisotropies just as geometric multigrid does. However, in AMG, due to its higher flexibility in creating the coarser levels, this problem is much less severe and easy to cure in the cases considered here, at least in terms of robust convergence. The convergence behavior of geometric multigrid in such cases is considered in [292]. $\gg$

### A.8.5.2 Convection–diffusion problems

So far, we have only considered symmetric problems. However, as mentioned earlier, RAMG05 does not make use of the symmetry and can formally also be applied to non-symmetric problems. Practical experience has shown that, generally, the nonsymmetry by itself does not necessarily cause particular problems for AMG. Other properties of the given system typically influence the performance of RAMG05 to a much larger extent, for instance, whether or not the underlying matrices are (approximately) weakly diagonally dominant. If this is strongly violated, there is no guarantee that the method converges.

We are not going to discuss nonsymmetric problems in detail here but rather present results for a typical example from a class of nonsymmetric problems for which AMG provides robust and fast convergence. This is the class of convection-dominant equations

$$-\varepsilon \Delta u + a(x, y)u_x + b(x, y)u_y = f(x, y) \tag{A.8.25}$$

with some small $\varepsilon > 0$, discretized by standard first-order upwind differences. Note that the resulting discretization matrices are off-diagonally negative.

Generally, for such equations, AMG converges very quickly, in particular, if the characteristics are straight lines. Heuristically, this is an immediate consequence of AMG's coarsening strategy. Since strong connections are only in the upstream direction, interpolation to any F-point $i$ will typically use relatively many of them for interpolation and, consequently be rather accurate in the sense of AMG. Thus, the reason for fast convergence is essentially algebraic, and not a consequence of smoothing in the usual sense.

When the characteristics change over the region, the simple directionality of the strong connections on coarser grids is lost, and AMG will exhibit a more typical convergence behavior. As an example, consider a worst case problem given by selecting

$$a(x, y) = -\sin(\pi x)\cos(\pi y) \quad \text{and} \quad b(x, y) = \sin(\pi y)\cos(\pi x), \tag{A.8.26}$$

$f(x, y) \equiv 1$ and $u = \sin(\pi x) + \sin(13\pi x) + \sin(\pi y) + \sin(13\pi y)$ on the boundary of the unit square. Finally, we set $\varepsilon = 10^{-5}$.

The major difficulty with this particular example is that $a$ and $b$ are chosen to yield *closed characteristics* and a stagnation point in the center of the domain. Consequently, (A.8.25) becomes more and more singular for $\varepsilon \longrightarrow 0$. For $\varepsilon = 0$, the continuous problem is no longer well defined: any function which is constant along the characteristic curves, solves the homogeneous equation. According to the results shown in [292], geometric multigrid approaches have serious difficulties with this example: covergence becomes very slow and mesh dependent.

**Figure A.36.** (a) Solution contours; (b) standard coarsening pattern.

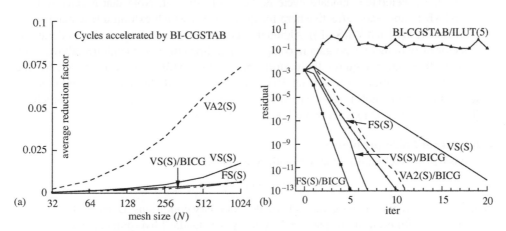

**Figure A.37.** (a) Average reduction factors; (b) convergence histories for $N = 512$.

Figure A.36 depicts the coarsening strategy performed by AMG. Since strong connectivity is in the circular direction only, AMG attempts to not coarsen in the radial direction (within the limits imposed by the grid).

Figure A.37(a) shows average reduction factors of some AMG cycles if used as a preconditioner for BI-CGSTAB [397]. In all cases, instead of CF-relaxation, we employed symmetric Gauss–Seidel relaxation for smoothing (which usually gives faster convergence for convection dominant problems). The figure shows rapid convergence in all cases. In particular, the cycles using standard coarsening, on average reduce the residual by approximately two orders of magnitude per BI-CGSTAB iteration. (Note that each BI-CGSTAB iteration involves the performance of *two* AMG cycles.) The VA2(S)-cycle still converges very fast (better than 0.1 reduction per iteration) but exhibits a relatively significant $h$-dependency. Figure A.37(b) shows the convergence histories of cycles with and without acceleration.

**Table A.9.** Complexities and computing times ($N = 512$).

| Method | Complexities | | Setup time | Times (sec)/Pentium II, 300 MHz | | | |
| | $c_A$ | $c_G$ | | Stand-alone | | BI-CGSTAB | |
| | | | | Cycle | $\varepsilon_0 = 10^{-10}$ | Cycle | $\varepsilon_0 = 10^{-10}$ |
|---|---|---|---|---|---|---|---|
| AMG1R5 | 3.94 | 2.11 | 9.72 | 2.94 | 180.2 (58) | | |
| VS(S) | 3.33 | 1.92 | 12.0 | 3.48 | 88.6 (22) | 7.52 | 64.8 (7) |
| FS(S) | " | " | " | 8.34 | 95.4 (10) | 17.3 | 98.4 (5) |
| VA2(S) | 2.58 | 1.68 | 9.62 | 2.87 | 107.1 (34) | 6.22 | 78.0 (11) |
| VS(S-1F,0.02) | 5.14 | 1.94 | 89.3 | 4.93 | 138.6 (10) | 10.4 | 141.2 (5) |

Table A.9 shows detailed performance measurements for $N = 512$. The accelerated VS(S) cycle requires seven iterations to reduce the residual by ten orders of magnitude, the accelerated FS(S)- and VS(S-1F)-cycles only require five cycles. In terms of total cost, however, the accelerated standard cycle is the most efficient. Note that acceleration by BI-CGSTAB is not really effective here for those cycles which exhibit a fast stand-alone convergence (FS(S) and VS(S-1F)). Although acceleration reduces the number of iterations by a factor of two, there is no gain in computational time since, as mentioned above, each BI-CGSTAB iteration requires the performance of two AMG cycles. However, for the other cycles, acceleration *is* beneficial. Regarding the relatively high memory requirement observed in the table, note that we have a similar "fan out" effect (in the upstream direction) as described in the previous section (cf. Remark A.8.11).

### A.8.5.3 Indefinite problems

We consider the Helmholtz equation (with constant $c$)

$$-\Delta u - cu = f(x, y) \quad (c \geq 0) \tag{A.8.27}$$

on the unit square with homogeneous Dirichlet boundary conditions. Discretization is on a regular mesh with *fixed* mesh size $h = 1/N$ using the standard five-point stencil,

$$\frac{1}{h^2} \begin{bmatrix} & -1 & \\ -1 & 4 - ch^2 & -1 \\ & -1 & \end{bmatrix}. \tag{A.8.28}$$

The corresponding discretization matrix, $A_c$, is nonsingular as long as $c$ does not equal any of the eigenvalues

$$\lambda_{n,m} = \frac{2}{h^2}(2 - \cos n\pi h - \cos m\pi h) \quad (n, m = 1, 2, \ldots, N - 1) \tag{A.8.29}$$

of the corresponding discrete Laplace operator, $A_0$. If $c = \lambda_{n,m} (= \lambda_{m,n})$, $A_c$ is singular and its nullspace is spanned by the eigenfunctions

$$\phi_{n,m} = \sin(n\pi x)\sin(m\pi y) \quad \text{and} \quad \phi_{m,n} = \sin(m\pi x)\sin(n\pi y) \tag{A.8.30}$$

of $A_0$ corresponding to the eigenvalue $\lambda_{n,m}$.

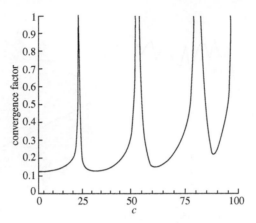

**Figure A.38.** Convergence factor of stand-alone VS(S)-cycle as a function of $c$ ($h = 1/256$).

$A_c$ is positive definite as long as $c$ is smaller than the first eigenvalue of $A_0$, that is, if $c < \lambda_{1,1}$. However, according to Section A.4.2.1 (see Example A.4.1), we have to expect a performance degradation of AMG if $c$ approaches $\lambda_{1,1}$. This is demonstrated in Fig. A.38 where the convergence factor of the VS(S)-cycle (used stand-alone) is depicted as a function of $c$ (for $h = 1/256$). Indeed, if $c$ approaches $\lambda_{1,1} = 2\pi^2 + O(h^2)$, AMG's convergence factor tends to one. Example A.4.1 showed that, in order to avoid this, interpolation in AMG necessarily would have to approximate $\phi_{1,1}$ increasingly better if $c \to \lambda_{1,1}$.

If $c > \lambda_{1,1}$, $A_c$ is no longer positive definite. Nevertheless, Fig. A.38 shows that AMG converges at a slightly reduced rate as long as $c$ remains comfortably between the first two eigenvalues (the second eigenvalue being $\lambda_{1,2} = \lambda_{2,1} \approx 49.4$). Increasing $c$ further, we see that AMG converges as long as $c$ remains comfortably between consecutive eigenvalues, but that this convergence becomes poorer and poorer. By the time $c$ approaches the sixth eigenvalue ($c \approx 150$), AMG diverges, even for $c$ in the "valleys" between the eigenvalues. This degradation occurs because the Gauss–Seidel relaxation, although it still has good smoothing properties (on the finer levels), diverges for all (smooth) eigenfrequencies $\phi_{n,m}$ with $\lambda_{n,m} < c$. Consequently, as in usual geometric multigrid, the overall method will still converge as long as the coarsest level used is fine enough to represent these smooth eigenfrequencies sufficiently well (and a direct solver is used on that coarsest level). That is, the size of the coarsest level limits the convergence of AMG when $c$ becomes larger: the more variables are represented on the coarsest level, the higher the value of $c$ for which AMG converges. (In the above computations, we used five AMG levels resulting in a coarsest grid containing 500 variables.)

If we use the VS(S)-cycle as a preconditioner for BI-CGSTAB rather than stand-alone, we obtain reasonable convergence for much larger values of $c$. This is demonstrated in Fig. A.39(a) which shows the average reduction factor of VS(S)/BI-CGSTAB as a function of $c$ in solving the homogeneous problem with the first approximation being constant to one. (The values shown are the average reduction factors *per BI-CGSTAB iteration*, each of which requires two AMG iterations, observed in reducing the residual by eight orders

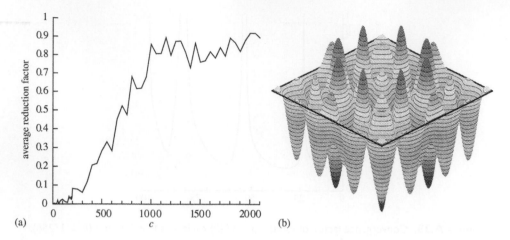

**Figure A.39.** (a) Average reduction factor of the VS(S)/BI-CGSTAB-cycle as a function of $c$ ($h = 1/256$); (b) solution of (A.8.27) for $f(x, y) \equiv 1$ and $c = 1000$.

of magnitude.) The figure shows that acceptable convergence is achieved up to $c \approx 1000$ (close to the 40th eigenvalue of $A_0$). Figure A.39(b) shows the solution of (A.8.27) for $f(x, y) \equiv 1$ and $c = 1000$.

For even larger values of $c$ the average reduction factor per iteration is only 0.8–0.9. Finally, for $c > 2100$, AMG breaks down since some diagonal entries in the coarsest level Galerkin operator become negative. As before, we used five levels for all computations. Decreasing the number of levels (i.e. increasing the number of variables on the coarsest level) gives convergence for even larger values of $c$. In any case, however, the size of $c$ permitted will remain limited. We will not discuss this problem any further since it is well-known that the efficient solution of Helmholtz equation with very large values of $c$ requires different algorithmical approaches [79].

## A.9 AGGREGATION-BASED ALGEBRAIC MULTIGRID

In this section, we consider a particularly simple limiting case of the AMG approach discussed in this appendix, namely, the case that interpolation is defined such that each F-variable interpolates from *exactly* one C-variable. That is, although each F-variable $i$ may have more than one connection to the set of C-variables, the sets of interpolatory variables, $P_i$, are restricted to contain exactly one C-variable each. According to Section A.4.2, the corresponding interpolation weight should equal one if the $i$th row sum of the given matrix is zero. To simplify interpolation further, let us *always* define this weight to be one even if the $i$th row sum of the given matrix is not zero.

Consequently, the total number of variables can be subdivided into "aggregates" $I_k$ where $k \in C$ and $I_k$ contains (apart from $k$ itself) all indices $i$ corresponding to F-variables which interpolate from variable $k$ (see Fig. A.40). With this notation, the computation of

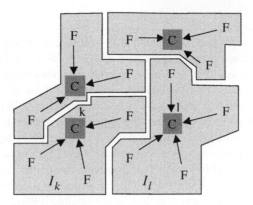

**Figure A.40.** Subdivision of fine-level variables into aggregates. The arrows indicate which C-variable an F-variable interpolates from.

the Galerkin coarse-level operator now becomes very simple. One easily sees that

$$I_h^H A_h I_H^h = \left( a_{kl}^H \right) \quad \text{where } a_{kl}^H = \sum_{i \in I_k} \sum_{j \in I_l} a_{ij}^h \ (k, l \in C), \tag{A.9.1}$$

that is, the coefficient $a_{kl}^H$ is just the sum of all crosscouplings between $I_k$ and $I_l$.

Obviously, regarding the coefficients $a_{kl}^H$, the particular role of the variables $k$ and $l$ (as being C-variables) is not distinguished from the other variables. In fact, the Galerkin operator merely depends *on the definition of the aggregates*. Consequently, we might as well associate each aggregate $I_k$ with some "new" coarse-level variable which has no direct relation to the C-variable $k$. The above interpolation is nothing else than *piecewise constant interpolation* from these new coarse-level variables to the associated aggregates.

This leads to the so-called *aggregation-type* AMG approaches [51, 398, 399] which had originally been developed the other way around: coarsening is defined by building aggregates (rather than constructing C/F-splittings), a new coarse-level variable is associated with each aggregate and interpolation is defined to be piecewise constant. The above description indicates that the aggregation approach can be regarded as a limiting case of the approach considered in this appendix (which started from the interpretation that the coarse-level variables form a subset of the fine-level ones; see Remark A.2.2).

Clearly, for a given subdivision into aggregates to be reasonable, all variables in the same aggregate should strongly depend on each other. Otherwise, piecewise constant interpolation makes no real sense. Since directionality of strength of connectivity plays no role in this approach, strength of connectivity is most naturally defined in a symmetric way. More precisely, one usually defines two variables $i$ and $j$ to be strongly connected to each other if $a_{ij}^2 / a_{ii} a_{jj}$ exceeds a certain size.

Unfortunately, an immediate implementation of this simple coarsening and interpolation approach leads to rather inefficient AMG cycles, even if used as a preconditioner. Convergence will be very slow and not at all robust (see Section A.4.2.1 (Variant 4)). In particular, V-cycle convergence will exhibit a *strong h-dependency* if applied to differential problems. In fact, if regarded as a limiting case of the approach considered in this appendix, the aggregation approach just *forces* worst case situations as discussed in Section A.6 (see Example A.6.1). By the definition of the approach, remedies to avoid such worst-case situations in practice (by a proper distribution of the C- and F-variables) as discussed in Section A.6, cannot be realized here. Finally, piecewise constant interpolation cannot account for any potential oscillatory behavior of the error and, thus, is not suitable if there are strong positive couplings.

Consequently, the basic idea of aggregation-based AMG needs certain improvements in order to become practical. In the following sections, we sketch two possibilities introduced in [51] and [398, 399], respectively. Since we just want to highlight the main ideas, we restrict ourselves to simple but characteristic (Poisson-like) problems.

At this point, for completeness, we also mention [106], where the concept of aggregation was introduced for the first time.

### A.9.1 Rescaling of the Galerkin Operator

It has been demonstrated [51] that the coarse-grid correction of smooth error, and by this the overall convergence, can often be substantially improved by using "overinterpolation", that is, by multiplying the actual correction (corresponding to piecewise constant interpolation) by some factor $\alpha > 1$. Equivalently, this means that the coarse-level Galerkin operator is rescaled by $1/\alpha$,

$$I_h^H A_h I_H^h \longrightarrow \frac{1}{\alpha} I_h^H A_h I_H^h.$$

To motivate this approach, let us consider the simplest case that $A_h$ is derived from discretizing $-u''$ on the unit interval with mesh size $h$, i.e. the rows of $A_h$ correspond to the difference stencil

$$\frac{1}{h^2}[-1 \quad 2 \quad -1]_h$$

with Dirichlet boundary conditions. Let us assume any error, $e^h$, to be given which satisfies the homogeneous boundary conditions. If no rescaling is done ($\alpha = 1$), the variational principle (last statement in Corollary A.2.1) tells us that the two-level correction, $I_H^h e^H$, is optimal in the sense that it minimizes $\|e^h - I_H^h e^H\|_1$ w.r.t. all possible corrections in $\mathcal{R}(I_H^h)$. Because of (A.3.13) this means that $I_H^h e^H$ minimizes

$$\|v^h\|_1^2 = (A_h v^h, v^h)_E = \frac{1}{2h^2} \sum_{i,j}{}' (v_i^h - v_j^h)^2 + \sum_i s_i (v_i^h)^2 \qquad (A.9.2)$$

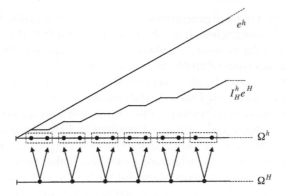

**Figure A.41.** Optimal approximation $I_H^h e^H$ of $e^h$ w.r.t. the energy norm.

where $v^h = e^h - I_H^h e^H$. (The prime indicates that summation is only over neighboring variables $i$ and $j$.) This, in turn, means that, away from the boundary (where we have $s_i = 0$), the Euclidean norm of the *slope* of $v^h$ is minimal. At the boundary itself we have $s_i \neq 0$, and $v^h$ equals zero.

The result of this minimization is illustrated in Fig. A.41 (see also [51, 67]), assuming the aggregates to be built by joining pairs of neighboring variables (marked by dashed boxes). We here consider a smooth error $e^h$ in the neighborhood of the left boundary of the unit interval. On each aggregate, interpolation is constant and the slope of $I_H^h e^H$ necessarily vanishes. On the remaining intervals, the Euclidean norm of the slope of $v^h$ becomes minimal if the slope of $I_H^h e^H$ equals that of $e^h$. Consequently, $I_H^h e^H$ has, on the average, *only half the slope of* $e^h$.

> This simple argument illustrates that the optimal approximation of $e^h$ by elements in $\mathcal{R}(I_H^h)$ w.r.t. the *energy norm* does not approximate to the actual *values* of $e^h$. Multiplying the resulting approximation by a factor of $\alpha = 2$ gives a much more effective correction in this sense. Note that subsequent smoothing smooths out the "wiggles", but does not improve the quality of the correction.

**Remark A.9.1** Note that the above aggregation approach for the model problem $-u''$ coincides with the approach considered in Example A.6.1, where we have shown that the Galerkin operator is off by a factor of two. In fact, we could have used this result immediately to show that a rescaling of the Galerkin operator by $\alpha \approx 2$ makes sense. However, the above considerations show the *origin* of the problem more clearly, namely, the inability of piecewise constant interpolation to approximate the *values* of smooth error if approximation is based on the energy norm. Piecewise *linear* (second-order) interpolation would not exhibit this problem (see the next section). $\gg$

The main argument carries over to the Poisson equation in 2D and 3D, assuming a uniform grid and the aggregates to be built by $2 \times 2$ and $2 \times 2 \times 2$ blocks of neighboring variables, respectively. For more general problems and/or different grids, the optimal weight is no longer $\alpha = 2$. Nevertheless, it has been demonstrated [51] that a slightly reduced value of $\alpha = 1.8$ (in order to avoid "overshooting") yields substantially improved V-cycle convergence for various types of problems, at least if the cycle is used as a preconditioner and if the number of coarser levels is kept fixed (in [51] four levels are always used). Smoothing is done by symmetric Gauss–Seidel relaxation sweeps.

Clearly, the robustness and efficiency of this (very simple and easy to program) approach are somewhat limited since a good value of $\alpha$ depends on various aspects such as the concrete problem, the type of mesh and, in particular, the size of the aggregates. For instance, if the aggregates are composed of three neighboring variables (rather than two) in each spatial direction, the same arguments as above show that the best weight would be $\alpha \approx 3$ in the case of Poisson's equation. If the size of the aggregates varies over the domain, it becomes difficult to define a good value for $\alpha$.

### A.9.2 Smoothed Aggregation

Another approach to accelerate aggregation-based AMG has been developed [398–400]. Here, piecewise constant interpolation is only considered as a first-guess interpolation which is improved by some *smoothing process* ("smoothed aggregation") before the Galerkin operator is computed. In [398, 399], this smoothing is done by applying one $\omega$-Jacobi relaxation step.

The operator corresponding to piecewise constant interpolation is denoted by $\tilde{I}_H^h$. Then the final interpolation operator used is defined by

$$I_H^h = (I_h - \omega D_h^{-1} A_h^f) \tilde{I}_H^h$$

where $D_h = diag(A_h^f)$ and $A_h^f$ is derived from the original matrix $A_h$ by adding all weak connections to the diagonal ("filtered matrix"). That is, given some coarse-level vector $e^H$, $e^h = I_H^h e^H$ is defined by applying one $\omega$-Jacobi relaxation step to the homogeneous equations $A_h^f v^h = 0$ starting with the first approximation $\tilde{I}_H^h e^H$. (Note that this process will increase the "radius" of interpolation and, hence, destroy the simplicity of the basic approach. Note also that Jacobi relaxation here serves a quite different purpose from the Jacobi F-relaxation in Section A.5.1.3.)

To illustrate this process, we again consider the 1D case of $-u''$ and assume the aggregates to consist of three neighboring variables (corresponding to the typical size of aggregates used in [398, 399] in each spatial direction). Note first that, since all connections are strong, we have $A_h^f = A_h$. Figure A.42 depicts both the piecewise constant interpolation (dashed line) and the smoothed interpolation obtained after the application of one Jacobi-step with $\omega = 2/3$ (solid line). Obviously, the smoothed interpolation just corresponds to *linear* interpolation if the coarse-level variables are regarded as the fine-level analogs of those variables sitting in the center of the aggregates.

In order to see this explicitly, we use the notation as introduced in the figure and note that the result of piecewise constant interpolation is

$$\tilde{e}_{i,k-1} = e_{k-1}, \quad \tilde{e}_{i,k} = e_k \quad \text{and} \quad \tilde{e}_{i,k+1} = e_{k+1} \qquad (i = -1, 0, 1). \qquad \text{(A.9.3)}$$

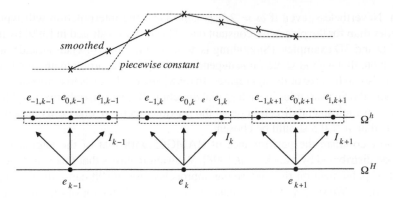

**Figure A.42.** Piecewise constant versus smoothed interpolation.

The application of one $\omega$-Jacobi step as formally described above, using $\tilde{e}$ as the first approximation, means that the final interpolated values within the aggregate $I_k$ (and analogously in all other aggregates) are computed as follows:

$$e_{-1,k} = \tilde{e}_{-1,k} + \omega(\overline{e}_{-1,k} - \tilde{e}_{-1,k}) \quad \text{where } \overline{e}_{-1,k} = \tfrac{1}{2}(\tilde{e}_{1,k-1} + \tilde{e}_{0,k}),$$
$$e_{1,k} = \tilde{e}_{1,k} + \omega(\overline{e}_{1,k} - \tilde{e}_{1,k}) \quad \text{where } \overline{e}_{1,k} = \tfrac{1}{2}(\tilde{e}_{0,k} + \tilde{e}_{-1,k+1}),$$
$$e_{0,k} = \tilde{e}_{0,k} + \omega(\overline{e}_{0,k} - \tilde{e}_{0,k}) \quad \text{where } \overline{e}_{0,k} = \tfrac{1}{2}(\tilde{e}_{-1,k} + \tilde{e}_{1,k}).$$

Inserting (A.9.3), this gives

$$e_{-1,k} = \frac{\omega}{2}e_{k-1} + \left(1 - \frac{\omega}{2}\right)e_k, \quad e_{1,k} = \left(1 - \frac{\omega}{2}\right)e_k + \frac{\omega}{2}e_{k+1} \quad \text{and} \quad e_{0,k} = e_k.$$

The special choice $\omega = 2/3$ indeed leads to linear interpolation as pointed out above,

$$e_{-1,k} = \tfrac{1}{3}e_{k-1} + \tfrac{2}{3}e_k, \quad e_{1,k} = \tfrac{2}{3}e_k + \tfrac{1}{3}e_{k+1} \quad \text{and} \quad e_{0,k} = e_k.$$

---

**Remark A.9.2** Linear interpolation does not exhibit a scaling problem as described in the previous section for piecewise constant interpolation. In fact, for the above model case, one can easily compute the Galerkin operator to be

$$\frac{1}{(3h)^2}[-3 \quad 6 \quad -3]_{3h}$$

which, after proper scaling of the restriction operator by $1/3$, is seen to exactly correspond to the "natural" $3h$-discretization of $-u''$.                    $\gg$

---

Of course, in more general situations, relaxation of piecewise constant interpolation will not give exact linear interpolation any more and a good choice of $\omega$ depends on the situation. Nevertheless, even if $\omega = 2/3$ is kept fixed, the interpolation will typically be much better than for the piecewise constant one. This is demonstrated in [399] by means of various 2D and 3D examples. (Smoothing is done by a mixture of Gauss–Seidel and SOR sweeps.) Note that a good value for $\omega$ depends not only on the problem and the underlying mesh, but also on the size of the aggregates. In [398], the tendency is to compose aggregates of three neighboring variables in each spatial direction. If, instead, only *two* neighbors were aggregated in each spatial direction (as in the previous section), one easily sees by similar arguments that $\omega \approx 0.5$ should be chosen.

We now compare the performance of RAMG05 with that of the aggregation-based AMG code distributed by Vanek *et al.* [398]. We want to stress that this is *not* meant to be any kind of judgement on the underlying approaches; too much can still be improved in either approach. We simply want to point out the differing behavior of the interpolations as currently used in these two codes.

In general, both codes perform comparably if applied to relatively smooth (Poisson-like) problems. Sometimes RAMG05 is slightly faster and sometimes the aggregation-based code. A major advantage of aggregation-type AMG is that, typically, it needs less memory than RAMG05 (due to its very fast coarsening which causes a lower operator complexity $c_A$). On the other hand, the aggregation-based code seems to *require* acceleration by conjugate-gradient to maintain its efficiency and robustness in more complex situations. Since RAMG05 puts more effort into the construction of interpolation and performs a slower coarsening, its performance seems to depend to a lesser extent on aspects such as strong discontinuities. This is demonstrated in Table A.10 for three examples all of which exhibit strong anisotropies or discontinuites. The results clearly show that, at least for the kind of problems considered here, aggregation-based AMG behaves critically if used stand-alone (with total computing times being higher than those of RAMG05 by factors of 8.5 to 16). However, if used as a preconditioner, efficiency substantially increases.

## A.10 FURTHER DEVELOPMENTS AND CONCLUSIONS

The AMG approach described here has been seen to provide very robust and efficient methods for solving certain types of matrix equations, such as those arising in the numerical solution of (scalar) elliptic PDEs. This has been demonstrated by a variety of applications of different type, on structured as well as unstructured grids, in 2D and 3D. Although all applications were geometrically based and many of them were even defined on very simple grids, AMG did not make use of any information other than that contained in the given matrix. The only reason for also considering certain model problems on simple geometries was that they most easily allow the investigation of AMG's asymptotic behavior as well as its dependency on various specific aspects such as anisotropies, discontinuities, singular perturbations and the like. AMG's performance in geometrically complex situations is comparable as was demonstrated by some examples.

From a practical point of view, this is the main strength of AMG. It is applicable to complex geometric situations, independent of the spatial dimension, and it can be applied to solve certain problems which are out of the reach of geometric multigrid, in particular,

**Table A.10.** Complexities and computing times.

| | Complexities | | Times (sec)/IBM Power PC, 333 MHz | | | | |
| | | | | Stand-alone | | Conjugate gradient | |
| Method | $c_A$ | $c_G$ | Setup time | Cycle | $\varepsilon_0 = 10^{-10}$ | Cycle | $\varepsilon_0 = 10^{-10}$ |
|---|---|---|---|---|---|---|---|
| Example from Section A.8.4, $h = 1/512$ | | | | | | | |
| Aggregat. AMG | 1.56 | 1.31 | 10.7 | 2.04 | 245.4 (115) | 2.48 | 57.90 (19) |
| VS(S) | 2.52 | 1.79 | 4.86 | 1.74 | 27.49 (13) | 2.11 | 21.71 (8) |
| FS(S) | 2.52 | 1.79 | 4.86 | 3.30 | 21.42 (5) | 3.64 | 19.52 (4) |
| VA2(S) | 2.14 | 1.58 | 4.76 | 1.45 | 19.28 (10) | 1.81 | 19.25 (8) |
| Example from Section A.8.4 (one million cell case) | | | | | | | |
| Aggregat. AMG | 2.64 | 1.29 | 108. | 17.0 | 4634. (266) | 19.1 | 414.4 (16) |
| VS(S) | 2.85 | 1.56 | 43.8 | 10.5 | 401.2 (34) | 12.6 | 245.1 (16) |
| FS(S) | 2.85 | 1.56 | 43.8 | 17.5 | 288.6 (14) | 19.5 | 258.2 (11) |
| VA1(S) | 1.41 | 1.13 | 26.7 | 5.56 | 476.4 (81) | 7.67 | 210.7 (24) |
| Example from Section A.8.5, $h = 1/512$, $\alpha = 20°$ | | | | | | | |
| Aggregat. AMG | 1.22 | 1.13 | 11.2 | 2.10 | 1610. (762) | 2.37 | 191.5 (76) |
| VS(S) | 3.24 | 1.84 | 8.82 | 2.57 | 186.2 (69) | 3.03 | 81.66 (24) |
| FS(S) | 3.24 | 1.84 | 8.82 | 5.03 | 129.6 (24) | 5.43 | 74.12 (12) |
| VA2(S) | 1.72 | 1.41 | 6.09 | 1.54 | 103.3 (63) | 1.99 | 49.86 (22) |

problems with no geometric or continuous background at all (as long as the underlying matrix satisfies certain conditions). That is, AMG provides an attractive multilevel variant whenever geometric multigrid is either too difficult to apply or can not be used at all.

Clearly, AMG should not be regarded as a competitor for geometric multigrid. Whenever geometric multigrid *can be* applied efficiently, it will usually be superior. Instead, AMG should be regarded as an efficient alternative to standard numerical methods such as CG accelerated by typical (one-level) preconditioners. AMG not only converges much faster but its convergence is, to a large extent, also independent of the size of the given problem. Although designed to be used stand-alone, practical experience has clearly shown that one often can increase efficiency further by using AMG as a preconditioner. We have seen that cheaper (e.g. low-memory) AMG variants, used as a preconditioner, are often better than more sophisticated ones applied stand-alone.

Further developments and applications which are close to the original AMG ideas are contained in [105, 160, 166, 200, 222, 263, 265, 310, 436, 437]. Related approaches, but with a focus on different coarsening and interpolation strategies, are found in [147, 214]. AMG methods based on smoothed aggregation (see Section A.9.2) are an efficient alternative to standard AMG, at least if employed as a preconditioner rather than stand-alone. Applications of the (nonsmoothed) aggregation-type approach in computational fluid dynamics are found in [250, 309, 326, 413].

Many aspects have not been addressed in this introduction, for instance, the further improvement of interpolation. The focus of this introduction was on purely matrix-based approaches. However, as long as interpolation is defined merely on the basis of the algebraic information contained in the given matrix, its "quality" (assuming an adequate geometric problem to be given) is somewhat limited. Although, for all type of problems considered here, this could be compensated for by a proper arrangement of the algorithm, in general this limitation is the major reason for the fact that the two-level theory presented does not carry over to a V-cycle theory (proving V-cycle convergence factors which are independent of the size of the given problem). Generally, the more effort is put into the construction of the interpolation, the faster the convergence can be, but, unfortunately, the required numerical work may increase even faster. That is, the main problem in designing efficient AMG algorithms is the trade-off between convergence and numerical work, and keeping the balance between the two is the ultimate goal of any practical algorithm.

Moreover, there are still many applications for which algebraically defined interpolation, and hence the resulting AMG performance, are not yet satisfactory. For instance, one of the major current research activities in AMG aims at its generalization to efficiently treating *systems* of PDEs such as linear elasticity problems. Although AMG has successfully been applied to various cases [51, 90, 256, 334, 399], its development has not yet reached the state where a particular approach is well accepted. However, even for scalar applications, there are still questions about the best ways to define coarsening and interpolation, for instance, if the given matrix is symmetric positive definite, contains relatively large positive off-diagonal entries, and is far from being weakly diagonally dominant. In such cases, the performance of classical AMG may be only suboptimal.

It is often possible to avoid such situations by simplifying the given matrix before applying AMG [312]. One can also imagine situations where it would be advantageous (and easy) to provide AMG with some additional information on the problem at hand. For instance, information on the geometry (in terms of point locations) or more concrete descriptions on what an "algebraically smooth" error looks like (e.g. in form of some user-provided "test-vector(s)"). This additional information can be used to fit AMG's interpolation in order to approximate certain types of error components particularly well. Straightforward possibilities have already been indicated [334].

In the following, we briefly summarize a few more recent approaches to defining interpolation which aim at increasing the robustness in cases such as those mentioned above.

A new way to construct interpolation (AMGe, [90]) starts from the fact that an algebraically smooth error is nothing but an error which is slow to converge w.r.t. the relaxation process. Hence, an algebraically smooth error, generally, corresponds to the eigenvectors of $A$ belonging to the smallest eigenvalues. Instead of defining interpolation by directly exploiting (A.3.11), the goal in [90] is to define interpolation so that the smaller the associated eigenvalue is, the better eigenvectors are interpolated. To satisfy this by explicitly computing eigenvectors is, of course, much too expensive. However, in the case of finite element methods, assuming the element stiffness matrices to be known, one can derive measures (related to measures used in classical multigrid theory) whose minimization allows the determination of *local* representations of algebraically smooth error components in the above sense. The added robustness has been demonstrated [90] by means of certain model applications. However, the approach is still in its infancy. Significant development work still has to be done to link the processes of coarsening and interpolation definition in order to obtain an optimal algorithm. However, it is an interesting new approach which has the potential to lead to more generally applicable AMG approaches.

Other algebraic approaches, designed for the solution of equations derived from finite element discretizations, have been considered [256, 409]. Both approaches are aggregation based and the coarse-space basis functions are defined so that their energy is minimized in some sense. (In the finite element context it is natural to define interpolation implicitly by constructing the coarse-space basis functions.) This does not require the element stiffness matrices to be known, but leads to a *global* (constraint) minimization problem the exact solution of which would be very expensive. However, iterative solution processes are proposed in both papers to obtain approximate solutions, indicating that the extra work (invested in the setup phase) is acceptable. While [409] concentrates on scalar applications, an extension to systems of PDEs (from linear elasticity) is a major aspect in [256]. Special attention is paid to the correct treatment of zero-energy modes (e.g. rigid-body modes in case of linear elasticity): such modes should be contained in the span of the coarse-space basis functions, at least away from Dirichlet boundaries. (Note that, for typical scalar problems, this corresponds to the requirement that constants should be interpolated exactly away from Dirichlet boundaries, see Remark A.4.1.) It is interesting that the approach in [256] can be regarded as an extension of the earlier work [399] on smoothed aggregation: if only one iteration step is performed to approximately solve the energy minimization problem, the resulting method coincides with the smoothed aggregation approach. In contrast to the latter, however, further iterations will *not* increase the support of the basis functions (i.e., the radius of interpolation). Some test examples [256] indicate the advantages of this new interpolation in terms of convergence speed. Unfortunately, however, this benefit is essentially offset by the expense of the minimization steps.

There are various other papers with focus on the development of multigrid methods to solve finite element problems on unstructured grids. Although some of them are also based on algorithmical components which are, more or less, algebraically defined, most of them are not meant to be algebraic multigrid solvers in the sense considered in this appendix. We therefore do not want to discuss such approaches further but rather refer, for example, to [104] and the references given therein.

In the approach of [407], $A$ is not assumed to be symmetric, and interpolation and restriction are constructed separately. Interpolation, for instance, is constructed so that a smooth error, $S_h e^h$, is interpolated particularly well w.r.t. the *Euclidean* norm, $\|.\|_E$. More precisely, the attempt is to make

$$\left\| S_h e^h - I_H^h e^H \right\|_E,$$

where $e^H$ denotes the straight injection of $S_h e^h$ to the coarse level, as small as possible (cf. (A.5.1)). In [407], this leads to certain local minimizations which are used to find, for each variable, pairs of neighboring variables which would allow a good interpolation in the above sense, and, at the same time, compute the corresponding weights (of both the interpolation and the restriction). Based on this information, a C/F-splitting is constructed which allows each F-variable to interpolate from one of the pairs found before. A heuristic algorithm is used to minimize the total number of C-variables.

In this context, we want to point out that, although standard AMG has been developed within the variational framework, it has successfully been applied to a large number of nonsymmetric problems without any modification. This can be explained heuristically, but no theoretical justification is available at this time. In the context of smoothed aggregation-based AMG, a theoretical analysis can be found in [168].

An important aspect which has not been addressed in this appendix is the parallelization of AMG. An efficient parallelization of AMG is rather complicated and requires certain algorithmical modifications in order to limit the communication cost without significantly sacrificing convergence. Most parallelization approaches investigated up to now either refer to simple aggregation-based variants [326] or use straightforward domain decomposition techniques (such as Schwarz' alternating method) for parallelization. A parallelization strategy which stays very close to the standard AMG approach has been presented [223]. Results for complex 3D problems demonstrate that this approach scales reasonably well on distributed memory computers as long as the number of unknowns per processor is not too small. The method discussed in [407] is also available in parallel. There are several further ongoing parallelization activities, for instance, at the University of Bonn and the Lawrence Livermore National Laboratory [110] and the Los Alamos National Laboratory, but no results have been published to date.

It is beyond the scope of this introduction to discuss the variety of hierarchical algebraic approaches which are not really related to the multigrid idea in the sense that these approaches are not based on the fundamental multigrid principles of smoothing and coarse-level correction. There is actually a rapid and very interesting ongoing development of such approaches. For completeness, we include some references. Various approaches based on approximate block Gauss elimination ("Schur-complement" methods) are found in [8, 10–12, 115, 282–284, 313, 406]. Multilevel structures have also been introduced into ILU-type preconditioners [338]. Recently, some hybrid methods have been developed which use ideas both from ILU and from multigrid [21–23, 314–316]. For a further discussion, see [408].

Summarizing, the development of hierarchically operating algebraic methods to efficiently tackle the solution of large sparse, unstructured systems of equations is currently one of the most active fields of research in numerical analysis. Many different methods have been investigated but, so far, none of them is really able to efficiently deal with *all* practically relevant problems. All methods seem to have their range of applicability but all of them may fail to be efficient in certain other applications. Hence, the development in this exciting area of research is expected to continue for a while.

# Appendix B

# SUBSPACE CORRECTION METHODS AND MULTIGRID THEORY

**Peter Oswald**

Bell Laboratories, Lucent Technologies, RM 2C-403, 600 Mountain Avenue,
Murray Hill, NJ 07974, USA

This is an introduction to the modern theory of iterative subspace correction methods for solving symmetric positive definite variational problems in a Hilbert space. The basics of stable space splittings and the convergence properties of additive and multiplicative Schwarz methods are given in the form of a discussion rather than a rigorous mathematical treatment. The standard applications to multigrid algorithms and domain decomposition schemes are covered. The examples are based on finite difference discretizations of Poisson's equation, and are adapted to the main material of the book which is oriented towards a more practically oriented user of multigrid methods in large-scale applications in the engineering sciences.

## B.1 INTRODUCTION

This monograph concentrates on the efficient parallel implementation of multigrid methods for PDE discretizations and emphasizes *quantitative* multigrid theory. The present appendix is complementary, and provides a bridge to the state-of-the-art *qualitative* multigrid theory. By *qualitative* we mean the analysis of optimality of multigrid algorithms and other methods used in scientific computing in the asymptotic range, i.e. if the mesh parameter $h$ in the discretization of the problem tends to 0 and, consequently, the number of equations tends to infinity. Even though we believe that the distinction between *quantitative* and *qualitative* multigrid theories is more of a philosophical nature, both appear to have their merits and shortcomings. Using the qualitative theory, the optimal $O(N)$ operation count of a FMG-cycle to solve a finite difference or finite element discretization of a second-order elliptic PDE problem within discretization error can be rigorously justified under conditions that are much more general than those required by the quantitative theory. Nonuniform grids and nested refinement can easily be treated. However, no specification of the size of the constants in the O($N$) estimate can be given. Neither approach can give a satisfactory treatment of the robustness problem at large, they do not provide reliable performance estimates for problems with strongly varying coefficients or predominantly nonsymmetric or indefinite behavior such as convection–diffusion problems or Helmholtz equations. Some recent developments in the qualitative theory have led to a unified treatment of seemingly different types of iterative solution methods for operator equations, including multigrid algorithms, domain decomposition methods, fictitious domain methods, but also "old-fashioned" block-iterative solvers. A simplified version of this modern theory of *subspace correction methods* will be outlined below, together with examples for multigrid and domain decomposition algorithms.

The basic idea of subspace correction methods consists in the following (in later places, we will replace the *matrix notation* used at this moment by an *operator notation* which will make it easier to see connections with other concepts, e.g. from applied Fourier analysis). Given a linear system

$$Lu = f \tag{B.1.1}$$

of large dimension $N$, in a subspace correction method we use a finite number of *auxiliary problems*

$$\tilde{L}_j \tilde{u}_j = \tilde{f}_j, \quad j = 1, \dots, J, \tag{B.1.2}$$

of usually smaller dimension $N_j$ ($N_1 + \cdots + N_J \geq N$). Note that in some applications the $\tilde{L}_j$ are just diagonal submatrices of $L$ and the $\tilde{u}_j$ subvectors of $u$ which has led to the synonym *subproblems* or *subspace problems* for (B.1.2). However, the main implicit assumption is that any of the $\tilde{L}_j$ is invertible, and that the auxiliary problems (B.1.2) can be solved fast, typically, in O($N_j$) operations. Finally, simple *prolongation matrices* $P_j$ of dimension $N \times N_j$ and *restriction matrices* $R_j$ of dimension $N_j \times N$ are necessary to link the subproblems (B.1.2) to the original system (B.1.1). In analogy with classical iterative methods such as Jacobi and Gauss–Seidel iterations, we can now define the two prototypes of algorithms for the solution of (B.1.2) based on the subproblems (B.1.2) and the given set of prolongation and restriction matrices.

**Additive subspace (AS) correction method**    $u^{k+1} = AS(\omega, u^k, L, f; \tilde{L}_j, P_j, R_j)$

(1)  Residual computation
        Compute $r^k = f - Lu^k$.
(2)  Restriction and solution of independent subspace problems
        For $j = 1, \dots, J$, compute $\tilde{v}_j^k = \tilde{L}_j^{-1} R_j r^k$.
(3)  Prolongation and update
        Compute $u^{k+1} = u^k + \omega \sum_{j=1}^J P_j \tilde{v}_j^k$.

The error iteration matrix for this AS iteration becomes

$$M_{\text{AS}} = I - \omega BL, \qquad B = \sum_{j=1}^J P_j \tilde{L}_j^{-1} R_j, \tag{B.1.3}$$

where $B$ can be considered as *preconditioner* for $L$ associated with the given choice of subspace problems (B.1.2), more precisely, with the choice of $\{\tilde{L}_j, P_j, R_j : j = 1, \dots, J\}$. The relaxation parameter $\omega$ is introduced for convenience, it could be replaced by individual relaxation parameters $\omega_j$, $j = 1, \dots, J$, and can be interpreted as a way of correctly *scaling* the subspace problems (B.1.2) with respect to the original system (B.1.1). Formally, the multiplication by $B$ looks very suitable for parallelization, however, the parallel efficiency truly depends on the choices for $\tilde{L}_j$, $P_j$, $R_j$. Obviously, the iteration AS as detailed above generalizes the $\omega$-Jacobi relaxation (for details, see Section B.3).

As it may be anticipated from its description, the AS method is usually not very fast. A more efficient way to use the subproblems to compose an iterative method for (B.1.1) seems to be the following analog of a SOR relaxation.

**Multiplicative subspace (MS) correction method**    $u^{k+1} = MS(\omega, u^k, L, f; \tilde{L}_j, P_j, R_j)$

(1) Initialization
      Set $v^1 = u^k$.
(2) Loop through subspace problems for $j = 1, \dots, J$,
      Compute $r^j = f - Lv^j$.
      Restrict and solve a subspace problem $\tilde{v}_j^j = \tilde{L}_j^{-1} R_j r^j$.
      Compute prolongation and update $v^{j+1} = v^j + \omega P_j \tilde{v}_j^j$.
(3) Exit
      Set $u^{k+1} = v^{J+1}$.

The error iteration matrix for the MS method possesses a product representation as follows:

$$M_{\text{MS}} = (I - \omega P_J \tilde{L}_J^{-1} R_J L) \cdots (I - \omega P_1 \tilde{L}_1^{-1} R_1 L). \tag{B.1.4}$$

At first glance, the sequential nature of the MS iteration and the computation of residuals involving $L$ in each step of the inner loop seem to make the algorithm less attractive for parallel implementations. However, a closer look at the implementation of the MS method reveals that such a statement is again dependent on the particular setting. The MS algorithm can be modified in several important directions, most importantly, the ordering of the subspace problems now matters (this is analogous with differences between GS-LEX and GS-RB) and subspaces can be used repeatedly in one loop. Some of these variations will be mentioned in Section B.3.

In Sections B.2 and B.3, we will present the abstract theory of subspace correction methods for the case of *symmetric positive definite systems* (B.1.1). This "soft" theory requires only knowledge about the basics of Hilbert spaces and classical numerical linear algebra. We provide examples, partly with finite element background, that should help readers to understand the concepts and the results. In Section B.4, we examplify the application of this theory to multigrid methods by deriving a qualitative V-cycle convergence result for the finite difference discretization of the Poisson equation. Domain decomposition methods, the other mainstream application of the theory of subspace correction methods, are considered in Section B.5.

Since this appendix is, in a certain sense, complementary to the main contents of this book, the presentation is kept on an informal level. The reader interested in more mathematical details and recent developments is recommended to consult the literature cited below. The same comment applies to the absence of information on the history of the theory presented below. Since the idea of transforming and splitting a large problem into a number of similar (sub-)problems is so obvious, analogous algorithms and attempts to formalize and treat them have been around in most areas of applied and numerical mathematics, e.g., in applied harmonic analysis and optimization. We specifically recommend the survey articles and books [54, 102, 117, 134, 176, 180, 298, 362, 425, 435] for further reading.

## B.2 SPACE SPLITTINGS

In this section, we change our notation slightly. Let $V$ denote a finite-dimensional Hilbert space, with scalar product $(u, v)_V$ and norm $\|u\|_V = \sqrt{(u, u)_V}, u, v \in V$. For simplicity, everything is assumed to be real-valued. A function $\ell \equiv \ell(u, v)$ with arguments $u, v \in V$ and values in $\mathbb{R}$ is called $V$-elliptic if it is linear in each argument (thus, representing a *bilinear form* on $V$) and satisfies the inequalities

$$|\ell(u, v)| \leq \bar{\gamma} \|u\|_V \|v\|_V, \quad \gamma \|u\|_V^2 \leq \ell(u, u) \qquad \forall\, u, v \in V.$$

The first inequality is also called the *continuity* or *boundedness* of the bilinear form $\ell$, the second the *stability* or *coercivity* of $\ell$. The best possible values of the positive constants $0 < \gamma \leq \bar{\gamma} < \infty$ represent the *ellipticity constants* of $\ell$. If in addition $\ell(u, v) = \ell(v, u)$ for all $u, v \in V$ then $\ell$ is called *symmetric*.

**Example B.2.1**  This example is related to the study of linear systems (B.1.1). Set $V = \mathbb{R}^N$, where the Euclidean scalar product

$$(u, v) = v^T u = \sum_{n=1}^{N} v_n u_n, \quad u = (u_1, \ldots, u_N)^T, \quad v = (v_1, \ldots, v_N)^T,$$

defines the Hilbert space structure (to make the notation simpler, we will use $(\cdot, \cdot)$ whenever the Euclidean scalar product in some $\mathbb{R}^N$ is meant). Set

$$\ell(u, v) = (Lu, v) = \sum_{m,n=1}^{N} l_{m,n} u_n v_m.$$

Obviously, this is a symmetric $\mathbb{R}^N$-elliptic bilinear form if and only if $L$ is symmetric positive definite. In this case the ellipticity constants $\gamma, \bar{\gamma}$ coincide with the smallest and largest eigenvalues of $L$, respectively.                                                        $\triangle$

**Example B.2.2**  Another example of importance arises if elliptic boundary value problems are treated by *Galerkin methods*, in particular, by *finite element methods* (FEM). We outline the details for the homogeneous Dirichlet problem for the Poisson equation and linear finite elements on a two-dimensional domain $\Omega$. Recall that the underlying continuous problem takes the form

$$-\Delta u(x, y) = f^\Omega(x, y), \quad (x, y) \in \Omega,$$
$$u(x, y) = 0, \quad (x, y) \in \Gamma, \tag{B.2.1}$$

where $f^\Omega$ is a given function. Suppose that (B.2.1) possesses a (sufficiently smooth) solution $u(x, y)$. Formally, by multiplying the differential equation by any (sufficiently smooth) function $v(x, y)$ that vanishes on the boundary $\Gamma$, integrating over $\Omega$, and applying Green's formula, we have

$$\int_\Omega f^\Omega(x, y) v(x, y) \, dx \, dy = -\int_\Omega \Delta u(x, y) v(x, y) \, dx \, dy$$
$$= \int_\Omega u_x(x, y) v_x(x, y) + u_y(x, y) v_y(x, y) \, dx \, dy$$
$$- \int_\Gamma u_n(x, y) v(x, y) \, d\Gamma(x, y)$$
$$= \int_\Omega \nabla u(x, y) \cdot \nabla v(x, y) \, dx \, dy$$

where $\nabla$ denotes the gradient operator. The integral with respect to $\Gamma$ has been dropped since $v(x, y) = 0$ on $\Gamma$. Thus, if we denote

$$\ell(u, v) = (u, v)_1 \equiv \int_\Omega \nabla u(x, y) \cdot \nabla v(x, y) \, dx \, dy,$$
$$f(v) = (f^\Omega, v)_0 \equiv \int_\Omega f^\Omega(x, y) v(x, y) \, dx \, dy,$$

then necessarily

$$\ell(u, v) = f(v) \tag{B.2.2}$$

for all sufficiently smooth $v(x, y)$ vanishing on $\Gamma$. The derivation of this so-called *variational formulation* (B.2.2) associated with (B.2.1) can be made mathematically precise, if we introduce the concept of *weak solutions* of (B.2.1) in the Sobolev space $H_0^1(\Omega)$ [89, 179]. For the purpose of this informal introduction into subspace correction methods, it suffices to switch immediately to Galerkin methods based on (B.2.2). Let us take any finite-dimensional space $V$ (dim $V = N$) of functions on $\Omega$ that vanish on $\Gamma$ and such that both $(\cdot, \cdot)_1$ and $(\cdot, \cdot)_0$ make sense as scalar product on $V$ (this essentially reduces to requiring that $\|v\|_k^2 = (v, v)_k = 0$ and $v \in V$ implies $v = 0$ for either $k = 0$ or $k = 1$). With this assumption, it is clear that $\ell(u, v)$ when considered as a bilinear form on that $V$ is symmetric and $V$-elliptic with respect to either of the two scalar products (if $V$ is equipped with the scalar product $(\cdot, \cdot)_1$ then the ellipticity constants are simply $\gamma = \bar{\gamma} = 1$, in the other case their ratio $\kappa(\ell) = \bar{\gamma}/\gamma$ is typically very large). In the following, we will sometimes use the more explicit notation $\{V; (\cdot, \cdot)_V\}$ to indicate which scalar product is meant. For instance, if $\ell$ is a symmetric $V$-elliptic form then $\{V; \ell\}$ itself is a possible choice.

As a consequence, the finite-dimensional variational problem of finding $u \in V$ such that

$$\ell(u, v) = f(v) \quad \forall v \in V \tag{B.2.3}$$

has a unique solution which is a minimizer of the *energy functional* $J(u) = \ell(u, u) - 2f(u)$ associated with (B.2.3). This solution is, roughly speaking, a projection of the exact solution of (B.2.1) into the $N$-dimensional space $V$. There is a well-understood procedure to estimate the discretization error associated with this *Galerkin projection* which we will not go into. To find the solution $u \in V$ of it computationally, typically a basis $\Phi = \{\varphi^i,\ i = 1, \ldots, N\}$ in $V$ will be chosen, and (B.2.3) turns into an equivalent system of linear equations (B.1.1), where coefficient matrix and right-hand side are given by

$$L = (l_{m,n} = \ell(\varphi^n, \varphi^m))_{m,n=1}^N, \qquad f = (f_m = f(\varphi^m))^T. \tag{B.2.4}$$

The matrix $L$ is necessarily symmetric positive definite: its properties are sensitive to both the bilinear form $\ell$ and the choice of the basis $\Phi$ in $V$.

Here is the finite element example we wish to discuss. Assume for simplicity that $\Omega$ is a polygonal domain and equipped with a quasi-uniform and regular triangulation $\mathcal{T}$ (this means that all triangles are well-shaped (i.e. the smallest angle is bounded from below by a fixed value) and have approximately the same diameter $\approx h$, and that neighboring triangles share a common vertex or a common edge). Figure B.1(a) shows a typical triangulation. The space $V = V(\mathcal{T})$ of *linear finite elements* associated with $\mathcal{T}$ consists of all continuous functions on $\Omega$ that vanish on $\Gamma$ and are *linear* (i.e. take the form $u(x, y) = a + bx + cy$ for some $a, b, c \in \mathbb{R}$) when restricted to any of the triangles in $\mathcal{T}$. Any function $u \in V(\mathcal{T})$ is uniquely determined by its values $u^i$ at the interior vertices or *nodal points* $P^i = (x^i, y^i)$ of $\mathcal{T}$, and can be recovered by linear interpolation on each triangle (clearly, values at boundary vertices are set to 0). The dimension $N$ of $V(\mathcal{T})$ coincides with the number of nodal points, and due to the assumptions on $\mathcal{T}$, satisfies $N \approx h^{-2}$. The standard basis functions $\varphi^i$, known as the nodal bases or simply hat functions, are defined as the Lagrange functions for this local, piecewise linear interpolation scheme, i.e. $\varphi^i \in V(\mathcal{T})$, $i = 1, \ldots, N$, is given

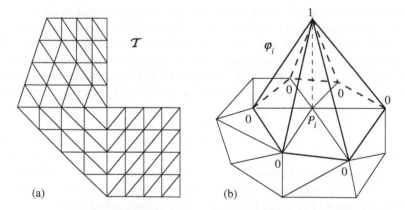

**Figure B.1.** (a) Triangulation; (b) linear finite element nodal basis function.

by requiring

$$\varphi^i(x^s, y^s) = \begin{cases} 1, & s = i, \\ 0, & s \neq i, \end{cases} \quad s = 1, \ldots, N.$$

The support of any $\varphi^i$ is very small: it consists of the union of triangles adjacent to $P^i$. Figure B.1(b) shows a typical nodal basis function. Let us mention in passing that under some assumptions on $\Omega$ and $f^\Omega$ (which are weaker than the corresponding conditions for finite difference schemes), this choice of $V$ as finite-dimensional space in the Galerkin formulation (B.2.3) leads to a discretization error of $O(h^2)$ in the $\| \cdot \|_0$ norm and of $O(h)$ in the $\| \cdot \|_1$ norm, respectively. The associated matrix $L$ is sparse, with $\approx N \approx h^{-2}$ nonzero entries, but has a deteriorating condition number $\kappa(L) \approx h^{-2}$ if $h \to 0$. A drawback of the finite element approach is that the computation of the scalar products in the formula for $l_{m,n}$ and $f_m$ (see (B.2.4)) requires numerical integration, and generally leads to more work in the assembly process of the linear system. △

**Example B.2.3** If $\Omega$ is a rectangle (or composed of several rectangles), partitions $\mathcal{R}$ into rectangles can be used instead of triangulations, and a completely similar setup leads to the space of *bilinear finite elements* $V(\mathcal{R})$ with analogous properties of the associated Galerkin formulation (B.2.3) resp. (B.2.4). Clearly, the restriction of any $u \in V(\mathcal{R})$ to any subrectangle of the partition will be a bilinear function: $u(x, y) = a + bx + cy + dxy$. △

There are many other choices (higher order finite element and spectral element spaces, wavelet spaces, linear combinations of radial basis functions or Gaussians) that appear in connection with data approximation and the solution of various operator equations and can be used within a Galerkin scheme. However, for the purpose of this appendix we will restrict ourselves to the above examples.

We will now introduce the notion of *stable space splittings* which allows us to give a unified treatment of subspace correction methods as methods based on properly representing Hilbert spaces by sums of other Hilbert spaces. The notation is chosen such that the analogy with the introduction becomes obvious. Fix the $N$-dimensional Hilbert space $V$, and consider the problem (B.2.3) where $\ell$ is a symmetric $V$-elliptic bilinear form. For $j = 1, \ldots, J$, let $\tilde{V}_j$ be a $N_j$-dimensional Hilbert space and $\tilde{\ell}_j$ symmetric $\tilde{V}_j$-elliptic bilinear forms. We do not assume that $\tilde{V}_j \subset V$, instead we require that a link between $\tilde{V}_j$ and $V$ is established by linear mappings (embeddings or *prolongations*) $P_j : \tilde{V}_j \to V$.

**Definition B.2.1** We call the formal decomposition

$$\{V; \ell\} \cong \sum_{j=1}^{J} P_j \{\tilde{V}_j; \tilde{\ell}_j\} \tag{B.2.5}$$

a stable space splitting of $\{V; \ell\}$ using the spaces $\{\tilde{V}_j; \tilde{\ell}_j\}$ and the embeddings $P_j$, $j = 1, \ldots, J$, if any $v \in V$ admits at least one representation

$$v = \sum_{j=1}^{J} P_j \tilde{v}_j, \quad \tilde{v}_j \in \tilde{V}_j, \quad j = 1, \ldots, J \tag{B.2.6}$$

and

$$\||v\|| = \inf \left( \sum_{j=1}^{J} (\tilde{\ell}_j \tilde{v}_j, \tilde{v}_j) \right)^{1/2}, \tag{B.2.7}$$

satisfies a two-sided inequality

$$\eta \ell(v, v) \le \||v\||^2 \le \bar{\eta} \ell(v, v) \quad \forall \, v \in V \tag{B.2.8}$$

with positive constants $\eta$, $\bar{\eta}$. The infimum in (B.2.7) is taken with respect to all admissible representations (B.2.6). The optimal constants $\eta$, $\bar{\eta}$ in (B.2.8) are called lower and upper stability constants of the splitting (B.2.5), respectively, their ratio $\kappa = \bar{\eta}/\eta$ is the condition of the splitting.

Note that in the finite-dimensional setting described here, the assumption (B.2.6) automatically implies (B.2.8). The definition can be extended to countably many spaces $\tilde{V}_j$ ($J = \infty$), and $V$ as well as $\tilde{V}_j$ could be separable Hilbert spaces. Then (B.2.8) becomes a real assumption. This extension is useful to connect the discrete theory outlined here with general recipes from approximation theory and the theory of function spaces. The importance of Definition B.2.1 will become clear in Section B.3, let us just mention that keeping the size of $\kappa$ small and independent of $J$ will be critical for the fast convergence of subspace correction methods. We now present some examples of stable splittings which illustrate the flexibility of the abstract concepts and are a preparation for the subsequent sections of this appendix.

**Example B.2.4** This example is related to classical *block-iterative solvers* for (B.1.1). Consider the situation of Example B.2.1. Split the index set $\Lambda = \{1, \dots, N\}$ into pairwise disjoint nonempty sets $\Lambda_j$ and set $N_j = \#\Lambda_j$. Without loss of generality, assume $\Lambda_1 = \{1, \dots, N_1\}$, $\Lambda_2 = \{N_1 + 1, \dots, N_1 + N_2\}$, and so on. By $\tilde{L}_j$ we denote the (symmetric positive definite) submatrices of $L$ of dimension $N_j$ corresponding to $\Lambda_j$. Set

$$\tilde{V}_j = \mathbb{R}^{N_j}, \qquad \tilde{\ell}_j(\tilde{v}_j, \tilde{v}_j) = (\tilde{L}_j \tilde{v}_j, \tilde{v}_j),$$

and introduce the prolongations $P_j \colon \mathbb{R}^{N_j} \to \mathbb{R}^N$ by

$$P_j \colon \tilde{v}_j \equiv (\tilde{x}_1, \dots, \tilde{x}_{N_j})^T \to v \equiv (x_1, \dots, x_N)^T, \quad x_l = \begin{cases} \tilde{x}_{l-N_1-\dots-N_{j-1}}, & l \in \Lambda_j \\ 0, & l \notin \Lambda_j, \end{cases}$$

$j = 1, \dots, J$. This gives a stable space splitting since (B.2.6) holds for exactly one choice of $\tilde{v}_j$:

$$\tilde{v}_j = R_j v, \quad R_j \colon (x_1, \dots, x_N)^T \to (x_{N_1+\dots+N_{j-1}+1}, \dots, x_{N_1+\dots+N_j})^T,$$
$$j = 1, \dots, J.$$

Thus, the infimum in (B.2.7) can be dropped, and it is easy to see that

$$\||v|\|^2 = \sum_{j=1}^{J}(\tilde{L}_j R_j v, R_j v) = (\tilde{L}v, v)$$

where $\tilde{L} = \mathrm{diag}(\tilde{L}_1, \ldots, \tilde{L}_J)$ is a block-diagonal matrix of dimension $N$ consisting of the diagonal submatrices of $L$ corresponding to the index sets $\Lambda_j$. The stability constants of the splitting and its condition coincide with the extremal eigenvalues and the spectral condition number of the matrix $\tilde{L}^{-1}L$, respectively.

An extremal case occurs if we choose one-element sets $\Lambda_j = \{j\}$, $j = 1, \ldots, N$. Then $N_j = 1$ and $J = N$. The $\tilde{L}_j$ are of size 1 and given by the diagonal entry $l_{j,j}$ of $L$. Thus, $\tilde{L} = \mathrm{diag}(L)$, and the space splitting has to do with *diagonal preconditioning*. In the general case, the splitting is associated with *block-diagonal preconditioning*. Since the other, trivial extremal case would be to choose $J = 1$ and $\Lambda_1 = \Lambda$ (then $\tilde{L}^{-1}L$ is just the identity matrix) there arises the interesting design problem of finding the right balance between the size of the submatrices $N_j$ and their overall number $J$. Figure B.2 shows the grid points associated with several choices of index sets $\Lambda_j$, $j = 1, 2$, for the Model Problem I, the five-point discretization of the Dirichlet problem for Laplace's equation on a uniform square grid. In each case, we could have further split the two index sets. While the first two examples are related to smoothers with red–black-ordering and line smoothing (the condition of the associated splitting is $\approx h^{-2}$ and of the same order as for diagonal preconditioning and the condition number of $L$ itself), the last one relates to the domain decomposition approach which will be discussed in Section B.5. In the latter case, the condition of the associated splitting is $\approx h^{-1}H^{-1}$ where $H > h$ is the stepsize parameter of the underlying choice of $\Lambda_j$. The reader is encouraged to establish these bounds, the extremal vectors that give the order of the constants $\eta$ and $\bar{\eta}$ in (B.2.8) are unit vectors, on the one hand, and vectors associated with the grid values of "smooth" functions such as $\varphi_h^{1,1}$, on the other.

There are many useful generalizations of this example: the sets $\Lambda_j$ may overlap and the matrices $\tilde{L}_j$ need not be the corresponding diagonal submatrices of $L$. The examples below (even though they are cast in a different language) are of this type.          $\triangle$

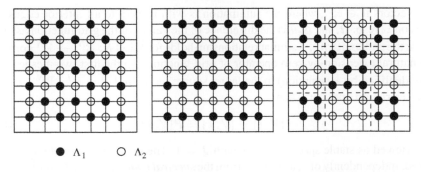

$\bullet$ $\Lambda_1$          $\bigcirc$ $\Lambda_2$

**Figure B.2.** Choices of $\{\Lambda_j\}$ for the five-point discretization.

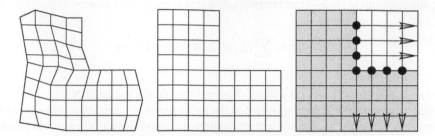

**Figure B.3.** Meshes for the fictitious space example.

**Example B.2.5**   Here is an example related to the *fictitious space method*. Consider three finite difference meshes of mesh size $\approx h$ on three different domains $\Omega \approx \hat{\Omega} \subset \tilde{\Omega}$ as shown schematically in Fig. B.3. The first mesh which is not a square partition is assumed to be a slight distortion of the square partition of the L-shaped domain $\hat{\Omega}$ shown in the center. We assume that there is a discrete one-to-one mapping between the grid points of the two meshes which is close to a linear mapping restricted to this set (such meshes are sometimes called *topologically equivalent*). The third mesh is a square mesh of a unit square $\tilde{\Omega}$ which contains the mesh of the L-shaped domain (the further details shown in the right-hand side picture will be explained below).

Suppose that we have to solve a linear problem (B.1.1) associated with a finite difference or finite element discretization of (B.2.1) on the first domain and mesh (although this has not been detailed in this monograph, there are standard methods to derive finite difference approximations on unstructured meshes). By $\hat{L}$ and $\tilde{L}$ we will denote the stiffness matrices of the finite difference discretization of (B.2.1) with respect to the meshes on $\hat{\Omega}$ and $\tilde{\Omega}$, respectively. Note that $\hat{L}$ is a submatrix of $\tilde{L}$. Set $V = \hat{V} = \mathbb{R}^N$ and $\tilde{V} = \mathbb{R}^{\tilde{N}}$ where $N$ and $\tilde{N} > N$ denote the number of interior grid points of the meshes for $\hat{\Omega}$ and $\tilde{\Omega}$, respectively. Define the bilinear forms $\ell$, $\hat{\ell}$, and $\tilde{\ell}$ as in Example B.2.1. With proper assumptions on the discretization on the first mesh and on its one-to-one mapping onto the second mesh, we have

$$\ell(u, u) \approx \hat{\ell}(u, u) = h^{-2} \sum_e |\Delta_e u|^2 \tag{B.2.9}$$

for all vectors $u \in \mathbb{R}^N$, where the summation is with respect to all interior edges of the meshes on $\Omega$ resp. $\hat{\Omega}$, and $\Delta_e u$ is the difference of the values of the grid function associated with $u$ at the endpoints of $e$. A similar relationship holds for $\tilde{\ell}$. Let $R : \mathbb{R}^N \to \mathbb{R}^{\tilde{N}}$ correspond to the natural extension-by-zero operator of grid functions on the first two meshes (which can be identified by assumption) to the larger square mesh, and $P = R^T : \mathbb{R}^{\tilde{N}} \to \mathbb{R}^N$ the natural restriction operator. Then

$$\{V; \ell\} \cong \{\hat{V}; \hat{\ell}\} \cong P\{\tilde{V}; \tilde{\ell}\} \tag{B.2.10}$$

can be viewed as stable space splittings with $J = 1$. The condition of the first splitting is bounded, independently of $h$, as follows from the *spectral equivalence* of $L$ and $\hat{L}$ expressed by (B.2.9).

Let us show that the condition of the second splitting is of the order $\approx h^{-1}$ as $h \to 0$ (this behavior can be improved if a better $R$ and $P = R^T$ based on discrete harmonic extension are used). Consider any $\tilde{u} \in \mathbb{R}^{\tilde{N}}$ such that $u = P\tilde{u}$. By (B.2.9) we have

$$\ell(u, u) \approx \hat{\ell}(u, u) = h^{-2} \sum_{e \subset \hat{\Omega}} |\Delta_e u|^2 \leq h^{-2} \sum_{e \subset \tilde{\Omega}} |\Delta_e Ru|^2 = \tilde{\ell}(Ru, Ru),$$

since $Ru = u$ on $\hat{\Omega}$ and $Ru = 0$ on $\tilde{\Omega} \backslash \hat{\Omega}$ by definition of $R$. Thus, since $PRu = u$ we obtain

$$\|\|u\|\|^2 = \inf_{\tilde{u} : u = P\tilde{u}} \tilde{\ell}(\tilde{u}, \tilde{u}) \leq \tilde{\ell}(Ru, Ru) = \hat{\ell}(u, u) \leq \bar{\eta}\ell(u, u)$$

for some $\bar{\eta} > 0$.

On the other hand, fix an arbitrary $u \in \mathbb{R}^N$ and consider any $\tilde{u} \in \mathbb{R}^{\tilde{N}}$ such that $u = P\tilde{u}$. Then

$$\hat{\ell}(u, u) = \tilde{\ell}(Ru, Ru) = h^{-2} \sum_{e \subset \hat{\Omega}} |\Delta_e Ru|^2$$

$$\leq h^{-2} \left( \sum_{e \subset \hat{\Omega}} |\Delta_e \tilde{u}|^2 + 3 \sum_{\tilde{P}^s \in \partial \hat{\Omega}} |\tilde{u}(x^s, y^s)|^2 \right)$$

$$\leq \tilde{\ell}(\tilde{u}, \tilde{u}) + 3h^{-2} \|\tilde{u}\|^2_{\partial \hat{\Omega}}.$$

The second term in the last expression (the Euclidean norm of the subvector of $\tilde{u}$ corresponding to the grid points $\tilde{P}^s$ on the boundary $\partial \hat{\Omega}$ of the L-shaped domain $\hat{\Omega}$) is bounded from above by $Ch^{-1}\tilde{\ell}(\tilde{u}, \tilde{u})$. For our example, this discrete Poincare-type estimate can be verified as follows. Observe that we can connect each of those $\tilde{P}^s$ with a grid point $\tilde{P}^{s'}$ on the boundary of $\tilde{\Omega}$ on a separate set $E^s$ of $\leq Ch^{-1}$ interior edges from the mesh on $\tilde{\Omega}$ (see the dashed lines in Fig. B.3). Since $\tilde{u}(x^{s'}, y^{s'}) = 0$, we have

$$|\tilde{u}(x^s, y^s)|^2 \leq \left( \sum_{e \in E^s} |\Delta_e \tilde{u}| \right)^2 \leq Ch^{-1} \sum_{e \in E^s} |\Delta_e \tilde{u}|^2.$$

Summation with respect to $s$ (recall that the sets $E^s$ are pairwise disjoint) gives the above bound. Altogether, we have

$$\hat{\ell}(u, u) \leq (1 + 3Ch^{-1})\tilde{\ell}(\tilde{u}, \tilde{u})$$

for all $\tilde{u}$ satisfying $u = P\tilde{u}$. It remains to take the infimum with respect to $\tilde{u}$ which yields

$$\ell(u, u) \approx \hat{\ell}(u, u) \leq 1/\eta \|\|u\|\|^2 \quad (\eta = (1 + Ch^{-1})^{-1} \approx h).$$

Thus, we have proved the upper bound $O(h^{-1})$ for the condition, that it cannot be improved is clear from looking at unit vectors $u$, on the one hand, and a $u$ obtained from the grid function $\varphi_h^{1,1}$ by restricting it to $\hat{\Omega}$, on the other. $\triangle$

**Example B.2.6** This example introduces to the *multilevel splittings of finite element spaces* which are central for the applications to multigrid theory. A more complete account of the underlying concepts and their roots in spline approximation and function space theory, has been given in [298]. The standard setting is to start with a sequence of partitions $\{\mathcal{T}_j\}$ of a polyhedral domain in $\mathbb{R}^d$ obtained by some sort of regular refinement, and such that a fixed type of finite element construction leads to an increasing sequence

$$V_1 \subset V_2 \subset \cdots \subset V_j \subset \cdots$$

of finite element spaces on these partitions. Not all finite element constructions share this property but there are a number of worked examples. For instance, linear finite element spaces on triangulations and tetrahedral partitions in two and three dimensions, respectively, which are of importance for the numerical solution of second-order elliptic boundary value problems satisfy the above nestedness assumption. For simplicity, let $\Omega$ be the unit square. Consider a sequence of uniform triangulations $\mathcal{T}_j$ of diameter $\approx 2^j$, $j = 1, 2, \ldots$, as shown in Fig. B.4. Note that $\mathcal{T}_{j+1}$ is obtained from $\mathcal{T}_j$ by subdividing all triangles into four (equal) subtriangles. This procedure is called *regular dyadic refinement*. Note that the triangulation shown in Fig. B.1(a) is an example of a triangulation on a polygonal domain obtained by two regular dyadic refinements from an initial, coarse triangulation into five triangles. More general refinement procedures are possible (bisection algorithms, nested local refinement) but will not be discussed here.

Let $\tilde{V}_j = V(\mathcal{T}_j)$ be the corresponding finite element spaces of Example B.2.2, and set $\tilde{\ell}_j(\tilde{u}_j, \tilde{v}_j) = 2^{2j}(\tilde{u}_j, \tilde{v}_j)_0$. Note that $\tilde{V}_j$ is a proper subspace of $\tilde{V}_{j+1}$, with scalar products that are identical up to a constant scaling factor, and that the dimensions $N_j = \dim \tilde{V}_j \approx 2^{2j}$ grow exponentially with $j$. Denote by $I_j^{j+1} : \tilde{V}_j \to \tilde{V}_{j+1}$ the natural embedding operators, and let $I_j^J = I_{J-1}^J \cdots I_j^{j+1}$, $j < J$, be their iterates. Assume now that we have to solve (B.2.2) with respect to the space $V_J = V(\mathcal{T}_J)$. Therefore, we set $\ell_J(u_J, v_J) = (u_J, v_J)_1$, $u_J, v_J \in V_J$. $\triangle$

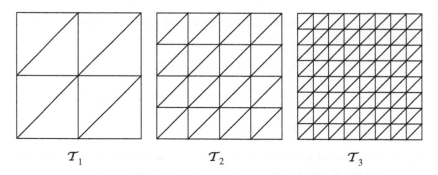

Figure B.4. Dyadically refined triangulations for Example B.2.6.

**Theorem B.2.1** *The space splitting*

$$\{V_J; \ell_J\} \cong \sum_{j=1}^{J} I_j^J \{\tilde{V}_j; \tilde{\ell}_j\} \tag{B.2.11}$$

*is stable, with stability constants* $\eta_J, \bar{\eta}_J$ *and condition* $\kappa_J$ *that remain bounded, independently of* $J$:

$$0 < \eta \leq \eta_J \leq \bar{\eta}_J \leq \bar{\eta} < \infty, \qquad \kappa_J \leq \kappa = \bar{\eta}/\eta. \tag{B.2.12}$$

Proofs of this result can be found in the literature cited in Section B.1. There are many variations connected with it. For instance, if

$$\tilde{u}_j = \sum_{i=1}^{N_j} \tilde{u}_j^i \varphi_j^i$$

is the unique representation of $\tilde{u}_j$ with respect to the nodal basis $\Phi_j$ in the finite element space $\tilde{V}_j$ then

$$\tilde{\ell}_j(\tilde{u}_j, \tilde{u}_j) \approx 2^{2j} \sum_{i=1}^{N_j} (\tilde{u}_j^i)^2 \|\varphi_j^i\|_0^2 \approx \sum_{i=1}^{N_j} (\tilde{u}_j^i)^2. \tag{B.2.13}$$

To see this so-called $L_2$-*stability of the nodal basis* $\Phi_j$, use the fact that for a function $u$ which is linear on a triangle $\Delta$ and takes values $\alpha, \beta, \gamma$ at its three vertices, we have

$$\int_{\Delta} u(x, y)^2 \, dx \, dy \approx |\Delta|(\alpha^2 + \beta^2 + \gamma^2)$$

with constants independent of $\Delta$ (as usual, $|\Delta|$ denotes the area of $\Delta$). Application to the piecewise linear function $\tilde{u}_j$ and each triangle in $\mathcal{T}_j$ leads to (B.2.13). This relationship allows us to conclude that the splitting into one-dimensional spaces $\tilde{V}_j^i$ associated with the basis functions $\varphi_j^i$

$$\{V_J; \ell_J\} \cong \sum_{j=1}^{J} \sum_{i=1}^{N_j} I_j^J \{\tilde{V}_j^i; \tilde{\ell}_j^i\} \tag{B.2.14}$$

is also stable, with stability constants and condition again satisfying (B.2.12) (with possibly different values for $\eta, \bar{\eta}$). The only requirement on $\tilde{\ell}_j^i$ is

$$\tilde{\ell}_j^i(\varphi_j^i, \varphi_j^i) \approx 1, \tag{B.2.15}$$

uniformly in $j$ and $i$, which is satisfied for choosing it as the restriction of either $\ell_J$ or $\tilde{\ell}_j$ to $\tilde{V}_j^i$. To obtain the necessary estimates for the triple bar norm associated with (B.2.14), take the corresponding result for the splitting (B.2.11), substitute (B.2.13), together with the scaling assumption (B.2.15) on $\tilde{\ell}_j^i$ (note that $\tilde{u}_j^i \in \tilde{V}_j^i$ means that $\tilde{u}_j^i$ can be written as a scalar multiple of the basis function $\varphi_j^i$).

We conclude with some general statements on stable space splittings. First, let us mention the following equivalent formulation of (B.2.2) if a stable splitting (B.2.5) is available. Define linear operators $T_j \colon V \to \tilde{V}_j$ and elements $\tilde{\phi}_j \in \tilde{V}_j$, $j = 1, \ldots, J$, by requiring

$$\tilde{\ell}_j(T_j u, \tilde{v}_j) = \ell(u, P_j \tilde{v}_j) \quad \forall \tilde{v}_j \in \tilde{V}_j, \tag{B.2.16}$$

and

$$\tilde{\ell}_j(\tilde{\phi}_j, \tilde{v}_j) = f(P_j \tilde{v}_j) \quad \forall \tilde{v}_j \in \tilde{V}_j. \tag{B.2.17}$$

For any given $u \in V$, these are well-defined Galerkin formulations on the spaces $\tilde{V}_j$. The operator

$$\mathcal{P} = \sum_{j=1}^{J} P_j T_j \colon V \to V \tag{B.2.18}$$

is called the *additive Schwarz operator* associated with (B.2.5). Also, define $\phi = \sum_{j=1}^{J} P_j \tilde{\phi}_j$.

**Theorem B.2.2**  *Assume that the space splitting (B.2.5) is stable. Then $\mathcal{P}$ is symmetric positive definite with respect to $\{V; \ell\}$, its spectral condition number coincides with the condition of the splitting, and its extremal eigenvalues with the values $\bar{\eta}^{-1}$ and $\eta^{-1}$. The operator equation*

$$\mathcal{P}u = \phi, \tag{B.2.19}$$

*has the same solution as the variational problem (B.2.2).*

For the elementary proof, see [298, Section 4.1]. In some cases, $\mathcal{P}$ can be written explicitly. For instance, in Example B.2.4, we have $\mathcal{P} = \tilde{L}^{-1} L$. The additive Schwarz operator associated with the splitting (B.2.14) takes the form

$$\mathcal{P}u = \sum_{j=1}^{J} \sum_{i=1}^{N_j} \frac{\ell(u, \varphi_j^i)}{\tilde{\ell}_j^i(\varphi_j^i, \varphi_j^i)} \varphi_j^i. \tag{B.2.20}$$

To see this, note that the problems (B.2.16) are one-dimensional and, therefore, can be solved explicitly, and that the prolongations are given by natural embeddings $I_j^J$ which are omitted in the above formula.

The formula (B.2.20) has a very familiar appearance, it reminds us of the Fourier series representation (with the difference that the system

$$\Phi = \bigcup_{j=1}^{J} \Phi_j = \{\phi_j^i \colon i = 1, \ldots, N_j, \ j = 1, \ldots, J\} \tag{B.2.21}$$

is neither orthogonal nor a basis in $V$). This connection is very useful, especially for proving the stability of certain space splittings by using known results from applied harmonic analysis and function space theory [298] but also to see the benefits and drawbacks of emerging wavelet algorithms for solving PDE discretizations [116].

The last remark is about verifying the stability of a given splitting. The *upper bound* for $\||v|\|^2$ requires us to find a *good decomposition* of arbitrary elements $u \in V$ with components $\tilde{v}_j \in \tilde{V}_j$ such that $v = \sum_j P_j \tilde{v}_j$. If there is only one admissible representation (B.2.6) then we have no choice but to consider this decomposition (thus, to "guess" a good set of auxiliary spaces $\tilde{V}_j$ is the important part of proving anything about the splitting). Otherwise, we have some choice in (B.2.6), and suitable decompositions are constructed by using various projections onto the spaces $\hat{V}_j$. For instance, when deriving Theorem B.2.1, one often relies on the $L_2$-orthoprojection operators $Q_j : L_2(\Omega) \to \tilde{V}_j$ given by

$$(Q_j u, \tilde{v}_j)_0 = (u, \tilde{v}_j)_0 \quad \forall \tilde{v}_j \in \tilde{V}_j, \tag{B.2.22}$$

$j \geq 1$. For them, the two-sided inequality

$$(u_J, u_J)_1 \approx \|Q_1 u_J\|_0^2 + \sum_{j=2}^{J} 2^{2j} \|Q_j u_J - Q_{j-1} u_J\|_0^2 \quad \forall u_J \in V_J, \tag{B.2.23}$$

can be proved (e.g., using approximation-theoretic and elliptic regularity results), again with constants uniformly bounded with respect to $J$.

Let us show that (B.2.23) implies Theorem B.2.1. By setting $\tilde{u}_j = Q_j u_J - Q_{j-1} u_J$ for $j = 2, \ldots, J$, and $\tilde{u}_1 = Q_1 u_J$, we have $\tilde{u}_j \in \tilde{V}_j$ and $u_J = \sum_{j=1}^{J} \tilde{u}_j$. This implies the upper bound for the stability of the splitting (B.2.11):

$$\||u_J|\|^2 = \inf_{\tilde{v}_j \in \tilde{V}_j : u_J = \sum_{j=1}^{J} \tilde{v}_j} \sum_{j=1}^{J} 2^{2j} \|\tilde{v}_j\|_0^2$$

$$\leq \|Q_1 u_J\|_0^2 + \sum_{j=2}^{J} 2^{2j} \|Q_j u_J - Q_{j-1} u_J\|_0^2$$

$$\leq C(u_J, u_J)_1 = C\ell_J(u_J, u_J).$$

On the other hand, for an arbitrary decomposition (B.2.6), using the fact that the spaces $\tilde{V}_j$ form an increasing sequence and that the $Q_j$ are linear projections, we have

$$Q_j u_J = Q_j \left( \sum_{l=1}^{J} \tilde{v}_l \right) = \sum_{l=1}^{j} \tilde{v}_l + \sum_{l=j+1}^{J} Q_j \tilde{v}_l$$

and

$$\|Q_j u_J - Q_{j-1} u_J\|_0^2 = \|\tilde{v}_j + \sum_{l=j+1}^{J} Q_j \tilde{v}_l - \sum_{l=j}^{J} Q_{j-1} \tilde{v}_l\|_0^2$$

$$\leq \left( 2 \sum_{l=j}^{J} \|\tilde{v}_l\|_0 \right)^2 = 4 \left( \sum_{l=j}^{J} 2^{-l/2} (2^{l/2} \|\tilde{v}_l\|_0) \right)^2$$

$$\leq 4 \left( \sum_{l=j}^{J} 2^{-l} \right) \left( \sum_{l=j}^{J} 2^l \|\tilde{v}_l\|_0^2 \right) \leq 2^{-j+3} \sum_{l=j}^{J} 2^l \|\tilde{v}_l\|_0^2.$$

An analogous estimation works for $\| Q_1 u_J \|_0^2$. Thus,

$$
\ell_J(u_J, u_J) \leq C \left( \| Q_1 u_J \|_0^2 + \sum_{j=2}^{J} 2^{2j} \| Q_j u_J - Q_{j-1} u_J \|_0^2 \right)
$$

$$
\leq C \left( \sum_{j=1}^{J} 2^{2j} 2^{-j+3} \sum_{l=j}^{J} 2^l \| \tilde{v}_l \|_0^2 \right) = C \left( \sum_{l=1}^{J} 2^l \| \tilde{v}_l \|_0^2 \left( \sum_{j=1}^{l} 2^{j+3} \right) \right)
$$

$$
\leq 16 C \left( \sum_{l=1}^{J} 2^{2l} \| \tilde{v}_l \|_0^2 \right).
$$

Taking the infimum with respect to all admissible representations (B.2.6), we see the lower bound for the stability of the splitting (B.2.11).

Speaking in practical terms, the orthoprojections $Q_j$ are still too involved (the solution of (B.2.22) is not straightforward), and one would like to replace them by more explicit constructions. Finite element interpolation operators $I_j u : C(\bar{\Omega}) \to \tilde{V}_j$ defined by requiring the interpolation condition $(I_j u)(x^i, y^i) = u(x^i, y^i)$ at all (boundary and interior) vertices of $\mathcal{T}_j$ come to mind but do not necessarily lead to the "optimal" decomposition to prove Theorem B.2.1, more recently, *quasi-interpolants* have been proposed. A simple set of quasi-interpolant operators which could be used for the above linear finite element spaces is given by

$$
\tilde{Q}_j u = \sum_{i=1}^{N_j} \frac{(u, \varphi_j^i)_0}{(1, \varphi_j^i)_0} \varphi_j^i, \quad j \geq 1. \tag{B.2.24}
$$

Although these $\tilde{Q}_j$ are not projections onto $\tilde{V}_j$, they at least reproduce constant functions locally in the interior of the triangulation $\mathcal{T}_j$ which is often enough (*local preservation of polynomials of a certain degree* is one of the characteristics of quasi-interpolant operators). More importantly, the $\tilde{Q}_j$ are well-defined and uniformly bounded with respect to $L_2(\Omega)$, and they can be computed by fast algorithms.

The typical method to establish the *lower bound* in the stability requirement (B.2.8) of a splitting (B.2.5) is the proof of so-called *strengthened Cauchy–Schwarz inequalities* [425, 435]. The simplest version is as follows: introduce a matrix $E = ((\varepsilon_{j,l}))_{j,l=1}^J$ where the entries are defined as the positive constants for which

$$
\ell(P_j \tilde{v}_j, P_l \tilde{v}_l)^2 \leq \varepsilon_{j,l} \tilde{\ell}_j(\tilde{v}_j, \tilde{v}_j) \tilde{\ell}_l(\tilde{v}_l, \tilde{v}_l) \quad \forall \tilde{v}_j \in \tilde{V}_j, \ \tilde{v}_l \in \tilde{V}_l. \tag{B.2.25}
$$

Without loss of generality, we may assume that $E$ is symmetric. Let $\lambda_{\max}(E)$ denote the largest eigenvalue of the matrix $E$.

**Lemma B.2.1** *For an arbitrary space splitting (B.2.5) we have*

$$
\ell(u, u) \leq \lambda_{\max}(E) \|| u |\|^2 \quad \forall u \in V, \tag{B.2.26}
$$

*where the matrix $E$ is determined from the strengthened Cauchy–Schwarz inequalities (B.2.25) as described above.*

The proof of this lemma is straightforward. Considering an arbitrary admissible representation (B.2.6), we obtain

$$
\ell(u, u) = \ell\left( \sum_{j=1}^{J} P_j \tilde{u}_j, \sum_{j=1}^{J} P_j \tilde{u}_j \right) = \sum_{j,l=1}^{J} \ell(P_j \tilde{u}_j, P_l \tilde{u}_l)
$$

$$
\leq \sum_{j,l=1}^{J} \varepsilon_{j,l} \tilde{\ell}_j(\tilde{v}_j, \tilde{v}_j)^{1/2} \tilde{\ell}_l(\tilde{v}_l, \tilde{v}_l)^{1/2} \leq \lambda_{\max}(E) \sum_{j=1}^{J} \tilde{\ell}_j(\tilde{v}_j, \tilde{v}_j).
$$

Since the representation was arbitrary, we arrive at (B.2.26).

By properly scaling the auxiliary bilinear forms $\tilde{\ell}_j$ we can ensure that $\varepsilon_{j,j} = 1$. As a consequence, we have $0 \leq \varepsilon_{j,l} \leq 1$ for all nondiagonal entries. This implies $1 \leq \lambda_{\max}(E) \leq J$, both extremes are possible. Going through the above examples, we see that $\eta = 1$ in Example B.2.4 because in this case $E$ can be chosen as the identity matrix. For examples where $J$ is small (such as Example B.2.5), we can use the trivial bound $\lambda_{\max}(E) \leq J \max \varepsilon_{j,j}$. A nontrivial situation arises in Example B.2.6. By carefully applying Green's formula on each triangle of the underlying triangulations, one can show that

$$
(\tilde{v}_j, \tilde{v}_l)_1 \leq C 2^{|j-l|/2} (2^j \|\tilde{v}_j\|_0)(2^l \|\tilde{v}_l\|_0) \quad \forall \, \tilde{v}_j \in \tilde{V}_j, \ \tilde{v}_l \in \tilde{V}_l, \ j, l \geq 1. \quad \text{(B.2.27)}
$$

Thus, we can choose $E = ((C 2^{|j-l|/2}))_{j,l=1}^{J}$, and because of the exponential decay of the $\varepsilon_{j,l}$ away from the diagonal, we obtain $\lambda_{\max}(E) \leq C$ for some absolute constant $C$, independently of $J$.

**Example B.2.7** We conclude with an appendix to Example B.2.6. Depending on the application, it may happen that the same spaces $V$ and $\tilde{V}_j$ are equipped with different choices of bilinear forms. For example, if the Poisson problem (B.2.1) is modified by adding a source term $q \cdot u(x, y)$, where, for simplicity, $q > 0$ is a constant, then the appropriate bilinear form takes the form

$$
\ell^q(u, v) = (u, v)_1 + q(u, v)_0 \quad \forall \, u, v \in V.
$$

For large $q$, one should definitely take into consideration the term associated with the $L_2$-scalar product. Therefore, if we again take the finite element spaces of Example B.2.6 then the following results are of interest. $\triangle$

**Lemma B.2.2** (a) *The splitting*

$$
\{V_J; (\cdot, \cdot)_0\} \cong \sum_{j=1}^{J} \{\tilde{V}_j; (\cdot, \cdot)_0\} \quad \text{(B.2.28)}
$$

*is stable and has condition $\kappa = J$. The following stability bounds for (B.2.28) are sharp:*

$$J^{-1}\|u_J\|_0^2 \leq \|\|u_J\|\|^2 \leq \|u_J\|_0^2 \quad \forall u_J \in V_J. \tag{B.2.29}$$

**(b)** *The splitting*

$$\{V_J; \ell^q\} \cong \sum_{j=J_0}^{J} \{\tilde{V}_j; 2^{2j}(\cdot, \cdot)_0\} \tag{B.2.30}$$

*is stable with condition $\kappa = O(1)$, independently of $q \geq 0$ and $J$ if $J_0 = J_0(q)$ is chosen according to the following rules: if $q \leq 1$ or $q \geq 2^{2J}$ then $J_0 = 1$ or $J_0 = J$, respectively, while in the intermediate range $1 < q < 2^{2J}$ the choice $J_0 = [\log_2 q/2]+1$ is appropriate.*

*Proof.* Since by definition of the orthoprojections $Q_j$ we have

$$\|u_J\|_0^2 = \|Q_{J_0}u_J\|_0^2 + \sum_{j=J_0+1}^{J} 2^{2j}\|Q_j u_J - Q_{j-1}u_J\|_0^2, \quad 1 \leq J_0 \leq J, \tag{B.2.31}$$

the upper bound in (B.2.29) is obvious (set $J_0 = 1$ and look at the definition of $\|\|u_J\|\|$ for the splitting (B.2.28)). The lower bound follows from Lemma B.2.1 in conjunction with the trivial estimate $\lambda_{\max}(E) \leq J$. That the bounds cannot be improved follows by considering special $u_J$. For example, take any $u_J \neq 0$ which belongs to the $L_2$-orthogonal complement space $W_J = V_J \ominus V_{J-1}$. By applying $Q_J - Q_{J-1}$ to any admissible representation $u_J = \sum_{j=1}^{J} \tilde{v}_j$, we obtain $u_J = (Q_J - Q_{J-1})\tilde{v}_J$, and since $Q_J - Q_{J-1}$ is an orthoprojection onto $W_J$, we obtain

$$\|u_J\|_0^2 = \|(Q_J - Q_{J-1})\tilde{v}_J\|_0^2 \leq \|\tilde{v}_J\|_0^2 \leq \sum_{j=1}^{J} \|\tilde{v}_j\|_0^2$$

which leads to $\|u_J\|_0^2 \leq \|\|u_J\|\|^2$ for such $u_J$, and to the sharpness of the upper bound. Concerning the lower bound, pick $u_J = \varphi_1^1 \in \tilde{V}_1 \subset V_J$, and look at the admissible representation given by $\tilde{v}_j = J^{-1}u_J$, $j = 1, \ldots, J$. Then

$$\|\|u_J\|\| \leq \sum_{j=1}^{J} J^{-2}\|u_J\|_0^2 = J^{-1}\|u_J\|_0^2.$$

This establishes the sharpness of the lower bound.

As to the stability of (B.2.30), we will concentrate on the intermediate range $1 < q < 2^{2J}$ (the reader will be able to deal with the remaining cases). By definition of $J_0$, we have

$2^{2(J_0-1)} \leq q < 2^{2J_0}$. Thus, by (B.2.23) and (B.2.31) we can estimate

$$
\ell^q(u_J, u_J)
$$

$$
\geq c \left( \|Q_1 u_J\|_0^2 + \sum_{j=2}^{J} 2^{2j} \|Q_j u_J - Q_{j-1} u_J\|_0^2 \right)
$$

$$
+ q \left( \|Q_{J_0} u_J\|_0^2 + \sum_{j=J_0+1}^{J} \|Q_j u_J - Q_{j-1} u_J\|_0^2 \right)
$$

$$
\geq c \left( 2^{2J_0} \left( \|Q_{J_0} u_J\|_0^2 + 2^{-2J_0} \|Q_1 u_J\|_0^2 + \sum_{j=2}^{J_0} 2^{2(j-J_0)} \|Q_j u_J - Q_{j-1} u_J\|_0^2 \right) \right.
$$

$$
\left. + \sum_{j=J_0+1}^{J} \left( 2^{2j} + 2^{2J_0} \right) \|Q_j u_J - Q_{j-1} u_J\|_0^2 \right)
$$

$$
\geq c \left( 2^{2J_0} \|Q_{J_0} u_J\|_0^2 + \sum_{j=J_0+1}^{J} 2^{2j} \|Q_j u_J - Q_{j-1} u_J\|_0^2 \right).
$$

This gives the upper stability estimate for the splitting (B.2.30), with a constant $\bar{\eta} \leq C$.

For the lower estimate, we complement the Cauchy–Schwarz inequalities (B.2.27) by their trivial counterparts for the $L_2$-scalar product

$$
(\tilde{v}_j, \tilde{v}_l)_0 \leq 2^{-(j+l)} (2^j \|\tilde{v}_j\|_0)(2^l \|\tilde{v}_l\|_0) \quad \forall \, \tilde{v}_j \in \tilde{V}_j, \, \tilde{v}_l \in \tilde{V}_l.
$$

Multiplying here by $q \approx 2^{2J_0}$, and adding the result to (B.2.27) we obtain

$$
\ell^q(\tilde{v}_j, \tilde{v}_l) \leq C (2^{|j-l|/2} + 2^{-(j+l-2J_0)})(2^j \|\tilde{v}_j\|_0)(2^l \|\tilde{v}_l\|_0) \quad \forall \, \tilde{v}_j \in \tilde{V}_j, \, \tilde{v}_l \in \tilde{V}_l,
$$

for all $J_0 \leq j, l \leq J$. Obviously, in this range of $j, l$, the first term $2^{|j-l|/2}$ dominates the second, therefore, again $\lambda_{\max}(E) \leq C$, independently on $q$, and $J$. Applying Lemma B.2.1 concludes the proof of Lemma B.2.2. $\qquad \square$

## B.3 CONVERGENCE THEORY

After this extended introduction to the concept of stable space splittings, we now derive the convergence theory for the associated subspace correction methods. Let us briefly link the notation of the previous section to the AS and MS methods as defined in the introduction. All we have to do is to fix basis systems in the spaces involved, and to identify elements of these spaces with coefficient vectors, and operators between them with matrices. Even though this might temporarily lead to some confusion, we will use the same notation for elements and vectors as well as for operators and matrices, respectively. Thus, $P_j$ will denote an operator from $\tilde{V}_j$ into $V$, and, at the same time, a rectangular $N \times N_j$ matrix representing this operator with respect to the bases chosen in $\tilde{V}_j$ and $V$, respectively. Assuming that (B.2.5) is stable, we will use the notation $L$ and $\tilde{L}_j$ for the matrices associated with $\ell$

(i.e. $\ell(u, u) = (Lu, u)$) and $\tilde{\ell}$. Thus, the matrix representation of the operators $T_j$ can be derived from (B.2.16):

$$(\tilde{L}_j T_j u, \tilde{v}_j) = \tilde{\ell}_j(T_j u, \tilde{v}_j) = \ell(u, P_j \tilde{v}_j) = (Lu, P_j \tilde{v}_j) = (P_j^T Lu, \tilde{v}_j).$$

This gives $T_j = \tilde{L}_j^{-1} P_j^T L$, and

$$\mathcal{P} = \sum_{j=1}^{J} P_j \tilde{L}_j^{-1} P_j^T L \equiv BL, \qquad B = \sum_{j=1}^{J} P_j \tilde{L}_j^{-1} P_j^T. \tag{B.3.1}$$

Thus, if we fix the restriction operators as the adjoint operators (transposed matrices) of the prolongations, i.e.

$$R_j = P_j^T, \quad j = 1, \ldots, J, \tag{B.3.2}$$

then everything falls into place. The *additive subspace correction method* defined in Section B.1 is simply the *extrapolated Richardson method* (or $\omega$-Richardson relaxation) applied to the reformulation (B.2.19) of the original variational problem (B.2.2). The assumption (B.3.2) is more or less natural since we are restricted to symmetric positive definite $L$ and $\tilde{L}_j$; it ensures that the preconditioner $B$ is also symmetric positive definite. As a by-product, the *preconditioned conjugate gradient* (PCG) *method* with preconditioner $B$ can be applied for solving (B.1.1), and the design of stable space splittings for $\{V; \ell\}$ with small condition can be viewed as a method of constructing good preconditioners $B$ in a systematic way. We will give this PCG-method the descriptive name AS-CG.

Here is another useful representation which leads to a unified treatment of AS and MS methods in terms of classical iterative methods for a so-called *extended semidefinite problem*. Set $\tilde{L}_{j,l} = \tilde{L}_j^{-1} P_j^T L P_l$, $j, l = 1, \ldots, J$, and $\tilde{N} = \sum_{j=1}^{J} N_j$. Define the $\tilde{N} \times \tilde{N}$ matrix $\tilde{\mathcal{P}}$ as a $J \times J$ block matrix whose entries are the $N_j \times N_l$ matrices $\tilde{L}_{j,l}$. We will use the notation

$$\tilde{\mathcal{P}} = \tilde{\mathcal{L}} + \tilde{\mathcal{D}} + \tilde{\mathcal{U}} \tag{B.3.3}$$

for the standard decomposition of the block matrix into lower triangular, diagonal, and upper triangular block matrices. Let $\tilde{v} = (\tilde{v}_1, \ldots, \tilde{v}_J)^T$ be the corresponding block representation of $\mathbb{R}^{\tilde{N}}$-vectors. Set $\tilde{\phi} = (\tilde{\phi}_1, \ldots, \tilde{\phi}_J)^T$, where the $\tilde{\phi}_j$ are determined in (B.2.17).

**Lemma B.3.1** *Assume that a stable space splitting (B.2.5) is given, and that (B.3.2) holds.*
**(a)** *If $\tilde{u}$ is a solution of the semidefinite problem*

$$\tilde{\mathcal{P}}\tilde{u} = \tilde{\phi}, \tag{B.3.4}$$

*then $u = \tilde{P}\tilde{u} \equiv \sum_{j=1}^{J} P_j \tilde{u}_j$ is the (unique) solution of (B.2.2) and its reformulation (B.2.19).*
**(b)** *For any fixed $\tilde{N} \times \tilde{N}$ matrix $\tilde{\mathcal{B}}$, we consider the linear iteration*

$$\tilde{u}^{k+1} = \tilde{u}^k + \tilde{\mathcal{B}}(\tilde{\phi} - \tilde{\mathcal{P}}\tilde{u}^k), \quad k \geq 0, \tag{B.3.5}$$

*with a starting vector $\tilde{u}^0$ given. The iteration (B.3.5) generates an iteration in $V$ by the formula $u^k = \tilde{P}\tilde{u}^k$, $k \geq 0$. If $\tilde{\mathcal{B}} = \tilde{\mathcal{B}}_{AS} \equiv \omega\tilde{\mathcal{I}}$, where $\tilde{\mathcal{I}}$ is the $\tilde{N} \times \tilde{N}$ identity matrix, then (B.3.5) generates the additive subspace correction method AS associated with the splitting. Analogously, if $\tilde{\mathcal{B}} = \tilde{\mathcal{B}}_{MS} \equiv (\omega^{-1}\tilde{\mathcal{I}} + \tilde{\mathcal{L}})^{-1}$, then (B.3.5) generates the multiplicative subspace correction method MS associated with the splitting.*

*Proof.* Part (a) can be seen from applying $\tilde{P}$ to both sides of (B.3.4) resulting in

$$\tilde{P}\tilde{\mathcal{P}}\tilde{u} = \sum_{j=1}^{J}\sum_{l=1}^{J} P_j\tilde{L}_j^{-1}P_j^T L P_l\tilde{u}_l = \mathcal{P}\tilde{P}\tilde{u} = \tilde{P}\tilde{\phi} = \phi.$$

Now compare with Theorem B.2.2.

Analogously, applying $\tilde{P}$ to both sides of the iteration (B.3.5) and using the relationship $u^k = \tilde{P}\tilde{u}^k$, we obtain

$$u^{k+1} = u^k + \tilde{P}\tilde{\mathcal{B}}(\tilde{L}_1^{-1}P_1^T, \ldots, \tilde{L}_1^{-1}P_1^T)^T r^k \qquad (B.3.6)$$

Here the explicit form of the $\tilde{L}_{j,l}$ and $\tilde{\phi}_j = \tilde{L}_j^{-1}P_j^T f$ has been utilized (for the latter formula, compare (B.2.17)). Thus, setting $\tilde{\mathcal{B}} = \tilde{\mathcal{B}}_{AS}$, we immediately arrive at $u^{k+1} = u^k + B_{AS}Lr^k$ which coincides with the AS iteration.

To see the result for the MS method, some algebraic transformations are necessary. For convenience, denote $B_j = \omega P_j\tilde{L}_j^{-1}P_j^T$. Consider one loop of the MS method. Denote $u = u^k$, $r = f - Lu^k$. By induction, we obtain

$$v^2 = u + B_1 r,$$
$$v^3 = u + ((B_1 + B_2) - B_2 L B_1)r,$$
$$v^4 = u + ((B_1 + B_2 + B_3) - (B_3 L B_1 + B_2 L B_1 + B_3 L B_2) + B_3 L B_2 L B_1)r,$$

$$\cdots$$

$$v^{J+1} = u + \left(\sum_{j=1}^{J} B_j - \sum_{1 \leq l < j \leq J} B_j L B_l + \cdots + (-1)^{J+1} B_J L B_{J-1} L \ldots B_2 L B_1\right)r.$$

Repeatedly using the formula $B_j L B_l = \omega^2 P_j\tilde{L}_{j,l}\tilde{L}_l^{-1}P_l^T$, we obtain

$$B_{j_s} L B_{j_{s-1}} L \cdots B_{j_2} L B_{j_1} = \omega^s P_{j_s}(\tilde{L}_{j_s,j_{s-1}} \cdots \tilde{L}_{j_3,j_2}\tilde{L}_{j_2,j_1})\tilde{L}_{j_1}^{-1}P_{j_1}^T,$$

for all $1 \leq j_1 < \cdots < j_s \leq J$. After substitution into the previous representation for $v^{J+1}$, and returning to the notation of the MS iteration ($v^{J+1} = u^{k+1}$, $u = u^k$, $r = r^k$), we have

$$u^{k+1} = u^k + \omega\left(\sum_{j=1}^{J}\sum_{l=1}^{j-1} P_j(I - \omega\tilde{L}_{j,l}\right.$$

$$\left. + \omega^2\sum_{m:l<m<j} \tilde{L}_{j,m}\tilde{L}_{m,l} - \cdots + (-\omega)^{j-l}\tilde{L}_{j,j-1}\cdots\tilde{L}_{l+1,l})\tilde{L}_l^{-1}P_l^T\right)r^k.$$

This has to be compared with (B.3.6) for

$$\tilde{B} = \tilde{B}_{MS} = \omega(\tilde{I} + \omega\tilde{L})^{-1} = \omega(\tilde{I} - \omega\tilde{L} + \cdots + (-1)^{J-1}\tilde{L}^{J-1}).$$

By computing the entries of the powers of the lower triangular block matrix $\tilde{L}$, it is not hard to see from the latter expressions that the two iterations coincide. This gives the statement of part (b) for the MS method. Lemma B.3.1 is proved. $\qquad\square$

Thus, the AS and MS methods can be viewed as $\omega$-Richardson relaxation and Richardson-SOR iteration applied to the block matrix $\tilde{P}$. Clearly, assuming that all diagonal blocks $\tilde{L}_{j,j}$ are invertible, we could have chosen $\tilde{B} = \omega\tilde{D}^{-1}$ to define the counterpart of the $\omega$-Jacobi relaxation.

By using this formalism, the convergence theory of subspace correction methods reduces to standard derivations as known for the case of block-Jacobi and block-SOR (see [167] or [298, Theorem 18]). Slightly different derivations can be found in [54, 425, 435]. To measure convergence behavior, we will use the energy norm $\|e^k\|_\ell = \sqrt{\ell(e^k, e^k)} = \sqrt{(Le^k, e^k)}$ of the error $e^k = u - u^k$ after $k$ iterations (consequently, $e^0$ denotes the error of the starting guess). The convergence rate is defined below as the energy norm of the corresponding error iteration matrix.

**Theorem B.3.1** *Let $V$ be a finite-dimensional Hilbert space, and $\ell$ a symmetric $V$-elliptic bilinear form on $V$. Assume that (B.2.5) is a stable space splitting (with stability constants $\eta, \bar{\eta}$ and condition $\kappa$) given by the auxiliary spaces $\{\tilde{V}_j; \tilde{\ell}_j\}$ and the prolongations $P_j$, $j = 1, \ldots, J$. The restrictions $R_j$ are defined by (B.3.2).*

*(i) The additive subspace correction method AS converges if and only if $0 < \omega < 2/\eta$. The optimal convergence rate is achieved for $\omega^* = 2\eta\bar{\eta}/(\eta + \bar{\eta})$, and equals*

$$\rho_{AS}^* = \min_{0<\omega<2\eta} \lambda_{\max}(M_{AS}) = 1 - \frac{2}{1+\kappa}. \tag{B.3.7}$$

*Correspondingly, the AS-CG method converges with the guaranteed error bound*

$$\|e^k\|_\ell \le 2\left(1 - \frac{2}{1+\sqrt{\kappa}}\right)^k \|e^0\|_\ell, \quad k \ge 1. \tag{B.3.8}$$

*(ii) Assume that, in addition to the above, strengthened Cauchy–Schwarz inequalities (B.2.25) hold such that $E$ is symmetric and $\varepsilon_{j,j} = 1$ for all $j = 1, \ldots, J$. Then the multiplicative algorithm subspace correction method MS converges for $0 < \omega < 2$. The analogously defined optimal convergence rate can be estimated by*

$$(\rho_{MS}^*)^2 \le 1 - \frac{1}{\bar{\eta}(2\lambda_{\max}(E) + 1)}. \tag{B.3.9}$$

*Without assuming (B.2.25), one has*

$$(\rho_{MS}^*)^2 \le 1 - \frac{1}{\log_2(4(J+1)) \cdot \kappa}. \tag{B.3.10}$$

*The results in (ii) remain valid for any reordering of the spaces in the splitting.*

Intuitively, it might seem that the multiplicative algorithm MS should perform better than AS which can be observed in many applications, and parallels the experience with Jacobi- and Gauss–Seidel methods for specific classes of linear systems. However, this practical observation is not reflected by the upper estimates of Theorem B.3.1. In fact, it has been shown [297] that the logarithmic factor in (B.3.10) cannot be removed. The counterexamples of [297] are based on some exotic Toeplitz matrices $L$ and the splittings mentioned in Example B.2.4.

There are numerous modifications of the multiplicative algorithm, and refined theories which serve some applications better, and lead to sharper estimates under special circumstances. We mention the *symmetric multiplicative subspace correction method* (SMS), which is the abstract counterpart of the SSOR-method. It combines two steps of the MS method into one and, is therefore, formally twice as expensive. The iteration operator takes the form

$$M_{\text{SMS}} = (I - \omega P_J T_J) \cdots (I - \omega P_1 T_1)(I - \omega P_1 T_1) \cdots (I - \omega P_J T_J),$$

and can be viewed as the MS method applied to the splitting

$$\{V; \ell\} \cong \{\tilde{V}_J; \tilde{\ell}_J\} + \cdots + \{\tilde{V}_1; \tilde{\ell}_1\} + \{\tilde{V}_1; \tilde{\ell}_1\} + \cdots + \{\tilde{V}_J; \tilde{\ell}_J\}.$$

In analogy to the situation with SOR and SSOR, one has $\rho_{\text{SMS}}^* = (\rho_{\text{MS}}^*)^2$, i.e. the convergence theory of the SMS method is covered by Theorem B.3.1 (ii). An advantage is that we can now write $M_{\text{SMS}} = I - BL$, where $B$ is symmetric. Thus, the application of the PCG-method is possible, this time with a *multiplicative preconditioner* rather than with the *additive preconditioner* associated with $\mathcal{P}$. The use of subspace correction methods as preconditioners in standard Krylov type iterative methods is becoming good practice and can considerably increase the robustness of a solver in comparison with applying subspace correction methods or Krylov space methods on their own.

A more general multiplicative version, the *variable symmetric multiplicative algorithm* has been proposed and analyzed by Bramble *et al.* [54, Algorithm III]. In a multigrid environment, the general recommendation is to allow for more subspace correction steps on the spaces $\tilde{V}_j$ corresponding to coarse meshes, and, therefore, to the inexpensive subproblems in (B.2.16). The benefit is that weaker assumptions suffice to state optimal convergence estimates, without significantly increasing the arithmetic complexity of the iteration in the asymptotical range ($J \to \infty$). Speaking in terms of subspace splittings, each auxiliary space $\{\tilde{V}_j; \tilde{\ell}_j\}$ appears $\nu_j$ times, with $\nu_j$ not necessarily a fixed number (the MS and SMS methods correspond to $\nu_j = 1$ and $\nu_j = 2$, respectively). Note that these modifications are important for the multiplicative algorithms but have no immediate impact on the additive method (usually, the estimates for $\kappa$ cannot be improved this way).

The reader is encouraged to look at the examples of Section B.1, and to interpret the results on the conditioning of the splittings introduced so far as indicators for the convergence rates of the associated additive and multiplicative subspace correction methods. It is also recommended that readers derive the corresponding matrix representations and estimate the complexity of the implementation. As should be clear from the abstract theory, the ultimative goal is to obtain stable splittings with small condition number (hopefully, independently of discretization and problem parameters) and a reasonable overall operation count for the components $\tilde{L}_j$, $P_j$, and $R_j$ $(= P_j^T)$, $j = 1, \ldots, J$, involved in the algorithms. Roughly speaking, the best we can hope for are so-called *asymptotically optimal algorithms*, the convergence rate of which stays well away from 1, and such that the computational work per iteration grows only linearly with the problem size. The examples of multigrid algorithms and domain decomposition methods discussed below are of this type. Relying mostly on the results for Example B.2.6, we will give examples of such asymptotically optimal algorithms for finite difference discretizations.

## B.4 MULTIGRID APPLICATIONS

In this section we will derive a V-cycle multigrid method for the five-point discretization of the Poisson problem and justify its asymptotic optimality by interpreting it as a special instance of a multiplicative subspace correction method and using the general theory for the latter. The result is *qualitative*: other than saying that the convergence factor of the method satisfies $\rho_h \leq \rho^* < 1$ (independently of the mesh parameter $h$), no concrete values of $\rho_h$ resp. of the upper bound $\rho^*$ can be predicted.

Consider the unit square $\Omega$ and the sequence of uniform grids $\mathcal{V}_j = \Omega_{2^{-j}, 2^{-j}}$, $j \geq 1$. Thus, $\mathcal{V}_j$ is the set of all interior vertices of the triangulation $\mathcal{T}_j$ (compare Figs B.4 and B.5). We will simultaneously speak about vectors in $\mathbb{R}^{N_j}$ and grid functions on $\mathcal{V}_j$ assuming that the connection between vector indices and grid points in $\mathcal{V}_j$ is clear (e.g., given by the ordering discussed in Chapter 1). All grid functions are extended to the boundary of $\Omega$ by assuming zero values at the boundary grid points. Thus, grid functions on $\Omega_j$ can be identified with finite element functions in $V(\mathcal{T}_j)$. Let

$$L_j u_j = f_j \tag{B.4.1}$$

be the linear system corresponding to the standard five-point finite difference discretization of (B.2.1) with respect to the grid $\mathcal{V}_j$, $j \geq 1$. For, simplicity, we will call (B.4.1) *the finite difference method* (FDM) *problem of level $j$*.

Our concern will be to construct a multigrid algorithm for the solution of any of these systems, say, of the FDM problem of level $J$. Thus, we set $V_J = \mathbb{R}^{N_J}$ and $\ell_J(u_J, v_J) = (L_J u_J, v_J)$, as before. The simple key to making a "qualified" guess for a suitable space

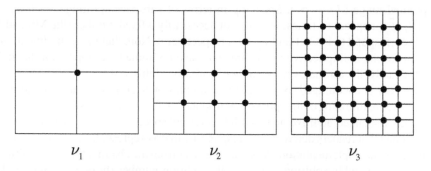

$$\nu_1 \qquad\qquad \nu_2 \qquad\qquad \nu_3$$

**Figure B.5.** Grids $\mathcal{V}_j$, $j \leq 3$, for FDM problems.

splitting is the following observation: the FDM matrices $L_j$ of level $j$ and the Galerkin stiffness matrix (B.2.4) of the bilinear form

$$\ell(u, v) = \int_\Omega \nabla u \cdot \nabla v \, dx \, dy$$

with respect to the finite element nodal basis $\Phi_j$ in $\tilde{V}_j$ coincide up to the forefactor $2^{2j}$ $(= h^{-2})$ in $L_j$. Just compute the few different values of $\ell(\varphi_j^n, \varphi_j^m)$ in (B.2.4) (only basis functions with nontrivial intersection of supports have to be considered). This is incidental, and does not generalize to more general domains, grids, differential operators or to the 3D case. Note, however, that *spectral equivalence* of the FDM problem of level $J$ with a corresponding FEM problem would be enough to derive useful results in essentially the same way as detailed below.

All notation is explained in Example B.2.6 or will be introduced below. Denote the transfer operator between finite element functions in $V(\mathcal{T}_j)$ and vectors in $V_j$ (grid functions on $\mathcal{V}_j$) by $\tilde{I}_j$, and set $\tilde{I}_j^J = \tilde{I}_J I_j^J$ for all $1 \le j \le J$. With the proper enumeration of the nodal basis functions in $\Phi_j$, the matrix representation of $\tilde{I}_j$ is the identity matrix, and we will omit this $\tilde{I}_j$ in future. Note that the prolongations $P_j = \tilde{I}_j^J$ have a simple meaning. Given any linear finite element function $\tilde{u}_j \in \tilde{V}_j$, the vector $P_j \tilde{u}_j$ represents the values of $\tilde{u}_j$ on the finest grid $\mathcal{V}_J$. In other words, $P_j$ corresponds to consecutive *linear interpolation* of grid functions from $\mathcal{V}_l$ onto $\mathcal{V}_{l+1}$ along the edges of the triangulation $\mathcal{T}_l, l = j, \ldots, J-1$. As a consequence of Theorem B.2.1 and (B.2.14), we have.

**Theorem B.4.1** *The following splittings are stable, with uniformly bounded stability constants and condition if $J \to \infty$ (compare (B.2.12)):*

$$\{V_J; \ell_J\} \cong \sum_{j=1}^{J} \tilde{I}_j^J \{\tilde{V}_j; \tilde{\ell}_j\} \cong \sum_{j=1}^{J} \sum_{i=1}^{N_j} \tilde{I}_j^J \{\tilde{V}_j^i; \tilde{\ell}_j^i\}, \tag{B.4.2}$$

*where $\tilde{\ell}_j(\tilde{u}_j, \tilde{v}_j) = 2^{2j} (\tilde{u}_j, \tilde{v}_j)_0$, and $\tilde{\ell}_j^i$ satisfies (B.2.15). The MS method associated with the second splitting represents a V-cycle multigrid method for solving the FDM discretization (B.4.1) of level $J$ while the AS method leads to a multilevel preconditioner. Both methods can be implemented with $O(N_J)$ operations per iteration and converge at rates $\le \rho < 1$, where $\rho$ does not depend on $J$.*

According to the material of Section B.3, the additive and multiplicative subspace correction methods based on the splittings in (B.4.2) should possess convergence rates

$$\rho_{J,\text{AS}}^* \le \rho_1^* < 1, \qquad \rho_{J,\text{MS}}^* \le \rho_2^* < 1.$$

Recall that the stronger estimate (B.3.9) of Theorem B.3.1 can be applied since strengthened Cauchy–Schwarz inequalities are available for the underlying finite element splitting. Provided that the relaxation parameter is well chosen, the iteration count to reach a given error reduction should therefore not grow with $J$ in any significant way. Alternatively,

PCG-methods such as AS-CG can be used, thereby avoiding the problem of choosing an appropriate $\omega$.

Let us derive the details of the algorithms using the second splitting of (B.4.2). We will show that the MS method (applied in reverse ordering) is indeed equivalent to a standard *V-cycle multigrid method*, with one Jacobi relaxation as the presmoothing step and no postsmoothing step. The AS method is simpler but still reveals the structure of a V-cycle. Recall that $\tilde{I}_j^J = \tilde{I}_J I_{J-1}^J \cdots I_j^{j+1}$ and that the matrix representations of the $\tilde{I}_j$ are identity matrices and can be omitted. The stencil notation of the finite element restriction operators $I_{j+1}^j = (I_j^{j+1})^T$ (with respect to the finite element nodal bases) is as follows:

$$I_{j+1}^j : \begin{bmatrix} & 1/2 & 1/2 \\ 1/2 & 1 & 1/2 \\ 1/2 & 1/2 & \end{bmatrix}.$$

These restrictions are intermediate to the FW and HW restriction operators discussed in Section 2.3.3. Finally, the scaling of the $\tilde{\ell}_j^i$ is fixed by setting

$$\tilde{\ell}_j^i(\varphi_j^i, \varphi_j^i) = (\varphi_j^i, \varphi_j^i)_1 = 4.$$

The inversion of $\tilde{L}_j^i$ on the one-dimensional $\tilde{V}_j^i$ corresponds to a scalar multiplication by $1/4$.

We start with the AS method. According to (B.1.3) and (B.3.1), it suffices to describe the matrix–vector multiplication for the preconditioner $B_J$ associated with the splitting. From (B.3.1) (compare also (B.2.20)) we conclude that

$$B_J = 2^{-2J-2} \sum_{j=1}^J I_{J-1}^J \cdots I_j^{j+1} (I_j^{j+1})^T \cdots (I_{J-1}^J)^T.$$

The factor $2^{-2J}$ comes from the forefactor $2^{2J}$ in the splitting while an additional $1/4$ comes from the inversion of the $\tilde{L}_j^i$ (see the above scaling for $\tilde{\ell}_j^i$). We will incorporate a factor $1/2$ into each $I_{J-1}^J$. Thus, we set

$$\hat{I}_j^{j+1} = 2^{-1} I_j^{j+1}, \qquad \hat{I}_j^J = \tilde{I}_J \hat{I}_{J-1}^J \cdots \hat{I}_j^{j+1}, \qquad \hat{L}_j = \text{diag}(L_j) = 2^{2j+2} I.$$
$$\tag{B.4.3}$$

The second splitting in (B.4.2) can be written in the equivalent form

$$\{V_J; \ell_J\} \cong \sum_{j=1}^J \hat{I}_j^J \{\tilde{V}_j; \hat{\ell}_j\}, \tag{B.4.4}$$

where $\hat{\ell}_j(\tilde{u}_j, \tilde{v}_j) = (\hat{L}_j \tilde{u}_j, \tilde{v}_j) = 2^{2j+2}(\tilde{u}_j, \tilde{v}_j)$ for all $\tilde{u}_j, \tilde{v}_j \in \tilde{V}_j$, $j = 1, \ldots, J$. As a result, we can simplify the formula for $B_J$ to

$$B_J = \sum_{j=1}^J \hat{I}_{J-1}^J \cdots \hat{I}_j^{j+1} \hat{L}_j^{-1} (\hat{I}_j^{j+1})^T \cdots (\hat{I}_{J-1}^J)^T,$$

which can be written in a recursive way:

$$B_1 = \hat{L}_1^{-1}, \qquad B_{j+1} = \hat{I}_j^{j+1} B_j (\hat{I}_j^{j+1})^T + \hat{L}_{j+1}^{-1}, \quad j = 1, \ldots, J-1. \qquad \text{(B.4.5)}$$

Before we can interpret (B.4.5) as a simplified multigrid V-cycle, we will look at the MS method for the second splitting in (B.4.2) or, what is the same, for (B.4.4). To obtain an efficient algorithm, reverse ordering is the correct choice (i.e. the inner loop of the MS iteration will start with the $J$th subproblem and end with the first one). Set

$$K_j = \omega \hat{I}_{J-1}^J \cdots \hat{I}_j^{j+1} \hat{L}_j^{-1} (\hat{I}_j^{j+1})^T \cdots (\hat{I}_{J-1}^J)^T, \quad j = 1, \ldots, J,$$

and show by induction that the matrices

$$L_j = (\hat{I}_j^{j+1})^T \cdots (\hat{I}_{J-1}^J)^T L_J \hat{I}_{J-1}^J \cdots \hat{I}_j^{j+1} \qquad \text{(B.4.6)}$$

indeed coincide with the FDM matrices $L_j$ of level $j = 1, \ldots, J$, respectively (for general $L_J$ based on five-point FDM discretizations, these matrices are usually different from the corresponding FDM discretization on the coarser grid $\mathcal{V}_j$ but still preserve the five-point stencil property [54, Section 7]). The induction step can be performed in stencil notation. With the stencils for

$$L_{j+1}: 2^{2j+2} \begin{bmatrix} & -1 & \\ -1 & 4 & -1 \\ & -1 & \end{bmatrix}, \qquad \hat{I}_{j+1}^j: \begin{bmatrix} & 1/4 & 1/4 \\ 1/4 & 1/8 & 1/4 \\ 1/4 & 1/4 & \end{bmatrix},$$

at hand, one computes

$$L_{j+1}\hat{I}_{j+1}^j: 2^{2j+2} \begin{bmatrix} & & -1/4 & -1/4 & \\ & -1/2 & 1/4 & 1/2 & -1/4 \\ -1/4 & 1/4 & 1 & 1/4 & -1/4 \\ -1/4 & 1/2 & 1/4 & -1/2 & \\ & -1/4 & -1/4 & & \end{bmatrix}$$

and

$$\hat{I}_j^{j+1} L_{j+1}\hat{I}_{j+1}^j: 2^{2j+2} \begin{bmatrix} & -1/4 & \\ -1/4 & 1 & -1/4 \\ & -1/4 & \end{bmatrix} = 2^{2j} \begin{bmatrix} & -1 & \\ -1 & 4 & -1 \\ & -1 & \end{bmatrix}.$$

The last stencil is with respect to $\mathcal{V}_j$ while the others are with respect to $\mathcal{V}_{j+1}$.

Using the above notation, the inner loop of one (reverse) MS step takes the form

$$z^j = z^{j+1} + K_j(f_J - L_J z^{j+1}), \quad j = J, \ldots, 1, \ (z^{J+1} = u^k, \ u^{k+1} = z^1).$$

Thus, the defect iteration of the MS method is given by

$$r^{k+1} = (I - L_J K_1)(I - L_J K_2) \cdots (I - L_J K_J) r^k. \qquad \text{(B.4.7)}$$

Let us look at the defect iteration of a V(1,0)-cycle using the $L_j$ as coarse grid stiffness matrices and $S_j = I - \omega \hat{L}_j^{-1} L_j$ as (pre-)smoothing. This is nothing but $\omega$-Jacobi relaxation used as a smoother on all levels. Denote the error propagation matrix for this V-cycle for the FDM problem of level $j$ by

$$M_j = I - C_j L_j, \quad j = 1, \ldots, J,$$

By Theorem 2.4.1 with $\nu_1 = 1$, $\nu_2 = 0$, and $\gamma = 1$, we have

$$M_{j+1} = (I - \hat{I}_j^{j+1} C_j (\hat{I}_j^{j+1})^T L_{j+1})(I - \omega \hat{L}_{j+1}^{-1} L_{j+1})$$
$$= I - (\omega \hat{L}_{j+1}^{-1} + \hat{I}_j^{j+1} C_j (\hat{I}_j^{j+1})^T (I - \omega L_{j+1} \hat{L}_{j+1}^{-1})) L_{j+1}.$$

To start, set formally $M_0 = 0$ or, directly, $M_1 = S_1 = I - \omega \hat{L}_1^{-1} L_1$. In our particular case, $\hat{L}_1 = L_1$. From this, a recurrence for $C_j$ can be derived:

$$C_{j+1} = \omega \hat{L}_{j+1}^{-1} + \hat{I}_j^{j+1} C_j (\hat{I}_j^{j+1})^T - \omega \hat{I}_j^{j+1} C_j (\hat{I}_j^{j+1})^T L_{j+1} \hat{L}_{j+1}^{-1},$$
$$j = 1, \ldots, J - 1, \tag{B.4.8}$$

where $C_0 = 0$ resp. $C_1 = \omega \hat{L}_1^{-1}$. In this relation, multiply by $\hat{I}_{J-1}^J \ldots \hat{I}_{j+1}^{j+2}$ from the left and by $(\hat{I}_{j+1}^{j+2})^T \ldots (\hat{I}_{J-1}^J)^T$ from the right, and recall the above expressions for $K_j$ and $L_j$:

$$\hat{K}_{j+1} \equiv \hat{I}_{J-1}^J \cdots \hat{I}_{j+1}^{j+2} C_{j+1} (\hat{I}_{j+1}^{j+2})^T \cdots (\hat{I}_{J-1}^J)^T = K_{j+1} + \hat{K}_j - \hat{K}_j L_J K_{j+1}.$$

Obviously, $\hat{K}_1 = K_1$ and $\hat{K}_J = C_J$. From this relation, we see that

$$I - L_J \hat{K}_{j+1} = (I - L_J \hat{K}_j)(I - L_J K_{j+1}), \quad j = 1, \ldots, J - 1,$$

which results in

$$I - L_J C_J = I - L_J \hat{K}_J = (I - L_J K_1)(I - L_J K_2) \cdots (I - L_J K_J).$$

We have shown that the defect iteration of the MS method (B.4.7) and of the above V(1,0) multigrid cycle are identical, which implies that the two iterations are also identical. More importantly, according to Theorem B.4.1 we have proved the optimality of this algorithm (that each iteration requires only $O(N_J) = O(2^{2J})$ operations was shown for general multigrid cycles). Thus, for any $0 < \omega < 2$, convergence is guaranteed and the convergence rate will be bounded away from 1, independently of $J$. The same holds true for the AS method (with small enough $\omega$) and the AS-CG algorithm. To a certain extent, we have obtained a strong result, since it guarantees optimality for the simplest multigrid V-cycle algorithm making the optimality of more advanced V-cycle and W-cycle algorithms highly probable (one could argue that the AS method is a yet simpler V-cycle method, see the following remarks).

The only, but important, difference between the AS and MS methods in the multigrid context can be seen from comparing the recursions for $B_j$ (B.4.5) and for $C_j$ (B.4.8). In a multiplicative algorithm, additional smoothing operations involving the coarse grid matrices $L_j$ on all levels are incorporated, whereas in the additive method the matrices $L_j$ ($j < J$) are not even required. The recursion for $C_j$ "degenerates" to the recursion for $B_j$ if we set $L_j = 0$, $j = 1, \ldots, J$. Thus, both algorithms can be implemented in essentially the same way. This observation is helpful if a multigrid method is used as preconditioner for $L_J$ in a Krylov space iteration. The reader is recommended to derive the details for the SMS method which leads (in contrast to the above MS method) to a symmetric multigrid preconditioner $C_j$.

The reader is encouraged to check the few changes that are necessary to adapt the above considerations to Example B.2.7. This example reveals one possibility of modifying the standard multigrid V-cycle to obtain a robust solution method for the linear problems that arise at each time step when parabolic problems such as the heat equation are solved by implicit schemes with variable time steps.

As can be concluded from the above derivation, the abstract theory of subspace correction methods covers only a certain part of multigrid theory. In particular, the *coarse grid matrices $L_j$ have to satisfy* (B.4.6), i.e. they are defined from $L_J$ by *Galerkin projection* and depend on the set of prolongation/restriction operators. If we change the above interpolation scheme inherited from the natural embeddings of the linear finite element spaces to FW (bilinear interpolation) than the associated Galerkin coarse grid matrices (B.4.6) would be defined by compact nine-point stencils, and depend on the difference $J - j$. The matrices $\tilde{L}_j$ resp. the bilinear forms $\tilde{\ell}_j$ which describe the auxiliary problems on the spaces $\tilde{V}_j$ of the splitting are essentially responsible for the smoothers. Here, we have restricted our attention to *symmetric positive definite smoothers* and, of course, to *symmetric positive definite problems* (B.1.1) from the very beginning. Extensions to cover a broader spectrum of multigrid applications are discussed in [54], see also [180, 425, 435]. For treatments which emphasize *multilevel preconditioning*, (i.e. the AS method in a multigrid context) in connection with finite element and wavelet space decompositions for operator equations, see [116, 117, 298].

## B.5 A DOMAIN DECOMPOSITION EXAMPLE

We will sketch some of the basic algorithmic ideas and the convergence theory, again using the Poisson equation (B.2.1) on the unit square $\Omega$ discretized by a five-point FDM scheme or, equivalently, by linear finite elements. For simplicity, we fix a grid $V_J$ of dyadic stepsize $h = 2^{-J}$ as our computational grid $V$ and, correspondingly, $T = T_J$ as the triangulation of the finite element space. Consider the linear system (B.1.1), where $L = L_J$ is the FDM matrix of level $J$.

The basic idea of a domain decomposition method is illustrated in Fig. B.6, where (a) shows a decomposition into four *nonoverlapping domains* and an *interface* $\Gamma$ while (b) shows a decomposition into two *overlapping domains*. On each of the domains, local problems are defined, e.g. by restricting the partial differential equation to the subdomain and complementing it by some boundary conditions. Solving (in parallel) the local problems and gluing them together leads to an approximation of the global problem on $\Omega$. Obviously, this procedure defines a preconditioner (i.e. an approximative inverse) for $L$, and represents one

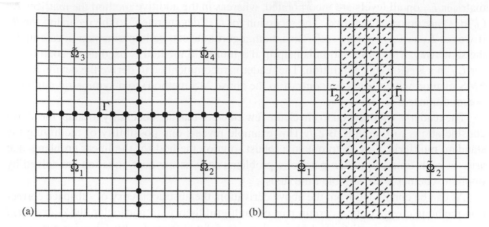

**Figure B.6.** (a) Nonoverlapping; and (b) overlapping domain decompositions.

step of an iterative domain decomposition method. Since it is based on defining subproblems, it should fit into the framework of subspace correction methods and allow for the same modifications as the abstract methods (e.g. CG-accelerations and multiplicative versions are possible).

The reader can imagine that in realistic applications much more general subdomain patterns than shown in Fig. B.6 can arise, and that the design of suitable decompositions is subject to many side conditions (e.g. the physical nature of the underlying problem, load balancing, and minimization of communication are typical issues). Decompositions into *strips*, where any grid point belongs to at most two subdomains, are somewhat easier to handle, and reduce essentially to the situation of two subdomains (such as shown for the overlapping case in Fig. B.6(b)). Interior vertices, as in Fig. B.6(a), where more than two subdomains touch each other, cause theoretical and practical problems. For both basic versions, subdomains are denoted by $\tilde{\Omega}_m$, $m = 1, \ldots, M$. We introduce the subgrids $\tilde{V}_m$ as the part of $\mathcal{V}$ interior to $\tilde{\Omega}_m$, analogously, $\tilde{T}_m$ denotes the restriction of $T$ to $\tilde{\Omega}_m$. The sets of all grid functions on $\mathcal{V}$ and $\tilde{V}_m$ (or, equivalently, linear finite element functions on $T$ and $\tilde{V}_m$) will be denoted by $V$ and $\tilde{V}_m$, respectively. To avoid confusion with the notation used in the previous subsection, we will not make any notational difference between spaces, matrices, and operators for grid functions and finite element functions of different levels $j = 1, \ldots, J$, assuming that the reader is aware of the identification process and the formal differences. In particular, we will consistently use $V_j$, $V_j^i$, $\ell_j$, $\ell_j^i$ $\tilde{V}_j$, $\tilde{V}_j^i$, $\tilde{\ell}_j$, $\tilde{\ell}_j^i$ for the spaces and bilinear forms defined above. The same applies to the prolongations $\tilde{I}_j^J$.

As auxiliary problems on $\tilde{V}_m$ we will consider five-point FDM discretizations of the same Poisson problem (B.2.1) with respect to the domains $\tilde{\Omega}_m$ instead of $\Omega$. In particular, homogeneous Dirichlet boundary conditions are assumed along $\partial\tilde{\Omega}_m$ (there are a lot of variations such as imposing Neumann or Robin boundary conditions which have been used successfully [102, 362] but we will not discuss them here). Thus, $\tilde{L}_m$ is the submatrix of $L$

associated with the grid points in $\mathcal{V}_m$, the associated bilinear form will be denoted by $\tilde{\ell}_m$. In the nonoverlapping case, where

$$\mathcal{V}_\Gamma \equiv \mathcal{V} \Big\backslash \bigcup_{m=1}^{M} \tilde{\mathcal{V}}_m \neq \emptyset,$$

we also need to create an auxiliary problem for the unknowns associated with the interface $\Gamma$. This so-called *interface problem* should approximate the *Schur complement matrix*

$$S_\Gamma = L_\Gamma - \sum_{m=1}^{M} L_{m,\Gamma}^T \tilde{L}_m^{-1} L_{m,\Gamma} \tag{B.5.1}$$

which represents the stiffness matrix for the reduced problem with respect to $V_\Gamma$, the set of grid functions on $\mathcal{V}_\Gamma$ (the finite element counterpart of $V_\Gamma$ is the trace space of $V$ onto the interface which consists of linear spline functions interpolating the grid functions defined on $\mathcal{V}_\Gamma$). In (B.5.1), the notation comes from rewriting the linear system $Lx = f$ in a block form related to the subgrids $\tilde{\mathcal{V}}_m$:

$$\begin{pmatrix} \tilde{L}_1 & \cdots & 0 & L_{1,\Gamma} \\ \vdots & \ddots & \vdots & \vdots \\ 0 & \cdots & \tilde{L}_M & L_{M,\Gamma} \\ L_{1,\Gamma}^T & \cdots & L_{M,\Gamma}^T & L_\Gamma \end{pmatrix} \begin{pmatrix} \tilde{x}_1 \\ \vdots \\ \tilde{x}_M \\ x_\Gamma \end{pmatrix} = \begin{pmatrix} \tilde{f}_1 \\ \vdots \\ \tilde{f}_M \\ f_\Gamma \end{pmatrix}.$$

Thus, $\tilde{x}_m = \tilde{L}_m^{-1}(\tilde{f}_m - L_{m,\Gamma} x_\Gamma), m = 1, \ldots, M,$ and

$$S_\Gamma x_\Gamma = f_\Gamma - \sum_{m=1}^{M} L_{m,\Gamma}^T \tilde{L}_m^{-1} \tilde{f}_m$$

represents the reduced problem for determining the grid values $x_\Gamma$ on $\mathcal{V}_\Gamma$. The solution of (B.1.1) can formally be written as

$$x_\Gamma = S_\Gamma^{-1} \left( f_\Gamma - \sum_{m=1}^{M} L_{m,\Gamma}^T \tilde{L}_m^{-1} \tilde{f}_m \right), \quad \tilde{x}_m = \tilde{L}_m^{-1}(\tilde{f}_m - L_{m,\Gamma} x_\Gamma), \quad m = 1, \ldots, M.$$

Since $S_\Gamma$ represents a dense matrix, the explicit computation and storage of which should be avoided, we look for an approximate substitute $\tilde{S}_\Gamma$ the inverse of which is easy to compute, i.e. we look for a symmetric positive definite preconditioner $B_\Gamma = \tilde{S}_\Gamma^{-1} \approx S_\Gamma^{-1}$. We introduce the associated bilinear forms by

$$s_\Gamma(x_\Gamma, y_\Gamma) = (S_\Gamma x_\Gamma, y_\Gamma), \qquad \tilde{s}_\Gamma(x_\Gamma, y_\Gamma) = (\tilde{S}_\Gamma x_\Gamma, y_\Gamma).$$

As the above formulas reveal, the extension $P_\Gamma$ of grid functions on $\mathcal{V}_\Gamma$ to $V$ needs special attention.

We will briefly discuss choices for the components and the stability question of the resulting splitting

$$\{V; \ell\} \cong \sum_{m=1}^{M} P_m\{\tilde{V}_m; \tilde{\ell}_m\} + P_\Gamma\{V_\Gamma; \tilde{s}_\Gamma\}, \tag{B.5.2}$$

where the rectangular matrices $P_m$ correspond to the extension-by-zero of grid functions on $\tilde{V}_m$ to $V$ (consequently, $P_m^T$ represents the natural restriction of grid functions on $V$ to $\tilde{V}_m$). For interpretation of later results, note that the choices

$$\tilde{s}_\Gamma = S_\Gamma, \qquad P_\Gamma^T = (-L_{1,\Gamma}^T \tilde{L}_1^{-1}, \ldots, -L_{M,\Gamma}^T \tilde{L}_M^{-1}, I_\Gamma), \tag{B.5.3}$$

would give rise to a *tight splitting* in (B.5.2), with $\eta = \bar{\eta} = \kappa = 1$ ($I_\Gamma$ is the identity matrix in the subspace $V_\Gamma$). This fact is expressed by the identity

$$(Lu, u) = \sum_{m=1}^{M} (\tilde{L}_m \tilde{u}_m, \tilde{u}_m) + (S_\Gamma u_\Gamma, u_\Gamma), \qquad u = \sum_{m=1}^{M} P_m \tilde{u}_m + P_\Gamma u_\Gamma. \tag{B.5.4}$$

Clearly,

$$L^{-1} = \sum_{m=1}^{M} P_m \tilde{L}_m^{-1} P_m^T + P_\Gamma S_\Gamma^{-1} P_\Gamma^T. \tag{B.5.5}$$

In the case of overlapping domain decompositions, the introduction of a special interface problem can be avoided, and one directly looks at

$$\{V; \ell\} \cong \sum_{m=1}^{M} P_m\{\tilde{V}_m; \tilde{\ell}_m\}. \tag{B.5.6}$$

As we will see, in both cases the results may depend on the number of domains $M$. For obtaining $M$-independent convergence results, a so-called *coarse grid problem* has to be included into the definition of $B_\Gamma$ resp. into the splitting (B.5.6). Another question is the systematic replacement of $\tilde{L}_m^{-1}$ by *inexact solves*, both for the solution of the subproblems associated with the subdomains $\tilde{\Omega}_m$, and in the application of $P_\Gamma$, see (B.5.3). This becomes particularly important if the dimension of the subproblems $\tilde{N}_m = \dim \tilde{V}_m$ is large, and makes the use of direct solvers prohibitive.

We will now derive a realization of the above concepts by applying Theorem B.2.1 resp. Theorem B.4.1 following essentially [296, 298]. Although the assumptions of this derivation are a little restrictive, the results are typical and can be used as a guideline in other, more realistic situations. In addition, since the only thing we will do is to regroup the one-dimensional spaces $V_j^i$ forming the multigrid splittings of Theorem B.4.1 with respect to the subdomains $\Omega_m$ and the interface $\Gamma$, the resulting domain decomposition algorithms could be viewed as a specific way to parallelize a multigrid method. This provides another link between the basic theme of this monograph and domain decomposition methods.

Fix some integer $J^* = 1, \ldots, J - 1$, and let the domains $\tilde{\Omega}_m, m = 1, \ldots, 2^{2J^*}$, form a uniform partition of the unit square $\Omega$ into squares of side length $H = 2^{-J^*}$. Fig. B.6(a) corresponds to the case $J^* = 1$. The *interface* $\Gamma$ consists of the horizontal and vertical grid lines associated with $\mathcal{V}_{J^*}$. To start with, let us assume that the linear systems with the coefficient matrices $\tilde{L}_m$ can be solved by a *direct method*, i.e. we assume that $\tilde{L}_m^{-1}$ is available (e.g. in the form of a $LU$-factorization). This means, that of all components in the representation (B.5.5) only $S_\Gamma^{-1}$ needs an easy replacement (in other words, we look for a preconditioner for $S_\Gamma$). We will provide this preconditioner by regrouping the components of the multilevel splittings mentioned above. From the definition of $S_\Gamma$, we have

$$(S_\Gamma u_\Gamma, u_\Gamma) = \inf_{u \,:\, u_\Gamma = u|_\Gamma} \ell(u, u),$$

we leave this as an exercise to the reader. From Theorem B.4.1 we have

$$\ell(u, u) \approx \inf_{u = I_{J^*}^J u_{J^*} + \sum_{j=J^*+1}^J I_j^J \sum_i u_j^i} \ell_{J^*}(u_{J^*}, u_{J^*})$$

$$+ \sum_{j=J^*+1}^J \sum_i \ell_j^i(u_j^i, u_j^i), \quad u \in V. \tag{B.5.7}$$

To prove (B.5.7), use the stability estimate for the first splitting in (B.4.2) with $J$ replaced by $J^*$ to substitute back $\ell_{J^*}(u_{J^*}, u_{J^*})$ for the components with $j < J^*$ in the second splitting of (B.4.2). Together this gives

$$(S_\Gamma u_\Gamma, u_\Gamma) \approx \inf_{u_\Gamma = (I_{J^*}^J u_{J^*} + \sum_{j=J^*+1}^J \sum_i I_j^J u_j^i)|_\Gamma} \ell_{J^*}(u_{J^*}, u_{J^*}) + \sum_{j=J^*+1}^J \sum_i \ell_j^i(u_j^i, u_j^i).$$

Since $\ell_j^i(u_j^i, u_j^i) \geq 0$, the infimum will not change if we omit all those terms for which $\varphi_j^i|_\Gamma = 0$ (for $j > J^*$ this is equivalent to $\text{supp}\,\varphi_j^i \subset \tilde{\Omega}_m$ for some $m$). For convenience, for each $j = J^*, \ldots, J$ we denote by $\hat{V}_j \subset V_j$ the set of all

$$\hat{u}_j = \sum_{i:\varphi_j^i|_\Gamma \neq 0} u_j^i \equiv \sum_{i:\varphi_j^i|_\Gamma \neq 0} c_j^i \varphi_j^i.$$

Note that $\hat{V}_{J^*} = V_{J^*}$. Obviously, any such $\hat{u}_j$ is uniquely determined by its values on $\Gamma$ (more precisely, by the grid values $c_j^i$ at the points in $\mathcal{V}_j^\Gamma = \mathcal{V}_j \cap \Gamma$), and can be recovered from its trace $\hat{u}_j|_\Gamma$ by the discrete *extension-by-zero operator* $E_j : V_j^\Gamma \equiv V_j|_\Gamma \to \hat{V}_j \subset V_j$ *of level $j$* defined by

$$E_j u_j^\Gamma = \begin{cases} u_j^\Gamma & \text{on } \mathcal{V}_j^\Gamma \\ 0 & \text{on } \mathcal{V}_j \backslash \mathcal{V}_j^\Gamma \end{cases}, \quad u_j^\Gamma \in V_j^\Gamma.$$

As before, in all these definitions we identify grid functions on $\mathcal{V}_j$ and $\mathcal{V}_j^\Gamma$ with the corresponding linear finite element functions on $\Omega$ and $\Gamma$, respectively. Observe finally that

$$2^{2j} \|\hat{u}_j\|_0^2 \approx \sum_{i:\varphi_j^i|_\Gamma \neq 0} (c_j^i)^2 \approx 2^j \|\hat{u}_j|_\Gamma\|_{0,\Gamma}^2, \quad \hat{u}_j \in \hat{V}_j \tag{B.5.8}$$

(the notation $(\cdot, \cdot)_{0,\Gamma}$ resp. $\|\cdot\|_{0,\Gamma}$ stands for the scalar product resp. the norm in $L_2(\Gamma)$). Taking all this into consideration, we can continue with

$$
\begin{aligned}
(S_\Gamma u_\Gamma, u_\Gamma) &\approx \inf_{u_\Gamma = (I^J_{J*} u_{J*} + \sum^J_{j=J*+1} \sum_i I^J_j u^i_j)|_\Gamma} \ell_{J*}(u_{J*}, u_{J*}) + \sum^J_{j=J*+1} \sum_{i:\varphi^i_j|_\Gamma \neq 0} \ell^i_j(u^i_j, u^i_j) \\
&\approx \inf_{u_\Gamma = (I^J_{J*} u_{J*} + \sum^J_{j=J*+1} I^J_j \hat{u}_j)|_\Gamma} \ell_{J*}(u_{J*}, u_{J*}) + \sum^J_{j=J*+1} 2^{2j} \|\hat{u}_j\|^2_0 \\
&\approx \inf_{u_\Gamma = u_{J*}|_\Gamma + \sum^J_{j=J*+1} u^\Gamma_j} \ell_{J*}(u_{J*}, u_{J*}) + \sum^J_{j=J*+1} 2^j \|u^\Gamma_j\|^2_0.
\end{aligned}
$$

The constants in the above two-sided inequalities are independent of $J^*$ and $J$.

The last relationship simply represents the stability assertion of a splitting for the Schur complement problem $\{V_\Gamma; s_\Gamma\}$ with respect to the hierarchy of spaces $V^\Gamma_{J*} \subset \cdots \subset V^\Gamma_J = V_\Gamma$. To follow the mathematical formalities, introduce $\ell^\Gamma_j(u^\Gamma_j, v^\Gamma_j) = 2^j(u^\Gamma_j, v^\Gamma_j)_{0,\Gamma}$ as the auxiliary scalar products on $V^\Gamma_j$, $j = J^* + 1, \ldots, J$, and denote the natural restriction of $I^J_j$ to the interface $\Gamma$ by $I^{J,\Gamma}_j$. Formally, we can write $I^{J,\Gamma}_j = \hat{I}^{J,\Gamma}_j E_j : V^\Gamma_j \to V^\Gamma$, where $\hat{I}^{J,\Gamma}_j = |_\Gamma \circ I^J_j$. This proves the following theorem.

**Theorem B.5.1**  *Under the above restrictions on $\{\Omega_m\}$, the space splitting*

$$
\{V_\Gamma; s_\Gamma\} \cong \hat{I}^{J,\Gamma}_{J*} \{V_{J*}; \ell_{J*}\} + \sum^J_{j=J*+1} I^{J,\Gamma}_j \{V^\Gamma_j; \ell^\Gamma_j\} \tag{B.5.9}
$$

*for the Schur complement problem governed by $S_\Gamma$ is stable, with stability constants and condition that remain bounded, independently of $J^*$ and $J$.*

It is straightforward to realize that the resulting AS and MS methods based on (B.5.9) represent modified multigrid V-cycles for the levels $J^*, \ldots, J$ if the bilinear forms $\ell^\Gamma_j(\cdot, \cdot)$ are discretized using the $L_2(\Gamma)$-stability of the basis $\{\varphi^i_j|_\Gamma\}$ expressed by the second relation in (B.5.8). The first modification in comparison with the V-cycles of Section B.5 consists in the *coarse grid problem* associated with $\{V_{J*}; \ell_{J*}\}$ which requires the solution of a FDM discretization of level $J^*$. The second difference is that the prolongation/restriction operations are now performed only with respect to the values on $\Gamma$. Therefore, the operation count of the preconditioning step (without multiplication by $S_\Gamma$ and costs for solving the coarse grid problem) will be proportional to the number of unknowns on $\Gamma$ which is $\approx 2^{J+J^*}$.

The coarse grid problem which arose naturally in the above derivation from the components with $j \leq J^*$ of the multilevel splitting (B.4.2) represents a bottleneck in the parallelization of a domain decomposition code. Historically, the first algorithms that used decompositions with many domains did not include a coarse grid problem, at the cost of reduced convergence rates. In our derivation, the *no-coarse-grid-problem* case can be mimicked as follows: instead of starting from (B.5.7), we could have dropped all components with $j < J^*$ in (B.4.2), and considered the reduced splitting

$$\{V; \ell\} \cong \sum_{j=J^*}^{J} I_j^J \{V_j; \ell_j\} \cong \sum_{j=J^*}^{J} \sum_i I_j^J \{V_j^i; \ell_j^i\} \tag{B.5.10}$$

as the starting point. This modification leads to a deterioration of the upper stability constant from $\approx 1$ for the splittings in (B.4.2) to $\approx 2^{2J^*}$ for (B.5.10). Indeed, going back to the finite element interpretation, for any $u \in V = V_J$, by definition of the triple bar norm for (B.4.2), there are $v_j \in V_j$ such that

$$u = \sum_{j=1}^{J} v_j, \qquad \sum_{j=1}^{J} 2^{2j} \|v_j\|_0^2 \le C\ell(u, u).$$

To simplify notation, we have dropped the natural embeddings $I_j^J$. Setting $u_{J^*} = \sum_{j=1}^{J^*} v_j$, we have by an application of the Cauchy–Schwarz inequality

$$\|u_{J^*}\|_0^2 \le \left( \sum_{j=1}^{J^*} 2^{-j}(2^j \|v_j\|_0) \right)^2 \le \sum_{j=1}^{J^*} 2^{2j} \|v_j\|_0^2,$$

which results in

$$u = u_{J^*} + \sum_{j=J^*+1}^{J} v_j,$$

$$2^{2J^*} \|u_{J^*}\|_0^2 + \sum_{j=J^*+1}^{J} 2^{2j} \|v_j\|_0^2 \le 2^{2J^*} \sum_{j=1}^{J} 2^{2j} \|v_j\|_0^2 \le C 2^{2J^*} \ell(u, u).$$

Thus, the deterioration is no more than by a factor $\approx 2^{2J^*}$. To see that this factor can be attained, consider a function from $V_1$ such as $u = \varphi_1^1$ which has norms $\ell(u, u) \approx \|u\|_0^2 \approx 1$ but is not well represented with respect to the functions $v_j$, $j \ge J^*$, allowed in the reduced splittings. The lower bound will remain $\approx 1$. The reader will easily verify these facts. If we now proceed as before, we will arrive at a splitting of the form

$$\{V_\Gamma; s_\Gamma\} \cong \sum_{j=J^*}^{J} I_j^{J,\Gamma} \{V_j^\Gamma; \ell_j^\Gamma\}. \tag{B.5.11}$$

This splitting does not involve a coarse grid problem, in exchange it inherits the worse condition number $\kappa \approx 2^{2J^*} = H^{-2}$ from (B.5.10).

If $J^*$ and $J$ increase, the dimension of the interface problem may become fairly large. For this (and other) reasons, many attempts have been made to further enhance parallelization. A very popular idea is to extend the domain decomposition principle to the interface problem, and to decompose $\Gamma$ into "subdomains" of its own. The first thing which comes to mind is a decomposition

$$\Gamma = \sum_{m,n} \Gamma_{m,n}, \qquad \Gamma_{m,n} = \partial\Omega_m \cap \partial\Omega_n.$$

where the summation extends over all $m, n$ for which $\Gamma_{m,n} \neq \emptyset$. In our example, the $\Gamma_{m,n}$ are edges associated with the grid $\mathcal{V}_{J*}$ which leads to the name *edge spaces* for the sets of grid functions $V_{m,n}^\Gamma = V_\Gamma|_{\Gamma_{m,n}}$ to be introduced as additional auxiliary spaces. The appealing part of this choice is that potential subproblems associated with these local interfaces are truly one-dimensional and all similar to each other. Problems should be expected at the interior vertices of the domain decomposition which has triggered the introduction of additional *vertex spaces*. The reader who has followed our considerations to this point will be able to introduce local problems on the respective $\Gamma$-components by further regrouping the subspaces $V_j^i$ associated with $\Gamma$ appearing in the above derivation of Theorem B.5.1. This will lead to potentially better parallelizable $S_\Gamma$-preconditioners (compare [362, p. 140]). See [102] for a more comprehensive and systematic discussion of the interface problems arising in connection with nonoverlapping domain decompositions, and [362] for numerical support. We have left out many other aspects such as the definition of infinite-dimensional trace spaces, the construction of approximate harmonic extension operators (replacements for $P_\Gamma$), and the connection with boundary integral equations and boundary element methods.

As mentioned before, it is often prohibitive to solve the subproblems $\tilde{L}_m \tilde{u}_m = \tilde{f}_m$, $m = 1, \ldots, M$, by a direct method (or by an iterative method within machine accuracy). Instead, one would like to replace the action of $\tilde{L}_m^{-1}$ by a simpler preconditioner and use *inexact solves*. However, this is by no means a trivial task since the $\tilde{L}_m^{-1}$ enter both $S_\Gamma$ and $P_\Gamma$ in a complicated way (see [102, Section 5] and [362, Section 4.4]). Some specific proposals, however, can easily be found if one reviews our derivation for Theorem B.5.1 carefully. Let us begin with an rearrangement of the splitting associated with (B.5.7):

$$\{V; \ell\} \cong I_{J*}^J\{V_{J*}; \ell_{J*}\} + \sum_{m=1}^{2^{2J*}} \left( \sum_{j=J*+1}^J \sum_{i \,:\, \mathrm{supp}\,\varphi_j^i \subset \tilde{\Omega}_m} I_j^J\{V_j^i, \ell_j^i\} \right)$$

$$+ \sum_{j=J*+1}^J \sum_{i \,:\, \varphi_j^i|_\Gamma \neq 0} I_j^J\{V_j^i; \ell_j^i\}$$

$$\cong I_{J*}^J\{V_{J*}; \ell_{J*}\} + \sum_{m=1}^{2^{2J*}} \left( \sum_{j=J*+1}^J \sum_{i \,:\, \mathrm{supp}\,\varphi_j^i \subset \tilde{\Omega}_m} I_j^J\{V_j^i, \ell_j^i\} \right)$$

$$+ \sum_{j=J*+1}^J I_j^J E_j\{V_j^\Gamma; \ell_j^\Gamma\}.$$

The stability constants of these splittings are uniformly bounded, independently of $J*$ and $J$. The replacement of the last group of components is admissible due to the properties of the extension operators $E_j$ as discussed above. This last group (considered together with the coarse grid problem) is the exact counterpart of the splitting (B.5.9). The groups of components associated with the subdomains $\tilde{\Omega}_m$ represent a replacement of $\{\tilde{V}_m; \tilde{\ell}_m\}$ by a local multigrid splitting. If the AS method associated with the above splitting are considered then this results in a replacement of $\tilde{L}_m^{-1}$ by the corresponding local multilevel preconditioner based on an application of Theorem B.4.1 on $\tilde{\Omega}_m$. Analogously, $P_\Gamma S_\Gamma^{-1} P_\Gamma^T$ is replaced by some multilevel preconditioner associated with the values on $\Gamma$ which is similar in structure to the above preconditioner for $S_\Gamma$ but also involves the extension operators $E_j$ and their transposes $E_j^T$. As a result, the exact solution of subproblems with $\tilde{L}_m$, i.e. the multiplication by $\tilde{L}_m^{-1}$, is avoided by replacing it with one iteration step of a multilevel preconditioned iterative method for the subproblem on $\tilde{\Omega}$. The reader is encouraged to work out the details.

After this discussion of the nonoverlapping case, we will present the standard result for the *overlapping case* in an analogous setting. In addition to $1 \leq J^* < J$, let us fix another integer $\hat{J}$ such that $J^* \leq \hat{J} \leq J$. Set $\delta = 2^{-\hat{J}}$, and define the $\tilde{\Omega}_m$, $m = 1, \ldots, 2^{2J^*}$, by extending the dyadic squares of side length $H = 2^{-J^*}$ used above by a corridor of width $\delta$ in both coordinate directions in the interior of $\Omega$. Thus, any two neighboring $\tilde{\Omega}_m$ overlap in a small strip of width $2\delta$. All other specifications are the same as in the nonoverlapping case.

**Theorem B.5.2**  *For the overlapping decomposition $\{\Omega_m\}$ just defined, the stability constants and condition of the space splitting*

$$\{V; \ell\} \cong I_{J^*}^J \{V_{J^*}; \ell_{J^*}\} + \sum_{m=1}^{2^{2J^*}} P_m \{\tilde{V}_m; \tilde{\ell}_m\} \tag{B.5.12}$$

*behave like*

$$0 < c \leq \eta \leq \bar{\eta} \leq C 2^{\hat{J} - J^*} = C \frac{H}{\delta}, \qquad \kappa \approx 2^{\hat{J} - J^*} = \frac{H}{\delta}. \tag{B.5.13}$$

*The constants in these estimates are independent of $J^*$, $\hat{J}$, and $J$.*

Before we sketch the proof of Theorem B.5.2, we will comment on its practical consequences. From (B.5.13) we see that only *sufficient overlap*, i.e. when the overlap parameter $\delta$ becomes proportional to $H$, and the inclusion of the coarse grid problem lead to the optimal preconditioning effect ($\kappa = O(1)$). Clearly, this means more work per local problem (if $\delta = H$ then a local problem is up to nine times larger, and the solution of all subproblems would take at least tenfold the time needed for the subproblems associated with a comparable nonoverlapping domain partition). However, as a practical observation, already a small overlap $\delta \approx 2h \ldots 4h$ often leads to reasonably good convergence rates, at little extra cost. For the splitting (B.5.6) which does not contain a coarse grid problem, the condition number may further increase, at most by a factor $\approx H^{-2}$. In an overlapping environment, the replacement of the direct solves (involving $\tilde{L}_m^{-1}$) by inexact solves on the subdomains is not an obstacle. Any spectral equivalent replacement $B_m \approx \hat{L}_m^{-1}$ would suffice. A drawback is the increased amount of data communication between neighboring subdomains.

To avoid unnecessary technicalities, let us sketch the argument for the finite element version of Theorem B.5.2. We will again omit the mappings $I_j^J$. The proof of (B.5.13) relies on two essential observations. First,

$$\{\tilde{V}_m; \tilde{\ell}_m\} \cong \sum_{j=J^*}^{J} \sum_{i \, : \, \mathrm{supp}\, \varphi_j^i \subset \tilde{\Omega}_m} \{V_j^i; \ell_j^i\} \tag{B.5.14}$$

is stable with $0 < c \le \eta \le \bar{\eta} \le C < \infty$ with $c, C$ independent of all parameters. For the domains $\tilde{\Omega}_m$ under consideration, this is a rather standard consequence of the basic results of Theorem B.4.1 which gives the same result for the domain $\Omega$. The reduction is by observing that (B.5.14) can be viewed as the *trivial localization* of the splitting

$$\{V_J; \ell_J\} \cong \sum_{j=1}^{J} \sum_i \{V_j^i; \ell_j^i\} \tag{B.5.15}$$

to the subdomain $\tilde{\Omega}_m$, where trivial means that all components of the splitting with support at least partially outside $\tilde{\Omega}_m$ are omitted. One should be aware that trivial localization of multilevel splittings to a general subdomain may lead to very poorly conditioned splittings (the above subdomains are among the "nice" ones in this respect). Since, by the same Theorem B.4.1,

$$\{V_{J^*}; \ell_{J^*}\} \cong \sum_{j=1}^{J^*} \sum_i \{V_j^i; \ell_j^i\} \tag{B.5.16}$$

with uniform bounds for the stability constants, we can substitute these splittings for the components of the splitting (B.5.12). This results in the splitting

$$\{V; \ell\} \cong \sum_{j=1}^{J^*} \sum_i \{V_j^i; \ell_j^i\} + \sum_{j=J^*}^{J} \sum_{i \, : \, \mathrm{supp}\, \varphi_j^i \subset \tilde{\Omega}_m} \{V_j^i; \ell_j^i\}, \tag{B.5.17}$$

which should have essentially the same stability constants and condition number as (B.5.12). These simple manipulations with stable splittings have been introduced and analyzed in [298, p. 82–83] under the names *refinement* and *clustering* of stable splittings.

Thus, it suffices to find estimates for the stability constants of (B.5.17). This can be done by comparing the triple bar norms of the splittings (B.5.17) and (B.5.15) with each other. Let us denote them by $\||u\||_{\mathrm{mod}}$ and $\||u\||$, respectively. Analogous notation is introduced for the stability constants. The differences between the two splittings are as follows. On the one hand, some of the components $\{V_j^i; \ell_j^i\}$ occur several times (but no more than five times) in (B.5.17). On the other, (B.5.17) represents a *subsplitting* of (B.5.15), i.e. all components in (B.5.17) are also contained in (B.5.15), the latter splitting contains some more components for $J^* < j < \hat{J}$ associated with the interface $\Gamma$ which is defined as in the nonoverlapping case. This immediately gives

$$5\||u\||_{\text{mod}}^2 \geq \||u\||^2 \geq \eta \ell(u, u) \implies \eta_{\text{mod}} \geq \frac{\eta}{5}.$$

However, in the other direction, we can only prove

$$\||u\||_{\text{mod}}^2 \leq C2^{\hat{J}-J^*}\||u\||^2 \leq C2^{\hat{J}-J^*}\bar{\eta}\ell(u, u).$$

Although this is technically involved, we will try to convey the idea. Take any close-to-optimal decomposition of $u \in V$ with respect to the splitting (B.5.15),

$$u = \sum_{j=1}^{J} v_j \equiv \sum_{j=1}^{J}\sum_i c_j^i \varphi_j^i : \quad \sum_{j=1}^{J}\sum_i 2^{2j}\|v_j\|_0^2 \leq C\||u\||^2,$$

and modify it such that it matches the decompositions admissible in the splitting (B.5.17). The only problematic terms are those for which $\varphi_j^i|_\Gamma \neq 0$ and $J* < j < \hat{J}$ (there is nothing to prove in the cases of sufficient overlap $\hat{J} = J^*$ or $\hat{J} = J^* + 1$). Summing all these terms with the same $j$, we define functions $\hat{v}_j \in \hat{V}_j$, $J^* < j < \hat{J}$, associated with $\Gamma$ (see the definition before (B.5.8)). Obviously,

$$\|\hat{v}_j\|_0^2, \|v_j - \hat{v}_j\|_0^2 \leq C\|v_j\|_0^2.$$

Setting $\hat{u}_{J*+1} = 0$, we will recursively define

$$\hat{w}_j = \hat{v}_j + \hat{u}_j, \quad \hat{u}_{j+1} = E_{j+1}(\hat{w}_j)|_\Gamma, \quad u_{j+1} = \hat{w}_j - \hat{u}_{j+1},$$
$$j = J^* + 1, \dots, \hat{J} - 1.$$

Note that the functions $u_j \in V_j$ as well as $\hat{u}_{\hat{J}} \in V_{\hat{J}}$ are linear combinations of terms admissible in (B.5.17), and that $\hat{v}_{J*+1} + \cdots + \hat{v}_{\hat{J}-1} = u_{J^*+2} + \cdots + u_{\hat{J}} + \hat{u}_{\hat{J}}$. Thus,

$$u = \sum_{j=1}^{J^*} v_j + \sum_{j=J^*}^{\hat{J}-1}(v_j - \hat{v}_j + u_j) + (v_{\hat{J}} + u_{\hat{J}} + \hat{u}_{\hat{J}}) + \sum_{j=\hat{J}}^{J} v_j \equiv \sum_{j=1}^{J} w_j$$

is an admissible decomposition in the definition of the triple bar norm associated with (B.5.17) which yields

$$\||u\||_{\text{mod}}^2 \leq \sum_{j=1}^{J} 2^{2j}\|w_j\|_0^2 \leq C \sum_{j=1}^{J} 2^{2j}\|v_j\|_0^2 + \sum_{j=J^*+1}^{\hat{J}} 2^{2j}\|u_j\|_0^2 + 2^{2\hat{J}}\|\hat{u}_{\hat{J}}\|_0^2.$$

If we can show that

$$\sum_{j=J^*+2}^{\hat{J}} 2^{2j}\|u_j\|_0^2 + 2^{2\hat{J}}\|u_{\hat{J}}\|_0^2 \leq C2^{\hat{J}-J^*}\||u\||^2,$$

then things fall into place. By definition of the recursion we have

$$
\begin{aligned}
\|u_{j+1}\|_0^2 &= \|\hat{w}_j - E_{j+1}\hat{w}_j|_\Gamma\|_0^2 \approx \|\hat{w}_j\|_0^2 \approx 2^{-j}\|\hat{w}_j|_\Gamma\|_{0,\Gamma}^2 \\
&= 2^{-j}\|(\hat{v}_{J^*+1} + \cdots + \hat{v}_j)|_\Gamma\|_{0,\Gamma}^2 \\
&\leq 2^{-j}\left(\sum_{l=J^*+1}^{j}\|\hat{v}_l|_\Gamma\|_{0,\Gamma}\right)^2 \leq C2^{-j}\left(\sum_{l=J^*+1}^{j} 2^{l/2}\|\hat{v}_l\|_0\right)^2 \\
&\leq C2^{-j}\left(\sum_{l=J^*+1}^{j} 2^{-l/2}(2^l\|\hat{v}_l\|_0)\right)^2 \leq C2^{-J^*-j}\sum_{l=J^*+1}^{j} 2^{2l}\|\hat{v}_l\|_0^2.
\end{aligned}
$$

This yields

$$
\sum_{j=J^*+2}^{\hat{j}} 2^{2j}\|u_j\|_0^2 \leq C2^{-J^*}\sum_{l=J^*+1}^{\hat{j}-1} 2^{2l}\|\hat{v}_l\|_0^2 \sum_{j=l}^{\hat{j}-1} 2^j \leq C2^{\hat{j}-J^*}\sum_{l=J^*+1}^{\hat{j}-1} 2^{2l}\|\hat{v}_l\|_0^2
$$

$$
\leq C2^{\hat{j}-J^*}\sum_{l=1}^{J} 2^{2l}\|v_l\|_0^2 \leq C2^{\hat{j}-J^*}\|u\|^2.
$$

Since $\hat{u}_{\hat{j}} = E_{\hat{j}}(\hat{w}_{\hat{j}-1}|_\Gamma)$ and, thus, $\|\hat{u}_{\hat{j}}\|_0^2 \leq C\|\hat{w}_{\hat{j}-1}\|_0^2$, the estimate for the last term is the same. This finishes the derivation of the upper bound

$$
\bar{\eta}_{\text{mod}} \leq C2^{\hat{j}-J^*}\bar{\eta},
$$

and Theorem B.5.2, (B.5.13), follows from the uniform bounds for $\eta, \bar{\eta}$ obtained in Theorem B.4.1.

# Appendix C

# RECENT DEVELOPMENTS IN MULTIGRID EFFICIENCY IN COMPUTATIONAL FLUID DYNAMICS

## A. Brandt

Department of Applied Mathematics and Computer Science,
The Weizmann Institute of Science,
Rehovot 76100, Israel

Very high efficiency has been attained by multigrid solvers for some types of problems, such as general uniformly elliptic problems. Our objective is to attain such an optimal performance for general fluid dynamics problems. A set of obstacles to achieving that goal is tabled below, along with a list of possible ways for overcoming each obstacle, their current state of development and references.

The table includes staggered and nonstaggered, conservative and nonconservative discretizations of viscous and inviscid, incompressible and compressible flows at various Mach numbers, as well as a simple (algebraic) turbulence model and comments on chemically reacting flows. The listing of associated computational barriers involves: nonalignment of streamlines or sonic characteristics with the grids; recirculating flows; stagnation points; discretization and relaxation on and near shocks and boundaries; far-field artificial boundary conditions; small-scale features not visible on some of the coarse grids; large grid aspect ratios; boundary layer resolution; and grid adaption.

## C.1 INTRODUCTION

The table below does *not* refer to a vast literature on multigrid methods in CFD (see for example [203]), in which enormous improvements over previous (single-grid) techniques have been achieved, but without adopting the systematic top multigrid efficiency (TME) approach. This approach insists on obtaining basically the same ideal efficiency to *every* problem, by a very systematic study of each type of difficulty, through a carefully chosen

573

sequence of model problems. Several fundamental techniques are typically absent in the multigrid codes that have not adopted the TME strategy. Most important, those codes fail to decompose the solution process into separate treatments of each factor of the PDE principal determinant, and therefore do not identify, let alone treat, the separate obstacles associated with each such factor. Indeed, depending on flow conditions, each of those factors may have different ellipticity measures (some are uniformly elliptic, others are nonelliptic at some or all of the relevant scales) and/or different set of characteristic surfaces, requiring different combinations of relaxation and coarsening procedures.

The table deals only with *steady-state* flows and their *direct* multigrid solvers, i.e. not through pseudotime marching. Time-accurate solvers for genuine *time-dependent* flow problems are in principle simpler to develop than their steady-state counterparts. Using semi-implicit or fully implicit discretizations, large and adaptable time steps can be used, and parallel processing across space *and time* is feasible. The resulting system of equations (i.e. the system to be solved at each time step) is much easier than the steady-state system because it has better ellipticity measures (due to the time term; cf. Section 2.8.2), it does not involve the difficulties associated with recirculations, and it comes with a good first approximation (from the previous time step). A simple multigrid "F-cycle" at each time step can solve the equations much below the discretization errors of that step [78]. It is thus believed that fully efficient multigrid methods for the steady-state equations will also yield fully efficient and highly parallelizable methods for time-accurate integrations.

## C.2 TABLE OF DIFFICULTIES AND POSSIBLE SOLUTIONS

Throughout the table, wherever appropriate, pointers to the subject and sections in the body of the book are provided.

When no comment is made in the "Status" column of the table it usually means that the discussed "Possible Solution" is not known to have been tried.

| Difficulty | Possible Solutions | Status |
|---|---|---|
| ⊙ **Uniformly elliptic scalar equation on uniform grids in general domains** | Multigrid cycles, guided by local mode analysis + FMG (see Chapters 2 and 4) | TME demonstrated 1971 [56, 58] and rigorously proved [68, 69] |
| ⊙ **Nonlinearity** | (1) FAS cycles (Section 5.3) <br> (2) Continuation processes (to obtain good initial approximations), integrated into one-shot FMG algorithm (Section 10.2) | (1) Demonstrated 1975 [58, 369]. <br> (2) Described in Sec. 8.3.2 of [66] |
| ⊙ **Fluid dynamics: general** | Basic ideas are reviewed in Sec. 2 of [500]; see also Sections 8.6–8.9 above. | |
| ⊙ **Nonscalar PDE systems** (see Chapter 8) | (1) General rules for the order of the intergrid transfer operators are given in Sec. 4.3 of [66] with some more details in Sec. 3.3 [69] <br><br> (2) A general approach to the design of relaxation is based on the operator principal matrix $L$ and on the factors of det $L$ Secs. 3.4 and 3.7 in [66]. In this approach a distribution matrix $M$ and a weighting (or "preconditioning") matrix $P$ are constructed so that $PLM$ is triangular, containing the factors of det $L$ on the main diagonal (separated from each other as much as possible, to avoid the complication with "product operators" discussed next). This (if necessary—together with the technique described next), leads to decomposing relaxation into simple schemes for the (scalar) factors of det $L$. Many specific examples are given below <br><br> (3) For systems of PDE which are of mixed type (elliptic–hyperbolic) another possibility is to sometimes introduce new unknowns in terms of which elliptic and hyperbolic parts are separated | TME demonstrated in a number of cases (see many details below). TME *proved* for uniformly elliptic systems [68, 69] <br><br><br><br> TME demonstrated for incompressible and compressible cases [381–384] |

| Difficulty | Possible Solutions | Status |
|---|---|---|
| • *Product operator*: an equation $LU = f$, where $L = L_2 L_1$. Assume a relaxation process for $L_j$ is given, with the amplification factor $\mu_j(\theta)$ and the smoothing factor $\bar{\mu}_j$, ($j = 1, 2$) | Two alternative approaches:<br><br>(1) Throughout the multigrid algorithm (not just in the relaxation sweeps), introduce an explicit new unknown function $V$, replacing the equation with the pair of equations $L_1 U - V = 0$ and $L_2 V = f$. The resulting smoothing factor is $\bar{\mu} = \max(\bar{\mu}_1, \bar{\mu}_2)$. See Section 8.3<br><br>(2) Use $V$ only as an auxiliary function in relaxation. That is: starting with $v = L_1 u$, where $u$ is the current approximation to $U$, perform $v_2$ sweeps on the equation $L_2 V = f$, yielding a new value $\tilde{v}$. Then perform $v_1$ sweeps on the equation $L_1 u = \tilde{v}$. The resulting amplification factor is $\mu(\theta) = \mu_1(\theta)^{v_1} + [1 - \mu_1(\theta)^{v_1}]\hat{L}_1(\theta)^{-1}\mu_2(\theta)^{v_2}\hat{L}_1(\theta)$, where the Fourier symbols are defined by $\hat{L}_j(\theta) = e^{-i\theta \cdot x/h} L_j e^{i\theta \cdot x/h}$. Hence in scalar cases $\bar{\mu} < \bar{\mu}_1^{v_1} + \bar{\mu}_2^{v_2}$. | (1) TME demonstrated for $L = \Delta^2$ [77, 239] |
| ⊙ **Smoothing for special CFD systems** | $M$ = distribution operator<br>$P$ = preconditioner (see discussion above) | |
| • Cauchy–Riemann on staggered grid<br><br>$L = \begin{pmatrix} \partial_x & \partial_y \\ \partial_y & -\partial_x \end{pmatrix}$ | Two alternatives:<br>(1) $M = L$, $P = I$<br>(2) $P = L$, $M = I$ | (1) TME demonstrated [72, 130]<br>(2) TME validated |
| • Stokes on staggered grid<br><br>$L = \begin{pmatrix} -\Delta & 0 & \partial_x \\ 0 & -\Delta & \partial_y \\ \partial_x & \partial_y & 0 \end{pmatrix}$ | (1) See Section 8.7<br><br>$M = \begin{pmatrix} 1 & 0 & -\partial_x \\ 0 & 1 & -\partial_y \\ 0 & 0 & -\Delta \end{pmatrix}$,   $P = I$<br><br>(2)<br><br>$P = \begin{pmatrix} 1 & 0 & 0 \\ 0 & 1 & 0 \\ \partial_x & \partial_y & -\Delta \end{pmatrix}$,   $M = I$ | (1) TME demonstrated [72, 130]<br><br>(2) TME validated |

- Stokes, nonstaggered

(1) Quasi-elliptic discretization

$$L = \begin{pmatrix} -\Delta & 0 & \partial_x^{2h} \\ 0 & -\Delta & \partial_y^{2h} \\ \partial_x^{2h} & \partial_y^{2h} & 0 \end{pmatrix}$$

with averaging of the resulting pressure

(2) $h$-elliptic discretization, e.g.

$$L = \begin{pmatrix} -\Delta & 0 & \partial_x^{2h} \\ 0 & -\Delta & \partial_y^{2h} \\ \partial_x^{2h} & \partial_y^{2h} & -\omega h^2 \Delta \end{pmatrix}$$

Analogous to the staggered case; e.g.

$$M = \begin{pmatrix} 1 & 0 & -\partial_x^{2h} \\ 0 & 1 & -\partial_y^{2h} \\ 0 & 0 & -\Delta \end{pmatrix}$$

No modifications of the FMG algorithm is required, even in the quasi-elliptic case (as explained in Sec. 18.6 of [66]). In generalization to Navier–Stokes, pressure averaging is required of coarse-level results before their interpolation to the next finer level (whenever the coarse-level employs the quasi-elliptic discretization)

(1) In a quasi-elliptic approach, TME demonstrated (Sec. 18.6 of [66], and [83]

(2) TME demonstrated (see Section 8.8)

- Nonconservative incompressible Euler, whose principal operator in 2D is

$$L = \begin{pmatrix} u \cdot \nabla & 0 & \partial_x \\ 0 & u \cdot \nabla & \partial_y \\ \partial_x & \partial_y & 0 \end{pmatrix}$$

(similarly 3D), on staggered grid, second- (or higher) order discretization

(1) Employ cycle index $\gamma = 2^p$, where $p$ is the order of discretization, with

$$M = \begin{pmatrix} 1 & 0 & -\partial_x \\ 0 & 1 & -\partial_y \\ 0 & 0 & u \cdot \nabla \end{pmatrix}$$

(2) With the same $M$, for each of the momentum equations employ a relaxation scheme which is fast *converging* for the advection operator $u \cdot \nabla$ (i.e. converging fast not only for high frequency, but also for smooth characteristic components; see discussion of advection below)

(3) Use canonical variable $(u, v, P)$ on staggered grid, where $P = (u^2 + v^2)/2 + p$. Upwind only $P$, use central discretization for $(u, v)$. Relaxation is marching for $P$, and weighted (preconditioning) for $(u, v)$

(1) TME for first-order discretization using W-cycles shown in [72, 130]

(2) TME demonstrated for 2D entering flows with second-order discretization [86] and for recirculating flows with first-order discretization [87]

(3) TME in [381–383]

- Low-Reynolds incompressible Navier–Stokes, staggered or not

Fully analogous to Stokes solvers: just replace $\Delta$ in $L$ by $Q = -R^{-1}\Delta + u \cdot \nabla$. See also Sec. 8.7

TME demonstrated 1978 [72, 130]

| Difficulty | Possible Solutions | Status |
|---|---|---|
| • High-Reynolds incompressible Navier–Stokes, staggered | Fully analogous to incompressible Euler (outside boundary layers: see discussion on such layers below); just replace $u \cdot \nabla$ everywhere with $Q$. See also Sec. 8.8 <br><br> $$M = \begin{pmatrix} 1 & 0 & 0 & 0 & 0 & -\rho(u\cdot\nabla)\partial_1 \\ 0 & 1 & 0 & 0 & 0 & -\rho(u\cdot\nabla)\partial_2 \\ 0 & 0 & 1 & 0 & 0 & -\rho(u\cdot\nabla)\partial_3 \\ 0 & 0 & 0 & 1 & 0 & -\rho^2\Delta \\ 0 & 0 & 0 & 0 & 1 & -p\Delta \\ 0 & 0 & 0 & 0 & 0 & \rho^2(u\cdot\nabla)^2 \end{pmatrix}$$ <br><br> The advection and full potential operators are each relaxed by one of the approaches described for them below (in the section on nonelliptic operators. The *semicoarsening* described there would then be used as an *inner* multigrid cycle for *relaxing* one factor of the determinant, to be distinguished from the *outer* multigrid cycle, which can use *full coarsening*) | TME demonstrated for first-order discretization on staggered [72, 130] and nonstaggered grids (Sec. 19.5 in [66]), and for second-order staggered discretization [86] |
| • *Compressible Euler*, nonconservative, on staggered grid: the subprincipal operator on $(u_1, u_2, u_3, \rho, \varepsilon, p)$ is <br><br> $$L = \begin{pmatrix} \rho u\cdot\nabla & 0 & 0 & 0 & 0 & \partial_1 \\ 0 & \rho u\cdot\nabla & 0 & 0 & 0 & \partial_2 \\ 0 & 0 & \rho u\cdot\nabla & 0 & 0 & \partial_3 \\ \rho^2\partial_1 & \rho^2\partial_2 & \rho^2\partial_3 & \rho u\cdot\nabla & 0 & 0 \\ p\partial_1 & p\partial_2 & p\partial_3 & 0 & \rho u\cdot\nabla & 0 \\ 0 & 0 & 0 & -\dfrac{\partial p}{\partial \rho} & -\dfrac{\partial p}{\partial \varepsilon} & 1 \end{pmatrix}$$ <br><br> $\det L = \rho^5(u \cdot \nabla)^3((u \cdot \nabla)^2 - a^2\Delta)$, <br> $a = (\partial p/\partial \rho + (p/\rho^2)(\partial p/\partial \varepsilon))^{1/2}$ is the sound speed, <br> $\rho, \varepsilon, p$ defined at cell centers, <br> $u_i$ – at center of cell faces perpendicular to the $i$th coordinate | | |
| • 2D Compressible Euler, nonconservative and conservative, staggered grid, using canonical variables $(u, v, S, H)$. Structured and unstructured grids | Use $(u, v)$ at cell edges, $H$ at middle of cell, $S$ at vertices. Upwind only $S$ at momentum equations. Relax $S$, $H$ by marching. $(u, v)$ by a weighting relaxation. Crocco's form is used here to define relaxation | TME in [382–384] |
| • 2D/3D incompressible and compressible Euler: canonical variables in which velocities are replaced by vector potential representation. Nonstaggered structured and unstructured grid | All variables at cell nodes. Relax hyperbolic quantities using marching. Relax vector potential using point Gauss–Seidel | TME achieved (unpublished) for interior and exterior flows in 2D, interior in 3D |

- *Compressible Navier–Stokes*, nonconservative. The subprincipal operator on $(u_1, u_2, u_3, \rho, \varepsilon, p)$ is

$$L_s = \begin{pmatrix}
Q_\mu - \bar\lambda\partial_{11} & -\bar\lambda\partial_{12} & -\bar\lambda\partial_{13} & 0 & 0 & \partial_1 \\
-\bar\lambda\partial_{21} & Q_\mu - \bar\lambda\partial_{22} & -\bar\lambda\partial_{23} & 0 & 0 & \partial_2 \\
-\bar\lambda\partial_{31} & -\bar\lambda\partial_{32} & Q_\mu - \bar\lambda\partial_{33} & 0 & 0 & \partial_3 \\
\rho^2\partial_1 & \rho^2\partial_2 & \rho^2\partial_3 & Q_0 & 0 & 0 \\
p\partial_1 & p\partial_2 & p\partial_3 & 0 & Q_\kappa & \frac{\partial p}{\partial\varepsilon} \\
0 & 0 & 0 & -\frac{\partial p}{\partial\rho} & -\frac{\partial p}{\partial\varepsilon} & 1
\end{pmatrix}$$

where $Q_\alpha = -\alpha\Delta + \rho u \cdot \nabla$, $\kappa = k/c_v$ (coefficient of thermal conductivity divided by the specific heat at constant volume), $\bar\lambda = \lambda + \mu$, $\lambda = (2/3)\mu$, $\det L_s = Q_\mu^2 \det L_c$, where $L_c$ is the "core operator"

$$L_c = \begin{pmatrix}
Q_0 & 0 & -\rho^2\Delta \\
0 & Q_\kappa & -p\Delta \\
-\frac{\partial p}{\partial\rho} & \frac{\partial p}{\partial\varepsilon} & Q_{\mu+\bar\lambda}
\end{pmatrix}$$

At standard conditions of laminar air flow the Prandtl number $\gamma\mu/\kappa \approx 0.72$; for turbulence $\gamma\mu/\kappa \approx 0.9$, with $\gamma = c_p/c_v = 1.4$

- Nonconservative *nonstaggered* Euler and NS

(1) Where $\bar\lambda, \mu, \kappa \ll \rho h|u|$ relax as in Euler above

(2) Otherwise use

$$M = \begin{pmatrix}
1 & 0 & 0 & 0 & 0 & -\partial_1 \\
0 & 1 & 0 & 0 & 0 & -\partial_2 \\
0 & 0 & 1 & 0 & 0 & -\partial_3 \\
0 & 0 & 0 & 1 & 0 & 0 \\
0 & 0 & 0 & 0 & 1 & 0 \\
\bar\lambda\partial_1 & \bar\lambda\partial_2 & \bar\lambda\partial_3 & 0 & 0 & Q_{\mu+\bar\lambda}
\end{pmatrix}$$

relaxing each $Q_\mu$ by one of the approaches described for the advection–diffusion below, and $L_c$ by procedures discussed for it below (in the section on nonelliptic operators)

(1) Probably similar to the staggered (cf. transition from staggered to nonstaggered in Stokes)

(2) In the 2D incompressible case: premultiply $L$ by a projection operator $P$, obtaining a Poisson equation for the pressure. Solve pressure equation with multigrid and the advection equation by marching downstream

TME demonstrated for 2D incompressible Euler [325] in the cases of channel (with bump) and airfoil flows

| Difficulty | Possible Solutions | Status |
|---|---|---|
| • *Conservative* discretization of any of the above systems | Apply a prefactor $P$ such that $PL$ has principally the above nonconservative form. See, however, the difficulty associated with FDA *factorizability* (discussed in the section on nonelliptic operators), which may arise with such $PL$ operators. See also Section 8.9 | Mentioned in Sec. 3.4 of [66], but not tested |
| ⊙ **Nonelliptic operators**<br>More precisely: small ellipticity measures at some (e.g. large) scales. The main operators of interest here are<br>(1) The advection operator (or, similarly, the convection–diffusion operator at large Reynolds numbers)<br>(2) The *near-sonic* full potential operator or more generally the core operator $L_c$<br>(3) The *supersonic* full potential operator or $L_{c'}$. (See below a separate discussion of anisotropies caused by the discretization) | The distributive Gauss–Seidel (DGS) relaxation of the full flow equations allows a specific individual treatment for each of these cases, taking into account its particular set of characteristics, as detailed below | |
| • Grid *aligned* with the characteristics: pointwise relaxation has only *semi*smoothing capability | Block (e.g. line or plane) relaxation schemes and/or semicoarsening, possibly in alternating directions, guided by mode analyses [58, 61] (see Sections 5.1 and 5.2); or ILU relaxation [211, 366] (see Section 7.5) | TME demonstrated in many cases |
| • Distinguishing different regimes (open vs. closed characteristics) | Running separately the relaxation subroutine of a given nonelliptic factor can<br>(1) Separately check its convergence properties<br>(2) Produce a scalar $\sigma \approx 1$ at regions of open characteristics and $\sigma \ll 1$ on closed characteristics (as in separated flow zones) | |

- *Nonaligned* grids, with *open* characteristics (e.g. entering flow): The main difficulty is the shorter distance (along the characteristics) for which a coarser grid still approximates some *smooth* solution components (characteristic components with intermediate cross-characteristic smoothness) [63, 85]

Three possible approaches, all guided by *half-space* two-level FMG mode analysis, using for simplicity the first differential approximation (FDA) to the discrete operator (see [63] and Sec. 7.5 in [66])

(1) Downstream-ordered relaxation marching: suitable only for the advection factor; sometimes still requires W-cycles, and not very good for massively parallel processing. In the case of an $O(h^p)$ discretization which is not purely upstreamed, relaxation should involve a predictor–corrector downstream marching. If the predictor order is $q$, the corrector should be applied at least $p/q$ times. See also Section 7.2

(2) Similarly, with downstream-ordered ILU relaxation: suitable for the advection operator (in 2D and 3D) *and* for the nearsonic full potential operator in 2D (*not* in 3D). See also Section 7.5

(3) Semicoarsening with controlled artificial dissipation at coarse levels (to match the target-grid numerical dissipation): suitable for all operators in 2D and 3D, and for massively parallel processing. See also Section 5.1

(4) Cycle index $= 2^{p/m}$, where $p$ is the order of discretization and $m$ is the order of the differential factor. (Only suitable for the advection operator, for which $m = 1$; especially attractive for $p = 2$ in 3D; not requiring ordered relaxation, but still disadvantageous for massively parallel processing because of the high cycle index)

(1) TME demonstrated in [86] and in recent calculations, both for the advection operator by itself and as part of the incompressible Euler system

(3) TME has been shown for the sonic full potential operator [74–75]

(4) For $p = 1$, TME has been shown on various occasions. Should be tried for $p = 2$

| Difficulty | Possible Solutions | Status |
|---|---|---|
| • The mixed *convection–diffusion* operator with order $p$ approximation, having natural viscosity $\nu$ and artificial viscosity $\alpha h^p$ | Treatment as elliptic operator on levels where $\nu \gtrsim (2^p \cdot 4 - 5)\alpha h^p$ and as the nonelliptic advection operator otherwise | Not precisely tried |
| • *Closed* characteristics (recirculating flows). Here *uniformity* of viscosity (including numerical viscosity) is important for accuracy, while the viscosity *size* is less important (except at resolved boundary layers, discussed below). In fact, a uniform $(h)$ artificial viscosity can yield higher order approximations. Full convergence may also be less important here (since in reality, too, steady state may take exceedingly long to establish) | Using the above-mentioned scalar $\sigma$, form a $\sigma$-dependent convergence test, to distinguish between slowness of open and closed characteristics (and possibly ignore the latter). Also based on $\sigma$, at recirculation regions use uniform (explicit) $O(h)$ numerical viscosity, with continuation from large to small viscosity integrated into the FMG algorithm. The cycles can employ one of the following three options. (See also Sections 7.2–7.3). | TME cycles by methods (1) and (2) were shown in [87] and [430] respectively. Method (3), which should be best for massive parallelization, has not been implemented |
| | (1) DCW method (using defect corrections within W-cycles), with suitable over-weighting of residuals [87]. Suitable only for $O(h)$ discretizations | |
| | (2) Effectively downstream relaxation ordering (using alternate direction sweeps) and doubling of transferred residuals (for $O(h)$ discretization) [430] | |
| | (3) Semicoarsening, generally similar to [74] | |
| • Full potential operator $(\mathbf{u} \cdot \boldsymbol{\nabla})^2 - a^2 \Delta$, $M_0 = |\mathbf{u}|/a \le 0.7$ (uniformly elliptic) | Any classical algorithm is suitable (see for example Section 5.3.6), but the algorithm of the next case is also adequate | TME well established |
| • Full potential $0.7 \le M_0 \le 1.4$ | Relaxation marching downstream (for transition to the supersonic case below) together with semicoarsening in the characteristic (cross-stream) direction | TME shown for the case $M = 1$ [74]. Other cases have not yet been implemented |
| • Full potential $1.4 \le M_0$ (uniformly hyperbolic, with the stream as the time-like direction, and with $O(1)$ "Courant number") | Marching in the stream direction, possibly with a predictor–corrector procedure. For full massive parallelization, however, wave methods (extending standing wave methods [76, 246] should be used | |

- The "core operator"

$$L_c = \begin{bmatrix} Q_0 & 0 & -\rho^2\Delta \\ 0 & Q_\kappa & -p\Delta \\ -\partial p/\partial\rho & -\partial p/\partial\varepsilon & Q_{\mu+\bar\lambda} \end{bmatrix}$$

should be relaxed as part of relaxing the compressible NS system, in the case that $\rho|u|h \le \max(\bar\lambda, \mu, \kappa)$. In the case of alignment between the grid and the flow, with mesh size $h_1$ and $h_2$ in the stream and cross-stream directions, respectively, and $h_2 \le h_1$ (e.g. in boundary layers), the case where $L_c$ need be relaxed is when $\rho|u|h_2^2 \le h_1 \max(\bar\lambda, \mu, \kappa)$. In aerodynamics, $\bar\lambda$, $\mu$ and $\kappa$ are comparable, so the case of interest is $|u|h_2^2 \le \nu h_1$, where $\nu = \mu/\rho$.

Best relaxation scheme depends on the flow parameters. For example:

(1) If $\kappa \ll \rho|u|h$, then $Q_\kappa \approx Q_0$ (in principal terms) and one can use DGS with

$$M = \begin{pmatrix} 1 & 0 & \rho^2\Delta \\ 0 & 1 & p\Delta \\ 0 & 0 & Q_0 \end{pmatrix}$$

resulting in the need to relax the first two equations each on an advection operator (see methods above), and the third equation on the operator $Q_0 Q_{\mu+\bar\lambda} - \rho^2 a^2\Delta$. In the case of interest the principal part of the latter is $((\mu + \bar\lambda)Q_0 + \rho^2 a^2)\Delta$, so it can be relaxed by the general method for relaxing a product operator (see $L = L_2 L_1$ above)

(2) In the aerodynamics and aligned case of interest, the term $Q_{\mu+\bar\lambda}$ in $L_c$ is not the principal term. Therefore relaxation can easily be conducted with the weighting (preconditioning) matrix

$$P = \begin{pmatrix} 1 & 0 & 0 \\ -p & \rho^2 & 0 \\ 0 & 0 & 1 \end{pmatrix}$$

and the distribution matrix

$$M = \begin{pmatrix} 0 & 1 & 0 \\ 0 & -p_\rho/p_\varepsilon & 1 \\ 1 & 0 & 0 \end{pmatrix}$$

yielding $PLM$ whose principal part is its main diagonal, on which separately appear the Laplace operator $\Delta$, the convection–diffusion operator $Q_{\bar\kappa}$ where $\bar\kappa = p_\rho\rho^2/(2pp_\varepsilon) = 1.25\kappa$ (for air), and a free function

| Difficulty | Possible Solutions | Status |
|---|---|---|
| • *FDA factorizability question*: the decomposition of a system relaxation into its scalar factors depends on the equality of the different occurrences of the advection–diffusion operator $Q$ (or $Q_{\mu+\tilde{\lambda}}$) appearing in $PL$, the prefactoring by $P$ of a conservative discretization $L$. However, for smooth-characteristic convergence in relaxing a nonelliptic discrete operator, important is not only the differential operator it approximates, but also its FDA terms in noncharacteristic directions; e.g. the cross-stream *numerical* viscosity of $Q$. These may not be the same in the different occurrences of $Q$, putting the factorization into question | (1) Examining several examples of conservative discretization of transonic flows, the FDA terms in various occurrences of $Q_{\mu+\tilde{\lambda}}$ turn out to be sufficiently close to each other (e.g. only 4% discrepancy) to allow full efficiency of the proposed relaxation schemes | (1) Further examination is needed |
| | (2) Conservative schemes may be designed so that the various FDAs of $Q_{\mu+\tilde{\lambda}}$ are identical, or at least sufficiently close so that the scheme is still factorizable | (2) Some "genuinely multidimensional upwind" schemes yield factorizable schemes, e.g. in the subsonic case in the control-volume structured-grid context [358]. Further studies are in progress |
| | (3) A general practical approach is a defect correction relaxation: the residuals are calculated by the given $PL$ system and fed into a DGS relaxation scheme whose driving factors may have different discretizations (as long as their numerical viscosities are not larger than those in the $PL$ system) | |
| • High-order discretization (away from shocks) | (1) "*Double discretization*" *schemes*: use high-order only in calculating residuals transferred to the coarse grid, *not* in relaxation (unless the high-order scheme is preferable also for high frequency modes). See also Section 7.8 | Introduced 1978 [72]. Successfully implemented in various *elliptic* cases (see description and refs in Sec. 10.2 of [66]). Methods for nonelliptic have not been tested beyond second-order |
| | (2) However, in relaxing *nonelliptic* factors (e.g. downstream relaxation marching for convection operator) the high-order must be used (e.g. by a predictor–corrector downstream relaxation). See also Section 7.4 | *Comment*: high-order approximations on unstructured grids are very expensive |

⊙ **Algebraic turbulence models**

These employ the (compressible or incompressible) Navier–Stokes equations, adding to the laminar viscosity $\mu_\ell$ (and similarly to $\lambda_\ell$ and $\kappa_\ell$) a turbulent viscosity $\mu_t$ (similarly $\lambda_t$, $\kappa_t$), which is defined in terms of geometric functions (such as the distance to the wall), flow-dependent boundary-layer-wide (BLW) parameters (such as the boundary layer thickness, the maximum and minimum total velocity across the layer, and the flow wall friction) and in terms of the local total vorticity $\omega = |\mathrm{curl}\, \boldsymbol{u}|$. E.g., the two-layer Baldwin–Lomax model [17], is defined in two regions as follows:

(1) *Outer layer* Here $\mu_t$ is defined only in terms of distance from the boundary and BLW parameters.

(2) *Inner layer* Closer to the wall, $\mu_t = \rho \ell^2 \omega$, where $\ell$ depends on the distance to the wall and on BLW parameters. In the 2D incompressible case, and neglecting the laminar viscosity, the resulting principal operator, on the vector of unknowns $(\omega, u, v, p)$, is

$$L = \begin{pmatrix} -\omega & (u_y - v_x)\partial_y & -(u_y - v_x)\partial_x & 0 \\ -A & Q_\mu & 0 & \partial_x \\ -B & 0 & Q_\mu & \partial_y \\ 0 & \partial_x & \partial_y & 0 \end{pmatrix}$$

where $A = \rho \ell^2 [2u_x \partial_x + (u_y + v_x)\partial_y]$, $B = \rho \ell^2 [2v_y \partial_y + (u_y + v_x)\partial_x]$, so that $\det L = \Delta \{\omega \partial_s - 2\rho \ell^2 (u_y - v_x) \times [v_x \partial_{xx} - u_y \partial_{yy} + (v_y - u_x)\partial_{xy}]\}$

*In the outer layer* the principal operator, hence also relaxation, are exactly as for the laminar case, with $\mu = \mu_t + \mu_\ell$. The BLW parameters are held unchanged during relaxation at scales finer than the boundary-layer width. Only at a suitable coarser level, where the cross-layer mesh size approaches the layer width, is the dependence of the BLW parameters on the flow relaxed, together with the flow equations themselves, by applying box relaxation near the boundary (cf. the section on boundary relaxation)

*In the inner layer*, suppose for example that the coordinate along the wall is $x$, and $u_y \gg \max\{|u_x|, |v_x|, |v_y|\}$. Then the principal operator takes the form

$$L = \begin{pmatrix} -\mu & \mu\partial_y & -\mu\partial_x & 0 \\ -\mu\partial_y & Q_\mu & 0 & \partial_x \\ -\mu\partial_x & 0 & Q_\mu & \partial_y \\ 0 & \partial_x & \partial_y & 0 \end{pmatrix}$$

A suitable distribution operator then is

$$M = \begin{pmatrix} 1 & \partial_y & -\partial_x & 0 \\ 0 & 1 & 0 & -\partial_x \\ 0 & 0 & 1 & -\partial_y \\ 0 & \mu\partial_x & -\mu\partial_y & Q_\mu \end{pmatrix}$$

yielding $LM$ with principal terms only on the main diagonal, where there appear the operators $\Delta$ (for the continuity equation ghost function) and $\boldsymbol{u} \cdot \nabla - 2\mu\partial_{yy}$ (for each of the momentum equations). The latter is nonelliptic, and its characteristics would often be aligned with the grid (cf. the section on nonelliptic operators)

TME should first be demonstrated for a simple turbulence model, such as the one described here

| Difficulty | Possible Solutions | Status |
|---|---|---|
| ⊙ **Chemically reacting flows**<br>*These feature three types of difficulty*<br>(1) A set of $N$ continuity equations, where $N$, the number of chemical species, may be quite large.<br>(2) The nonlinear source terms in these equations may be very stiff<br>(3) Some densities at some (few) points may become negative upon the coarse-to-fine FAS interpolation | At any grid point where any source term is *principal* (meaning: its production rate of species $i$ per unit volume is large compared with max $(h^{-2}D_i, h^{-1}\rho_i|u|)$, where $D_i$ is the local diffusion coefficient of species $i$) it should be included in the principal matrix operator $L$. As a result, at each grid point the weighted-distributive relaxation step (local inversion of the principal terms of *PLM*) may involve the inversion of a matrix of size up to $N \times N$. This would correspond to the (relaxation part of the) point-implicit method [94]. Fortunately, this will usually happen only on some coarser multigrid levels and/or at some restricted zones, thus requiring only a relatively small amount of work<br><br>Nonlinearity is treated by an FAS in which, instead of the fine-to-coarse transfers of densities $\rho_1, \ldots, \rho_N$ and the coarse-to-fine interpolation of the changes $\delta\rho_1, \ldots, \delta\rho_N$, transferred are the functions $f_1(\rho_1), \ldots, f_N(\rho_N)$ and interpolated are $\delta f_1(\rho_1), \ldots, \delta f_N(\rho_N)$, where $f_i(\rho_i)$ are properly chosen functions; e.g. $f_i(\rho_i) = \log \rho_i$, so that after interpolation $\rho_i = \exp\left(f_i(\rho_i^{\text{OLD}}) + \delta f_i(\rho_i)\right) > 0$. Furthermore, the continuity-equation residual restriction should be conservative (strictly full weighting) | TME should first be demonstrated for a simple model case; e.g. a 2D incompressible inviscid flow with two reacting species |
| ⊙ **Shocks**<br>• *Shock displacement question*: a (small) displacement should result from global solution changes that occur on coarse levels of the cycle. How can one obtain an accurate displacement, when those levels are too coarse to resolve it? | An accurate shock displacement is obtained if the fine-to-coarse residual transfer is conservative (e.g. "FW") and the coarse-to-fine correction interpolation is followed by local relaxation passes near the shock | Full efficiency shown [356] |
| • Relaxation near strong shocks | Add extra relaxation passes, using general robust schemes (e.g. box Kaczmarz), until all local residuals drop to their level away from shocks | TME shown in a quasi-1D case [502] |

- Poor $h$-ellipticity of high-resolution schemes

Construction of new, genuinely multidimensional upwind schemes

Developed in the context of unstructured triangular grids [357]

⊙ **Boundary related difficulties**

- *Discretization* near boundaries

For best multigrid efficiency, use Cartesian coordinates throughout the domain, with boundary-fitted local grid patches, regarded as finer multigrid local levels [58, 66]. Only crude (e.g. first-order) discretization is then needed near the boundaries on the Cartesian grids

Ruge and Brandt have devised a near-general-boundary discretization for incompressible Euler on staggered Cartesian grid (unpublished)

- *Relaxation at and near boundaries:* Difficulties:

  (1) There is no smoothing analysis when the boundaries are not aligned with the grid

  (2) The fine-to-coarse residual weighting near boundaries is generally very imprecise, hence the residuals should be reduced there *more* than in the interior

  (3) Larger residuals are created near boundaries upon coarse-to-fine interpolations (of solution or corrections)

A general-type robust relaxation scheme, e.g. *box Kaczmarz*, throughout a zone that is the width of several meshes and is near the boundary. The box size in each direction should be several mesh sizes and the boxes should have substantial overlap. One can afford several passes of such a relaxation for each full interior sweep since the zone width is $O(h^{1-\varepsilon})$, with $0 \le \varepsilon < 1$. In particular, add near-boundary relaxation passes after the FMG interpolation (allowing the latter to be of lower order near the boundary). The local relaxation passes should continue until all local residuals have dropped well below their global average magnitude. See examples and discussions in Sections 5.5, 5.6, 8.2.6 and 8.4.2

For uniformly elliptic equations it has been proved [68, 69] and demonstrated computationally (for cases of reentrant corners [14]) that the interior efficiency as predicted by mode analysis (implying TME) can always be obtained. TME has been demonstrated for incompressible Euler on staggered Cartesian grids for a variety of boundary conditions [501]

- *Boundary layers* (if they need be resolved. See also section on grid adaptation below)

Resolved by boundary-fitted local grid patches, with local semirefinements: finer levels, in narrower layers near the boundary, have smaller cross-layer mesh sizes, allowing the physical cross-stream viscosity to dominate over the numerical one (cf. Section 9.4). Additional terms in the governing equations (Navier–Stokes instead of Euler, or turbulent modelling etc.) may be used in these patches. Downstream marching relaxation and cross-stream semicoarsening should feature in the multigrid cycles. A "λ-FMG" type of algorithm (see Sec. 9.6

The local refinement techniques for Poisson's equation, with TME, are demonstrated in [14]

| Difficulty | Possible Solutions | Status |
|---|---|---|
| • *Far-field* artificial boundary conditions: requiring in some cases nonlocal *absorbing boundary conditions* (ABC) for some wave factor or radiation boundary conditions (RBC) | in [66] and also Section 9.4 above) should be employed, so that coarse FMG stages already include local semirefinements at the boundary, thus effectively incorporating into the FMG stages a process of continuation in $R_e$<br><br>Increasingly coarser grids covering increasingly larger domains. The *size* of each domain is based on accuracy-to-work optimization criteria (similar to those in [58, §8, 66, §9.5]), implying also a natural criterion for the largest domain needed. On *interior* boundaries (boundaries of a grid residing in the interior of the next coarser grid) the solution is interpolated from the coarser grid. On such boundaries, if ABC is needed, only high-frequency components need be absorbed, for which the ABC are *local*, and can be enforced as part of the relaxation process (of the corresponding wave factor) | Details of the algorithm have been worked out, and TME (or its equivalent accuracy-to-work relation) has recently been demonstrated for the 2D Poisson equation in the unbounded plane (cf. [70, §4]). Techniques for nonelliptic cases have not been systematically studied. For indefinite cases with RBC, TME has been obtained [79] |
| • *Small-scale singularities* invisible on the next coarser grid, such as small "islands" or "holes" in the domain (e.g. an airplane smaller than the mesh size of some coarser grid) or small boundary conditions (BC) features (e.g. small regions of Neumann BC and otherwise Dirichlet BC) | Local relaxation passes around the singularities after return from the next coarser grid, together with either one of the following three devices:<br>(a) Enlarging the singularity on the coarser grid<br>(b) Modifying the interior coarse-grid equation near the singularity<br>(c) If the coarse grid equations are not modified, then the convergence is slow, but slow to converge are just few very special components. Hence slowness can be eliminated by recombining iterants (see below) | TME shown in elliptic cases [82] |
| ⊙ **Grid-induced slow convergence** | One can avoid many of the following maladies by using suitable multigrid structures (described below under "grid adaptation") | |
| • Large aspect ratios | Either of the following:<br>(1) Block (part-line or part-plane) relaxation, analyzed by mode analysis (see [58] and Sections 5.1–5.2) | TME has been shown in a variety of elliptic cases |

(2) Semicoarsening [5] (often natural, since the large aspect ratio is created in the first place by semirefinements) with relaxation "semismoothing" analysis (see Sections 5.1–5.2; Secs. 2.1 and 3.2 in [61]; and Sec. 3.3 in [66])

(3) Combinations of block relaxation in some directions and semicoarsening in others (cf. Section 5.2.3)

• Expanding grids

⊙ **Grid adaptation**

Relaxation marching in the direction of increasing mesh size [412]; or distributive relaxation (Sec. 6 in [60])

Use local patches of multigrid levels in creating any desired local refinement, aspect ratio, boundary fitting or even flow fitting. Base refinement criteria on the fine-to-coarse multigrid correction ($\tau$). Adaptation can be integrated into the $\lambda$-FMG algorithm together with proper (e.g. Reynolds number) continuations. See Section 9.4

Introduced in Secs. 7–9 of [58] and Sec. 9 of [66], but tried only for Poisson equation near singularities [14]

⊙ **Stagnation point** causing an instability in the coarse-grid correction and problems with some of the relaxation methods described above

Coarse-grid numerical viscosity depending on the *average* (e.g. "FW") of the fine-grid numerical viscosity, not on its injected value (Sec. 4.5 in [87]). Relaxation near stagnation should be based on full Newton linearization, not on the operator principal matrix $L$ generally used elsewhere

⊙ **A small number of slowly converging components** may arise in many situations, especially when the "Possible Solutions" described in many of the sections above are not *fully* implemented. That "small number" would often slowly but unboundedly increase with decreasing mesh size

A general method to expel a few slow components is by *recombining iterants*, or equivalently, using the multigrid cycle as a preconditioner for *Krylov subspace acceleration* [410]. To inexpensively expel a larger number of slow components (without executing many multigrid cycles and storing many fine-grid iterants), iterants may also be recombined *at various coarse levels* of the multigrid cycle [82, 410, 294]; see Section 7.8

TME shown in an example [87]

# REFERENCES

[1] Aftosmis, M., Melton, J. and Berger, M., Adaptation and surface modeling for Cartesian mesh methods. *AIAA Paper* 95-1725, 1995.

[2] Aftosmis, M., Berger, M. and Melton, J., Robust and efficient Cartesian mesh generation for component-based geometry. *AIAA Paper* 97-0196, 1997.

[3] Alcouffe, R.E., Brandt, A., Dendy, J.E. and Painter, J.W., The multi-grid method for the diffusion equation with strongly discontinuous coefficients. *SIAM J. Sci. Comput.* **2**, 430–454, 1981.

[4] Allmaras, S., Analysis of semi-implicit preconditioners for multigrid solution of the 2-D compressible Navier–Stokes equations. *AIAA Paper* 95-1651-CP, 1995.

[5] Arlinger, B., Multigrid technique applied to lifting transonic flow using full potential equation. SAAB-SCANIA Rep. L-0-1 B439, 1978.

[6] Astrakhantsev, G.P., An iterative method of solving elliptic net problems. *USSR Comp. Math. Math. Phys.* **11** (2), 171–182, 1971.

[7] Auzinger, W. and Stetter, H.J., Defect corrections and multigrid iterations. *Multigrid Methods*, Lecture Notes in Mathematics 960 (eds W. Hackbusch and U. Trottenberg). 327–351. Springer, Berlin, 1982.

[8] Axelsson, O., The method of diagonal compensation of reduced matrix entries and multilevel iteration. *J. Comp. Appl. Math.* **38**, 31–43, 1991.

[9] Axelsson, O. and Barker, V.A., *Finite Element Solution of Boundary Value Problems*. Academic Press, Orlando, Florida, 1984.

[10] Axelsson, O. and Neytcheva, M., Algebraic multilevel iteration method for Stieltjes matrices. *Numer. Linear Algebr. Appli.* **1** (3), 213–236, 1994.

[11] Axelsson, O. and Vassilevski, P.S., Algebraic multilevel preconditioning methods I. *Num. Math.* **56**, 157–177, 1989.

[12] Axelsson, O. and Vassilevski, P.S., Algebraic multilevel preconditioning methods II. *SIAM J. Numer. Anal.* **27**, 1569–1590, 1990.

[13] Babuska, I., The $p$ and $h-p$ versions of the finite element method: The state of the art. *Finite Elements, Theory and Applications* (eds Dwoyer, Hussaini and Voigt). Springer, New York, 1988.

[14] Bai, D. and Brandt, A., Local mesh refinement multilevel techniques. *SIAM J. Sci. Comput.* **8**, 109–134, 1987.

[15] Bai, X.S. and Fuchs, L., A multi-grid method for calculation of turbulence and combustion, *Multigrid Methods IV*, Proceedings of the Fourth European Multigrid Conference (eds P.M. Hemker and P. Wesseling). 35–43. Birkhäuser, Basle, 1994.

[16] Bakhvalov, N.S., On the convergence of a relaxation method with natural constraints on the elliptic operator. *USSR Comp. Math. Math. Phys.* **6**, 101–135, 1966.

[17] Baldwin, B. and Lomax, H., Thin-layer approximation and algebraic model for separated turbulent flows. *AIAA Paper* 78–257, 1978.

[18] Bank, R.E., Dupont, T. and Yserentant, H., The hierarchical basis multigrid method. *Numer. Math.* **52**, 427–458, 1988.

[19] Bank, R.E. and Mittelmann, H.D., Continuation and multi-grid for nonlinear elliptic systems. *Multigrid Methods II*, Lecture Notes in Mathematics 1228 (eds W. Hackbusch and U. Trottenberg). 23–37. Springer, Berlin, 1986.

[20] Bank, R.E., PLTMG: A Software Package for Solving Elliptic Partial Differential Equations. User's Guide 7.0, *Frontiers Appl. Math.* **15**. SIAM, Philadelphia, 1994.

[21] Bank, R.E. and Smith, R.K., The incomplete factorization multigraph algorithm. *SIAM J. Sci. Comput.* **20**, 1349–1364, 1999.

[22] Bank, R.E. and Smith, R.K., The hierarchical basis multigraph algorithm. *SIAM J. Sci. Comput.* (submitted).

[23] Bank, R.E. and Wagner, Ch., Multilevel ILU decomposition. *Numer. Math.* **82**, 543–576, 1999.

[24] Barcus, M., Peric M. and Scheuerer, G., A control volume based full multigrid procedure of two-dimensional, laminar, incompressible flows. *Proceedings of the 7th GAMM Conference on Numerical Methods Fluid Mechanics*. Vieweg, Braunschweig (Germany), 1987.

[25] Barinka, A., Barsch, T., Charton, P., Cohen, A., Dahlke, S., Dahmen, W. and Urban, K., Adaptive wavelet schemes for elliptic problems – implementation and numerical experiments. (Revised version, November 1999) IGPM Rep. 173, RWTH Aachen, 1999.

[26] Barret, R., Berry, M., Chan, T.F., Demmel, J., Donato, J., Dongarra, J., Eijkhout, V., Pozo, R., Romine, C. and van der Vorst, H., *Templates for the Solution of Linear Systems: Building Blocks for Iterative Methods*. SIAM, Philadelphia, 1994.

[27] Barros, S.R.M., The Poisson equation on the unit disk: a multigrid solver using polar coordinates. *Appl. Math. Comput.* **25**, 123–135, 1988.

[28] Barros, S.R.M., Multigrid methods for two- and three-dimensional Poisson-type equations on the sphere. *J. Comp. Phys.* **92**, 313–348, 1991.

[29] Bastian, P., Parallele adaptive Mehrgitterverfahren. Ph.D. thesis (in German), University of Heidelberg, Germany, 1994.

[30] Bastian, P., Birken, K., Johannsen, K., Lang, S., Neuss, N., Rentz-Reichert, H. and Wieners, C., UG – a flexible toolbox for solving partial differential equations. *Comp. Visual. Sci.* **1**, 27–40, 1997.

[31] Bastian, P., Load balancing for adaptive multigrid methods. *SIAM J. Sci. Comput.* **19** (4), 1303–1321, 1998.

[32] Bastian, P., Hackbusch, W. and Wittum, G., Additive and multiplicative multi-grid – A comparison. *Computing* **60**, 345–364, 1998.

[33] Batchelor, G.K., *An Introduction to Fluid Dynamics*. Cambridge University Press, Cambridge, 1967.

[34] Batycky, R., Blunt, M. and Thiele, M., A 3D Field-scale Streamline-based Reservoir Simulator. *SPE Reservoir Engineering*, November 1997.

[35] Baumgardner, J.R. and Frederickson, P.O., Icosahedral discretization of the two-sphere. *SIAM J. Numer. Anal.* **22**, 1107–1115, 1985.

[36] Bell, J.B., Shubin, G.R. and Stephens, A.B., A segmentation approach to grid generation using biharmonics. *J. Comp. Phys.* **47**, 463–472, 1982.

[37] Bernardi, Ch and Maday, Y., Spectral methods. *Handbook of Numerical Analysis Vol. 5: Techniques of Scientific Computing Part 2.* (eds P.G. Ciarlet and J.L. Lions). 209–485. North Holland, Amsterdam, 1997.

[38] Berger, M. and Oliger, J., Adaptive mesh refinement for hyperbolic partial differential equations. *J. Comp. Phys.* **53**, 484–512, 1984.

[39] Berger, M.J. and Jameson, A., Automatic adaptive grid refinement for the Euler equations. *AIAA J.* **23**, 561–568, 1985.

[40] Bernert, K., $\tau$-extrapolation – theoretical foundation, numerical experiment and application to Navier–Stokes equations. *SIAM J. Sci. Comput.* **18** (2), 460–478, 1997.

[41] Bey, J., *Finite-Volumen- und Mehrgitter-Verfahren für elliptische Randwertprobleme.* Teubner Verlag, Stuttgart, 1998.

[42] Bleecke, H.M., Eisfeld, B., Heinrich, R., Ritzdorf, H., Fritz, W., Leicher, S. and Aumann, P., Benchmarks and Large Scale Examples. *Portable Parallelization of Industrial Aerodynamic Applications (POPINDA), Results of a BMBF project.* Notes on Numerical Fluid Mechanics 71. 89–103. Vieweg, Braunschweig, Germany, 1999.

[43] Börgers, C., Mehrgitterverfahren für eine Mehrstellendiskretisierung der Poissongleichung und für eine zweidimensionale singulär gestörte Aufgabe. Master's thesis, University of Bonn, West Germany, 1981.

[44] Bolstad, J.H. and Keller, H.B., A multigrid continuation method for elliptic problems with turning points. *SIAM J. Sci. Comput.* **7**, 1081–1104, 1986.

[45] Bornemann, F. and Yserentant, H., A basic norm equivalence for the theory of multilevel methods. *Numer. Math.* **64**, 455–476, 1993.

[46] Bornemann, F. and Deuflhard, P., The cascadic multigrid method for elliptic problems. *Numer. Math.* **75**, 135–152, 1996.

[47] Bornemann, F. and Krause, R., Classical and cascadic multigrid – a methodological comparison. *Domain Decomposition Methods in Science and Engineering* (eds P. Bjorstad, M. Espedal and D. Keyes). 64–71. DD-Press, Bergen, Norway, 1998.

[48] Braess, D., The contraction number of a multigrid method for solving the Poisson equation. *Numer. Math.* **37**, 387–404, 1981.

[49] Braess, D., The convergence rate of a multigrid method with Gauss – Seidel relaxation for the Poisson equation. *Multigrid Methods.* Lecture Notes in Mathematics 960 (eds. W. Hackbusch and U. Trottenberg). 220–312. Springer, Berlin, 1982.

[50] Braess, D., *Finite Elements. Theory, Fast Solvers and Applications in Solid Mechanics.* Cambridge University Press, Cambridge, 1997.

[51] Braess, D., Towards algebraic multigrid for elliptic problems of second order. *Computing* **55**, 379–393, 1995.

[52] Braess, D. and Hackbusch, W., A new convergence proof for the multigrid method including the V-Cycle. *SIAM J. Numer. Anal.* **20**, 967–975, 1983.

[53] Braess, D. and Sarazin, R., An efficient smoother for the Stokes problem. *Appl. Num. Math.* **23**, 3–20, 1997.

[54] Bramble, J.H., *Multigrid Methods.* Pitman Research Notes in Mathematics Series 294. Longman, Harlow, 1993.

[55] Bramble, J.H., Pasciak, J.E. and Xu, J., Parallel multilevel preconditioners. *Math. Comp.* **55**, 1–22, 1990.

[56] Brandt, A., Multi-level adaptive technique (MLAT) for fast numerical solution to boundary value problems. *Proceedings of the 3rd International Conference on Numerical Methods in Fluid Mechanics*, Lecture Notes in Physics 18 (eds H. Cabannes and R. Temam). 82–89. Springer, Berlin, 1973.

[57] Brandt, A., Multi-level adaptive techniques (MLAT). I. The multigrid method. Research Rep. RC 6026, IBM T.J. Watson Research Center, Yorktown Heights, NY, 1976.

[58] Brandt, A., Multi-level adaptive solutions to boundary-value problems. *Math. Comput.* **31**, 333–390, 1977.

[59] Brandt, A., Multi-level adaptive techniques (MLAT) for partial differential equations: ideas and software. *Mathematical Software III* (ed. J.R. Rice). 277–318. Academic Press, New York, 1977.

[60] Brandt, A., Multi-level adaptive techniques (MLAT) for singular perturbation problems. *Numerical Analysis of Singular Perturbation Problems* (eds P.W. Hemker and J.J.H. Miller). 53–142. Academic Press, New York, 1979.

[61] Brandt, A., Stages in developing multigrid solutions. *Proceedings of the 2nd International Congress on Numerical Methods for Engineers* (eds E. Absi, R. Glowinski, P. Lascaux and H. Veysseyre). 23–43. Dunod, Paris, 1980.

[62] Brandt, A., Numerical stability and fast solutions to boundary value problems. *Boundary and Interior Layers – Computational and Asymptotic Methods* (ed. J.J.H. Miller). 29–49. Boole Press, Dublin, 1980.

[63] Brandt, A., Multigrid solvers for non-elliptic and singular-perturbation steady-state problems. Weizmann Institute of Science, Rehovot, Israel, December 1981.

[64] Brandt, A., Guide to multigrid development. *Multigrid Methods*, Lecture Notes in Mathematics 960 (eds W. Hackbusch and U. Trottenberg). 220–312. Springer, Berlin, 1982.

[65] Brandt, A., Multigrid solvers on parallel computers. *Elliptic Problem Solvers* (ed. M.H. Schultz). 39–84. Academic Press, New York, 1981.

[66] Brandt, A., Multigrid techniques: 1984 guide with applications to fluid dynamics. GMD-Studie Nr. 85, Sankt Augustin, West Germany, 1984.

[67] Brandt, A., Algebraic multigrid theory: the symmetric case. *Appl. Math. Comp.* **19**, 23–56, 1986.

[68] Brandt, A., Rigorous local mode analysis of multigrid. Preliminary Proceedings of the 4th Copper Mountain Conference on Multigrid Methods, Copper Mountain, Colorado, 55–133, April 1989.

[69] Brandt, A., Rigorous quantitative analysis of multigrid: I. Constant coefficients two level cycle with $L_2$ norm. *SIAM J. Numer. Anal.* **31**, 1695–1730, 1994.

[70] Brandt, A., The Gauss Center research in scientific computation. *Electron Trans. Num. An.* **6**, 1–34, 1997.

[71] Brandt, A., General highly accurate algebraic coarsening schemes. Proceedings of the 9th Copper Mountain Conference on Multigrid Methods, Copper Mountain, April 1999.

[72] Brandt, A. and Dinar, N., Multigrid solutions to flow problems. *Numerical Methods for Partial Differential Equations* (ed. S. Parter). 53–147. Academic Press, New York, 1979.

[73] Brandt, A. and Diskin, B., Multigrid solvers on decomposed domains. *In:* Quarteroni, A., *et al.* (eds.), Domain decomposition methods in science and Eng. *Contemp. Math.* **157**, 135–155, 1994.

[74] Brandt, A. and Diskin, B., Multigrid solvers for the non-aligned sonic flow: the constant coefficient case. *Comput. Fluids* **28**, 511–549, 1999.

[75] Brandt, A. and Diskin, B., Multigrid solvers for non-aligned sonic flows. *SIAM J. Sci. Comput.* **21**, 473–501, 1999.

[76] Brandt, A. and Diskin, B., Efficient multigrid solvers for the linearized transonic full-potential operator. (submitted).

[77] Brandt, A. and Dym, J., Effective boundary treatment for the biharmonic Dirichlet problem, *Proceedings of the 7th Copper Mountain Conference on Multigrid Methods*, NASA Conference Publication 3339 (eds. N.D. Melson *et al.*). 97–108. 1996.

[78] Brandt, A. and Greenwald, J., Parabolic multigrid revisited. *Multigrid Methods III*, Proceedings of the 3rd International Conference on Multigrid Methods (eds W. Hackbusch and U. Trottenberg). 143–154. Birkhäuser, Basle, 1991.

[79] Brandt, A. and Livshits, I., Wave-ray multigrid methods for standing wave equations. *Electr. Trans. Num. Analysis* **6**, 162–181, 1997.

[80] Brandt, A., McCormick, S.F. and Ruge, J., Algebraic multigrid (AMG) for automatic multigrid solution with application to geodetic computations. Institute for Computational Studies, Fort Collins, Colorado, 1982.

[81] Brandt, A., McCormick, S.F. and Ruge, J., Algebraic multigrid (AMG) for sparse matrix equations. *Sparsity and its Applications* (ed. D.J. Evans). 257–284. Cambridge University Press, Cambridge, 1984.

[82] Brandt, A. and Mikulinsky, V., Recombining iterants in multigrid algorithms and problems with small islands. *SIAM J. Sci. Comput.* **16**, 20–28, 1995.

[83] Brandt, A. and Ta'asan, S., Multigrid solutions to quasi-elliptic schemes. *Progress and Super-computing in Computational Fluid Dynamics* (eds E.M. Murman and S.S. Abarbanel). 235–255, Birkhäuser, Boston, 1985.

[84] Brandt, A. and Ta'asan, S., Multigrid method for nearly singular and slightly indefinite problems. *Multigrid Methods II*, Lecture Notes in Mathematics 1228 (eds W. Hackbusch and U. Trottenberg). 99–121. Springer Berlin, 1986.

[85] Brandt, A. and Yavneh, I., Inadequacy of first-order upwind difference scheme for some recirculating flows. *J. Comp. Phys.* **93**, 128–143, 1991.

[86] Brandt, A. and Yavneh, I., On multigrid solution of high-Reynolds incompressible entering flows. *J. Comp. Phys.* **101**, 151–164, 1992.

[87] Brandt, A. and Yavneh, I., Accelerated multigrid convergence and high-Reynolds recirculating flows. *SIAM J. Sci. Comput.* **14**, 607–626, 1993.

[88] Brenner, S.C., Convergence of the multigrid V-cycle algorithm for second order boundary values problems without full elliptic regularity. Preprint 1999:07, Department of Mathematics, University of South Carolina, 1999.

[89] Brenner, S.C. and Scott, L.R., *The Mathematical Theory of Finite Element Methods*. Springer, Berlin, 1994.

[90] Brezina, M., Cleary, A.J., Falgout, R.D., Henson, V.E., Jones, J.E., Manteuffel, T.A., McCormick, S.F. and Ruge, J.W., Algebraic Multigrid Based on Element Interpolation (AMGe). LLNL Techn. Rep. UCRL-JC-131752, submitted to *SIAM J. Sci. Comput.*

[91] Briggs, W.L., *A Multigrid Tutorial*. SIAM, Philadelphia, 1987.

[92] Bulgak, H. and Zenger, C. (eds), *Error Control and Adaptivity in Scientific Computing*, NATO Science Series, Series C: Mathematics and Physical Sciences, Vol. 536. Kluwer, Dordrecht, 1999.

[93] Buneman, O., A compact non-iterative Poisson solver. Res. Rep. 294, Stanford University Institute of Plasma Research, 1969.

[94] Bussing, T.R.A. and Murman, E.M., Finite-volume method for the calculation of compressible chemically reacting flows. *AIAA J.* **26**, 1070–1078, 1988.

[95] Buzbee, B.L., Golub, G.H. and Nielson, C.W., On direct methods for solving Poisson's equation. *SIAM J. Numer. Anal.* **7**, 627–656, 1970.

[96] Catalano, L.A. and Deconinck, H., Two-dimensional optimization of smoothing properties of multi-stage schemes applied to hyperbolic equations. *Multigrid Methods: Special Topics and Applications II*, GMD-Studien 189 (eds W. Hackbusch and U. Trottenberg). 43–56. GMD, Sankt Augustin, West Germany, 1991.

[97] Chakravarthy, S. and Osher, S., A new class of high accuracy TVD schemes for hyperbolic conservation laws. *AIAA Paper* 85-0363, 1985.

[98] Concus, P., Golub, G. and O'Leary, D., A generalized conjugate gradient method for the numerical solution of elliptic partial differential equations. *Sparse Matrix Computations* (eds J. Bunch and D. Rose). 309–332. Academic Press, New York, 1976.

[99] Canu, J. and Linden, J., Multigrid solution of the 2D Euler equations: A comparison of Osher's and Dick's flux difference splitting schemes. Arbeitspapiere der GMD Nr. 693, GMD, Sankt Augustin, Germany, 1992.

[100] Canu, J. and Ritzdorf, H., Adaptive, block-structured multigrid on local memory machines. Proceedings of 9th GAMM Seminar, Kiel, 1993.

[101] Chan, T.F. and Keller, H.B., Arc-length continuation and multi-grid techniques for nonlinear elliptic eigenvalue problems. *SIAM J. Sci. Comput.* **3**, 173–194, 1982.

[102] Chan, T. and Mathew, T., Domain decomposition algorithms. *Acta Numerica 94* **3**, 61–143, 1994.

[103] Chan, T.F., Go, S. and Zikatanov, L., Multilevel methods for elliptic problems on unstructured grids. VKI Lecture Series 1997-02, van Karman Institute, Rhode St.G. Belgium, 1997.

[104] Chan, T., Zikatanov, L. and Xu, J., An agglomeration multigrid method for unstructured grids. Proceedings of the 10th International Conference on Domain Decomposition Methods, Boulder, Colorado, August 1997, (eds J. Mandel, C. Farhat and X.-C. Cai). *Cont. Math.* **218**, 67–81, 1998.

[105] Chang, Q., Wong, Y.S. and Fu, H., On the algebraic multigrid method. *J. Comp. Phys.* **125**, 279–292, 1996.

[106] Chatelin, F. and Miranker, W.L., Acceleration by aggregation of successive approximation methods. *LAA* **43**, 17–47, 1982.

[107] Chen, L. and Armfield, S., A simplified marker and cell method for unsteady flows on non-staggered grids. *Int. J. Num. Meth. Fluids* **21**, 15–34, 1995.

[108] Cimmino, G., La ricerca scientifica ser. II 1. *Pubbliz. dell'Inst. pre le Appl. del Calculo* **34**, 326–333, 1938.

[109] Cleary, A.J., Falgout, R.D., Henson, V.E., Jones, J.E., Manteuffel, T.A., McCormick, S.F., Miranda, G.N. and Ruge, J.W., Robustness and scalability of algebraic multigrid. *SIAM J. Sci. Comput.* (special issue on the Fifth Copper Mountain Conference on Iterative Methods, 1998), **21**, 1886–1908, 2000.

[110] Cleary, A.J., Falgout, R.D., Henson, V.E. and Jones, J.E., Coarse-grid selection for parallel algebraic multigrid. *Proceedings of the 5th International Symposium on Solving Irregularly Structured Problems in Parallel*, Lecture Notes in Computer Science 1457. 104–115. Springer, New York, 1998.

[111] Collatz, L., *The Numerical Treatment of Differential Equations*. Springer, Berlin, 1966.

[112] Cooley, J.W. and Tukey, J.W., An algorithm for the machine calculation of complex Fourier series. *Math. Comp.* **19**, 297–301, 1965.

[113] Courant, R., Friedrichs, K.O. and Lewy, H., Über die partiellen Differenzengleichungen der mathematischen Physik. *Math. Annal.* **100**, 32–74, 1928.

[114] Dahmen, W. and Kunoth, A., Multilevel preconditioning. *Numer. Math.* **63**, 315–344, 1992.

[115] Dahmen, W. and Elsner, L., *Algebraic multigrid methods and the Schur complement*, Notes on Numerical Fluid Mechanics 23. Vieweg, Braunschweig, 1988.

[116] Dahmen, W., Kurdila, A. and Oswald, P. (eds), *Multiscale Wavelet Methods for Partial Differential Equations*. Academic Press, San Diego, 1997.

[117] Dahmen, W., Wavelet and multiscale methods for operator equations, *Acta Numerica* **6**, 55–228, 1997.

[118] Darmofal, D.L. and Siu, K., A robust multigrid algorithm for the Euler equations with local preconditioning and semi-coarsening. *J. Comp. Phys.* **151**, 728–756, 1999.

[119] Decker, D.W. and Keller, H.B., Multiple limit point bifurcation, *J. Math. Anal.* **75**, 417–430, 1980.

[120] Dendy (Jr.), J.E., Black box multigrid, *J. Comp. Phys.* **48**, 366–386, 1982.

[121] Dendy (Jr.), J.E., Black box multigrid for nonsymmetric problems, *Appl. Math. Comp.* **13**, 261–284, 1983.

[122] Dendy, J.E., McCormick, S.F., Ruge, J.W., Russel, T.F. and Schaffer, S., Multigrid methods for three-dimensional petroleum reservoir simulation. Paper SPE 18409, Tenth Symposium on Reservoir Simulation, Houston, Texas, 1989.

[123] Deuflhard, P., Cascadic conjugate gradient methods for elliptic partial differential equations. Algorithm and numerical results. *Domain Decomposition Methods in Scientific and Engineering Computation* (eds D.E. Keyes and J. Xu). 29–42. AMS, Providence, 1994.

[124] Dick, E., Multigrid solution of steady Euler equations based on polynomial flux-difference splitting. *Int. J. Num. Meth. Heat Fluid Flow*, **1**, 51–62, 1991.

[125] Dick, E. and Linden, J., A multigrid method for steady incompressible Navier–Stokes equations based on flux difference splitting. *Int. J. Num. Meth. Fluids* **14**, 1311–1323, 1992.

[126] Dick, E. and Riemslagh, K., Multi-staging of Jacobi relaxation to improve smoothing properties of multigrid methods for steady Euler equations. *J. Comp. Appl. Math.* **50**, 241–254, 1994.

[127] Dick, E. and Riemslagh, K., Multi-staging of Jacobi relaxation to improve smoothing properties of multigrid methods for steady Euler equations, II. *J. Comp. Appl. Math.* **59**, 339–348, 1995.

[128] Dick, E. and Steelant, J., Coupled solution of the steady compressible Navier–Stokes equations and the $k-\varepsilon$ turbulence equations with a multigrid method. *Appl. Numer. Math.* **23**, 49–61, 1997.

[129] Dick, E., Riemslagh, K. and Vierendeels, J. (eds), *Multigrid Methods VI*, Proceedings of the 6th European Multigrid Conference, Lecture Notes in Computational Science and Engineering 14. Springer, Berlin, 2000.

[130] Dinar, N., Fast methods for the numerical solutions of boundary value problems. Ph.D. thesis, Weizmann Institute of Science, Rehovot, Israel, 1979.

[131] Dryja, M., Smith, B.F. and Widlund, O.B., Schwarz analysis of iterative substructuring algorithms for elliptic problems in three dimensions. *SIAM J. Numer. Anal.* **31** (6), 1662–1694, 1994.

[132] Dryja, M. and Widlund, O.B., Domain decomposition algorithms with small overlap. *SIAM J. Sci. Comput.* **15**, 604–620, 1994.

[133] Dryja, M. and Widlund, O.B., Schwarz methods of Neumann–Neumann type for three-dimensional elliptic finite element problems. *Comm. Pure Appl. Math.* **48** (2), 121–155, 1995.

[134] D'yakonov, E.G., *Minimization of the Computational Work – Asymptotically Optimal Algorithms for Elliptic Equations* (in Russian). Nauka, Moscow, 1989.

[135] Eiseman, P.R., Grid generation for fluid mechanics computations. *Ann. Rev. Fluid Mech.* **17**, 487–522, 1985.

[136] Elman, H.R., Ernst, O.G. and O'Leary, D.P., A multigrid method enhanced by Krylov subspace iteration for discrete Helmholtz equations, University of Maryland, Comp. Science Dept., Preprint CS-TR-4029, 1999.

[137] Eriksson, K., Estep, D., Hansbo, P. and Johnson, C., *Introduction to Adaptive Methods for Differential Equations* (ed. A. Iserles). 105–158. Cambridge University Press, Cambrige, 1995.

[138] Farhat, C. and Roux, F.-X., Implicit parallel processing in structural mechanics. *Comput. Mech. Adv.* **2** (1), 1–124, 1994.

[139] Fedorenko, R.P., A relaxation method for solving elliptic difference equations. *USSR Comp. Math. Math. Phys.* **1** (5), 1092–1096, 1962.

[140] Fedorenko, R.P., The speed of convergence of one iterative process. *USSR Comp. Math. Math. Phys.* **4** (3), 227–235, 1964.

[141] Foerster, H., Stüben K. and Trottenberg, U., Non-standard multigrid techniques using checkered relaxation and intermediate grids. *Elliptic Problem Solvers* (ed. M.H. Schultz). 285–300. Academic Press, New York, 1981.

[142] Frederickson, P.O., Fast approximate inversion of large sparse linear systems. Mathematics Rep. 7-75, Department of Mathematical Sciences, Lakehead University, Ontario, 1975.

[143] Frederickson, P.O. and McBryan, O.A., Parallel superconvergent multigrid. *Multigrid Methods: Theory, Applications and Supercomputing* (ed. S. McCormick). 195–210. Marcel Dekker, New York, 1988.

[144] Frederickson, P.O. and McBryan, O.A., Recent developments for the PSMG multiscale method. *Multigrid Methods III*, Proceedings of the 3rd International Conference on Multigrid Methods (eds W. Hackbusch and U. Trottenberg). 21–40. Birkhäuser, Basle, 1991.

[145] Frohn-Schauf, C., Flux-Splitting-Methoden und Mehrgitterverfahren für hyperbolische Systeme mit Beispielen aus der Strömungsmechanik. Ph.D. thesis (in German), University of Düsseldorf, Germany, 1992.

[146] Fromm, J.E., A method for reducing dispersion in convective difference schemes. *J. Comp. Phys.* **69**, 176–189, 1968.

[147] Fuhrmann, J., A modular algebraic multilevel method. Tech. Rep. Preprint 203, Weierstrass-Institut für Angewandte Analysis Stochastik, Berlin, 1995.

[148] Fuchs, L., Multigrid systems for incompressible flows. *Efficient Solution of Elliptic Systems*, Notes on Numerical Fluid Mechanics 10 (ed. W. Hackbusch). 38–51. Vieweg, Braunschweig, 1984.

[149] Fuchs, L. and Zhao, H.S., Solution of three-dimensional viscous incompressible flows by a multi-grid method. *Int. J. Num. Meth. Fluids* **4**, 539–555, 1984.

[150] Garabedian, P.R., *Partial Differential Equations*. Wiley, New York, 1964.

[151] Gärtel, U., Parallel multigrid solver for 3D anisotropic elliptic problems. *Hypercube and Distributed Computers* (eds F. André and J.P. Versus). 37–47. North Holland, Amsterdam, 1989.

[152] Gärtel, U., Joppich, W., Schüller, A., Schwichtenberg, H., Trottenberg, U. and Winter, G., Two strategies in parallel computing: Porting existing software versus developing new parallel algorithms – two examples. *Fut. Generation Comp. Systems* **10**, 257–262, 1994.

[153] Gärtel, U., Krechel, A., Niestegge, A. and Plum, H.-J., Parallel multigrid solution of 2D and 3D anisotropic elliptic equations: standard and nonstandard smoothing. *Multigrid Methods III*, Proceedings of the 3rd International Conference on Multigrid Methods (eds W. Hackbusch and U. Trottenberg). 191–209. Birkhäuser, Basle, 1991.

[154] Gear, C.W., *Numerical Initial Value Problems for Ordinary Differential Equations*. Prentice Hall, Englewood Cliffs, 1971.

[155] Gerhold, T., Friedrich, O., Evans, J. and Galle, M., Calculation of complex three-dimensional configurations employing the DLR-TAU-code. *AIAA Paper* 97-0167, 1997.

[156] Gerlinger, P. and Brüggemann, D., An implicit multigrid scheme for the compressible Navier–Stokes equations with low-Reynolds-number closure. *Trans. ASME/J. Fluids Eng.* **120**, 257–262, 1998.

[157] Ghia, U., Ghia, K.N. and Shin, C.T., High-Re solutions for incompressible flow using the Navier–Stokes equations and a multigrid method. *J. Comp. Phys.* **48**, 387–411, 1982.

[158] Godunov, S.K., Finite difference method for numerical computation of discontinuous solutions of the equations of fluid dynamics. *Mat. Sbornik* **47**, 271–306, 1959.

[159] Golub, G. and van Loan, C., *Matrix Computations*, 2nd edn. The John Hopkins University Press, Baltimore, 1989.

[160] Grauschopf, T., Griebel, M. and Regler, H., Additive multilevel-preconditioners based on bilinear interpolation, matrix dependent geometric coarsening and algebraic multigrid coarsening for second order elliptic PDEs. *Appl. Numer. Math.* **23**, 63–96, 1997.

[161] Greenbaum, A., A multigrid method for multiprocessors. *Appl. Math. Comp.* **19**, 23–45, 1986.

[162] Greenbaum, A., *Iterative Methods for Solving Linear Systems*. SIAM, Philadelphia, 1997.

[163] Griebel, M., Multilevel algorithms considered as iterative methods on semidefinite systems. *SIAM J. Sci. Comput.* **15**, 547–565, 1994.

[164] Griebel, M., *Multilevelmethoden als Iterationsverfahren über Erzeugendensystemen*. Teubner Verlag, Stuttgart, 1994.

[165] Griebel, M. and Thurner, V., The efficient solution of fluid dynamics problems by the combination technique. *Int. J. Num. Meth. Heat Fluid Flow* **5**, 51–269, 1995.

[166] Griebel, M., Neunhoeffer, T. and Regler, H., Algebraic multigrid methods for the solution of the Navier–Stokes equations in complicated geometries. SFB-Bericht Nr. 342/01/96 A, Institüt für Informatik, Technische Universität München, Munich, 1996.

[167] Griebel, M. and Oswald, P., Remarks on the abstract theory of additive and multiplicative Schwarz methods. *Numer. Math.* **70**, 163–180, 1995.

[168] Guillard, H. and Vanek, P., An aggregation multigrid solver for convection-diffusion problems on unstructured meshes. Rep. 130, Center for Computational Mathematics, University of Denver, 1998.

[169] Hackbusch, W., Ein iteratives Verfahren zur schnellen Auflösung elliptischer Randwertprobleme. Rep. 76-12, Institute for Applied Mathematics, University of Cologne, West Germany, 1976.

[170] Hackbusch, W., On the convergence of a multi-grid iteration applied to finite element equations. Rep. 77-8, Institute for Applied Mathematics, University of Cologne, West Germany, 1977.

[171] Hackbusch, W., Convergence of multi-grid iterations applied to difference equations. *Math. Comp.* **34**, 425–440, 1980.

[172] Hackbusch, W., On the convergence of multi-grid iterations. *Beiträge Numer. Math.* **9**, 213–239, 1981.

[173] Hackbusch, W., Multi-grid convergence theory. *Multigrid Methods*, Lecture Notes in Mathematics 960 (eds W . Hackbusch and U. Trottenberg). 177–219. Springer, Berlin, 1982.

[174] Hackbusch, W. and Trottenberg, U., (eds), *Multigrid Methods*, Lecture Notes in Mathematics 960. Springer, Berlin, 1982.

[175] Hackbusch, W., Parabolic multigrid methods. *Computing Methods in Applied Sciences and Engineering VI*. Proceedings of the 6th International Symposium on Computational Methods in Applied Sciences and Engineering (eds R. Glowinski and J.R. Lions). 20–45. North Holland, Amsterdam, 1984.

[176] Hackbusch, W., *Multi-grid Methods and Applications*. Springer, Berlin, 1985.

[177] Hackbusch, W. and Trottenberg, U. (eds), *Multigrid Methods II*, Lecture Notes in Mathematics 1228. Springer, Berlin 1986.

[178] Hackbusch, W. and Trottenberg, U. (eds), *Multigrid Methods III*, Proceedings of the 3rd International Conference on Multigrid Methods, International Series on Numerical Mathematics 98, Birkhäuser, Basle, 1991.

[179] Hackbusch, W., *Theory and Numerical Treatment of Elliptic Differential Equations*. Springer, New York, 1992.

[180] Hackbusch, W., *Iterative Solution of Large Sparse Systems of Equations*. Springer, New York, 1994.

[181] Hackbusch, W. and Wittum, G. (eds), *Multigrid Methods V*, Proceedings of the 5th European Multigrid Conference, Lecture Notes in Computational Science and Engineering 3. Springer, Berlin, 1998.

[182] Harlow, F.H. and Welch, J.E., Numerical calculation of time-dependent viscous incompressible flow of fluid with free surface. *Phys. Fluids* **8**, 2182–2189, 1965.

[183] Harten, A., High resolution schemes for hyperbolic conservation laws, *J. Comp. Phys.* **49**, 357–393, 1983.

[184] Hemker, P.W., A note on defect correction processes with an approximate inverse of deficient rank. *J. Comp. Appl. Math.* **8**, 137–139, 1982.

[185] Hemker, P.W., Mixed defect correction iteration for the accurate solution of the convection diffusion equation. *Multigrid Methods*, Lecture Notes in Mathematics 960 (eds W. Hackbusch and U. Trottenberg). 485–501. Springer, Berlin, 1982.

[186] Hemker, P.W., Mixed defect correction iteration for the solution of a singular perturbation problem. *Defect Correction Methods: Theory and Applications*, Computing Suppl. 5 (eds K. Böhmer and H.J. Stetter). 123–145. Springer, Berlin, 1984.

[187] Hemker, P.W., On the order of prolongations and restrictions in multigrid procedures. *J. Comp. Appl. Math.* **32**, 423–429, 1990.

[188] Hemker, P.W. and Spekreijse, S.P., Multiple grid and Osher's scheme for the efficient solution of the steady Euler equations. *Appl. Num. Math.* **2**, 475–493, 1986.

[189] Hemker, P.W. and Wesseling, P. (eds), *Multigrid Methods IV*, Proceedings of the 4th European Multigrid Conference. Birkhäuser, Basle, 1994.

[190] Hempel, R., The MPI Standard for Message Passing. *HPCN Conference München*, Lecture Notes in Computer Science 796 (eds W. Gentzsch and U. Harms). 247–252. Springer, Berlin, 1994.

[191] Hempel, R. and Ritzdorf, H., The GMD communications subroutine library for grid-oriented problems. Arbeitspapiere der GMD 589, Sankt Augustin, West Germany, 1991.

[192] Hempel, R. and Schüller, A., Experiments with parallel multigrid algorithms using the SUPRENUM communications subroutine library. GMD Arbeitspapier 141, GMD, Sankt Augustin, West Germany, 1988.

[193] Hentschel, R. and Hirschel, E.H., Self adaptive flow computations on structured grids. *Computational Fluid Dynamics '94* (eds S. Wagner, E.H. Hirschel and J. Periaux). 242–249. Wiley, Chichester, 1994.

[194] Hestenes, M.R. and Stiefel, E., Methods of conjugate gradients for solving linear systems. *J. Res. Nat. Bur. Stand. Sect. B* **49**, 409–436, 1952.

[195] Hirsch, C., *Numerical Computation of Internal and External Flows*. Vol. 1, Wiley, Chichester, 1989.

[196] Hirsch, C., *Numerical Computation of Internal and External Flows*. Vol. 2, Wiley, Chichester, 1990.

[197] Hoppe, H. and Mühlenbein, H., Parallel adaptive full-multigrid methods on message-based multiprocessors. *Parallel Comput.* **3**, 269–287, 1986.

[198] Horton, G., The time-parallel multigrid method. *Comm. Appl. Num. Meth.* **8**, 585–596, 1992.

[199] Horton, G. and Vandewalle, S., A space-time multigrid method for parabolic pdes, *SIAM J. Sci. Comput.* **16** (4), 848–864, 1995.

[200] Huang, W.Z., Convergence of algebraic multigrid methods for symmetric positive definite matrices with weak diagonal dominance. *Appl. Math. Comp.* **46**, 145–164, 1991.

[201] Huang, Y., A multigrid method for solution of vorticity-velocity form of 3-d Navier–Stokes equations. *Fifth Copper Mountain Conference on Multigrid Methods*, Vol. 2, 23–60, University of Colorado, Denver 1991.

[202] Jameson, A., Schmidt, W. and Turkel, E., Numerical simulation of the Euler equations by finite volume methods using Runge–Kutta time stepping schemes. *AIAA Paper* 81-1259, 1981.

[203] Jameson, A., Solution of the Euler equations for two dimensional transonic flow by a multigrid method, *Appl. Math. Comp.* **13**, 327–356, 1983.

[204] Jameson, A., Transonic flow calculations for aircraft, *Numerical Methods in Fluid Mechanics*, Lecture Notes in Mathematics 1127 (ed. F. Brezzi). 156–242. Springer, Berlin, 1985.

[205] Jameson, A., Analysis and design of numerical schemes for gas dynamics, 1: artificial diffusion, upwind biasing, limiters and their effect on accuracy and multigrid convergence. *Int. J. Comp. Fluid Dyn.* **4**, 171–218, 1995.

[206] Joppich, W. and Mijalković, S., *Multigrid Methods for Process Simulation*. Springer, Vienna, 1993.

[207] Kaczmarz, S., Angenäherte Auflösung von Systemen linearer Gleichungen. *Bulle. Acad. Pol. Sci. Lett. A* **15**, 355–357, 1937.

[208] Kaspar, W. and Remke, R., Die numerische Behandlung der Poisson-Gleichung auf einem Gebiet mit einspringenden Ecken. *Computing* **22**, 141–151, 1979.

[209] Keller, H.B., Numerical solution of bifurcation and nonlinear eigenvalue problems. *Applications of Bifurcation Theory* (ed. P. Rabinowitz). 359–384. Academic Press, New York, 1977.

[210] Kershaw, D., The incomplete Cholesky-conjugate gradient method for the iterative solution of systems of linear equations. *J. Comp. Phys.* **26**, 43–65, 1978.

[211] Kettler, R., Analysis and comparison of relaxation schemes in robust multigrid and preconditioned conjugate gradient methods. *Multigrid Methods*, Lecture Notes in Mathematics 960 (eds W. Hackbusch and U. Trottenberg). 502–534. Springer, Berlin, 1982.

[212] Kettler, R. and Wesseling, P., Aspects of multigrid methods for problems in three dimensions. *Appl. Math. Comp.* **19**, 159–168, 1986.

[213] Khalil, M., Analysis of linear multigrid methods for elliptic differential equations with discontinuous and anisotropic coefficients. Ph.D. thesis, Delft University of Technology, Delft, The Netherlands, 1989.

[214] Kickinger, F., Algebraic multi-grid for discrete elliptic second order problems. Institutsbericht 513, Institut für Mathematik, Universität Linz, Austria, 1997.

[215] Kim, J. and Moin, P., Application of a fractional-step method to incompressible Navier–Stokes equations. *J. Comp. Phys.* **59**, 308–323, 1985.

[216] Kim, Y. and Chung, T., Finite-element analysis of turbulent diffusion flames. *AIAA J.* **27**, 330–339, 1988.

[217] Klawonn, A. and Widlund, O.B., FETI and Neumann–Neumann Iterative Substructuring Methods: Connections and New Results. Technical Rep. 796, Computer Science Department, Courant Institute for Mathematical Sciences, New York, 1999.

[218] Knapek, S., Matrix-dependent multigrid homogenization for diffusion problems. *SIAM J. Sci. Comput.* **20** (2), 515–533, 1998.

[219] Kolp, O. and Mierendorff, H., Efficient multigrid algorithms for locally constrainted parallel systems. *Appl. Math. Comp.* **19**, 169–200, 1986.

[220] Koren, B., Defect correction and multigrid for an efficient and accurate computation of airfoil flows. *J. Comp. Phys.* **183**, 193–206, 1988.

[221] Krechel, A., Plum, H-J. and Stüben, K., Parallelization and vectorization aspects of the solution of tridiagonal linear systems. *Parallel Comput.* **14**, 31–49, 1990.

[222] Krechel, A. and Stüben, K., Operator dependent interpolation in algebraic multigrid. *Multigrid Methods V*, Proceedings of the 5th European Multigrid Conference, Lecture Notes in Computational Science and Engineering 3 (eds W. Hackbusch and G. Wittum). 189–211. Springer, Berlin, 1998.

[223] Krechel, A. and Stüben, K., Parallel algebraic multigrid based on subdomain blocking. GMD-Report 71, 1999. Submitted to Parallel Comput.

[224] Kroll, N., Eisfeld, B. and Bleecke, H.M., FLOWer. *Portable Parallelization of Industrial Aerodynamic Applications (POPINDA), Results of a BMBF Project*, Notes on Numerical Fluid Mechanics 71 (ed. A. Schüller). 58–71. Vieweg, Braunschweig, Germany, 1999.

[225] Kroll, N., Radespiel, R. and Rossow, C.-C., Accurate flow solvers for 3d applications on structured meshes. VKI Lecture Series on CFD 1994, AGARD R-807, 4.1-4.59, 1995.

[226] Kronsjö, L. and Dahlquist, G., On the design of nested iterations for elliptic difference equations. *BIT* **11**, 63–71, 1972.

[227] Kronsjö, L., A note on the "nested iteration" method. *BIT* **15**, 107–110, 1975.

[228] Laasonen, P., On the discretization error of the Dirichlet problem in a plane region with corners. *Ann. Acad. Sci. Fenn. Ser. AI Math. Dissertat.* **408**, 1–16, 1967.

[229] Lax, P., *Hyperbolic Systems of Conservation Laws and the Mathematical Theory of Shock Waves*, Regional Conference Series in Applied Mathematics 11. SIAM, Philadelphia, 1973.

[230] Lax, P. and Wendroff, B., Systems of conservation laws. *Comm. Pure Appl. Math.* **13**, 217–237, 1960.

[231] van Leer, B., Flux-vector splitting for the Euler equations. *Proceedings of the 8th International Conference on Numerical Methods in Fluid Dynamics*, Lecture Notes in Physics 170 (ed. E. Krause). 507–512, Springer, 1982.

[232] van Leer, B., On the relation between the upwind-differencing schemes of Godunov, Engquist-Osher and Roe. *SIAM. J. Sci. Comput.* **5** (1), 1–20, 1984.

[233] van Leer, B., Upwind-difference methods for aerodynamic problems governed by the Euler equations. *Large Scale Computations in Fluid Mechanics* Lectures in Applied Mathematics, II (eds B. Enquist, S. Osher and R. Somerville). 327–336. American Mathematical Society, Providence, Rhode Island, 1985.

[234] van Leer, B., Lee W.-T. and Roe, P.L., Characteristic time-stepping or local preconditioning of the Euler equations. *AIAA Paper* 91-1552, 1991.

[235] Lemke, M., Multilevel Verfahren mit selbst-adaptiven Gitterverfeinerungen für Parallelrechner mit verteiltem Speicher. Ph.D. thesis (in German), University of Düsseldorf, Germany, 1993.

[236] Leonard, B.P., A stable and accurate convective modelling procedure based on quadratic upstream interpolation. *Comp. Meth. Appl. Mech. Eng.* **19**, 59–98, 1979.

[237] Le Tallec, P., Domain decomposition methods in computational mechanics. *In:* J. Tinsley Oden (ed.), *Computational Mechanics Advances*, **1** (2), 121–220. North Holland, Amsterdam, 1994.

[238] Lin, F.B. and Sotiropoulos, F., Strongly-coupled multigrid method for 3-D incompressible flows using near-wall turbulence closures. *Trans. ASME J. Fluids Eng.* **119**, 314–324, 1997.

[239] Linden, J., A multigrid method for solving the biharmonic equation on rectangular domains. Arbeitpapiere der GMD, Sankt Augustin, West Germany, 1985.

[240]  Linden, J., Mehrgitterverfahren für das erste Randwertproblem der biharmonischen Gleichung und Anwendung auf ein inkompressibles Strömungsproblem. Ph.D. thesis (in German), University of Bonn, West Germany, 1985.

[241]  Linden, J., Lonsdale, G., Ritzdorf, H. and Schüller, A., Scalability aspects of parallel multigrid. *Fut. Generation Comp. Systems* **10**, 429–439, 1994.

[242]  Linden, J., Lonsdale, G., Steckel, B. and Stüben, K., Multigrid for the steady-state incompressible Navier–Stokes equations; a survey. Arbeitspapiere der GMD 322, Sankt Augustin, West Germany, 1988.

[243]  Linden, J., Steckel, B. and Stüben, K., Parallel multigrid solution of the Navier–Stokes equations on general 2D-domains. *Parallel Comput.* **7**, 461–475, 1988.

[244]  Linden, J. and Stüben, K., Multigrid methods: An overview with emphasis on grid generation processes. *Numerical Grid Generation in Computational Fluid Dynamics.* (eds J. Häuser and C. Taylor). 483–509. Pineridge Press, Swansea, 1986.

[245]  Liu, F. and Zheng, X., A strongly coupled time-marching method for solving the Navier-Stokes and $k-\omega$ turbulence model equations with multigrid. *J. Comp. Phys.* **128**, 289–300, 1996.

[246]  Livshits, I., Multigrid solvers for wave equations. Ph.D. thesis, Bar-Ilan University, Ramat Gan, Israel, 1995.

[247]  Lonsdale, G., Solution of a rotating Navier–Stokes problem by a nonlinear multigrid algorithm. *J. Comp. Phys.* **74**, 177–190, 1988.

[248]  Lonsdale, G., Ritzdorf, H., Stüben, K., The $L_i SS$ package. Arbeitspapiere der GMD 745, Sankt Augustin, West Germany, 1993.

[249]  Lonsdale, G. and Schüller, A., Multigrid efficiency for complex flow simulations on distributed memory machines. *Parallel Comput.* **19**, 23–32, 1993.

[250]  Lonsdale, R.D., An algebraic multigrid solver for the Navier–Stokes equations on unstructured meshes. *Int. J. Num. Meth. Heat Fluid Flow* **3**, 3–14, 1993.

[251]  Machenhauer, B., The Spectral Method. *Numerical Methods Used in Atmospheric Models.* GARP: Global atmospheric research programme, Geneva, World Meteorological Organisation Publication Series **17**, 121–275, 1979.

[252]  Hart, L., McCormick, S.F., O'Gallagher, A. and Thomas, J., The fast adaptive composite-grid method (FAC): algorithms for advanced computers. *Appl. Math. Comp.* **19**, 103–125, 1986.

[253]  Mandel, J. and Lett, G.S., Domain decomposition preconditioning for $p$-version finite elements with high aspect ratios. *Appl. Num. Math.* **8**, 411–425, 1991.

[254]  Mandel, J. and Brezina, M., Balancing domain decomposition for problems with large jumps in coefficients. *Math. Comp.* **65**, 1387–1401, 1996.

[255]  Mandel, J. and Tezaur, R., Convergence of a substructuring method with Lagrange multipliers. *Numer. Math.* **73**, 473–487, 1996.

[256]  Mandel, J., Brezina, M. and Vanek, P., Energy optimization of algebraic multigrid bases. *Computing* **62**, 205–228, 1999.

[257]  Mavripilis, D.J., Multigrid strategies for viscous flow solvers on anisotropic unstructured meshes. *J. Comp. Phys.* **145**, 141–165, 1998.

[258]  Mavripilis, D.J. and Venkatakrishnan, V., Agglomeration multigrid for two-dimensional viscous flows. *Comp. Fluids* **24**, 553–570, 1995.

[259]  Mavripilis, D.J. and Venkatakrishnan, V., A 3D agglomeration multigrid solver for the Reynolds averaged Navier–Stokes equations on unstructured meshes. *Int. J. Num. Meth. Fluids* **23**, 527–544, 1996.

[260]  Mavripilis, D.J. and Venkatakrishnan, V., A unified multigrid solver for the Navier–Stokes equations on mixed element meshes, *Int. J. Comp. Fluid Dyn.* **8**, 247–263, 1998.

[261] McBryan, O.A., Frederickson, P.O., Linden, J., Schüller, A., Solchenbach, K., Stüben, K., Thole, C.A. and Trottenberg, U., Multigrid methods on parallel computers—a survey of recent developments. *Impact Comput. Sci. Eng.* **3**, 1–75, 1991.

[262] McCormick, S.F., *Multilevel Adaptive Methods for Partial Differential Equations,* Frontiers in Applied Mathematics 6. SIAM, Philadelphia, 1989.

[263] McCormick, S. and Ruge, J., Algebraic multigrid methods applied to problems in computational structural mechanics. *State-of-the-Art Surveys on Computational Mechanics.* 237–270. ASME, New York, 1989.

[264] Meijerink, J. and van der Vorst, H., An iterative solution method for linear systems of which the coefficient matrix is a symmetric M-matrix. *Math. Comp.* **31**, 148–162, 1977.

[265] Mertens, R., De Gersem, H., Belmans, R., Hameyer, K., Lahaye, D., Vandewalle, S. and Roose, D., An algebraic multigrid method for solving very large electromagnetic systems. *IEEE Trans. Magn.* **34**, 3327–3330, 1998.

[266] van der Maarel, E., A local grid refinement method for the Euler equations. Ph.D. thesis, University of Amsterdam, Amsterdam, The Netherlands, 1993.

[267] Michielse P.H. and van der Vorst, H.A., Data transport in Wang's partition method. *Parallel Comput.* **7**, 87–95, 1988.

[268] Mierendorff, H., Parallelization of multigrid methods with local refinements for a class of nonshared memory systems. *Multigrid Methods: Theory, Applications and Supercomputing* (ed. S.F. McCormick). 449–465. Marcel Dekker, New York, 1988.

[269] Mittelmann, H.D., Multi-grid methods for simple bifurcation problems. *Multigrid Methods*, Lecture Notes in Mathematics 960 (eds W. Hackbusch and U. Trottenberg). 558–575. Springer, Berlin, 1982.

[270] Molenaar, J., Multigrid methods for semiconductor device simulation. CWI Tract 100, CWI, Amsterdam, 1993.

[271] Morano, E. and Dervieux, A., Steady relaxation methods for unstructured multigrid Euler and Navier–Stokes solutions. *Comp. Fluid Dyn.* **5**, 137–167, 1995.

[272] Morano, E., Mavripilis, D.J. and Venkatakrishnan, V., Coarsening strategies for unstructured multigrid techniques with application to anisotropic problems, *7th Copper Mountain Conference on Multigrid Methods*, NASA Conference Publication 3339 (eds N.D. Melson, T.A. Manteuffel, S.F. McCormick and C.C. Douglas). 591–606. NASA, Hampton, 1996.

[273] Moulton, J.D., Dendy, J.E. and Hyman, J.M., The black box multigrid numerical homogenization algorithm. *J. Comp. Phys.* **141**, 1–29, 1998.

[274] Mulder, W.A., Multigrid relaxtion for the Euler equations. *J. Comp. Phys.* **60**, 235–252, 1985.

[275] Mulder, W.A., A new multigrid approach to convection problems. *J. Comp. Phys.* **83**, 303–323, 1989.

[276] Mulder, W.A., A high resolution Euler solver based on multigrid, semi-coarsening, and defect correction. *J. Comp. Phys.* **100**, 91–104, 1992.

[277] Naik, N.H. and van Rosendale, J., The improved robustness of multigrid elliptic solvers based on multiple semicoarsened grids. *SIAM J. Numer. Anal.* **30**, 215–229, 1993.

[278] Nicolaides, R.A., On multiple grid and related techniques for solving discrete elliptic systems. *J. Comp. Phys.* **19**, 418–431, 1975.

[279] Nicolaides, R.A., On the $l^2$ convergence of an algorithm for solving finite element equation. *Math. Comp.* **31**, 892–906, 1977.

[280] Niestegge, A. and Witsch, K., Analysis of a multigrid Stokes solver. *Appl. Math. Comp.* **35**, 291–303, 1990.

[281] Nonino, C. and del Guidice, S., An improved procedure for finite-element methods in laminar and turbulent flow. *Numerical Methods in Laminar and Turbulent Flow* I (ed. C. Taylor). 597–608. Pineridge Press, Swansea, 1985.

[282] Notay, Y., An efficient algebraic multilevel preconditioner robust with respect to anisotropies. *Algebraic Multilevel Iteration Methods with Applications* (eds O. Axelsson and B. Polman). 111–228. University of Nijmegen, The Netherlands, 1996.

[283] Notay, Y., Using approximate inverses in algebraic multilevel methods. *Numer. Math.* **80**, 397–417, 1998.

[284] Notay, Y., Optimal V-cycle algebraic multilevel preconditioning. *Numer. Lin. Alg. Appl.* **5**, 441–459, 1998.

[285] Oertel, K.-D., Praktische und theoretische Untersuchungen zur ILU-Glättung bei Mehrgitterverfahren, Master's thesis (in German), University of Bonn, West Germany, 1988.

[286] Ortega, J.M. and Rheinboldt, W.C., *Iterative solution of Nonlinear Equations in Several Variables*. Academic Press, New York, 1970.

[287] Osher, S. and Chakravarthy, S., Upwind schemes and boundary conditions with applications to Euler equations in general geometries. *J. Comp. Phys.* **50**, 447–481, 1983.

[288] Osher, S. and Solomon, F., Upwind difference schemes for hyperbolic systems of conservation laws. *Math. Comp.* **38**, 339–374, 1982.

[289] Oosterlee, C.W. and Wesseling, P., On the robustness of a multiple semi-coarsened grid method. *Z. Angew. Math. Mech.* **75** (4), 251–257, 1995.

[290] Oosterlee, C.W. and Ritzdorf, H., Flux difference splitting for three-dimensional steady incompressible Navier–Stokes equations in curvilinear coordinates. *Int. J. Num. Meth. Fluids*, **23**, 347–366, 1996.

[291] Oosterlee, C.W., A GMRES-based plane smoother in multigrid to solve 3D anisotropic fluid flow problems. *J. Comp. Phys.* **130**, 41–53, 1997.

[292] Oosterlee, C.W. and Washio, T., An evaluation of parallel multigrid as a solver and as a preconditioner for singularly perturbed problems. *SIAM J. Sci. Comput.* **19**, 87–110, 1998.

[293] Oosterlee, C.W., Gaspar, F.J., Washio, T. and Wienands, R., Multigrid line smoothers for higher order upwind discretizations of convection-dominated problems. *J. Comp. Phys.* **139**, 274–307, 1998.

[294] Oosterlee, C.W. and Washio, T., Krylov subspace acceleration of nonlinear multigrid with application to recirculating flow. *SIAM J. Sci. Comput.*, **21**, 1670–1690, 2000.

[295] Oswald, P., On function spaces related to finite element approximation theory. *Z. Anal. Anwendungen* **9**, 43–64, 1990.

[296] Oswald, P., Stable subspace splittings for Sobolev spaces and some domain decomposition algorithms. *Proceedings of the 7th Symposium on Domain Decomposition Methods*, Contemporary Mathematics 180 (eds D. Keyes and J. Xu). 87–98. AMS, Providence, Rhode Island, 1994.

[297] Oswald, P., On the convergence rate of SOR: a worst case estimate. *Computing* **52**, 245–255, 1994.

[298] Oswald, P., *Multilevel Finite Element Approximation. Theory and Applications*. Teubner Verlag, Stuttgart, 1994.

[299] Patankar, S.V., *Numerical Heat Transfer and Fluid Flow*. Hemisphere, Washington DC, 1980.

[300] Patankar, S.V., A calculation procedure for two-dimensional elliptic situations. *Num. Heat Transfer* **4**, 409–425, 1981.

[301] Patankar, S.V. and Spalding, D., A calculation procedure for heat, mass and momentum transfer in three-dimensional parabolic flows. *Int. J. Heat Mass Transfer* **15**, 1787–1806, 1972.

[302] Peric, M., A finite volume method for the prediction of three-dimensional fluid flow in complex ducts. Ph.D. thesis, Imperial College, London, UK, 1985.

[303] Pierce, N.A. and Giles, M.B., Preconditioned multigrid methods for compressible flow calculations on stretched meshes, *J. Comp. Phys.* **136**, 425–445, 1997.

[304] Pierce, N.A., Giles, M.B., Jameson, A. and Martinelli, L., Accelerating three-dimensional Navier–Stokes calculations. *AIAA Paper* 97–1953, 1997.

[305] Pierce, N.A. and Alonso, J.J., Efficient computation of unsteady viscous flows by an implicit preconditioned multigrid method. *AIAA J.* **36**, 401–408, 1998.

[306] Quarteroni, A. and Valli, A., *Domain Decomposition Methods for Partial Differential Equations.* Oxford Science, Oxford, UK, 1999.

[307] Radespiel, R. and Swanson, R.C., Progress with multigrid schemes for hypersonic flow problems. *J. Comp. Phys.* **116**, 103–122, 1995.

[308] Rannacher, R., Error control in finite element computations, an introduction to error estimation and mesh-size adaptation. *Error Control and Adaptivity in Scientific Computing*, NATO Science Series, Series C: Mathematics and Physical Sciences, Vol. 536 (eds H. Bulgak and C. Zenger). 247–278. Kluwer, Dordrecht, 1999.

[309] Raw, M., A coupled algebraic multigrid method for the 3D Navier–Stokes equations. Rep.: Advanced Scientific Computing Ltd., Waterloo, Ontario, Canada.

[310] Regler, H., Anwendungen von AMG auf das Plazierungsproblem beim Layoutentwurf und auf die numerische Simulation von Strömungen. Ph.D. thesis (in German), Technische Universität München, Munich, Germany, 1997.

[311] Reichelt, C., Theoretische und numerische Aspekte von Mehrgitterverfahren zur Lösung der Diffusionsgleichung mit stückweise stetig differenzierbaren Koeffizienten. Master's thesis, University of Bonn, Germany, 1995.

[312] Reitzinger, S., Algebraic multigrid and element preconditioning I. SFB Rep. 98-15, University of Linz, Austria, 1998.

[313] Reusken, A.A., Multigrid with matrix-dependent transfer operators for a singular perturbation problem. *Computing* **50** (3), 199–211, 1993.

[314] Reusken, A.A., A multigrid method based on incomplete Gaussian elimination. Rep. RANA 95-13, Eindhoven University of Technology, The Netherlands, 1995.

[315] Reusken, A.A., Approximate cyclic reduction preconditioning. *Multigrid Methods V*, Proceedings of the 5th European Multigrid Conference, Lecture Notes in Computational Science and Engineering 3 (eds W. Hackbusch and G. Wittum). 243–259. Springer, Berlin, 1998.

[316] Reusken, A.A., On the approximate cyclic reduction preconditioner. Rep. 144, Institut für Geometrie und Praktische Mathematik, RWTH Aachen, 1997. *SIAM J. Sci. Comput.* **21**, 565–590, 1999.

[317] Rhie, C.M. and Chow, W.L., Numerical study of the turbulent flow past an airfoil with trailing edge separation. *AIAA J.* **21**, 1525–1532, 1983.

[318] Riemslagh, K. and Dick, E., Multi-stage Jacobi relaxation in multigrid methods for the steady Euler equations. *Int. J. Comp. Fluid Dyn.* **4**, 343–361, 1995.

[319] Ries, M., Trottenberg, U. and Winter, G., A note on MGR methods. *Linear Algebra Appl.* **49**, 1–26, 1983.

[320] Ritzdorf, H., Lokal verfeinerte Mehrgitter-Methoden für Gebiete mit einspringenden Ecken. Master's thesis (in German), University of Bonn, West Germany, 1984.

[321] Ritzdorf, H., Self-adaptive local refinements supported by the CLIC-3D Library. *Portable Parallelization of Industrial Aerodynamic Applications (POPINDA), Results of a BMBF Project,*

Notes on Numerical Fluid Mechanics 71 (ed. A. Schüller). 190–199, Vieweg, Braunschweig, 1999.

[322]   Ritzdorf, H., Schüller, A., Steckel, B. and Stüben, K., $L_i SS$ – an environment for the parallel multigrid solution of partial differential equations on general 2D domains. *Parallel Comput.* **20**, 1559–1570, 1994.

[323]   Ritzdorf, H. and Stüben, K., Adaptive multigrid on distributed memory computers. *Multigrid Methods IV*, Proceedings of the 4th European Multigrid Conference (eds P.W. Hemker and P. Wessling). 77–96. Birkhäuser, Basle, 1994.

[324]   Rixen, D. and Farhat, C., A simple and efficient extension of a class of substructure based preconditioners to heterogeneous structural mechanics problems. *Int. J. Numer. Meth. Eng.* **44**, 489–516, 1999.

[325]   Roberts, T.W., Sidilkover, D. and Swanson, R.C., Textbook multigrid efficiency for the steady Euler equations. *AIAA Paper* 97–1949, 1997.

[326]   Robinson, G., Parallel computational fluid dynamics on unstructured meshes using algebraic multigrid. *Parallel Computational Fluid Dynamics 92* (eds R.B. Pelz, A. Ecer and J. Häuser). Elsevier Science, Amsterdam, 1993.

[327]   Rodi, W., Majumdar, S. and Schönung, B., Finite volume methods for two-dimensional incompressible flow problems with complex boundaries. *Comp. Meth. Appl. Mech. Eng.* **75**, 369–392, 1989.

[328]   Roe, P.L., Approximate Riemann solvers, parameter vectors and difference schemes. *J. Comput. Phys.* **43**, 357–372, 1981.

[329]   Rood, R.B., Numerical advection algorithms and their role in atmospheric transport and chemistry models. *Rev. Geophys.* **25**, 71–100, 1987.

[330]   Rosenfeld, M., Kwak, D. and Vinokur, M., A fractional step solution method for the unsteady incompressible Navier–Stokes equations in generalized coordinate systems. *J. Comp. Phys.* **94**, 102–137, 1991.

[331]   Rosenfeld, M. and Kwak, D., Multigrid acceleration of a fractional-step solver in generalized curvilinear coordinate systems. *AIAA Paper* 92-0185, 1992.

[332]   Rüde, U., *Mathematical and Computational Techniques for Multilevel Adaptive Methods*, Frontiers in Applied Mathematics, Vol. 13. SIAM, Philadelphia, 1993.

[333]   Ruge, J.W. and Stüben, K., Efficient solution of finite difference and finite element equations by algebraic multigrid (AMG). *Multigrid Methods for Integral and Differential Equations*, Institute of Mathematics and its Applications Conferences Series 3 (eds D.J. Paddon and H. Holstein). 169–212. Clarendon Press, Oxford, 1985.

[334]   Ruge, J.W. and Stüben, K., Algebraic Multigrid (AMG). *Multigrid Methods, Frontiers in Applied Mathematics* (ed. S.F. McCormick). 73–130. SIAM, Philadelphia, 1987.

[335]   Saad, Y. and Schultz, M.H., GMRES: A generalized minimal residual algorithm for solving nonsymmetric linear systems. *SIAM J. Sci. Comput.* **7**, 856–869, 1986.

[336]   Saad, Y., A flexible inner–outer preconditioned GMRES algorithm. *SIAM J. Sci. Comput.* **14**, 461–469, 1993.

[337]   Saad, Y., *Iterative Methods for Sparse Linear Systems*. PWS Publishing, Boston, USA, 1996.

[338]   Saad, Y., ILUM: a multi-elimination ILU preconditioner for general sparse matrices. *SIAM J. Sci. Comput.* **17**, 830–847, 1996.

[339]   Sbosny, H. and Witsch, K., Parallel multigrid using domain decomposition. *Proceedings of the 6th GAMM Seminar*, Notes on Numerical Fluid Mechanics (ed. W. Hackbusch). **31**, 200–215, 1991.

[340] Schröder, J., Zur Lösung von Potentialaufgaben mit Hilfe des Differenzenverfahrens. *Z. Angew. Math. Mech.* **34**, 241–253, 1954.

[341] Schröder, J., *Operator Inequalities*, Mathematics in Science and Engineering 147, Academic Press, 1980.

[342] Schröder, J. and Trottenberg, U., Reduktionsverfahren für Differenzengleichungen bei Randwertaufgaben I. *Numer. Math.* **22**, 37–68, 1973.

[343] Schröder, J., Trottenberg, U. and Reutersberg, H., Reduktionsverfahren für Differenzengleichungen bei Randwertaufgaben II. *Numer. Math.* **26**, 429–459, 1976.

[344] Schröder, J., Trottenberg, U. and Witsch, K., *On Fast Poisson Solvers and Applications*, Lecture Notes in Mathematics 631. 153–187, Springer, Berlin, 1978.

[345] Schüller, A., Anwendung von Mehrgittermethoden auf das Beispiel einer ebenen, reibungsfreien und kompressiblen Unterschallströmung am Beispiel des Kreisprofils. Master's thesis (in German), University of Bonn, West Germany, 1983.

[346] Schüller, A., *Mehrgitterverfahren für Schalenprobleme* Ph.D. thesis (in German), University of Bonn. GMD-Bericht Nr. 171, Oldenbourg Verlag, Munich, 1988.

[347] Schüller, A., A multigrid algorithm for the incompressible Navier–Stokes equations. *Numerical Treatment of Navier–Stokes Equations*, Notes on Numerical Fluid Mechanics 30 (eds W. Hackbusch and R. Rannacher). 124–133. Vieweg, Braunschweig, 1990.

[348] Schüller, A. (ed.) *Portable Parallelization of Industrial Aerodynamic Applications (POPINDA), Results of a BMBF Project*. Notes on Numerical Fluid Mechanics 71. Vieweg, Braunschweig, 1999.

[349] Schwarz, W., Dreidimensionale Netzgenerierung für ein Finites Volumenverfahren, MBB, Bericht Nr. MBB/LKE122/S/PUB/206, Ottobrunn, West Germany, 1985.

[350] Schwichtenberg, H., *Erweiterungsmöglichkeiten des Programmpaketes MG01 auf nichtlineare Aufgaben*. Master's thesis (in German), University of Bonn, West Germany, 1985.

[351] Shaidurov, V.V., *Multigrid Methods for Finite Elements*. Kluwer, Dordrecht, 1995.

[352] Shaw, G.J. and Sivaloganathan, S., On the smoothing of the SIMPLE pressure correction algorithm. *Int. J. Num. Meth. Fluids* **8**, 441–462, 1988.

[353] Sheffer, S.G., Martinelli, L. and Jameson, A., An efficient multigrid algorithm for compressible reactive flows. *J. Comp. Phys.* **144**, 484–516, 1998.

[354] Shieh, C.F., Three-dimensional grid generation using elliptic equations with direct grid distribution control. *AIAA J.* **22**, 361–364, 1984.

[355] Shortley, G.H. and Weller, R., Numerical solution of Laplace's equation. *J. Appl. Phys.* **9**, 334–348, 1938.

[356] Sidilkover, D., Higher order accurate method for recovering shock location. Ph.D. thesis, Weizmann Institute of Science, Rehovot, Israel, 1989.

[357] Sidilkover, D., A genuinely multidimensional upwind scheme and efficient multigrid solver for the compressible Euler equations. ICASE Rep. 97-84, 1994.

[358] Sidilkover, D., Some approaches towards constructing optimally efficient multigrid solvers for the inviscid flow equations. ICASE Rep. 97-39. *Comput. Fluids* **28**, 551–571, 1999.

[359] Sidilkover, D. and Brandt, A., Multigrid solution to steady-state two-dimensional conservation laws. *SIAM J. Numer. Anal.* **30**, 249–274, 1993.

[360] Sivaloganathan, S., The use of local mode analysis in the design and comparison of multigrid methods. *Comput. Phys. Comm.* **65**, 246–252, 1991.

[361] Smith, R.M. and Hutton, A.G., The numerical treatment of advection: A performance comparison of current methods. *Num. Heat Transfer* **5**, 439–461, 1982.

[362] Smith, B., Bjorstad, P. and Gropp, W., *Domain Decomposition – Parallel Multilevel Methods for Elliptic Partial Differential Equations*. Cambridge University Press, Cambridge, 1996.

[363] Soare, M., *Application of Finite Difference Equations to Shell Analysis*. Pergamon Press, Oxford, 1967.

[364] Solchenbach, K. and Trottenberg, U., On the multigrid acceleration approach in computational fluid dynamics. Arbeitspapiere der GMD 251, Sankt Augustin, West Germany, 1987.

[365] Sonar, Th., Strong and weak norm refinement indicators based on the finite element residual for compressible flow. *Impact Comput. Sci. Eng.* 111–127, 1993.

[366] Sonneveld, P., Wesseling, P. and de Zeeuw, P.M., Multigrid and conjugate gradient methods as convergence acceleration techniques. *Multigrid Methods for Integral and Differential Equations* (eds D.J. Paddon and H. Holstein). 117–168. Clarendon Press, Oxford, 1985.

[367] Southwell, R.V., Stress calculation in frameworks by the method of systematic relaxation of constraints I, II. *Proc. Roy. Soc. London Ser. A*, **151**, 56–95, 1935.

[368] Southwell, R.V., *Relaxation Methods in Theoretical Physics*. Clarendon Press, Oxford, 1946.

[369] South, J.C. and Brandt, A., Application of a multi-level grid method to transonic flow calculations. *Transonic Flow Problems in Turbo Machinery* (eds T.C. Adam and M.F. Platzer). 180–207. Hemisphere, Washington, 1977.

[370] Sparis, P.D., A method for generating boundary-orthogonal curvilinear coordinate systems using the biharmonic equation. *J. Comp. Phys.* **61**, 445–462, 1985.

[371] Spekreijse, S.P., Multigrid solution of second order discretizations of hyperbolic conservation laws. *Math. Comp.* **49**, 135–155, 1987.

[372] Spekreijse, S.P., Multigrid solution of the steady Euler equations. CWI Tract 46, Centrum voor Wiskunde en Informatica, Amsterdam, 1988.

[373] Stevenson, R., On the validity of local mode analysis of multi-grid methods. Ph.D. thesis University of Utrecht, The Netherlands, 1990.

[374] Stiefel, E., Über einige Methoden der Relaxationsrechnung. *Z. Angew. Math. Phys.* **3**, 1–33, 1952.

[375] Stüben, K., Approximation und Fehlerabschätzung für Potentiale ebener kompressibler Unterschallströmungen. Ph.D. thesis (in German), University of Cologne, West Germany, 1977.

[376] Stüben, K., Algebraic multigrid (AMG): Experiences and comparisons. *Appl. Math. Comp.* **13**, 419–452, 1983.

[377] Stüben, K., A review of algebraic multigrid. GMD-Report 69, 1999. To appear in *J. Comput. Appl. Math.* 2000.

[378] Stüben, K. and Trottenberg, U., Multigrid methods: fundamental algorithms, model problem analysis and applications. *Multigrid Methods*, Lecture Notes in Mathematics 960 (eds W. Hackbusch and U. Trottenberg). 1–176. Springer, Berlin, 1982.

[379] Stüben, K. and Trottenberg, U., *On the Construction of Fast Solvers for Elliptic Equations*, Computational Fluid Dynamics, Lecture Series 1982-04. von Karman Institute for Fluid Dynamics, Rhode-Saint-Genese, Belgium, 1982.

[380] Swarztrauber, P., The direct solution of the discrete Poisson equation on the surface of a sphere. *J. Comp. Phys.* **15**, 46–54, 1974.

[381] Ta'asan, S., Canonical forms of multidimensional steady state inviscid flows. ICASE Rep. 93-34, 1993.

[382] Ta'asan, S., Optimal multigrid method for inviscid flows. *Proceedings of the European Multigrid Conference*, EMG93, Amsterdam, 309–320, July 1993.

[383] Ta'asan, S., Canonical-variables multigrid method for steady-state Euler equations. *Proceedings of the 14th International Conference on Numerical Methods for Fluid Dynamics*, Bangalore, India, 1994.

[384] Ta'asan, S., Essentially optimal multigrid method for steady state Euler equations. *AIAA Paper* 95-0209, 1995.

[385] Tai, C.-H., Sheu, J.-H. and van Leer, B., Optimal multistage schemes for Euler equations with residual smoothing. *AIAA J.* **33**, 1008–1016, 1995.

[386] Thiebes, H.-J., Mehrgitterverfahren und Reduktionsverfahren für indefinite elliptische Randwertaufgaben. Ph.D. thesis (in German), University of Bonn, West Germany, 1983.

[387] Thiele, M., Batycky, R. and Blunt, M., A streamline-based 3D field-scale compositional reservoir simulator. Paper SPE 38889 presented at the 1997 SPE Annual Technical Conference and Exhibition, San Antonio, Texas, 5–8 October, 1997.

[388] Thole, C.A. and Trottenberg, U., Basic smoothing procedures for the multigrid treatment of elliptic 3-D operators. *Appl. Math. Comp.* **19**, 333–345, 1986.

[389] Thole, C.A. and Trottenberg, U., A short note on standard multigrid algorithms for 3D problems. *Appl. Math. Comp.* **27** (2), 101–111, 1988.

[390] Thompson, M.C. and Ferziger, J.H., An adaptive multigrid technique for the incompressible Navier–Stokes equations. *J. Comp. Phys.* **82**, 94–121, 1989.

[391] Thompson, J.F., Warsi, Z.U.A. and Mastin, C.W., Numerical Grid Generation. North Holland, Amsterdam, 1985.

[392] Trompert, R., Local uniform grid refinement for time dependent partial differential equations. CWI, Tract 107, Centrum voor Wiskunde en Informatica, Amsterdam, 1995.

[393] Turek, S., *Efficient Solvers for Incompressible Flow Problems*. Lecture Notes in Computer Science and Engineering 6. Springer, Berlin, 1999.

[394] Turek, S. and Schäfer, M., Benchmark computations of laminar flow around cylinder. *Flow Simulation with High-Performance Computers II*, Notes on Numerical Fluid Mechanics 52 (ed. E.H. Hirschel). 547–566. Vieweg, Braunschweig, 1996.

[395] Tveito, A. and Winther, R., *Introduction to Partial Differential Equations. A Computational Approach*. Texts in Applied Mathematics. Springer, Berlin, 1998.

[396] Vandewalle, S., *Parallel Multigrid Waveform Relaxation for Parabolic Problems*. Teubner Verlag, Stuttgart, 1993.

[397] van der Vorst, H., BICGSTAB: A fast and smoothly converging variant of Bi-CG for the solution of non-symmetric linear systems. *SIAM J. Sci. Comput.* **13**, 631–644, 1992.

[398] Vanek, P., Mandel, J. and Brezina, M., Algebraic multigrid on unstructured meshes. University of Colorado at Denver, UCD/CCM Rep. 34, 1994.

[399] Vanek, P., Mandel, J. and Brezina, M., Algebraic multigrid by smoothed aggregation for second and fourth order elliptic problems. *Computing* **56**, 179–196, 1996.

[400] Vanek, P., Brezina, M. and Mandel, J., Convergence of algebraic multigrid based on smoothed aggregation. UCD/CCM Rep. 126, 1998. Submitted to *Numer. Math.*

[401] Vanka, S.P., Block-implicit multigrid solution of Navier–Stokes equations in primitive variables. *J. Comp. Phys.* **65**, 138–156, 1986.

[402] Vanka, S.P., A calculation procedure for three-dimensional steady recirculation flows using multigrid methods. *Comm. Meth. Appl. Mech. Eng.* **55**, 321–338, 1986.

[403] Varga, R.S., *Matrix Iterative Analysis*. Prentice-Hall, Englewood Cliffs, 1962.

[404] Vierendeels, J., Riemslagh, K. and Dick, E., Smoothers for viscous flows on high aspect ratio grids, *Computational Fluid Dynamics '98*, Proceedings of the ECCOMAS Conference,

Vol 2 (eds K.D. Papailiou, D. Tsahalis, J. Périaux, Hirsh Ch. and M. Pandolfi). 258–265, Wiley, Chichester, 1998.

[405] Vinokur, M., An analysis of finite-difference and finite-volume formulations of conservation laws. *J. Comp. Phys.* **81**, 1–52, 1989.

[406] Wagner, C., Kinzelbach, W. and Wittum, G., Schur-complement multigrid – a robust method for groundwater flow and transport problems. *Numer. Math.* **75**, 523–545, 1997.

[407] Wagner, C., On the algebraic construction of multilevel transfer operators, *Computing*, **65**, 73–95, 2000.

[408] Wagner, C., Introduction to algebraic multigrid, Course notes of an algebraic multigrid course at the University of Heidelberg in the winter semester 1998/99. http://www.iwr.uni-heidelberg.de/~Christian.Wagner, 1999.

[409] Wan, W.L., Chan, T.F. and Smith, B., An energy minimization interpolation for robust multigrid methods. Department of Mathematics, University of California, Los Angeles, UCLA CAM Report 98-6, 1998.

[410] Washio, T. and Oosterlee, C.W., Krylov subspace acceleration for nonlinear multigrid schemes. *Electr. Trans. Num. Anal.* **6**, 271–290, 1997.

[411] Washio, T. and Oosterlee, C.W., Flexible multiple semicoarsening for three-dimensional singularly perturbed problems. *SIAM J. Sci. Comput.* **19**, 1646–1666, 1998.

[412] Washio, T. and Oosterlee, C.W., Error analysis for a potential problem on locally refined grids. To appear in *Num. Math.*

[413] Webster, R., An algebraic multigrid solver for Navier–Stokes problems in the discrete second order approximation. *Int. J. Num. Meth. Fluids* **22**, 1103–1123, 1996.

[414] Wesseling, P., A robust and efficient multigrid method. *Multigrid Methods*, Lecture Notes in Mathematics 960 (eds W. Hackbusch and U. Trottenberg). 614–630. Springer, Berlin, 1982.

[415] Wesseling, P., *An Introduction to Multigrid Methods*. Wiley, Chichester, 1992.

[416] Wesseling, P. and Oosterlee, C.W., Geometric multigrid with applications to computational fluid dynamics. To appear in *J. Comp. Appl. Math.* 2000.

[417] Wesseling, P., Segal, A. and Kassels, C.G.M., Computing flows on general three-dimensional nonsmooth staggered grids. *J. Comp. Phys.* **149**, 333–362, 1999.

[418] Widlund, O., DD methods for elliptic partial differential equations. *Error Control and Adaptivity in Scientific Computing*, NATO Science Series, Series C: Mathematics and Physical Sciences, Vol. 536 (eds H. Bulgak and C. Zenger). 325–354. Kluwer, Dordrecht, 1999.

[419] Wittum, G., Distributive Iterationen für indefinite Systeme als Glätter in Mehrgitterverfahren am Beispiel der Stokes- und Navier–Stokes-Gleichungen mit Schwerpunkt auf unvollständigen Zerlegungen. Ph.D. thesis (in German), University of Kiel, West Germany, 1986.

[420] Wittum, G., Linear iterations as smoothers in multigrid methods: theory with applications to incomplete decompositions. *Impact Comput. Sci. Eng.* **1**, 180–215, 1989.

[421] Wittum, G., On the robustness of ILU smoothing. *SIAM J. Sci. Comput.* **10**, 699–717, 1989.

[422] Wittum, G., Multi-grid methods for Stokes and Navier–Stokes equations with transforming smoothers: algorithms and numerical results. *Numer. Math.* **54**, 543–563, 1989.

[423] Wittum, G., On the convergence of multi-grid methods with transforming smoothers. *Num. Math.* **57**, 15–38, 1990.

[424] Wu, J., Ritzdorf, H., Oosterlee, C.W., Steckel, B. and Schüller, A., Adaptive parallel multigrid solution of 2D incompressible Navier–Stokes equations. *Int. J. Num. Methods Fluids* **24**, 875–892, 1997.

[425] Xu, J., Iterative methods by space decomposition and subspace correction. *SIAM Rev.* **34**, 581–613, 1992.

[426] Yanenko, N.N. and Shokin, Y.I., On the correctness of first differential approximation of difference schemes. *Dokl. Acad. Nauk. SSSR.* **182**, 776–778, 1968.

[427] Yavneh, I., Multigrid smoothing factors for red-black Gauss-Seidel relaxation applied to a class of elliptic operators. *SIAM J. Numer. Anal.* **32**, 1126–1138, 1995.

[428] Yavneh, I., On red-black SOR smoothing in multigrid. *SIAM J. Sci. Comput.* **17**, 180–192, 1996.

[429] Yavneh, I., Coarse-grid correction for nonelliptic and singular perturbation problems. *SIAM J. Sci. Comput.* **19**, 1682–1699, 1998.

[430] Yavneh, I., Venner, C.H. and Brandt, A., Fast multigrid solution of the advection problem with closed characteristics. *SIAM J. Sci. Comput.* **19**, 111–125, 1998.

[431] Young, D., *Iterative Solution of Large Linear Systems.* Academic Press, New York, 1971.

[432] Yserentant, H., Hierarchical bases of finite element spaces in the discretization of nonsymmetric problems. *Computing* **35**, 39–49, 1985.

[433] Yserentant, H., On the multi-level splitting of finite element spaces. *Numer. Math.* **49**, 379–412, 1986.

[434] Yserentant, H., Hierarchical bases give conjugate gradient type methods a multigrid speed of convergence. *Appl. Math. Comp.* **19**, 347–357, 1986.

[435] Yserentant, H., Old and new convergence proofs for multigrid methods. *Acta Numerica* **2**, 285–326, 1993.

[436] Zaslavsky, L., An adaptive algebraic multigrid for multigroup neutron diffusion reactor core calculations. *Appl. Math. Comp.* **53**, 13–26, 1993.

[437] Zaslavsky, L., An adaptive algebraic multigrid for reactor critical calculations. *SIAM J. Sci. Comput.* **16**, 840–847, 1995.

[438] de Zeeuw, P.M. and van Asselt, E.J., The convergence rate of multi-level algorithms applied to the convection-diffusion equation. *SIAM J. Sci. Comput.* **6**, 492–503, 1985.

[439] de Zeeuw, P.M., Matrix-dependent prolongations and restrictions in a blackbox multigrid solver. *J. Comp. Appl. Math.* **33**, 1–27, 1990.

[440] de Zeeuw, P.M., Incomplete line LU as smoother and as preconditioner. *Incomplete decompsitions (ILU) – Algorithms, Theory and Applications* (eds G. Wittum and W. Hackbusch). 215–224. Vieweg, Braunschweig, 1993.

[441] de Zeeuw, D. and Powell, K.G., An adaptively-refined Cartesian mesh solver for the Euler equations. *J. Comp. Phys.* **104**, 56–68, 1993.

[442] Zeng, S. and Wesseling, P., An ILU smoother for the incompressible Navier–Stokes equations in general coordinates. *Int. J. Num. Methods Fluids* **20**, 59–74, 1995.

[443] Zenger, C., Sparse grids. Parallel algorithms for partial differential equations. *Proceedings of the 6th GAMM-Seminar*, Notes on Numeical Fluid Mechanics 31. (ed. W. Hackbusch). 297–301. Vieweg, Braunschweig, 1991.

[444] Zijlema, M., Computational modeling of turbulent flow in general domains. Ph.D. thesis, Delft University of Technology, Delft, The Netherlands, 1996.

[445] Debus, B., Ansatz spezieller Mehrgitterkomponenten für ein zweidimensionales, singulär gestörtes Modellproblem: Grobgitter- und Glättungsoperatoren. Master's thesis, University of Bonn, West Germany, 1985.

[446] Griebel, M. and Zumbusch, G.W., Parallel multigrid in an adaptive PDE solver based on hashing and space-filling curves. *Parallel Comput.* **25**, 827–843, 1999.

[447] Kroll, N., Direkte Anwendung von Mehrgittertechniken auf parabolische Anfangsrandwertaufgaben. Master's thesis, University of Bonn, West Germany, 1981.

[448] Ritzdorf, H., The High-level Communications Library CLIC. *Portable Parallellization of Industrial Aerodynamic Applications (POPINDA), Results of a BMBF Project*, Notes on Numerical Fluid Mechanics 71 (ed. A. Schüller). 33–50. Vieweg, Braunschweig, 1999.

[449] Thole, C.-A., Mayer, S. and Supalov, A., Fast solution of MSC/NASTRAN sparse matrix problems using a multilevel approach. *Electr. Trans. Num. Anal.* **6**, 246–254, 1997.

[450] Trottenberg, U. and Witsch, K., Zur Kondition diskreter elliptischer Randwertaufgaben. GMD-Studie Nr. 60, Sankt Augustin, West Germany, 1981.

[451] Brandt, A., The Weizmann Institute Research on Multilevel Computation: 1988 Report, in *Proc. 4th Copper Mountain Conf. on Multigrid Methods* (Mandel, J. et al., eds), SIAM, 1989, pp. 13–53.

[452] Thomas, J.L., Diskin, B. and Brandt, A., "Textbook Multigrid Efficiency for the Incompressible Navier–Stokes Equations: High Reynolds Number Wakes and Boundary Layers", ICASE Report 99–51, December 1999. Computer & fluids, to appear.

[453] Thomas, J.L., Diskin, B. and Brandt, A., "General Framework for Achieving Textbook Multigrid Efficiency: Quasi-1-D Euler Example", *Frontiers of Computational Fluid Dynamics – 2000*, Half Moon Bay, CA, June 2000.

# INDEX

Page numbers in **bold** refer to major references, those in *italic* refer to figures or tables.